Strawberries

Production, Postharvest Management and Protection

Strawberries

Production, Postharvest Management and Protection

Edited by
Radha Mohan Sharma, Rakesh Yamdagni,
Anil Kumar Dubey, and Vikramaditya Pandey

CRC Press is an imprint of the
Taylor & Francis Group, an **informa** business

CRC Press
Taylor & Francis Group
6000 Broken Sound Parkway NW, Suite 300
Boca Raton, FL 33487-2742

© 2019 by Taylor & Francis Group, LLC
CRC Press is an imprint of Taylor & Francis Group, an Informa business

No claim to original U.S. Government works

Printed on acid-free paper

International Standard Book Number-13: 978-1-4987-9609-5 (Hardback)

This book contains information obtained from authentic and highly regarded sources. Reasonable efforts have been made to publish reliable data and information, but the author and publisher cannot assume responsibility for the validity of all materials or the consequences of their use. The authors and publishers have attempted to trace the copyright holders of all material reproduced in this publication and apologize to copyright holders if permission to publish in this form has not been obtained. If any copyright material has not been acknowledged please write and let us know so we may rectify in any future reprint.

Except as permitted under U.S. Copyright Law, no part of this book may be reprinted, reproduced, transmitted, or utilized in any form by any electronic, mechanical, or other means, now known or hereafter invented, including photocopying, microfilming, and recording, or in any information storage or retrieval system, without written permission from the publishers.

For permission to photocopy or use material electronically from this work, please access www.copyright.com (http://www.copyright.com/) or contact the Copyright Clearance Center, Inc. (CCC), 222 Rosewood Drive, Danvers, MA 01923, 978-750-8400. CCC is a not-for-profit organization that provides licenses and registration for a variety of users. For organizations that have been granted a photocopy license by the CCC, a separate system of payment has been arranged.

Trademark Notice: Product or corporate names may be trademarks or registered trademarks, and are used only for identification and explanation without intent to infringe.

Library of Congress Cataloging-in-Publication Data

Names: Sharma, R. M. (Radha Mohan), editor.
Title: Strawberries : production, postharvest management and protection / editors: R M Sharma, Rakesh Yamdagni, A K Dubey, Vikramaditya Pandey.
Description: Boca Raton, FL : CRC Press, Taylor & Francis Group, 2019. | Includes index.
Identifiers: LCCN 2019010555| ISBN 9781498796095 (hardback : alk. paper) | ISBN 9781498796125 (ebook)
Subjects: LCSH: Strawberries.
Classification: LCC SB385 .S744 2019 | DDC 634/.75--dc23
LC record available at https://lccn.loc.gov/2019010555

Visit the Taylor & Francis Web site at
http://www.taylorandfrancis.com

and the CRC Press Web site at
http://www.crcpress.com

Printed and bound in Great Britain by
TJ International Ltd, Padstow, Cornwall

Contents

Foreword ix
Preface xi
Editors xiii
Contributors xv
Introduction xix

Chapter 1   Introduction 1

V. Pandey, R. M. Sharma, R. Yamdagni, A. K. Dubey, and Tushar Uttamrao Jadhav

Chapter 2   Composition, Quality and Uses 23

Monika Sood and Julie Dogra Bandral

Chapter 3   Structures and Functions 31

R.M. Sharma, Rajesh Kumar Shukla, and Nav Prem Singh

Chapter 4   Origin, Taxonomy and Distribution 37

A.K. Dubey

Chapter 5   Breeding and Improvement 49

R.K. Salgotra, Manmohan Sharma, and Anil Kumar Singh

Chapter 6   Varieties 81

V.K. Tripathi, R.M. Sharma, and Sanjeev Kumar

Chapter 7   Tissue Culture 121

Anil Kumar Singh, Gyanendra Kumar Rai, Sreshti Bagati, and Sanjeev Kumar

Chapter 8   Markers and Genetic Mapping 141

Era Vaidya Malhotra and Madhvi Soni

Chapter 9   Climatic Requirements 161

R. Yamdagni and A.D. Sharma

Chapter 10   Soil 169

P.K. Rai

Chapter 11   Propagation 179

Nimisha Sharma, Laxuman Sharma, and B.P. Singh

v

Chapter 12 Planting 193

Biswajit Das

Chapter 13 Nutrition 209

Arti Sharma and Bindiya Sharma

Chapter 14 Water Management 229

V. Pandey, R.M. Sharma, Dinesh Kumar, S.D. Sharma, and S.K. Jena

Chapter 15 Weed Management 269

Manpreet Kour and A.S. Charak

Chapter 16 Mulching 289

S.K. Singh, Prashant Kalal, and Pramod Kumar

Chapter 17 Flowering 301

Kiran Kour, Bikramjit Singh, and Tanjeet Singh Chahal

Chapter 18 Pollination 321

M.S. Khan, Poonam Srivastava, and R.M. Srivastava

Chapter 19 Fruit Development 335

Sangita Yadav, Sandeep Kumar, Seema Sangwan, and Shiv K. Yadav

Chapter 20 Use of Plant Bio-regulators 349

V.K. Tripathi, Sanjeev Kumar, and Vishal Dubey

Chapter 21 Special Cultural Practices 359

O.P. Awasthi and Sunil Kumar

Chapter 22 Protected Cultivation 365

K.K. Pramanick and Poonam Kashyap

Chapter 23 Soilless Culture 373

Mahital Jamwal and Nirmal Sharma

Chapter 24 Harvesting 399

Sushil Sharma and Kuldeep Singh

Contents | vii

Chapter 25 Yield and Varietal Performance 403

Sanjeev Kumar, V.K. Tripathi, and Parshant Bakshi

Chapter 26 Post-Harvest Handling and Storage 411

B.V.C. Mahajan and Alemwati Pongener

Chapter 27 Value Addition.. 431

Julie Dogra Bandral and Monika Sood

Chapter 28 Physiological Disorders 445

A.K. Goswami, S.K. Singh, and Satyabrata Pradhan

Chapter 29 Integrated Disease Management 453

M.K. Pandey, B.K. Pandey, and A.K. Tiwari

Chapter 30 Integrated Insect Pest Management 487

Uma Shankar and D. P. Abrol

Chapter 31 Pesticide Residues 507

Tarun Verma, Beena Kumari, and Kaushik Banerjee

Chapter 32 Production Economics and Marketing 519

Sudhakar Dwivedi and Pawan Kumar Sharma

Chapter 33 Challenges, Potential and Future Strategies.. 531

K. Kumar

Index.. 539

Foreword

Strawberries find mention in several Roman and European literatures. Although, these have been in cultivation before the beginning of Christian era but their scientific cultivation at large scale is only about two centuries old. The genus *Fragaria*, to which strawberries belong, is a group of low perennial creeping herbs distributed in wild in the temperate and sub-tropical regions of world and forms a polyploidy series ranging from diploid to octoploid with basic chromosome number of x=7. In 1712, a French Army Officer A. F. Frezier introduced large fruited, female strawberry (*F. chiloensis*) plants from Chile to France which easily got hybridized with other octoploid species, *F. virginiana*, to give rise to the present day large fruited cultivated strawberry *Fragaria* x *ananassa* Duch. In the subsequent decades, breeding efforts led to development of cultivars for specific purposes. Development of high impact production technologies could be largely attributed for its large scale cultivation across the world both under open and protected conditions.

Attractive fruits of unique taste with high nutritive value and amenability to round the year cultivation under varied management conditions have made strawberries as choice crop for growers. With short duration of cropping and precocious bearing, strawberries are ideal for high return and attractive profit per unit area. The consumption of fruits as dessert and several processed as well as value added products have made strawberry crop popular among growers and consumers alike throughout world. Their utility in vertical or terrace gardens in metropolitan cities across the world is now a reality and, therefore, strawberries have become an integral part of "Balcony Gardens"- a modern trend in urban Horticulture. The adaptability of strawberries to selective tropical and sub-tropical climatic regions under protected conditions has made it an ideal choice in crop diversification programmes especially for inter-space utilization in young orchards for runner or fruit production. Thus, it could be a candidate crop for increasing the income of farmers. Besides, multicoloured and double flowering genotypes of strawberries have great ornamental value.

Strawberries are ready for harvesting at four to five weeks after blossoming with duration between first and full bloom to be 10 to 12 days. Large variations in productivity of strawberries ranging from as low as 6 t/ha to as high as 83 t/ha have been reported in about 120 to 150 days of cropping period. With average price of strawberry fruits at US $ 200/q, a gross return up to US $ 15000, US $ 14200 and US $ 13800/ha during 1st, 2nd and 3rd year, respectively could be expected. The cost: benefit ratio may vary greatly depending upon variety, season, region and productivity ranging from 1:1 to 1:1.2 depending on fixed and variable factors of production and price situation in a particular locality and season. This could be reason of strawberry cultivation gaining immense popularity. As a result, several commercial growers have come forward to invest in high-tech production of strawberries such as soilless culture, vertical gardening, protected cultivation etc. There is vast scientific literature available on strawberry but scattered in the form of scientific research papers and reports. There was a long felt need to compile the huge scientific information available on various aspects of strawberries in most comprehensive style and a single volume like the present one "The Strawberries: Production, Post-harvest Management and Protection" for the benefit of end users.

I am confident that the present book will fulfill the long cherished demand for a most up to date and comprehensive scientific literature on strawberries. I wish that the book will be very useful to all the stakeholders on strawberries. I complement the authors of different chapters and editors for compiling vast scientific information on strawberries in such a user friendly manner.

Prof. Roderick Drew
Griffith University, Queensland, Australia
&
Ex-President,
International Society for Horticultural Science (ISHS), Belgium

Preface

Strawberries have a peculiar history of evolution. Systematic exploration and collection of parental genotypes to the development of present-day, everbearing cultivars spans over five centuries. But considering evolution in nature on a time scale, this is a very short phase. And yet within a period of less than five centuries, strawberries have gained immense popularity among the masses across the globe. The nutritional qualities of the fruit, such as vitamins, minerals and antioxidants, along with its versatile, uses in processed products, have greatly contributed to its enhanced popularity across the large section of consumers, traders and processors. The strawberry crop is of very short duration and versatile and can be grown both indoors in protected conditions, as well as outdoors in suitable climatic conditions. The special feature of this soft fruit is that unlike many other fruit crops it is very short (almost a creeper), and hence amenable to diverse management techniques. Unlike most other fruit crops, which are perennial and tall with a long gestation period, strawberries are of very short duration crop. The duration from planting to fruit harvest (commencement) is about 75–80 days. Not only this, the strawberries are ideal for modern techniques of scientific management such as rooftop gardening, vertical gardening, soilless culture and hi-tech protected cultivation. In addition, it also fits well in crop diversification especially as an intercrop in perennial fruit crop orchards, and makes possible the concept of interspace utilization for cultivating "fruits within fruits" ensuring enhanced return per unit of land cultivated.

As such, there has been growing demand for it throughout the year, and various special stores and supermarkets have appropriately chipped in to ensure the supply of strawberry for an extended duration to the consumers. This feat has been due to the concerted efforts of scientists, extension functionaries and need-based support from policy implementing agencies. But the demand for increased fruit production is increasing day by day due to enhanced consumer awareness. This provides the opportunity to further take up cutting edge research for developing varieties not only with increased productivity, but also tolerant to several biotic and abiotic stresses matching with hi-tech production to maintain pace with the growing demand. Further increased awareness facilitates competitive prices to different classes of stakeholders, and most importantly, assured increased income to growers. The availability of objective-specific cultivars has attracted several fruit growers to opt for strawberries as their choice of crop, leading to extension of this crop to non-traditional areas such as subtropical regions of India. The beginners are in the search of basic technical know-how, and the intelligently equipped growers are trying to adopt most advanced production systems. There exists a wide gap in the technological needs of strawberry growers and other stakeholders in various parts of the world.

To cater to the needs of students, teachers, extension functionaries, farmers and traders for scientific crop management techniques on strawberry, a modest attempt was made by the first two editors of this book, who wrote *Modern Strawberry Cultivation* during the year 2000. But considering the rapid advancements in the strawberry sector, as is also evident from the facts and figures presented in forthcoming chapters of this book, there has been a long-felt need to collect and compile the available scientific information worldwide to update the knowledge of one and all concerned with strawberry industry. We have attempted to bring all the latest available scientific information together, covering basic aspects to the most advanced production technologies, into a single volume. We have emphasized keeping the language and style of presentation simple for the easy understanding of the readers. Efforts have also been made to make the compilation free of errors but if there are a few, we humbly apologize and request suggestions to make needed improvements in revised editions in future.

Maximum credit in shaping this book goes to those who have conducted research under the most difficult situations and generated the vast scientific information on various aspects of strawberry. We are really indebted to one and all of them, whose works and publications have been referred to with due credit in the references section. We are highly thankful to all the authors who have readily agreed and contributed various chapters of topical interest. We wish to thankfully acknowledge the support of our office for granting the requisite permission for revising and compiling the book in present form. It would not have been possible for us to complete this project without loving inspiration from Rinki, Bobby, Chinki, Lovely, Shivangi and Mitthi. The sincere help extended by Mr Daya Shankar, ICAR-IARI, New Delhi on various aspects of computer assistance is thankfully acknowledged. Lastly, we would like to thank Mr Anand Srivastava for drawing some figures incorporated in this manuscript.

We hope that this compilation – *Strawberries: Production, Postharvest Management and Protection* – will be of immense use to all stakeholders including scientists, extension functionaries, students, growers, traders and policy makers alike, and facilitate production and utilization of strawberries in various agroclimatic regions of the world.

R.M. Sharma, R. Yamdagni, A.K. Dubey, and V. Pandey

Editors

Prof Radha Mohan Sharma did his doctorate in Pomology at the Dr Y.S. Parmar University of Horticulture and Forestry, Solan, HP (India) in 1996, and became Assistant Professor at SKUAST J&K in 1999. Prof. Sharma has headed the Regional Horticultural Research Sub-Station, Doda for five years, and served at SKUAST-Jammu for 13 years in various capacities. He has also served at RVSKVV, Gwalior, MP (India) as Professor (Hort.) for one year. Presently, Prof Sharma is working as Principal Scientist in the Division of Fruits & Horticultural Technology, ICAR-Indian Agricultural Research Institute, New Delhi (India). Prof Sharma has about 14 years research experience of temperate fruits, and has exclusively dealt with the research on strawberry, walnut, apple and pear. Presently, he is engaged with the research on citrus improvement and rootstock research. Establishment of field gene bank of temperate fruits, identification of elite clones of walnut and *Ambri* apple, varietal evaluation and standardization of propagation technique of strawberry in subtropical areas and development of two acid lime varieties are among his credits. Prof Sharma has 13 years experience of teaching undergraduate and postgraduate students, and he has guided 5 MSc (Fruit Science) students. He has published three books, 12 book chapters and 50 research papers in national and international journals. He is a Fellow of Horticultural Society of India.

Prof Rakesh Yamdagni did his Doctorate in Horticulture at the Government Agriculture College, Kanpur under Agra University (now C. S. Azad University of Agriculture and Technology, Kanpur) in 1969, and worked at the Punjab Agriculture University, Hisar (now Chaudhary Charan Singh Haryana Agricultural University, Hisar) as Assistant Professor in 1967. Prof Yamdagni served at Haryana Agriculture University, Hisar for more than 30 years in the capacities of Head, Department of Horticulture, Head, Department of Forestry, Regional Director of Research for Dryland Agriculture and Director of Extension Education. He also worked as Vice Chancellor, Narendra Dev University of Agriculture and Technology, Faizabad (UP) for three years. Prof Yamdagni has published one book, more than 100 research papers and 150 articles. He guided 13 PhD and 9 MSc students during his academic carrier.

Dr Anil Kumar Dubey did his Doctorate in Horticulture at the C. S. Azad University of Agriculture and Technology, Kanpur in 1996, and joined Agriculture Research Services in 1995. Presently, Dr Dubey is working as Principal Scientist at the Division of Fruits & Horticultural Technology, ICAR-Indian Agricultural Research Institute, New Delhi. Dr Dubey has made an outstanding contribution in the fields of mango and citrus improvement and their rootstock standardization. He has developed nine varieties of fruit crops including mango (four) and citrus (five), standardized softwood grafting for the Khasi mandarin, identified salt tolerant rootstocks for citrus fruits and standardized the gamma irradiation doses and ovule age for maximum recovery of haploids in citrus. He has guided seven MSc and three PhD students. In addition, he has published 87 research papers in national and international journals, 3 books, 3 practical manuals, 5 technical bulletins, 27 popular articles and 17 e-publications. He is the recipient of Rajiv Gandhi Gyan Vigyan Maulik Pusthak Lekhan Purshkar (Pratham Purshkar) from Ministry of Home Affairs, Govt of India and Dr Shyam Singh Best Scientist Award for Citrus from ISC. He is a Fellow of the Horticultural Society of India.

Dr Vikramaditya Pandey did his Doctorate degree in Horticulture at the Banaras Hindu University, Varanasi (UP) in 1998, and joined the Agricultural Research Service in 1996. He has served at Central Horticultural Experiment Station, Bhubaneswar (Odisha), National Research Centre for Banana, Trichy (TN) and Indian Institute of Vegetable Research, Varanasi (UP) in various capacities. Presently, Dr Pandey is working as Principal Scientist (Horticulture) at the Horticultural Science Division of the Indian Council of Agricultural Research, and looking after the national coordination and monitoring of various research and development issues on horticultural crops. He was involved in evolving three varieties, one each of mango, custard apple and chilli. He has worked on various aspects of improvement, production, integrated cropping system and natural resource management in horticultural crops with special reference to fruits and vegetables. Dr Pandey has published about 40 research papers, 4 books, 10 book chapters, 6 technical bulletins, 21 extension folders and 24 popular articles. He has visited four central Asian countries to explore the possibilities of developing collaborative research projects on crops and aspects of mutual interest of partner countries.

Contributors

D. P. Abrol
Faculty of Agriculture
Sher-e-Kashmir University of Agricultural
    Sciences and Technology of Jammu
Jammu, India

O. P. Awasthi
Division of Fruits & Horticultural Technology
ICAR – Indian Agricultural Research Institute
Pusa, India

Sreshti Bagati
School of Biotechnology
Sher-e-Kashmir University of Agricultural
    Sciences and Technology, Jammu
Jammu, India

Julie Dogra Bandral
Division of Food Science & Technology,
    Faculty of Agriculture
Sher-e-Kashmir University of Agricultural
    Sciences and Technology of Jammu
Jammu, India

Parshant Bakshi
Advanced Centre for Horticulture Research
Sher-e-Kashmir University of Agricultural
    Sciences and Technology of Jammu
Jammu, India

Kaushik Banerjee
National Referral Laboratory ICAR-National
    Research Centre for Grapes
Pune, India

Tanjeet Singh Chahal
Punjab Agricultural University Fruit Research
    Station, Jallowal-Lesriwal
Jalandhar, India

A. S. Charak
Krishi Vigyan Kendra, Doda
Sher-e-Kashmir University of Agricultural
    Sciences and Technology of Jammu
Jammu, India

Biswajit Das
ICAR – Research Complex for NEH Region
Tripura Centre
Lembucherra, India

A. K. Dubey
Division of Fruits & Horticultural Technology
ICAR – Indian Agricultural Research Institute
Pusa, India

Vishal Dubey
Agriculture Marketing and Agriculture Foreign
    Trade Department
Jhansi, India

Sudhakar Dwivedi
Division of Agriculture Economics &
    Statistics, Faculty of Agriculture
Sher-e-Kashmir University of Agricultural
    Sciences and Technology of Jammu
Jammu, India

A. K. Goswami
Division of Fruits & Horticultural Technology
ICAR – Indian Agricultural Research Institute
Pusa, India

Tushar Uttamrao Jadhav
Pune, India

Mahital Jamwal
Regional Horticultural Research Substation
Sher-e-Kashmir University of Agricultural
    Sciences and Technology of Jammu
Doda, India

S. K. Jena
ICAR – Indian Institute of Water Management
Bhubaneswar, India

Prashant Kalal
Division of Fruits & Horticultural Technology
ICAR – Indian Agricultural Research Institute
Pusa, India

Poonam Kashyap
ICAR – Indian Institute of Farming Systems
    Research
Modipuram, India

M. S. Khan
Department of Entomology
Gobind Ballabh Pant University of Agriculture
    & Technology
Pantnagar, India

Kiran Kour
Advanced Centre for Horticulture Research
Sher-e-Kashmir University of Agricultural
    Sciences and Technology of Jammu
Jammu, India

Manpreet Kour
Division of Instructional Livestock Complex
    Centre, Faculty of Veterinary Sciences &
    Animal Husbandry
Sher-e-Kashmir University of Agricultural
    Sciences and Technology of Jammu
Jammu, India

Dinesh Kumar
Division of Crop Production
ICAR – Central Institute for Sub-Tropical
    Horticulture
Lucknow, India

K. Kumar
Department of Fruit Science
Dr Y.S. Parmar University of Horticulture &
    Forestry
Nauni-Solan, India

Pramod Kumar
Department of Fruit Science
Dr Y. S. Parmar University of Horticulture &
    Forestry
Nauni-Solan, India

Sandeep Kumar
Division of Germplasm Evaluation
ICAR – National Bureau of Plant Genetic
    Resources
New Delhi, India

Sanjeev Kumar*
Uttar Pradesh Council of Agricultural Research
Lucknow, India

Sanjeev Kumar**
Division of Crop Improvement
Indian Institute of Sugarcane Research
Lucknow, India

Sunil Kumar
Division of Fruits & Horticultural Technology
ICAR – Indian Agricultural Research Institute
Pusa, India

Beena Kumari
Department of Entomology
CCS Haryana Agriculture University
Hisar, India

B.V.C. Mahajan
Punjab Horticultural Post-Harvest Technology
    Centre
Punjab Agricultural University
Ludhiana, India

Era Vaidya Malhotra
Tissue Culture and Cryopreservation Unit
ICAR – National Bureau of Plant Genetic
    Resources
New Delhi, India

B. K. Pandey
Horticultural Science Division
Indian Council of Agricultural Research,
    Krishi Anusandhan Bhawan-II
Pusa, India

M. K. Pandey
Division of Plant Breeding & Genetics, Faculty
    of Agriculture
Sher-e-Kashmir University of Agricultural
    Sciences and Technology of Jammu
Jammu, India

V. Pandey
Horticultural Science Division
Indian Council of Agricultural Research,
    Krishi Anusandhan Bhawan-II
Pusa, India

Contributors

Alemwati Pongener
ICAR – National Research Centre on Litchi
Muzaffarpur, India

Satyabrata Pradhan
Division of Fruits & Horticultural Technology
ICAR – Indian Agricultural Research
Institute
Pusa, India

K. K. Pramanick
ICAR – Indian Agricultural Research Institute,
Regional Station
Shimla, India

Gyanendra Kumar Rai
School of Biotechnology
Sher-e-Kashmir University of Agricultural
Sciences and Technology of Jammu
Jammu, India

P. K. Rai
Advanced Centre for Horticulture Research
Sher-e-Kashmir University of Agricultural
Sciences and Technology of Jammu
Jammu, India

R. K. Salgotra
School of Biotechnology
Sher-e-Kashmir University of Agricultural
Sciences and Technology of Jammu
Jammu, India

Seema Sangwan
Department of Microbiology
College of Basic Sciences and Humanities,
CCS Haryana Agriculture Univeristy
Hisar, India

Uma Shankar
Division of Entomology, Faculty of Agriculture
Sher-e-Kashmir University of Agricultural
Sciences and Technology of Jammu
Jammu, India

Arti Sharma
Division of Fruit Science, Faculty of
Agriculture
Sher-e-Kashmir University of Agricultural
Sciences and Technology of Jammu
Jammu, India

A. D. Sharma
Germplasm Conservation Division
ICAR – National Bureau of Plant Genetic
Resources
New Delhi, India

Bindiya Sharma
Zimplistic India Pvt Ltd
Gurugram, India

Laxuman Sharma
Department of Horticulture
Sikkim University
Gangtok, India

Manmohan Sharma
School of Biotechnology
Sher-e-Kashmir University of Agricultural
Sciences and Technology of Jammu
Jammu, India

Nimisha Sharma
Division of Fruits & Horticultural Technology
ICAR – Indian Agricultural Research Institute
Pusa, India

Nirmal Sharma
Division of Fruit Science, Faculty of
Agriculture
Sher-e-Kashmir University of Agricultural
Sciences and Technology of Jammu
Jammu, India

Pawan Kumar Sharma
Krishi Vigyan Kendra, Poonch
Sher-e-Kashmir University of Agricultural
Sciences and Technology of Jammu
Poonch, India

R. M. Sharma
Division of Fruits & Horticultural Technology
ICAR – Indian Agricultural Research Institute
Pusa, India

Sushil Sharma
Division of Agricultural Engineering, Faculty
of Agriculture
Sher-e-Kashmir University of Agricultural
Sciences and Technology of Jammu
Jammu, India

S. D. Sharma
Department of Fruit Science, College of
  Horticulture and Forestry
Dr Y.S. Parmar University of Horticulture and
  Forestry
Neri, Hamirpur, India

Rajesh Kumar Shukla
Department of Horticulture
Govind Ballabh Pant University of Agriculture
  & Technology
Pantnagar, India

Anil Kumar Singh
School of Biotechnology
Sher-e-Kashmir University of Agricultural
  Sciences and Technology of Jammu
Jammu, India

Bikramjit Singh
Krishi Vigyan Kendra, Pathankot
Punjab Agriculture University
Punjab, India

B. P. Singh
ICAR – National Research Centre on Plant
  Biotechnology
Pusa, India

Kuldeep Singh
Division of Fruits & Horticultural Technology
ICAR – Indian Agricultural Research Institute
Pusa, India

Nav Prem Singh
Department of Fruit Science
Punjab Agricultural University
Ludhiana, India

S. K. Singh
Division of Fruits & Horticultural Technology
ICAR – Indian Agricultural Research Institute
Pusa, India

Madhvi Soni
Division of Plant Pathology
ICAR – Indian Agricultural Research Institute
Pusa, India

Monika Sood
Division of Food Science & Technology,
  Faculty of Agriculture
Sher-e-Kashmir University of Agricultural
  Sciences and Technology of Jammu
Jammu, India

Poonam Srivastava
Department of Entomology
Govind Ballabh Pant University of Agriculture
  & Technology
Pantnagar, India

R. M. Srivastava
Department of Entomology
Govind Ballabh Pant University of Agriculture
  & Technology
Pantnagar, India

A. K. Tiwari
Department of Plant Pathology
Govind Ballabh Pant University of Agriculture
  & Technology
Pantnagar, India

V. K. Tripathi
Department of Horticulture
Chandra Shekhar Azad University of
  Agriculture and Technology
Kanpur, India

Tarun Verma
Department of Entomology
CCS Haryana Agriculture University
Hisar, India

Sangita Yadav
Division of Seed Science and Technology
ICAR – Indian Agricultural Research Institute
Pusa, India

Shiv K. Yadav
Division of Seed Science and Technology
ICAR – Indian Agricultural Research Institute
Pusa, India

R. Yamdagni
Gurugram, India

*Chapter 6, 20 & 25
**Chapter 7

Introduction

Strawberries are known to human civilization since pre-historic times but became popular only after introduction in to Europe in the 16th century. They are widely adopted to variable agro-climates across the temperate to sub-tropical regions of the world. Strawberries are essentially creepers and polyploids with basic chromosome number (x=) as seven. The present day strawberry cultivars, *Fragaria* x *ananassa* Duch, have evolved in nature due to hybridization between *Fragaria chiloensis* (diploid) with *F. virginiana* (octaploid). Subsequently, a large number of trait specific varieties amenable to open and protected cultivation have been evolved.

Standardization of protected cultivation technologies has enabled round the year production of strawberries. The nutrient rich fruits with a unique taste are put to varied uses both for fresh and process market. Precocious cropping, high returns per unit area, short duration, feasibility of round the year production and cultural factors cropping have made strawberry an attractive crop. Recent research results have shown a great promise for integration of strawberry cultivation into vertical gardening. Strawberries are an essential element of 'Valentines Day' across the world. Doubling farmers' income can easily be achieved by shifting to crops such as strawberries but this requires scientific knowledge. This book is a valuable contribution in that direction. There is a need to identify best varieties, reduce cost of cultivation, address issues of food safety and explore export.

I feel great pleasure in introducing the book "The Strawberries: Production, Post-harvest Management and Protection" to the stakeholders. It covers most recent and comprehensive scientific information available on various aspects of production and utilization of strawberries. I am confident that the information provided in the book will fulfill the needs of stakeholders and prove very useful to one and all concerned with production and utilization of strawberries across the world. I complement the authors of different chapters and editors for putting in great efforts in compiling latest scientific information on strawberries in a most comprehensive and concise volume.

Dr. N. K. Krishna Kumar
Regional Representative for South and Central Asia
Bioversity International (CGIAR)
NASC Complex, DPS Marg, New Delhi
&
Ex-DDG (Hort. Sci.),
Indian Council of Agricultural Research
Pusa, New Delhi

1 Introduction

V. Pandey, R. M. Sharma, R. Yamdagni,
A. K. Dubey, and Tushar Uttamrao Jadhav

CONTENTS

1.1 Historical Background 1
1.2 Botany and Taxonomic Status 2
1.3 Evolution and Distribution 3
1.4 Crop Improvement and Varietal Wealth 5
1.5 Growing Environments 6
1.6 Production Technology 7
1.7 Protected Cultivation 11
1.8 Flowering and Fruiting 12
1.9 Harvesting, Handling and Processing 13
1.10 Yield 14
1.11 Economics 14
1.12 Plant Protection 14
1.13 Epilogue 15
References 16

Excellent features for fruits such as an attractive appearance, unique taste, high nutritive value, year-round availability even when other fresh fruits are scarce in the market, precocious bearing, short duration of cropping, high return per unit area with an attractive profit margin and varied uses as fresh fruit for direct consumption and processing into several value-added products (Sharma et al., 2013; Nile and Park, 2014) have made strawberry a most sought-after fruit crop among growers and consumers alike throughout world. In addition to its food value, multicoloured and double-flowering genotypes of strawberries evolved through intergeneric and interspecific crosses have great ornamental value with good fruit set. Consequently, strawberries have become an integral part of "balcony gardens" – a modern trend in urban horticulture (Olbricht et al., 2014). Their utility in vertical or terrace gardens in metropolitan cities across the world is now a reality and not just imaginary. In addition, the adaptability of strawberries to selective tropical and subtropical climatic regions has made it an ideal choice in crop diversification programmes, especially if used to fill gaps in young fruit orchards of various types.

1.1 HISTORICAL BACKGROUND

Strawberries are believed to have been in cultivation by Romans even before the beginning of the Christian era. A wild species in the genus *Fragaria*, specifically *Fragaria vesca*, is said to have been observed growing abundantly throughout the Northern Hemisphere long before the development of present-day cultivated strawberries. In the United States, the cultivation of strawberries started in the early 18th century. Strawberry had been observed growing in different parts of the world, and this heart-shaped fruit of love was mentioned by the Roman poets Virgil and Ovid in the first centuries BC and AD, and in England, gardeners have been cultivating strawberries since the 16th century AD (Boriss et al., 2006). In most other European countries too, strawberries have had been in cultivation from some time in the 18th century. The first documented botanical illustration of a strawberry plant appeared as a figure in herbaries in 1454 and the English word "strawberry" comes from the Anglo-Saxon *"streoberie"*, and was not spelt in the modern fashion until 1538.

The genus *Fragaria* was first summarized in the pre-Linnaean literature by C. Bauhin (1623). The authors of the first *Cataloguedu Jardin du Roi* called it merely *"Fraisieretranger"*, or foreign strawberry. Robins, who wrote the *Catalogue* in 1624, thought it to have come from Pannonie, a region lying between the Danube River to the north and the Illyria to the south in Europe, and gave the "foreign strawberry" the botanical name *Fragaria major pannonica*. In 1629, Parkinson called it the "Bohemia strawberry" or *Fragaria major sterilisseubohemica*. By the end of the 15th century, the two strawberries cultivated in gardens were the wood strawberry, *F. vesca*, and the musky strawberry, *Fragaria moschata*, both characterized by small, distinctly flavoured fruits. In all, the early botanists of the 16th century had named three European species: *F. vesca* and its subspecies *Fragaria vesca semperflorens*; *F. moschata*; and *Fragaria viridis*, the green strawberry.

1.2 BOTANY AND TAXONOMIC STATUS

Earlier, Linnaeus (1738) described the strawberry under only a monotypic genus *Fragaria flagellis-reptan* in *Hortus Cliffortianus* but later in *Species Plantarum* (Linneaus, 1753), he described three species including varieties, although several European species earlier not known properly but now known were omitted, and one belonging to *Potentilla* was included (Staudt, 1962).

Duchesne (1766), in his classic book *L'Histoire Naturelle des Fraisiers*, which was revised in 1788, quoted the beginning of strawberry cultivation, and he is credited with publishing the best early taxonomic literature on strawberries (Hedrick, 1919; Staudt, 1962). He maintained the strawberry collection at the Royal Botanic Gardens, having living collections documented from various regions and countries of Europe and the Americas. Later on, he distributed samples to Linnaeus in Sweden.

Taxonomically, strawberries belong to Division: Magnoliophyta; Class: Magnoliopsida; Order: Rosales; Family: Rosaceae; Subfamily: Rosoideae; Tribe: Potentilleae; Subtribe: Fragariinae; Genus: *Fragaria* L. of the Plant Kingdom. The genus *Fragaria* is a group of low perennial creeping herbs distributed in the wild in the temperate and subtropical regions of world (Anon, 1956), which forms a polyploidy series (Potter et al., 2007), from diploid to octoploid with basic chromosome number x = 7. It has a close taxonomic relative under the genus *Potentilla*.

Three flower types exist among octoploid species: pistillate, staminate and hermaphrodite or complete. Plants may also be polygamodioecious, with the first few flowers on predominately pistillate plants producing fruits (Hancock and Bringhurst, 1979). Most modern cultivars have only hermaphrodite flowers. The blossoms are composed of many pistils, each with its own style and stigma attached to a receptacle that develops into a fleshy "fruit" after fertilization of pistils. The true fruit are the nutlets (achenes), each containing one seed, which are found on the surface of the receptacle and result from the fertilization and development of the pistils.

Flowers are white (occasionally tinged with pink), having five petals (*Fragaria iinumae* has 6–8 petals). In some species, male and female flowers are readily distinguished, but in others (e.g., gynodioecious *F. vesca* subsp. *bracteata*), the female flowers have anthers, and are very similar to the bisexual flower (Ashmanet al., 2012). The berries are more diverse, and are mostly used to differentiate species and varieties.

The present *Fragaria* taxonomy (Chapter 4) includes 20 named wild species, three described naturally occurring hybrid species and two commercially important and cultivated hybrid species. The wild species are distributed in the north temperate and Holarctic zones (Staudt, 1989, 1999a, b; Rousseau-Gueutin et al., 2008).

Staudt (1999a, b) has monographed the American and European *Fragaria* subspecies, besides revising the Asian species (Staudt, 1999a, b, 2003, 2005; Staudt and Dickore, 2001). The general distribution of specific ploidy levels within certain continents is used to infer the history and evolution of *Fragaria* species (Staudt, 1999a, b) as detailed in Table 1.1.

TABLE 1.1
Ploidy Level in *Fragaria* and Their Distribution

Ploidy Level	Distribution
2× (diploid)	Central Asia and Far East except ssp. *vesca* occur in both Eurasia and America
4× (tetraploid)	East and South East Asia
6× (hexaploid)	Europe
8×(octoploid)	North and South America (only one species is endemic in the Far East to South Kuriles)

1.3 EVOLUTION AND DISTRIBUTION

In 1712, a French army officer, A.F. Frezier who was returning from Chile saw large fruited strawberry (*Fragaria chiloensis*) at Concepcion and brought five plants to France. He distributed two plants to the cargo master of the ship, one to the King's garden in Paris, one to his superior at Brest and only one kept for himself. An interesting fact is that all the plants brought by Frezier from Chile to France were female, which were easily hybridized with other the octoploid species, *Fragaria virginiana*, to give rise to the present-day large-fruited cultivated strawberry *Fragaria × ananassa* Duch. So, the present-day cultivated strawberry *Fragaria × ananassa* Duch. is considered to have been derived from that hybridization.

The most significant event in the history of the evolution of the strawberry could be attributed to the introduction of *F. virginiana* from eastern North America to Europe during the 16th century. Later on, widespread popularity of the strawberry has arisen during the past six decades as strawberry breeders have utilized the available fruit diversity in breeding and genetic improvement programmes to evolve cultivars for specific purposes (Sharma and Yamdagni, 2000). As per estimate (faostat.fao.org), the total world production of strawberry during 2014 was 8,114,373 tonnes. The leading strawberry-producing countries (Figure 1.1) are spread over Asia, North America, Europe and Africa. China (38.36%) is the leading strawberry producer followed by the United States (16.90%), Mexico (5.66%), Turkey (4.60%), Egypt (3.49%), Spain (3.60%), South Korea (2.59%), Poland (2.50%), Russia (2.33%), Germany (2.08%) and Japan (2.02%).

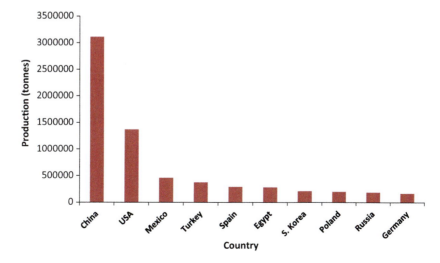

FIGURE 1.1 Top 10 strawberry-producing countries of the world.

In India, strawberry cultivation gained momentum in the late 1960s in Himachal Pradesh and the hills of the then Uttar Pradesh (now Uttarakhand). Presently, strawberry is grown in Maharashtra, Jammu and Kashmir, Himachal Pradesh and Uttarakhand and in the plains of Northern India. In India, it is cultivated over 1000 ha with an annual production of 5000 tonnes (Anon, 2017). Maharashtra is the leading producer of strawberry, and its plantations are mainly concentrated in Mahabaleshwar (hilly regions), which contributes 85% of the total strawberry production of India (Figures 1.2 and 1.3). Nowadays, its cultivation is extending in the Pune, Nashik and Sangli areas of Maharashtra. The "Mahabaleshwar Strawberry" has been granted GI status during 2010 (Joshi, 2016).

China is regarded as one of the richest countries in wild *Fragaria* germplasm resources, and since the 1980s, 103 wild accessions of *Fragaria* genotypes have been collected throughout China and conserved. Of about 20 recognized *Fragaria* species in the world, 11 species, including eight diploids, *F. vesca*, *Fragaria nilgerrensis*, *Fragaria pentaphylla*, *Fragaria gracilis*, *Fragaria nubicola*, *F. viridis*, *Fragaria daltoniana* and *Fragaria mandschurica*, and three tetraploid species, *Fragaria orientalis*, *Fragaria moupinensis* and *Fragaria corymbosa*, have been reported to be distributed in China (Jia Jun et al., 2006, 2014).

During the summers of 1996, 1998 and 1999, prospecting explorations were carried out in southern Chile, and plants, runners and seeds of native *Fragaria* germplasm were collected from the natural ecosystems, as well as from small family orchards, where the white-fruited "Chilean

FIGURE 1.2 Strawberry production in Mahabaleshwar area of Maharashtra (India).

FIGURE 1.3 Vertical strawberry productions in Maharashtra (India).

Introduction 5

strawberry" (*F. chiloensis*) was in cultivation. Within the past two decades, over 20 trips to collect native octoploid strawberry germplasm in North and South America were taken to expand the genetic diversity of *ex situ* collections (Hancock et al., 2002; Gambardella et al., 2002). In the course of the three trips, differences in plant morphological characteristics were found and three different types have been characterized *in situ* (see Chapter 4).

1.4 CROP IMPROVEMENT AND VARIETAL WEALTH

Factors such as precocity of cropping, varied uses of fruits, high profit and versatility of adoptability has made strawberry a choice crop for farmers, especially in areas near large cities. Strawberry has attracted the attention of breeders and other crop improvement scientists over the years due to its economic importance and tremendous inter- and intraspecific genetic variability. As a result, intensive breeding for evolving varieties for specific purposes have been taken up in Europe and the United States. Genetic improvement in strawberry has, in most cases, been achieved by selection from natural seedling populations and subsequent clonal propagation, although it has resulted in a narrow genetic base. Conventional breeding has proved a powerful crop improvement tool, and concerted efforts have resulted in identification of a large number of varieties. Superior varieties have been maintained and clonally multiplied and are being grown in different parts of the world.

Breeding of strawberries was started in the early 1800s, particularly in Great Britain and perhaps the most important early selection of garden strawberry was made by Michael Keen (1806), who developed Keen's Seedling, which dominated English strawberry acreage in the late-19th century, and is in the pedigree of many modern-day cultivars. The British breeder Thomas A. Knight, Charles Hovey of Massachusetts, Arthur Howard, Harlow Rockhill and Albert Etter in California were some of breeders who actually laid the foundation for genetic improvement of the strawberry (see Chapter 5).

Until about 1930, breeding was mostly taken up by private breeders and involved both intraspecific crosses and hybridization of European species with wild species. By the middle of the 20th century, emphasis in strawberry breeding was shifted to many public institutions in other European countries, such as Scotland, Germany, and the Netherlands (Hancock, 1999). Many public institutions in the United States, including the University of California, the University of Florida and the United States Department of Agriculture, and East Malling in the UK started active strawberry breeding programmes in the first half of the 20th century. In the latter half of the 20th century, many private companies such as Driscoll Strawberry Associates, Plant Sciences and California Giant in the United States, Edward Vinson in the UK and Planasa in Spain dominated strawberry breeding, and have significantly contributed to genetic improvement for evolving improved strawberry cultivars worldwide. Subsequent advances in breeding methodologies, over the years, were exploited in enhancing the size and yield of the berries. Thorough understanding of genetics of the crop and inheritance of characters has facilitated crop improvement programmes and has led to development of varieties for specific purposes.

In 1994 and 1995, the main European collection was completed with 900 strawberry cultivars. In 1998, the European COST action 836 (organization of the integrated research in berries: a model for strawberry of quality, in accordance with environment rules and consumer requirements) was started to coordinate the scientific action in 20 partner countries. The overview report of the working group showed that the collections maintained by participating institutes have changed significantly within few years. In 18 institutes, 2747 cultivar accessions and 418 accessions of wild species were preserved. Nearly half of the 928 listed cultivars were grown only at one site and four important cultivars seem to have been lost. Subsequently, 106 cultivars were selected to be maintained in a core collection based on historical significance (Geibel, 2002). The first strawberry hybrid "Hudson" was developed in the United States in 1780.

Climate-specific varieties have extended the availability of strawberries year-round. Besides this, the cultivars for soilless culture have revolutionized the world strawberry industry. Farmers now have a wide option of varieties resistant to pests and diseases, adapted to desired climate, productive over a longer period and producing high-quality fruits (see Chapter 6). In the past few years, DNA marker (morphological, biochemical and molecular) technologies and several PCR-based techniques for multilocus analysis such as RFLP, RAPDs, CAPS, SCAR, AFLPs, ISSRs and SSRs have been standardized, and are widely used for varietal identification, diversity analysis, true-to-type assessment, development of genetic linkage maps and Quantitative Trait Loci (QTL) mapping in strawberry; each method having its own strengths and limitations.

1.5 GROWING ENVIRONMENTS

Strawberry thrives best in a temperate climate but it can flourish in varied climatic conditions from tropical highlands to subarctic regions. It is widely cultivated in plains as well as in hills up to an elevation of 3000 metres above mean sea level in humid or dry regions. It can be grown under protected as well as open conditions in temperate and subtropical climatic conditions with maximum temperature of 22°–25°C during the day and 7°–13°C during night. But in countries such as India, where strawberry-growing areas are temperate and rainfed, it has become acclimatized to subtropical areas with remarkable adaptability and acceptability by growers due to the higher yield and early availability of fruit (December–March) compared with those produced in temperate areas (April–June). The adaptation factors affecting growth, flowering, fruiting, yield and quality are photoperiod, light intensity, chilling and heat units, soil type, temperature and soil water balance (Sherman and Beckman, 2003; Maltoni et al., 2006) and the environment. Environmental factors modify plant responses through the production of endogenous hormones and biochemical compounds in the plants (Hughes et al., 1969; Atkinson et al., 2005).

Strawberry is a short-day plant, which requires exposure to about 10 days of less than eight hours' sunshine for flower initiation. In typical temperate (extreme cool) areas, its plants remain dormant during winter, and low temperatures are required to break their dormancy. With the commencement of spring, the plants resume growth and begin flowering. In varieties grown in subtropical climate, plants start growth after planting (October) and fruiting from December to March.

High growth rates of strawberries are maintained at day temperatures of 22°–23°C. Root temperatures of 18°–20°C tend to increase the growth of aerial parts of plants, while top-root ratio increases with high root temperature (Shoemaker, 1954). The optimum day/night temperatures for leaf and petiole growth is 25°/12°C, while for roots and fruits, it is 18°/12°C. High temperatures are more detrimental to photosynthesis and productivity than moderate or low temperatures. Net CO_2 assimilation rate (A) is markedly reduced at high temperature. Maintenance of higher A at high temperature by an individual genotype indicates its tolerance to heat stress (Kadir et al., 2006). Frost damage is of primary concern during spring as growers take advantage of method of cultivation to produce an early crop (Ourecky and Reich, 1976). Varieties suitable for growing in subtropical areas are more prone to frost injury due to peak flowering during December to January, hence there is need for artificial protection.

Strawberry cultivars are commonly classified on the basis of photoperiod requirement for flower initiation. June-bearers are facultative short-day plants (Darrow, 1936), and initiate flowers in late summer and autumn in northern climates. Everbearers are long-day plants, which initiate flowers during long days under favourable temperatures (Darrow and Waldo, 1934), whereas, day-neutral cultivars can initiate flowers independently of photoperiod (Bringhurst and Voth, 1980; Nicoll and Galletta, 1987).

The critical day length is about 14–15 hours and the optimal temperature for flower induction is around 15°–18°C. For an optimal flower induction, the plants should have at least three weeks (preferably five weeks or more) of short days (Uleberg et al., 2016).

Runners are initiated only when the day length is 12 hours or longer with temperatures above 10°C; however, optimum temperatures for growth are organ specific. Good-quality strawberry production essentially requires good exposure to sunlight, which facilitates advanced berry ripening

Introduction 7

due to higher absorption of solar heat. Besides light duration, light quality significantly influences the growth and development of the strawberry crop. Higher irradiance with LED light is an effective method to improve strawberry yield (Hidaka et al., 2013). The supplementation of ambient light has been reported to improve the fruit quality of strawberry (Choi et al., 2015).

The strawberry can be grown on any type of soil-poor sand to heavy clay, if provided with proper moisture, organic matter and drainage system. The best soil for strawberry production is deep, well-drained sandy loam, well supplied with humus (over 2% organic matter). High organic matter content is highly desirable for strawberry production (Barney et al., 1992). Soil should be porous and rich in humus with enough depth of at least 60 cm. Strawberry roots are generally distributed within 15–40 cm deep soil profile and spread up about 30–100 cm (Mann and Ball, 1927; Hansen, 1931). Strawberry ripens somewhat earlier on sandy than on clay soils.

Heavy clay soils, which are usually poorly drained, encourage disease development and impede precisely timed field operations. Plants established in low-lying muck or organic soils are more vulnerable to frost injury. Strawberries require well-drained soils, so raised beds are used for sites with shallow soils (Barney et al., 1992), and they prefer slightly acidic soils at 5.5–6.5 pH. Soils with higher than 8 pH can adversely affect certain nutrients, especially iron levels, in strawberry plants. For each unit increase in salinity above the threshold value (1 mS/cm), fruit yields decrease by 33%. Saline soils are not appropriate since strawberries have a low tolerance to salts. To overcome soil and water-related constraints, nowadays soilless culture of strawberry is slowly getting momentum. Soil biosolarization, a new technique that combines soil biofumigation and soil solarization, has been observed to be a good practice for soil preparation for cultivating strawberry. In this technique, soil is biofumigated by amendment with chicken manure (25 t/ha), then solarized for 30 days by covering with a clear plastic mulch. This practice can tremendously increase the fruit yield of strawberry.

1.6 PRODUCTION TECHNOLOGY

Strawberry propagation through seeds is not a commercial practice, and is limited to raising hybrid progenies. Stratification of seeds for 11 days at 1°C (Borgman, 1950) or for 60 days at 2.5°C or treatment with 96% H_2SO_4 (Jonkers, 1985) or GA_3 and thiourea (Iyer et al., 1970) has been observed to help enhance germination.

Strawberry plants are commercially grown from the plantlets that are proliferated at the runner nodes of the adult plants. Initially, the runner plants produce few roots, thereafter making a fibrous root system and, once the runner plant is big enough, the stolon is separated from the mother plant, and runner is planted elsewhere. Micropropagation has successfully been exploited for multiplication of strawberries on a large scale.

Use of Chlormequat, SADH, CCC and 6-benzyl-aminopurine (BA) treatments has been observed to increase the number of crowns per plant (Sach et al., 1972; Dwivedi et al., 1999), while application of GA_3 has been observed to enhance the production of runners in many cultivars of strawberries (Sharma et al., 2005; Singh and Tripathi, 2010).

There are four types of quality planting material of strawberries, namely, plug plants, dormant runners, freshly dug runners and test tube (tissue-cultured) plants. "Plug/tray/balled plants" can be procured as such for direct planting or can be raised by planting runner tips in plastic/polystyrene root trainer trays in batches during mid-July to the last week of August (Lieten, 1998). Both fresh plug plants or cold stored/frozen plug plants are used for planting. Planting of strawberry basically consists of three systems, that is, the hill system, matted row and spaced row. The matted row system and annual hill system are the main planting systems commonly followed (Black et al., 2008). Apart from this, single hedge row, double hedge row, single hill row or double hill row etc. can also be followed depending upon requirements and the choice of the growers. In hill system, runners are planted in single or double rows on 15–25 cm raised beds at 30–35 cm spacing within the rows and 20–35 cm between the rows under double row planting. The spacing can be adjusted depending upon the vigour of different cultivars.

Plant spacing of 30 cm×30 cm or 40 cm×30 cm has been observed to be better, but this varies with cultivars (Poling et al., 2005; Nenadovic et al., 2006). The planting densities vary from 40,000–70,000 plants per ha (Duralija et al., 2006; Ahmad, 2009). A proper planting system along with spacing and plant density is optimized for maximization of production of good-quality strawberry fruits. Variations or modifications in the planting methods in matted row or hill row system may be done to adjust according to the requirements of site and agro-climatic conditions. Density also depends on the type of strawberry varieties, day-neutral varieties produce less runners, whereas June-bearing varieties are prolific runner producers. Accordingly, the spacing and plant density are adjusted (Janssen, 2005).

According to Dane et al. (2016), the intercropping of strawberries with legumes may be profitable. Intercropping systems reduce usage of pesticide and nitrogenous fertilizers (Pelzer et al., 2012), and are thus more appealing to the consumers. Intercropped plants serve as biological and physical barriers against pests and diseases (La Mondia et al., 2002). Consequently, reduced pesticide usage, minimized environmental load, meeting the demand of consumers for healthy food etc. are the possible outcomes of judicious intercropping practices. An intercropping system may increase the water use efficiency, but water is needed more to support the yield (Kumar and Dey, 2011). In addition, strawberry has also been observed to be a good intercrop in litchi (Figure 1.4) and young apple (Figure 1.5) orchards. Therefore, interspaces of young orchards, both in temperate or subtropical climatic conditions, could beneficially be made use of in improving yield efficiency, crop diversification and enhanced income of stakeholders. In Ratnagiri area of Maharashtra, India

FIGURE 1.4 Strawberry production as an intercrop in litchi orchards.

FIGURE 1.5 Strawberry as an intercrop in young apple orchard for runner production.

Introduction

low-cost soilless culture has also been developed, which can be shifted to other intended locations, if required (Figure 1.6).

Strawberries return a greater profit from water management practices than most other crops. Efficient water management practice is also useful in utilizing the applied nutrients in the soil, managing abiotic stresses; especially frost damage of the blossoms (Mitra, 1991), hastening the growth of runners and increasing the size of berries and yields of late varieties. An efficient irrigation management system is not only of primary importance for the profitability and sustainability of field strawberry production but it also influences the fruit yield, water use efficiency and diffuse pollution of ground and surface water. In the recent past, adoption of more efficient water management practices such as subsurface drip irrigation (SDI) and use of plastic mulches has greatly improved the water use efficiency in commercial strawberry production.

Irrigation is essential for high yield and mitigating the adverse effects of frost for consistent production of strawberries. Irrigation water with significant salt content, combined with poor soil structure, may also cause development of unacceptable salinity levels in soils. In low-rainfall areas, supplementary moisture is necessary to optimize fruit production. An average of 30 cm of irrigation water is required over the growing season. Water applications may also be needed for mitigating the adverse effects of spring frost and cooling the summer crop when temperatures are above average.

The depth of irrigation varies according to growth stage. In the initial crop stage (after planting), the majority of the roots of strawberries are concentrated within the top 10–12 cm, so the irrigation water should not percolate below 12 cm soil profile. But at the full plant growth stage, the active root depth of strawberry extends approximately up to 25 cm top soil profile. Accordingly, in an ideal irrigation design with an efficient scheduling system, irrigation water should not be allowed to percolate below 10–12 cm in the beginning nor below 25 cm soil profile at full growth. Soil moisture meters/probes are very useful in verifying the correct irrigation schedule and helpful in avoiding deep percolation of water (Espejo et al., 2009).

The duration of each irrigation event varies in different soil textural classes. For lighter soils, the duration is generally less but the frequency is more, whereas, for heavier soils, the duration is more but the frequency of irrigation is less.

The strawberry plants need 1.9–3.2 cm of water per week during the growing season. The amount of water to apply during each irrigation varies with the soil type. In general, 1.25–2.5 cm of water on sandy soil, 1–3.9 cm on loam and 1.5–2 cm on clay soils are used. In the first year, the strawberry plants are capable of drawing water from 18 inches (46 cm) depth (Harrold, 1955) and form extensive and deep root systems (Rom and Dana, 1960). The optimum growth of strawberry plants occurs under the soil moisture conditions where tension does not exceed 1.0 atmosphere. Thus, frequent irrigation, rather than a few heavy ones, is favoured for strawberries (Mortensen, 1934).

FIGURE 1.6 Low cost soilless culture initiated in Maharashtra (India).

Application of 8235 m^3/ha and 10,000 m^3/ha irrigation water to Sabrina and Antilla cultivars of strawberry in Spain resulted in 1027–1084 g/plant and 731–754 g/plant fruit yield, respectively. The land productivity varied from 74.7–78.9 t/ha and 48.7–50.3 t/ha in Sabrina and Antilla cultivars, respectively. Similarly, the water productivity ranged from 16.5–18.3 kg/m^3 and from 13.8–14.3 kg/m^3 water applied in Sabrina and Antilla cultivars, respectively (Lozano et al., 2016). A wide range of consumptive water use from 300 mm (Trout and Gartung, 2004) to 797 mm (Strand, 2008) has been reported in strawberry.

Under adverse situations, waste water with proper treatment can also be utilized for irrigating strawberries.

As the strawberry may be cropped for one year only or for up to three years or more, manuring varies according to the duration of time the plantation is to be retained. It is preferable to apply 50–75 tonnes of farmyard manure per hectare and mix well into the soil before planting strawberry runners. Deep litter or broiler house poultry manure and spent mushroom compost are often available, and useful, but as they contain more nitrogen, phosphorous and potash per tonne than does the farmyard manure, they should not be used at the same rate. Five–7.5 tonnes of deep litter or broiler house manure are equivalent to 50 tonnes per hectare of farmyard manure. It is essential to apply bulky organic manures regularly every three or four years if an annual crop practice on the same site is to be followed (Hughes et al., 1969).

Under most conditions, 454–680 kg of a fertilizer, 2–3% of nitrogen, 6–8% phosphorous and 6–8% of potash should be sufficient for strawberries. On lands where legumes have been grown, a part of the nitrogen may be left out. It should be kept in mind that large applications of commercial fertilizers are profitable only when used on soils in good physical condition and well supplied with humus. Calcium ammonium nitrate should be applied safely due to nitrate toxicity in strawberries (Jackson, 1972). The use of 120 kg/ha of N as slow release fertilizer applied two-thirds in autumn and one-third in spring is good for strawberries.

The effect of foliar fertilization on strawberry yield depends, to a great extent, on the individual cultivar and mulching materials to be used (Karp and Starast, 2002). Foliar spraying with urea is not as common in strawberry as in many other perennial crops but may be a good approach to improve plant N status if soil N availability and/or root uptake are insufficient to meet short-term plant N demand.

Injected solution of fertilizers or fertigation has positive effects on the yield and proportion of the first and second grade of berries (Koszanskiet al., 2005). Therefore, muriate of potash (KCl) fertilizer, which has a low price and high solubility, can be safely used as source of potassium for strawberry plants (Ibrahim et al., 2004). Gutal et al. (2005) studied scheduling of nitrogen through drip irrigation by applying four levels of recommended dose of nitrogen (50, 75, 100 and 125% RDF), and compared with the recommended dose (120:60:60 kg/ha) as band placement.

Mulching in strawberry influences plant growth, yield and fruit quality (Sharma and Singh, 1999). Mulching helps in keeping the fruits clean and protects it from getting in contact with the soil and thereby soil-borne pathogens that cause fruit rot. Mulching also moderates the hydrothermal regime and increases water use efficiency (Verma and Acharya, 1996). Mulching with organic or inorganic materials covers soils and forms a physical barrier to limit soil water evaporation, control weeds, maintain a good soil structure and protect crop from soil contamination. Natural mulches are those derived from animal and plant materials. Oat, hay, rye, paddy straw or wheat straw are the easily available and ideal materials because these are light in weight and loose, and don't smother the tender and herbaceous strawberry plants. With judicious application, these are beneficial and help in maintaining soil organic matter and tilth (Tindall et al., 1991) and provide food and shelter for earthworms and other desirable soil macro- and microorganisms (Doran, 1980). Burning of leftover paddy and wheat straw in the field after harvesting the main crop is an environmental/ecological issue in certain regions, especially in paddy-wheat growing belts of India such as Punjab, Haryana and Uttar Pradesh, and as such, the leftover crop stubble can suitably be utilized for mulching in strawberry.

Introduction **11**

Plastic mulches have beneficially been utilized in strawberry cultivation, especially under protected conditions.

Strawberry is a poor competitor with weeds for light, nutrients, moisture etc. and most weeds that invade strawberry fields are annuals. During stand establishment, little mallow (*Malva parviflora*), bur clover (*Medicago polymorpha*), sweet clover (*Melilotus officinalis*) and filaree (*Erodium cicutarium*) are common weeds as these survive even after fumigation. After planting, grasses and broadleaved weeds with windblown seeds, including thistle (*Onopordum acanthium*) and common groundsel *(Senecio vulgaris)*, become problems. In certain sites, perennial weeds such as field bindweed and bermuda grass or yellow nut sedge (*Cyperus esculentus*) may require control especially in fields, where the crop is carried over into second year of production (Fennimore et al., 2010). Season long uncontrolled weed growth can reduce the strawberry productivity by 51%, and for every 100 g/m^2 increase in weed biomass, fruit yield decreases by 6% (Prittis and Kelly, 2004). Moreover, harvesting fruit by hand presents a challenge in fields infested by weeds. In the *pick your own berries* system of harvesting, weed-free fields are important for repeated sales and customer satisfaction. A weed control programme integrates knowledge of how weeds enter the field (prevention of infestation), culture practices including mulching, and chemical control measures. The primary goal of weed management is to optimize yield by minimizing weed competition.

Judicious use of plant bioregulators improves the growth and yield of strawberries both for production of plant materials and fruits. Quality plant materials are prerequisites for successful strawberry production at a commercial scale. Plant bioregulators have been successfully used to enhance the runner production of strawberry cultivars throughout the world. In many areas, such as in India, constraints of availability of suitable land, undulated topography, inadequate soil fertility and availability of irrigation water in temperate hilly regions; the availability of sufficient quality planting materials is a limiting factor in strawberry production. Use of GA_3 (50–100 ppm) has been observed very encouraging in enhancing the runner production in several cultivars (Franciosoi et al., 1985; Singh and Tripathi, 2010). Similarly, enhanced growth of strawberry plant has also been reported with increasing NAA level up to 35 ppm (Mir et al., 2004).

There is improvement in fruit productivity and quality of strawberry with the application of 50–100 ppm GA_3 (Singh and Tripathi, 2010; Tripathi and Shukla, 2010). Treatment of strawberry plants with BA is well known for its effects on runner growth (Waithaka and Dana, 1978). The use of BA (45–50 ppm) has been reported to enhance the regeneration capacity of the Rabunda (Elizalde et al., 1979), Gorella and Chandler cultivars of strawberry (Sharma et al., 2005).

1.7 PROTECTED CULTIVATION

Most strawberries are grown in the open field, but plastic tunnels and greenhouses have become popular for extending the season and/or off-season fruit production in many parts of Europe, North Africa, Israel, Korea and Japan, particularly in areas with mild winters or runner production for new plantations. Greenhouses are frequently used to hasten ripening, but they can also be used for strawberry production during fall and winter seasons. Besides, this facility is very much useful for protecting the crops against rains too. Although the percentage of malformed or small fruits produced are increased but this problem can be alleviated by improved pollination (Lopez-Galarza et al., 1997). In the UK, by combining glasshouses, plastic tunnels and field culture, the supply of strawberries is maintained during most of the months of the year (Takeda, 1999). In Europe, the production of strawberry for fresh market in forced and protected conditions, which was initially aimed at enhancing the earliness of June-bearing short-day varieties, is increasing. Nowadays, the objective is to have year-round availability of fruits, forcing and protecting the strawberry crop against adverse weather conditions (Neri et al., 2012) with better control of diseases (Andriolo et al., 2002; Fernandes-Jr et al., 2002). Protected cultivation involves construction of either polyethylene/plastic greenhouses or tunnels/row covers of plastic film over the wire-hoops.

1.8 FLOWERING AND FRUITING

Flowering or onset of reproductive cycle is a very complex process and dependent on a series of events during the growth of strawberry. Plants of some strawberry species, including the parent species of the present-day cultivated strawberry, are dioecious. They have male (staminate) or female (pistillate) flowers on separate plants. In these instances, female plants must be interplanted with male or perfect-flowered plants to facilitate pollination, effective fertilization and production of fruits. However, even in species where the flowers with different sexes are borne on separate plants, some male plants bear few perfect flowers and produce a little fruit. Since the mid-19th century, practically all cultivars have originated through hybridization and human selection. These have been chosen with perfect flowers, hoping to effect self-fertilization. Occasionally, sterile or partially sterile plants are also found, which produce no fruit or no late fruit even though the blossoms appear bisexual. Flowering in strawberries is assumed to be controlled by both promoting and inhibiting processes. Under flower-inducing conditions, certain recognizable biochemical or physiological changes occur such as the photoperiodic control of flowering (Thompson and Guttridge, 1960) through flower inhibitor system produced in the leaves. Decline in the level of auxins after receiving the required number of photoinductive cycles (12–15) is a consequence of the changes from vegetative to floral states (Moore and Hough, 1962). Flowering in strawberries is governed by cultivars such as June-bearer or short-day types, everbearers or long-day types, day-neutrals, remontant types and exceptionally a few being amphiphotoperiodic types; nutrition, internal level and external application of plant bioregulators, photoperiod and temperature.

In early days, the old cultivars with pistillate flowers were common, requiring pollen from staminate flowers for fruit setting (Darrow, 1927, 1937). The flowers in most of the present-day, modern cultivars are hermaphrodite (perfect flowered) ensuring self-fertility although few varieties may have either pistillate flowers only or flowers with few stamens or flowers with stamens that fail to produce pollen, thus practically self-sterile (Eaton and Smith, 1962; Hyams, 1962). For getting maximum size and perfect shape of the fruit, pollination of all the pistils of a flower is necessary. In perfect-flowered cultivars, natural sterility is the primary cause for low fruit set in late-borne flowers; however, if the first flowers develop into nubbins (irregularly shaped berries), yet late-borne flowers produce good berries, the poor fruit development is probably due to inadequate pollination (Darrow, 1966). Pollination results in fertilization of the seed embryo in the ovule that develops rapidly to form well-shaped fruit. Exclusion of insects from strawberry flowers gives decreased yield due to lesser fruit set and more malformed or misshaped fruits (Katayama, 1987; Dag et al., 1994). Under protected cultivation, malformation of the early set fruits in some varieties might be due to poor supply of viable pollen grains (Hughes et al., 1969). Owing to inadequate pollination, growth-promoting substances are not synthesized due to absence of fertilized seed in the achenes and the receptacle in its area fails to grow, subsequently leading to malformed and misshaped fruits (Hughes, 1961, 1962). The percentage of malformed fruits in strawberry is increased due to insufficient pollination (Kruistum et al., 2006).

In Srinagar area of Jammu & Kashmir, India, the main insect visitors to strawberry flowers are honey bees (*Apis cerana)* and *Lasioglossum* sp. (Abrol, 1989; Partap et al., 2000). In addition to pollination, the pollinators such as the bumble bees, *Bombus terrestris*, and honey bees, *Apis mellifera*, can vector a microbial control agent to reduce the incidence of grey mould pathogen, *Botrytis cinerea*, in greenhouse strawberries, resulting in higher yields (Mommaerts et al., 2011; Hokkanen et al., 2015). Recently, methods of artificial pollination and instruments such as contact vibration devices and ultrasonic pollination devices have been developed for effective pollination in strawberries (Shimizu et al., 2015).

Fruit growth of strawberry may be sigmoid or double-sigmoid (Coombe, 1976; Rosati, 1993) depending upon the variety and fruit part used for the measurement of growth. Ozark Beauty cultivar has been observed to show both types of growth patterns under different light conditions and sample frequencies (Mudge et al., 1981; Veluthambi et al., 1985). In strawberry, the time taken for

Introduction 13

the berry to become fully ripe is closely related to temperature (Perkins-Veazie and Huber, 1987), and can vary from 20–60 days, probably reflecting an overall effect of temperature on metabolic rate as determined by enzyme activities. Strawberry ripeness are generally divided into four stages, namely, green, white, pink (or turning) and red.

Despite their non-climacteric nature (Perkins-Veazie, 1988), strawberry fruits undergo distinct changes during ripening. Fruits continue to increase in size during the ripening process (Abeles and Takeda, 1990), which is marked by simultaneous changes in sensory attribute in normal fruits. However, "White Carter" (a mutant cultivar of *Fragaria* × *ananassa*) lacks red pigmentation (anthocyanin) even when fully ripe. This cultivar shows otherwise normal ripening behaviour with respect to rest of the sensory attributes.

1.9 HARVESTING, HANDLING AND PROCESSING

Strawberry is ready for harvesting at four or five weeks after blossoming. The duration of time between first bloom and full bloom may be 10–12 days. A great increase in the number of ripe fruit occurs over the first four–six days of harvest. The strawberry fruits can attain full colour in storage even if detached at the white or pink stage but being a non-climacteric fruit, changes in texture, sugars and acidity fail to develop properly after harvesting. The ripe berries for fresh market and processing are picked every third and fourth day, respectively. Berries must be picked during the cooler part of the day to minimize the ill effect of field heat and it should be done by pinching and twisting, not by pulling. The stem is pinched off about 7–8 mm above the cap. The caps from berries need to be removed, if harvested for processing. Generally, 15–25 pickers are required to pick the fruit of one hectare of strawberry, which also depends on how fast the berries ripen, the crop volume and picker's experience. The harvested berries are arranged in plastic punnets, which are placed in corrugated fibre trays and kept in a cool place to reduce water loss from the berries (Sharma and Yamdagni, 2000).

Strawberry fruits are highly susceptible to mechanical injury, physiological disorders, fungal attack and water loss. The resultant short post-harvest life limits fresh market potential and consumer acceptability. The date of harvest is generally determined on the basis of colour of berry surface. All berries should be harvested at near to the fully ripe (more than three-quarters red coloured) stage as sugar content and eating quality do not improve after harvest. Appearance (colour, size, shape and freedom from defects), firmness, flavour (soluble solids, titratable acidity and flavour volatiles), and nutritional value (vitamin C) are all important quality characteristics. A minimum 7°B soluble solids and/or a maximum 0.8% titratable acidity are reliable indices of fruit maturity for harvest (Kader, 1999).

The most common method to precool strawberry fruits is forced-air cooling, which is the most widely adaptable and fastest method for small-scale operations. Since the fruits are extremely perishable, each berry should be examined properly and graded accordingly, as even minimum infection by *Botrytis cinerea* in the beginning results in rotting of the whole lot subsequently. Post-harvest treatment of fruits with $CaCl_2$ and many other such chemicals improves the shelf-life of fruits. Chitosan – an edible active coatings (EACs) chemical compound – is very useful in extending the shelf life of strawberries. Similarly, hot water treatment, modified atmosphere storage and many such other interventions facilitate successful post-harvest handling of fruits. Strawberry fruits have a maximum storage life of five–seven days at 32°F and 95% relative humidity.

The fruits are amenable to processing in to several value-added products. The processing and value addition not only help in minimizing post-harvest losses but make strawberry cultivation a highly profitable enterprise. Strawberry jam is very popular and the fruits are also utilized (Haffner, 2002) in preparation of products such as juices, squashes, wines, pickles, chutneys, spread, yoghurts and ice creams etc. In the case of ice creams, fresh strawberries are used as toppings or blended as small pieces into semi-processed or finished ice cream products. The bulk of strawberries can be frozen.

1.10 YIELD

Although, strawberry cultivation is basically taken up for fruit production, quality plant production also forms an integral part of plantations. Accordingly, strawberry productivity could be assessed both in terms of runner and fruit production. Yield is a polygenic character and greatly influenced by genetic constitution of varieties and environmental factors including crop management practices. It varies from 14.3 t/ha–78.9 t/ha depending on the cultivars and growing conditions and systems (Lozano et al., 2016; Kumar et al., 2012).

It varies from 14.3 t/ha (cv. Marmalada) in Russia and 15.55 t/ha (cv. Chandler) to 30.9 t/ha (cv. Oso Grande) in India to as high as 43.83 t/ha (cv. Redgauntlet) in Lithuania. Land productivity has been observed to vary from 74.7–78.9 t/ha and 48.7–50.3 t/ha in Sabrina and Antilla cultivars, respectively on one hand and the water productivity from 16.5–18.3 kg/m^3 and from 13.8–14.3 kg/m^3 water applied in Sabrina and Antilla cultivars, respectively (Lozano et al., 2016). Similarly, drip irrigation at 100% of crop ET (1.0 "V" volume of water) has been found to give significantly higher fruit yield (57.07q/ha) when compared with surface irrigation (47.17q/ha). The fruit yield improves further (78.63q/ha) due to increased efficiency of drip irrigation when combined with black polyethylene mulch (Kumar et al., 2005).

Similarly, great variations in terms of land productivity ranging from 6 t/ha (Stevens et al., 2011) to as high as 83 t/ha (Serrano et al., 1992) have been reported. Even in terms of water productivity, there is wide variation ranging from 4 kg/m^3 (Kirschbaum et al., 2003) to as much as 14.67 kg/m^3 of water used (Serrano et al., 1992). Interestingly, great variations in amount of water applied has also been reported ranging from 4071 m^3/ha–15,2014 m^3/ha (Garcia Morillo et al., 2015), which influence the productivity of strawberries.

1.11 ECONOMICS

The strawberry is a short-duration crop, which, if planted in September–October starts giving yield in February–March. A new plantation continues to give yield up to third year, thereafter it needs to be replanted. The average yield per year of strawberry was worked out as 67 quintals per hectare. The yield decreases from 70 quintals per hectare in the first year to 66 quintals in the second year to 64 quintals per hectare in the third year. The average price of strawberry was US$200 per quintal. Thus, gross return works out to be US$15,000, US$14,200 and US$13,800 per hectare for the first, second and third year, respectively. The cost:benefit ratio may vary greatly depending upon variety, season, region and productivity. On an average, it could range from 1:1 to 1:1.2 depending on fixed and variable factors of production and price situation in a particular locality and season.

1.12 PLANT PROTECTION

Strawberry suffers due to several pests and diseases causing considerable losses in quality and quantity of potential fruit productivity (Abrol and Kumar, 2009; Pandey et al., 2012). The entire insect-pest complexes are responsible for causing 15–40% damage (Sharma et al., 2008), resulting into huge losses. These are sucking and lepidopteran insect pests, soil arthropods, diseases, nematodes, birds and mammals. The insect pests comprise aphids, mites, whiteflies, bugs, thrips, cutworms, foliage defoliators, leaf folders, bud weevils, crown borers, white grubs, termites etc. Others include nematodes, snails and slugs and mammals including domestic animals. An effective IPM combining cultural (modifying crop production procedures), physical control (exclusion and hand-picking), biological (use of predators, parasites and pathogens), sanitation, resistant varieties and application of safe pesticides greatly help in management of these pests. There are many new molecules available that are substantially less toxic, selective and more eco-friendly than the conventional pesticides, which can suitably be made use of in pest management programmes (Pandey et al., 2012).

Introduction   **15**

Biological control utilizing natural enemies to suppress insect-pests is now a new thrust area to overcome the ill effects of pesticides.

Strawberry suffers due to several diseases such as powdery mildew (*Sphaerotheca macularis* f. sp. *fragariae*), anthracnose (*Colletotrichum acutatum, Colletotrichum fragariae* and *Colletotrichum gloeosporioides*), verticillium wilt (*Verticillium* spp.), black root rot, red stele (syn. red core, red root rot, brown stele or black stele) caused by *Phytophthora fragariae*, leather rot and crown rot (*Phytophthora cactorum*), rhizopus rot (*Botrytis cinerea* and *Rhizopus stolonifer*), Fusarium wilt (*Fusarium oxysporum* f. sp. *Fragariae*); leaf spots, scorch and blights; and vector-borne viruses such as strawberry crinkle virus (SCV) and strawberry mild yellow edge virus (SMYEV). An integrated approach comprising tolerant cultivars, healthy planting materials, modification of growing conditions, soil solarization, judicious spray of pesticides, monitoring and management of vectors and pesticide residues forms the core principle of successful disease management in strawberries.

Strawberry fruits are consumed fresh; therefore, it is desirable to decontaminate them from pesticide residues to reduce health hazards. A number of insecticides including synthetic pyrethroids (fenvalerate, deltamethrin etc.), organophosphates (dimethoate, chlorpyriphos etc.) and other granular pesticides are applied for management of the quoted insect, mite and nematode pests. Similarly, for the management of diseases, carbendazim, copper oxychloride, mancozeb, sulphur and thiram are generally used. The use of such substances has a great toxicological significance due to the fact that they may cause health hazards. Therefore, many developed countries have started addressing this issue and fixed some maximum residues limits (MRL) usually at 0.01 or 0.005 mg/kg (Gebara et al., 2011).

The strategies to minimize the residues of pesticides in strawberries include judicious pesticide applications and needful regulations such as use of relatively easily degradable pesticides; integration of alternative methods of managing pests and diseases such as bio-agents and bio-pesticides, phytosanitation, crop rotation, resistant varieties and good agricultural practices (GAP); surveillance of pesticide residues in soil and water being utilized for production; regulation on dumping of industrial effluents in river, canals, or any other water bodies used for strawberry production and assessing health costs while deciding pesticide production and use policy of the country under reference.

The processing of food commodities generally implies the transformation of the perishable fresh commodity to a value-added product with a greater shelf life, and is closer to being ready for table use.

Besides pests and diseases, the strawberry crop suffers heavily due to several physiological disorders, viz., malformed berries and nubbins or button berries, albinism, phyllody, June yellows, sunscald, frost injury, tip burn, leaf burn, bronzing and dried calyx disorder. Selection of suitable cultivars, macro- and microclimatic and soil considerations, standard production technologies including GAP and nutrition helps alleviate the problems associated with physiological disorders.

1.13 EPILOGUE

Although strawberry is mainly a crop for temperate climates, but areas near major cities in northern India such as Delhi, Gurugram, Ambala, Chandigarh, Ludhiana and Jammu and other similar places throughout the world have witnessed immense popularity of strawberries among growers and consumers alike. But due to severe hot summers, strawberry cultivation is a risky and costly enterprise in the subtropical zone. Overall, non-availability of quality planting materials at cheaper rates and high investments for irrigation infrastructure, mulching materials, protected structures, availability of labourers and packing materials are the most serious limitations in the area expansion under this crop. Planting materials and berry-packing materials are most costly production inputs in traditional strawberry production.

Nevertheless, in spite of several constraints, strawberry sector has made progress by leaps and bounds during the past six decades and has gained immense popularity among growers, traders, processing industry and consumers mainly because of its typical nutritional values, taste, flavour and varied uses in fresh form and processed food products. As a result, several commercial growers have come forward to invest in high-tech production of strawberries such as soilless culture, vertical gardening, protected cultivation etc. In addition, strawberry is slowly becoming popular in crop diversification both for fruit production and runner production and successfully being integrated with perennial fruit production system.

From the foregoing description, it is evident that there is vast scientific literature available on strawberry but scattered in the form of scientific research papers and reports. There was a long-felt need to compile the huge scientific information available on genetic resources, breeding and crop improvement, crop production and protection technologies under different conditions, post-harvest handling and processing of strawberries. And, going by the popular proverb "necessity is the mother of all inventions", the present book, *Strawberries: Production, Postharvest Management and Protection*, is an outcome of sincere efforts made by all the authors and editors in compiling the worldwide available literature on technologies for strawberry production for the benefit of end users in the most comprehensive style.

However, in an era of such fast-growing scientific advancement and versatility of information technology, it is really a difficult task to say that the information provided in this compilation is complete in all aspects at this date. So the scientific information presented in the book is subject to revision with the advancement of scientific knowledge and our understanding of the subject in the future. Information pertaining to strawberries has been appropriately elaborated in the forthcoming chapters, supported by appropriate tables and figures to make the subject interesting and clearly understandable for students, teachers, researchers, extension workers and policy makers.

REFERENCES

Abeles, F.B. and Takeda, F. 1990. Cellulase activity and ethylene in ripening strawberry and apple fruits. *Scientia Horticulturae*, 42: 269–275.

Abrol, D.P. 1989. Studies on ecology and behaviour of insect pollinators frequenting strawberry blossoms and their impact on yield and fruit quality. *Tropical-Ecology*, 30(1): 96–100.

Abrol, D.P. and Kumar, A. 2009. Foraging activity of *Apis* species on strawberry blossoms as influenced by pesticides. *Pakistan Entomologist*, 31(1): 57–65.

Ahmad, M.F. 2009. Effect of planting density on growth and yield of strawberry. *Indian Journal of Horticulture*,66(1): 132–134.

Andriolo, J.L., Bonini, J.V. and Boemo, M.P. 2002. Acumulacao de materiaseca e rendimento de frutos de morangueiro cultivado emsubstrato com different essolucoes nutritivas. *Horticultura Brasileira*, 20: 24–27.

Anon. 1956. *Wealth of India*. Vol. IV, CSIR, New Delhi, India: 57–60.

Anon. 2017. Release of all India 2016–17 (Third Advance Estimates) of Area and Production of Horticultural Crops. National Horticulture Board, Govt. of India.

Ashman, T.L., Spigler, R.B., Goldberg, M.T. and Govindarajulu, R. 2012. *Fragaria*: a polyploid lineage for understanding sex chromosome evolution. In R. Navajas-Pérez (ed.). *New Insights on Plant Sex Chromosomes*, Nova Science, Hauppauge, New York: pp. 67–90.

Atkinson, C.J., Nestby, R., Ford, Y.Y. and Dodds, P.A.A. 2005. Enhancing beneficial antioxidants in fruits: a plant physiological perspective. *BioFactors*, 23(4): 229–234.

Barney, D.L., Davis, B.B. and Fellman, J.K. 1992. *Strawberry Production: Overview*. The Alternative Agricultural Enterprises publication series was supported by a grant from the Northwest Area Foundation, St. Paul, Minnesota: p.4.

Black, B., Pace, M. and Goodspeed, J. 2008. *Strawberries in the Garden*. Utah State University Cooperative Extension (http://extension.usu.edu/files/publications/publication/H orticulture_Fruit_2008-06pr.pdf)

Borgman, H.H. 1950. Propagation. In R.M. Sharma and R. Yamdagni (eds.). *Modern Strawberry Cultivation*, Kalyani Publishers, Ludhiana, India: pp. 58–62.

Boriss, H., Brunke, H. and Kreith, M. 2006. Commodity profile: Strawberries. Agricultural Issues Center, University of California, Davis, California. Website http://aic.ucdavis.edu/profiles/Strawberries-2006.

Bringhurst, R.S. and Voth, V. 1980. Six new strawberry varieties released. *California Agriculture*, 34: 12–15.

Introduction

Choi, H.G., Moon, B.Y. and Kang, N.J. 2015. Effects of LED light on the production of strawberry during cultivation in a plastic greenhouse and in a growth chamber. *Scientia Horticulturae*, 189: 22–31.

Coombe, B.G. 1976. The development of fleshy fruits. *Annual Review of Plant Physiology*, 27: 207–228.

Dag, A., Dotan, S. and Abdul Razek, A. 1994. Honeybee pollination of strawberry in greenhouses. *Hassadeh*, 74(10): 1068–1070.

Dane, S., Laugale, V., Lepse, L. and Sterne, D. 2016. Possibility of strawberry cultivation in intercropping with legumes: a review. *Acta Horticulturae*, 1137: 83–86.

Darrow, G.M. 1927. Sterility and fertility in the strawberry. *Journal of Agriculture Research*, 34: 393–411.

Darrow, G.M. 1936. Interrelation of temperature and photoperiodism in the production of fruit-buds and runners in the strawberry. *Proceedings of American Society of Horticultural Science*, 34: 360–363.

Darrow, G.M. 1937. *Strawberry Improvement*. USDA Yearbook:pp. 445–495.

Darrow, G.M. 1966. *The Strawberry*. Holt, Rinehart and Winston, New York, Chicago, and San Francisco: p.447.

Darrow, G.M. and Waldo, G.F. 1934. Responses of strawberry varieties and species to the duration of the daily light period. *USDA Technical Bulletin*, 453.

Doran, J.W. 1980. Microbial changes associated with residue management with reduced tillage. *Soil Science Society of America Journal*, 44: 518–524.

Duchesne, A.N. 1766. Histoire naturelle des fraisiers. Didot le jeune, Paris, France. Website http://reader. digitale-sammlungen.de/ resolve/display/bsb10301393.html

Duralija, B., Cmelik, Z., Druzic Orlic, J. and Milicevic, T. 2006. The effect of planting system on the yield of strawberry grown out-of-season. *Acta Horticulturae*,708: 89–92.

Dwivedi, M.P., Negi, K.S., Jindal,K.K. and Rana, H.S. 1999. Effect of bio-regulators on vegetative growth of strawberry. *Scientific Horticulturae*,6: 79–84.

Eaton, G.W. and Smith, M.V. 1962. Fruit pollination. Publication: Ontario Department of Agriculture No. 172.

Elizalde, M.M., De, B., Guitman, M.R. and Biain-de-Elizalde, M.M. 1979. Vegetative propagation in everbearing strawberry as influenced by a morphactin, GA_3 and BA. *Journal of the American Society for Horticultural Science*, 104(2): 162–164.

Espejo, A.J., Vanderlinden, K., Infante, M.J. and Muriel, J.L. 2009. Empleo de sensores de humedad para caracterizar la dinámica de flujo de agua de riego en fresa. *Vida Rural*. 15/Septiembre.

Fennimore, S.A., Daugovish, O. and Smith, R.F. 2010. In IPM Pest Management Guidelines: Strawberry Statewide IPM Program, Agriculture and Natural Resources, University of California.

Fernandes, Jr,F., Furlani, P.R., Ribeiro, I.J.A. and Carvalho, C.R.L. 2002. Producao de frutos e estolhos do morangueiroem diferentessistemas de cultivoemambienteprotegido. *Bragantia*, 61: 25–34.

Franciosoi, R., Salas, P., Yamashio, E. and Duorte, O. 1985. Effect of gibberellic acid on runner formation and yield of strawberry. *Proceeding Tropical Region American Society for Horticulture Science*, 24: 127–129.

Gambardella, M., Infante, R., Lopez Aranda, J.M., Faedi, W. and Roudeillac, P. 2002. Collection of wild and cultivated native *Fragaria* in southern Chile. *Acta Horticulturae*, 567(1): 61–63.

Garcia Morillo, J.G., Martin, M., Camacho, E., Rodriguez Diaz, J.A. and Montesino, P. 2015. Toward precision irrigation for intensive strawberry cultivation. *Agricultural Water Management*, 151: 43–51.

Gebara, A.B., Ciscato, C.H.P., Monteiro, S.H. and Souza, G.S. 2011. Pesticide residues in some commodities: dietary risk for children. *Bulletin of Environmental Contamination and Toxicology*, 86: 506–510.

Geibel, M. 2002. Genetic resources in strawberries in Europe. *Acta Horticulturae*, 567(1): 73–75.

Gutal, G.B., Barai, V.N., Mane, T.A., Purkar, J.K. and Bote, N.L. 2005. Scheduling of irrigation for strawberry through drip. *Journal of Maharashtra Agricultural Universities*, 30(2): 215–216.

Haffner, K. 2002. Postharvest quality and processing of strawberries. *Acta Horticulturae*, 567(2): 715–722.

Hancock, J.F. 1999. Strawberries. Crop production science in horticulture series, No 11. CABI, Wallingford, UK.

Hancock Jr, J.F. and Bringhurst, R.S. 1979. Ecological differentiation in perennial, octoploid species of *Fragaria*. *American Journal of Botany*, 66: 367–375.

Hancock, J.F., Hokanson, S.C., Finn, C.E. and Hummer, K.E. 2002. Introduction of supercore collection of wild octoploid strawberries. *Acta Horticulturae*, 567(1): 77–79.

Hansen, H.C. 1931. Comparison of root and top development in varieties of strawberry. *American Journal of Botany*, 18: 658–673.

Harrold, L.L. 1955. Evapotranspiration rates for different crops. *Journal of American Society of Agricultural Engineers*, 36: 669–672.

Hedrick, U.P. 1919. *Sturtevant's Edible Plants of the World*. Dover, New York (formerly published in 1919 by JB Lyon Co, Albany as *Sturtevant's Notes on Edible Plants*).

Hidaka, K., Dan, K., Imamura, H., Miyoshi, Y., Takayama, T., Sameshima, K., Kitano, M. and Okimura, M. 2013. Effect of supplemental lighting from different light sources on growth and yield of strawberry. *Environmental Control Biology*, 51(1): 41–47.

Hokkanen, H.M., Menzler-Hokkanen, I. and Lahdenpera, M.L. 2015. Managing bees for delivering biological control agents and improved pollination in berry and fruit cultivation. *Sustainable Agriculture Research*, 4(3): 89–102.

https://humanelivingnet.net.

https://commons.wikimedia.org/.../File: Linnaeus_Hortus_Cliffortianus_frontispiece_cr.

Hughes, H.M. 1961. Preliminary studies on the insect pollination of soft fruits. *Experimental Horticulture*, 6: 44.

Hughes, H.M. 1962. Pollination studies. Progress report. M.A.A.F., N.A.A.S., Rep. Rept. Efford Exp. Hort. Stat. UK.

Hughes, H.M., Duggan, J.B. and Banwell, M.G. 1969. Effect of potassium and magnesium levels on performance indicators of *strawberry* plants grown in hydroponic. Strawberry Bull. HMSO 10, 6d, Min. Agric. Fish Food, UK.

Hyams, E. 1962. *Strawberry Growing Complete: A system of Procuring Fruit throughout the Year.* Faber and Faber, London, UK.

Ibrahim, A., AbdelLatif, T., Gawish, S. and Elnagar, E. 2004. Response of strawberry plants to fertigation with different K-KCl/K-KNO$_3$ combination ratios. *Arab Universities Journal of Agricultural Sciences*, 12(1): 469–480.

Iyer, C.P.A., Chacko, E.K. and Subramaniam, M.D. 1970. Ethrel for breaking dormancy of strawberry seeds. *Current Science*, 39: 271–272.

Jackson, D.C. 1972. Nitrate toxicity in strawberries. *Agrochemophysica*, 4: 2–45.

Janssen, D. 2005. Growing Strawberries. University of Nebraska-Lincoln. Extension. http://lancaster.unl.edu

Jonkers, H. 1985. Accelerated flowering of strawberry seedlings. *Euphytica*, 1: 41–46.

Joshi, H. 2016. Mahabaleshwar strawberry gets GI status (14 May 2010) *Business Standard*. Retrieved 27 January 2016.

Kader, A.A. 1999. Fruit maturity, ripening, and quality relationships. *ActaHorticulturae*, 485: 203–208.

Kadir, S., Sidhu, G. and Al Khatib, K. 2006. Strawberry (*Fragaria × ananassa* Duch.) growth and productivity as affected by temperature. *Hort Science*, 41(6): 1423–1430.

Karp, K. and Starast, M. 2002. Effects of spring time foliar fertilization on strawberry yield in Estonia. *Acta Horticulturae*, 594: 501–505.

Katayama, E. 1987. Utilization of honey bees as pollinators for strawberries in plastic greenhouse in Tochigi prefecture. *Honeybee Science*, 8(4): 147–150.

Kirschbaum, D.S., Correa, M., Borguez, A.M., Larson, K.D. and Dejon, T.M. 2003. Water requirement and water use efficiency of fresh and waiting bed strawberry plants. *IV International Symposium on Irrigation of Horticultural Crops*, 664: 347–352.

Koszanski, Z., Friedrich, S., Podsiado, C., Rumasz-Rudnicka, E. and Karczmarczyk, S. 2005. The influence of irrigation and mineral fertilization on morphology and anatomy, some physiological processes and yielding of strawberry. *Woda Srodowisko Obszary Wiejskie*, 5(2): 145–155.

Kruistum, G.V., Blom, G., Meurs, B. and Evenhuis, B. 2006. Malformation of strawberry fruit (*Fragaria* x *ananassa*) in glasshouse production in spring. *Acta Horticulture*, 708: 413–416.

Kumar, S. and Dey, P. 2011. Effect of different mulches and irrigation methods on root growth, nutrient uptake, water-use efficiency and yield of strawberry. *Scientia Horticulturae (Amsterdam)*,127(3): 318–324.

Kumar, S., Sharma, I.P. and Raina, J.N. 2005. Effect of levels and application methods of irrigation and mulch materials on strawberry production in North-West Himalayas. *Journal of the Indian Society of Soil Science*, 53(1): 60–65.

Kumar, P.S., Chaudhary, V.K. and Bhagawati, R. 2012. Influence of mulching and irrigation level on water-use efficiency, plant growth and quality of strawberry (*Fragaria × ananassa*). *Indian Journal of Agricultural Sciences*, 82(2): 127–133.

La Mondia, J.A., Elmer, W.H., Mervosh, T.L. and Cowles, R.S. 2002. Integrated management of strawberry pests by rotation and intercropping. *Crop Protection*, 21: 837–846.

Jia Jun, L., Han Ping, D., Chang Hua, T., Ming Qin, D., Mi Zhen, Z. and Ya Ming, Q. 2006. Studies on the taxonomy of the strawberry (*Fragaria*) species distributed in China. *Acta Horticulturae Sinica*, 33(1): 1–5.

Lei, J.J., Xue, L., Dai, H.P. and Deng, M.Q. 2014. Taxonomy of Chinese *Fragaria* species. *Acta Horticulturae*, 1049: 289–294.

Lieten, F. 1998. Recent advances in strawberry plug transplant technology. *Acta Horticulturae*, 513: 383–388.

Introduction

Linnaeus, C. 1738. Classes plantarum, seu systema plantarum omnia a fruictificatione desumpta ... secundum classes, ordines et nomina generic cum clave cuiusvis methodi et synonymis genericis. Conrad Wishoff, Leiden, the Netherlands.

Linnaeus, C. 1753. *Species Plantarum*, 1st edn. Impensis Laurentii Salvii, Holmiae.

Lopez-Galarza, S., Maroto, J.V., San Baautista, A. and Alagarda, J. 1997. Performance of waiting-bed strawberry plants with different number of crowns in winter plantings. *Acta Horticulturae*, 439: 439–448.

Lozano, D., Ruiz, N. and Gavilan, P. 2016. Consumptive water use and irrigation performance of strawberries. *Agricultural Water Management*, 169: 44–51.

Mann, C.E.T. and Ball, E. 1927. Studies in the root and shoot growth of the strawberries. *Journal of Pomology and Horticultural Science*, 5: 149–169.

Maltoni, M.L., Magnani, S., Baruzzi, G. and Faedi, W. 2006. Studio sulla variabilita delle principali caratteristiche qualitative e nutrizionali della fragola. *Rivista di Frutticoltura e di ortofloricoltura*, 68(4): 56–60.

Mir, M.M., Barche, S., Kirad, K.S. and Singh, D.B. 2004. Studies on the response of plant growth regulator on growth, yield and quality of strawberry (*Fragaria x ananassa* Duch.) cv. Sweet Charley. *Plant Archives*, 4(2): 331–333.

Mitra, S.K. 1991. Strawberry. In: T.K. Bose, D.S. Rathore and S.K. Mitra (eds.). *Temperate Fruits*, Hort. Allied Pub., Calcutta, India: pp. 549–596.

Mommaerts, V., Put, K. and Smagghe, G. 2011. Bombus terrestris as pollinator-and-vector to suppress Botrytis cinerea in greenhouse strawberry. *Pest Management Science*, 67(9): 1069–1075.

Moore, J.N. and Hough, L.F. 1962. Relationships between auxin levels, time of floral induction and vegetative growth of the strawberry. *Proceedings of the American Society for Horticultural Science*, 81: 255–264.

Mortensen, F. 1934. Runner plant production in southwest Texas. *Proceedings of the American Society for Horticultural Science*, 32: 424–428.

Mudge, K.W., Narayanan, K.R. and Poovaiah, B.W. 1981. Control of strawberry fruit set and development with auxins. *Journal of American Society of Horticultural Science*, 106: 80–84.

Nile, S.H. and Park, S.W. 2014. Edible berries: bioactive components and their effect on human health. *Nutrition*, 30: 134–144.

Nenadovic, M.E., Milivojevic, J. and Urovic, D. 2006. The influence of planting distance on the fruit characteristic of newly introduced strawberry cultivars. *Vocarstvo*, 40(2): 123–132.

Neri, D., Baruzzi, G., Massetani, F. and Faedi, W. 2012. Strawberry production in forced and protected culture in Europe as a response to climate change. *Canadian Journal of Plant Sciences*, 92: 1021–1036.

Nicoll, M.F. and Galletta, G.J. 1987. Variation in growth and flowering habits of June bearing and everbearing strawberries. *Journal of the American Society for Horticultural Science*, 112, 872–880.

Olbricht, K., Pohlheim, F., Eppendorfer, A., Vogt, F. and Rietze, E. 2014. Strawberries as balcony fruit. *Acta Horticulturae*, 1049: 215–218.

Ourecky, D.K. and Reich, J.E. 1976. Frost tolerance in strawberry cultivars. *Hort Science*, 11: 413–414.

Pandey, M.K., Shankar, U. and Sharma, R.M. 2012. Sustainable strawberry production in sub-tropical plains. In: D.P. Abrol and U. Shankar (eds.).*Ecologically Based Integrated Pest Management*, New India Publishing Agency, New Delhi, India: pp. 787–820.

Partap, U., Matsuka, M., Verma, L.R., Wongsiri, S., Shrestha, K.K. and Partap, U. 2000. Pollination of strawberry by the Asian hive bee, *Apis cerana*. Asian bees and beekeeping: progress of research and development. *Proceedings of the Fourth Asian Apicultural Association International Conference, Kathmandu, Nepal*, 23–28 March: pp. 178–182.

Pelzer, E., Bazot, M., Makowski, D., Corre-Hellou, G., Naudin, C., Al Rifai, M., Baranger, E., Bedoussac, L., Biarnès, V., Boucheny, P. 2012. Pea-wheat intercrops in low-input conditions combine high economic performances and low environmental impact. *European Journal of Agronomy*, 40: 39–53.

Perkins-Veazie, P. 1988. Development and use of an in vitro system to study the ripening physiology of strawberry fruit. Ph.D. Thesis, University of Florida, Gainesville, Florida.

Perkins-Veazie, P. and Huber, D.J. 1987. Growth and ripening of strawberry fruit under field conditions. *Proceeding of Florida State Horticultural Society*, 100: 89–93.

Poling, E.B., Krewer, G. and Smith, J.P. 2005. Southeast Regional Strawberry Plasticulture Production Guide. 1–21. (https://www.uaex.edu/farmranch/crops-commercialhorticulture/docs/Guide%20to%20 Strawberry%20Plasticulture.pdf)

Potter, D., Eriksson, T., Evans, R.C., Oh, S., Smedmark, J.E., Morgan, D.R., Kerr, M., Robertson, K.R., Arsenault, M., Dickinson, T.A. and Campbell, C.S. 2007. Phylogeny and classification of Rosaceae. *Plant Systematics and Evolution*, 266: 5–43.

Prittis, M.P. and Kelly, M.J. 2004. Weed competition in a mature matted row strawberry planting. *Weed HortScience*, 39(5): 1050–1052.

Rom, R.C. and Dana, M.N. 1960. Strawberry root growth studies in fine sandy soil. *Journal of the American Society for Horticultural Science*, 75: 367–372.

Rosati, P. 1993. Recent trends in strawberry production and research: an overview. *Acta Horticulturae*, 348: 23–44.

Rousseau Gueutin, M., Lerceteau Kohler, E., Barrot, L., Sargent, D.J., Monfort, A., Simpson, D., Arus, P., Guerin, G. and Denoyes Rothan, B. 2008. Comparative genetic mapping between octoploid and diploid *Fragaria* species reveals a high level of collinearity between their genomes and the essentially disomic behavior of the cultivated octoploid strawberry. *Genetics*, 179: 2045–2060.

Sach, M., Iszak, E. and Geisenberg, C. 1972. Effect of chlormequat and SADH on runner development and fruiting behavior of summer planted strawberry. *Hort Science*, 7: 364–386.

Serrano, L., Carbonell, X., Save, R., Marpha, O. and Penuelas, J. 1992. Effects of irrigation regimes on the yield and water use of strawberry. *Irrigation Science*, 13: 45–48.

Sharma, R.R. and Singh, S.K. 1999. Strawberry cultivation: a highly remunerative farming enterprise. *Agro India*, 3: 20–22.

Sharma, R.M. and Yamdagni, R. 2000. *Modern Strawberry Cultivation*. Kalyani Publishers, Ludhiana, India: p.172.

Sharma, R.M., Khajuria, A.K. and Kher, R. 2005. Chemical manipulation in the regeneration capacity of strawberry cultivars under Jammu plains. *Indian Journal of Horticulture*, 62(2): 190–192.

Sharma, R.M., Kher, R., Dogra, J., Sood, M., Shankar, U. and Verma, V.S. 2008. Sustainable Strawberry Production in North Indian Plains. Abstracts published in 3rd Indian Horticulture Congress, November 6–9, 2008: 168–169.

Sharma, R.M., Singh, A.K., Sharma, S., Masoodi, F.A. and Shankar, U. 2013. Strawberry regeneration and assessment of runner quality in subtropical plains. *Journal of Applied Horticulture*, 15: 191–193.

Sherman, W.B. and Beckman, T.G. 2003. The climatic adaptation in fruit crops. *Acta Horticulturae*, 622: 411–428.

Shimizu, H., Hoshi, T., Nakamura, K. and Park, J.E. 2015. Development of a non-contact ultrasonic pollination device. *Environmental Control in Biology*, 53(2): 85–88.

Singh, V.K. and Tripathi, V.K. 2010. Efficacy of GA3, boric acid and zinc sulphate on growth, flowering, yield and quality of strawberry cv. Chandler. *Progressive Agriculture*, 10(2): 345–348.

Staudt, G. 1962. Taxonomic studies in the genus *Fragaria*: typification of *Fragaria* species known at the time of Linnaeus. *Canadian Journal of Botany*, 40: 869–886.

Staudt, G. 1989. The species of *Fragaria*, their taxonomy and geographical distribution. *Acta Horticulturae*, 265: 23–33.

Staudt, G. 1999a. Notes on Asiatic *Fragaria* species: *Fragaria nilgerrensis* Schltdl. Ex J. Gay. *Botanische Jahrbücher für Systematik*, 121(3): 297–310.

Staudt, G. 1999b. Systematics and geographic distribution of the American strawberry species: Taxonomic studies in the genus *Fragaria* (Rosaceae: Potentilleae). SERBIULA (sistema Librum 2.0). 81.

Staudt, G. 2003. Notes on Asiatic *Fragaria* species: III. *Fragaria orientalis* Losinsk. and *Fragaria mandshurica* spec. nov. *Botanische Jahrbücher für Systematik*, 124(4): 397–419.

Staudt, G. 2005. Notes on Asiatic *Fragaria* species: IV. *Fragaria iinumae*. *Botanische Jahrbucher fur Systematik*, 126(2): 163–175.

Staudt, G. and Dickore, W.B. 2001. Notes on Asiatic *Fragaria* species: *Fragaria pentaphylla* Losinsk. and *Fragaria tibetica* spec. nov. *Botanische Jahrbucher fur Systematik*, 123: 341–354.

Stevens, M.D., Black, B.L., Lee-Cox, J.D. and Feuz, D. 2011. Horticultural and economic considerations in the sustainability of three cold-climate strawberry production system. *HortScience*, 46: 445–451.

Strand, L.L. 2008. Integrated pest management for strawberries, vol. 3351, UCANR Publication.

Takeda, F. 1999. "Out-of-season" greenhouse strawberry production in soilless substrate. *Advances in Strawberry Research*, 18: 4–15.

Thompson, P.A. and Guttridge, C.G. 1960. The role of leaves as inhibitors of flower induction in strawberry. *Annals of Botany*, 24: 482–490.

Tindall, J.A., Beverly, R.B. and Radcliffe, D.E. 1991. Mulch effect on soil properties and tomato growth using micro-irrigation. *Agronomy Journal*, 83: 1028–1034.

Tripathi, V.K. and Shukla, P.K. 2010. Influence of plant bio-regulators, boric acid and zinc sulphate on yield and fruit characters of strawberry cv. Chandler. *Progressive Horticulture*, 42(2): 186–188.

Trout, T.J. and Gartung, J. 2004. Irrigation water requirements for strawberries. *Acta Horticulturae*, 664: 665–671.

Uleberg, E., Sonsteby, A., Jaakola, L. and Martinussen, I. 2016. Climatic effects on production and quality of berries – a review from Norway. *Acta Horticulturae*, 1117: 259–262.

Veluthambi, K., Rhee, J.K., Mizrahi, Y. and Poovaiah, B.W. 1985. Correlation between lack of receptacle growth in response to auxin accumulation of a specific polypeptide in a strawberry (*Fragaria ananassa* Duch.) variant genotype. *Plant Cell Physiology*, 26: 317–324.

Verma, M.L. and Acharya, C.L. 1996. Water stress indices of wheat in relation to soil water conservation practices and nitrogen. *Journal of Indian Society Soil Science*, 44: 368–375.

Waithaka, K. and Dana, M.N. 1978. Effect of growth substances on strawberry growth. *Journal of the American Society for Horticultural Science*, 103(5): 627–628.

2 Composition, Quality and Uses

Monika Sood and Julie Dogra Bandral

CONTENTS

2.1 Composition 23
2.2 Fruit Quality 23
    2.2.1 Genetic Diversity with Respect to Fruit Quality 25
2.3 Uses 26
    2.3.1 Medicinal Uses 26
    2.3.2 Other Uses 28
References 28

2.1 COMPOSITION

The fruits of strawberries are very popular in the world for being sweet, juicy, tasty and attractively red, with captivating flavour and rich in aesthetic values. The soluble solid content of the fruits comprise sugars, acids and other substances dissolved in the fruit juice. About 80–90% of the soluble solid content consists of sugars, including reducing sugars, fructose and glucose (80–90% in 1:1 ratio), with small amounts of non-reducing sugars, such as sucrose (Forney and Breen, 1985; Kader, 1991). Citric acid is the primary organic acid in strawberry found at all the stages of fruit growth (Kim et al., 1993). The contents of acids in fruit juice largely determine pH, contribute to colour stability and regulate the activity of oxidative enzymes. Strawberry fruit contains an appreciable amount of ascorbic acid, which increases as the fruits attain maturity and ripening (Spayd and Morris, 1981). Strawberry fruits are a rich source of several phytochemicals, minerals and vitamins too. The general composition of fresh strawberry fruits is given in Table 2.1.

2.2 FRUIT QUALITY

The International Organization for Standardization defines quality as "the degree to which a set of inherent characteristics fulfils requirements" (ISO 9000, 2000). Requirements also include consumer expectations. Fully ripe fruits of strawberries have the highest ratings for flavour, sweetness, juiciness, total soluble solids content and ascorbic acid. On the other hand, one-half to three-quarters ripe strawberries have the highest ratings for crispness and firmness (Munbodh and Aumjaud, 2004). Basically, the quality of fresh fruits is a combination of attributes, properties or characteristics that give each commodity value in terms of human food (Kader, 1999), including size and sensory attributes as well as nutritional value and health benefits (Dhaliwal and Singh, 1983; Joolka and Badiyala, 1983). The colour, firmness, flavours and nutritive value of strawberries are related to their composition at harvest, and compositional changes during post-harvest handling. There are large genotypic variations in the physical and chemical composition of strawberry fruits. Thus, there is huge potential to develop new cultivars that have good eating and processing qualities and resistance to physical damage for maintaining their firmness during the course of post-harvest handling. The berry length of strawberry fruits range from 1.66–3.86 cm and breadth from 1.66 cm–2.55 cm depending on cultivars, growing conditions and crop management (Dhaliwal and Singh, 1983) practices. Strawberry cultivars with excellent eating quality have significantly higher total sugar

TABLE 2.1
General Composition of Fresh Strawberry Fruit

Nutrient	Content (per 100 g)
Water (g)	91.57
Energy (kcal)	30
Protein (g)	0.61
Fat (g)	0.37
Carbohydrate (g)	7.02
Total dietary fibre (g)	2.3
Ash (g)	0.43
Minerals	
Calcium (mg)	14
Iron (mg)	0.38
Magnesium (mg)	10
Phosphorus (mg)	19
Potassium (mg)	166
Sodium (mg)	1
Zinc (mg)	0.13
Copper (mg)	0.05
Manganese (mg)	0.29
Selenium (µg)	0.7
Vitamins	
Vitamin C (mg)	56.7
Thiamine (mg)	0.02
Riboflavin (mg)	0.07
Niacin (mg)	0.23
Pantothenic acid (mg)	0.34
Vitamin B_6 (mg)	0.06
Folate (µg)	18
Vitamin A (IU)	27
Vitamin E (mg)	0.14
Total phenolic (mg)	58–210
Total anthocyanin (mg)	55.15
Pelargonidin 3-glucoside	88% of total anthocyanin
Cyanidin 3-glucoside	12% of total anthocyanin

Source: Green (1971), Kader (1991), Kim et al. (1993), USDA (1998).

contents, total sugar/organic acid ratio, sucrose/organic acid ratio and lower total organic acid content than those with poor eating quality. The cultivars that retain their eating quality throughout the harvest season have a significantly lower coefficient of variation (cv) with respect to total sugar content, sucrose/total sugar ratio, glucose/total sugar ratio and citric acid/organic acid ratio than cultivars with variable eating quality (Sone et al., 2000). Strawberry flavour involves perception of the tastes and aroma of many chemical constituents. More sugars and relatively more acids are required for good flavour. High acids and low sugars produce a tart strawberry, while high sugars and low acids result in a bland taste. When both sugars and acids are low, the result is a tasteless strawberry fruit or one with flat taste. Strawberry fruits with high pH (3.27–3.86) help to stabilize attractive colour, while high titratable acidity (0.58–1.35 g/100 g) and TSS/acid ratio (8.52–13.79) contribute the flavour of fruits. High TSS (8.0–11.5 °Brix) makes strawberry fruits ideal for making juice concentrate.

Composition, Quality and Uses

2.2.1 GENETIC DIVERSITY WITH RESPECT TO FRUIT QUALITY

Varietal differences with respect to fruit quality of strawberry were studied by Sharma et al. (1981) and their resultant observations revealed that varieties such as Tioga and Torrey were better in appearance, uniformity, fruit firmness and taste than other cultivars under observation. Strawberry fruit quality is determined by the combination of various constituents. Neither TSS nor acidity can be considered to be solely responsible for better dessert quality. Fruit quality parameters are greatly influenced by the growing conditions too. Hassan et al. (2001) obtained a wide range in fruit TSS (5.2–9.87 °Brix) and titratable acid (0.47–1.28%) content, suggesting Chandler, Ofra and Oso Grande to be the most promising cultivars for warm areas.

Various chemical characteristics such as pH, soluble solids, titratable acidity and volatiles are related to overall quality assessment of fresh strawberry fruit (Alavoine and Crochon, 1989; Douillard and Guichard, 1990). The sugar content of strawberry comprises monosaccharides (glucose and fructose) and disaccharides (sucrose and maltose).

The genetic constitution of a variety has a greater effect on the flavour and chemical composition than the cultivation techniques. Shamaila et al. (1992) reported cv. Totem with deep red colour to be a popular and preferred cultivar by the processing industry. In the sensory attributes, cv. Redcrest has been rated lowest in overall quality because of high sourness and limited sweetness. Minimum pH and high titratable acidity are responsible for its intense sourness. However, high variation among strawberry varieties reveals the potential of available genetic resources to edit for different industrial purposes (Hakala et al., 2002). The quality attributes of some cultivars are summarized in Table 2.2.

In strawberry cultivars, much softer fruit have maximum water-soluble pectin content, so the measurement of fruit pectin methyl esterase (PME) activity could be a criterion in preliminary screening of fruit maturity characteristics (Lefever et al., 2004). As strawberry fruit ripens, the anthocyanin content is generally reported to increase with accompanied decrease in firmness and chlorophyll content (Spayd and Morris, 1981). Most consumers prefer strawberries that do not lose too much juice when sliced. Firmness influences the susceptibility of strawberry fruits to physical damage and consequently their suitability for shipping. Huber (1984) concluded that both polyuronides and hemicelluloses might be important in influencing the texture of strawberry fruits. Morphological defects, such as appearance of the fruits, may originate in the field; fruit malformation is the most serious of them, caused by low temperature, poor pollination or infestations with pests. Minor blemishes, which would not distract from eating quality, are acceptable, but more

TABLE 2.2
Quality Attributes of Some Strawberry Cultivars

Cultivars	Sugars (g/100 g)	Vitamin C (mg/100 g)	TSS (° Brix)	Titratable Acidity (%)	References
Senga Sengana, Jonsok, Korona, Polka, Honeoye and Bounty	5.4–11	32.4–84.7	–	–	Hakala et al. (2002)
Marquise	–	21.37	8.6	–	Munbodh and Aumjaud (2004)
Campinero	–	85	2.7	–	Beatriz et al. (2002)
Diver		40	7.1		
Aromas	–	68.4	7.9	0.69	Yommi et al.
Camarosa	–	83.87	9.4	0.94	(2003)
Gaviota	–	83.25	8.9	0.65	
Selva	–	67.05	8.6	0.75	
Senga Sengana	7.6	83	8.7	0.95	Skupien (2003)

serious defects can influence appearance, firmness, loss of water and susceptibility to decay, and the presence of such defects may either render strawberry fruits unmarketable or could be a cause for sale at reduced prices.

Volatile compounds are essential not only to the aroma but also for the overall flavour of strawberries. The characteristic strawberry aroma is due to presence of a complex network of volatile compounds (Testoni and Lovati, 2004). Seven volatile compounds – ethyl hexanoate, methyl hexanoate, methyl butanoate, ethyl heptanoate, ethyl propionate, ethyl butanoate and 2-hexenylacetate – contribute to much of the aroma associated with ripe strawberry fruit (Guichard et al., 1991). The flavour of the fruit is attributed to the presence of volatile esters. Salicylic acid has been frequently identified in distilled juices but it is not certain whether it pre-exists, as has often been claimed, in the form of methyl salicylate in the juice (Anonymous, 1956). Ripe fruits contain slightly more lipids than the unripe fruits, and contain more oleic acid and less linoleic acid (Couture et al., 1988).

Strawberry possesses remarkable nutritional quality, correlated especially to its high levels of vitamin C, folate and phenolic constituents (Giampieri et al., 2012). Strawberries contain fat-soluble vitamins (i.e. vitamin A and tocopherol) and carotenoids (i.e. lutein and zeaxanthin), besides high vitamin C content. Together with vitamin C, folate plays a crucial role in contributing to the nutritional quality of strawberry fruit and it is one of the richest natural sources of this essential biochemical compound, which has a vital role in health promotion and disease prevention (Tulipani et al., 2008, 2009). Moreover, both the constituents of strawberry fruits, viz., dietary fibre and fructose contents, may contribute significantly to regulate level of blood sugars by retarding digestion, while the fibre content may control calorie intake by its satiating effect on hunger. Finally, strawberry fruits have also been observed to be a good source of micronutrients such as manganese, potassium, iodine, magnesium, copper, iron and phosphorus.

2.3 USES

2.3.1 MEDICINAL USES

Strawberry fruits, like many other berries, are considered to be rich in phytonutrients, namely amino acids (alanine, cystine, glycine, glutamic acid, arginine, aspartic acid, threonine, tryptophan, valine, histidine etc.), phenolics (caffeic acid, ellagic acid, flavonols etc.), vitamins (Vitamin B, C and E), minerals (Mg, Mn, B, K etc.) and dietary fibre, which are essential parts of the human diet for healthy living (Reiger, 2002). Strawberry fruit is considered a good source of micronutrients, such as minerals, vitamin C, folate and phenolic substances, most of which are natural antioxidants and contribute to the high nutritional quality of the fruit. In particular, the phenolic compounds in strawberry fruits are best known for their antioxidant and anti-inflammatory properties. The fruits are considered to directly and indirectly have antimicrobial, anti-allergy and antihypertensive properties, as well as the capacity to inhibit the activities of some physiological enzymes and receptor properties (Giampieri et al., 2013). In addition to traditional nutrients, strawberries are among the richest dietary sources of phytochemicals, mainly represented by phenolic compounds, a large and heterogeneous group of biologically active non-nutrients, showing many non-essential functions in plants and huge biological potentialities in human beings (Hakkinen and Torronen, 2000). The major class of phenolic compounds in strawberry is represented by flavonoids (mainly anthocyanins, with flavonols and flavanols contributing at minor level), followed by hydrolyzable tannins (ellagitannins and gallotannins) as the second most abundant class, and phenolic acids (hydroxybenzoic acids and hydroxycinnamic acids) together with condensed tannins (proanthocyanidins) being the minor constituents. The flavonols identified are derivatives of quercetin and kaempferol, quercetin derivates being the most abundant, while phenolic acids occur as derivatives of hydroxycinnamic acid (i.e. caffeic acid) and hydroxybenzoic acid (i.e. gallic acid). The quercetin and kaempferol contents in strawberry cultivars have been observed to range from 0.7–2.6 mg/100 g and 0.9–2.2 mg/100 g, respectively, in fresh fruit (Mikkonen et al., 2002). Strawberry fruits contain polyphenols (tannins), chlorogenic acid, D-catechin and p-coumaric acid (Wesche-Ebling and Montgomery, 1990). The flavonoids present in strawberry fruits include anthocyanins, flavonols, catechins and proanthocyanidins. Quantitatively, anthocyanins are the most important group. A high content of ellagic acid (as ellagitannins) is a special feature of berries from the Rosaceae family including strawberry. Ellagic acid can be liberated from ellagitannins by efficient acid hydrolysis (Torronen and Matta, 2002). Ellagic acid has been reported to have anti-HIV, anti-carcinogenic, anti-haemorrhage and antioxidant properties (Reiger, 2002).

As regards anthocyanins, more than 25 different anthocyanin pigments have been observed in strawberry fruits across different varieties and selections (Lopes da Silva et al., 2007). Pelargonidin-3-glucoside is the major anthocyanin (60–80%) in strawberry fruits, independent from genetic and environmental factors and the presence of cyanidin-3-glucoside also seems to be constant in strawberries, although only in smaller proportions (Bridle and Garcia, 1997). Ellagitannins together with anthocyanins are the most abundant phenolic compounds in strawberry (Aaby et al., 2005; Kahkonen et al., 2001). Chemically, they are different combinations of hexahydroxydiphenic acid with glucose, with a wide range of structures such as monomers (i.e. ellagic acid glycosides), oligomers (i.e. sanguiin H-6, the most typical ETs in strawberry) and complex polymers.

Strawberry fruits, a rich source of phytochemicals and vitamins, have been highly ranked among dietary sources of polyphenols and antioxidant capacity depending on cultivars, cultivation practices, storage and processing methods, viz., freezing versus dry heat, have been associated with maximum retention of bioactive compounds in strawberry fruits in several studies. Nutritional epidemiology shows an inverse association between strawberry consumption and incidence of hypertension or serum C-reactive protein. Strawberry phenolics show a wide range of biological activities in the prevention of inflammation, oxidative stress, cardiovascular disease (CVD), certain types of cancers, type 2 diabetes and obesity. The addition of berries in the human diet can positively affect the risk factors for oxidation variety of stress and age-related diseases (Basu et al., 2010; Mazza, 2007; Youdim et al., 2000). Further studies are needed to define the optimal dose and duration of strawberry intake in influencing the levels of biomarkers or pathways related to chronic diseases (Basu et al., 2014). The ellagitannin content is actually related to decreased rates of cancer death. In one study, conducted among a group of over 1000 elderly people, strawberries topped a list of eight foods most linked to lower rates of cancer deaths. Those eating the most strawberries were three times less prone to develop cancer than those who eat few or no strawberries (Kahkonen et al., 2001).

The total soluble phenols decrease as strawberries ripen probably due to synthesis of anthocyanins (Anonymous, 1956). Anthocyanins are considered to have pharmacological effects, such as lowering the atherogenic index and decreasing triglyceride and free fatty acid levels. Moreover, anthocyanin inhibits the growth of cancer cells (Mori and Sakurai, 1996). The anthocyanins in strawberry not only provide red colour to its pulp, they also serve as potent antioxidants that have repeatedly been observed to help protect cell structures in the body and to prevent oxygen damage in all the organ systems of the body. The phenol content in strawberry fruits makes them a unique heart protective, an anti-cancer and anti-inflammatory fruit, all rolled into one. The anti-inflammatory properties of strawberries include the ability of phenols in this fruit to minimize the activity of enzyme cyclo-oxygenase, or COX (Kahkonen et al., 2001; Olsson et al., 2006; Boivin et al., 2007).

Strawberry fruits contain 1% achenes on a fresh weight basis but they contribute to about 11% of total phenolics and 14% of total antioxidant activity in strawberries. Ellagic acid, ellagic acid glycosides and ellagitannins are the main contributors to the antioxidant activities of achenes. The major anthocyanin in pulp is pelargonidin-3-glucoside, whereas, achenes contain nearly equal amounts of cyanidin-3-glucoside and pelargonidin-3-glucoside (Aaby et al., 2005; Yoshida and Tamura, 2005).

Certainly, antioxidation is one possible and relevant mechanism, since strawberry consumption significantly decreases oxidative stress, by decreasing malondialdehyde formation, protecting LDL from oxidation and protecting mononuclear blood cells against increased DNA damage (Azzini et al., 2010; Tulipani et al., 2011). Relatively long-term consumption of moderate amounts of berries increases HDL cholesterol, reduces blood pressure and results in favourable changes in platelet functions, thus playing a vital role in the reduction of CVD risks at normal intake levels (Erlund et al., 2008; Giongo et al., 2010).

Moreover, the strawberry has also been recently investigated for its potential contribution to the dietary management of hyperglycaemia linked to type 2 diabetes and related complications of hypertension.

Regarding cancer prevention, some of the known chemopreventive agents present in berries include vitamins (A, C and E and folic acid), minerals (Ca and Se), dietary fibre, carotenoids, phytosterols (beta-sitosterol and stigmasterol), triterpene esters and various phenolic compounds (Duthie, 2007; Seeram, 2008). Strawberry extracts also seem to modulate cell signalling in cancer cells by inhibiting proliferation of several types of cancer cells (Zhang et al., 2008), inducing cell cycle arrest and apoptosis (Boivin et al., 2007; Seeram et al., 2006) and suppressing tumour angiogenesis (Atalay et al., 2003).

However, strawberry consumption also has some limitations. Strawberries contain measurable amounts of oxalates. High oxalates concentrated in body fluids may crystallize and cause health problems. For this reason, individuals with already existing and untreated kidney or gallbladder problems may avoid eating strawberries. Strawberries contain goitrogens, which can interfere with the functioning of the thyroid gland. Cooking may help to inactivate the goitrogenic compounds found in food. However, it is inconclusive as to what extent the goitrogenic compounds are inactivated by cooking or exactly how much risk is involved with the consumption of strawberries by individuals with preexisting and untreated thyroid problems (Meyers et al., 2003).

2.3.2 OTHER USES

Large quantities of strawberry are quick frozen, sliced or preserved whole with sugar in the United States. Strawberries may be made into preserves, jams, jellies, juice, nectar, wine, puree, instant pickle, sweet 'n' sour pickle, sauce, squash, crush, spread and syrup, and may also be canned. Crushed strawberries and strawberry syrup are used in soda fountains in beverages and ice creams.

REFERENCES

Aaby, K., Skrede, G. and Wrolstad, R.E. 2005. Phenolic composition and antioxidant activities in flesh and achenes of strawberries (*Fragaria ananassa*). *Journal of Agricultural and Food Chemistry*, 53(10): 4032–4040.

Alavoine, F. and Crochon, M. 1989. Taste quality of strawberry. *Acta Horticulturae*, 265: 449–452.

Anonymous. 1956. *Wealth of India*. New Delhi, India: CSIR, Vol. IV: pp. 57–60.

Atalay, M., Gordillo, G., Roy, S., Rovin, B., Bagchi, D., Bagchi, M. and Sen, C.K. 2003. Anti-angiogenic property of edible berry in a model of hemangioma. *FEBS Letters*, 544: 252–257.

Azzini, F., Intorre, P., Vitaglione, P., Napolitano, A., Foddai, M.S., Durazzo, A. and Maiani, G. 2010. Absorption of strawberry phytochemicals and antioxidant status changes in human. *Journal of Berry Research*, 2: 81–89.

Basu, A., Rhone, M. and Lyons, T.J. 2010. Berries: emerging impact on cardiovascular health. *Nutrition Reviews*, 68: 168–177.

Basu, A., Nguyen, A., Betts, Nancy M. and Lyons, Timothy J. 2014. Strawberry as a functional food: An evidence-based review. *Critical Reviews in Food Science and Nutrition*, 54: 790–806.

Beatriz, R.C., O do Nascimento, J.R., Genovese, M.I. and Franco, M.L. 2002. Influence of cultivar on quality parameters and chemical composition of strawberry fruits grown in Brazil. *Journal of Agricultural and Food Chemistry*, 50(9): 2581–2586.

Boivin, D., Blanchett, M., Barrette, S., Mogharbi, A. and Beliveau, R. 2007. Inhibition of cancer cell proliferation and suppression of TNF-induced activation of NF kappa B by edible berry juice. *Anti Cancer Research*, 27(2): 937–948.

Bridle, P. and Garcia-Viguera, C. 1997. Analysis of anthocianins in strawberries and elderberries. A comparison of capillary zone electrophoresis and HPCL. *Food Chemistry*, 59: 299–304.

Composition, Quality and Uses

Couture, R., Willemot, C., Avezard, C., Castaigne, F. and Gosselin, A. 1988. Improved extraction of lipids from strawberry. *Phytochemistry*, 7: 2033–2036.

Dhaliwal, G.S. and Singh, K. 1983. Evaluation of strawberry cultivars under Ludhiana conditions. *Haryana Journal of Horticultural Sciences*, 12(1–2): 36–40.

Douillard, D. and Guichard, E. 1990. The aroma of strawberry (*Fragaria ananassa*): characterization of some cultivars and influence of freezing. *Journal of Science of Food and Agriculture*, 50: 517–531.

Duthie, S.J. 2007. Berry phytochemicals, genomic stability and cancer: evidence for chemoprotection at several stages in the carcinogenic process. *Molecular Nutrition and Food Research*, 51: 665–674.

Erlund, I., Koli, R., Alfthan, G., Marniemi, J., Puukka, P., Mustonen, P. and Jula, A. 2008. Favourable effects of berry consumption on platelet function, blood pressure, and HDL cholesterol. *American Journal of Clinical Nutrition*, 87: 323–331.

Forney, C.F. and Breen, P.J. 1985. Growth of strawberry fruit and sugar uptake of fruit discs at different inflorescence positions. *Scientia Horticulturae*, 27: 55–62.

Giampieria, F., Alvarez-Suareza, J.M., Mazzonia, L., Romandinia, S., Bompadreb, S., Diamantic, J., Capocasac, F., et al. 2013. The potential impact of strawberry on human health. *Natural Product Research*, 27(4–5): 448–455.

Giampieri, F., Tulipani, S., Alvarez-Suarez, J.M., Quiles, J.L., Mezzetti, B. and Battino, M. 2012. The strawberry: composition, nutritional quality, and impact on human health. *Nutrition*, 28: 9–19.

Giongo, L., Bozza, E., Caciagli, P., Valente, E., Pasquazzo, M.T., Pedrolli, C. and Costa, A. 2010. Short-term blueberry intake enhances biological antioxidant potential and modulates inflammation markers in over-weight and obese children. *Journal of Berry Research*, 1: 147–158.

Green, A. 1971. Soft fruits. In: A.C. Hulme (ed.). *The Biochemistry of Fruits and their Products*, Academic Press, New York, vol. 2: 375–410.

Guichard, E., Issanchou, S., Descourvieres, A. and Etievant, P. 1991. Pectin concentration, molecular weight and degree of esterification: Influence on volatile composition and sensory characteristics of *strawberry* jam. *Journal of Food Science*, 56: 1621–1627.

Hakala, M., Tahvonen, R., Huspalahti, R. and Lapvetelainen, A. 2002. Quality factors of Finnish strawberries. *Acta Horticulturae*, 567: 727–730.

Hakkinen, S.H. and Torronen, A.R. 2000. Content of flavonols and selected phenolic acids in strawberries and Vaccinium species: influence of cultivar, cultivation site and technique. *Food Research International*, 33: 517–524.

Hassan, G.I., Godara, A.K., Kumar, J. and Huchche, A.D. 2001. Evaluation of different strawberry cultivars under Haryana conditions. *Haryana Journal of Horticultural Sciences*, 30(1–2): 41–43.

Huber, D.J. 1984. Strawberry fruit softening: The potential roles of polyuronides and hemicelluloses. *Journal of Food Science*, 49: 1310–1315.

ISO 9000. 2000. *Quality Management Systems-fundamentals and Vocabulary*. Switzerland: ISO.

Joolka, N.K. and Badiyala, S.D. 1983. Studies on the comparative performance of strawberry cultivars. *Haryana Journal of Horticultural Sciences*, 12(3–4): 173–177.

Kader, A.A. 1991. Quality and its maintenance in relation to post harvest physiology of strawberry. In: Luby, J.J. and Dale, A. (eds.). *The Strawberry into the 21st Century*, Timber Press, Portland, Oregon: 145–152.

Kader, A.A. 1999. Fruit maturity, ripening and quality relationships. *Acta Horticulturae*, 485: 203–208.

Kahkonen, M.P., Hopia, A.L. and Heinonen, M. 2001. Berry phenolics and their antioxidant activity. *Journal of Agricultural and Food Chemistry*, 49(8): 4076–4082.

Kim, J.K., Moon, K.D. and Sohn, T.H. 1993. Effect of PE film thickness on MA storage of strawberry. *Journal of Korean Society of Food Science and Nutrition*, 22: 78–84.

Lefever, G., Vieuille, M., Delage, N.D., Harlingue, A., Monteclerc, J.D. and Bompeix, G. 2004. Characterization of cell wall enzyme activities, pectin composition, and technological criteria of strawberry cultivars (*Fragaria × ananassa* Duch). *Journal of Food Science*, 69(4): 221–226.

Lopes da Silva, F., Escribano-Bailon, M.T., Perez Alonso, J.J., Rivas-Gonzalo, J. and Santos-Buelga, C. 2007. Anthocyanin pigments in strawberry. *LWT – Food Science and Technology*, 40: 374–382.

Mazza, G.J. 2007. Anthocyanins and heart health. *Annali dell'Istituto Superiore di Sanita*, 43: 369–374.

Meyers, K.J., Watkins, C.B., Pritts, M.P. and Lui, R.H. 2003. Antioxidant and antiproliferative activities of strawberries. *Journal of Agricultural and Food Chemistry*, 51(23): 6887–6892.

Mikkonen, T.P., Hukkanen, A.T., Maatta, K.R., Kokko, H.I., Torronen, A.R., Karenlampi, S.O. and Karjalainen, R.O. 2002. Flavonoid content in strawberry cultivars. *Acta Horticulturae*, 567(2): 815–818.

Mori, T. and Sakurai, M. 1996. Riboflavin affects anthocyanin synthesis in nitrogen culture using strawberry suspended cells. *Journal of Food Science*, 61(4): 698–702.

Munbodh, R.S. and Aumjaud, B.E. 2004. Quality attributes of Marquise *strawberry* variety. *Revue-Agricole-et-Sucriere-de-l'-Ile-Maurice*, 83(2/3): 11–17.

Olsson, M.E., Anderson, C.S., Oredsson, S., Berglund, R.H. and Gustavsson, K.E. 2006. Antioxidant levels and inhibition of cancer cell proliferation in vitro by extracts from organically and conventionally cultivated strawberries. *Journal of Agricultural and Food Chemistry*, 54(4): 1248–1255.

Reiger, M. 2002. http://www.uga.edu/fruit/strawbry.htm

Shamaila, M., Baumann, T.E., Eaton, G.W., Powrie, W.D. and Skura, B.J. 1992. Quality attributes of strawberry cultivars grown in British Columbia. *Journal of Food Science*, 57(3): 696–699.

Sharma, R.L., Badiyala, S.D. and Lakhanpal, S.C. 1981. Varietal differences in fruit quality of strawberry (*Fragaria ananassa*). *Haryana Journal of Horticultural Sciences*, 10(3–4): 193–195.

Skupien, K. 2003. Biological value of selected cultivars of strawberry fruit (*Fragaria ananassa* Duch.). *Folia Horticulturae*, 15(2): 167–172.

Sone, K., Mochizuki, T. and Noguchi, Y. 2000. Relationship between stability of eating quality of strawberry cultivars and their sugar and organic acid contents. *Journal of the Japanese Society for Horticultural Science*, 69(6): 736–743.

Spayd, S.E. and Morris, J.R. 1981. Effects of immature fruit and holding on strawberry puree and color stability. *Journal of the American Society of Horticultural Science*, 106: 211–216.

Seeram, N.P. 2008. Berry fruits for cancer prevention: current status and future prospects. *Journal of Agricultural and Food Chemistry*, 56: 630–635.

Seeram, N.P., Adams, L.S., Zhang, Y., Lee, R., Sand, D., Scheuller, H.S. and Heber, D. 2006. Blackberry, black raspberry, blueberry, cranberry, red raspberry, and strawberry extracts inhibit growth and stimulate apoptosis of human cancer cells in vitro. *Journal of Agricultural and Food Chemistry*, 54: 9329–9339.

Testoni, A. and Lovati, F. 2004. La qualita delle fragile in rapport alle aspettative dei consumatori e alle innovazioni di prodotto. *Rivista di Frutticoltura-e-di ortofloricoltura*, 66(4): 47–53.

Torronen, R. and Matta, K. 2002. Bioactive substances and health benefits of strawberries. *Acta Horticulturae*, 567(2): 797–803.

Tulipani, S., Mezzetti, B., Capocasa, F., Bompadre, S., Beekwilder, J., Ric de Vos, C.H. and Battino, M. 2008. Antioxidants, phenolic compounds, and nutritional quality of different strawberry genotypes. *Journal of Agricultural and Food Chemistry*, 56: 696–704.

Tulipani, S., Mezzetti, B. and Battino, M. 2009. Impact of strawberries on human health: insight into marginally discussed bioactive compounds for the Mediterranean diet. *Public Health Nutrition*, 12: 1656–1662.

Tulipani, S., Alvarez-Suarez, J.M., Busco, F., Bompadre, S., Mezzetti, B. and Battino, M. 2011. Strawberries improve plasma antioxidant status and erythrocyte resistance to oxidative hemolysis. *Food Chemistry*, 128: 1880–1886.

USDA. 1998. Nutrient database for standard reference, Release 12. Available at: http://www.nal.usda.gov/fnic/foodcomp. Google Scholar.

Wesche-Ebling, P. and Montgomery, M.W. 1990. Strawberry polyphenoloxidase: Anthocyanin degradation. *Journal of Food Science*, 55: 731–735.

Yommi, A.K., Borquez, S.L., Quipildor, S.L. and Kirschbaum, D.S. 2003. Fruit quality evaluation of strawberry cultivars grown in Argentina. *Acta Horticulturae*, 628(2): 871–878.

Yoshida, Y. and Tamura, H. 2005. Variation in concentration and composition of anthocyanins among strawberry cultivars. *Journal of the Japanese Society for Horticultural Science*, 74(1): 36–41.

Youdim, K.A., Martin, A. and Joseph, J.A. 2000. Incorporation of the edelberry anthocyanins by endothelial cells increases protection against oxidative stress. *Free Radical Biology and Medicine*, 29: 51–60.

Zhang, Y., Seeram, N.P., Lee, R., Feng, L. and Heber, D. 2008. Isolation and identification of strawberry phenolics with antioxidant and human cancer cell anti-proliferative properties. *Journal of Agricultural and Food Chemistry*, 56: 670–675.

3 Structures and Functions

R.M. Sharma, Rajesh Kumar Shukla, and Nav Prem Singh

CONTENTS

3.1 Roots 32
3.2 Crown 32
3.3 Runners 33
3.4 Leaves 33
3.5 Flower 33
3.6 Fruit 34
References 35

The taste of wild strawberries was recognized many centuries ago before they were domesticated and cultivated. It is believed that the Romans and Greeks used to consume strawberries and there is evidence for them having been collected even earlier, probably during prehistoric era. The present-day cultivated strawberries, *Fragaria × ananassa* Duch. are hybrid progenies of *F. chiloensis* and *F. virginiana*, first recorded in France during the 18th century (Sharma and Yamdagni, 2000). Subsequent efforts in genetic improvement have resulted in the evolution of genotypes with a large size and high-quality fruits, which are nowadays cultivated on a large scale in open fields or in protected conditions. In fact, it is grown practically across the length and breadth of the globe under temperate to mild subtropical climatic conditions and even in mild-climate highland areas of many tropical regions. Strawberry fruits are highly prized for their universal appeal to human beings in physical appearance, aroma, and taste. Besides, they are rich sources of vitamins and nutrient elements.

There is some disagreement about the nomenclature of "strawberry". The English word "strawberry" is considered to have been taken from the Old English *"streawberige"* or the Anglo-Saxon *"streoberie"* (or *"stroeberrie"*). Regardless, it was not spelt in the modern fashion until about AD 1538. It is believed that various types of straw that were used traditionally as mulch to suppress weeds in field and keep the berries (fruits) free from contamination of soil gave rise to the name of strawberry. However, some workers are of the opinion that the straw-like appearance of the runners led to its current English name (Pritts and Handley, 1998). However, according to Buczacki (1994), "the name strawberry does not seem to have anything to do with the practice of putting straw mulch around the plants and long predates their cultivation. It probably comes from 'strayberry', for the runners cause young plants to stray from the parents". The genus name, *Fragaria*, derived from the Latin word for strawberry, *"fraga"*. And, *"fraga"* itself is a derivative of *"fragum"*, which means "fragrant" and accurately characterized the olfactory sensation that harvested fruits of strawberry have (Pritts and Handley, 1998).

The strawberry plant is a herbaceous perennial that produces roots, crowns, leaves, stolons (runners), flowers and fruits (berries) in patterns, determined by its genetic makeup (Figure 3.1). Annually, leaves and roots develop at higher points on the crown. Thus, the plant tends to grow out of the ground and develops poor root-soil contact with age (Anon, 2006). Being short statured (<30 cm tall), the strawberry plant creeps on the ground and can be grown for many years (Shoemakers, 1955). Nevertheless, it is increasingly being cropped for less than a year and, at most, not more than three or four years continually at the same place. This is because of the fact that the young plants tend to have highly branched (multiple crowned) plants and bear large and better-quality fruits than older plants. The structural features of a strawberry plant are described in this chapter.

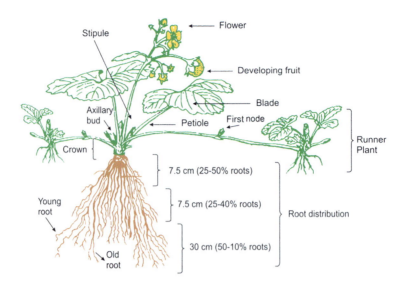

FIGURE 3.1 Structural features of a strawberry flower. (Modified from Yannai, 1988. Training programme on production, planning, cultivation techniques and post-harvest handling. Ministry of Agriculture, State of Israel.)

3.1 ROOTS

The underground part (roots) of strawberry supports the plant and supplies water and mineral nutrients from the soil that are essential for growth and berry production. The roots develop from the crown (Figure 3.1) mainly during late summer and autumn (subtropical region) or spring (temperate region). The length and number of roots formed depends upon soil conditions and planting density. The strawberry plant has a shallow root system, with 80–90% in the top 15 cm of clay, and 50% in the top 15 cm of well-drained sandy loam soils, which result in sensitivity to the levels of salinity and water stresses in the soil. These issues must be addressed, while choosing and preparing the site for planting. The strawberry roots can penetrate the soil up to 30 cm deep with half of the roots located in the lower 15 cm in light sandy soils. Strawberry's root system consists basically of two types of roots, viz., primary roots and feeder roots.

Primary roots supply water and nutrients to the aerial parts and last for more than one season. There are about 20–35 roots/plant, measuring 10–15 cm in length. In subsequent years, more primary roots are regenerated on the crown. With proper management, plants can produce new healthy primary roots above the old ones, thus helping plants that had a poorly developed root system in the past to recover. The new roots are initiated by plants at succeeding higher levels on the crown and when exposed to low temperature or moisture stress may die, while younger plants live.

The feeder roots emerge from the primary roots and these absorb water and nutrients, and survive only for a few days, and are replaced regularly. Feeder roots are highly sensitive to abiotic stresses such as drought and extremes of temperatures. Placing a 2.5 cm thick layer of soil over the plant bed after fruit harvest encourages the development of new roots and renders the plants less vulnerable to these stresses.

The roots also serve as storage sites for starch reserves during winter, which is utilized for vigorous growth, profuse flowering and large berry production (Almadhagi et al., 2014).

3.2 CROWN

The strawberry plant has a compressed and thickened, modified stem referred to as "crown", which has a growing point at the upper end and roots at its base. New leaves and flower clusters emerge from "fleshy buds" in the crown during winter (in subtropical regions) and early spring (in temperate

Structures and Functions

regions). Branch crowns are smaller crowns that emerge from the main crown. Plants can have one or two additional flower clusters on each branch crown. The short and cool days of autumn and the early spring promote the development of the branch crown, and every branch crown adds to the yield of the main crown by producing its own "flower cluster", botanically referred to as "inflorescence". Branch crowns and main crowns become structurally identical. It is very much essential to have enough growth of plants (three or four crowns/plant) for fruiting however; excessive vegetative growth (more than six crowns/plant) of plants may decrease the size of berries (www.hort.cornell.edu). Although early flowering is positively correlated with the size of crown (Fridiaa et al., 2016), the size of crown has been reported a reliable criterion for selecting quality planting materials (Diengngan et al., 2016) too. Therefore, a balance has to be struck in allowing strawberry plantation for vegetative growth for plant (runner/crown) production or commercial fruit production.

3.3 RUNNERS

Strawberry plants produce runners (stolons or daughter plants), which are utilized for clonal propagation of cultivars (Figure 3.1) at commercial scale. Strawberry cultivars vary widely in their ability to produce runners, with day-neutral types typically producing few runners due to their longer fruiting season than short-day types. Runner development is a natural process, which arises from the crown, creating a runner train. More than one runner train may arise from one plant, and one runner train may produce several daughter plants (up to 15 plants) in ideal growing conditions. Temperature plays very crucial role in runner production, and 30–35°C has been observed to be ideal for proliferation of runners. Higher temperature (>40°C) can significantly reduce the production of number of runners per plant. Providing partial shade (50–75%) to the runner bed can moderate the adverse effects of high temperature, particularly in subtropical areas (Sharma et al., 2013). Runner trains on every alternate node form roots, and give rise to the development of new daughter plants. The daughter plants become independent from the mother plant at 30–40 days after rooting.

3.4 LEAVES

Strawberry plants have compound leaves with the leaf blade divided into three separate leaflets (trifoliate) with broad serrations. Leaves are borne on petioles (leafstalks) along the crown and are arranged in spiral fashion around the crown. Leaf size and pattern of serrations are variety specific. The average life of leaf is 1–3 months. By and large, older leaves die during the winter and are replaced by new leaves in the autumn (subtropical regions) or spring (temperate regions). Although leaf emergence takes place throughout the season, with most production occurring during long days, the process may be slowed down under unfavourably extreme temperature conditions (>40°C or <0°C). A well-established leaf canopy system is important to provide energy for flower bud initiation. The death of original leaves in freshly dug plants may be a setback to the plants established, and can delay the fruiting, hence the number of leaves and total leaf area before flower bud initiation are directly related to fruit yield. The leaves with long petiole length are more prone to leaf death following planting than the leaves with short petiole length. Photosynthates synthesized by leaves are translocated into developing fruits and other parts of the plant to support and sustain their further growth. Late in the autumn, sugar is converted to starch in root tissue for winter storage. The strawberry leaves have large numbers of stomata, so they have very high water requirements, particularly during summer (April–June).

3.5 FLOWER

Basically, strawberry inflorescence is a modified stem terminating in a primary blossom. Inflorescence develops terminally and the continued vegetative extension of the crown is taken over thereafter by the uppermost lateral bud, which resumes dominance over lower laterals and displaces the inflorescence to one side. The stolons develop laterally from axillary buds. The strawberry inflorescence is a cyme, with one primary flower subtended by up to 14 low-order flowers in the order of secondary and tertiary flowers. Four types of flowers (primary, secondary, tertiary and quaternary) emerge in the inflorescence of strawberry. The primary or terminal flower produces the largest fruit. In modern cultivars, secondary flowers develop terminally on each of the two or three (or sometimes four) branches beneath the primary flower on the main floral axis. Thereafter, the structure becomes strictly dichasial, with two tertiary flowers forming on each secondary branch, and two quaternaries on every tertiary, and so on (Anderson and Guttridge, 1982).

> **Dichasial cyme or dichasium:** This is a more normal cymose-type inflorescence, in which two lateral branches develop on two sides of the terminal apical (oldest) flowers. The lateral branches may again branch, similarly, after the manner of *biparous* cymose branching. Typical examples are found in jasmine (*Jasminum*), *Nyctanthus*, *Clerodendron* and pink (*Dianthus chinensis* of the family Caryophyllaceae) flowers (Gangulee et al., 1988).

The flowers are white, 2.5–3.75 cm across with 5–10 green sepals, five oval petals, numerous styles and two–three dozen stamens arranged in three whorls. Flowers are complete, usually self-fertile and composed of numerous pistils. The pistils are borne on a roundish or conical flower-supporting stem called the "receptacle", which becomes enlarged after maturity. After fertilization, each pistil develops into a single-seeded fruit (achene). These true fruits are distributed in a Fibonacci spiral pattern (Darnell, 2003) on the outside of the receptacle (edible fleshy tissue). Many *Fragaria* species are dioecious. However, hermaphroditism is also present and ranges in expression from plants with entire cymes that are completely self-fertile to plants where individual flowers are occasionally self-fertile (Hancock, 1999). Almost all the cultivars of strawberries nowadays are completely hermaphroditic, although occasionally pistillate cultivars (e.g., Pegasus, Orleans) are not uncommon.

> The achene is a one-seeded, superior and mature ovary but the pericarp and seed coat are free from each other. More commonly, the achenes are observed to form aggregate fruits. The flowers with apocarous multiple gynoecium are apt to give rise to a number of fruits as there are a number of free ovaries each capable of giving rise to one fruit. But as all these fruits develop from a single flower, they together from an aggregate fruit (better called fruitlets). Sometimes these fruits coalesce together, giving the appearance of a single fruit. while in many cases, the fruits remain free from one another forming a bunch or etaerio of fruitlets. The etaerio of achenes are very commonly found as aggregate fruits and may be seen in *Narvelia* and *Clematis* of the family Ranunculaceae, in which the fruitlets are provided with feathery persistent styles. The rosehip shows a number of achenes within the cup-shaped thalamus, while in strawberries, the minute achenes are buried as small brown specks on a fleshy edible thalamus. On the other hand, the spongy thalamus of the lotus (*Nelumbo lutea*) is also an aggregate of achenes, the fruitlets being sunk into thalamus (Gangulee et al., 1988).

3.6 FRUIT

Botanically, the red fruit called the "berry" is an enlarged flower stem (receptacle) with many seeds (achenes) imbedded in the outer surface. The fruits or berries are in fact swollen receptacles with the true fruits (achenes) on their surface. The berry size is positively related with flower size and the number of floral parts, particularly the number of carpels (Janick and Eggert, 1968; Moore et al., 1970), but can also be modified by a number of environmental factors (Guttridge, 1985).

A medium-sized fruit of most of the varieties of strawberry has about 20–500 achenes. Actually, what looks like seeds really are the "true fruits", properly referred to as achenes. Inside the dry ovary wall of each achene is a real seed (ovule) with the potential of becoming a unique strawberry plant (seedling) (www.hort.cornell.edu). The fruit is an aggregate of achenes or one-seeded fruitlets. It has a receptacle, which accumulates sugars and vitamins and ripens like a true fleshy fruit. Each achene contains a single ovule and can therefore be considered an individual fruit. Besides having a large berry size and more achenes, primary berries are first to ripen, followed by secondary and tertiary berries. Quaternary berries are the smallest (generally unmarketable) and ripen last. From blossoming to ripening, berry development takes 20–35 days, depending on the growing conditions.

REFERENCES

Almadhagi, I.A.U., Hasan, S.M.Z., Ahmad, A. and Yusoff, W.A. 2014. The starch status during growth and development of strawberry plant under tropical climatic conditions. *Acta Horticulturae*, 1024: 115–120.

Anderson, H.M. and Guttridge, C.G. 1982. Strawberry truss morphology and the fate of higher order flower buds. *Crop Research*, 22: 105–122.

Anon. 2006. Midwest Strawberry Production Guide. Bulletin 926, The Ohio State University Extension: pp. 11–18.

Buczacki, S. 1994. *Best Soft Fruit*. Hamlyn an imprint of Reed Consumer Books Limited, London, UK.

Darnell, R.L. 2003. Strawberry growth and development. In: N.F. Childers (ed.). *The Strawberry*. Dr. Norman F. Childers Publications, Gainesville, Florida, USA: pp. 3–10.

Diengngan, S., Mahadevamma, M. and Murthy, B.N. 2016. Efficacy of *in vitro* propagation and crown sizes on the performance of strawberry (*F. x ananassa* Duch.) cv. Festival under field condition. *Journal of Agricultural Science & Technology*, 18 (1): 255–264.

Fridiaa, A., Winardiantika, V., Lee, Y.H., Choi, I.V., Yoon, C.S. and Yeoung, Y.R. 2016. Influence of crown size on plant growth, flowering and yield of day-neutral strawberry cultivars. *Acta Horticulturae*, 1117: 347–353.

Gangulee, H.C., Das, K.S. and Datta, C. 1988. *The Fruit, College Botany Vol. I* (Revised by Sen, S.), Books and Allied Pvt. Ltd, Calcutta 700009, India.

Guttridge, C.G. 1985. *Fragaria ananassa*. In: A.H. Halevy (ed.) *CRC Handbook of Flowering. Volume III*. CRC Press, Inc., Boca Raton, Florida, USA: pp. 16–33.

Hancock, J.F. 1999. *Strawberries*. CABI Publishing, New York, USA.

Janick, J. and Eggert, D.A. 1968. Factors affecting fruit size in the strawberry. *Proceedings of the American Society for Horticultural Science*, 93: 311–316.

Moore, J.N., Brown, G.R. and Brown, E.D. 1970. Comparison of factors influencing fruit size in large-fruited and small-fruited clones of strawberry. *Journal of the American Society for Horticultural Science*, 95: 827–831.

Poling, E.B., *Strawberry Plant Structure and Growth Habit*. http://www.hort.cornell.edu/expo/proceedings/2012/Berries/Berry%20Plant%20Structure%20Poling.pdf

Pritts, M. and Handley, D. 1998. *Strawberry Production Guide*. Northeast, Midwest and Eastern Canada (NRAES-88).

Sharma, R.M. and Yamdagni, R. 2000. *Modern Strawberry Cultivation*. Kalyani Publishers, Ludhiana Punjab, India: p. 172.

Sharma, R.M., Singh, A.K., Sharma, S., Masoodi, F.A. and Shankar, U. 2013. Strawberry regeneration and assessment of runners quality in subtropical plains. *Journal of Applied Horticulture*, 15(30): 191–194.

Shoemakers, J.S. 1955. *Small Fruit Culture*. 3rd edn. McGraw Hill, London, UK.

Yannai, B. 1988. Training programme on production, planning, cultivation techniques and post-harvest handling. Ministry of Agriculture, State of Israel.

4 Origin, Taxonomy and Distribution

A.K. Dubey

The genus *Fragaria* is a group of low perennial creeping herbs distributed in the wild state in the temperate and subtropical regions of the world (Anon, 1956). The cultivated strawberry *Fragaria* × *ananassa* Duch. has been derived from the hybridization of two North American species, *F. chiloensis* (L.) Duch and *F. virginiana* Duch, which was first developed in France in the 18th century. Duchesne in his classical book *L'Histoire Naturelle des Fraisiers* (1766), which was revised in 1788, discussed the beginning of the cultivated strawberry. In 1712, a French army officer, A.F. Frezier, on his way back from Chile to France saw the large-fruited strawberry (*F. chiloensis*) at Concepcion, and brought five plants with him. On arrival in France, of the total five collected plants, he distributed two plants to the cargo master of the ship, one to the King's garden in Paris, one to his superior at Brest and kept one plant himself. The introduction of *F. chiloensis* to France was one of the two important steps in the origin of cultivated strawberries. The second step was that Frezier brought only female plants to France, which were easily hybridized with another octoploid species, *F. virginiana*, to give rise to the large-fruited, cultivated strawberry *F.* × *ananassa* Duch.

The genus *Fragaria*, to which strawberries also belong, is a member of the family Rosaceae and subfamily Rosoideae (Potter et al., 2007), which forms a polyploidy series, from diploid to octoploid with basic chromosome number of x = 7. The genus *Potentilla* is closely related to *Fragaria*. The genus *Fragaria* was first summarized in pre-Linnaean literature by C. Bauhin (1623). The authors of the first *Catalogue du Jardin du Roi* called it merely *"Fraisieretranger"*, or foreign strawberry. The Robins, who wrote the catalogue in 1624, thought it to have come from Pannonia, a region of Europe between the Danube river to the north and Illyria to the south, and gave their "foreign strawberry" the botanical name *F. major pannonica*. In 1629, Parkinson called it the "Bohemia strawberry" or *F. major sterilisseubohemica*. By the end of AD 1500, two strawberries cultivated in gardens were the wood strawberry, *F. vesca*, and the musky strawberry, *F. moschata*, both characterized by small, distinctly flavoured berries. Botanists of the early 17th century named three European species: *F. vesca* and its subspecies *F. vesca semperflorens*; *F. moschata* and *F. viridis*, the green strawberry. The single most important significance of the 17th century in strawberry history was the introduction of *F. virginiana* from eastern North America to Europe, as later on it contributed in to the evolution of today's modern, big-fruited strawberries. Guide la Brosse, head of the Jardin du Roi at Paris, in the year 1636, included a *F. americana* magno fructurubro in his *Catalogue of the Garden*. In 1738, Linnaeus identified it in the garden of George Clifford, a wealthy banker who had a large botanical collection in Amsterdam.

The first strawberry hybrid "Hudson" was developed in the United States during 1780. The English word "strawberry" derives from the Anglo-Saxon word *"streoberie"*, not spelt in the modern way until during 1538. The first documented botanical illustration of a strawberry plant is considered to have appeared as an illustration in herbaries during 1454. In *Hortus Cliffortianus*, Linnaeus (1738) described it as monotypic genus containing *F. flagellisreptan*; but in *Species Plantarum*, he (Linnaeus, 1753), described three species including varieties, although several European species were omitted, but are now known, and the one belonging to genus *Potentilla* was included (Staudt, 1962). Duchesne (1766) has been credited for publishing the best early taxonomic treatment of strawberries (Hedrick, 1919; Staudt, 1962). Duchesne maintained a collection of strawberries at the

38   Strawberries

Royal Botanical Garden, had living collections documented from various regions and countries of Europe and the Americas and supplied samples to Linnaeus in Sweden.

The present *Fragaria* taxonomy includes 20 named wild species, three described naturally occurring hybrid species and two commercially important cultivated hybrid species. The wild *Fragaria* species have been found to be distributed in the north temperate and Holarctic zones (Staudt, 1989, 1999a, b; Rousseau-Gueutin et al., 2008). American and European *Fragaria* subspecies have been monographed by Staudt (1999a, b), who also revised the status of Asian species (Staudt, 1999a, b, 2003, 2005; Staudt and Dickore, 2001). Chinese and mid-Asian species were studied (Lei et al., 2005). But considering global diversity further collection and comparison are required to understand the taxonomy of genus *Fragaria* comprehensively. The distribution of specific ploidy levels within certain continents (see Table 1.1) has been used to infer the history and evolution of these species (Staudt, 1999a, b).

During the summers of 1996, 1998 and 1999, prospecting trips were made to southern Chile to collect native *Fragaria* germplasm. Plants, runners and seeds were collected from natural ecosystems as well as from small family orchards where the white-fruited "Chilean strawberry" (*F. chiloensis*) was cultivated. In the course of the three trips, differences in plant morphological characteristics were found. A first characterization *in situ* showed three different types:

1. The first corresponded to wild plants that grew on the coast, very close to the shore, where they colonized beaches and sand dunes of the eastern shore of Chiloe Island. These plants had small, bright dark green, coriaceous, smooth leaves; vigorous and at times very long runners with perfect flowers and small red berries.
2. A second type was found normally in the lower elevations of mountain ranges in the regions of Araucania and the Lakes District. In contrast with the first type, the plants were large with greyish green leaves having abundant trichomes. Flowers were pistillate, staminate or hermaphrodite and fruits small-to-medium and red.
3. The third type corresponded to the white cultivated strawberry, which is mainly found in small orchards of the coastal mountain ranges in the region of Bio Bio, near Concepcion city. Morphologically, this plant resembled the second type. It is vigorous, with hairy leaves, runners and petioles. It was distinguished by its characteristic large, white or pale pink fruits. The achenes were dark, and they contrasted sharply with the light fruit peel. These strawberries were very sweet with intense aroma and, therefore, highly prized in the local markets (Gambardella et al., 2002).

Strawberries as *Fragaria* species exist as a natural polyploidy series ranging from diploid ($2n = 2x = 14$) to decaploid ($2n = 10 = 70$). Diploid *Fragaria* species are endemic to *boreal* Eurasia and North America*. *F. vesca* is a native of the regions from the west of the Urals throughout northern Europe and across the North American continent. But according to the flora of these regions (Hulten, 1968), this diploid species is not native to Siberia, Sakhalin, Hokkaido, Japan, Kamchatka, or to the Kurile, Aleutian, or Hawaiian archipelagos, and it has been introduced in many parts of these areas.

Diploid strawberry species are reported to occur on many of these islands of and surrounding Japan, in Hokkaido, Sakhalin and the greater and lesser Kuriles (Makino, 1940). Diploid and tetraploid species are native to China, Siberia and the Russian Far East. Wild pentaploids ($2n = 5x = 35$) have been observed to have natural occurrence in California (*F. bringhurstii*) and China (Lei et al., 2005). These strawberries exist in colonies with other ploidy levels nearby. The only known wild hexaploid ($2n = 6x = 42$) species, *F. moschata* (musk strawberry), is native to Europe as far east as Lake Baikal (Hancock, 1999). Wild octoploid species have been reported to be distributed from Unalaska eastward in the Aleutian Islands (Hulten, 1968), completely across the North American

---

* The **boreal forest**, also known as *Taiga*, a Turkish word that mean coniferous forests.

Origin, Taxonomy and Distribution 39

continent, on the Hawaiian Islands and in Chile, South America (Staudt, 1999a, b). Wild (*Fragaria iturupensis*) decaploids (2n = 2x = 70) are native to the Kurile Islands (Hummer and Hanckock, 2009) and the old Cascades in western North America (Hummer unpublished).

Over the past 50 years, strawberry breeders have increasingly relied on a narrow base of domesticated strawberry germplasm to evolve new varieties. The progenitors of the modern cultivated strawberry exist in native populations in North and South America, which have many useful traits of commercial importance for modern breeding purposes. Within two decades over 20 collection trips have been made to collect native octoploid strawberry germplasm in North and South America for expanding the genetic diversity of *ex situ* collections (Hancock et al., 2002).

Five wild diploid strawberry lines, identified as *F. mandschurica*, were collected from Jilin, Heilongjiang and Inner Mongolia (Nei Menggu) in northeast China. Morphological and RAPD analysis highlighted differences between this species and another wild species *F. vesca*. Close relationships were noted *between F. mandschurica* and tetraploid *F. orientalis* and hexaploid *F. moschata*. It is suggested that *F. mandschurica* is most likely the primary species of both of these two species (Lei et al., 2001).

A natural strawberry genotype, "Jilin 4", was collected from the Changbai Mountains located in Gongzhuling Region, Jilin Province, Northeast China. It was identified as a pentaploid (2n = 5x = 35). Of the several genotypes evaluated, Jilin 4 was observed to be the most vigorous among representatives of all 15 wild species of *Fragaria* and the other more than 60 cultivars of *F.* × *ananassa* in a common strawberry plantation with size of flowers and anthers being similar to other cultivars but it produced abundant runners, viable pollen and sterile pistils. It appeared tolerant to drought and cold, but showed symptoms of virus infection. There are three possible origins of Jilin 4 (Lei et al., 2005).

The occurrence of *F. viridis* and sterile *F. vesca* × *F. viridis* hybrids at the northern edge of the distribution of southern Finland, can be explained by high soil Ca, high Ca:K ratio, a lowest temperature of less than −25°C and after a mean −3.5°C temperature for less than five days' lag before persisting snow. A high Ca: K ratio may be associated with the tolerance to freezing injury. Weakness and sterility appeared in experimental *F. vesca* × *viridis* hybrids. A hybrid swarm declined on an uninhabited islet since 1930. This can be ascribed to leaching due to acid rains from top soil, lacking in the supply of sufficient Ca minerals (Ahokas, 2002).

There are three wild diploid *Fragaria* species, viz., *F. iinumae, F. nipponica, F. yezoensis* and a botanical variety, *F. nipponica* var. *yakusimensis* in Japan. Of these, *F. iinumae* Makino is distributed in areas ranging from the lowlands of Hokkaido in the north to the mountains of main island Honshu (300–2000 m elevation) with heavy snowfalls along the Sea of Japan. The southernmost limit is Mount Daisen in Tottori Prefecture. *F. nipponica* Makino is distributed in the mountainous regions (500–2000 m elevation) on the Pacific side in central part of Honshu. *F. nipponica* var. *yakusimensis* Masamune is found in isolated areas on Yakushima (1800 m elevation), a small island, located south of Kyushu. While *F. yezoensis* Hara grows along the eastern shores of Hokkaido. Native *Fragaria* of Japan can though be seen found growing in diverse environments and habitats (Oda, 2002).

China is regarded as one of the richest country in wild strawberry resources, and since the 1980s, 103 accessions of wild strawberry genotypes have been collected from various regions of China and conserved. Of about 20 recognized *Fragaria* species in the world, 11 are distributed in China, which includes eight diploid species such as *F. vesca, F. nilgerrensis, F. pentaphylla, F. gracilis, F. nubicola, F. viridis, F. daltoniana* and *F. mandschurica* and three tetraploid species such as *F. orientalis, F. moupinensis* and *F. corymbosa* (Lei et al., 2006, 2014). The Changbai Mountains of China have a wide distribution of wild strawberry germplasm resources, and during the past 10 years, a total of 20 wild strawberry genotypes have been collected from this region and classified into three species, viz., diploid *F. mandschitrica* (*F. mandschurica*), tetraploid *F. orientalis* and *F. corymbosa*. The *Fragaria orientalis* var. *concolor* distributed in Changbai Mountains and earlier regarded as a variety of *F. orientalis* was identified as *F. corymbosa*. This is thought to be the first report that *F. corymbosa* is distributed in Changbai Mountains (Dai et al., 2007). The taxonomic description and distribution of strawberry species are summarized in Table 4.1.

TABLE 4.1
Status and Occurrence of *Fragaria* Species

Species	Ploidy Level	Subspecies	Characteristics	Distribution
F. bucharica Losinsk	2x		Self-incompatible with sympodial runners, a characteristic that distinguishes it from *F. nubicola* (Hook. f.) Lindl. Ex Lacaita, also found in the Himalayas	Western Himalayas
		bucharica	Bigger bractlets	Tajikistan to Afghanistan, Pakistan, and Himachal Pradesh in India (Staudt, 2006)
		darvasica	Smaller bractlets	Himalayan region
F. chinensis Losinska	2x		Leaves are trifoliate, nearly sessile, extended sparse appressed hairs, runners monopodial branching, fruits pale red-to-red, small with no flavour.	Western and southern China
F. daltoniana J. Gay	2x		Slender plants, self-compatible with sympodial runners, leaf trifoliate, hairs on petioles, runners and peduncles, ovoid to cylindrical large berries with spongy tasteless pulp, shiny and coriaceous leaves, similar to *F. chiloensis.*	Himalayan region (India to Myanmar including Tibet).
F. iinumae Makino	2x		Entirely different from other diploid species because of its glaucous leaves resembling somewhat with *F. virginana* ssp. *glauca*, flowers always greater than five (six–eight) with six–nine petals, berries elongated with small calyx, spongy and tasteless pulp, sympodial runners, crown as rosettes, but they sometimes rise above the ground in "tufts", making this species conspicuous. Above ground shoots die back through the crown and roots remain alive.	Lowlands in the north to the mountains of the main island Honshu in areas of heavy snow along the Sea of Japan of Hokkaido
F. mandshurica Staudt	2x		Differs from *F. vesca* but corresponds in some characters with *F. orientalis* (4x) distributed in eastern Siberia, outer Mongolia, Manchuria and Korea, hence considered the diploid ancestor of *F. orientalis*. Leaf trifoliate, sympodially branched runners with hairs (also on petioles), flowers hermaphrodite, self-incompatible, berries subglobose to obovoid, highly acidic with yellow achenes placed in shallow pits. Some accessions may blossom again in winter.	East banks of Lake Baikal, Mongolia and South Korea and spreads to northeastern China

(Continued)

Origin, Taxonomy and Distribution

TABLE 4.1 (CONTINUED)
Status and Occurrence of *Fragaria* Species

Species	Ploidy Level	Subspecies	Characteristics	Distribution
F. nilgerrensis Schlect. (Nilgiri strawberry)	2x		Robust plants, self-compatible, roundish leaflets, trifoliate leaves with hairs, sympodial branching runners, sub-globose berries, taste like banana, white and pale yellow tint with many achenes, much lesser cold tolerance. The leaf morphology of the tetraploid *F. moupinensis*, resembles that of *F. nilgerrensis* (Darrow, 1966).	Southeastern Asia
		Nilgerrensis and subsp. (Staudt, 1999a).	Fruits white to cream	Northwestern and southwestern India, east Himalaya, northeastern Myanmar, northern Vietnam, southwest and central China
		hayatae	Pink to red peel, creamy cortex (Staudt, 1999a) known for high anthocyanin in all plant parts including the berries (Staudt, 1989).	Recorded only in Taiwan
F. nipponica Makino	2x		Now includes the submerged species *F. yezoensis*, self-incompatible.	Japan
		nipponica		Honshu and Hokkaido in Japan, and Sakhalin and the Kuriles in Russia
		yakusimensis		Yakushima Islands of Japan
		chejuensis		Island of Cheju-do off the Korean mainland
F. nubicola Lindl	2x		Leaves are pinnately quinquefoliate, rarely trifoliate, obovate with sharp sawtooth, petioles, runners and peduncles covered with appressed hairs, fruit globose or elliptic. Monopodial branching of stolon, which is the only distinguishing feature separating it from *F. bucharica*. This species was observed to form accessory leaflets probably associated with the time of year.	Southern slopes of the Himalayas to southeast Tibet, and in southwest China
F. pentaphylla Losinsk	2x		Leaves are pinnately quinquefoliate, rarely trifoliate, long petioles, nearly glabrous above and purple red beneath, petioles and runners covered with hairs, self-incompatible, contains accessory leaflets, closely related to a tetraploid species, *F. tibetica*, which also has monopodial runners and a white-fruited form, *F. tibetica* f. *alba*. The two species are distinguished from each other by heteroecy, tetraploidy, larger pollen grains, and larger achenes found in *F. tibetica*. Fruits may be white or red. It is highly resistant to leaf spot disease.	North China

(Continued)

TABLE 4.1 (CONTINUED)
Status and Occurrence of *Fragaria* Species

Species	Ploidy Level	Subspecies	Characteristics	Distribution
F. vesca L.	2x		Self-compatible, runner production sympodial	Largest native range including Europe, Asia west of the Urals, disjunct in North America
		vesca	Relatively small flowers, erect inflorescence mostly exceeding the leaves, subglobose to somewhat ovoid berries with good flavour, reflexed calyx, easily separable from the berries.	Endemic across Europe to Siberia, as far as Lake Baikal
		americana	Berries more or less elongated, somewhat pointed with always reflexed calyx. Differs from other subspecies by its slender morphological structure.	Distributed in many parts of the USA from Virginia to South Dakota, North Dakota, Missouri, Nebraska and Wyoming. Also found in Ontario and British Columbia, Canada
		californica	Subglobose to elongated berries, with reflexed calyx, achenes superficial or in shallow pits.	California
		bracteata	Gynodioecious (contains both hermaphrodite and occasionally female sex forms), flowers significantly larger, spreading to appressed calyx to the berries, achenes set in deep pits.	Extends to rocky mountains southwards to Mexico, after the end of Pleistocene glaciations migrated northwards to Pine Pass where meets now the subspecies *americana*, migrated to the west after glaciations from an eastern refuge along two routes (along northern route to the Peach River Valley in the Yukon territory and along a more southern route to North and South Dakota). Around the coastal and Cascades mountain ranges from British Columbia through Washington and Oregon, and the Sierra Nevada in California. Its distribution extends into Mexico, where it is referred to as *F. mexicana* Schltdl (Staudt, 1999b).

(Continued)

Origin, Taxonomy and Distribution

TABLE 4.1 (CONTINUED)
Status and Occurrence of *Fragaria* Species

Species	Ploidy Level	Subspecies	Characteristics	Distribution
F. viridis Duch.	2x		Leaves trifoliate, petioles covered with hairs, monopodial branching, inflorescence taller than leaves. Distinguishable from *F. vesca* by its large flowers and fewer-flowered inflorescence (4–10 flowers/inflorescence), hermaphrodite but self-incompatible, wine red peel, cortex and pith yellowish-green, large berries and scapes lie along the ground at the time of ripening, usually appressed berries, achenes set in deep pits, calyx not easily detached from the berries.	Eastern Europe extending to Lake Baikal and the Altai mountains and Asia. Distributed in Xinjiang, China
F. × bifera Duch.	2x		Morphological features of this hybrid species are mostly intermediate and include the stolon branching system and leaf colour.	In regions where *F. vesca* and *F. viridis* distributions overlap, including Russia, Germany, France, Finland, and Italy. Hybridization has occurred, resulting in the hybrid species *Fragaria × bifera*.
F. corymbosa Losinsk	4x		Second tetraploid species known only as male plant, somewhat oval leaflets and derived from diploid *F. chinensis* (Staudt, 2003). Leaves are pinnately quinquefoliate or trifoliate, obovate, petioles, covered with long but thin spreading hairs, monopodial branching, flowers small, fruit red, pulp pinkish-white, tasteless and slightly acidic. It is susceptible to high temperature.	Russian Far East, and north, west and central China
F. gracilis A. Los.	4x		Plants are extremely dwarf, leaves trifoliate and almost sessile, sparse spreading hairs on petioles, runners and peduncle, dioecious, few flowers/inflorescence, fruits red, small and tasteless. It is not resistant to high temperature during summer.	Northwestern China
F. moupinensis (French.) Card	4x		Leaves are pinnately quinquefoliate or trifoliate, nearly sessile, petioles light red to red, thickly spreading hairs on petioles, runners and peduncle, branching monopodial, 2–3 flowers/inflorescence and dioecious, leaflets serrate elongated and oval, berries orange red with tasteless spongy pulp, achenes placed in deep pits and derived from diploid *F. chinensis* (Staudt, 2003).	Northern and southwestern China

(Continued)

TABLE 4.1 (CONTINUED)
Status and Occurrence of *Fragaria* Species

Species	Ploidy Level	Subspecies	Characteristics	Distribution
F. orientalis Losinsk	4x		Leaves are trifoliate, with thick spreading hairs sometimes purplish red beneath, nearly sessile or only central leaflets have short peduncle, runners have sympodial branching. Coarser leaves and teeth, flowers larger than *F. mandschurica* with sexual dimorphism in size, trioecious (male, female and hermaphrodite plants, due to the monofactorial tetrasomic inheritance of sex of the quadruple nullipex type 1) (Staudt, 1967), berries large, obovoid and very much acidic, it can be distinguished from *F. mandshurica* by the size of its pollen grains, a characteristic related to the number of chromosomes. Female plants often blossom again in autumn.	Eastern Siberia, Outer Mongolia, Manchuria and northeastern China
F. tibetica Staudt & Dickore	4x		Diploid *F. pentaphylla* seems to be the putative ancestor of this species (Staudt and Dickore, 2001), dioecious. Leaves pinnately quinquefoliate or trifoliate, nearly sessile, appressed or ascending hairs, monopodial branching, few flowers per inflorescence, fruits orange red to light red.	Southwestern China
F. bringhurstii Staudt	5x (9x)			California
F. sp. *novb*				China
F. moschata Duch.	6x		Only hexaploid species, originated from the cross between *F. vesca* × *F. nubicola* in which unreduced gametes chromosome numbers frequently occur and give rise to offspring with 28, 35, 42 and even higher chromosome numbers, growth vigorous, somewhat rhombic leaflets, inflorescences mostly exceeding the leaves, flowers large, dioecious or trioecious, scapes lie along the ground due to the weight of berries, berries with strong musklike taste and smell with reflexed calyx, fruit only has colour on the peel, while the cortex and pith are yellowish-white.	Europe extending as far as the Ural Mountains, Siberia
F. chiloensis (L.) Miller	8x		Also called beach strawberry	Western North America, Hawaii, Chile

(Continued)

Origin, Taxonomy and Distribution

TABLE 4.1 (CONTINUED)

Status and Occurrence of *Fragaria* Species

Species	Ploidy Level	Subspecies	Characteristics	Distribution
		pacifica	Small red berries	From Aleutian Islands Southwards to California near San Francisco
		lucida	Berries of different sizes, shapes and colours with sweet taste, achenes quite large and dark brown, calyx large, usually clasping the berries.	From Queen Charlotte Islands to San Luis Obispo County, California
		chiloensis f. *patagonica*	White-fruited landrace of *F. chiloensis* was first domesticated by the Mapuche Indians. This forma has larger flower and fruit structures than other *F. chiloensis* subspecies. This large, white-fruited landrace with hairy petioles was imported from Chile to Europe in the early 18th century. It is the maternal progenitor of the cultivated strawberry.	Coastal mountains, the central valley in Chile and in the Andes in southern Chile with the southern limit of its distribution in Argentina
		sandwichensis		Mountainous regions of Hawaii and Maui
F. virginiana Miller	8x		Called as scarlet strawberry	North America
		virginiana		Eastern North America and spreads to British Columbia in the west
		glauca	Distinguished from other subspecies by the smooth leaf surface and the dark to light bluish (glaucous) leaves.	
		grayana		Northwestern Texas, to Nebraska, Iowa, and Illinois. It is also found in Louisiana, Alabama, Indiana, Ohio, Virginia, and New York.
		platypetala	The leaves of *F. virginiana* subsp. *platypetala* are also blue green but only slightly (Staudt, 1999b).	British Columbia, extending southward to Washington, Oregon and northern California
F. × *ananassa* Duch. ex Lamarck	8x			Cultivated worldwide
		cuneifolia	Suspected as a natural hybrid of *F. chiloensis* subsp. *pacifica* or *Fragaria chiloensis* subsp. *lucida* and *F. virginiana* subsp. *platypetala*. Hybrid has smaller leaves, flowers, and fruits.	Coastal regions of British Columbia (Vancouver Island) south to Fort Bragg and Point Arena lighthouse in California

(Continued)

TABLE 4.1 (CONTINUED)
Status and Occurrence of *Fragaria* Species

Species	Ploidy Level	Subspecies	Characteristics	Distribution
F. iturupensis Staudt	10x		More primitive than *F. virginiana* subsp. glauca. Oblate fruit and erect inflorescence, flavour components of this polyploidy population resemble *F. vesca*.	Eastern slopes of Mt. Atsonupurion Iturup, the second island in the southern section of the greater Kuril Island archipelago.
F. virginiana	10x	*platypetala* Miller		Oregon, United States
F. x vescana Bauer & A. Bauer	10x			Cultivated in Europe

Source: Lei et al. (2005, 2014), Staudt (1989, 2008, 2009).

REFERENCES

Ahokas, H. 2002. Factors controlling the north-edge distribution of *Fragaria viridis* and its *F. vesca* hybrids in South Finland. *Acta Horticulturae*, 567(1): 385–388.

Anon. 1956. *Wealth of India*, vol. IV. CSIR, New Delhi, India: pp. 57–60.

Dai, H., Lei, J.J. and Deng, M.Q. 2007. Investigation and studies on classification of wild *Fragaria* spp. distributed in Changbai Mountains. *Acta Horticulturae Sinica*, 34(1): 63–66.

Darrow, G. 1966. *The Strawberry: History, Breeding and Physiology*. Holt, Rinehart and Winston, New York, New York and Chicago, Illinois: p. 447.

Hancock, J.F. 1999. *Strawberries*. Crop production science in horticulture series, No 11. CABI, Wallingford, UK.

Hancock, J.F., Hokanson, S.C., Finn, C.E. and Hummer, K.E. 2002. Introduction of supercore collection of wild octoploid strawberries. *Acta Horticulturae*, 567(1): 77–79.

Hedrick, U.P. 1919. *Sturtevant's Edible Plants of the World*. Dover, New York (formerly published in 1919 by J. B. Lyon Co, Albany as *Sturtevant's Notes on Edible Plants*).

Hulten, E. 1968. *Flora of Alaska and Neighboring Territories: A Manual of the Vascular Plants*. Stanford University Press, Stanford, California.

Hummer, K.E. and Hancock, J.H. 2009. Strawberry genomics: botanical history, cultivation, traditional breeding, and new technologies, Chapter 11. In: K.M. Folta and S.E. Gardiner (eds.). *Plant Genetics and Genomics of Crops and Models, Vol 6: Genetics and Genomics of Rosaceae*. Springer, Germany: pp. 413–435.

Gambardella, M., Infante, R., Lopez Aranda, J.M., Faedi, W. and Roudeillac, P. 2002. Collection of wild and cultivated native *Fragaria* in southern Chile. *Acta Horticulturae*, 567(1): 61–63.

JiaJun, L., HanPing, D., ChangHua, T., MingQin, D., MiZhen, Z. and YaMing, Q.I. 2006. Studies on the taxonomy of the strawberry (*Fragaria*) species distributed in China. *Acta Horticulturae Sinica*, 33(1): 1–5.

Lei, J.J., Mochizuki, T. and Deng, M.Q. 2001. Studies on the diploid strawberry species *Fragaria mandschurica* Staudt. *Journal of Fruit Science*, 18(6): 337–340.

Lei, J.J., Xue, L., Dai, H.P. and Deng, M.Q. 2014. Taxonomy of Chinese *Fragaria* species. *Acta Horticulturae*, 1049: 289–294.

Lei, J., Li, Y., Du, U. Dai, El.and Deng, M. 2005. A natural pentaploid strawberry genotype from the Changhai Mountains in Northeast China. *HortScience*, 40(5): 1194–1195.

Makino, T. 1940. *Makino's New Illustrated Flora of Japan*. Hokuryukan, Tokyo, Japan: pp. 270–271.

Oda, Y. 2002. Photosynthetic characteristics and geographical distribution of diploid *Fragaria* species native in Japan. *Acta Horticulturae*, 567(1): 381–384.

Potter, D., Eriksson, T., Evans, R.C., Oh, S., Smedmark, J.E., Morgan, D.R., Kerr, M., Robertson, K.R., Arsenault, M., Dickinson, T.A. and Campbell, C.S. 2007. Phylogeny and classification of Rosaceae. *Plant Systematics and Evolution*, 266: 5–43.

Rousseau Gueutin, M., Lerceteau Kohler, E., Barrot, L., Sargent, D.J., Monfort, A., Simpson, D., Arus, P., Guerin, G. and Denoyes Rothan, B. 2008. Comparative genetic mapping between octoploid and diploid *Fragaria* species reveals a high level of collinearity between their genomes and the essentially disomic behavior of the cultivated octoploid strawberry. *Genetics*, 179: 2045–2060.

Staudt, G. 1962. Taxonomic studies in the genus *Fragaria*: typification of *Fragaria* species known at the time of Linnaeus. *Canadian Journal of Botany*, 40: 869–886.

Staudt, G. 1967. The genetics and evolution of heteroecy in the genus Fragaria. I. Investigations on *Fragaria orientalis*. *Zeitschrift für Pflanzenzuchtuny*, 58: 245–277.

Staudt, G. 1989. The species of *Fragaria*, their taxonomy and geographical distribution. *Acta Horticulturae*, 265: 23–33.

Staudt, G. 1999a. Notes on Asiatic *Fragaria* species: *Fragaria nilgerrensis* Schltdl. Ex J. Gay. *Botanische Jahrbücher für Systematik*, 121(3): 297–310.

Staudt, G. 1999b. Systematics and geographic distribution of the American Strawberry species: taxonomic studies in the genus *Fragaria* (Rosaceae: Potentilleae). In: *Botany*, vol 81. University of California Press, Berkeley, California: p. 162.

Staudt, G. 2003. Notes on Asiatic *Fragaria* species: III. *Fragaria orientalis* Losinsk. and *Fragaria mandshurica* spec. *nov. Botanische Jahrbücher für Systematik*, 124(4): 397–419.

Staudt, G. 2005. Notes on Asiatic *Fragaria* species: IV. *Fragaria iinumae*. *Botanische Jahrbücher für Systematik*, 126(2): 163–175.

Staudt, G. 2006. Himalayan species of *Fragaria* (Rosaceae). *Botanische Jahrbücher für Systematik*, 126(4): 483–508.

Staudt, G. 2008. Notes on Asiatic *Fragaria* species V: *F. nipponica* and *F. iturupensis*. *Botanische Jahrbücher für Systematik*, 127(3): 317–341.

Staudt, G. 2009. Strawberry biogeography, genetics and systematic. *Acta Horticulturae*, 842: 71–84.

Staudt, G. and Dickore, W.B. 2001. Notes on Asiatic *Fragaria* species: *Fragaria pentaphylla* Losinsk. and *Fragaria tibetica* spec. *nov. Botanische Jahrbücher für Systematik*, 123: 341–354.

5 Breeding and Improvement

R.K. Salgotra, Manmohan Sharma, and Anil Kumar Singh

CONTENTS

5.1 History and Evolution 50
5.2 Genetics 52
5.3 Quantitative Genetics...55
    5.3.1 Coefficient of Variation, Heritability and Genetic Advance 55
    5.3.2 Combining Ability 56
    5.3.3 Correlations 58
5.4 Cytogenetics 58
    5.4.1 Chromosome Number..61
    5.4.2 Chromosome Doubling..61
5.5 Floral Biology 62
    5.5.1 Crossing Techniques 62
5.6 Breeding Methods 62
    5.6.1 Introductions and Selection 63
    5.6.2 Hybridization 63
        5.6.2.1 Interspecific Hybridization 64
        5.6.2.2 Intergeneric Hybridization 65
    5.6.3 Mutagenesis 66
    5.6.4 Apomixis 66
5.7 Breeding Objectives...66
    5.7.1 Adaptability to Soil Factors 66
    5.7.2 Environmental Factors 67
    5.7.3 Special Plant Characters 67
        5.7.3.1 Stolon Production..67
        5.7.3.2 Variegation 68
        5.7.3.3 Fruiting Habit..68
        5.7.3.4 Ripening Time and Fruit Development 68
        5.7.3.5 Fruit Size, Yield and Quality 68
        5.7.3.6 Pest Resistance..69
        5.7.3.7 Disease Resistance 70
5.8 Molecular Markers and Genome Mapping...72
References...72

Most fruit crops have been under cultivation since Neolithic times and as a result of continuous selection, natural or artificial, have evolved to such a great extent that they show little resemblance to their wild progenitors. Strawberries (*Fragaria* spp.) are one of the most modern and extensively investigated horticultural crops grown widely in different parts of the world. Strawberries, like other berries, represent one of the most important sources of bioactive compounds with antioxidant activity (Battino et al., 2009). With understanding of their health benefits at the consumer level, the importance of strawberries has also increased (Alvarez-Suarez et al., 2011, 2014; Giampieri et al., 2012; Tulipani et al., 2014).

Strawberries have drawn the attention of scientists involved in their breeding for crop improvement over the years due to their economic importance and the availability of large inter- and intraspecific genetic variability. Genetic improvement in strawberries has, in most cases, been achieved by selection from natural seedling populations, leading to development of many varieties. But intense selection pressures and subsequent maintenance of unique genotypes by clonal propagation have resulted in a narrow genetic base. Conventional breeding has proved a powerful tool for crop improvement because it is evolutionary in nature and progress can be achieved by selecting improved individuals, which continually serve as parents for subsequent cycles of breeding. Strawberries are often characterized by a combination of high heterozygosity, polyploidy, sterility and incompatibility. Therefore, breeding efforts for genetic improvement of this crop is restrained (Janick and Moore, 1996) due to:

1. the inability of the sexual system to incorporate variation from non-related species
2. the reliance on natural variation
3. the linkage of desirable and undesirable characters
4. the limitation of selection in detecting infrequent recombinations

The complex octoploid nature ($2n = 8x = 56$) of the cultivated strawberries has made genetic, cytogenetic and molecular studies very difficult compared with the diploid *F. vesca* (Stewart, 2011), so their usefulness as a model plant for studies of genetic mechanisms is minimized. Nevertheless, the available biological diversity of wild *Fragaria* species has been regularly used as a source of novel genetic variation that can be introgressed into the cultivated strawberry through breeding (Hancock et al., 1999; Chambers et al., 2013). The biological diversity of *Fragaria*, especially with respect to its sexual system and polyploidy, also makes it a very attractive biological entity for studies and understanding of ecological and evolutionary genomics.

5.1 HISTORY AND EVOLUTION

Present-day cultivated strawberries are believed to have originated in European gardens during the 18th century when female clones of Frutillar (*F. chiloensis)* from Chile were interplanted with clones of Scarlet strawberry (*F. virginiana*) that had previously been introduced from eastern North America during early 17th century. The resultant hybrid progenies as chance seedlings had combined traits of both the parents such as the hardiness and productivity of *F. virginiana* and the relatively large fruit size of *F. chiloensis*. Systematic breeding work for genetic improvement of strawberry was started during the early parts of 18th century in Europe, particularly in Great Britain. The most important early selection of garden strawberries is believed to have been made by Michael Keen during the year 1806, referred to as Keen's Seedling, which dominated English strawberry acreage during the late-19th century, and contributed to the pedigree of many modern cultivars. Duchesne is considered to be first strawberry researcher, credited to have collected specimens, classified into species, studied sexual compatibility among the different types, and documented the need for pollination during second half of the 18th century. British breeder Thomas A. Knight started the first systematic breeding program during 1817 and attempted intercrossing of the octoploid ecotypes, leading to generation of lines that served as germplasm for breeding and development of many modern cultivars. Charles Hovey of Massachusetts, in 1836 evolved the first cultivar from controlled hybridization between a European cultivar and a native *F. virginiana* selection in North America, which he named "Hovey". Arthur Howard began systematic breeding during the 1880s and, after his death, the variety "Howard 17", introduced during 1907, is in the pedigree of many important varieties evolved later during the 20th century. Harlow Rockhill was the first breeder of everbearing strawberries. The everbearing strawberry "Progressive" was introduced during 1912.

Albert Etter, during the early parts of the 20th century, carried out an active breeding programme in California using native *F. chiloensis* clones, and varieties evolved by him (especially "Ettersburg 80") constitute the genetic background of many modern cultivars. When European *Fragaria × ananassa* clones were brought to North America during the middle of the 19th century,

Breeding and Improvement

breeding work (Hancock et al., 2008) increased dramatically both in the public and private sectors. The parental species, *F. virginiana* and *F. chiloensis*, also got naturally hybridized in northwestern North America (Luby et al., 1992; Salamone et al., 2013). The first Japanese strawberry cultivar, "Fukuba", was selected during 1899 from seedlings of "General Chanzy". Most of the Japanese forcing strawberry cultivars have Fukuba in their pedigree.

Until about 1930, most of the breeding work was done by breeders in private sector and involved both intraspecific crosses and hybridization of European material with wild species. By the middle of the 20th century, the emphasis in strawberry breeding had shifted to institutions in the public sector in several European countries, such as Scotland, Germany and the Netherlands (Hancock, 1999). Many institutions in the public sector in the United States, including the University of California, the University of Florida and the US Department of Agriculture, and East Malling in the UK started active strawberry breeding programmes during the first half of the 20th century. In the latter half of the 20th century, many private companies, such as Driscoll Strawberry Associates, Plant Sciences and California Giant in the United States, Edward Vinson in the UK and Planasa in Spain, dominated strawberry breeding and have actively contributed many strawberry cultivars worldwide. Advances in breeding methodologies, over the years, have been exploited in enhancing the size and yield of the berry.

Research on strawberry breeding and genetics in northwest Europe is actively being undertaken by public as well as private institutions (Simpson, 2014). In Netherlands, the Fresh Forward breeding programme is focused on improving June-bearing and everbearing cultivars for northern Europe. Research on Genetics at Plant Research International (PRI) includes developing a linkage map for *Fragaria × ananassa*, molecular markers for resistance to *Phytophthora cactorum* and molecular studies on fruit quality, including firmness, flavour volatiles and antioxidants. Research at the University of Helsinki has been focused on identifying the genes for everbearing flowering in the wild diploid strawberry, *F. vesca*. In Germany, Hansabred GmbH & Co. aims to develop commercial cultivars for northern Europe with an emphasis on improving flavour and disease resistance. Research at the Julius Kuhn-Institut includes studies on the inheritance of flavour volatiles and the potential of aroma analysis as a selection tool. The breeding programme in Norway is being managed by Graminor. The research programme at Bioforsk includes investigating the genetics of *Phytophthora cactorum* in *F. vesca*. In the UK, the national programme on breeding in strawberries is based at East Malling Research Station, and there are private programmes of Edward Vinson Ltd., Driscoll's Genetics Ltd. and Angus Soft Fruits. Research on the genetics of strawberries in the UK is currently aimed at developing molecular markers for resistance to *Verticillium dahliae* and *Podosphaera aphanis* and improved water use efficiency.

Strawberry breeding at the National Institute for Research in Agriculture in Uruguay has focused on developing cultivars adapted to the different environments and local production systems (Vicente et al., 2009). In Italy, Marche Polytecnic University Ancona (IT) breeding program is specifically focused to evolve and evaluate new strawberry genetic material with increased adaptability to the mid-Adriatic areas and improve fruit quality and nutritional qualities (Capocasa et al., 2009).

The objectives of the Spanish strawberry breeding programme are aimed at evolving early and productive cultivars for fresh market, suitable for shipping, with high sensory qualities and neutraceutical properties that would thrive in production systems without treatments with methyl bromide and well adapted to Huelva's agro-environmental conditions.

In Japan, research institutions in both the public and private sectors are actively involved to breed everbearing strawberry varieties, in addition to June-bearing varieties (Takahashi et al., 2003). American or European cultivars have rarely been used as breeding material in recent years in Japanese strawberry breeding programmes. The present Japanese strawberry breeders are, however, concerned about the increase in inbreeding coefficient in Japanese strawberry cultivars, and are suggesting that foreign gene sources should be used more for improvement of the cultivars for forcing (Morishita, 2014), that is, protected or offseason cultivation.

5.2 GENETICS

To understand the possibilities and limitations of strawberry breeding, the results of studies on inheritance of different traits need to be understood. Professor A. Millardet at Bordeaux, France, was really the first modern strawberry geneticist. Millardet (1894) studied the cross *F. vesca alba*× *F. chiloensis* in which he obtained one sterile plant similar to *vesca* and three plants like *chiloensis*, the pollen parent. In all, he obtained four cases of paternal plants resembling the *chiloensis* in all general characters. Strasburger (1909) sectioned flowers of the *F. virginiana*×*F. moschata* at 12-hour intervals after pollination to determine if actual fertilization of the egg by the pollen took place. The union of egg and pollen typical of sexual reproduction, indicated that the so-called paternals were true hybrids and the characters of the pollen parent were remarkably dominant in the seedlings where *F. moschata* was a parent.

Richardson (1914, 1918, 1920 and 1923) published a series of notes on genetics of *Fragaria* that gave insights into the inheritance of different traits in diploid (*F. vesca*) and octoploid species. Thereafter, the genetics of number of simply inherited traits have been studied in *F. vesca* with limited reports in octoploid types. Strasburger (1909) studied the inheritance of runnered and runnerless, three-leaflet and one-leaflet, white- and red-fruited forms in everbearing wood strawberry, *Fragaria vesca semperflorens* (Table 5.1) and reported the inheritance in the diploid strawberry to be similar to that in other plants.

Mangelsdorf and East (1927) studied numerous diploid×diploid crosses and concluded that they were entirely interfertile; however, the diploids under study were merely varieties of *F. vesca*. They crossed pink-flowered genotypes with white-flowered ones and observed pink flowers in F_1 as dominant. In the F_2, they obtained 128 pink to 46 white, close to a 3:1 ratio. Ichijima (1930) studied the genetics of *Fragaria* and reported that crosses of *F. nilgerrensis* with other diploid species were dwarf, and crosses of the musk strawberry, *F. moschata* (2n=42) with diploids gave seedlings with the *F. moschata* chromosome number.

East (1930) also observed similar Mendelian inheritance in *F. vesca* (2n=14)× *F. virginiana* (2n=56) cross. Majority of the studies conducted afterwards have reported quantitative genetic control of most of the traits in octoploid genotypes/varieties with exception of inheritance of genes conferring resistance to disease causing microbes such as *Phytophthora fragariae* causing red stele disease (Van de Weg, 1997) and *Rca2* locus conferring resistance to pathotype 2 of *Colletotrichum acutatum*, a causal agent of anthracnose in strawberry (Denoyes-Rothan et al., 2005).

Yarnell (1929) studied tetraploid obtained by crossing two diploids, *F. bracteata*× *F. rosea*. The tetraploid so obtained was selfed and produced tetraploids. These tetraploids were crossed with the diploids *F. vesca*, *F. viridis* and *F. bracteata*. All resultant seedlings were triploids. Yarnell (1931) categorized the diploids into four groups:

1. *F. vesca* and its varieties, *F. bracteata*, *F. californica*, *F. americana*, *F. rosea*, and *F. mexicana*
2. *F. viridis* (= *collina*)
3. *F. nilgerrensis*
4. *F. vesca* type from China

TABLE 5.1
Study of Mendelian Inheritance for Different Traits in Strawberry

Cross Combination	F_1	F_2 ratio	References
Runnered/Runnerless	Runnered	Runnered: Runnerless (3:1)	Strasburger (1909)
Three-leaflet/one-leaflet	Three-leaflet	Three-leaflet:one-leaflet (3:1)	Strasburger (1909)
White-flowered/pink-flowered	White-flowered	White-flowered:pink-flowered (3:1)	Strasburger (1909)

Breeding and Improvement

F. viridis produced seedlings that were vigorous in crosses with the other diploids, but were only partially fertile. *F. nilgerrensis* produced only dwarf hybrids in all crosses except those with *F. viridis*, and even these hybrids produced no flowers. True hybrids could not be obtained in crosses of *F. moschata* with diploid species, but some maternals were obtained. The octoploids, *F. virginiana* and *F. chiloensis*, set no fruit when used as female parents in crosses with *F. vesca*, but, when used as male parents on *F. vesca*, they produced both maternal and partially fertile plants. *F. chiloensis*×*F. vesca* var. *bracteata* produced three sterile seedlings. *F. virginiana* and *F. chiloensis* both gave vigorous sterile hybrids with *F. moschata* pollen, but fruit set could not be obtained when they were used as pollen parents. All types of crosses gave at least some completely fertile plants, which were like the mother parent. Increased pairing of chromosomes took place with the increase in temperature.

Harland and King (1957) studied the inheritance of mildew resistance in strawberry. Susceptibility to mildew was observed to be due to two dominant genes in the diploid *F. vesca*. In the F_2 population of a cross involving resistant × susceptible genotypes, a 15:1 ratio of susceptible to resistant was obtained. American octoploid strawberry genotypes derived from East Asian *F. vesca* types instead of American *F. vesca* types. The octoploids *F. virginiana* and *F. chiloensis* have been reported to derive from different ancestors (Staudt, 1959).

The tetraploids *F. orientalis* and *F. moupinensis* might have originated through hybridization between the two diploids with overlapping habitats followed by chromosome doubling, or by the union of an unreduced chromosome complement (gamete) with a double unreduced chromosome complement (gamete) from two diploids (Islam, 1960). Islam (1960) studied the accidental haploid plant obtained from the cross *F. vesca*×*F. ananassa* and compared it with the parent *F. vesca*. It was observed to be smaller in all its parts. Ellis (1962) reviewed crosses made between *F. vesca* and *Potentilla* by different strawberry workers. He obtained hybrids from the Sans Rivale strawberry × *Potentilla fruiticosa* and × *Potentilla palustris* with 5x and 6x surviving seedlings, and 3x seedlings in *F. vesca* (4x) × *Potentilla fruiticosa*.

Tom Sjulin and Adam Dale carried out pedigree analysis of North American strawberry varieties during the 1980s. The breeding systems of different species of *Fragaria* are given in Table 5.2.

Strawberry genetics and crop improvement have been reviewed by several research workers (Galetta and Mass, 1990; Hancock, 2005; Flachowsky et al., 2011). Bringhurst (1990) ascertained that *F. chiloensis* was the female parent of all F_1 hybrids between *F. chiloensis* and *F. vesca*, as evidenced by the following two reasons:

- First that female *F. chiloensis* is heterogametic, segregating 1:1 for sex and the distribution of the hybrids as to sex fits at 1:1 ratio.
- Second, *F. vesca* ssp. *californica* (Staudt, 1962) of coastal California is normally self-pollinating (Arulsekar et al., 1981).

The cultivated strawberry has a narrow germplasm base, although the progenitors of the cultivated species have extensive geographical range. A high collinearity between the strawberry genomes and *F. vesca* has been reported by several research workers (Hancock et al., 2005; Rousseau-Gueutin et al., 2008; Sargent et al., 2009; Zorrilla-Fontanesi et al., 2011). This, together with the availability of a comprehensive genome sequence and annotated gene predictions for the diploid species (Shulaev et al., 2011), will greatly facilitate the independent genetic investigations in the cultivated strawberry.

Shaw and Larson (2008) evaluated the genetic improvement of cultivars released from the University of California during 1945–66 and during 1993–2004. Average fruit yield, fruit size, commercial fruit appearance and fruit firmness were 47–140% greater for modern cultivars as compared to early generation of cultivars with the highest increase being observed in fruit yield.

TABLE 5.2
Breeding System in Different *Fragaria* Species

Ploidy Level	Species	Self-Compatible (SC)	Self Incompatible (SI)	Gynodioecious (GD)	Dioecious (D)	Trioecious (T)
2x	*F. vesca*			GD		
	F. viridis		SI			
	Fragaria× hagenbachiana					
	F. nubicola		SI			
	Fragaria spec. Nov.		SI			
	F. daltoniana	SC				
	F. nilgerrensis	SC				
	F. mandschurica		SI			
	F. nipponica		SI			
	F. iinumae					
	F. yezoensis		SI			
4x	*F. moupinensis*				D	
	F. corymbosa				D	
	F. orientalis					T
6x	*F. moschata*					T
8x	*F. iturupensis*	SC				T
	F. virginiana				D	T
	F. chiloensis	SC			D	
	Fragaria× ananassa	SC (cultivated)			D	
5x, 6x, 9x	*Fragaria× briughurstii*				D	

Source: Staudt (1989).

Breeding and Improvement

5.3 QUANTITATIVE GENETICS

Most of the plant and fruit characters of the cultivated strawberry are inherited quantitatively (Galletta and Maas, 1990). The potential for application of quantitative genetic methods for the improvements of strawberry varieties is well recognized, but has rarely been utilized. Comprehensive quantitative studies of various quantitative characters in strawberry have been conducted in different strawberry growing countries (Bedard et al., 1971; Gooding et al., 1975; Shaw et al., 1989). The study of quantitative genetics of floral traits in the two sex morphs of gynodioecious *F. virginiana* revealed significant genetic variation underlying all floral traits, although narrow-sense heritabilities were, in most cases, much lower than broad-sense ones (Ashman, 1999a). Moreover, the sex morphs differed significantly in their heritabilities for shared traits, such as stamen length, and showed a tendency towards differing significantly in others, such as carpel number and petal length. In addition, correlations between the sex morphs for these traits were significantly greater than 0, but less than 1, and it indicated that greater sexual dimorphism could evolve in this population of *F. virginiana*, even if selection on these traits was not divergent. However, strong developmental integration of floral traits (e.g., stamen length and petal length) and high-levels of non-additive genetic variance may represent barriers to the evolution of complete sexual dimorphism.

In strawberry, the inheritance patterns of key aromatic compounds in segregating populations of strawberry and *F. virginiana* exhibit quantitative inheritance, typical of polygenic traits (Carrasco et al., 2005). Honjo et al. (2011) evaluated varietal differences in everbearing strawberry and selection indicators of flowering pattern by investigating the number of inflorescence and individual flowers together with their flowering dates from spring to autumn. Highly significant ($P \leq 0.001$) differences among varieties in these traits indicated a genetic basis for the strength of everbearing genotypes. The results suggested that parentage were involved in the strength of everbearing genotypes. The use of early-flowering June-bearer genotypes has been suggested to facilitate the breeding of strong everbearing varieties.

Which genetic factors control strawberry fruit aroma is complex. Cluster analysis grouped the volatiles into distinct chemically related families revealing a complex metabolic network underlying volatile production in strawberry fruit. Seventy QTLs covering 48 different volatiles have been detected, with several of them being stable over time and mapped as major QTLs (Zorrilla-Fontanesi et al., 2012).

5.3.1 COEFFICIENT OF VARIATION, HERITABILITY AND GENETIC ADVANCE

Consistency in the performance of selection in the succeeding generations depends upon the magnitude of heritable variation present in relation to the observed variation. Literature covering most of the reports on the coefficient of variation (CV), heritability (h^2) and genetic advance (GA) in strawberry for different characters is summarized in Table 5.3.

Narrow-sense heritability for plant growth traits has been reported low to moderate with little contribution of non-additive variance (Shaw, 1988). Substantial amounts of additive variance for yield and fruit size have been observed, although the relative proportions of additive and dominance variance have been suggested to vary widely across testing environments and propagule types (Pringle and Shaw, 1998; Shaw, 1989; Shaw et al., 1989; Shaw and Larson, 2005). Greater dominance variance is typically found for soluble solid and titratable acid contents (Shaw et al., 1987).

Whitaker et al. (2012) studied the genetics of 12 commercially important traits including fruit quality traits (soluble solids content and titratable acidity), other fruit and yield traits (early and total marketable yields, proportion of total cull fruit, proportion of misshapen fruit, proportion of water-damaged fruit and shape score), and vegetative traits (plant height and total runners). Narrow-sense heritability varied from low to moderate ($h^2 = 0.13 \pm 0.07$ to 0.32 ± 0.09) except for shape score ($h^2 = 0.06 \pm 0.04$) and total average weight ($h^2 = 0.52 \pm 0.07$). Broad-sense heritability was larger ($h^2 = 0.18 \pm 0.03$ to 0.53 ± 0.04), and for more than half the traits, over 50% of the total genetic

TABLE 5.3

Characters Showing High Genotypic Coefficient Variation (GCV), Phenotypic Coefficient Variation (PCV), Heritability (h^2) and Genetic Advance (GA)

S. No.	Character	References
1.	Fruit firmness and fruit skin toughness	Mori (2000)
2.	Fruit volume, number of flowering trusses per plant, number of fruits per plant, fruit weight, percentage of plant flowering and fruit length	Verma et al. (2002), Verma et al. (2003)
3.	Plant height, petiole length, flower number and flower disk diameter	Verma et al. (2003)
4.	Fruit size, fruit arising node length and its diameter ratio	Nathewet et al. (2007)
5.	Soluble solid content, titratable acidity, total number of runners, plant height, total market yield, early market yield, average fruit weight, total number of unmarketable (cull) fruit, total number of fruit, total misshapen fruit and total water-damaged fruits	Whitaker et al. (2012)
6.	Plant height, plant spread, number of leaves per plant, leaf area index, number of flowers per plant, number of fruits per plant, total soluble solids, titratable acidity, reducing sugar, total sugars, ascorbic acid, fruit length, fruit diameter, fresh fruit weight, fruit volume, dry fruit weight and fruit yield per plant	Mishra et al. (2015)

variation was observed to be non-additive. Large genetic and genotypic correlations were found for some traits, most notably between soluble solids content and early marketable yield (–0.68 ± 0.22). Genetic gains for this pair of traits based on a Monte Carlo simulation illustrated the tradeoff between these two traits, showing that about 27% increase in early yield could be obtained through selection but at the expense of about 8% decrease in soluble solids. However, it was suggested that moderate gains could be achieved in both traits using appropriate index coefficients.

Evaluation of strawberry genotypes (Mishra et al., 2015) to study genetic variation and the relationship between yield and its components showed significant variance among genotypes of all traits. The phenotypic coefficient of variation (PCV) for all the characters was slightly higher than genotypic coefficient of variation (GCV), which signified the presence of environmental influence to some degree in the phenotypic expression of characters. Dry fruit weight had maximum PCV (52.47) and GCV (48.26). The estimates of narrow-sense heritability (h^2) were observed to be lower than those of broad-sense heritability (h^2) for all the characters. Maximum genetic advance was recorded for fruit yield/plant (76.84), whereas genetic advance as per cent of mean was maximum for dry fruit weight (84.09). Maximum heritability (h^2, 98.44) was coupled with higher GA (76.84) estimated for fruit yield/plant, indicating the controlled character by additive genes, and therefore further improvement could be brought by selection. Fruit yield was significantly and positively associated with most of the characters except number of leaves per plant, titratable acidity, and ascorbic acid at both genotypic and phenotypic levels; therefore, these are important prerequisites to formulating a successful genetic improvement programme of strawberries.

5.3.2 Combining Ability

Knowledge of the breeding value (general and specific combining ability) of the parental forms used in crossbreeding programmes as well as the fundamental knowledge on the genetic determinants of the intensity of quantitative traits at the population level accelerates and increases the likelihood of achieving the intended purpose. Estimates of combining ability are essential in identifying superior parents and to study the heterotic effects in breeding programmes. It is well known that determination of general combining ability (GCA) gives an indication about the performance of the parents. In crossbreeding, the use of parental forms with high GCA effects or additive effects of genes (Griffing, 1956a; Griffing, 1956b) for the desired functional traits allows us to expect, with high probability, that their off-springs will be characterized by high values for the intended

Breeding and Improvement

traits (Vicente et al., 2009). GCA of the parents for a trait determines their overall usefulness, with respect to that trait, for creating new cultivars. The most valuable hybrid progeny in terms of quantitative traits can be obtained by crossing parents that exhibit favourable GCA effects for these traits (Masny et al., 2005, 2008, 2009; Yashiro et al., 2002).

Specific combining ability (SCA) is attributed primarily to deviations from the additive scheme caused by dominance and epistasis (Rojas and Sprague, 1952). Conversely, knowledge of the SCA of a pair of parental forms, which is a genetic interaction of both parents as a result of non-additive gene action (dominance and epistasis), showed in their offspring (Griffing, 1956a; Griffing, 1956b), in respect of the trait under consideration, allows for an examination of the differences between the value of the revealed trait and the value expected on the basis of the sum of GCA effects for those parental forms (Bestfleisch et al., 2014). The quotient of the mean square deviations for GCA and SCA provides for the determination of which genetic effects (additive or non-additive) have a predominant share in determining a given trait in the tested progeny.

The cultivars having good GCA for different characters to be used in strawberry breeding programme are summarized in Table 5.4.

Shaw et al. (2010) screened strawberry genotypes for resistance to *Verticillium dahlia* Kleb., an important soilborne pathogen of strawberry, and observed that selection reduced genetic variation for the resistance score, and GCV decreased in the breeding population from 34.4–11.6% from 1994–2008. Genotypic scores suggested that the pattern change in GCV might not be due to scarcity of variation, but rather due to limitations in the detection test.

Masny and Zurawicz (2014) estimated GCA and SCA effects of 10 strawberry cultivars for severity of *Verticillium* wilt and suggested that genetic, additive and non-additive effects were involved in the inheritance of strawberry for tolerance to *Verticillium* wilt. A significant negative GCA effect for the severity of *Verticillium* wilt in plants was found in cultivars such as Selvik, Filon and Sonata, indicating genetic transmission of tolerance from parents to the offspring. On the other hand, a significant and positive GCA effect for the severity of *Verticillium* wilt in cultivar Figaro indicated transmission of relatively high susceptibility to the wilt from this parent to its off-springs. Only in one hybrid family, Albion× Charlotte, the SCA effect was observed to be significantly positive for severity of *Verticillium* wilt in plants, whereas two other hybrid families, such as Selvik× Salsa and Sonata× Albion showed significantly negative effects for SCA.

TABLE 5.4
List of Cultivars with High GCA for Specific Characters

S.No.	Character	Variety	References
1.	Early ripening	Selection 2589 and Selection 1890	Rutkowski et al. (2006)
2.	Number of inflorescence	Selection 3387 and Selection 1387	Gawronski and Zebrowska (2005)
3.	Number of flowers per inflorescence	Selection 3387 and Selection 1387	Gawronski and Zebrowska (2005)
4.	Number of crowns	Selection 3387 and Selection 1387	Gawronski and Zebrowska (2005)
5.	Number of fruits per plant	Selection 3387 and Korona	Gawronski and Zebrowska (2005)
6.	Fruit size	Selva	Masny et al. (2005)
7.	Average fruit weight	Pandora and 160 Ms	Gawronski and Zebrowska (2005)
8.	Fruit yield	Selva, Selection 3387, 160 Ms and Senga Sengana	Masny et al. (2005)
9.	Fruit firmness	Selva	Masny et al. (2005)
10.	Fruit colour	Capitola	Masny et al. (2005)
11.	Resistance to grey mould	Selva	Masny et al. (2005)
12.	Verticillium wilt	Selvik, Filon and Sonata	Masny and Zurawicz (2014)
13.	leaf spot, leaf scorch and powdery mildew	Salsa, Figaro and San Andreas	Masny et al. (2016)

Masny et al. (2016) studied the general and specific combining abilities (GCA and SCA) of 13 dessert strawberry cultivars with respect to major leaf diseases including strawberry leaf spot (*Mycosphaerella fragariae* (Tul.) Lindau), leaf scorch (*Diplocarpon earliana* (Ell. & Ev.) Wolf) and, powdery mildew (*Sphaerotheca macularis* (Wallr.) U. Braun). The most significant and positive GCA effects were estimated for cv. "Salsa" for low plant susceptibility (tolerance) to all three diseases assessed. High GCA effects were also shown for cvs "Figaro" and "San Andreas"-for low plant susceptibility to leaf spot and leaf scorch, and for cv. Diamante for low plant susceptibility to leaf spot and powdery mildew. Lower breeding values for the estimated traits were observed with cv. Camarosa for low susceptibility to leaf spot; with cvs Palomar and Granda for low plant susceptibility to leaf scorch, and with cvs Monterey, Portola and Charlotte for low plant susceptibility to powdery mildew. The lowest GCA effects for low plant susceptibility to all the three leaf diseases tested were observed with cv. Albion. The Aromas× Elianny combination, showed significantly negative SCA effects for the susceptibility of plants to both strawberry leaf spot and leaf scorch and observed highly usefulness for breeding new, resistant varieties.

5.3.3 CORRELATIONS

Yield is a complex character and depends upon the expression of a number of components known as yield components. Correlation studies indicate a magnitude of association between any pair of characters. Knowledge of the association of yield components with each other and with other characters is helpful in the improvement of the complex characters, particularly yield, for which direct selection is not much effective.

Many studies have correlated vegetative and reproductive trait performance (Nicoll and Galletta, 1987; Lal and Seth, 1981; Hancock et al., 1982). There are significant positive correlations, usually between yield and fruit number, inflorescence number, leaf number and crown number. The berry yield has been found to have negative correlations with plant and root size, leaf area, petiole length, and runner number (Nicoll and Galletta, 1987).

Within genotypes, yield per plant mainly depends on number of fruits in either of the two-culture system. Potential yield differences within genotypes are apparently established before or during flower bud differentiation. Vegetative variables have strong correlations with yield, when the genotypes are grown as matted rows. When grown singly with less interplant competition, reproductive variables can be correlated with yield among genotypes. In some genotypes, runner production and fruiting compete for food assimilates. Genotypic yield variability suggests that genotypes with similar yield can have different routes to yield. Literature on correlation studies in strawberry is summarized in Table 5.5.

5.4 CYTOGENETICS

Cytogenetic investigations on fruit plants and their wild relatives leading to a knowledge of their interrelationships and the sterility barriers have not only provided a better understanding of the evolutionary processes but also have enabled the artificial introgression of desirable characters over natural isolation barriers. The induction of mutations and polyploidy has been proved to be of immediate benefit in fruit breeding. Investigations carried out on general cytology, induced polyploidy, interspecific hybridization and genome analysis in strawberry have opened new avenues of improvement, and successfully contributed to the development of radically new and better genotypes.

The earliest reports on *Fragaria* cytogenetics published by Ichijima (1926), focusses on chromosome counting and observations of any abnormal chromosomal behaviour (Figure 5.1).

Ichijima (1926) arranged *Fragaria* into four groups: diploids, tetraploids, hexaploids and octoploids, thereafter many studies to karyotype diploid *Fragaria* species have been made by several researchers (Iwastubo and Naruhashi, 1989; Iwastubo and Naruhashi, 1991). Longly (1926) first reported the chromosome number of *F. vesca*, *F. moschata*, *F. chiloensis*, *F. virginiana* and many

TABLE 5.5

Characters Showing Positive Correlation with Other Characters of Strawberry

S.No.	Character	Characters with Positive Correlation (r)	References
1.	Number of fruits per plant	Fruit volume, fruit weight, flower disk diameter, number of runners per plant, fruit length and fruit width	Verma et al. (2003)
2.	Fruit volume	Number of plantlets, days to runner formation, length of leaf petiole, flower diameter and fruit weight	Verma et al. (2002)
3.	Fruit size	Fruit arising node length/total peduncle length ratio and fruit arising node/lower node diameter ratio	Nathewet et al. (2007)
4.	Fruit yield	Number of berries per plant, harvest duration, berry weight, berry length and canopy spread	Das et al. (2006)
5	First flowering date and the number of leaves emerging between spring and summer inflorescences	Number of inflorescence and individual flowers	Honjo et al. (2011)
6.	Fruit yield	Days to runner formation, inflorescences number, fruit number, fruit size and number of achenes	Lal and Seth (1981)

varieties and hybrids. He considered *F. vesca* to represent the primitive form from which the others evolved, and held that dioeciousness (male and female flowers on separate plants) in the strawberry was associated with polyploidy. He also confirmed Millardet's results in obtaining maternal and paternal forms in crosses.

Kihara (1930) and Fedorova (1934) indicated that three of the genomes of *Fragaria × ananassa* were homologous with those of *F. moschata*. Fedorova obtained a tetraploid from *F. vesca × F. moschata* indicating the homology of the genomes of *F. vesca* with those of *F. moschata*. They observed that three of the *Fragaria × ananassa* genomes seemed closely related to that of *F. vesca*. Homology of the genomes of *F. moschata* with those of *F. vesca* was concluded by Schiemann (1944) from a tetraploid hybrid (*F. vesca × F. moschata*), which produced normal pollen grains and had always 14 bivalents in Meiosis-I. Thus, the homology of the three diploid species *F. vesca*, *F. viridis* and *F. nipponica* could be established.

Senanayake and Bringhurst (1967) proposed the genomic constitution of AAA'A'BBBB (2A2A'4B), which was generally accepted. Later Bringhurst (1990) suggested that the genomic structure of the octoploids (*Fragaria × ananassa*, *F. chiloensis*, and *F. virginiana*) may be modified to AAA'A'BBB'B' (2A2A'2B2B') based on the cytological and genetic (Arulsekar et al., 1981) evidences. In an attempt to determine the probability of polysomic inheritance in strawberry cultivars, Mok and Evans (1971), while studying the diakinesis of nine eastern North American cultivars, observed multivalent pairing in each of the nine cultivars during both the years, varying from quadrivalents and hexavalents to occasional octovalents. The authors concluded that tetrasomic inheritance was likely to be important in the cultivated strawberry. However, they noted that the rest of the meiotic cycles appeared normal.

Byrne and Jelenkovic (1976), while studying chromosome pairing of nine cultivars and 32 S_1 seedlings of *Fragaria × ananassa*, reported all chromosome pairing as bivalents, indicating cytological diploidization. Five cells with apparent multivalents were interpreted as pseudo multivalents because of their end-to-end or side-to-side associations, rather than the typical ring and chain multivalent associations. A completely sterile seedling was observed to have complete bivalent pairing at pachytene, but desynapsis to an almost completely unpaired condition by diplotene. Pentaploid

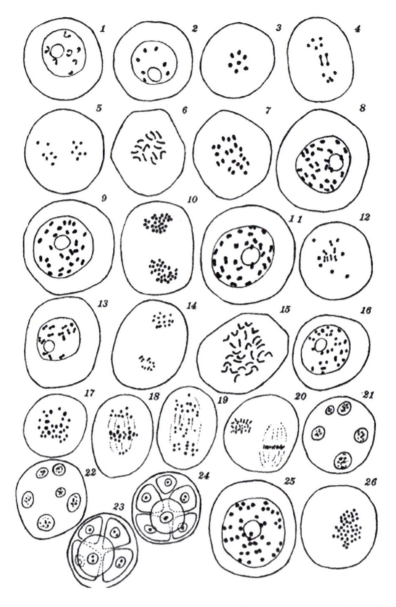

FIGURE 5.1 Original observations on mitosis and meiosis in *Fragaria* species (Ichijima, 1926).

hybrids between *Fragaria × ananassa* and the unrelated diploid species *F. nubicola* averaged 11.6 bivalents per Pollen Mother Cell (PMC) and frequent multivalent associations, indicating a residual homology between ancestral genomes of the octoploid strawberry.

In a group of sterile garden strawberry (8x), hautboy strawberry (6x) hybrids and their parent clones, meiosis has been reported essentially normal in the parent clones. The heptaploid (7x, 2n = 49) sterile hybrids exhibited several abnormalities including presence of univalents, trivalents, tetravalents and pentavalents at diakinesis; cytomixis during Prophase I; chromosome alignment off the spindles and premature chromosome movement to the poles in Metaphase I and Anaphase I; laggards at Anaphase I; and chromosome ejection into the cytoplasm at Telophase I, which may be repeated during the second meiotic division, causing abnormal spore numbers and pollen sterility (Berezenko, 1981).

Breeding and Improvement

Improved staining methods allowed clear karyotype analysis of higher ploidy *Fragaria* species, including octoploid *F. chiloensis* (Nathewet et al., 2007; Nathewet et al., 2009a). Fluorescent *in situ* hybridization (FISH) has been used to study 5S and 18S-25S rDNA in diploid and polyploid species (Liu and Davis, 2011). The 18S-25S rDNA sites are commonly located in the Nucleolus Organizing Region (NOR) of satellite (SAT) chromosomes. The 18S-25S rDNA sites were observed in one or two satellite chromosomes in some diploid *Fragaria* species (Liu and Davis, 2011). These sites can be used as chromosome-specific markers for higher resolution karyotyping.

5.4.1 CHROMOSOME NUMBER

The ability to count chromosomes has long been a valuable tool for plant breeders and cytogeneticists, as chromosome number plays important role in the systematics and taxonomy of strawberries in differentiating various species. This work is very difficult with strawberries since the chromosomes are extremely small and frequently have only a few cell divisions visible in a single root tip. Meiotic analysis are sometimes more successful but require plants at flowering stage. Cultivated large-fruited strawberries and the species *F. chiloensis*, *F. virginiana* and *F. ovalis*, from which they have been derived, intercross freely and their hybrids, with certain exceptions, produce fertile seedlings. But the cultivated varieties and the octoploid species, from which they have been derived, do not cross readily with diploid, tetraploid and hexaploid species, and seedlings of such crosses are nearly or entirely sterile. The barrier to hybridization has largely been due to differences in chromosome number, which are placed under the four chromosome number groups – 2x, 4x, 6x and 8x (Darrow, 1937).

5.4.2 CHROMOSOME DOUBLING

Chromosome doubling of germinating seedlings, especially of diploids, has been quite successful. Morrow and Darrow (1941) obtained tetraploid *F. vesca* by treating germinating seedlings with 0.2% colchicine solution for 24 hours. The seedling of Dorsett with 16-ploid was obtained by treating seedlings for five and six hours. Hull's 16-ploids resulting from colchicine treatment of seeds were chimeral. Ellis (1958) reported that 78% of the surviving treated seedlings of *F. vesca* were tetraploid with the use of 1% colchicine for 24 hours.

Ellis (1958) in his study summarized that *F. moschata* had been derived from the allopolyploid of *F. vesca* (4x) × *F. nubicola* (2x). He gave two methods to obtain octoploids with some chromosomes of *F. vesca*:

1. doubling of chromosomes of *Fragaria vesca* (2x) × *F. moschata* (6x) hybrids
2. doubling of chromosomes of *F. vesca* (4x) *Fragaria* × *moschata* (6x) hybrids and then back-crossing to *F. moschata*

And the cross *F. vesca* (4x) and *F. moschata* (6x) gave only vigorous pentaploids, which were all fertile males.

The ploidy level of species is helpful in explaining the phylogenetic relationships. Species with lower chromosome numbers are in most cases ancestral to the species with higher numbers. Cultivated strawberries (*Fragaria* × *ananassa*) are octoploid (2n = 8x = 56). Thirteen diploid species (2n = 2x = 14) are known in *Fragaria* in two centres of diversity, namely Central Asia and the Far East, four tetraploid species (2n = 4x = 28) confined to East and South East Asia, one hexaploid species (2n = 6x = 42) native to Europe and three octoploid (2n = 8x = 56) species (Staudt, 2008). Artificial triploid (2n = 3x = 21), tetraploid, pentaploid, octoploid, decaploid, 16-ploid and 32-ploid plants have also been developed and cultivated. Chromosome counts and flow cytometry revealed that *F. iturupensis* included natural decaploid genotypes (Hummer et al., 2009)

5.5 FLORAL BIOLOGY

Three flower types exist among octoploid species:

1. pistillate, which is devoid of anthers (female)
2. staminate, with nonfunctional pistils (male)
3. hermaphrodite (perfect) or complete

Plants may also be polygamodioecious, with the first few flowers on predominantly pistillate plants producing fruits (Hancock and Bringhurst, 1979). Most modern cultivars have hermaphrodite flowers only and, therefore, must be emasculated for controlled cross pollinations (Janick and Moore, 1996). Strawberry blossom is composed of many pistils, each with its own style and stigma attached to a receptacle which on fertilization of pistils develops into a fleshy "fruit". The true fruit are the nutlets (achenes) containing one seed each, which are found on the surface of the receptacle and result from the fertilization and development of the pistils. However, there is little information on the genetic variation of floral traits in species with separate genders. Gynodioecious (co-occurrence of female and hermaphrodites) species have a breeding system intermediate between hermaphroditism and complete separation of the sexes (dioecy), and thus can provide insights into the genetic architecture underlying floral phenotypes with respect to both primary (stamens and carpels) and secondary (petals) sexual traits (Ashman, 1999b).

In some species, male and female flowers can be readily distinguished, but in others (e.g., gynodioecious *F. vesca* subsp. *bracteata*), the female flowers have anthers, and resemble bisexual ones closely (Ashman et al., 2012). The mature receptacles ("fruits") are more diverse, and can be used to differentiate species.

5.5.1 CROSSING TECHNIQUES

Being a highly heterozygous species, selfing or inbreeding in strawberry usually poses many problems in successful breeding programmes. The pistil is allowed to be fully exposed through emasculation, which should be done one or three days before anthesis to prevent selfing. When emasculating, the thumbnail is generally used to remove the stamens, corolla and calyx at one operation. When working with small flowers, such as *F. vesca*, sharp pointed scalpels or tweezers sharpened to a cutting edge are used to remove the flower parts. If humidity is low, anthers can dehisce a day or two before anthesis. The emasculated flowers must be bagged to avoid the transfer of undesired pollen on them.

Pollen should be collected before one or two days of anthesis and thereafter kept in vials or paper-lined Petri dishes to dehisce and transferred to stigmas with a small hairbrush. Pollen viability can be extended for several years if kept at 4°C under low humidity. After 25–30 days of pollination, the fruits ripen under normal temperature conditions (18–25°C) and seeds can be separated from the pulp at that time.

Seeds should be stored at 1–4°C. To hasten the germination, the seeds should be subjected to the process of after ripening for 2.5–3 months (Bringhurst and Voth, 1957) followed by exposing the seeds to light and scarification for 10–15 minutes in concentrated H_2SO_4 (Janick and Moore, 1996). Seeds are sown in seed trays containing suitable sterilized medium for 40 days, and thereafter transplanted into soil in pots for another 45–55 days before being placed in the field.

5.6 BREEDING METHODS

As most of the characters segregate in F_2 and advanced generations due to highly heterozygous conditions of strawberry, selection pressure must be exerted simultaneously for important traits. To

Breeding and Improvement 63

avoid the chances of inbreeding and loss of vigour, a wide genetic base of parentage is desirable. New and superior strawberry cultivars may be evolved through introduction, selection, hybridization and mutagenesis.

5.6.1 INTRODUCTIONS AND SELECTION

Commercial strawberry cultivars that originated from introductions were basically developed in three different ways:

1. directly by increase *en masse* from the introduced stock(s)
2. selection made from introductions
3. hybrid offspring obtained from the introduced germplasm crosses with the local adapted varieties

Although most introductions did not play a significant role in early strawberry improvement work, moreover, modern strawberry varieties had descended from the crossing of *F. virginiana* with *F. chiloensis.* Various introductions such as Noble, Red Chilean, White Chilean, Ambato, Methven Scarlet, New Jersey Scarlet, Crescent Old Scarlet, Crimson Cone, Royal Sovereign etc. have enriched the genetic pool of strawberry and have extensively been used as sources of resistance genes for biotic or abiotic stresses (Darrow, 1937).

5.6.2 HYBRIDIZATION

Breeding for heterosis by means of inbreeding and hybridization has been found effective for *F. ananassa.* Selection and hybridization of the most useful inbred lines from different varieties gave much better results in most cases than the hybridization of varieties. The fruiting of seedlings could be accelerated by a year by scarification of freshly harvested seed with concentrated sulfuric acid and early spring sowing of the scarified seed under protected cultivation, with spring pricking out of the seedlings. The breeding process could also be accelerated by testing seedlings under artificial conditions of frost and infection and selecting resistant seedlings before planting them in the nursery.

Since strawberry is a highly heterozygous outbred fruit crop, two breeding schemes are most fruitful, viz., interspecific hybridization and intergeneric hybridization. Wide crosses within *Fragaria* and among *Fragaria* and closely related species (chiefly *Potentilla)* have been made for basically two reasons:

1. One was to introduce exotic genes into the garden strawberry.
2. The other was to produce polyhaploid individuals that could be doubled (chromosome) to yield isogenic lines.

The development of a superior strawberry cultivar begins with the production of hybrids exhibiting a high degree of heterosis for fruit yield, quality and stability. Cultivars may range from one- or two-way $(P_1 \times P_2)$ crosses to complex cross populations involving more than two parents. Depending on the breeding objectives, many strawberry breeding programmes exploit the full array of crosses, e.g., two-way $(P_1 \times P_2)$, three way $(P_1 \times P_2) \times P_3$; four-way $(P_1 \times P_2 \times P_3) \times P_4$; double crosses $(P_1 \times P_2) \times (P_3 \times P_4))$, backcrosses $(P_1 \times P_2) \times P_1$, and complex crosses involving many parents. Selection pressure must be exerted simultaneously for important traits in each segregating population. Hybridization in breeding programmes has evolved a number of modern cultivated strawberry varieties, such as Headliner, Armore, Dunlop, Catskill, Tannessee Beauty, Midway Redglow, Cambridge Regent, Cambridge Sentry, Cavalier etc.

5.6.2.1 Interspecific Hybridization

Direct crosses between most species result in low fertility but the chromosome numbers at the lower ploidy levels can be easily doubled using colchicine and unreduced gametes, which are reasonably common and result in the production of fertile seedlings (Bringhurst and Gill, 1970) as illustrated by the following:

F. vesca (2x)
↓
Colchicine treatment
F. vesca (4x) × *F. ananassa* (8x)
↓
6x hybrids × *F. ananassa* (8x)
(Unreduced gamete from 6x hybrids)
↓
10x hybrids (have small proportion of fertile seedlings)

This makes it possible to artificially transfer genes across the ploidy levels. Three approaches for transferring genes between the lower ploidy levels and octaploidy have been tested:

1. Genomically balanced decaploids (2n = 10x = 70) are used as bridges. These are produced using pentaploid or 10-ploid intermediates (Bringhurst and Gill, 1970).
2. It involves backcrossing nonaploids (2n = 9x = 63) to octoploids (Bringhurst and Gill, 1970). The nonaploids are generated by crossing decaploids with octaploids or by hybridization of 16-ploid plants with diploids.
3. This approach utilizes synthetic octaploids produced by chromosome doubling and redoubling lower ploidy levels (Darrow, 1966; Evans and Jones, 1967).

Natural hybrids with different ploidy levels have been detected in a coastal area of California where *F. vesca* and *F. chiloensis* were grown continuously with the following three options:

i. *F. vesca* (2x) × *F. chiloensis* (8x)
↓
5x hybrid (partially fertile)
ii. *F. vesca* (2x) × *F. chiloensis* (8x)
↓
6x hybrid (partially fertile)
(unreduced gamete from *F. vesca*)
iii. *F. vesca* (2x) × *F. chiloensis* (8x)
↓
9x hybrid (partially fertile)
(unreduced gamete from *F. chiloensis*)

When these three hybrids were pollinated with *F. chiloensis*, many of the resulting seedlings had chromosome numbers that could come only from unreduced gametes. The pentaploid yielded mostly nine-ploid seedlings. The hexaploid yielded more than 50% 10x seedlings, and 9x yielded aneuploid with chromosome numbers of 57 and 60 and a few with 2n = 56 and 2n = 63 euploids. Due to the relatively high fertility of nine-ploid hybrids, they may serve as a bridge for introgression of *F. vesca* genes into *F. chiloensis*. Presumably the same pathway could be utilized with cultivated strawberry.

The incorporation of these characters into cultivated strawberries is often complicated by the presence of some deleterious characters, viz., poor size, colour and quality of fruits, and susceptibility to leaf diseases. So the native octaploid species had to be used in the breeding programme for three–four generations of seedlings from outcross or backcross to the cultivated species to obtain

Breeding and Improvement

TABLE 5.6

Interspecies Crossability in the Genus *Fragaria*

	Male Parent								
Seed Parent	VE	VI	NL	NIP	OR	MO	VI	CH	AN
				Diploid					
F. vesca (VE)	1	1,3	1,3	3	3	3	2,3	2,3	2,3
F. viridis (VI)	1,4	1	4	1	3	2,3			3
F. nilgerrensis (NL)		4	4	1	3				
F. nipponica (NIP)		4	1	4	4	3			
				Tetraploid					
F. orientalis (OR)	2,3	2,3				1	2,3		
				Hexaploid					
F. moschata (MO)	3	2,3			2,3	1	3	3	3
				Octoploid					
F. virginiana (VI)	2,3	3				2,3	1	1	1
F. chiloensis (CH)	2,3	3				2,3	1	1	1
F. ananassa (AN)	2,3	3				2,3	1	1	1

Source: Hancock *et al.*, 1991

*1: Full seed set, viable F_1 plants, 2: Partial seed set, viable F_1 plants, 3: No viable F_1 plants or sterile F_1, 4: No seed set.

commercial progeny (Bringhurst and Voth, 1984). In addition, large populations are required to provide segregation among various characters involved in a practical breeding programme (Janick and Moore, 1996). The interspecies crossability in the genus *Fragaria* is presented in Table 5.6.

Some diploid species do intercross readily, but most hybrids have reduced fertility. The two tetraploids are said to be interfertile and their hybrids fully fertile. The three octaploids are entirely interfertile.

Darrow and Scott (1947) studied the ploidy levels of different species of strawberry. They produced decaploids by crossing *F. vesca* (4x) with cultivated octaploids to obtain hexaploids, then these hexaploids were crossed with octaploids to select out the 10-ploids. Ten-ploids were then intercrossed and several hundred seedlings raised from which commercial types could be selected.

The lower ploidy levels (2x, 4x and 6x) of *Fragaria* species possesses desirable characteristics such as unique flavours, vigour, disease and pest resistance, and adaptability to a wide range of habitats, which can be effectively used in strawberry breeding programmes (Harbut and Sullivan, 2004).

Marta et al. (2004) produced 35 hybrids between *Fragaria vesca* and *Fragaria ananassa*, out of which only 14% germinated and yielded short-lived plants due to less-developed endosperms and small embryos. Lei et al. (2005) studied the botanical and fruiting traits, meiosis and pollen of the interspecific hexaploid hybrid HH-1(6x) obtained from the cross between the cultivated strawberry Honeoye (8x) and the natural wild species Heilongjiang 1 (5x), and described the origins of various ploidy levels. They also obtained various hexaploid, heptaploid and octaploid seedlings from backcrossing the hexaploid hybrid HH-1 (6x) with Honeoye and Hokowase.

5.6.2.2 Intergeneric Hybridization

Hybridization between *Fragaria* and its related genus *Potentilla* is also possible but most hybrid progenies are sterile or rarely survive. However, two fertile colchicine-induced amphidiploids have been produced, which could be used to transfer genes between genera. These are:

1. a 14-ploid hybrid between octoploid *Fragaria × ananassa* and hexaploid *Potentilla polustris* (Ellis, 1962)
2. a decaploid hybrid between *F. chiloensis* and *Potentilla glandulosa* (Senanayake and Bringhurst, 1967)

Marta et al. (2004) studied the crosses between *Duchesnea indica* and *Fragaria × ananassa* and observed many putative hybrids when *D. indica* was used as female parent but very few hybrids were obtained when *D. indica* was used as the male parent due to pollen-pistil self-incompatibility reaction.

5.6.3 MUTAGENESIS

Mutations are the ultimate source of all genetic variability. Mutation breeding still provides a useful adjunct to standard breeding plans, and has a secure place in the fruit breeder's repertoire of techniques. Currently, mutation breeding is used to complement other breeding approaches, rather than as a standalone approach to strawberry improvement, and of late has begun to make a significant impact.

5.6.4 APOMIXIS

Apomixis is another form of asexual reproduction in which a seed is produced without the actual fusion of male and female gametes. Several variations of apomixis occur, but the essential component of all types is that of female gamete, either reduced or unreduced, being formed and developing into a seed. A new individual is formed without genetic contribution from the male. Apomixis is often confused with sexual reproduction because a seed is formed in both cases and extensive research may be required to determine if the progeny is produced sexually or apomictically. Sukhareva et al. (2002) developed an apomictic progeny between *F. vesca × F. ananassa* to enrich the genetic pool of multispecific decaploid hybrids of *Fragaria*. The resultant apomictic progenies with limited genetic diversity have been selected for further studies.

5.7 BREEDING OBJECTIVES

Strawberry breeding programmes must have well-defined objectives, keeping economical and biological viability in view. Since adversities that limit strawberry yield significantly differ from one production area to another, this leads to the inference that the objectives of the strawberry breeding programmes have to be different depending on the conditions that affect strawberry production. Usually, strawberry breeders attempt to improve different traits of economic importance (say 10–15). Of the various breeding objectives, fruit yield is the most important and biologically very complex indeed. Other breeding objectives that tend to stabilize strawberry production include resistance to diseases, insect pests, and tolerance to adverse environmental factors such as drought, cold, high temperature and fruit quality including nutritional value in accordance with consumers' demand.

5.7.1 ADAPTABILITY TO SOIL FACTORS

The soil habitants of different *Fragaria* species do differ. For example *F. chiloensis* grows better on sandy, well-drained soil than *F. virginiana* or *F. vesca* (Hancock and Bringhurst, 1979). The thick leathery leaved, deep rooted *F. chiloensis* is considered more drought tolerant than *F. virginiana* (Darrow, 1966). *F. orientalis*, *F. vesca* subsp. *americana* and *F. virginiana* subsp. *glauca* have been identified as having high resistance against drought (Goncharova and Dobrenkova, 1981).

The clones of *F. vesca* have much lower NaCl tolerance than *F. chiloensis*, the mean tolerance values being about 65000 ppm with little interpopulation variation (Hancock and Bringhurst, 1979). A list of drought- and salt-tolerant varieties of strawberries is given in Table 5.7.

Cultivated strawberries appear to differ in their relative sensitivity to changes in soil pH. The strawberry cultivar "British Sovereign" responded to limiting pH (4.5–5) when grown on low-lying delta soils, whereas Marshall did not (Boyce and Matlock, 1966). Wild strawberries have been observed growing vigorously on acidic, neutral and alkaline soils.

Breeding and Improvement

TABLE 5.7

List of Abiotic Stress-Tolerant Cultivars

Stress	Cultivars	References
Drought area	Dorsett, Fairfax, Catskill, Premier, Dunlap Blakemore Missionary, Surecrop, Ettersburg 121, Marshall, Ettersburg 80, Festival naya	Darrow (1966), Darrow and Deway (1934), Goncharova and Dobrenkova (1981)
Salt-affected areas	Festival naya Fletcher	Goncharova and Dobrenkova (1981), Dziadczyk et al. (2005)
Low temperature	Earlidawn, Howard 17	Scott and Lawrance (1975)
High temperature	Blakemore Missionary, *Fragaria virginiana*	Darrow (1966)

5.7.2 ENVIRONMENTAL FACTORS

The chilling requirement of strawberries varies with the cultivars. *F. chiloensis* is important in deriving successful cultivars for winter production. Day neutrality in everbearing strawberry cultivars (Aptos, Brighton, Tillikum, Fern, Mark, Yolo, Muir, Selva, Tribute and Tristar) was derived by cv. Bringhurst from high-elevation populations of *F. virginiana* ssp. *glauca* (Bringhurst and Voth, 1978; 1984). The primary source for the long-day trait in everbearing plants used in breeding programmes in the United States came from "Pan American" (Sjulin and Dale, 1987a), which arose as a mutant in a field of short-day cv. Bismarck during 1898. Examples of these are Geneva, Deep Red, Fort Laramie and Red Giant. Powers (1954) released "Ogallala" and "Arapahoe" cultivars, which were the combinations of *F. virginiana* ssp. *glauca* and cultivars containing the genes of cv. Pan American.

The flowers of two selections of *F. virginiana* from North Dakota have been found especially resistant to frost injury during a period of low temperatures (Darrow and Scott, 1947). Flowers of a selection of *F. virginiana* ssp. *glauca* from Cheyenne, Wyoming also have considerable resistance to frost. Some of the progeny from crosses between North Dakota selections and Midland cultivated strawberry have been reported to be as frost tolerant as the *F. virginiana* parent suggesting that frost tolerance is highly heritable.

Sjulin and Dale (1987b) summarized that the level of cold hardiness corresponds to the regions of origin and ancestry of these cultivars. The most hardy material was observed from Minnesota, Northern and Northern everbearing clusters, which are largely derived from "Dunlap" and "Premier" and have large ancestral contributions from the pure *F. virginiana* ssp. *virginiana* genotypes "New Jersey Scarlet" and "The Native Lowa", the *F. virginiana* ssp. *virginiana* derivative "Methven Scarlet" or *F. virginiana* ssp. *glauca*. High-latitude collections of *F. chiloensis* from Chile and Alaska are also useful for cold hardiness. The list of temperature stress tolerant cultivars is given in Table 5.7.

5.7.3 SPECIAL PLANT CHARACTERS

5.7.3.1 Stolon Production

Rates of stolon production are highly variable among cultivars, and substantial variations have been observed in wild collected clones of *F. vesca*, *F. chiloensis* and *F. virginiana* (Holler and Abrahamson, 1977; Honcock and Bringhurst, 1979). The nonrunning trait is highly heritable in *Fragaria* × *ananassa* and is closely correlated with the everbearing character. With few exceptions when nonrunning plants are selfed, their seedlings produce both running and nonrunning plants. Inbreeding to the S_3 generations gives a significant increase in the nonrunning seedlings among progenies, but no homozygous nonrunning genotypes have been obtained (Corbett and Meader, 1965).

In *F. vesca*, the runnering versus nonrunnering character is inherited with runnering being dominant and segregating in 3:1 ratio appearing in the F_2 population (Brown and Wareing, 1965). The early fruiting and adequate stolon production can be combined in an everbearing genotype (Simpson and Sharp, 1988).

5.7.3.2 Variegation

Variegation (June yellow or transient yellow) is an abnormal yellow-green mottled appearance of the leaves of strawberry plants. Notable examples of the delayed appearance of the condition are cultivars such as Auchincruive Climax, Blakemore, Tufts and Goldsmith (Z5A), which had gained extensive prominence before variegation appeared and weakened the plants. The parents carrying this disorder include Auchincruive Climax, Blakemore, Dixieland, Howard-17, Klonmore, Madame Moutot and Vermilion.

5.7.3.3 Fruiting Habit

Three types of fruiting habits are now recognized in strawberry plants:

1. single crop or June fruiting habit, which is the characteristic of most of the cultivars
2. everbearing two or more times/season and summer flowering governed by long photoperiods
3. day neutral

Powers (1954) intercrossed and selfed three everbearing (EB) and seven non-everbearing (NEB) parents in all the combinations. Some progenies segregated in a 3:1 (EB:NEB) ratio, some 3:5 and some in a ratio of 9:7. He postulated dominant and recessive genes for everbearing, with two or more complementary dominant genes of unequal potency and at least four recessive genes that govern expression of the character. In diploids of wild type *F. vesca*, everbearing was observed to be governed by a single recessive gene and acted independently of single recessive gene for non running. Bringhurst and Voth (1978) have argued that day neutrality is regulated by a single dominant allele.

5.7.3.4 Ripening Time and Fruit Development

The time of full bloom, time of ripening and rate of fruit development among strawberry seedlings are inherited quantitatively without heterosis (Peterson, 1953).

5.7.3.5 Fruit Size, Yield and Quality

Size of fruit is inherited quantitatively with six–eight allelic pairs controlling fruit expansion (Sherman et al., 1966; Spangelo et al., 1971). Baker (1952) observed heterosis in fruit size in some progenies of cross cultivars. Darrow (1966) suggested that *F. nilgerrensis* and *F. daltoniana* may possess useful genes for increasing fruit size.

Crown number per plant is often the factor most strongly associated with yield, which can be increased through either high-levels of stolon production or branch crown formation. In matted rows, rates of early stolon production are most strongly correlated with yield, while in hills, branch crown number per plant is of overriding importance. "Guardian" and "Honeoye" are the best genotypes for all purposes (Hancock et al., 1982). Crown production per plant can be increased either by shifting to the everbearing habit or by selecting prolonged fruiting in short-day types. *F. virginiana* ssp. *glauca* has been a rich source for both types of genes (Powers, 1954; Bringhurst and Voth, 1981). Clones of *F. chiloensis* also show a wide range of fruiting seasons, although none is everbearing (Bringhurst et al., 1977). *F. vesca* contains both extended fruiting and everbearing types (Brown and Wareing, 1965; Hancock and Bringhurst, 1978). Hancock et al. (1989) found *F. chiloensis* to have 10–40% higher photosynthetic rates than *F. ananassa*.

Breeding and Improvement 69

A strawberry breeding programme emphasizing improvement in the quality of fruit was taken up at Ohio State University, Ohio. Good fruit quality includes good taste, a good balance of sweetness and acidity, firmness, large berry size with long conical regular shape, a light red skin colour, uniformity between berries and longer shelf life. To select parents for hybridizations, a cross-section of cultivars were surveyed for desired fruit quality traits. Fruit firmness is a function of both flesh and skin toughness. Qurecky and Bourne (1968) found positive correlation between skin toughness and flesh firmness. Satisfactory handling quality depends on a combination of firm flesh and tough skin, which are greatly influenced by temperature and humidity during fruit development. Hansche et al. (1968) found fruit firmness to be highly heritable but negatively correlated with *Verticillium* wilt resistance.

The importance of strawberry, with respect to human health benefits, is also emerging with recognition at the consumer level (Alvarez-Suarez et al., 2011, 2014; Giampieri et al., 2012; Tulipani et al., 2014). Berries play an important role in the human diet due to their chemical composition, which is naturally enriched with many nutritional and bioactive compounds, such as minerals, vitamin C, folate, and phenolic substances (Giampieri et al., 2012). Several breeding studies have established that vitamin C content is quantitatively controlled (Anstey and Wilcox, 1950; Lundergan and Moore, 1975). Anstey and Wilcox (1950) noted a high correlation between the mean vitamin C content of fruits of a progeny and the mean of the parents. High vitamin C content of the fruit was partially dominant.

Fruit flavour is probably the most important attribute of a cultivar. According to Shaw (1988), the soluble solid and acid contents of fruits are genetically controlled with varying levels of additive and dominance control. Internal and external colour of the fruits are both highly heritable and quantitatively inherited (Lundergan and Moore, 1975).

Scalzo et al. (2005) studied nutritional and quality attributes of a few cultivars of strawberries such as antioxidant activity, total phenolic content, fruit firmness, colour, soluble solid content and titratable acidity, observed a high range of variation with respect to the characters under observation and suggested introgressing the characters in the desired cultivars through breeding.

Carrasco et al. (2005) studied the inheritance of fruit aroma character in F_2 segregating populations of strawberries and estimated maximum heritability for some of the biochemical compounds contributing to quality traits such as 2-methyl ethyl butanoate and 3-hexen-1-ol (Z), whereas heritability was low for furaneol, ethyl butanoate, butyl acetate, methyl butanoate, ethyl hexanoate and methyl hexanoate.

Similarly, Mezzetti et al. (2005) studied various quantitative and qualitative traits and observed significant and positive correlation between total antioxidant capacity as well as fruit yield and quality of different genotypes of strawberries, viz., "Sveva", "Cilady", "AN94.268.51", "Maya", "Onda", "Queen Elisa" and "Patty". They also observed higher nutritive values in Sveva, Cilady, AN94.268.51 and Maya cultivars.

Diamanti et al. (2012) screened 80 strawberry offspring, originated from eight inter- and intraspecific crosses of *Fragaria* spp., with respect to total antioxidant capacity, anthocyanin content and total phenolic content and succeeded in identifications and selection of new genetic material with improved combination of these nutritional and quality components of fruit. Progenies from strawberry crosses with *F. virginiana* ssp. *glauca* as a common parent showed a significant increase in their fruit quality and nutritional quality, which confirmed the potential of wild *Fragaria* genotypes to improve several desired characters of cultivated strawberries. The list of objectives specific cultivars is given in Table 5.8.

5.7.3.6 Pest Resistance

The genetics of resistance to the strawberry aphid has been observed to be controlled by gene(s) with more than a single locus, having partial dominance or additive gene action. Some clones of native *F. chiloensis* have been observed to be resistant to aphids (Shanks and Barritt, 1974; Crock et al., 1982). On crossing the genotype "Del Norte" with *Fragaria × ananassa* cultivars, the level

TABLE 5.8

List of Objective Specific Cultivars for Different Characters

Objective	Cultivar	References
Mechanical harvesting	Puget Beauty, Olympus, Totem, Red Chief,	Barritt (1976), MacIntyre and Gooding
Easy to cap	Mimak, Primek and Providence	(1978), Dale et al. (1987)
Long pedicels (>40 mm)	Saladin, Silver Jubilee and Troubadour	
Long necked fruit	Veeglow, Puget Beauty,	
	V 6747 R-6, V 72551-1 and SCR 165-L III	
Skin toughness	NY 844, Albritton, Apollo, Tennessee Shipper,	Ourecky and Bourne (1968), Hancock et
	Tioga Cardinal, Gorella and Holiday	al. (1996)
Vitamin C/High ascorbic	Marshall, Clarke, Catskill, Suwannee,	Hansen and Waldo (1944), Darrow (1966)
acid	Marshall, Sparkle, Fairpeake, Progressive	
	and *F. chiloensis*	
Flavoured processing type	Earlibelle, Veegem, Rainier and Shuksan	Sistrunk and Moore (1971), Stanley (1987)
Red-fleshed colour in the	Earlibelle, Cardinal, Veegam, Vontage, Rainer,	Sistrunk and Moore (1980)
frozen pack	Shuksan, Totem, Tyee, Senga Sengana and	
	Bounty	

of resistance in hybrids was intermediate to both the parents. Resistance of these hybrids could be transmitted through backcrossing, and seedlings could be recovered that are as resistant as Del Norte (Shanks and Barritt, 1974; Barritt and Shanks, 1980).

Resistance in *Fragaria× ananassa* to two-spotted spider mites appears to be highly heritable and quantitative (Barritt and Shanks, 1981). The continuous segregation for resistance in terms of plant injury ratings in relation to parental means, has been observed to be intermediate in progenies. A loss of resistance has been observed when resistant selections were selfed, but the resistance was reported when selfed selections were outcrossed to less-related parents (Chaplin et al., 1968, 1970).

Simpson et al. (2002) studied the inheritance of resistance to blossom weevil caused by *Anthonomus rubi* in the cultivated strawberries, *Fragaria× ananassa* involving seven cultivars and three improved breeding lines and observed a significant role of additive genetic variation in the inheritance of resistance to *A. rubi*. They also suggested "Idea" and "Alice" cultivars to be promising for use in strawberry breeding programmes.

5.7.3.7 Disease Resistance

Inheritance of resistance in strawberries to some of the important diseases such as *Alternaria* leaf spot, *Ramularia* leaf spot, leaf scorch, powdery mildew, *Verticillium* wilt, red stele root rot, anthracnose, crown rot, grey mould and *Fusarium* wilt has been studied well, with a fairly good degree of success. Resistance has been observed to be both quantitative and additive (Wilhelm, 1955), with a high estimate of heritability, partially dominant over susceptibility and governed by several genes (Bringhurst et al., 1968). *Verticillium* wilt resistance is negatively correlated with yield (Bringhurst et al., 1968). A list of strawberry varieties resistant/tolerant to some important diseases is furnished in Table 5.9.

The germplasm base for resistance to *Alternaria* leaf spot appears to be large, although region specific. It is not known whether this variation was due to environmental or pathogen variability. The susceptibility to the pathogen and sensitivity to toxins have been reported to be inherited as a single locus, with two alleles expressing incomplete dominance when heterozygous. Homozygous susceptible seedlings were more susceptible to the pathogen and more sensitive to its toxin than their heterozygous parent (Yamamoto et al., 1985). Breeding for resistance to leaf spot is generally restricted to the races that occur locally. No parent has been found that is resistant to all races (Nemec, 1971), although "Drabreak" from Louisiana, has been observed to be the most resistant

Breeding and Improvement

TABLE 5.9

List of Diseases Resistance/Tolerant Varieties of Strawberry

Disease	Varieties/Parent	References
Leaf spot	Albritton, British Sovereign, Earlibelle, Elista, Himiko, Hood, Howard-17, Lassen, Lateglow, Persikovaya, Rosanne, Surecrop, Tangi, Blakemore, Massey, Missionary Tokuku No.1	Hancock et al. (1990)
Leaf scorch	Sunrise x selfed (97% res.) Albritton x selfed (72% res.) Redglow x Headliner (13% res.)	Nemec and Blake (1971)
Angular leaf spot	Arrowhead, Massey, Sioux, Grenedier, Wisconsin 214, US 4808, US 4809	Lewers et al., (2003)
Powdery mildew	Puget Beauty, Catskill, Dunlap, Atmore, Sparkle Solprins, Kent, Patty Tioga, Brington	Daubeny (1961), Davik and Honne (2005), Simpson (1987)
Verticillium wilt	Sierra, Climax x selfed, Marshall x selfed, Pearle de Prague x selfed, Redcrop x selfed, Temple x selfed Catskill, Robinson, Blakemore, Cambridge Favourite, Redgauntlet, Aberdeen Gola, Gam K 40	Wilhelm (1955), Sowik et al. (2004)
Anthracnose	Sequoia, *F. vesca* Strawberry Parental Line No. 1	Gooding et al. (1981)
Fusarium wilt	Yachigo, Daehal 1, Line 10-2, Senga Sengana, Kurume 38, Senga Giana, Terunoka WB-A15, WB-A15 and WB-B33 Strawberry Parental Line No. 1	Honda et al. (1981), Kim et al. (1982) Takahashi et al. (2003)
Crown root	*F. vesca*, Senga Sengana	Horn and Carver (1963)
Verticillium wilt	Selvik, Filon and Sonata	Masny et al. (2014)
Leaf spot, leaf scorch	Salsa	Masny et al. (2016)

cultivar in Illinois. In quantitative studies on nature of inheritance for resistance to powdery mildew, it has been observed (Hsu et al., 1969) that non-additive variance is more important than additive, and considerable epistatic variance is generally present. Segregation appeared to depend on two additive genes for resistance and one pair of epistatic genes for susceptibility. The resistance to *Verticillium* wilt is partially dominant, inherited quantitatively and is mainly additive with a high heritability estimate (Wilhelm, 1955). Olbricht et al. (2006) studied the cross breeding of several accessions of *F. chiloensis* with *F. ananassa* cultivars "Elsanta", "Honeoye" and "Surecrop", and observed a high level of resistance to *Verticillium* wilt when these cultivars were crossed with specific accessions of *F. chiloensis* ssp. *lucida* and *F. chiloensis* ssp. *pacifica*. The grey mould resistance is quantitative and inherited additively, at least in part. Fruit firmness may play an important role in resistance to various diseases (Barritt, 1979).

Mori and Kifamusa (2003) studied the genetic variation in resistance to anthrocnose (*Colletotrichum acutatum*) in different crosses involving susceptible, medium and resistant cultivars of strawberry. Narrow- and broad-sense heritabilities were observed in most of the crosses, suggesting the presence of additive effect. He also observed selection responses in the F_1 and F_2 populations, and suggested that the concentration of resistant genes could be achieved through recurrent selection.

Denoyes-Rothan et al. (2005) studied the inheritance of resistance to anthracnose in *Fragaria × ananassa* cv. Belrubi, and observed that a single dominant gene *Rca2*, along with number of minor genes, was involved in the control of resistance to the disease. They also observed that the intermediate level of resistance to anthracnose was controlled by minor genes having quantitative inheritance.

The symptoms of virus infection vary with the variety, plant growth stage and weather conditions. Infected plants of some varieties express no obvious symptoms, but yield poorly due to viral infection. According to Sjulin et al. (1986), the GCA was higher than the SCA, which indicates a high proportion of genetic variance to be additive. Virus tolerance may vary in different areas due to the presence of different races or aphid population pressures (Hancock et al., 1990).

The nematodes enter into strawberry roots to feed and remain alive even after transplanting of runners, affecting nursery stocks. Most of the root nematodes are more destructive in sandy than in clay soils. Black roots in many soils are due to root lesion nematode, since the black root fungi enter roots injured by this nematode (Chen and Rich, 1962). The "British Sovereign", "Delite", "Earlidawn", "Gem", "Marlate", "Northwest", "Our Own", "Pocahontas", "Quinault", "Raritan", "Redstar", "Sparkle" and "Tristar" cultivars are resistant to nematode-borne virus (tomato ringspot virus).

5.8 MOLECULAR MARKERS AND GENOME MAPPING

Future strawberry breeding will involve more genetics and less empirical breeding. Therefore, breeding strategies are likely to focus on marker-assisted breeding, genetic modifications and genomics-based approaches. One of the most important and promising tools for strawberry research and the strawberry industry is that of molecular markers (Hokanson et al., 2001; Folta et al., 2005; Sargent et al., 2009; Whitaker, 2011).

Genomics, the analysis of an organism's complete DNA sequence, has been one of the most transformative influences on biological studies. The genome sequencing of organisms provides an understanding of the functions of individual genes and their networks, for defining evolutionary relationships and processes, in addition to revealing previously unknown regulatory mechanisms that coordinate the activities of genes.

The genome sequence of *F. vesca* was a scientific milestone, representing the first published plant genome obtained solely with next-generation sequencing technology (Shulaev et al., 2011). The small genome size in *Fragaria* and the availability of a reference sequence also facilitate comparative genomics. Whole-genome sequencing of the cultivated strawberry and four wild relatives (Hirakawa et al., 2014) has been done, which accelerates the precision breeding of the crop using marker-assisted selection and other genomic selection approaches. The central thrust of many research programmes is to now connect genes and traits using a variety of genetic, physiological, and developmental tests (Folta et al., 2005). The information on role of molecular markers and genome mapping in strawberry improvement has been discussed separately in depth under Chapter 8.

REFERENCES

Alvarez-Suarez, J.M., Dekanski, D., Ristić, S., Radonjic, N.V., Petronijevic, N.D., Giampieri, F., Astolfi, P., González-Paramás, A.M., Santos-Buelga, C., Tulipani, S. and Quiles, J.L. 2011. Strawberry polyphenols attenuate ethanol-induced gastric lesions in rats by activation of antioxidant enzymes and attenuation of MDA increase. *PLoS One*, 6(10): 258–278.

Alvarez-Suarez, J.M., Giampieri, F., Tulipani, S., Casoli, T., Di Stefano, G., González-Paramás, A.M., Santos-Buelga, C., Busco, F., Quiles, J.L., Cordero, M.D. and Bompadre, S. 2014. One-month strawberry-rich anthocyanin supplementation ameliorates cardiovascular risk, oxidative stress markers and platelet activation in humans. *The Journal of Nutritional Biochemistry*, 25(3): 289–294.

Anstey, T.H. and Wilcox, A.N. 1950. The breeding value of selected inbred clones of strawberries with respect to their vitamin C content. Scientific Agriculture 30: 367–374.

Arulsekar, S., Bringhurst, R.S. and Voth, V. 1981. Inheritance of PGI and LAP isozymes in octoploid cultivated strawberries. *Journal of the American Society for Horticultural Science*, 16: 679–683.

Ashman, T.L. 1999a. Determinants of sex allocation in a gynodioecious wild strawberry: implications for the evolution of dioecy and sexual dimorphism. *Journal of Evolutionary Biology*, 12(4): 648–661.

Ashman, T.L. 1999b. Quantitative genetics of floral traits in a gynodioecious wild strawberry *Fragaria virginiana*: implications for the independent evolution of female and hermaphrodite floral phenotypes. *Heredity*, 83: 731–741.

Ashman, T.L., Spigler, R.B., Goldberg, M.T. and Govindarajulu, R. 2012. *Fragaria*: a polyploid lineage for understanding sex chromosome evolution. In R. Navajas-Pérez (ed.) *New Insights on Plant Sex Chromosomes*, Nova Science, Hauppauge, New York, New York: 67–90.

Baker, R.E. 1952. Inheritance of fruit characters in the strawberry: a study of several F_1 hybrid and inbred populations. *Journal of Heredity*, 43(1): 9–14.

Barritt, B.H. 1976. Evaluation of strawberry parent clones for easy calyx removal. *Journal of the American Society for Horticultural Science*, 101: 590–591.

Barritt, B.H. 1979. Breeding strawberries for fruit firmness. *Journal of the American Society for Horticultural Science*, 104(5): 663–665.

Barritt, B.H. and Shanks Jr, C.H. 1980. Breeding strawberries for resistance to the aphids *Chaetosiphon fragaefolii* and *C. thomasi*. *HortScience*, 15(3): 287–288.

Barritt, B.H. and Shanks Jr, C.H. 1981. Parent selection in breeding strawberries resistant to two spotted spider mites. *HortScience*, 16: 323–324.

Battino, M., Beekwilder, J., Denoyes-Rothan, B., Laimer, M., McDougall, G.J. and Mezzetti, B. 2009. Bioactive compounds in berries relevant to human health. *Nutrition Reviews*, 67(suppl 1): 145–150.

Bedard, P.R., Hsu, C.S., Spangelo, L.P.S., Fejer, S.O. and Rousselle, G.L. 1971. Genetic, phenotypic and environmental correlations among 28 fruit and plant characters in the cultivated strawberry. *Canadian Journal of Genetics and Cytology*, 13(3): 470–479.

Berezenko, N.P. 1981. Study of meiosis in the hautboy strawberry and sterile strawberry-hautboy hybrids (in Russian). *Tsitologiya i Genetika*, 15: 37–40.

Bestfleisch, M., Möhring, J., Hanke, M.V., Peil, A. and Flachowsky, H. 2014. A diallele crossing approach aimed on selection for ripening time and yield in breeding of new strawberry (*Fragaria* × *ananassa* Duch.) cultivars. *Plant Breeding*, 133(1): 115–120.

Boyce, B.R. and Matlock, D.L. 1966. Strawberry nutrition. In: N.F. Childers (Ed.) *Fruit nutrition*, Horticultural Publications, Rutgers, The State University, New Brunswick, New Jersey: pp. 518–548.

Bringhurst, R.S. 1990. Cytogenetics and evolution in American *Fragaria*. *HortScience*, 25(8): 879–881.

Bringhurst, R.S. and Gill, T. 1970. Origin of Fragaria polyploids. II. Unreduced and doubled unreduced gametes. *American Journal of Botany*, 57: 969–976.

Bringhurst, R.S. and Voth, V. 1957. Effect of stratification on strawberry seed germination. *Proceedings of the Society for Horticultural Science*, 70: 144–149

Bringhurst, R.S. and Voth, V. 1978. Origin and evolutionary potentiality of the day-neutral trait in octoploid *Fragaria*. *Genetics*, 90: 510.

Bringhurst, R.S. and Voth, V. 1984. Breeding octoploid strawberries. *Iowa State Journal of Research*, 58: 371–381.

Bringhurst, R.S., Hansche, P.E. and Voth, V. 1968. Inheritance of verticillium wilt resistance and the correlation of resistance with performance traits. *Proceedings of the Society for Horticultural Science*, 92: 369–375.

Bringhurst R.S., Hancock, J.F. and Voth, V. 1977. The beach strawberry, an important natural resource. *California Agriculture*, 31: 10.

Bringhurst, R.S. and Voth, V. 1979. Strawberry plant. United States Patent (19). Plant 4,487.

Brown, I. and Wareing, P.F. 1965. Evaluating genetic sources of fruit detachment traits in strawberry. *HortScience*, 10: 120–121.

Byrne, D. and Jelenkovic G. 1976. Cytological diploidization in the cultivated octoploid strawberry Fragaria × ananassa. *Canadian Journal of Genetics and Cytology*, 18: 653–659.

Capocasa, F., Bordi, M. and Mezzetti, B. 2009. Comparing frigo and fresh plant in not fumigated and heavy soil: the response of ten strawberry genotypes. *Proceedings of the VIth International Strawberry Symposium Acta Horticulturae*, 842: 129–133.

Carrasco, B., Hancock, J.F., Beaudry, R. and Retamales, J. 2005. Chemical composition and inheritance patterns of aroma in *Fragaria* x *ananassa* and *Fragaria virginiana* progenies. *Horticulture Science*, 40: 1649–1650.

Chambers, A., Carle, S., Njuguna, W., Chamala, S., Bassil, N., Whitaker, V.M., Barbazuk, W.B. and Folta, K.M. 2013. A genome-enabled, high-throughput, and multiplexed fingerprinting platform for strawberry (*Fragaria* L.). *Molecular Breeding*, 31(3): 615–629.

Chaplin, C.E., Stoltz, L.P. and Rodriguez, J.G. 1968. The inheritance of resistance to the two-spotted mite *Tetranychus urticae* Koch in strawberries. *Proceedings of the Society for Horticultural Science*, 92: 376–380.

Chaplin, C.E., Stoltz, L.P. and Rodriguez, J.G. 1970. Breeding behaviour of mite resistant strawberries. *Journal of the American Society for Horticultural Science*, 95: 330–333.

Chen, T.A. and Rich, A.E. 1962. The role of Pra-tylenchus penetrans in the development of strawberry black root rot. *Plant Disease Reporter*, 46: 839–843.

Corbett, E.G. and Meader, E.M. 1965. Non-stoloniferous strawberry. *Journal of Heredity*, 56(5): 237–241.

Crock, J.E., Shanks Jr., C.H. and Barritt, B.H. 1982. Resistance in *Fragaria chiloensis* and *F.* × *ananassa* to the aphids *Chaetosiphon fragaefolii* and *C. thomasi. HortScience*, 17: 959–960.

Dale, A., Reid, M., Wang, S., Roderick, I., Sullivan, A., van Schyndel, S., Morton, J., Mclaughlin, I., Hughes, B. and Vandenberg, A. 1987. Ontario Coordinated Berry Crop Trials Report. Horticultural Research Institute of Ontario, Horticultural Experimental Station, Simcoe, Ontario, Canada: p. 47.

Darrow, G.M. 1937. *Strawberry Improvement.* Better plants and animals II. U.S. Department of Agriculture: pp. 445–495.

Darrow, G.M. 1966. *The Strawberry: History, Breeding and Physiology.* Holt, Rinehart and Winston, New York, New York.

Darrow, G.M. and Dewey, G.W. 1934. Studies on the stomata of strawberry varieties and species. *Proceedings of the Society for Horticultural Science*, 32: 440–447.

Darrow, G.M. and Scott, D.H. 1947. Breeding for cold hardiness of strawberry flowers. *Proceedings of the Society for Horticultural Science*, 50: 239–242.

Das, A.K., Singh, B. and Sahoo, R.K. 2006. Correlation and path analysis in strawberry (*Fragaria ananassa* Duch.). *Indian Journal of Horticulture*, 63(1): 83–85.

Daubeny, H.A. 1961. Powdery mildew resistance in strawberry progenies. *Canadian Journal of Plant Science*, 41: 239–243.

Davik, J. and Honne, B.I. 2005. Genetic variance and breeding values for resistance to a wind-borne disease [*Sphaerotheca macularis* (Wallr. Ex Fr.)] in strawberry (*Fragaria* × *ananassa* Duch.) estimated by exposing mixed and spatial models and pedigree information. *Theoretical* and *Applied Genetics*, 111: 256–264.

Denoyes-Rothan, B., Guérin, G., Lerceteau-Kohler, E. and Risser, G. 2005. Inheritance of resistance to *Colletotrichum acutatum* in *Fragaria* × *ananassa. Phytopathology*, 95(4): 405–412.

Diamanti, J., Capocosa, F., Balducci, F., Battino, M., Hancock, J.F. and Mezzeti, B. 2012. Increasing strawberry fruit sensorial and nutritional quality using wild and cultivated germplasm. *PLoS One*, 7(10): e46470.

Dziadczyk, P., Kiszczak, W., Tyrka, M.M. and Daiufer, K. 2005. Evaluation of strawberry (*Fragaria ananassa* Duch.) cultivar's salt stress tolerance. *Journal of Food Agriculture and Environment*, 3(2): 275–281.

East, E.M. 1930. The origin of the plants of maternal type which occur in connection with interspecific hybridizations. *The Proceedings of the National Academy of Sciences*, 16: 377–380.

Ellis, J.R. 1958. Cytogenetic studies in the genera *Fragaria* and *Potentilla*. PhD Thesis. University College, London.

Ellis, J.R. 1962. *Fragaria potentilla* intergeneric hybridization and evolution in *Fragaria.* In *Proceedings of the Linnean Society of London*, Blackwell Publishing Ltd., 173(2): 99–106.

Evans, W.D. and Jones, J.K. 1967. Incompatibility in *Fragaria. Candian Journal of Genetics and Cytology*, 9: 831–836.

Fedorova, N. 1934. Polyploid interspecific hybridizing *Fragaria. Genetica* 16: 525–541.

Flachowsky, H., Hofer, M. and Hanke, M.V. 2011. Strawberry. *Fruit, Vegetable and Cereal Science and Biotechnology*, 5: 8–26.

Folta, K.M., Staton, M., Stewart, P.J., Jung, S., Bies, D.H., Jesdurai, C. and Main, D. 2005. Expressed sequence tags (ESTs) and simple sequence repeat (SSR) markers from octoploid strawberry (*Fragaria* × *ananassa*). *BMC Plant Biology*, 5(1): 12.

Galletta, G.J. and Maas, J.L. 1990. Strawberry genetics. *Horticulture Science*, 25(8): 871–879.

Gawronski, J. and Zebrowska, J. 2005. Ocenazdolnoscikombinacyjnejwybranych genotype owtruskawki [*Fragaria* x *ananassa* Duch.]. Czesc I. Liczbakoron, kwiatostanowikwiatow wkwiatostanie. AnnalesUniversitatisMariae Curie-Skłodowska. *Sectio EEE: Horticultura* (15): 1–8.

Giampieri, F., Romandini, S., Alvarez-Suarez, J.M., Quiles, J.L., Mezzetti, B. and Battino, M. 2012. The strawberry: composition, nutritional quality, and impact on human health. *Nutrition*, 28(1): 9–19.

Goncharova, E.A. and Dobrenkova, L.G. 1981. Growth processes and yield in strawberry under conditions of drought and salinity. *Trudy po Prikladnoi Botanike, Genetike i Selektsii* 70: 97–102 [Plant Breed. Abstract 54: 9189].

Gooding, H.J., Jennings, D.L. and Topham, P.B. 1975. A genotype-environment experiment on strawberries in Scotland. *Heredity*, 34: 105–115.

Gooding, H.J., McNicol, R.J. and MacIntyre, D. 1981. Methods of Screening Strawberries for Resistance to *Sphaerotheca maculons* (Wall ExFrier) and *Phytophthora cactorum* (Leb. and Cohn). *Journal of Horticultural Science*, 56(3): 239–245.

Griffing, B. 1956a. Concept of general and specific combining ability in relation to diallel crossing systems. *Australian Journal of Biological Sciences*, 9(4): 463–493.

Griffing, B. 1956b. A generalized treatment of the use of diallel crosses in quantitative inheritance. *Heredity*, 10(1): 31–50.

Hancock, J.F. 1999. Strawberries. Crop production science in horticulture series, No 11. CABI, Wallingford, UK.

Hancock, J.F. 2005. Contributions of domesticated plant studies to our understanding of plant evolution. *Annals of Botany*, 96: 953–963.

Hancock, J.F. and Bringhurst, R.S. 1978. Inter-population differentiation and adaptation in the perennial, diploid species *Fragaria vesca* L. *American Journal of Botany*, 65: 795–803.

Hancock Jr, J.F. and Bringhurst, R.S. 1979. Ecological differentiation in perennial, octoploid species of *Fragaria*. *American Journal of Botany*, 66: 367–375.

Hancock, J.F., Siefker, J., Schulte, N. and Pritts, M.P. 1982. The effect of plant spacing and runner removal on twelve strawberry cultivars. *Advances in Strawberry Production*, 1: 19–20.

Hancock, J.F., Flore, J.A. and Galletta, G.J. 1989. Variation in leaf photosynthetic rates and yield in strawberries. *Journal of Horticultural Science*, 64: 449–454.

Hancock, J.F., Maas, J.L., Shanks, C.H., Breen, P.J. and Luby, J.J. 1990. Strawberries (*Fragaria ssp.*). In: J.N. Moore and J.R. Ballington (eds.) *Genetic Resources in Temperate Fruit and Nut Crops*. Intl. Sot. Hort. Sci, Wageningen, the Netherlands: pp. 489–546.

Hancock, J.F., Maas, J.L., Shanks, C.H., Breen, P.J. and Luby, J.J. 1991. Strawberries (*Fragaria*). *Acta Horticulturae*, 290: 491–548.

Hancock, J.F., Lavín, A.R. and Retamales, J.B., 1999. Our southern strawberry heritage: *Fragaria chiloensis* of Chile. *HortScience*, 34(5): 814–816.

Hancock, J.F., Drake, C.A. and Callow, P.W. 2005. Genetic Improvement of the Chilean native strawberry, *Fragaria chiloensis*. *HortScience*, 40: 1644–1645.

Hancock, J.F., Scott, D.H. and Lawrence, F.J. 1996. Strawberries. In: *Fruit Breeding. Vol II. Vine and Small Fruits* (Janick J, Moore J.N. eds.). John Wiley and Sons, New York, New York: 419–470.

Hancock, J.F., Sjulin, T.M. and Lobos, G.A. 2008. Strawberries. In: J.F. Hancock (ed.) *Temperate Fruit Crop Breeding: Germplasm to Genomics*. Springer, New York, New York: 393–438.

Hancock, J.F. Weebadde, C. and Serce, S. 2008. Challenges faced by Day- neutral strawberry breeders in the continental climates of Eastern United States and Canada. *HortScience*, 43: 1635–1636.

Hansche, P.E., Bringhurst, R.S. and Voth, V. 1968. Estimates of genetic and environmental parameters in the strawberry. *Proceedings of the Society for Horticultural Science*, 92: 338–345.

Hansen, E. and Waldo, G.F. 1944. Ascorbic acid content of small fruits in relation to genetic and environmental factors. *Food Research*, 9: 453–461.

Harbut, R.M. and Sullivan, J.A. 2004. Breeding potential of lower-ploidy *Fragaria* species. *Journal-American Pomological Society*, 58: 37–41.

Harland, S.C. and King, E. 1957. Inheritance of mildew resistance in *Fragaria* with special reference to cytoplasmic effects. *Heredity*, 11: 287.

Hirakawa, H., Shirasawa, K., Miyatake, K., Nunome, T., Negoro, S., Ohyama, A. et al. 2014. Draft genome sequence of eggplant (*Solanum melongena* L.): the representative *Solanum* species indigenous to the Old World. *DNA Research*, 21: 649–660.

Hokanson, S.C., Lamboy, W.F., Szewc-McFadden, A.K. and McFerson, J.R. 2001. Microsatellite (SSR) variation in a collection of *Malus* (apple) species and hybrids. *Euphytica*, 118: 281–294.

Holler, L.C. and Abrahamson, W.G. 1977. Seed and vegetative reproduction in relation to density in *Fragaria virginiana* (Rosaceae). *American Journal of Botany*, 64: 1003–1007.

Honda, F., Matsuda, T., Morshita, M., Iwanaga, Y. and Fushihara, H. 1981. Studies on the new strawberry cultivar Terunoka. *Bull Veg Ornam Crop Research Stn C Crume*, 5: 1–13.

Honjo, M., Nunome, T., Kataoka, S., Yano, T., Yamazaki, H., Hamano, M., Yui, S. and Morishita, M. 2011. Strawberry cultivar identification based on hypervariable SSR markers. *Breeding Science*, 61: 420–425.

Horn, N.L. and Carver, R.G. 1963. A new crown rot of strawberry plants caused by *Colletotrichum fragariae*. *Phytopathology*, 53: 768–770.

Hsu, C.S., Watkins, R., Bolton, A.T. and Spangelo, L.P.S. 1969. Inheritance of resistance to powdery mildew in the cultivated strawberry. *Canadian Journal of Genetics and Cytology*, 11: 426–438.

Hummer, K., Nathewet, P. and Yanagi, T. 2009. Decaploidy in *Fragaria iturupensis* (Rosaceae). *American Journal of Botany*, 96(3): 713–716.

Ichijima, K. 1926. Cytological and genetic studies on *Fragaria*. *Genetics*, 11(6): 590.

Ichijima, K. 1930. Studies on the genetics of *Fragaria*. *Molecular and General Genetics*, 55(1): 300–347.

Islam, A.S. 1960. Possible role of unreduced gametes in the origin of polyploidy. *Fragaria*. *Biologia*, 6: 189–192.

Iwastubo, Y. and Naruhashi, N. 1989. Karyotypes of three species of *Fragaria* (Rosaceae). *Cytologia*, 54: 493–497.

Iwastubo, Y. and Naruhashi, N. 1991. Karyotypes of *Fragaria nubicola* and *F. daltoniana*. *Cytologia*, 56: 453–457.

Janick, J. and Moore, J.N. 1996. *Fruit Breeding, Tree and Tropical Fruits*, Vol. 1. John Wiley & Sons, Inc., New York, New York.

Kihara, H. 1930. Karyologishe studien an *Fragaria* mit besonderer Beruck-sichtinung der geschlechtchromosomen. *Cytologia*, 1: 345–357.

Kim, C.H., Seo, H.D., Cho, W.D. and Kim, S.B. 1982. Studies on the varietal resistance and chemical control to the wilt of strawberry caused by *Fusarium oxysporum*. *Korean Journal of Plant Protection*, 21: 61–67.

Lal, D. and Seth, J.N. 1981. Studies on combining ability in strawberry (*Fragaria* x *ananassa*): 1 Number of inflorescences, number of flowers, days to maturity, and number of fruits. *Canadian Journal of Genetics and Cytology*, 23: 373–378.

Lewers, K.S., Mass, J.L., Hokanson, S.C., Gouin, C. and Hartung, J.S. 2003. Inheritance of resistance in strawberry to bacterial angular leafspot disease caused by *Xanthomonas fragariae*. *Journal of the American Society for Horticultural Science*, 128: 209–212.

Lei, J., Li, Y., Du, U., Dai, E.L. and Deng, M. 2005. A natural pentaploid strawberry genotype from the Changhai Mountains in Northeast China. *Horticulture Science*, 40(5): 1194–1195.

Liu, B. and Davis T.M. 2011. Conservation and loss of ribosomal RNA gene sites in diploid and polyploid *Fragaria* (Rosaceae). *BMC Plant Biology*, 11: 157.

Longly, A.E. 1926. Chromosomes and their significance in strawberry classification. *Journal of Agricultural Research*, 32: 559–568.

Luby, J.J., Hancock, J.F. and Ballington, J.R. 1992. Collection of native strawberry germplasm in the Pacific Northwest and Northern Rocky Mountains of the United States. *HortScience*, 27(1): 12–17.

Lundergan, C.A. and Moore, J.N. 1975. Inheritance of ascorbic acid content and colour intensity in fruits of strawberry (*Fragaria* x *ananassa* Duch *Journal of the American Society for Horticultural Science*, 100: 633–635.

MacIntyre, D. and Gooding, H.J. 1978. The assessment of strawberries for decapping by machine. *Horticulture Research*, 18: 127–137.

Mangelsdorf, A.J. and East, E.M. 1927. Studies on the genetics of *Fragaria*. *Genetics*, 12(4): 307–339.

Marta, A.E., Camadro, E.L., Diaz-Ricci, J.C., et al. 2004. Breeding barriers between the cultivated strawberry *Fragaria* x *ananassa* and related wild germplasm. *Euphytica*, 136: 139.

Masny, A. and Zurawicz, E. 2014. Combining ability analysis in 10 strawberry genotypes used in breeding cultivars for tolerance to Verticillium wilt. *Journal of American Society of Horticulture Science*, 139(3): 275–281.

Masny, A., Madry, W. and Zurawicz, E. 2005. Combining ability analysis of fruit yield and fruit quality in ever-bearing strawberry cultivars using an incomplete diallele cross design. *Journal of Fruit and Ornamental Plant Research*, 13: 5–17.

Masny, A., Mądry, W. and Zurawicz, E. 2008. Combining ability for important horticultural traits in medium- and late-maturing strawberry cultivars. *Journal of Fruit and Ornamental Plant Research*, 16: 133–152.

Masny, A., Zurawicz, E. and Mądry, W. 2009. General combining ability of ten strawberry cultivars for ripening time, fruit quality and resistance to main leaf diseases under Polish conditions. In VI International Strawberry Symposium, 842: 601–604.

Masny, A., Mądry, W. and Żurawicz, E. 2014. Combining ability of selected dessert strawberry cultivars with different fruit ripening periods. *Acta Scientiarum Polonorum Hortorum Cultus*, 13: 67–78.

Masny, A., Pruski, K., Zurawicz, E. and Mądry, W. 2016. Breeding value of selected dessert strawberry (*Fragaria*× *ananassa* Duch.) cultivars for ripening time, fruit yield and quality. *Euphytica*, 207(2): 225–243.

Mezzetti, B., Scalzo, J., Capocasa, F., Paladrani, S. and Battino, M. 2005. Il miglioramento genetico per aumentare qualita e capacita antiossidante delle fragole. *Frutticoltura*, 4: 26–29.

Millardet, A. 1894. Note sur Hybridation sans Croisement ou Fausse Hybridation. *Memoires Societe des Sciences Physiques et Naturelles de Bordeaux*, 4: 347–372.

Mishra, P.K., Ram, R.B. and Kumar, N., 2015. Genetic variability, heritability, and genetic advance in strawberry (*Fragaria*× *ananassa* Duch.). *Turkish Journal of Agriculture and Forestry*, 39(3): 451–458.

Mok, D.W.S. and Evans, W.D. 1971. Chromosome associations at diakinesis in the cultivated strawberry. *Canadian Journal of Genetics and Cytology*, 13: 231–236.

Mori, T. 2000. Heritability and selection effectiveness for fruit firmness in strawberry. *Journal of the Japanese Society for Horticultural Science*, 69: 90–96.

Breeding and Improvement

Mori, T. and Kitamura, H. 2003. Comparison of fruit quality and earliness of strawberry between inoculated and non-inoculated progenies for resistance to anthracnose. *Journal of the Japanese Society for Horticultural Science*, 72(1): 64–68.

Morishita, M. 2014. The status of strawberry breeding and cultivation in Japan. *Acta Horticulturae*, 1049: 125–131.

Morrow, E.B. and Darrow, G.M. 1941. Inheritance of some characteristics in strawberry varieties. *Proceeding of American Society for Horticultural Science*, 39: 262–268.

Nathewet, P., Yanagi, T., Iwatsubo, Y., Sone, K., Taketa, S. and Okuda, N. 2007. Chromosome observation method at metaphase and pro-metaphase stages in diploid and octoploid strawberries. *Scientia Horticulturae*, 114: 133–137.

Nathewet, P., Yanagi, T., Hummer, K.E., Iwatsubo, Y. and Sone, K. 2009a. Karyotype analysis in wild diploid, tetraploid and hexaploid strawberries, *Fragaria* (Rosaceae). *Cytologia*, 74(3): 355–364.

Nathewet, P., Yanagi, T., Iwatsubo, Y., Sone, K., Tamura, T. and Okuda, N. 2009b. Improvement of staining method for observation of mitotic chromosomes in octoploid strawberry plants. *Scientia Horticulturae*, 120: 431–435.

Nemec, S. 1971. Studies on resistance of strawberry varieties and selections to *Mycosphaerella fragariae* in southern Illinois. *Plant Disease Reporter*, 55: 573–576.

Nemec, S. and Blake, R.C. 1971. Reaction of strawberry cultivars and their progenies to leaf scorch in southern Illinois. *Horticulture Science*, 6: 497–498.

Nicoll, M.F. and Galleta, G.J. 1987. Variations in growth and flowering habits of Junebearing and everbearing strawberries. *Journal of American Society for Horticultural Science*, 112: 872–880.

Olbricht, K., Ulrich, D. and Dathe B. 2006. Cross breeding with accessions of *Fragaria chiloensis* resulting in selections with outstanding disease resistance and fruit quality characteristics. *Acta Horticulturae*, 708: 507–509.

Ourecky, D.K. and Bourne, M.C. 1968. Measurement of strawberry texture with an Instron machine. *Proceeding of the American Society for Horticultural Science*, 93: 317–325.

Peterson, R.M. 1953. Breeding behavior of the strawberry with respect to time of blooming, time of ripening, and rate of fruit development. Doctorial Series Publ. 5366 (Minn. Expt. Sta.).

Powers, L. 1954. Inheritance of period of blooming in progenies of strawberries. *Proceeding of American Society for Horticultural Science*, 64: 293–298.

Pringle, G.J. and Shaw, D.V. 1998. Predicted and realized response of strawberry production traits to selection in differing environments and propagation systems. *Journal of the American Society for Horticultural Science*, 123(1): 61–66.

Qurecky, D.K. and Bourne, M.C. 1968. Measurement of strawberry texture with Instron machine. *Proceeding of American Society for Horticultural Science*, 93: 317–325.

Richardson, C.W. 1914. A preliminary note on the genetics of *Fragaria*. *Journal of Genetics*, 3(3): 171–177.

Richardson, C.W. 1918. A further note on the genetics of *Fragaria*. *Journal of Genetics*, 7(3): 167–170.

Richardson, C.W. 1920. Some notes on *Fragaria*. *Journal of Genetics*, 10(1): 39–46.

Richardson, C.W. 1923. Notes on *Fragaria*. *Journal of Genetics*, 13(2): 147–152.

Rojas, B.A. and Sprague, O.F. 1952. A comparison of variance components in corn yield trials: III. General and specific combining ability and their interaction with locations and years. *Agronomy Journal*, 44(9): 462–466.

Rousseau-Gueutin, M., Lerceteau-Kohler, E., Barrot, L., Sargent, D.J., Monfort, A., Simpson, D., Arus, P., Guérin, G. and Denoyes-Rothan, B. 2008. Comparative genetic mapping between octoploid and diploid *Fragaria* species reveals a high level of collinearity between their genomes and the essentially disomic behavior of the cultivated octoploid strawberry. *Genetics*, 179(4): 2045–2060.

Rutkowski, P.K., Kruczynska, D.K. and Zurawicz, E. 2006. Quality and shelf life of strawberry cultivars in Poland. *Acta Horticulture*, 708: 329–332.

Salamone, I., Govindarajulu, R., Falk, S., Parks, M., Liston, A. and Ashman, T.L. 2013. Bioclimatic, ecological, and phenotypic intermediacy and high genetic admixture in a natural hybrid of octoploid strawberries. *American Journal of Botany*, 100(5): 939–950.

Sargent, D.J., Fernandez-Fernandez, F., Ruiz-Roja, J.J., Sutherland, B.G., Passey, A., Whitehouse, A.B. and Simpson, D.W. 2009. A genetic linkage map of the cultivated strawberry (*Fragaria × ananassa*) and its comparison to the diploid *Fragaria* reference map. *Molecular Breeding*, 24(3): 293–303.

Scalzo, J., Politi, A., Pellegrini, N., Mezzetti, B. and Battino, M. 2005. Plant genotype affects total antioxidant capacity and phenolic contents in fruit. *Nutrition*, 21(2): 207–213.

Schiemann, E. 1944. Artkreuzungen bei *Fragaria*. Ill, Die *vesca*-bastarde, Flora, Jena, 37: 66–192.

Scott, D.H. and Lawrence, F.J. 1975. Strawberries. In: J. Janick and Moore J.N. (eds.). *Advances in Fruit Breeding*, Purdue University Press, West Lafayette, Indiana: pp. 71–79.

Senanayake, Y.D.A. and Bringhurst, R.S. 1967. Origin of *Fragaria* polyploids: I. Cytological analysis. *American Journal of Botany*, 54: 221–228.

Shanks Jr, C.H. and Barritt, B.H. 1974. *Fragaria chiloensis* clones resistant to the strawberry aphid. *Horticulture Science*, 9(3): 202–203.

Shaw, D.V. 1988. Genotypic variation and genotypic correlations for sugar and organic acids of strawberries. *Journal of the American Society for Horticultural Science*, 113: 770–774.

Shaw, D.V. 1989. Variation among heritability estimates for strawberries obtained by offspring-parent regressions with relatives raised in separate environments. *Euphytica*, 44: 157–162.

Shaw, D.V. and Larson, K.D. 2005. Genetic variation and response to selection for early season fruit production in California strawberry seedling (*Fragaria* × *ananassa* Duch.) populations. *Journal of the American Society for Horticultural Science*, 130(1): 41–45.

Shaw, D.V. and Larson, K.D. 2008. Performance of early-generation and modern strawberry cultivars from the University of California breeding programme in growing systems simulating traditional and modern horticulture. *The Journal of Horticultural Science and Biotechnology*, 83(5): 648–652.

Shaw, D.V., Bringhurst, R.S. and Voth, V. 1987. Genetic variations for quality traits in an advanced cycle breeding populations of strawberry. *Journal of the American Society for Horticultural Science*, 113: 451–456.

Shaw, D.V., Bringhurst, R.S. and Voth, V. 1989. Genetic parameters estimated for an advanced-cycle strawberry breeding population at two locations. *Journal of the American Society for Horticultural Science*, 114: 823–827.

Shaw, D.V., Gordon, T.R., Hansen, J. and Kirkpatrick, S.C. 2010. Relationship between the extent of colonization by *Verticillium dahliae* and symptom expression in strawberry (*Fragaria* × *ananassa*) genotypes resistant to *Verticillium* wilt. *Plant Pathology*, 59(2): 376–381.

Sherman, W.B., Janick, J. and Erickson, H.T. 1966. Inheritance of fruit size in strawberries. *Proceeding of American Society for Horticultural Science*, 89: 309–17.

Shulaev, V., Sargent, D.J., Crowhurst, R.N., Mocker, T.C., Folkerts, O., Delcher, A.L., Jaiswal, P., et al. 2011. The genome of woodland strawberry (*Fragaria vesca*). *Nature Genetics*, 43: 109–116.

Simpson, D.W. 1987. The inheritance of mildew resistance in everbearing and day-neutral strawberry seedlings. *Journal of Horticultural Science*, 62(3): 329–334.

Simpson, D.W. 2014. Strawberry breeding and genetics research in North West Europe. *Acta Horticulturae*, 1049: 107–111.

Simpson, D.W. and Sharp, D.S. 1988. The inheritance of fruit yield and stolon production in everbearing strawberries. *Euphytica*, 38(1): 65–74.

Simpson, D.W., Easterbrook, M.A. and Bell, J.A. 2002. The inheritance of resistance to the blossom weevil, *Anthonomus rubi*, in the cultivated strawberry, *Fragaria* × *ananassa*. *Plant Breeding*, 121(1): 72–75.

Sistrunk, W.A. and Moore, J.N. 1971. Strawberry quality studies in relation to new variety development. Arkansas, Agriculture Experimental Station Bulletin. 761.

Sistrunk, W.A. and Moore, J.N. 1980. Evaluating strawberry selections for mechanization and quality. *Oregon Agriculture Experimental Station Bulletin*, 645: 133–141.

Sjulin, T. and Dale, A. 1987. Genetic diversity of North American strawberry cultivars. *Journal of the American Society for Horticultural Science*, 112(2): 375–386.

Sjulin, T.M. and Dale, A. 1987. Genetic diversity in North American strawberry. *Journal of the American Society for Horticultural Science*, 114: 823–827.

Sjulin, T.M., Robbins, J. and Barritt, B.H. 1986. Selection for virus tolerance in strawberry. *Journal of the American Society for Horticultural Science*, 111: 458–464.

Sowik, I., Borkowska, B., Wawrzynczak, D. and Michalczuk, L. 2004. Morphological characteristics and photosynthetic activity of selected strawberry clone with enhanced resistance to Verticillium wilt. *Acta Horticulturae*, 649: 181–184.

Spangelo, L.P.S., Hsu, C.S., Fejer, S.O., Bedard, P.R. and Rouselle, G.L. 1971. Heritability and genetic variance components for 20 fruit and plant characters in the cultivated strawberry. *Canadian Journal of Genetics and Cytology*, 13: 443–456.

Stanley, R. 1987. Stability of strawberry cultivars for freezing. *Journal of Horticulture*, 62(4): 501–505.

Staudt, G. 1962. Taxonomic studies in the genus Fragariatypification of *Fragaria* species known at the time of Linnaeus. *Canadian Journal of Botany*, 40(6): 869–886.

Staudt, G. 1989. The species of *Fragaria*, their taxonomy and geographical distribution. *Acta Horticulturae*, 265: 23–33.

Staudt, G. 2008. Strawberry biogeography, genetics and systematics. *Acta Horticulturae*, 842: 71–84.

Staudt, J.G., Dow Chemical Co. 1959. Method of fracturing oil wells. U.S. Patent 2,881,837.

Breeding and Improvement

Stewart, P.J. 2011. *Fragaria* history and breeding. In: Folta KM, Cole C (eds.) *Genetics, Genomics and Breeding of Berries,* Science Publisher, Boca Raton, Florida: pp. 114–137.

Strasburger, E. 1909. Zeitpunkt der Bestimmung des Geschlechts, Apogamie, Parthenogenesis und Reduktionsteilung. *Jena*, 41–47.

Sukhareva, N.B., Baturin, S.O. and Trajkovski, K. 2002. Introduction to apomixis in Fragaria vescana. *Acta Horticulturae*, 567: 231–234.

Takahashi, H., Yoshida, Y., Kanda, H., Furuya, H. and Matsumoto, T. 2003. Breeding of Fusarium wilt resistant strawberry cultivar suitable for field culture in North Japan. *Acta Horticulturae*, 626: 113–118.

Tulipani, S., Armeni, T., Giampieri, F., Alvarez-Suarez, J.M., González-Paramás, A.M., Santos-Buelga, C., Busco, F., Principato, G., Bompadre, S., Quiles, J.L., Mezzetti, B. and Battino, M. 2014. Strawberry intake increases blood fluid, erythrocyte and mononuclear cell defenses against oxidative challenge. *Food Chemistry*, 156: 87–93.

Van de Weg, W.E. 1997. A gene-for-gene model to explain interactions between cultivars of strawberry and races of *Phytophthora fragariae* var *fragariae. Theoretical and Applied Genetics*, 94: 445–451.

Verma, S.K., Singh, R.K. and Arya, R.R. 2002. Variability and correlation studies in strawberry germplasm for quantitative traits. *Indian Journal of Horticulture*, 59(1): 39–43.

Verma, S.K., Singh, R.K. and Arya, R.R. 2003. Estimate of genetic variability, heritability and genetic advance in strawberry. *Journal of Applied Horticulture Lucknow* 5(2): 102–104.

Vicente, E., Giménez, G., Manzzioni A., Vilaro, F., González, M. and Cabot, M. 2009. Strawberry Breeding in Uruguay. *Acta Horticulture*, 842: 411–414.

Whitaker, V.M. 2011. Applications of molecular markers in strawberry. *Journal of Berry Research*, 1: 115–127.

Whitaker, V.M., Osprio, L.F. and Hasing, T. 2012. Estimation of genetic parameters for 12 fruit and vegetative traits in the University of Florida Strawberry breeding population. *Journal of the American Society for Horticultural Science*, 137(5): 316–324.

Wilhelm, S. 1955. Verticillium wilt of the strawberry with special reference to resistance. *Phytopathology*, 45: 387–391.

Yamamoto, M., Namiki, F., Nishimura, S. and Khmoto, K. 1985. Studies on host-specific AF-toxins produced by *Alternaria alternate* strawberry pathotype causing Alternaria black spot of strawberry, use of toxin for determining inheritance of disease reaction in strawberry cultivar Morioka. *Annual Phytopathology Society of Japan*, 51: 530–535.

Yarnell, S.H. 1929. Notes on the somatic chromosomes of the seven-chromosome group of *Fragaria. Genetics*, 14(1): 78.

Yarnell, S.H. 1931. A study of certain polyploid and aneuploid forms in *Fragaria. Genetics*, 16(5): 455.

Yashiro, K., Tomita, K. and Ezura, H. 2002. Is it possible to breed strawberry cultivars which confer firmness and sweetness? *Acta Horticulturae*, 567: 223–226.

Zorrilla-Fontanesi, Y., Cabeza, A., Torres, A.M., Botella, M.A., Valpuesta, V., Monfort, A., Sanchez-Sevilla, J.F. and Amaya, I. 2011. Development and bin mapping of strawberry genic-SSRs in diploid *Fragaria* and their transferability across the Rosoideae subfamily. *Molecular Breeding*, 27(2): 137–156.

Zorrilla-Fontanesi, Y., Rambla, J.L., Cabeza, A., Medina, J.J., Sanchez-Sevilla, J.F., Valpuesta, V., Botella, M.A., Granell, A. and Amaya, I. 2012: Genetic analysis of strawberry fruit aroma and identification of O-methyltransferaseFaOMT as the locus controlling natural variation in mesifurane content. *Plant Physiology*, 159(2): 851–870.

6 Varieties

V.K. Tripathi, R.M. Sharma, and Sanjeev Kumar

CONTENTS

Acadie (Glooscap × Guardian) 85
Adina (88-042-35 × Parker)..85
Adria (Gramda × Miss)...85
Aguedilla (Camarosa × 67.35)..86
Alessandra (Director P. Wallbaum × Pocahontas) 86
Alice 86
Alinta (88-011-30 × Chandler) 86
Amelia 86
Anaheim (Irvine × Cal 85.92-602)..86
Annapolis [K74-5(Micmac × Raritan) × Earligow]..86
Ariel 87
Baotong (Benihomalei × Camarosa)..87
Barak (Tamir × ARO 730)...87
Bell Raugex (77-36 × Morioka 19) × (Morioka 19)...87
Belrubi (Pocahontas × Redcoat) 87
Benicia (Palomar × Cal 0.18-601) 87
Blomidon [K72-4 × {Micmac × (Guardsman × Tioga)}]..87
Bob (1501 × 1242)...88
Bolero (includes Redgauntlet, Gorella, Cordinal and Selva)...88
Bountiful (Linn × Totem)..88
Brunswick (Cavendish × Honeoye) 88
Cabot (K87-5 × K86-19)..89
Calypso (Rapella × Selva)...89
Camarosa (Douglas × Cal 85.218-605) 89
Cambridge Prizewinner (Early Cambridge × Howard 17) 89
Cambridge Favourite [(Etter Seedling × Avant Tout) × Blackmore] 90
Cambridge Rival (Dorsette × Early Cambridge) 90
Cambridge Vigour...90
Capitola (Cal 75.121-101 × Rarker)..90
Capri (CIVRI-30 × R6R1-26)..90
Carisma (Oso Grande × Vilanova)...91
Carmine (Rose Linda × FL 93-53)...91
Catskill (Marshall × Premier) 91
Cavalier (Valentine × Sparkle) 91
Chambly (Sparkle × Honeoye) 92
Chandler (Douglas × Cal 72.361-105)...92
Cristina (Selection AN01.211.51)..92
Chunxing (Chunxiang × Haiguan Zaohony)..92
Chunxu (Chunxiang and a Polish cultivar) 93
Crusader [Auchincruive Sdlg. 50 × 5 {Sdlg. 6M17 (Temple selfed × Sdlg. 11 (Aberdeen
open)} × Cambridge Vigour] 93
Cuesta [Seascape × Cal 83.25-2 (Fern × Parker)]...93

Cupcake (102213 × 1581) 93
Dange (Venta × Redgauntlet) 93
Delia (Honeoye × ISF 80-52-1) 93
Dely (T2-6 × A20–17) 93
Donna (Fern seeding × Douglas) 94
Douglas (Tufts × 64-57-108 (Tioga × Sequoia)) 94
Dris Straw Eighteen (91J302 × 26H165) 94
Dris Straw Nineteen (Driscoll Atlantis × 43J313) 95
Dris Straw Twenty (2K297 × Driscoll Ojai) 95
Dris Straw Twenty One (13H377 × 587L48) 95
Dris Straw Twenty Two (Dris Straw Three × 50L206) 95
Dris Straw Twenty Three (1M16 × 87K286) 95
Dris Straw Twenty Four (3M44 × 50L174) 95
Dris Straw Twenty Five (18L33 × 192M122) 96
Dris Straw Twenty Six (18L33 × 193M68) 96
Dris Straw Twenty Seven (Dris Straw Eight × 10L297) 96
Dris Straw Twenty Eight (95L299 × 251M27) 96
Earlibrite (Rosa Linda × FL 90-38) 96
Earlidawn (Midland × Tennessee Shipper) 96
Early Bommel (New Dutch seedling No. 69-16, was introduced commercially as Early
Bommel) 97
Elegance (EM834 × EM1033) 97
Elista [(Jucunda × Self) × US-3763] 97
Elvira (Vola × Gorella) 97
Enzed Levin (Orion) (84-1-38 × 84-8-443) 97
Erie (Sparkle × Premier) 97
Etna (Marlate × Belrubi) 98
Evangeline (K 88-4 × NYUS 119) 98
Everest (Irvine × Evita) 98
Evita [Chandler × (Gorella × Brighton)] 98
Fengguan 1 (Toyonoka × Tochitome) 98
Fern (Tufts × Cal 69.63-103) 98
Ferrara (Md. US4083 × Holiday) 98
Festival (Rosa Linda × Oso Grande) 98
FL 05-107 or Winterstar™ (Florida Radiance × Earlibrite) 99
Flair (Flevo00-24-7 × Flevo 00-08-4) 99
Flavorfest (B759 × B786) 99
Florence [Tioga × (Redgauhtlet × (Weltguard ×
Gorella)] × (Providence × Self) 100
Florida Radiance (Winter Dawn × FL 99-35) 100
Galletta (NCH 87-22 × Earliglow) 100
Gardena (selected in progeny from the cross of Addie × Pajaro) 100
Gea [R66/60{(Senga Sengana × Valentine) × Surprise des Halls} × Md US38] 100
Gem (a bud mutation of Champion Everbearing; also called Brilliant and Superfection) 100
Ginza (1929 × 1902) 101
Glory (Nelson. PS-1269 × PS-2286) 101
Gorella (Juspa × US-3763) 101
Grenadier (Valentine × Fairfax) 101
Gurdsman (Claribel × Sparkle) 101
Haruyoi (Harunoka × Hokowase) 102
Herriot (NYUS299 × Winona) 102

Varieties 83

Honeoye (Vibrant × Holiday) 102
Hongjia.. 102
Hongshimei (Zhangji × Ducra)... 102
Hongxiutianxiang (Sel. 06-37-2) (Camarosa × Benihoppe)............................. 102
Huxley (also known as Ettersburg 80) 102
Independence [ORUS 850-48(Linn × ORUS 3727) × ORUS 750-1
(Totem × ORUS 3746)]... 103
Joly (T2-6 × A20-17) 103
Jewel (selected in 1971 from the progeny of NY 1221 × Holiday).................... 103
Jingyuxiang (Camarosa × Benihoppe) Syn. Sel. 06-37-8................................. 103
Jingyixiang (Camarosa × Benihoppe) Syn. Sel. 06-37-10................................ 103
Jingchunxiang {(01-12-15′ (advanced selection) ×
Kinuama)} Syn. Sel. 06-54-2 103
Jingquanxiang {(01-12-15 (advanced selection) ×
Benihoppe)} Syn. Sel. 06-56-6).. 103
Joliette (Jewel × SJ 85189) 104
Jukhyang (Redpearl × Maehyang) 104
Kaorino (Selected from hybrid lines, line 0028401 × Line 0023001)............... 104
Kentaro (Kita-ekubo × Toyonoka) 104
Kimberley (Gorella × Chandler).. 104
Kita-ekubo [59 Kou-13-37(Aito × Morioko 19) × Reiko].............................. 105
L'Authentique Orleans (AC-L'Acadie × Joliette)... 105
Laguna (Irvine × Cal 85.92-602).. 105
Lambada [Selected from the cross between IVT76013 (Silvetta × Holiday) × IVT74112
(Karina × Primella)].. 105
Lincoln (Donna × 85-24-1, derived from Cruz, Pajaro, Douglas and Holiday) 105
Laurel (Allstar × Cavendish)... 105
Louise (Ettersburg 80 selfed; also called Churchill).. 105
Maehyang (Trochinomine × Akihime) 106
Malwina (Unnamed seedling × Sophie) 106
Mari (Pocahontas × Lihama) 106
Marina (Cartuno × Camarosa) 106
Marvel (PS-1269 × PS-2280).. 106
Merit (PS-2880 × PS-4630) 106
Merton Princess 106
Milsei (Parker × Chandler) 106
Mira (Scoot × Honeoye) 107
Missionary... 107
MNUS 210 [Earligrow × MNUS 52] Syn. Winona TM 107
Mojave (Palomar × Cal 1.57-601) 107
Nabila (Ventanax Q6Q8-26) 107
Niigata S3 (Kei 812 (seed parent) × Asuka-Ruby)... 108
Ningyu (Sachinoka × Akihime) 108
Oka (K 75-13 × Kentville).. 108
Ontaria.. 109
Oso Grande [Parker (U.S. Plant Pat. No. 5263) × Cal 77.3-603] 109
Pajaro (Cal 63.7-101 × Sequoia).. 109
Palomar (Camino Real × Ventana).. 109
Pavana 110
Pegasus (Redgauntlet × Gorella) Syn. ES608.. 110
Pink Beauty and Pretty Beauty (Kinuama × Pink Panda)................................. 110

84 Strawberries

Pink Panda (*F. comarum* hybrid involving *F. chiloensis*) 110
Pinnacle (Laguna × ORUS 1267-250) 110
Planasa 02-32 (94-020 × 9719) 110
Premier (Crescent × Howard No 1) 111
Rania (Ventanax Q6Q8-26) 111
Rebecka [(Fern × *F. vesca*) × *F. ananassa* F 861502] 111
Red Coat (Sparkle × Valentine) 111
Red Rich (also called Red-Glo and Hagerstrom's Everbearing) 111
Redgauntlet (New Jersey 1051 × Auch. Climax) 111
Romina (Selection AN99, 78, 51 derived from the cross of 95.617.1 × Darselect) 111
Rosalyne [Fern × (SJ9661-1 × Pink Panda)] 112
Roseberry [(Raritan × K74-12) × (SJ9616-1 × Pink Panda)] 112
Rosie (Honeoye × Forli) 112
Roxana [Surprise des Halles (Market Surprise) × Senga Sengana] 112
Sable (Veestar × Cavendish) 112
Sabrina (9719 × 94-020) 112
Saint Pierre (Chandler × Jewel) 112
Saint-Laurent d'Orleans [L Acadie × (SJ 8916 × Pink Panda)] 113
Salla (Addie × Pajaro) 113
Santaclara (Carisma × Plasirfre) 113
Saulene (Shuksan × Senga Sengana) 113
Selva (Cal 70.3-117 × Cal 71.98-605) 113
Senga Sengana (Markee × Sieger) 113
Shimei 3 (183-2 × All Star) 113
Shimei 4 (Hokowase × Shimei 1) 113
Shimei-7 (Tochiotome × Allstar) 114
Sophie (NY 1261 × Holiday) 114
Sparkle (Fairfax × Aberdeen; also called Paymaster) 114
Strawberry Festival (Rosa Linda × Oso Grande) 114
Surecrop (Fairland × Maryland-U.S. 1972) 114
Summer Ruby (57K104 × 572I16) 114
Sveva (EM483 × 87.734.3) 114
Sweet Ann (4A28 × 10B131) 115
Sweet Charlie (FL 80-456 × Pajaro) 115
Talisman (New Jersey × Auch. Climax) 115
Tango (Rapella × Selva) 115
Templer (Seedling Auch. 11 × Cambridge Vigour) 115
Terunoka (Hokowase × Donner) 115
Tillamook (Cuesta × Puget Reliance) 115
Tioga [Fresno × Torrey (Lassen × Cal 42.8-16)] 116
Tochiotome (Kurume 49 × Tochinomine) 116
Torrey (Lassen × Cal 42.8.16) 116
Vibrant [(SDBL101 × EM881) × (Rosie × Onebor)] 116
Winterstar™ (Florida Radiance × Earlibrite) Syn. FL 05-107 116
Yamaska (Pandora × Bogota) 117
Yueli (Benihoppe × Sachinoka) 117
Yuezhu (Akihime × Toyonoka) 117
Zarina (1007 × 880) 117
Zuoheqingxiang 117
References 117

Right from the beginning of its domestication during the 16th century, the overall improvement in productivity and enhanced quality of fruits in strawberries have largely been due to the untiring efforts of plant breeders and amateurs in developing trait-specific varieties. Being genetically precocious, strawberry has become a choice crop for large, marginal and small farmers alike. Further, the technology for soilless culture has made it possible to cultivate strawberries across the geographical regions in controlled climatic conditions. Availability of climate-specific varieties has facilitated the culture and, therefore, availability of strawberries round the year. The growers' choice of strawberry varieties depends largely on traits such as resistance to pests and diseases, adaptability to specific climatic conditions, long reproductive phase for higher yield of better-quality fruits, etc. According to Spangelo (1962), the following characteristics should be considered in selecting the varieties:

- high yielding and stable berry size even after the third picking
- firm fruit and good appearance after shipping
- good dessert quality
- good processing quality such as bright red surface of fruit peel and flesh, easily removable hulls, fruits of uniform, medium size, ability to retain fruit shape, firm but tender texture, good flavour and minimum trimming waste
- suitable time of ripening
- good resistance of or tolerance to diseases and pests

In 1994 and 1995, the number of main European collection of strawberry cultivars was 900. In 1998, the European COST Action 836 "Integrated Research in Berries" was started to coordinate the scientific action in 20 partner countries. The overview report of working groups has shown that the collections maintained by participating institutes have changed significantly within few years. In 18 institutes, 2,747 cultivar accessions and 418 accessions of wild species are being conserved. Nearly half the 928 listed cultivars were grown only at one site and four important cultivars seemed to had been lost. Subsequently, based on historical significance, 106 cultivars were selected for further maintenance in a core collection (Geibel, 2002).

A detailed description of strawberry cultivars grown in different parts of the word is given in the following paragraphs.

ACADIE (GLOOSCAP × GUARDIAN)

June-bearing cultivar released in 1989. Fruits are large, firm with good storage life at room temperature and ideal for shipping. Resistance to leaf scorch, leaf blight and leaf spot (*Mycosphaerella fragariae*) and moderately resistant to races of the pathogen causing red stele disease (Khanizadeh et al., 2002a; Khanizadeh, 2004).

ADINA (88-042-35 × PARKER)

Evolved by B. J. Morrison in Knoxfield, and introduced in 1997. It is moderately productive, short day, semi-spreading, medium vigour; fruits very large, highly attractive, bright red, moderately glossy; seeds level with surface of the fruit. Primary fruits are wedge shaped and, secondary and tertiary ones conical. Flesh light-red with white core; comparatively more firm than Pajaro. Plant breeder's rights for this hybrid are licensed to Agriculture Victoria Services Pvt. Ltd. (Finn, 1999).

ADRIA (GRAMDA × MISS)

Evolved during 1995 in Italy, late in harvesting season with improved productivity of high-quality fruits having high antioxidant properties (Capocasa et al., 2004a, b).

AGUEDILLA (CAMAROSA × 67.35)

A short-day Spanish cultivar developed in 1998, which produces large-sized and wedge-shaped fruits during extra early, mid or late season with a low percentage of second-grade fruits (Lopez et al., 2005).

ALESSANDRA (DIRECTOR P. WALLBAUM × POCAHONTAS)

The plants are vigorous and bear only female flowers. The fruits are large, bright red and juicy and have pleasant flavour. It is particularly suitable for growing in hilly regions or under cover for autumn cropping.

ALICE

This is a medium-late (June) bearing, moderately vigorous and productive cultivar released during 1993 by Horticulture Research International (HRI), East Malling, UK. Fruits are firm, red, sweet and pleasant flavour. Resistant to fungi, *Verticillium dahliae* and *Phytophthora cactorum* (Simpson et al., 2002).

ALINTA (88-011-30 × CHANDLER)

It was evolved by B. J. Morrison during 1998 in Knoxfield, Australia. It is a day-neutral, semi-spreading, medium dense, vigorous and early cultivar. Fruits are very large, orange-red, very glossy; fruits set during early season are short and wedge shaped and tend to conical shape on second-set fruits; flesh is light-red throughout, similar to Selva in firmness. It continues flowering during hot weather and fruits are suitable for fresh market (Finn, 1999).

AMELIA

It is a short-day variety, developed in 1998, bearing medium–large, conical fruits with uniform red skin, firm texture; good favour; ripens late during June–July about 10 days after Elsanta. Plants are vigorous with moderate fruit yield and runner production, resistant to powdery mildew (*Podosphaera aphanis*, formerly *Sphaerotheca macularis*) and *Phytophthora* crown rot (*P. cactorum*) but moderately susceptible to *Verticillium* (*V. dahliae*) wilt (Kim, 2010).

ANAHEIM (IRVINE × CAL 85.92-602)

It has originated in California, South Coast Research and Extension Center (SCREC), near Irvine and introduced during 1992. It is more vigorous and erect than Chandler. Plants are short-day-type producing fruits over an extended period in a subtropical climate. Compared with Chandler, fruits are slightly smaller, more firm, with external colour lighter, more orange and less glossy; internal colour lighter and less aromatic. Achenes yellow–light-red and slightly extruded fruits have very good flavour, suitable for fresh market, processing and home/garden uses. It is moderately resistant to common leaf spot and powdery mildew and tolerant to two-spotted spider mite (Daubeny, 1994).

ANNAPOLIS [K74-5(MICMAC × RARITAN) × EARLIGOW]

It originated in Kernville, Nova Scotia Agriculture Canada Research Station, and was introduced during 1984. Plants are vigorous. Primary fruits are large; secondaries are medium size, medium firm, and light–medium red external and internal colour. Primaries are conical, secondaries are globose-conical, calyx moderately difficult to remove, early flowering and ripening, suitable for

Varieties

fresh market. It is winter hardy, susceptible to powdery mildew, moderately resistant to *Verticillium* wilt, and highly resistant to races A-4, A-6 and A-7 of the red stele (*P. fragariae*) (Daubeny, 1994).

ARIEL

It is a late-maturing variety suitable for Mediterranean areas. It was introduced during 2001 by Censorzio Italiano Vivaisti. Harvesting of fruits starts during early April and lasts up to mid-May. Fruits are firm, bright red, good to taste with high sugar content (Leis et al., 2002).

BAOTONG (BENIHOMALEI × CAMAROSA)

It has large fruits (weigh up to 44.7 g/fruit) with neat and flat surface and, moderately firm (0.58 kg/cm^2) texture, the fruit pulp has 10.3 °Brix total soluble solids (TSS) and is good to taste. In comparison to "Toyonoka" and "Benihoppe" varieties, "Baotong" is more resistant to powdery mildew and foliar diseases during vegetative growth. Further, it ripens comparatively earlier than the "Sweet Charlie" and "Darselect" cultivars and has better flavour as well as capacity for producing more runners (Duan et al., 2014).

BARAK (TAMIR × ARO 730)

It is a short-day variety adapted to the Sharon coastal plains of Israel for winter production and developed by Ministry of Agriculture and Rural Development, Agricultural Research Organization, Bet Dagan, Israel. Fruits are glossy dark red (colour index RHS45A), medium sized (19–24 g), uniform conical with large reflexed calyx, sweet, low acidity, early season. Fruit yield is 744–746 g/plant; moderately tolerant (Gasic and Preece, 2014) to *Botrytis* fruit-rot (*B. cinerea*) and powdery mildew (*S. macularis*).

BELL RAUGEX (77-36 × MORIOKA 19) × (MORIOKA 19)

This is a mid–late-season variety, moderately vigorous in warm regions although less so in cool climates; fruit yield ranges between those of Morioka 16 and Hokowase; fruits are very attractive, conical, bright red and glossy with good taste, weighing about 14 g with tough skin and firm flesh, giving excellent shelf life and transportability.

BELRUBI (POCAHONTAS × REDCOAT)

It originated in France. Fruits are large, long conical, juicy with moderate flavour (Figure 6.1).

BENICIA (PALOMAR × CAL 0.18-601)

It is a short-day variety developed by the University of California. It is adapted to arid, subtropical climates. Fruits red (Munsell 2.5 R 7/10); very large (33.7 g); medium–long conical, flesh orange-red (Munsell 7.5 R 5/13), outstanding flavour. Fruits mature earlier than Camarosa but later than Ventana and Palomar. Plants are moderately vigorous, medium stolon number, very high yielding (2.566 kg/plant); moderately resistant to powdery mildew and tolerant to two-spotted spider mite (*Tetranychus urticae*). But it is (Gasic and Preece, 2014) moderately susceptible to anthracnose/crown rot (*C. acutatum*), *Phytophthora* crown rot (*P. cactorum*), common leaf spot (*Ramularia tulasnei*), and very susceptible to *Verticillium* wilt (*V. dahliae*).

BLOMIDON [K72-4 × {MICMAC × (GUARDSMAN × TIOGA)}]

It originated in Kentville, Nova Scotia, Canada and was introduced during 1984. Plants are vigorous and moderate to high yielding, calyx moderately difficult to remove. Primary fruits are large

FIGURE 6.1 Berubi.

to very large and secondary ones are medium-size; firm, glossy, peel medium–deep red but white under calyx. Fruit pulp is medium-red with white fibres and tart flavour. Fruits mid–late to ripen. It is moderately resistant to *Botrytis* fruit rot and to A-6, the most common race of the red stele, and susceptible to June yellows (Daubeny, 1994)

BOB (1501 × 1242)

It is a day-neutral variety developed by Sweet Darling Sales, Inc., Aptos, California and adapted to areas spread over coastal central California. Fruits are red (RHS 45A), medium to large-sized (weight 20.7 g); conical, cylindrical and wedge shaped. Fruit pulp firm, orange-red (RHS41B); calyx medium-size clasping to reflexed surface and white (RHS 155C). It has early-season fruit maturity but is a low-yielding (153.4 g/plant) variety and susceptible to bruising and rain (Gasic and Preece, 2014) damage.

BOLERO (INCLUDES REDGAUNTLET, GORELLA, CORDINAL AND SELVA)

It is an everbearing cultivar which was released during 1996 by HRI, East Malling, England. It produces crop during August to October. Fruits are firm, bright and attractive, regular shape, glossy peel and good shelf life. Plants are vigorous and moderately resistant to *V. dahliae* and *P. cactorum* (Simpson et al., 2002).

BOUNTIFUL (LINN × TOTEM)

It originated in Corvallis and was introduced jointly by USDA Research Service, Oregon State University and Washington State University during 1993. Plants are vigorous, low spreading and high yielding. Fruits are of the same size or slightly smaller than cv. Totem; firm, conical with uniform bright red peel and pulp. It is of late maturity group with very less ripening period, and suitable for processing. It is susceptible to *Botrytis* rot and similar to cv. Totem for tolerance to aphid-borne virus complex in the Pacific Northwest but comparatively more susceptible than cv. Totem to complex races of the red stele (Daubeny, 1994) pathogen but relatively resistant to powdery mildew.

BRUNSWICK (CAVENDISH × HONEOYE)

This variety was introduced in Canada by the Atlantic Food and Horticulture Research Centre of Agriculture and Agri-Food Canada during 1999. Plants are vigorous, produce more runners and perform well in continental areas with cold winters. Fruits are large (13–15 g), uniform in colour and shape, ripen during early midseason, resistant to red stele root rot (*Phytophthora fragaria* var. *fragariae*), leaf scorch (*Diplocarpon earlianum*), leaf spot (*M. fragariae*) and powdery mildew (*S. macularis*), and moderately tolerant to fruit rot (*B. cinerea*) (Jamieson, 2002; Jamieson et al., 2004a).

CABOT (K87-5 × K86-19)

It was commercially introduced in Canada by Atlantic Food and Horticulture Research Centre of Agriculture and Agri-Food Canada during 1998. Plants are vigorous and produce few runners. Fruits are very large (20–24 g), peel somewhat rough with medium strength; pulp firm; ripening during late midseason. Plants are resistant to red stele root rot (*P. fragariae* var. fragariae), leaf scorch (*D. earliana*), leaf spot (*M. fragariariae*) and powdery mildew (*S. macularis*) but susceptible to *Botrytis* fruit rot (Finn, 1999; Jamieson, 2002; Jamieson et al., 2004c) diseases.

CALYPSO (RAPELLA × SELVA)

It originated at East Malling Research Station, England and was introduced during 1992. It is a day-neutral variety and produces abundant runners. Plants are smaller than Rapella, fruits are firm (primary fruits may be irregular), good flavour, quality better than standard Rapella, suitable for the fresh market. It is moderately resistant to *Verticillium* wilt and susceptible to powdery mildew (symptoms largely remain confined to leaves and seldom observed on fruit) and two-spotted spider mite (Daubeny, 1994).

CAMAROSA (DOUGLAS × CAL 85.218-605)

It is a short-day variety and originated in California. South Coast Research and Extension Center (SCREC), Irvine introduced it during 1992. Plants are vigorous and produce fruits over an extended period in subtropical arid climates. Camarosa is high yielding, early and produces large but less firm fruits than Chandler. It is suitable for the fresh market as well as processing. It is moderately susceptible to common leaf spot, relatively resistant to powdery mildew and tolerant to two-spotted spider mite (Daubeny, 1994) (Figure 6.2).

CAMBRIDGE PRIZEWINNER (EARLY CAMBRIDGE × HOWARD 17)

It is essentially an early variety with very attractive fruits but not a heavy cropper. This is rather shy in producing runners and further, needs good soil for satisfactory growth and fruiting. It is important to obtain certified runners to avoid disease problem (Hughes et al., 1969). Plants are medium sized, growth moderate–weak and compact according to locality. Fruits are conical, occasionally wedge shaped, bright red, size large at first harvest but gets drastically reduced in subsequent harvests.

FIGURE 6.2 Camarosa.

Pulp moderately firm, pale red with subacidic taste and moderate flavour. It is less susceptible to *Botrytis* rot but susceptible to *Verticilium* wilt and red core diseases.

CAMBRIDGE FAVOURITE [(ETTER SEEDLING × AVANT TOUT) × BLACKMORE]

It is an early–midseason variety. Plants fairly large, moderately open and vigorous. Fruits large and the size is maintained well over the picking season; round, plump, pinkish-scarlet turning light-red, colour uniform but somewhat dull. Flesh firm, moderately juicy, pale pink–pale red with moderate flavour. It is suitable for processing. It is susceptible to *Botrytis* rot during wet season but fairly tolerant to mildews and field resistant to a few strains of red core pathogen (*P. fragariae* Hickman) and *Verticillium* wilt, tolerant to aphid-transmitted viral diseases but a few isolated plants may be frequently infected with green petal virus. Streaky yellows are very occasionally seen on this variety. It profusely produces runners (Hughes et al., 1969).

CAMBRIDGE RIVAL (DORSETTE × EARLY CAMBRIDGE)

Plant tall, upright, fairly open, moderately vigorous with characteristic light yellow to green foliage. Fruits large and often wedged, conical, shiny, bright crimson, turning dark red when fully ripe. Tip of fruit may not ripen. Flesh firm, red and sweet. It is a moderately good cropper of large strawberry fruits with distinctly good flavour. It is field resistant to most strains of red core disease but susceptible to *Verticillium* wilt and least susceptible to *Botrytis* rot. Very dark fruit colour is a limiting factor in its marketing. It is shy in runner production and needs early planting, if crop from one-year-old plantation is required (Hughes et al., 1969).

CAMBRIDGE VIGOUR

It is a fairly heavy cropper right from one-year-old plantations with good fruit quality. Plants are large–very large, upright and spreading, and very dense. Fruits are fairly large but inclined to be very small towards the end of picking season, sharply conical, regular but sometimes a little wedged, very attractive orange-red to light-red, turning scarlet. Flesh firm, juicy, moderately sweet, good flavour. Fruits from maidens ripen easily, but fruits from older plants ripen during the later midseason. It has field resistance to some strains of red core, but is very susceptible to *Verticllium* wilt and mildew, and sensitive to drought. Attractive fruits that do not withstand transport well, and are inclined to be soft, if at all overripe, so need more frequent picking. Older plants produce too many small berries. It produces plenty of runners and may grow too vigorously on good soils (Hughes et al., 1969).

CAPITOLA (CAL 75.121-101 × RARKER)

It originated at Wolfskill Experimental Orchards, near Davis, University of California and was introduced during 1991. Plants are day-neutral, vigorous, runner production equal or better than cv. Selva; both mother and daughter plants tend to bear flower and fruit. Fruits are larger than cv. Selva but less firm, colour (internal and external) similar to Chandler, pleasant subacidic flavour and comparable with cv. Selva, suitable both for fresh market and processing. It is quite susceptible to common leaf spot, moderately susceptible to two-spotted spider mite and very tolerant to viruses including mild yellow edge occurring in California (Daubeny, 1994).

CAPRI (CIVRI-30 × R6R1-26)

It is a fully remontant strawberry variety developed by Consorzio Italiano Vivaisti–Societa ConsortileA R.L., Ferrara, Italy and adapted to European continental climates. Fruits are glossy red (RHS 46A), large (20–33 g/fruit), moderately uniform and conical; peel orange-red (RHS 33A, RHS 35A) and pulp white (RHS 155D), TSS ranges from 7.5–8.5 °Brix, very good flavour, long growing season (June to November). It is a high yielding (1–1.2 kg/plant) variety, resistant to many fungal diseases and fairly tolerant to mildews but moderately susceptible to leaf spot disease (Gasic and Preece, 2014).

CARISMA (OSO GRANDE × VILANOVA)

It is a short-day, highly productive cultivar, which was evolved in Spain through selection in 1998. Fruits are large (weight 25–27 g/fruit), attractive with good flavour, marketable yield varies from 750–900 g/plant, suitable for high elevations (Bartual et al., 2002).

CARMINE (ROSE LINDA × FL 93-53)

It was developed during 1995. It produces very early fruits (December–February) having firm texture and deep red peel, suitable for areas with mild winter climates (Chandler et al., 2004).

CATSKILL (MARSHALL × PREMIER)

This variety was introduced by the New York State Agricultural Experiment Station, Geneva during 1934. The large, attractive fruit is good for local market but is soft and does not survive long-distance transport. The berries become dark on holding, and they are unsatisfactory for processing. The plants are fairly resistant to *Verticillium* wilt, powdery mildew, leaf scorch and leaf spot (Spangelo, 1962) (Figure 6.3).

CAVALIER (VALENTINE × SPARKLE)

It was introduced by the Canada Department of Agriculture, Ottawa, during 1957. The berries are of good size during the early part of the season but some growers find that the size gets smaller too rapidly during the later period of crop growth and development. The fruits are attractive, much more firm than that of the Catskill, Premier or Dunlap cultivars, and good for both dessert and processing.

FIGURE 6.3 Catskill.

This very early-ripening variety is highly resistant to *Verticillium* wilt but is fairly susceptible to powdery mildew, leaf scorch and leaf spot (Spangelo, 1962).

CHAMBLY (SPARKLE × HONEOYE)

It is believed to be the first June-bearing variety evolved during 1982 in Canada. It gives high yields of firm, deep red fruits with raised neck, elevated calyx and uniformly well-coloured flesh. It is recommended for the fresh market and processing. Plants are medium in size, not too vigorous, leaflets are slightly folded, medium size, dark green, oval to circular and serrated. Petioles are moderately long and slightly drooping. Inflorescence are four–six in number and erect on long, moderately thick peduncles during blooming and become arched (semi-erect) as the fruits mature. Fruits are conical, medium in size (8–10 g) with a white raised (2–3 mm) neck. Skin is shiny and deep red at maturity and flesh is red throughout. Calyx does not separate from the fruit at harvesting (Khanizadeh, 2004).

CHANDLER (DOUGLAS × CAL 72.361-105)

It was released during 1983 in California. The fruits are of exceptionally high dessert quality with outstanding colour, flavour and texture. The fruits have good resistance to physical damage from rain. This is a high-yielding, short-day variety (Figure 6.4).

CRISTINA (SELECTION AN01.211.51)

It was derived from a cross (CN 95, 602, 8 × CN 95, 419, 4) and selected during 2002. It is a short-day and late-maturing variety with good adaptability to non-fumigated soils. The plant has high number of crowns and medium vigour. It is highly productive, very large fruits, which are regular and conical with round tip, with medium-sized calyx that has a medium-easy removal. Achenes are yellow, medium in size and at surface level. Fruits are bright red with high sensory qualities (Capocasa et al., 2016).

CHUNXING (CHUNXIANG × HAIGUAN ZAOHONY)

It is vigorous and productive (up to 37.5 t/ha) cultivar originated in China during 1990. Fruits are conical, uniform and large (weighing 30.3 g), mature in mid-May, have bright red skin and sunken seeds. Flesh is orange-red, fine and juicy, very good eating quality, soluble solids 11% and ascorbic acid 1273 µg/g (Hao et al., 2002).

FIGURE 6.4 Chandler.

CHUNXU (CHUNXIANG AND A POLISH CULTIVAR)

It is highly productive, yielding large fruits of high quality. Fruits weigh 15–36 g, glossy, bright red; flesh is soft, red, tender, fine and very juicy. It has strong aroma with 536.9 µg ascorbic acid per gram pulp and highly resistant to high temperature and drought (Ma et al., 2001).

CRUSADER [AUCHINCRUIVE SDLG. 50 × 5 {SDLG. 6M17 (TEMPLE SELFED × SDLG. 11 (ABERDEEN OPEN)} × CAMBRIDGE VIGOUR]

The plants are vigorous, very dense and leafy. Flowers are small on short trusses. It has field resistance to most strains of red core and may be suitable for soils where other varieties do not grow satisfactorily. It ripens a few days before Templar, and individual berries being generally larger than Templar's and more like Redgauntlet's but lighter in total crop. It is susceptible to *Verticillium* wilt. Very large, early and glossy fruits of this variety are very attractive, although yield is not so heavy as Favourite or Red Gauntlet. Foliage is susceptible to powdery mildew. Not susceptible to *Botrytis*. It is worth trial, both in the open and in protection, as an early variety for the dessert market (Hughes et al., 1969).

CUESTA [SEASCAPE × CAL 83.25-2 (FERN × PARKER)]

It has originated at University of California. This is a late- and short-day variety. Fruits are larger than Chandler with good flavour and suitable for the fresh market, processing and home garden uses. It is moderately susceptible to common leaf spot and powdery mildew but tolerant to two-spotted spider mites (Daubeny, 1994).

CUPCAKE (102213 × 1581)

It is a day-neutral variety developed by Sweet Darling Sales, Inc., Aptos, California and adapted to coastal central California. Fruits are glossy, light-red (RHS 45A); mid-large-sized (31.5 g fruit), globose, conical, cylindrical, orange-red (RHS 41B) peel and white (RHS155C) flesh, balanced flavour, early season. Very low yield (105.1 g/plant); upright, resistant to bruising and rain (Gasic and Preece, 2014).

DANGE (VENTA × REDGAUNTLET)

It is a medium-late cultivar having attractive and good fruit size with high and stable yield every year (Rugienius et al., 2004).

DELIA (HONEOYE × ISF 80-52-1)

It is a short-day variety, fruits with orange-red skin; large; conical; very uniform and firm with moderate flavour; ripens in May–June (one week before Elsanta). Plants with moderate vigour; moderate–high yield; good runner production; susceptible to powdery mildew and *Phytophthora* crown rot (Kim, 2010).

DELY (T2-6 × A20–17)

It is a short-day variety adapted to European continental climates developed by Consorzio Italiano Vivaisti–Societa Consortile A.R.L., Ferrara, Italy. Fruit are glossy red (RHS 45A), medium–large sized, moderately uniform conical, red (RHS 44A) to orange-red (RHS 39) peel and white (RHS 155D) flesh, very good TSS (9 °Brix) and flavour; early season, resistant to local leaf and root diseases (Gasic and Preece, 2014).

DONNA (FERN SEEDING × DOUGLAS)

It has originated at the Turner & Growers Mangere Research Station (TGMRS), near Auckland, New Zealand during 1993. This is a late and day-neutral variety, suitable for offseason production for the fresh market. Fruits are glossy red, 20–21 g; uniform conical to wedge shaped; flesh firm, orange-red; TSS 7.2 °Brix; calyx reflexed and medium sized; achenes level with surface; early maturity. Plants are medium to strongly vigorous; high stolon number (15–20 stolons/plant); medium density foliage; flowers level to above foliage; semi-erect fruiting in trusses; tolerant to powdery mildew (*S. macularis*); moderate–high (1.16 kg/plant) yield (Daubeny, 1994; Kim, 2012).

DOUGLAS (TUFTS × 64-57-108 (TIOGA × SEQUOIA))

Douglas, a short-day cultivar developed at UC Davis in 1979, was named for the late Malcolm B. Douglas, manager of the California Strawberry Advisory Board. This remained a leading strawberry variety of the United States (Figure 6.5), which was further used in a breeding programme to develop high-yielding varieties.

DRIS STRAW EIGHTEEN (91J302 × 26H165)

It is fully everbearing variety developed by Driscoll Strawberry Associates, Inc., Watsonville, CA, by M. D. Ferguson and adapted to coastal southern California. Fruits are dark red (RHS 46B), medium-size (27.2 g/fruit), uniform conical, peel medium red-orange (RHS N34B) with sweet, moderately acidic, fine textured and white (RHS 155C) flesh. Calyx large and spreading to reflexed surface. This variety is low yielding (335.2 g/plant), moderately resistant to wind and high pH. This variety is, however, moderately susceptible to powdery mildew, leather rot (*P. cactorum*), common leaf spot, leaf scorch (*D. earlianum*), leaf blight (*Phomopsis obscurans*), black root-rot disease complex (*Rhizoctonia fragariae, Coniothyrium fuckelii, Hainesia lythri, Idriellalunata, Pyrenochaeta* sp., *Pythium* spp.), angular leaf spot (*Xanthomonas fragariae*), two-spotted spider mite, cyclamen mite (*Tarsonemu spallidus*), strawberry foliar nematode (*Aphelencoides fragariae*), root lesion nematode (*Pratylenchus penetrans*), stem eelworm (*Ditylenchus dipsac*), strawberry blossom weevil (*Anthonomu srubi*), aphids (*Aphis* spp.), high temperatures, waterlogging and susceptible to *Botrytis* fruit rot, *Verticillium* wilt, red stele (*P. fragariae*), Lygus bug (*Lygus hesperus*), drought and salinity (Gasic and Preece, 2014).

FIGURE 6.5 Douglas.

Varieties 95

DRIS STRAW NINETEEN (DRISCOLL ATLANTIS × 43J313)

This is a day-neutral variety developed by Driscoll Strawberry Associates, Inc., Watsonville, California and adapted to Florida. Fruits are dark red (RHS 46A), medium sized (25.3 g/fruit), moderately uniform conical–biconical, firm, medium red (RHS 41B) peel and white (RHS 155B) flesh. This is moderately sweet, low acidity, medium texture, large raised clasping to reflexed calyx, very early season, low yielding (210 g/plant), moderately resistant to *Botrytis* fruit rot, powdery mildew, angular leaf spot, two-spotted spider mite but highly susceptible to *Lygus* bug (Gasic and Preece, 2014).

DRIS STRAW TWENTY (2K297 × DRISCOLL OJAI)

It is a short-day variety developed by Driscoll Strawberry Associates, Inc., Watsonville, California and adapted to coastal southern California. Fruits are glossy red (RHS 46A); large (25.8 g/fruit); uniform conical; firm, peel is red (RHS 44B, RHS 2B) and flesh white (RHS 155B); moderately sweet; low acidity; medium texture; large raised reflexed calyx; late-maturing, moderate yielder (544 g/plant), resistant to powdery mildew and angular leaf spot and moderately resistant to two-spotted spider mite (Gasic and Preece, 2014).

DRIS STRAW TWENTY ONE (13H377 × 587L48)

This is a fully everbearing variety developed by Driscoll Strawberry Associates, Inc., Watsonville, California. Fruits are dark red (RHS 46A), large (30.8 g/fruit), moderately uniformly biconical; moderately firm, peel red (RHS 44B, RHS 41B) and flesh white (RHS155B), mild flavour. Plants are (Gasic and Preece, 2014) moderate yielding (487 g/plant).

DRIS STRAW TWENTY TWO (DRIS STRAW THREE × 50L206)

It is a fully everbearing variety developed by Driscoll Strawberry Associates, Inc., Watsonville, California and adapted to coastal southern California. Fruits are dark red (RHS 46A); medium-size (30.6 g/fruit); uniform conical to cylindrical; firm, peel red (RHS 46A, RHS 43A) and flesh white (RHS 155D, RHS N155B), balanced flavour, medium-sized reflexed surface calyx, and low (285.1 g/plant) yielding (Gasic and Preece, 2014).

DRIS STRAW TWENTY THREE (1M16 × 87K286)

This is a short-day variety developed by Driscoll Strawberry Associates, Inc., Watsonville, California. Fruits are dark red-orange (RHS N34A); very large (29.2 g/fruit); uniform biconical; moderately firm, peel red (RHS 42B, RHS 39A) and flesh white (RHS 155B, RHS 155D) with balanced flavour. It is medium yielding (616.3 g/plant), moderately resistant to high temperatures but moderately susceptible to *Botrytis* fruit rot, powdery mildew, *Verticillium* wilt and susceptible to angular leaf spot and cold temperatures (Gasic and Preece, 2014).

DRIS STRAW TWENTY FOUR (3M44 × 50L174)

It is a partially everbearing variety developed by Driscoll Strawberry Associates, Inc., Watsonville, California. Fruits are glossy, dark red (RHS 46B), very large (28.4 g/fruit), moderately uniform conical, firm, peel red (RHS 44A, RHS 39B) and flesh white (RHS 155B), very early maturing. Plants are moderate yielding (682 g/plant), moderately resistant to high temperatures, high soil pH, cool weather/freezes but susceptible to *Botrytis* fruit rot, powdery mildew, *Verticillium* wilt and angular leaf spot (Gasic and Preece, 2014).

DRIS STRAW TWENTY FIVE (18L33 × 192M122)

It is a short-day variety developed by Driscoll Strawberry Associates, Inc., Watsonville, California and adapted to coastal southern California. Fruits are dark red (RHS 46B), very large (33.6 g/fruit), firm, peel orange-red (RHS 33B) and flesh white (RHS 155D), sweet with medium acidity and texture, early maturity. This is high yielding (1.208 kg/plant), moderately resistant to *Verticillium* wilt, angular leaf spot, wind, high pH salinity and moderately susceptible to high temperatures, susceptible to *Botrytis* fruit rot, powdery mildew, aphids and *Lygus* bug (Gasic and Preece, 2014).

DRIS STRAW TWENTY SIX (18L33 × 193M68)

This is a short-day variety developed by Driscoll Strawberry Associates, Inc., Watsonville, California. Fruits are dark red (RHS 46A), large (30 g/fruit), moderately uniform conical; moderately firm, peel light-red (RHS 39B) and flesh white (RHS 155A), balanced flavour, mid-sized, reflexed surface calyx. Plants very low yielding (94.3 g/plant), moderately resistant to high pH, high soil salt levels; moderately susceptible to powdery mildew, angular leaf spot, high temperatures, wind; susceptible to *Botrytis* fruit rot, *Verticillium* wilt, aphids and *Lygus* bug (Gasic and Preece, 2014).

DRIS STRAW TWENTY SEVEN (DRIS STRAW EIGHT × 10L297)

This is a short-day variety developed by Driscoll Strawberry Associates, Inc., Watsonville, California. Fruits are glossy dark red (RHS 46A), very large (30.3 g/fruit), moderately uniform conical, peel firm red (RHS 40C) and flesh white (155D), sweet, low acidity, reflexed calyx; very early maturing. Plants high yielding (1.063 kg/plant), moderately resistant to *Verticillium* wilt, wind, high temperatures, high pH, high soil salt levels, water logging but moderately susceptible to powdery mildew, common leaf spot, angular leaf spot, aphids and, susceptible to *Botrytis* fruit rot and *Lygus* bug (Gasic and Preece, 2014).

DRIS STRAW TWENTY EIGHT (95L299 × 251M27)

It is a partially everbearing variety developed by Driscoll Strawberry Associates, Inc., Watsonville, California and adapted to coastal central California. Fruits are dark red (RHS 46A), medium-size (21.1 g/fruit), uniformly conical, firm, peel red (RHS 43B, RHS 43D, RHS 49B) and white (155B) flesh, moderately sweet, medium texture, medium sized, midseason, high yielding (1.218 kg/plant), resistant to moderately resistant to *Botrytis* fruit rot but moderately susceptible to powdery mildew, *Verticillium* wilt, high temperatures and moderately susceptible or susceptible to wind (Gasic and Preece, 2014).

EARLIBRITE (ROSA LINDA × FL 90-38)

It is a short-day cultivar originated during 1996–97 in Florida, USA. It is comparatively less susceptible (Chandler et al., 2000a) to *B. cinerea* and powdery mildew (*S. macularis* f. sp. *fragariae*).

EARLIDAWN (MIDLAND × TENNESSEE SHIPPER)

The United States Department of Agriculture introduced this variety during 1956 at Beltsville, Maryland. The fruits are large, bright red and attractive, comparatively more firm than var. Catskill or Premier. The berries retain their good colour, texture and flavour when frozen. Earlidawn is productive and very early variety but susceptible to powdery mildew, and highly susceptible to Verticillium wilt (Spangelo, 1962).

EARLY BOMMEL (NEW DUTCH SEEDLING NO. 69-16, WAS INTRODUCED COMMERCIALLY AS EARLY BOMMEL)

It is very vigorous, early flowering and yields more than Glasa or Karina. In the glasshouse, it is very susceptible to powdery mildew (*S. macularis*), but it forces well, provided the glasshouse temperature is kept high enough from early January to check its vigorous growth. Its main advantage is that, it can be useful in extending the cropping season (Verwijs, 1980).

ELEGANCE (EM834 × EM1033)

It has a complex pedigree including three North American cultivars ("Cardinal", "Honeoye" and "Selva") and two cultivars from the UK ("Alice" and "Eros"), and is suited to both a protected or an outdoor production system using an annual or perennial crop cycle. The average fruit size is larger than "Elsanta" and the percentage of Class-1 fruits has been observed to be consistently higher than Elsanta, frequently exceeding 90%. The plants are vigorous with erect habit. The firm berries have good skin strength and are very attractive with a uniform conical shape, glossy skin finish and bright orange/red colour. The texture is firm and juicy and the flavour is best if the fruit is harvested at fully ripe stage. Elegance has moderate resistance to wilt (*V. dahliae*) but is susceptible (Simpson et al., 2014) to powdery mildew (*P. aphanis*).

ELISTA [(JUCUNDA × SELF) × US-3763]

Plants are moderately vigorous, rather upright, a little spreading in the third year with an open growth habit. Fruits are large–medium, round conical–conical, sometimes wedge shaped, orange-red becoming scarlet with red flesh. Flavour is good but a little acidic. It yields comparatively more heavily than Cambridge Favourite, and with good-size fruits in early harvests. Later fruit tends to remain small, particularly if the weather is dry. This is a late midseason cultivar and amenable to prolong the picking period by at least a week after the cv. Favourite. Fruits do not extend beyond leaf canopy but are well exposed. It shows some resistance to both *Botrytis* and mildew. It may be of interest for processing because of its deep-coloured fruit (Hughes et al., 1969).

ELVIRA (VOLA × GORELLA)

It is a productive, early–midseason variety, with large, glossy, orange-red fruits, which are firm and juicy and have good flavour. Although it is mainly suitable for growing in glasshouses or polyhouses, it can also be grown in the open conditions.

ENZED LEVIN (ORION) (84-1-38 × 84-8-443)

This selection involves Hecker Pajaro, Fern and Douglas in its derivations and originated at the TGMRS, New Zealand during 1993. This is a short-day and early-maturing variety. Fruits are large, wedged, having good shelf life and flavour, and suitable for fresh market use (Daubeny, 1994).

ERIE (SPARKLE × PREMIER)

It was introduced by the New York State Agricultural Experiment Station, Geneva, during 1951. The fruits are very large and attractive, but soft and not good for processing. This late-ripening variety is recommended for trial to the local fresh-fruit market because of the size rather than the quality of the berries (Spangelo, 1962).

ETNA (MARLATE × BELRUBI)

It is highly productive, medium-late ripening, vigorous and resistant to *Verticillium* wilt but not *R. frazaiac* or *R. solani*. It can be grown in the open.

EVANGELINE (K 88-4 × NYUS 119)

It was commercially introduced in Canada during 1999 by the Atlantic Food and Horticulture Research Centre of Agriculture and Agri. Food Canada. It is early in maturity; fruits are firm, more attractive, conical, dark and large. It is resistant to red stele root rot disease (*P. fragariae* var. *fragariae*), leaf scorch (*D. earliana*), moderately resistant to leaf spot (*M. fragariae*), powdery mildew (*S. macularis*) and fruit rot (*B. cinerea*) diseases (Jamieson, 2002; Jamieson et al., 2004b).

EVEREST (IRVINE × EVITA)

It is day-neutral variety, fruits with uniform red skin; medium to large in size; predominantly conical; having medium-red flesh; firm; centre is absent or weakly expressed hollow; high acidity; glossy smooth surface; achenes below surface; ripens early during June–October. Plants are medium–strong in vigour; good yield; medium-dense foliage; easy harvest; poor runner production; does not require chilling for flower induction; poor tolerance to high temperatures but good tolerance to powdery mildew (Kim, 2010).

EVITA [CHANDLER × (GORELLA × BRIGHTON)]

It has originated in Faversham Kent during 1993. This is a day-neutral variety, ripening season from July–October in southwest England and suitable for the fresh market (Daubeny, 1994).

FENGGUAN 1 (TOYONOKA × TOCHITOME)

It was developed during 1998 and grows healthy and sturdy with lot of stolons, higher production, highly resistant to diseases such as grey mould (*B. cinerea*), powdery mildew (*S. macularis*) and anthracnose (*Glomerella cingulata*). Berries are large, weighing 17.8 g on average with bright skin; flesh is little bit soft, rich flavour, very good eating quality containing 11.3 °Brix TSS (Lu et al., 2005).

FERN (TUFTS × CAL 69.63-103)

It originated in California. It is an early and day-neutral cultivar. Fruits are large–medium, conical and solid internally. Flesh is firm with excellent flavour and suitable for both fresh market and processing (Figure 6.6).

FERRARA (MD. US4083 × HOLIDAY)

It is moderately productive, medium late in maturity, suitable for growing in the open. It shows low susceptibility to *Verticillium* and *Rhizoctonia*. The berries are well amenable to freezing.

FESTIVAL (ROSA LINDA × OSO GRANDE)

It has medium-sized, uniform, broad-conical fruit; deep red externally and medium–dark red internally; moderately acidic but balanced flavour; very firm with excellent shipping quality; flexible skin is extremely resistant to rain damage. It is resistant to root rot caused by *P. cactorum*; moderately

Varieties 99

FIGURE 6.6 Fern.

FIGURE 6.7 Festival.

resistant to fruit rot caused by *C. acutatum* Simm. and/or *B. cinerea*) but susceptible to angular leaf spot (caused by *Xanthamonas* spp.) and crown rot (caused by *Colletotrichum gloeosporioides*) (Figure 6.7).

FL 05-107 OR WINTERSTAR™ (FLORIDA RADIANCE × EARLIBRITE)

It is a short-day variety developed by University of Florida. Fruits glossy, bright red (RHS 34B), medium-sized (30–40 g primaries, 10–30 g secondaries and tertiaries), uniform conical, light orange (RHS 32C) flesh, sweet, low acidity, medium–large and attractive calyx, early maturing. Plants moderate–high yielding (453 g–1.06 kg/plant), resistant to abrasion and fruit rot due to *C. acutatum* (Gasic and Preece, 2014).

FLAIR (FLEVO00-24-7 × FLEVO 00-08-4)

It is a short-day variety bearing fruits that are glossy red-orange (RHS 34A); medium sized; conical to round-cordate; moderately firm, orange-red (RHS 33A) flesh, balanced flavour; medium sized, raised and spreading calyx, early maturity (Gasic and Preece, 2014).

FLAVORFEST (B759 × B786)

It is a short-day variety, fruits bright red, large, round-conical, light-red flesh, excellent flavour with 6.4–9.8 °Brix TSS, high yielding (ranging from 341 g–1.114 kg/plant) averaging over

690 g/plant, large, resistant to fruit rot and crown rot (caused due to *C. accutatum*, *C. fragariae* CF63) and red stele (Race A-3, Rpf1), moderately resistant to Botrytis fruit rot, leaf blight, leaf scorch, moderately susceptible to powdery mildew, red stele (Race A-5) but susceptible to *C. fragariae* CG163 and *C. gloeosporoides* CG162 induced anthracnose crown rot (Gasic and Preece, 2014).

FLORENCE [TIOGA × (REDGAUHTLET × (WELTGUARD × GORELLA)] × (PROVIDENCE × SELF)

It is a medium-late, June-bearing and productive cultivar released by HRI, East Malling, England during 1997. Fruits are large, firm with deep red peel colour and sweet flavour. It is resistant to *S. macularis*, *V. dahliae*, *P. cactorum* and tolerant to *Otiorhynchus sulcatus* (Simpson et al., 2002).

FLORIDA RADIANCE (WINTER DAWN × FL 99-35)

It has large, uniform, conical fruits with glossy appearance; deep red externally and medium red internally; moderately acidic but balanced flavour; moderately firm with good shipping quality; flexible skin that is moderately resistant to rain damage. Low–medium vigour with very open canopy; very long fruit stems facilitate and improve picking efficiency. Moderately resistant to anthracnose fruit rot, *Botrytis* fruit rot, and *Colletotrichum* crown rot but moderately susceptible to angular leaf spot, with symptoms particularly noticeable on the calyx; and highly susceptible to *Phytophthora* root rot.

GALLETTA (NCH 87-22 × EARLIGLOW)

It is a short-day variety; fruits are dark purple-red and red-blend skin; as large as Camarosa (32 g primary, 20 g secondary and 10 g tertiary fruits); long conical to conical; orange-red flesh, ripens very early, achenes slightly below surface. Plants are medium in vigour; fruit yield at least as big as Chandler and Camarosa; medium canopy density; resistant to crown and fruit anthracnose (Kim, 2010).

GARDENA (SELECTED IN PROGENY FROM THE CROSS OF ADDIE × PAJARO)

This is moderately vigorous Italian variety, bears fairly large dark red fruits with a mean weight of 29 g and a uniform colouration. The flesh is firm and orange-red. It flowers early and ripens two days before Addie variety. It is adapted for cultivation in northern Italy.

GEA [R66/60{(SENGA SENGANA × VALENTINE) × SURPRISE DES HALLS} × MD US38]

It is moderately productive, early maturity, suitable for protected cultivation, slightly susceptible to *Verticillium* but very susceptible to *Rhizoctonia*.

GEM (A BUD MUTATION OF CHAMPION EVERBEARING; ALSO CALLED BRILLIANT AND SUPERFECTION)

This variety was introduced by the Heart-O-Michigan Farmers and Nurseries, Farwell, Michigan, during 1933. This is a productive, everbearing variety but suitable only for dessert. It is generally inferior to recommended June-bearing varieties, and is of value only for special markets (Spangelo, 1962).

GINZA (1929 × 1902)

It is a day-neutral variety, fruit red (RHS 45A), medium-sized (16.8 g), variable, conical, long conical, long-wedge, orange-red (RHS 41B) and white (RHS 155C) flesh, balanced flavour, early maturity, very low yielding (91.7 g/plant) and resistant to variations in weather (Gasic and Preece, 2014).

GLORY (NELSON. PS-1269 × PS-2286)

It is an everbearing variety having red-orange-red, medium size (21.4 g) fruits; variable in shape ranging from cylindrical, conical, cordiform to wedged; flesh red, good favour, 8.4°Brix soluble solids; large, spreading to reflexed calyx; achenes level to below surface. Plants are medium in vigour, stolon medium in number (8.3 stolons/plant); low–moderate yield (511 g/plant); moderately susceptible to two-spotted spider mite, flower-thrips (*Frankliniella* spp.), powdery mildew, *Botrytis* fruit rot and angular leaf spot (Kim, 2012).

GORELLA (JUSPA × US-3763)

This cultivar was raised in the Netherlands. Plants moderate in growth, sparse with coarse thick foliage. Fruits are very large, conical, crimson red, sometimes with a green tip and well exposed beneath the upright foliage. Flesh bright crimson, juicy with average flavour (Hughes et al., 1969) (Figure 6.8).

GRENADIER (VALENTINE × FAIRFAX)

The Canadian Department of Agriculture, Ottawa, introduced this variety during 1957. The fruit is about the same size or larger than cv. Premier. Very firm, dark red berries are not suitable for all markets but they are very good for dessert. They are also good for processing as frozen whole or sliced berries and as jam. This is very good quality, midseason variety, yields well but is fairly susceptible to powdery mildew (Spangelo, 1962).

GURDSMAN (CLARIBEL × SPARKLE)

It was introduced by the Canadian Department of Agriculture, Ottawa during 1957. The medium–large berries do not retain their size if the soil moisture is inadequate. They are very firm and attractive, and ripen late in the season. The berries are highly acidic if harvested before they are well coloured and ripe. This very high-yielding variety is resistant to powdery mildew but susceptible to *Verticillium* wilt (Spangelo, 1962).

FIGURE 6.8 Gorella.

HARUYOI (HARUNOKA × HOKOWASE)

This cultivar (registered as Norin-14) is used for protected cultivation in the south of Tahoku, Japan. Fruits weigh 11.5–14.4 g and are light orange-red. Growth of the plants is vigorous and there is no tendency to dwarfing during forcing.

HERRIOT (NYUS299 × WINONA)

It is a short-day variety, adapted to perennial matted row production in temperate climates. Fruits are large; bright red; wide conical; firmness moderate; flavour good. Plants are vigorous and spreading; high runner production, resistant to powdery mildew, tolerant to replant diseases (Kim, 2012).

HONEOYE (VIBRANT × HOLIDAY)

This cultivar was derived in Geneva (New York, USA), and exhibited good vigour, intermediate plant density, globous growth habit and early in ripening. Fruits are large, firm, conical, bright red, aromatic, glossy, sour and attractive, which are consumed fresh or used in processing. It is very susceptible to root rot and collar rot diseases caused by *P. cactorum, P. fragariae, Verticillium albo-atrum* and others, but not susceptible to (Ludvikova and Paprstein, 2003a) grey mould (*B. cinerea*) and mildew (*S. macularis*).

HONGJIA

It is a vigorous, high-yielding (34.6 t/ha) Japanese cultivar released during 1999, having a very low chilling requirement (150–200 hours), resistant to powdery mildew (*S. macularis*) and leaf spot diseases. Fruits are large, weighing 18.6 g on average, reaching up to 50 g, bright red skin, flesh red, fine, firm, crisp with sweet flavour, 8-13°Brix TSS contents, with good shipping and eating quality (Kong et al., 2003; Li and Ding, 2006).

HONGSHIMEI (ZHANGJI × DUCRA)

This was developed during 1998. Plants are highly tolerant to cold and high temp (2–32°C), resistant to powdery mildew (*S. macularis*). Fruits mature during late-May–early-June, large, weighing 45.7 g on average, but reaching over 100 g with bright red skin. Seeds are yellow and sunken. Flesh is light-red, fine; quite firm with a light white core, good eating quality with 10.5°Brix TSS (Gu et al., 2005).

HONGXIUTIANXIANG (SEL. 06-37-2) (CAMAROSA × BENIHOPPE)

This is a high-quality June-bearing cultivar with large, firm fruit that have good, sweet flavour, long conical or wedge shape and red glossy surface. "Hongxiutianxiang" is highly resistant to powdery mildew (Zhang et al., 2014).

HUXLEY (ALSO KNOWN AS ETTERSBURG 80)

Huxley was developed in 1910. It is a midseason–late cultivar and extremely drought tolerant. Plants are large, flat and spreading, dense and vigorous with coarse and glossy leaves. Fruits are medium–large, round, dark red and shiny; flesh firm, pale red and fairly juicy. Berries remain firm and may be picked less frequently than other varieties. It has moderate flavour (Hughes et al., 1969).

Varieties

INDEPENDENCE [ORUS 850-48(LINN × ORUS 3727) × ORUS 750-1 (TOTEM × ORUS 3746)]

It was formerly known as ORUS 1076-126 and was introduced during 1998. It is a very late, short-day type, vigorous cultivar tolerant to virus, winter, root weevil and wet soil. Besides, it is resistant to foliar and *Botrytis* fruit-rot diseases. Fruits are of average size similar to Totem, very firm, slight wedge and uneven, bright, uniform red externally and very good internally with tender skin, acidic but very pleasant flavour, mature very late (seven days after Red Crest) and are suitable for fresh market (Finn, 1999).

JOLY (T2-6 × A20-17)

It is a short-day variety. The fruits are glossy red (RHS 45B), large, uniform conical, firm, red (RHS 44B, RHS 40B) and flesh white (RHS155D), very good flavour (8.5 °Brix), early maturing, moderate to high yielding (800 g/plant), and resistant to local leaf and root diseases (Gasic and Preece, 2014).

JEWEL (SELECTED IN 1971 FROM THE PROGENY OF NY 1221 × HOLIDAY)

It is hardy, consistent in cropping with fruit yield similar to that of Guardian. Berries are large (11.3 g), bright red of attractive appearance, moderately firm, easily picked and high quality for both fresh and frozen uses. Flavour is assessed as consistently high. It is resistant to post-harvest fruit rot and recommended for cultivation in the northeast and midwest of the United States (Sanford et al., 1985).

JINGYUXIANG (CAMAROSA × BENIHOPPE) SYN. SEL. 06-37-8

This is a June-bearing cultivar. Berries are large and have good, sweet flavour, long conical or wedge shaped and red glossy surface. It is highly resistant to powdery mildew (Zhang et al., 2014).

JINGYIXIANG (CAMAROSA × BENIHOPPE) SYN. SEL. 06-37-10

This cultivar bears large, bright red, medium firmness and long conical berries, which develop balanced sugar–acid flavour. Plants are vigorous and highly resistant to powdery mildew (Zhang et al., 2014).

JINGCHUNXIANG {(01-12-15′ (ADVANCED SELECTION) × KINUAMA)} SYN. SEL. 06-54-2

This cultivar bears large, conical or wedge shaped, bright red, attractive fruits with firm texture and a mild acidic flavour. Plants are vigorous and resistant to powdery mildew (Zhang et al., 2014).

JINGQUANXIANG {(01-12-15 (ADVANCED SELECTION) × BENIHOPPE)} SYN. SEL. 06-56-6)

This cultivar bears large, conical or wedge-shaped, bright red, attractive fruits with medium firmness and excellent flavour. Fruits are suitable for pick-your-own markets. Plants are vigorous and high yielding, but occasionally susceptible to powdery mildew (Zhang et al., 2014).

JOLIETTE (JEWEL × SJ 85189)

It is a short-day and June-bearing cultivar, released during 1989 by S. Khanizadeh in Canada. It gives high yield of large, moderately firm, glossy fruits with good flavour. It is resistant to leaf spot, leaf scorch, powdery mildew, water stress, low winter temperature (<−30°C), post-harvest fruit rot and six North American Eastern (NAE) races of red stele, and recommended for the fresh market (Khanizadeh, 2004).

JUKHYANG (REDPEARL × MAEHYANG)

It is a June-bearing cultivar, vigorous and bears attractive, conical fruits. It is ready for harvest earlier than Redpearl and because of weak dormancy, it is suitable for forcing. The fruits have a good balance of sugar, acid and aroma; very firm with a good shelf life. It is resistant to powdery mildew. The marketable yield of Jukhyang is relatively higher than that of Redpearl because this cultivar bears comparatively a low proportion of abnormal fruits. Also, it takes less effort to remove the flowers because the number of flowers is fewer than that of Redpearl and Maehyang (Lee et al., 2016).

KAORINO (SELECTED FROM HYBRID LINES, LINE 0028401 × LINE 0023001)

The fruits are conical and large, with an orange-red skin colour and white flesh and core. The eating quality of fruits is superior with refined aroma, high TSS and low acidity. It shows high level of anthracnose resistance equal to Hokowase and Sanchigo (Kitamura et al., 2015).

KENTARO (KITA-EKUBO × TOYONOKA)

It is a short-day cultivar suitable for cold regions. Plants are weaker and grow more vigorous with medium growth habit. Fruits are conical, very glossy bright red skin, light orange flesh with small hollow core. They are resistant to *Fusarium* wilt (*Fusarium oxysporum* f. sp. *fragariae*) and powdery mildew (*S. macularis*), moderately resistance to *Verticillum* wilt (*V. dahliae*), and less susceptible to *Botrytis* rot (*B. cinerea*). Plants also require temperature lower than 5°C for about 1000 hours for breaking of the dormancy (Kawagishi et al., 2001).

KIMBERLEY (GORELLA × CHANDLER)

A very promising early variety from the Netherlands, bearing large sweet berries of good quality having fewer misshapen fruits. It is suitable to grow in open as well as in protected structures. The plants are highly tolerant for mildew, and therefore a good choice for protected cultivation. The variety is susceptible to *Colletotrichum*. The variety has a good yield, similar to Elsanta (Figure 6.9). In India, Kimberley was introduced first in Uttarakhand state.

FIGURE 6.9 Kimberley.

Varieties

KITA-EKUBO [59 KOU-13-37(AITO × MORIOKO 19) × REIKO]

It is a short-day variety suitable for semi-forcing culture. Fruits are large, conical, very glossy, bright red skin, light orange flesh, good flavour with longer shelf life and large hollow core. Plants are resistant to *Fusarium* wilt (*F. oxysporum* f. sp. *fragariae*), *Verticillium* wilt (*V. dahliae*), powdery mildew (*S. macularis*) but susceptible to *Botrytis* rot (*B. cinerea*). In this variety, the requirements of low temperature (<5°C) for breaking dormancy is very long (Konno et al., 2001).

L'AUTHENTIQUE ORLEANS (AC-L'ACADIE × JOLIETTE)

It is a June-bearing, midseason cultivar released during 1996 for eastern central Canada, with higher yield of large, firm, light red and tasty fruits, having a long shelf life of up to four days at room temperature with higher level of antioxidants (gallic acid, ascorbic acid, protocantechuic acid, catechin, p-hydroxybenzoic acid, epicatechin and ellagic acid) and shipping attributes (Khanizadeh et al., 2002b; Khanizadeh et al., 2003).

LAGUNA (IRVINE × CAL 85.92-602)

It was evolved at the University of California during 1992. This is a short-day variety, producing fruit over an extended period in subtropical arid climatic conditions. Fruits are firm, have very good flavour and suitable for fresh market, processing and home garden uses. It is moderately resistant to common leaf spot and powdery mildew, and comparatively more resistant than Chandler to two-spotted spider mite.

LAMBADA [SELECTED FROM THE CROSS BETWEEN IVT76013 (SILVETTA × HOLIDAY) × IVT74112 (KARINA × PRIMELLA)]

It is a Dutch cultivar. Fruits are shiny, large and with an attractive aroma and easy to pick. For crops grown in the open, early planting is essential. It is suitable for glasshouse culture. It matures later than Avanta, but slightly earlier than Elvira. It is not susceptible to infection by *P. fragariae* or *P. cactorum*.

LINCOLN (DONNA × 85-24-1, DERIVED FROM CRUZ, PAJARO, DOUGLAS AND HOLIDAY)

It is early, day-neutral and midseason variety. Fruits have excellent flavour and suitable for fresh market and processing uses.

LAUREL (ALLSTAR × CAVENDISH)

It is a short-day variety; fruits are medium-red surface; medium–large (15 g/fruit); more broad than long, short conical, cordate or globose; achenes slightly below fruit surface; firm; fresh, favour very good, sweet, aromatic; midseason, high vigour; medium–high runner numbers; medium yields in matted rows; resistant to red stele root rot (*P. fragariae*) but moderately susceptible to powdery mildew and *Botrytis* fruit rot (Kim, 2012).

LOUISE (ETTERSBURG 80 SELFED; also called CHURCHILL)

This variety was introduced by the Canadian Department of Agriculture, Ottawa during 1942. It produces medium to large and attractive berries, which matures late. Fruits are good for dessert. The plants are very susceptible to leaf scorch and leaf spot (Spangelo, 1962).

MAEHYANG (TROCHINOMINE × AKIHIME)

Plants of this variety are vigorous with erect growth and weak dormancy, resistant to powdery mildew, producing high yield of quality fruits. Fruits are red, long, conical with 15 g fruit weight (Kim et al., 2004).

MALWINA (UNNAMED SEEDLING × SOPHIE)

It is short-day variety, fruits are glossy red (RHS 45A, RHS 53A), large, conical, very good flavour, small, matures very late in the season, resistant to rain cracking, sunburn, *Verticillium* wilt, moderately resistant to *Botrytis* fruit rot, powdery mildew, susceptible to flower-thrips (*Frankliniella* spp.) and common leaf spot (Gasic and Preece, 2014).

MARI (POCAHONTAS × LIHAMA)

It is heavy yielder (130 kg/100 m²) of commercially viable fruit. Fruits are large (up to 13g/fruit), firm, slightly sour and bright red. Bushes are tall and, airy and moderately resistant to diseases such as *B. cinerea* and *S. morsuvae*.

MARINA (CARTUNO × CAMAROSA)

It is short-day, early, vigorous, compact cultivar in Spain. It has high productivity with fruits pleasant aroma and excellent flavour. It has low susceptibility to powdery mildew (*S. macularis*). Fruits are good for direct consumption as well as for processing (Lopez et al., 2004; Medina et al., 2006).

MARVEL (PS-1269 × PS-2280)

It is an everbearing, midseason variety, having glossy red fruits, medium–large (23.5 g); conical–cylindrical; flesh red, good favour, 8.6°Brix TSS; large and spreading calyx; achenes level to above surface; medium–strong vigour, medium stolon number (7.7 stolons/plant); low–moderate yield (615–618 g/plant); medium to dense foliage; prostrate fruiting trusses; moderately susceptible to two-spotted spider mite, flower-thrips, powdery mildew, *Botrytis* fruit rot, and angular leaf spot; susceptible to *Lygus* bugs (Kim, 2012).

MERIT (PS-2880 × PS-4630)

It is an everbearing variety; fruits glossy orange-red–red (RHS 34B), small (18.1 g), uniform conical, orange-red (RHS 34C) flesh, good to very good flavour (TSS 8.8°C Brix), late maturing, moderate yield (503 g/plant), and moderately resistant to bacterial angular leaf spot but moderately susceptible to two-spotted spider mite, flower-thrips, powdery mildew and *Botrytis* fruit rot (Gasic and Preece, 2014).

MERTON PRINCESS

It is a mid- to late-season, outstanding variety known for the size of fruit, but susceptible to *Botrytis* and red core and generally not recommended. Plants are large and dense. Fruits are very large, generally pointed, rather soft, salmon-pink with good flavour. Fruits should be picked carefully (Hughes et al., 1969).

MILSEI (PARKER × CHANDLER)

Milsei is a short-day, vigorous and productive cultivar, which has been patented since 1992. Fruits are very large, conical to wedge shaped, smooth, glossy and red exterior; interior orange-red, firm, juicy and sweet. Flavour is medium–strong, acidity higher than Chandler. It is suitable for the fresh market, having very good shipping characteristics (Finn, 1999).

MIRA (SCOOT × HONEOYE)

It was released during 1992 at Kenville, Canada. It is a midseason, vigorous cultivar, which is resistant to red stele root rot (*P. fragariae* var. *fragariae*) disease (Jamieson et al., 2001).

MISSIONARY

This is a very old variety of unknown parents which, was released in 1900 from Virginia (Figure 6.10). In India, it was in tested 25 years back in Jammu and Kashmir state.

MNUS 210 [EARLIGROW × MNUS 52] SYN. WINONA TM

It was developed in St. Paul, Minnesota during 1985. It is a late-season, June-bearing cultivar, which bears large fruits. This is resistant to multiple diseases and can tolerate cold winter and warm summer temperatures (Luby et al., 2001).

MOJAVE (PALOMAR × CAL 1.57-601)

It is a short-day variety, with dark red fruits (Munsell 5 R 3/7), which are very large (36.1 g), short-medium conical, very firm, red (Munsell 5 R6/11) flesh, outstanding flavour, very high yielding (2.271 kg/plant), moderately resistant to powdery mildew, *Verticillium* wilt, common leaf spot and tolerant to two-spotted spider mite and local strawberry viruses, moderately susceptible to anthracnose crown rot (*C. acutatum*) and highly susceptible to *Phytophthora* crown rot (Gasic and Preece, 2014).

NABILA (VENTANAX Q6Q8-26)

It is short-day variety, fruits red (RHS45A), large, uniform conical, red (RHS 41A) and orange (RHS29A, RHS 159C) flesh, very good flavour (TSS 7°Brix), medium-sized, spreading surface

FIGURE 6.10 Missionary.

FIGURE 6.11 Nabila.

calyx, very early season maturity, moderate to high yield (850–900 g/plant), and resistant to local leaf and root diseases (Gasic and Preece, 2014) (Figure 6.11).

NIIGATA S3 (KEI 812 (SEED PARENT) × ASUKA-RUBY)

It is early flowering variety, fruits have high TSS content and good colouration, maturing about 34 days earlier than Echigohime and nine days earlier than Asuka-Ruby. The marketable yield of Niigata S3 is 85% of Echigohime and 107% of "Asuka-Ruby", while the early yield was 145% of Echigohime, 85% of Asuka-Ruby. The shape of the fruit is long conical, and its skin colour medium-red. Its average TSS content is 11.4°Brix, which is higher than Echigohime and Asuka-Ruby. It does not bear apical overripe fruits. This new cultivar is adaptable to the climatic conditions of Niigata, as well as other regions that experience low winter temperatures and insolation (Hamato et al., 2014).

NINGYU (SACHINOKA × AKIHIME)

It was released from Jiangsu Academy of Agricultural Sciences during 2010. It is a new and early-fruiting variety bearing fruits having balanced aroma with anthracnose resistance. Compared with "Sweet Charlie", "Ningyu" has less plant height (7.34 cm) and leaf area (21.83 cm^2), thinner crown (22.42 cm) and lighter weight of leaf (0.18 g) and fruit (21.66 g) than Sweet Charlie. "Sweet Charlie's" fruit surface and flesh colour are more intense red (hue angle value smaller) and greater in pigment intensity (chroma value bigger) than Ningyu's. Ningyu has higher amount of fructose (148.98 mg/100 g FW), glucose (195.92 mg/100 g FW), reducing sugar (344.91 mg/100 g FW), ascorbic acid (68.9575 mg/100 g FW) and antioxidant capacity (1.433 mmol trolox/100 g FW) than "Sweet Charlie" (Wang et al., 2014).

OKA (K 75-13 × KENTVILLE)

This is a midseason, high-yielding, June-bearing cultivar evolved during 1982. It has excellent productivity, good fruit colour, flavour and firmness, and is recommended for fresh pick-your-own market. Plants are medium in size; leaflets are small–medium (5–8 cm long) size, flat, dark green, almost shiny, obovate to round, showing 20–22 acute teeth on margin. Leaf stalk is 10 cm long, pubescent, without bract. Inflorescence is medium to long size, stalks are held erect during blooming. Fruits are large, wedge shaped and moderately firm with moderately glossy and medium-red skin when fully ripe with light-red flesh (Khanizadeh, 2004).

ONTARIA

It was one of the leading commercial varieties in Ontario, where it usually gave a high yield. However, growers would like a variety to replace it because the fruit is soft, darkens in storage and is not outstanding for processing. Plants are vigorous; moderate–low yield, 381 g/plant; short pedicels; moderately susceptible to June yellows, two-spotted spider mite, thrips (*Frankliniella occidentalis*), *Botrytis* fruit rot and angular leaf spot but fairly resistant to *Verticillium* wilt, powdery mildew, leaf scorch and leaf spot. Virus-free stocks of the Ontaria strain have been reported to be not available (Spangelo, 1962; Kim, 2010).

OSO GRANDE [PARKER (U.S. PLANT PAT. NO. 5263) × CAL 77.3-603]

This invention relates to a new and distinctive short-day type strawberry cultivar designated as "Oso Grande" (hereinafter to be called "Oso" in this chapter), which is the result of a cross made in 1981. "Oso" first fruited at the South Coast Field Station in 1982, where it was selected and designated originally as Cal 81.43-603. It was tested later as advanced selection C43. It is a high-yielding cultivar to produce exceptionally large and firm berries with fine flavour (www.google.co.in/patents/USPP6578). Also suitable to grow in subtropical areas (Figure 6.12).

PAJARO (CAL 63.7-101 × SEQUOIA)

It was released during 1979. The fruit is of high dessert quality, large, symmetrical, attractive and firm. The fruit are quite susceptible to physical damage from rain, but since most of it is harvested during the dry season in California, the proportion of discarded fruits is generally very low.

PALOMAR (CAMINO REAL × VENTANA)

It is a short-day variety; fruit skin and flesh is similar to Ventana in colour; very large (30.2 g); short conical; calyx weakly attached; achenes yellow–dark red even with fruit surface. Plants are small, vigour similar to Camino Real but more compact than Ventana; extremely high yield (2031 g/plant); moderately resistant to powdery mildew, anthracnose crown rot and *Verticillium* wilt; moderately susceptible to *Phytophthora* crown rot and common leaf spot; conditional tolerance to two-spotted spider mite. It responds well to early winter planting (Kim, 2010).

FIGURE 6.12 Oso Grande.

PAVANA

It is a late variety that is vigorous with an open growth habit, long flower clusters and mid-harvesting date. Fruits are shiny, conical, very juicy, bright red; flesh is green and very firm. It is resistant to red core and crown rot, and less sensitive to *Verticillium* wilt and powdery mildew on the leaves (Meulenbrock et al., 2002).

PEGASUS (REDGAUNTLET × GORELLA) SYN. ES608

It is a self-fertile, vigorous and medium-late cultivar. It produces large, firm, pale red and conical fruits having excellent flavour, and ripens three to seven days later than Elsanta and seven–10 days before Pandora. Fruits are suitable for fresh consumption. Large fruit size is maintained until the second cropping year. It is resistant to *B. cinerea* and *P. fragariae* races 1, 2 and 3 (Ludvikova and Paprstein, 2003b).

PINK BEAUTY AND PRETTY BEAUTY (KINUAMA × PINK PANDA)

Pink Beauty and Pretty Beauty are two new ornamental and red-flowered edible strawberry cultivars with large fruits, which were released by Shenyang Agricultural University, China during 2010.

"Pink Beauty" had large and pink flowers with diameter ranging from 2.7–3.5 cm. Its fruit are sweet, red, conical and uniform with average primary fruit weight of 14.9 g. Its fruits are a bit sour.

Pretty Beauty has red flowers, larger (3.5–3.7 cm diameter) than "Pink Beauty". Its fruits are red, short wedged and uniform with 10.5 g average primary fruit weight. Both the new red-flowered cultivars could be grown for ornamental purposes with yield of 30–40 flowers/plant annually. Each flower generally lasts for five–seven days and the whole florescence could last for more than one month in the open field. Pink Beauty and Pretty Beauty bear better fruits having stronger leaf spot resistance than Pink Panda (Xue et al., 2014).

PINK PANDA (*F. COMARUM* HYBRID INVOLVING *F. CHILOENSIS*)

It is an intergeneric ornamental hybrid (*Fragaria × ananassa*) × *Potentilla palustris*) released in England during 1971 and considered to be the first red-flowered strawberry cultivar in the world but with either no fruit set or very small fruits.

Plants appear similar to the common strawberry but its flowers are pinkish-red to bright red with good decorative value. Fruits are small, red, weighing 5–10 g, with acidic flavour, 8% TSS content, titratable acidity 1.3% and poor eating quality (Dai et al., 2005).

PINNACLE (LAGUNA × ORUS 1267-250)

It is a short-day, June-bearing cultivar selected during 1996 in a USDA-ARS (U.S. Department of Agriculture, Agricultural Research Service) breeding programme and released in cooperation with the Oregon Agricultural Experiment Station (OAES), Washington and Idaho Agricultural Experiment Station. It is a high-yielding, large-fruited, midseason cultivar with very good fruit quality, and most suited for fresh markets and for processed products (Finn et al., 2004).

PLANASA 02-32 (94-020 × 9719)

It is a short-day variety adapted to protected cultivation. Fruits red (RHS 44A, RHS 44B), large (24–26 g), uniform conical, red-orange (RHS 33A, RHS 33B) flesh, moderately sweet (TSS 6.44 °Brix); and midseason maturity (Gasic and Preece, 2014).

Varieties   **111**

PREMIER (CRESCENT × HOWARD NO 1)

It was introduced by A. B. Howard, Belehrtown, Massachusetts, during 1909.

RANIA (VENTANAX Q6Q8-26)

It is a short-day variety bearing very large, uniform conic, firm and red fruits (RHS46A–46B) with orange-red (RHS 33A,RHS 35B) to whitish (RHS 155D) flesh having very good flavour and TSS content of 8°Brix. It has medium-size, spreading surface calyx, very early maturity and resistant to local leaf and root diseases (Gasic and Preece, 2014).

REBECKA [(FERN × *F. VESCA*) × *F. ANANASSA* F 861502]

This is a day-neutral cultivar, bearing flowers and fruits after the normal season and on the new runner plants. Fruits are medium–small (8–9 g), soft with nice aroma of wild wood strawberries, and more suitable for home gardens (Trajkovski, 2002).

RED COAT (SPARKLE × VALENTINE)

This is a very high-yielding variety, introduced by the Canadian Department of Agriculture, Ottawa during 1957. It retains good berry size throughout the harvest season and is one of the most outstanding varieties for the fresh market. Fruit are bright red, very firm and attractive, and maintain their appearance in storage. The plants are susceptible to *Verticillium* wilt and leaf spot but resistant to powdery mildew (Spangelo, 1962).

RED RICH (ALSO CALLED RED-GLO AND HAGERSTROM'S EVERBEARING)

This variety was introduced by M. Hagerstrom, Enfield, Minnesota during 1949. This is a most promising everbearing variety. Fruit are good for dessert and freezes as well. In some conditions, probably related to photoperiod and temperature, the plants develop very few stolons. Virus-free stocks were reported to be not available (Spangelo, 1962).

REDGAUNTLET (NEW JERSEY 1051 × AUCH. CLIMAX)

It is a heavy-yielding variety with good berry size. Plant growth is sometimes weak, particularly in poor soil conditions. Fruits are large to very large, round to spherical-wedged, broad, and blunt; often uneven with few seeds, attractive and scarlet initially and later turn dark crimson. Flesh is firm, fairly juicy, deep scarlet with pleasant flavour. The fruit is well exposed, so this variety is easy to pick, but the fruit stalk is sometimes tough. It is field resistant to some strains of red core and *Verticillium* wilt, and not susceptible to mildew or *Botrytis*. This variety has the ability to initiate trusses during summer, particularly if the first crop is protected during the spring, and then flowers during August, and will crop satisfactorily generally with some protection, from mid-September until November, depending on season. Occasionally, during very warm years, picking of the second crop can be done in the autumn from unprotected plants. This second crop does not appear to reduce the crop in the following year from normal, autumn-initiated trusses. This variety is less prolific in runner production than Favourite and responds well to warm climatic conditions and sheltered sites (Hughes et al., 1969).

ROMINA (SELECTION AN99, 78, 51 DERIVED FROM THE CROSS OF 95.617.1 × DARSELECT)

It is short-day cultivar selected during 2000 and has high adaptability to non-fumigated soil conditions. It is very early blooming and maturity (Faedi's Precocity Index 137). The plants have medium crown number, strong vigour and high uniformity. Fruits are conical or biconical with round tip. Calyx is medium, easy to remove; achenes are medium sized and yellow. Fruits are red, firm, sweet and tasty with good shelf life. The fruits have high contents of vitamin C, folates (Tulipani et al., 2008) and flavonoids, mainly quercetin glucuronide (Diamanti et al., 2014). The high concentration of flavonols and other phytochemicals could be contributing to a good colour stability of pasteurized puree prepared from the fruits of this variety (Capocasa et al., 2016).

ROSALYNE [FERN × (SJ9661-1 × PINK PANDA)]

Plants are moderately vigorous, long, hardy and attractive; flowering starts from mid-July and continues up to mid- or late-July; flowers are large (4.3 cm), dark pale to pink (red-purple group 65 A), more suitable for home garden for ornamental purposes. Fruits are small–medium in size, without neck, skin red and moderately glossy, flesh medium red, moderately firm with excellent flavour. Plants are resistant (Khanizadeh et al., 2003; Khanizadeh, 2004) to leaf spot (*M. fragariae*) and leaf scorch (*D. earliana*).

ROSEBERRY [(RARITAN × K74-12) × (SJ9616-1 × PINK PANDA)]

It is a seedling selection from the progeny of a multiple cross, winter hardy, day-neutral and produces attractive blooms during summer. This is resistant to leaf scorch (*D. earlianum*). It is mostly grown in home gardens for ornamental purposes (Khanizadeh et al., 2002c).

ROSIE (HONEOYE × FORLI)

This is an early, June-bearing, moderately vigorous variety, released during 1999 by HRI, East Malling, England. Fruits are large, conical, strong red with glossy skin and sweet aroma (Simpson et al., 2002).

ROXANA [SURPRISE DES HALLES (MARKET SURPRISE) × SENGA SENGANA]

It is mid-early maturity (between Surprise des Halles and Senga Sengana) variety and resistant to *S. macularis* and *B. cinerea*. The fruits are heart shaped, of uniform size, rather firm and glossy, bright red, having an acid-sweet taste and pleasant flavour. Yield is good with a large proportion of top-quality fruits suitable for fresh consumption, freezing or jam making.

SABLE (VEESTAR × CAVENDISH)

It was introduced in 1998 by Agriculture and Agri-Food Canada Strawberry Breeding Programme at Kentville. It is early, more productive, vigorous and resistant to *P. fragariae*. Fruits are large (10–12 g) with strong aroma, suitable to pick-your-own plantations, but not fit for distant shipping (Jamieson, 2002).

SABRINA (9719 × 94-020)

It is a short-day variety adapted to protected cultivation. Fruits are red (RHS 43A, RHS 43B), large (24–26 g), uniform conical, red (RHS 41A, RHS 41B) flesh, pleasant flavour with 6.9°Brix TSS), and very high (1.612 kg/plant) yielding (Gasic and Preece, 2014).

SAINT PIERRE (CHANDLER × JEWEL)

It is a vigorous cultivar having an upright growth and bears two–five inflorescences per crown. Fruits are very large, firm, light-red with good shelf life and free from darkening during storage.

Varieties 113

This is resistant (Khanizadeh et al., 2002e) to leaf scorch (*D. earlianum*), leaf blight (*Dedrophoma obscurance*) and leaf spot (*M. fragariae*).

SAINT-LAURENT D'ORLEANS [L ACADIE × (SJ 8916 × PINK PANDA)]

It is a medium-vigour cultivar (Khanizadeh et al., 2005) with firm, light-red and shiny fruits having long shelf life, resistant to leaf scorch (*D. earlianum*) and leaf spot (*M. fragariae*).

SALLA (ADDIE × PAJARO)

This is a vigorous and high-yielding Italian variety and bears large, uniformly vermilion fruits with 36.6 g weight and firm flesh of a fairly deep red but not entirely uniform colouration. It matures about four days after Addie and is adapted to cultivation in northern Italy.

SANTACLARA (CARISMA × PLASIRFRE)

It is a new short-day variety and has been released from the Spanish National strawberry breeding programme (INIA Agreements CC05-024-C3-1). It has a high level of adaptation to different mild agro-environments and is noted for high ascorbic acid content, its medium to highly balanced agronomic, post-harvest and fruit-quality traits, and comparable with excellent quality varieties for fresh-fruit market (Soria et al., 2014).

SAULENE (SHUKSAN × SENGA SENGANA)

It is an early-maturing and high-yielding cultivar producing fruits of good quality and excellent taste, which are highly suitable for fresh use (Rugienius et al., 2004).

SELVA (CAL 70.3-117 × CAL 71.98-605)

It is an everbearing variety introduced into Belgium during 1987. Fruits are either grown in the open with low plastic cloches placed all over during early September or in peat bags in a greenhouse. The yield/plant or per unit area is less than that of Rapella, but average fruit weight is greater.

SENGA SENGANA (MARKEE × SIEGER)

It was introduced in Europe during 1952. It produces a healthy plant, and its runners begin to emerge during August. Fruits are attractive, red with medium flesh firmness and good freezing quality.

SHIMEI 3 (183-2 × ALL STAR)

This variety was developed during 1990 and suitable for north and northeast China. It is vigorous, early-mid in maturity with high yield. Fruits are large, having good flavour and good eating quality (Yang et al., 2005).

SHIMEI 4 (HOKOWASE × SHIMEI 1)

This variety is early in maturity with low chilling requirements. Fruits are coniform; fruit weight is 36.7 g with good storage quality and higher yield. Flesh is fine and fragrant, and highly resistant to various diseases (Yang et al., 2004).

SHIMEI-7 (TOCHIOTOME × ALLSTAR)

It was released during 2002 to adapt to modern agriculture and the development of the suburban fruit-picking tourism industry. The fruits are conical, bright red, fruit surface smooth, with strong and glossy waxy layer. The fruit are uniform and large (sometimes as much as 57 g), but average weight remains about 33.6 g for primary fruit and 21.5 g for secondary fruit. The fruit flesh is orange and thin, the texture is dense, fiber is low, fragrance is intense, and quality is excellent with TSS content of 10.5°Brix and 0.851 kg/cm^2 fruit elastic force and 0.447 kg/cm^2 fruit flesh resistance, and tolerant to storage and transportation. The plants are vigorous, producing 358.6 g/plant, and are resistant to leather rot and *B. cinerea*. The dormancy period is about 300 hours, and it is suitable for semi-forcing culture and is a mid–early-maturity cultivar (Yang et al., 2014).

SOPHIE (NY 1261 × HOLIDAY)

It is a late and June-bearing, vigorous and very productive variety, released during 1997 by HRI, East Malling, England. Berries are attractive with regular shape, strong red with glossy skin. Plants do not require other pollenizing cultivars (Simpson et al., 2002).

SPARKLE (FAIRFAX × ABERDEEN; ALSO CALLED PAYMASTER)

This variety was introduced by the New Jersey Agriculture Experiment Station, New Brunswick, New Jersey during 1942. The medium-sized, firm, rather dark and attractive berries are good for dessert and acceptable for freezing. It is fairly high yielding in many areas and is moderately susceptible to *Verticillium* wilt, leaf scorch and leaf spot but resistant to powdery mildew (Spangelo, 1962).

STRAWBERRY FESTIVAL (ROSA LINDA × OSO GRANDE)

It is a short-day variety released during 1995 in Florida, USA. It bears attractive, flavourful, and firm fruits. Mean fruit weight and yield are similar to Sweet Charlie (Chandler et al., 2000b).

SURECROP (FAIRLAND × MARYLAND-U.S. 1972)

This variety was introduced by the United States Department of Agriculture, Beltsville and Maryland Agricultural Experiment Station, College Park, Maryland during 1956. The fruits retain their good size throughout the season. It is attractive, firm, good for dessert and satisfactory for freezing. The plants are resistant to certain races of red stele and fairly resistant to *Verticillium* wilt, leaf scorch and leaf spot (Spangelo, 1962).

SUMMER RUBY (57K104 × 572I16)

It is a short-day and midseason variety similar to Kent; bears large and very bright fruits having medium-red skin; medium-red and moderately firm flesh; good flesh favour; firm skin. Plants are vigorous; yield and runner production good; hardy but susceptible to anthracnose (Kim, 2012).

SVEVA (EM483 × 87.734.3)

It was developed during 1996 in Italy, which is late in maturity with improved productivity of high-quality fruits having high antioxidant values (Capocasa et al., 2004a,b).

Varieties

SWEET ANN (4A28 × 10B131)

It is an early maturing and day-neutral variety, fruit glossy red (Pantone 185C), large (33 g), moderately uniform conical, but some are wedge shaped, moderately firm, red (1788C) flesh, balanced flavour with 10.5 to 13.2°Brix soluble solids content (Gasic and Preece, 2014).

SWEET CHARLIE (FL 80-456 × PAJARO)

This is a short-day and early-bearing variety developed by Florida University, Florida during 1992. It is suitable in areas with warm winters, and good for forced cultivation. Sweet Charlie is resistant to anthracnose (*Glomerella cingulata*). Fruits are conical, large, weighing on average 17 g and are orange-red. Flesh texture is firm, orange colour with some white strings, high TSS (7°Brix), low titratable acidity (0.66%) and high ascorbic acid (526 μ/g) content (Zhang et al., 2006; Chandler, 2005).

TALISMAN (NEW JERSEY × AUCH. CLIMAX)

It was originated at the Scottish Horticultural Research Institute and was introduced during 1965. Plant are medium sized, very slow to develop in the spring, upright and dense. Fruits are conical to wedge shaped, medium at first, becoming smaller as the season advances, quite attractive, bright scarlet and turning dark scarlet when full mature. Flesh is firm, juicy, scarlet, fine textured with good flavour, grown both for dessert and processing markets but smaller berry size is now chief disadvantage. It is field resistant to red core and mildew. It profusely produces runners (Hughes et al., 1969).

TANGO (RAPELLA × SELVA)

It is a high-yielding and day-neutral variety, originated at the HRI, East Malling, England by D. Simpson during 1986. Fruits are firm with good shape and appearance; flavour is acidic and more suitable for fresh market. It is resistant to *Verticillium* wilt and some races of red stele but susceptible to powdery mildew (Daubeny, 1994).

TEMPLER (SEEDLING AUCH. 11 × CAMBRIDGE VIGOUR)

It was introduced during 1964. The plants are vigorous, very dense and leafy. Flowers are small on short trusses, hidden in the foliage which makes picking difficult. It bears less heavily than established varieties and may make too vigorous a plant. However, it has field resistance to most strains of red core and may be suitable for soils where other varieties do not grow satisfactorily, midseason maturity with quite good fruit flavour. It is susceptible to powdery mildew (Hughes et al., 1969).

TERUNOKA (HOKOWASE × DONNER)

It is suitable for forced culture with additional lighting particularly in west of the Kwanto district and northern parts of Tohoku, Japan. Fruits weigh 10–15 g, glossy, scarlet red and conical. The flesh is firm, making the fruit highly suitable for handling and transporting. The pollen retains a high viability at low temperatures, so when conditions are adverse, few malformed fruits are produced. It is highly resistant to *Sphaerotheca humili* (macularis) and *F. oxysporum* f. sp. *fragariae*.

TILLAMOOK (CUESTA × PUGET RELIANCE)

It is a short-day (June-bearing) cultivar selected during 1996 by the United States Department of Agriculture, Agricultural Research Service (USDA-ARS) breeding programme and released in cooperation with the OAES, Washington and Idaho Agricultural Experiment Station. It is a high-yielding, large-fruited, midseason cultivar with very good fruit quality, suitable for the fresh and processed markets. It was named after the Native American tribe who lived along the pacific Coast on Tillamook Bay and current site of the city of Tillamook, Oregon (Finn et al., 2004).

TIOGA [FRESNO × TORREY (LASSEN × CAL 42.8-16)]

It was released during 1964 and has a very short harvest season, with typical characteristics of successively reduced fruit size with advances in the crop season. As the fruits tend to "cap" easily and get separated without calyx, hence it is somewhat difficult to pick the fruits of this variety for the fresh market. The fruits are suitable for processing but not for the fresh market (Bringhurst and Voth, 1989).

TOCHIOTOME (KURUME 49 × TOCHINOMINE)

This cultivar was released during 1994 and is suitable for sunny greenhouses. If planted during late August, then fruits are ready for picking during late December. Fruits are large (20.5 g), with pretty, smooth, bright red skin with good eating and shipping qualities. Seeds are little bit sunken. Flesh contains 10.2 °Brix soluble solids, 0.91% titratable acidity and 733 µ/g ascorbic acid (Zhu et al., 2005).

TORREY (LASSEN × CAL 42.8.16)

It was originated at the California Agriculture Experiment Station during 1961. The fruits are attractive with large size, firm flesh and, cap easily.

VIBRANT [(SDBL101 × EM881) × (ROSIE × ONEBOR)]

One of the female parents (SDBL101) of this multicross hybrid is of unknown pedigree and Italian origin but was best adapted to southern Italy. In trials, the fruit yield in the main-season crop has been observed to be approximately 85% of Elsanta but with a higher proportion of large berries (>35 mm). The yield in 60-day trials has been slightly higher than that of Elsanta. Plants are moderately vigorous with erect growth and the flowers are held above the foliage on long peduncles, resulting in a very good fruit display which facilitates faster harvesting. Berries are glossy, strong red and conical. The flavour is pleasant, sweet with juicy texture, while mean Brix readings and shelf life are both similar to Elsanta. It has good resistance to powdery mildew (*P. aphanis*) but is moderately susceptible to wilt (*V. dahliae*). Although it is a short-day cultivar, it produces a second crop in the UK, although the timing and magnitude of this depend on the season, location and growing system. If grown without protection in southern England, the harvesting of first crop typically begins during late-May or early-June, and subsequently a second flush of flowers appears during August, which is harvested during September. The fruit quality on the second crop is equivalent to the first and the total yield from both the crops comes out to be equivalent to 135% of the yield from a single crop on "Elsanta" (Simpson et al., 2014).

WINTERSTAR™ (FLORIDA RADIANCE × EARLIBRITE) SYN. FL 05-107

The plants of this variety are moderately vigorous and compact, allowing for high-density plantings; medium–long fruit stems. Large, very uniform, conical–broadly conical, bright red fruits with glossy appearance that does not become overly dark at the end of the season. Pulp is medium–light-red; low-acid flavour giving a perception of greater sweetness; firm with shipping quality nearly

Varieties   **117**

comparable to "strawberry cv. "Festival"". Skin of this cultivar is tough particularly during late in the season; therefore, the variety is moderately resistant to rain damage. Besides, it is moderately resistant to anthracnose fruit rot and susceptible to *Botrytis* fruit rot; moderately susceptible to angular leaf spot; susceptible to *Phytophthora* root rot.

YAMASKA (PANDORA × BOGOTA)

It is a June-bearing (short-day), late-ripening and vigorous cultivar introduced during 1998. Fruits are very large, attractive dark red and glossy, and can be stored up to five days at room temperature. Flesh is fairly firm. The plant produces two–five inflorescences per crown and can tolerate (with 10 cm thick cover of straw mulch) winter temperatures below −30°C (Khanizadeh et al., 2002d; Khanizadeh et al., 1999; Finn, 1999).

YUELI (BENIHOPPE × SACHINOKA)

The plants grow vigorously, with relatively smooth stem. The edge of the leaf shows a ragged shape. The leaves are green and elliptic. The first node bears 9.45 flowers/inflorescence, dimorphic, heterostyly, petals white and arranged in a single layer. The fruits are early maturity, big, conical, uniform, bright red with luster, firm with excellent flavour. It is comparatively better than variety "Benihoppe" in resistance to powdery mildew, grey mould and anthracnose (Jiang et al., 2014).

YUEZHU (AKIHIME × TOYONOKA)

In this variety, the leaves are green and ovoid and edge of leaves show ragged shape. The plants grow vigorously (height 16.23 cm and spread 35.7 cm × 29.8 cm), with relatively smooth stem. The first node bears 16 flowers/inflorescence, dimorphic heterostyly, petals white and arranged in single layer. The fruit are early-ripening, short conical, uniform, bright red, shiny fruit with good flavour. The stolon is reddish-green and flourishing (Jiang et al., 2014).

ZARINA (1007 × 880)

It is a day-neutral variety with early maturity, fruits are red (RHS 45A), large (25.6 g), conical–long-conical, long wedged, firm, flesh orange-red (RHS 41B) to whitish (RHS 155C), balanced flavour (Gasic and Preece, 2014).

ZUOHEQINGXIANG

This is a Japanese cultivar suitable for sunny greenhouses. Fruits are large (35 g) and sometimes extra large (52.5 g), conical and bright. Fruit flesh is white, contains 10.2% TSS and 0.94% titratable acidity, firm, rich in subacidic and pleasant flavour, with good eating quality (Wang and Lei, 2004).

REFERENCES

Bartual, R., Marsal, J.I., Arjona, A., Lopez, A.J.M., Medina, J.J., Lopez, M.R. and Lopez, M.J. 2002. Carisma: a new Spanish strawberry cultivar. *Acta Horticulturae*, 567: 187–189.

Bringhurst, R.S. and Voth, V. 1989. California strawberry cultivars. *Fruit Varietal Journal*, 43: 12–19.

Capocasa, F., Scalzo, J., Murri, G. and Mezzetti, B. 2004a. "Adria" and "Sveva" – new late maturing strawberry cultivars. *Informatore Agrario*, 60(67): 51–52.

Capocasa, F., Scalzo, J., Murri, G., Baruzzi, G., Faedi, W. and Mezzetti, B. 2004b. "Adria" and "Sveva" – two new strawberry cultivars from the breeding programme at the Marche Polytechnic University. *Acta Horticulturae*, 649: 85–86.

Capocasa, F., Mezzetti, B., Williams, S., Hargreaves, R., Bernardini, D. and Danesi, L. 2016.Romina' and "Cristina": two new strawberry cultivars for the European and USA market. *Acta Horticulturae*, 1117: 71–75.

Chandler, C.K. 2005. "Sweet Charlie" strawberry. *Journal of American Pomological Society*, 59(2): 67.

Chandler, C.K., Legard, D.E., Crocker, T.E. and Sims, C.A. 2004. "Carmine" strawberry. *HortScience*, 39(6): 1496–1497.

Chandler, C.K., Legard, D.E., Dunigan, D.D., Crocker, T.E. and Sims, C.A. 2000a. "Earlibrite" strawberry. *HortScience*, 35(7): 1363–1365.

Chandler, C.K., Legard, D.E., Dunigan, D.D., Crocker, T.E. and Sims, C.A. 2000b. "Strawberry Festival" strawberry. *HortScience*, 35(7): 1366–1367.

Dai, H.P., Lei, J.J., Yan, Y.H. and Li, H.Y. 2005. "Pink Panda" a promising strawberry cultivar with red flowers and its cultural techniques. *China Fruits*, 6: 43–44.

Daubeny, H. 1994. Agricultural Canada Research Station, Vancouve, B.C. *HortScience*, 29(9): 960–961.

Diamanti, J., Mazzoni, L., Balducci, F., Cappelletti, R., Capocasa, F., Battino, M., Dobson, G., Stewart, D. and Mezzetti, B. 2014. Use of wild genotypes in breeding program increases strawberry fruit sensorial and nutritional quality. *Journal of Agriculture and Food Chemistry*, 62(18): 3944–3953.

Duan, L., Tian, F., Zhang, X., Liu, Y., Wang, C. and Xu, X. 2014. "Baotong" – a new, early and high quality strawberry cultivar. *Acta Horticulturae*, 1049: 253–254.

Finn, C.E. 1999. Register of new fruit and nut varieties, List 39, Strawberry. *HortScience*, 32(2): 197–201.

Finn, C.E., Yorgey, B., Strik, B.C. and Moore, P.P. 2004. "Tillamook" and "Pinnacle" strawberries. *HortScience*, 39(6): 1487–1489.

Gasic, Ksenija and Preece, John E. 2014. Register of new fruit and nut cultivars list 47. *HortScience*, 49(4): 396–421.

Geibel, M. 2002. Genetic resources in strawberries in Europe. *Acta Horticulturae*, 567(1): 73–75.

Gu, J., Jiang, Z.T., Huang, R.Q., Wang, C.H. and Wang, W.L. 2005. Breeding report on new strawberry cultivar "Hongshimei". *China Fruits* 3: 3–5.

Hamato, N., Kotake, O., Ono, N., Kurashima, H., Nakano, M., Iwamoto, Y. and Takahashi, Y. 2014. "Niigata S3" is a new strawberry cultivar suitable for forcing culture under low temperature and insolation conditions. *Breeding Science*, 64: 427–434.

Hao, B.C., Li, Y., Jia, Y.N., Li, M.C. and Chu, F.J. 2002. The breeding report for new strawberry variety Chunxing. *China Fruits* 2: 3–4.

Hughes, H.M., Duggan, J.B. and Banweil, M.G. 1969. Strawberry Bull. 95 HMSO, 10, 6d, Ministry of Agriculture Fish and Food, UK.

Jamieson, A.R. 2002. New strawberries from the AAFC breeding program at Kentville. *Acta Horticulturae*, 567(1): 153–156.

Jamieson, A.R., Nickerson, N.L., Forney, C.F., Sanford, K.R. and Craig, D.L. 2001. "Mira" strawberry. *HortScience*, 36(2): 389–391.

Jamieson, A.R., Nickerson, N.L., Sanderson, K.R., Prive, J.P., Tremblay, R.J.A. and Hendrickson, P. 2004a. "Brunswick" strawberry. *HortScience*, 39(7): 1781–1782.

Jamieson, A.R., Nickerson, N.L., Sanderson, K.R., Prive, J.P., Tremblay, R.J.A. and Hendrickson, P. 2004b. "Evangeline" strawberry. *HortScience*, 39(7): 1783–1784.

Jamieson, A.R., Nickerson, N.L., Forney, C.F., Sanford, K.R., Prive, J.P., Tremblay, R.J.A and Hendrickson, P. 2004c. "Cabot" strawberry. *HortScience*, 39(7): 1778–1780.

Jiang, G.H., Zhang, Y.C., Miao, L.X. and Yang, X.F. 2014. Strawberry breeding in Zhejiang of China. *Acta Horticulturae*, 1049: 183–186.

Kawagishi, K., Kato, S., Ubukata, M., Abe, T., Tachikawa, S., Inagawa, Y. and Fukukawa, E. 2001. "Kentaro" a new short-day strawberry variety for cold regions. *Bulletin of Hokkaido Prefectural Agricultural Experiment Stations* 81: 11–20.

Khanizadeh, S. 2004. New strawberry cultivars. *Agriculture and Agri-Food Canada Bulletin*: 2–11.

Khanizadeh, S., Carisse, O. and Buszard, D. 2002a. Yamaska and Acadie strawberries. Strawberry research to 2001. *Proceeding of the 5th North American Strawberry Conference*: pp. 38–40.

Khanizadeh, S., Cousineau, J., Buszard, D., Gauthier, L. and Hebert, C. 2002b. "L'Authentique Orleans": a new strawberry cultivar with high levels of antioxidants. *Acta Horticulturae*, 567(1): 175–176.

Khanizadeh, S., Cousineau, J., Deschenes, M. and Levasseur, A. 2002c. "Roseberry" and "Rosalyne": two new hardy, day-neutral, red-flowering strawberry cultivars. Strawberry research to 2001. *Proceeding of the 5th North American Strawberry Conference*: p. 42.

Khanizadeh, S., Cousineau, J., Deschenes, M. and Levasseur, A. 2002d. The Quebec day-neutral strawberry breeding program of Agriculture and Agri-Food Canada at St- Jean-sur-Richelieu: a progress report. Strawberry research to 2001. *Proceeding of the 5th North American Strawberry Conference*: p. 37.

Varieties **119**

Khanizadeh, S., Cousineau, J., Deschenes, M., Levasseur, A., Carisse, O., DeEll, J., Gauthier, L. and Sullivan, J.A. 2002e. "Saint-Pierre" strawberry. *HortScience*, 37(7): 1133–1134.

Khanizadeh, S., Carisse, O., Deschenes, M., Gauthier, L., Gosselin, A., Hebert, C., DeEll, J. and Buszard, D. 2003. "L' Authentique Orleans strawberry. *Acta Horticulturae*, 626: 37–38.

Khanizadeh, S., Deschenes, M., Levasseur, A., Carisse, O., Charles, M.T., Rekika, D., Tsao, R., Yang, R., DeEll, J., Gauthier, L., Gosselin, A., Sullivan, J.A. and Davidson, C. 2005. "Saint-Laurent d'Orleans" strawberry. *HortScience*, 40(7): 2195–2196.

Khanizadeh, S., Theriault, B., Carisse, O. and Buszard, D. 1999. Yamaska strawberry. *HortScience*, 34(7): 1286–1287.

Kim, S.L. 2010. Register of new fruit and nut cultivars, list 45. *HortScience*, 45(5): 748–752.

Kim, S.L. 2012. Register of new fruit and nut cultivars, list 46. *HortScience*, 47(5): 556–559.

Kim, T., Jang, W.S., Choi, J.H., Nam, M.H., Kim, W.S. and Lee, S.S. 2004. Breeding of strawberry "Maehyang" for forcing culture. *Korean Journal* of *Horticultural Science and Technology*, 22(4): 434–437.

Kitamura, H., Mori, T., Kohori, J., Yamada, S. and Shimizu, H. 2015. Breeding and extension of the new strawberry cultivar "Kaorino" with extremely early flowering and resistance to anthracnose. *Horticulture Research (Japan)*, 14(1): 89–95.

Kong, Z.L., Tong, Y.F., Zhang, G.Z., Shao, Z. and Lei, Q.H. 2003. Hongjia, a new strawberry variety suitable to be grown under protection. *South China Fruits*, 32(5): 61.

Konno, H., Inagawa, Y., Kawagishi, K., Sawada, K., Schiozawa, K., Kato, S. and Tachikawa, S. 2001. A new short-day strawberry varity "Kita-ekubo". *Bulletin of the Hokkaido Prefectural Agricultural Experiment Station*, 81: 1–10.

Lee, C.G., Jang, P.H., Seo, J.B., Shin, G.H. and Yang, W.M. 2016. A new strawberry "Jukhyang" with high sugar contentand firmness. *Acta Horticulturae*, 1117: 39–43

Leis, M., Castagnoli, G. and Martinelli, A. 2002. Pomological novelties: Ariel. *Rivista di Frutticoltura e di Ortofloricoltura*, 64(6): 47.

Li, S.J. and Ding, H. 2006. Hongjia, a new strawberry cultivar and its cultural techniques. *South China Fruits,* 1: 5–9.

Lopez, A.J.M., Soria, C., Miranda, L., Sanchez, S.J.F., Galvez, J., Villalba, R., Romero, F., Santos, B.D.L., Medina, J.J., Palacios, J,, Bardon, E., Arjona, A., Refoyo, A., Martinez, T.A., de Cal, A., Melgarejo, P. and Bartual, R. 2005. "Aguedilla" strawberry. *HortScience*, 40(7): 2197–2199.

Lopez, A.J.M., Soria, C., Sanchez, S.J.F., Galvez, J., Medina, J.J., Arjona, A., Marsal, J.I. and Bartual, R. 2004. Marina strawberry. *HortScience*, 39(7): 1776–1777.

Lu, P.F., Zhang, Q. and Yu, G.W. 2005. Breeding report for new strawberry cultivar "Fengguan 1". *China Fruits* 6: 15–17.

Luby, J.J., Wildung, D.K. and Galletta, G.J. 2001. MNUS 210 (Winona TM) strawberry. *HortScience*, 36(2): 392–394.

Ludvikova, J. and Paprstein, F. 2003a. Strawberry cultivar "Honeoye". *Vedecke Prace Ovocnarske*, 18: 169–171.

Ludvikova, J. and Paprstein, F. 2003b. Strawberry cultivar "Pegasus". *Vedecke Prace Ovocnarske*, 18: 173–174.

Ma, H.X., Duan, X.M., Dai, Z.L., Chen, X.L. and Wu, W.M. 2001. Breeding new strawberry varieties "Chunxu". *China Fruits*, 3: 3–5.

Medina, J.J., Bartual, R., Soria, C., Arjona, A., Miranda, L., Sanchez, S.J.F., Galvez, J., Clavero, I. and Lopez, A.J.M. 2006. Adaptation and agronomical characterization of "Medina" and "Marina" strawberry cutivars. *Acta Horticulturae*, 708: 73–75.

Meulenbrock, E.J., Lindeloof, C.P.J. and Kanne, H.J. 2002. "Pavana" a new strawberry cultivar from Plant Reseach International. *Acta Horticulturae*, 567(1): 183–185.

Rugienius, R., Sasnauskas, A. and Shikshnianas, T. 2004. "Saulene" and "Dange" – two recent Lithuanian strawberry cultivars. *Acta Horticulturae*, 649: 73–75.

Sanford, J.C., Ourecky, D.K. and Reich, J.E. 1985. "Jewel" strawberry. *HortScience*, 20: 1136–1137.

Simpson, D.W., Bell, J.A, Hammond, K.J. and Whitehouse, A.B. 2002. The latest strawberry cultivars from Horticultural Research International. *Acta Horticulturae*, 567(1): 165–168.

Simpson, D.W., Whitehouse, A.B., Johnson, A.W., McLeary, K.J. and Passey, A.J. 2014. "Elegance" and "Vibrant", two new strawberry cultivars for Programmed Cropping in Northern Europe. *Acta Horticulturae*, 1049: 259–261.

Soria, C., Medina, J.J., Dominguez, P., Miranda, L., Villalba, R., Refoyo, A. and Lopez-Aranda, J.M. 2014. "Santaclara", a new strawberry cultivar developed by the Spanish Public Breeding Program. *Acta Horticulturae*, 1049: 249–255.

Spangelo, L.P.S. 1962. Growing strawberries in eastern Canada – planting and harvesting. Canada Department of Agriculture Bulletin, 1179.

Trajkovski, K. 2002. Rebecka, a day-neutral *Fragaria x vescana* variety from Balsgard. *Acta Horticulturae*, 567(1): 177–178.

Tulipani, S., Mezzetti, B., Capocasa, F., Bompadre, S., Beekwilder, J., de Vos, C.H., Capanoglu, E., Bovy, A. and Battino, M. 2008. Antioxidants, phenolic compounds, and nutritional quality of different strawberry genotypes. *Journal of Agriculture and Food Chemistry*, 56(3): 696–704.

Verwijs, A.J. 1980. Is strawberry seedling No.69–16 an asset? *Grownten en Fruit*, 36: 58–59.

Wang, M.X. and Lei, J.J. 2004. Performance of the introduced new Japanese strawberry cultivar "Zuoheqingxiang". *China Fruits*, 3: 25–26.

Wang, J., Zhao, M.Z., Wang, Z.W., Yu, H.M. and Huang, W.Y. 2014. Ningyu: a new cultivar with high reducing sugar, ascorbic acid and antioxidant capacity. *Acta Horticulturae*, 1049: 231–235.

Xue, L., Lei, J.J., Dai, H.P. and Deng, M.Q. 2014.Two new ned-flowered strawberry cultivars "Pink Beauty" and "PrettyBeauty". *Acta Horticulturae*, 1094: 219–224.

Yang, L., Hao, B.C., Li, L., Wang, J.T., Jia, Y.Y. and Chu, F.J. 2004. A new early ripening strawberry variety "Shimei 4". *Acta Horticulturae Sinica*, 31(1): 135.

Yang, L., Li, L., Hao, B.C., Jia, Y.Y., Li, M.C. and Chu, F.J. 2005. A new high-yield and high quality strawberry "Shimei 3". *Acta Horticulturae Sinica*, 32(3): 560.

Yang, L., Li, L., Yang, L., Zhang, J., Li, H. and Hao, B. 2014. Strawberry cultivar "Shimei-7" for sightseeing and fruit-picking tourism. *Acta Horticulturae*, 1049: 247–248.

Zhang, Y.T., Wang, G.X., Dong, J. and Wang, P. 2006. "Sweet Charlie", a new promising strawberry cultivar and its cultural techniques. *China Fruits*, 1: 22–23.

Zhang, Y., Wang, G., Chang, L., Dong, J., Zhong, C. and Wang, L. 2014. Current status of strawberry production and research in China. *Acta Horticulturae*, 1049: 67–71.

Zhu, H., Dai, H.P., Tan, C.H. and Lei, J.J. 2005. The performance of new strawberry cultivar "Tochiotome" in a sunny greenhouse. *China Fruits*, 4: 19–20.

7 Tissue Culture

Anil Kumar Singh, Gyanendra Kumar Rai, Sreshti Bagati, and Sanjeev Kumar

CONTENTS

7.1 Tissue Culture Techniques...122
    7.1.1 Micropropagation 122
    7.1.2 Protoplast Culture..123
    7.1.3 Anther/Pollen Culture..125
    7.1.4 Meristem Culture and Micro-Cuttings..126
    7.1.5 Somatic Embryogenesis...127
7.2 Applications of Tissue Culture Techniques 127
    7.2.1 Somaclonal Variation...127
    7.2.2 Virus Elimination 129
    7.2.3 *In vitro* Selection for Biotic/Abiotic Stress Tolerance...130
    7.2.4 Genetic Transformation 131
        7.2.4.1 Mechanism of *Agrobacterium*-Mediated Gene Transfer..........................131
References...135

As per the International Treaty on Plant Genetic Resources, strawberry is a crop of global horticultural significance (Hummer, 2009). This calls for a fundamental shift in its overall production management that includes bringing new areas under cultivation and converting old plantations to improved practices coupled with use of superior varieties. To keep pace with this situation, it becomes imperative to adopt faster methods of propagation for large plantations. Tissue culture techniques could be helpful in addressing such problems, including development of trait-specific novel varieties and their large-scale commercial multiplication.

Tissue culture is a broad term that covers *in vitro* culture of plant cells, tissues and organ. But in a strict sense, it refers to the practice of *in vitro* culture of plant cells in the form of an unorganized mass termed as callus culture. Similarly, cell culture refers to *in vitro* culture of single or relatively small group of plant cells, such as suspension cultures. On the other hand, when the organized plant structures such as root tips, shoot tips, embryos, etc. are used for *in vitro* culture for regeneration of organized structures, it is called organ culture. Therefore, the term tissue culture in a broad sense refers to aseptic culture of plant cells, tissues and organs under *in vitro* conditions. The phenomenon of callusing and dedifferentiation of cells in plant systems are the cornerstone in tissue culture.

> *Callus is an unorganised mass of cells in various stages of lignifications. Callus initiates from the young cells at the cut-ends in the region of the vascular cambium, although cells of the cortex and the pith may also contribute to its formation (Hartmann and Kester, 1989).*

> *The phenomenon of mature cells to return to a meristamatic condition and develop a new growing point is referred to as dedifferentiation. Normally, dedifferentiation in mature plant cells leads to callus formation, which finally marks the beginning of new growth in mature plant cells and organs.*

Accordingly, in genetic improvement of strawberry, *in vitro* techniques could be very useful for assisting in the introgression of new traits into selected progenies in a comparatively reduced time frame and further regeneration at mass scale (Taji et al., 2002). The ability to regenerate strawberry plants is critical in successful application of *in vitro* methods (Cao and Hammerschlag, 2000). In classical breeding approaches, integration of an efficient *in vitro* regeneration system could accelerate the breeding process for evolving varieties/hybrids with desired traits. In addition, adoption of tissue culture techniques can improve nursery enterprises through large-scale micropropagation of elite strawberry varieties, leading to enhanced profit to growers.

Tissue culture techniques are also extremely useful in storage and long-term conservation of germplasm. There are more than 12,000 accessions of different species and cultivated lines of strawberry maintained at 37 locations in 27 countries across the globe. Half these conserved accessions represent advanced breeding lines of the cultivated hybrid strawberry (*Fragaria × ananassa*). Six countries, viz., the United States, Canada, Russia, Chile, Germany and Spain, are maintaining the major gene bank collections (at least 500 accessions). In addition, private companies have more than 15,000 accessions with proprietary status for their internal use only. Primary collections at national gene banks are being maintained in the form of living plants protected in containers in greenhouses, screen houses and tube structures, or as field gene banks. On the other hand, the secondary backup collections are maintained *in vitro*, while long-term backup collections of meristems are placed in cryogenic storage conditions. In addition, genetically diverse species are maintained as seed lots stored at −18°C or backed-up in cryogenic conditions (Hummer, 2009). This chapter discussion is made on developments in tissue culture of strawberry and their practical applications to various aspects of genetic improvement and maintenance breeding.

> ***Totipotency*** *is the ability of a plant cell to perform all the functions of development, which are characteristic of the zygote, [that is], its ability to develop into a complete plant. The concept of totipotency is implicit in the Cell Theory propounded by Schleiden and Schwann in 1938, and its experimental demonstration further attempted by Vochting in 1878 (Singh, 2011).*

7.1 TISSUE CULTURE TECHNIQUES

7.1.1 MICROPROPAGATION

The plants systems are bestowed with a unique natural phenomenon called *totipotency*. The presence of totipotency (Vasil and Hilderbrandt, 1965) in the apical meristem and the adjacent tissues in the shoot tip is the cornerstone for its potential application in micropropagation of plants in strawberry and several other crops at commercial scale. Micropropagation differs from all other conventional methods of multiplication as the maintenance of aseptic conditions is a prerequisite to achieve enhanced success. The technique of micropropagation is discussed in detail in Chapter 11. Micropropagation of strawberry plants was introduced in the 1970s (Boxus, 1974), and using this technique, some of the most important European nurseries started regenerating millions of strawberry plants per year. Micropropagation provided a more practical solution to the problems associated with the strongly pathogenic fungi in soil that are associated with strawberry plantations. In addition, tissue-cultured plants have been observed to produce comparatively more runners per mother plant in the field in a short span of time (Mohan et al., 2005). Micropropagation has also been widely used for commercial propagation of strawberry and in maintenance breeding programmes to regenerate a large number of plants in a short span of time (Zimmerman, 1981).

Micropropagated strawberry plants can be stored under refrigerated conditions (Mullin and Schlegel, 1976), so this is also a useful technique for storage of germplasm. For cryopreservation

Tissue Culture

of strawberry germplasm (Hofer, 2011) at The National German Strawberry Genebank, all the distinct cryopreservation protocols tested, such as Plant Vitrification Solution 2 (PVS2), Encapsulation Dehydration and Controlled Rate Cooling, have been observed to be equally effective for 23 different *Fragaria* accessions. But wild *Fragaria* species have been observed to have significantly better regrowth with the PVS2 vitrification protocol that includes 14 days alternate-temperature cold acclimation (16 h at −1°C and 8 h at 22°C). Similarly, shoot tips of strawberry (*F. vesca*×*F. ananassa*) cv. "Heilongjiang No.1" were successfully preserved *in vitro* using a slow-growth culture method which survived one year and exhibited an increase of 1.56 cm in shoot length (Hao et al., 2005). Complete new plants can be derived from tissues either from preexisting buds through (1) shoot proliferation after shoot morphogenesis via adventitious shoot regeneration, or (2) formation of somatic embryos.

The rooting and acclimatization, and fresh/dry weight of rooted plantlets of *in vitro* proliferated microshoots of "Elvira" and "Selva" cultivars have been recorded to be *at par* in four different media, viz., (1) MRA: Murashige and Skoog (MS) agar medium with 30 g/l sucrose supplemented with plant growth regulators (PGRs), (2) MRXA: MS liquid medium supplemented with 30 g/l sucrose and PGRs, (3) MSYA: MS agar medium supplemented with 15 g/l sucrose without PGRs, and (4) ENSHI: Enshi liquid medium supplemented with only inorganic salts. Observations showed that there was no requirement for growth regulators in MSYA and ENSHI media for root development in both the cultivars. ENSHI medium contained no organic substances but it enhanced photo-autotrophic development, which resulted in the development of more functional roots leading to better leaf and root formation at two months after the acclimatization period, and later on, these plantlets produced significantly more daughter plants (Hofer and Reed, 2011).

7.1.2 Protoplast Culture

The plant protoplast provides a unique single cell system to understand several aspects of biological functions and also provide opportunities for cell manipulations. For example, protoplasts could be fused together to create cell hybrids. Under normal circumstances, isolated protoplasts do not fuse together since they carry a negative charged surface causing them to repeal one another. The two most successful techniques for protoplast fusion are:

> *Protoplast is a cell devoid of cell wall and consists of cytoplasm and nucleus. The protoplasm is the physical basis of life and organized mass of the protoplasm is known as protoplast. Although, a plant cell is considered to be separable into two distinct parts, viz., protoplast and the cell wall but this separation, by no means, is complete, as there are very fine fibrils or threads of protoplasm which pass through the minute perforations of the cell wall and connect the protoplasm of the adjoining cells. These fibrils are referred to as "plasmodesmata", which establish the organic continuity of protoplasm in a living body.*

1. Polyethylene glycol (PEG)-mediated fusion: In this technique, the protoplast fusion is achieved through the addition of PEG (MW, 1500–6000) in the presence of a high concentration of calcium ions (50 mmol/l) at 35–37°C, and pH between 8.0 and 10.0. The technique involves treatment of protoplast mixture with 15–45% PEG for 45 min. As the protoplast mixture is gradually washed with high pH Ca^{2+} ions, the fusion takes place with frequency ranging of 1–20%.
2. Electrofusion: In this technique, the protoplast fusion is achieved by the application of short pulses of direct electrical current. It consists of low voltage, non-uniform pulses of alternating electrical current which brings the protoplast in close contact followed a high voltage (500–1000 V/cm) DC current for a few microseconds, which creates transient

disturbances in the organization of plasmalemma and leads to fusion of adjacent proto-plasts. Electrofusion is carried out manually in an electroporator. In a more advanced version, *microelectrofusion*, one-by-one selective protoplast pairs are transferred into a microfusion chamber with the help of controlled micromanipulator to facilitate fusion of only desired pairs of protoplasts. Each microfusion chamber contains one pair of protoplast and fusion is achieved by subjecting it to one or several negative DC pulses of 800–1000 V/cm for 50 microseconds after mutual *dielectrophoresis* (1 MHz; 65–80 V/cm).

In strawberry breeding, genetic variation can artificially be created by alternative methods such as gene transfer (directed variation by introduction of foreign gene(s)), somatic hybridization (combining genomes by fusion of somatic cells) and somaclonal variation (utilizing existing or induced variation in cells) to supplement the conventional breeding without inbreeding depression (Wallin et al., 1993). Several reports reiterate the usefulness of protoplast technology in strawberry improvement (Jungnickel, 1988; Galletta and Maas, 1990; Jain and Pehu, 1992). Therefore, there is need for exploitation of protoplast fusion technology for gene transfer, somatic hybridization and for somaclonal variation (Larkin and Scowcroft, 1981) in strawberry breeding because somatic hybridization through protoplasm fusion offers the possibility of exchange of genetic material between the diploid *F. vesca* and cultivated octoploid strawberry *Fragaria × ananassa*.

Efficient procedures available for protoplast isolation and culture of strawberry are a few (Davey et al., 2005). Protoplasts can either be isolated mechanically by cutting or breaking the cell wall, or by digesting the cell wall with the help of enzymes or by a combination of both mechanical and enzymatic separation (George, 1993). Isolated strawberry protoplasts can be cultured on artificial nutrient medium such as 8P medium (Glimelius et al., 1986) supplemented with either 1.0 mg/l 2,4-D + 0.5 mg/l BA (Nyman and Wallin, 1992), or 2.0 mg/l NAA + 0.5 mg/l Thidiazuron (Wallin et al., 1993). Shoot organogenesis is induced on MS medium supplemented with 2% sucrose and 2.0 mg/l BA + 0.2 mg/l NAA. Fong et al. (1889) observed partial success in protoplast isolation but could not achieve success in culturing the resultant fused tissues, whereas, Wallin (1997) successfully regenerated plants from calli originating from protoplast fusion between *F. ananassa × F. vesca*. Protoplasts of hygromycin resistant *F. ananassa* could also be fused with kanamycin resistant protoplasts of *F. ananassa* accessions. Protoplasts isolated from leaf and petiole (Nyman and Wallin, 1988, 1992) or from callus (Wallin, 1997) have been successfully cultured with yield of viable protoplasts to the tune of 3.3×10^6 protoplasts/g fresh weight (Nyman and Wallin, 1992).

Many attributes influence the ability of protoplasts and protoplast-derived cells to express their totipotency and to develop into complete plantlets, such as the source of tissue, nutrient medium use for culture and environmental factors etc. In quantitative terms, the DNA content of protoplast fusion derived plants may have varied ploidy levels but Infante and Rosati (1993) could isolate protoplasts and regenerate plantlets in wood strawberry (*F. vesca*) with normal growth under greenhouse conditions, which did not show any phenotypic variation when compared with the micropropagated original clone. On the other hand, there have been morphological dissimilarities in plants derived through protoplast fusion with that of parent plants having more than 56 chromosomes (2n), which could be attributed to somaclonal variation (Wallin, 1997).

The growth of strawberry cultivars *in vitro* has been observed to be significantly influenced due to quality of light, for example, exposure to cellophane film cover

> *Double haploid (DH) is the common term for a plant obtained through doubling of haploid (x) chromosome number. It is generally achieved through colchicine treatment. Double haploid plants are completely homozygous (2n). Generally, the anthers/pollen or ovary/ovules are cultured to produce haploids and subsequently homozygous double haploid plants are obtained through colchicine treatment.*

Tissue Culture

mediated variable colours of lights (blue, green, yellow, red and clear) during direct shoot regeneration from leaf discs or proliferation from shoot tips and, during rooting (Mohamed et al., 2017), has variable effect on growth and development of tissue-cultured plantlets. Maximum frequency (10 shoots/explant) of adventitious bud regeneration was recorded due to exposure to red and green but, maximum shoot proliferation was recorded due to green (15.3 shoots/explant) followed by red (14 shoots/explant) and blue (13 shoots/explant) lights. There was significant increase in fresh weight of plantlets, percent of emergence of trifoliate leaves, length of roots and number of roots due to exposure to white light followed by yellow or blue light during the rooting stage. Maximum total chlorophyll content was observed due to blue and yellow light whereas, the content of N, P and K in plant tissues was increased due to red, blue and green light, respectively.

7.1.3 ANTHER/POLLEN CULTURE

Anther/pollen culture involves aseptic culture of immature anthers/pollen to regenerate fertile haploid plants from microspores. The production of haploid plants through anther/pollen culture is widely used as an alternative to repeated cycles of inbreeding/backcrossing in genetic improvement programmes. Initially, the pollen may be isolated from the anthers by squeezing or float culturing of anthers in the medium followed by squeezing with a glass rod and filtering through a nylon mesh, and centrifugation at 500–800 rpm for 5 min. The pollen grains thus obtained in a pellet form are washed and suspended in a suitable nutrient medium at 10^3-10^4 pollen/ml for *suspension culture* or put on a shallow liquid medium in petridishes as *float cultures*.

In suspension culture, the use of liquid nutrient medium facilitates easy and extensive scaling-up of tissue culture by employing culture flasks (250 ml flasks with 50 ml nutrient medium) or bioreactors (1–1000 liter). Large culture systems, such as bioreactors, generally have provisions for aeration, stirring mechanisms for nutrient and cell mixing, control of pathogens and other contaminants replacement of used nutrient and/or cultured cells – all controlled by sophisticated systems. For this, the flower buds of appropriate developmental stage are collected and surface sterilized and carefully excised anthers/pollen are placed on a suitable culture medium. The pollen grains isolated from the anthers, when cultured *in vitro*, produce haploid embryos or callus tissues and the process is termed as *androgenesis*. Similarly, the culture of unfertilized ovaries gives rise to haploid plants from egg cells or the other haploid cells of the embryo sac termed as ovary culture through a process known as *gynogenesis*. Use of growth regulators is very important in gynogenesis and at higher concentrations, the growth regulators may induce callusing of somatic tissues. There are two distinct stages in ovary culture, *viz.*, *induction* in which the ovaries are floated on to a liquid medium supplemented with low concentrations of auxins and incubated in the dark and *regeneration* when the cultures are transferred onto an agar medium supplemented with a little higher concentration of auxins and incubated in light.

Young embryos are placed on to a suitable nutrient medium to obtain plantlets; the cultured embryos generally do not undergo complete development and germinate prematurely to give rise to new plantlets through precocious germination. Anther/pollen or ovary/ovule culture generally give rise to haploid plantlets with only one set of chromosomes (x). But through the technique of chromosome doubling, it could pave the way for immediate establishment of double haploids, having homozygosity for all useful loci. Further, isolated microspores, being unicellular, are very much amenable for protoplast isolation and applications aiming at transformation (Dhlamini et al., 2005; Germana, 2006) leading to production of targeted transgenic homozygous plants in a comparatively short period of time.

Haploid recovery in strawberry through aseptic anther culture was not successful in the beginning (Rosati et al., 1975; Niemirowicz-Szczytt et al., 1983) but later, haploid plants from anther cultures of "Chandler", "Honeoye" and "Redchief" varieties were developed (Owen and Miller, 1996). The maximum shoot regeneration (8%) was obtained on a semi-solid MS medium containing 2.0

mg/l IAA + 1.0 mg/l BA + 0.2 M glucose. Chromosome counts of root tip cells from *ex vitro* grown plants confirmed the haploid status of plants obtained from all the three varieties.

7.1.4 MERISTEM CULTURE AND MICRO-CUTTINGS

Meristem culture is used for obtaining plantlets free from pathogenic microorganisms such as viruses, viroides, mycoplasma and even fungi and bacteria. Care needs to be exercised to remove as small portion of apical meristem (explant) as possible so as to have little surrounding tissues and minimize the possibility of virus being carried over to excised plant tissue. The practice calls for sufficient expertise to dissect out the shoot apical meristem with only one or at most two leaf primordia (100–500 μm) and to be careful further in preventing desiccation and contamination with pathogenic microorganisms. In general, there is a positive relationship between length of explants and their survival and shoot development, but the chances of infections being carried over with the explants also increases with their increasing length. Therefore, in practice, a balance has to be struck in deciding the length of meristem tissue, success in shoot and root proliferation and degree of freedom from pathogenic microorganisms in overall success in meristem culture. In strawberry, tissue culture of meristematic tissues has successfully eliminated complexes of viruses such as crinkle virus, latent A virus, latent C virus and mottle virus (Singh, 2011).

On the basis of stages of development, plant tissues are classified into two groups, viz., meristematic and permanent tissues. Meristem, as it is commonly referred to, is a group of immature cells that have the characteristic feature of continuous cell division. The plants having vascular systems, are characterized by an open system of growth, which involves formation of new tissues/organs throughout its life. It is possible due to presence of certain organized regions where the cells are continuously in immature condition and can form new cells throughout. These meristems are dome shaped and generally measure 0.1 mm in diameter and 0.25 mm length.

There are five stages in meristem/tissue culture, viz., identification and preparation of explants, culture initiation, multiplication, rooting of shoots and acclimatization of plantlets and transfer to soil (Debergh and Maene, 1981). In the first stage, the mother plant is identified and prepared in such a way that it produces true-to-type, healthy plant parts suitable for collection of explants and culture initiation. The mother plants should be of known performance and mature but the parts selected as explants should be healthy and juvenile/young enough to undergo dedifferentiation easily. In the second stage, the explants are sterilized and established/inoculated *in vitro*, supplemented with suitable nutrient medium. The main feature is maintenance of aseptic conditions and elimination/control of contamination. Explants are generally cleaned by removal of their outer covering, followed by wiping with 70% ethanol. If required, a suitable antibiotic and/or fungicide (e.g., trimethoprim, carbendazim etc.) may be added in the nutrient medium to better control contaminants.

In third stage, a few or many shoots or somatic embryos are produced from each explant, which is achieved by enhanced proliferation of axillary shoot buds, induction of adventitious buds and somatic embryogenesis. At two to three weeks after culture initiation, the explants are shifted to a nutrient medium enriched with auxins and cytokinins to facilitate shoot multiplication. Improper handling of culture at this stage or anomalies as a result of excessive concentration or wrong selection of growth regulators, especially cytokinins or excessive subculturing, may lead to reduced or missing root regeneration, excessive or abnormal flowering, deformities of fruits etc. in plantlets of strawberries and can be overcome by suitably modifying the culture conditions/procedures.

In the fourth stage, cultures are shifted to a modified *in vitro* growth medium to facilitate rhizogenesis, that is, regeneration of roots. Normally, the nutrient medium is supplemented with a lower

Tissue Culture 127

concentration of salts (half-strength MS medium) with reduced sugar (usually 1 g/l). As strawberry is easy to root, normally addition of growth regulators in nutrient medium for facilitating regeneration of roots is not required, but 0.1 to 1.0 mg/l NAA or IBA may result in a robust root system. Shoots are generally transferred to a suitable nutrient medium for root initiation, such as agar medium, but nowadays vermiculite or potting mixtures are generally used for planting of plantlets for root initiation and the process is referred to as *in vivo* or *ex vitro* rooting.

In the fifth stage, satisfactorily rooted plantlets are taken out of the nutrient medium but still maintained under protected conditions *in vitro* with high humidity, low solar irradiance, low CO_2 levels and slightly elevated sugar content. The rooted plantlets are slowly shifted to the outside environment with due care to make them autotropic and get them acclimatized to the harsh outside environment. Provisions such as a shade net house, sprinkler/foggers, high humidity and low light intensity facilitate easy hardening and improve the establishment of plants.

In microcuttings of strawberry, shoots are multiplied and made to initiate root in the same culture medium (Debnath, 2006). The use of microcuttings for regeneration of both roots and shoots in an *in vitro* growth medium supplemented with cytokinin has been observed as a better alternative for micropropagation of strawberry than multiple shoot proliferation (using a growth medium supplemented with cytokinin) with subsequent regeneration of roots in the proliferated shootlets. Proliferation, elongation and root initiation in *in vitro*-derived strawberry shoots can be obtained in a growth medium supplemented with zeatin alone at low concentrations (1.0–2.0 μM), which facilitates production of two to three shoots per explant, with ~90% rooting of strawberry var. "Bounty" in a single growth medium. Besides, there is no requirement of auxins in the culture medium in this protocol, which further minimizes the cost of regeneration. In addition, the possibility for occurrence of somaclonal variation among the proliferated plants is also minimized by adoption of this protocol. The main advantage of this protocol is that all the shoot tips of the *in vitro* grown plantlets can be used for induction of shoots and regeneration of roots, whereas basal nodal segments with developed roots can subsequently be transferred to a suitable potting medium and further acclimatized in the greenhouse. In the microcutting protocol for micropropagation of strawberry, Stage II is generally eliminated, which facilitates an increased rate of both multiplication and rooting, which ultimately translates into faster rate of micropropagation.

7.1.5 SOMATIC EMBRYOGENESIS

The technique of somatic embryogenesis involves development of bipolar embryos *in vitro* from embryogenically competent somatic cells. Thus, unlike organogenesis (microshoots and roots develop in different media), somatic embryogenesis is an apparently one-step protocol wherein development of bipolar embryos (having both a shoot and a root pole), similar to the zygotic embryos, takes place. The initiation and development of embryos from somatic tissues was first observed in cultures of carrot tissues (Steward et al., 1958; Reinert, 1958) and later observed in strawberry cotyledons innoculated on MS medium supplemented with 22.6 μM 2,4-D+2.2 μM BA+500 mg/l casein hydrolysate where few of the embryogenic tissues developed into somatic embryos. Morphologically normal plants obtained from somatic embryos were transferred to MS medium containing 2.89 μM GA_3 or 2.22 μM BA+0.54 μM NAA but the maintenance of the embryogenic cultures was, however, unsuccessful (Wang et al., 1984). Somatic embryogenesis in 8% embryogenic calli in strawberry cv. "Clea" has been observed when the somatic embryos were cultured in MS medium supplemented with 4.88 μM BA+4.90 μM IBA (Donnoli et al., 2001).

7.2 APPLICATIONS OF TISSUE CULTURE TECHNIQUES

7.2.1 SOMACLONAL VARIATION

The occurrence of variation in plants regenerated from *in vitro* cultures is referred to as "somaclonal variation" and has frequently been observed in micropropagated strawberry (Larkin and Scowcroft,

1981; Graham, 2005) with respect to traits such as plant morphology and yield. Clonal fidelity is an important consideration in commercial micropropagation and as such, true-to-type propagules and genetic stability are prerequisites for the application of tissue culture techniques for propagation of strawberry *in vitro*. There are limitations related to heritable genetic changes in strawberry resulting due to micropropagation. Several discrete variations in morphological characters of micropropagated strawberry plants have been observed such as, leaf variegation consisting of a narrow white streak in the leaf blade (Swartz et al., 1981), leaf chlorosis (Swartz et al., 1981), and changes in patterns of growth including dwarfness, compact trusses, lack of runner production and female sterility (Swartz et al., 1981). Variations among micropropagated subclones of strawberry var. "Olympus" have been observed and attributed most likely as transient responses to the micropropagation environment rather than due to genetic constitution (Moore et al., 1991). Generally, micropropagated plants have comparatively enhanced vigour, profuse runner production and more fruit yield than runner-propagated plants (Swartz et al., 1981), but increase in yield varies among different cultivars (Cameron et al., 1985).

> *Chimera are plants composed of two or more genetically different types of tissues that grow adjacent to each other but separately in the same plant. Chimeras generally originate due to spontaneous or induced mutations in dividing cells of meristem. These tissues could be usually arranged in layers or sectors of the stem and, in principle, originate due to mutation in one of the shoot apices. The extent of stability of chimeras differs and is regulated by their structure and the genotype of plant. chimeras are of different types such as periclinal, mericlinal or sectoral.*

Maintenance of genetic stability throughout the process of micropropagation is regulated by various factors such as genetic constitution of variety, presence/absence of chimeral tissues, type and origin of explants, type of culture media, types/concentrations of growth regulators, culture conditions (temperature, light, etc.) and duration of tissue culture (Graham, 2005). Neither somatic embryogenesis nor shoot organogenesis is widely used in micropropagation of strawberries at commercial scale because there are strong probabilities of occurrence of somaclonal variations due to use of adventitious tissues as explants for regeneration of plantlets. Somaclonal variations can be differentiated from normal plantlets by their morphological, biochemical, physiological and genetic characters. Molecular markers are powerful tools in identification of genetic variation in somaclonal population with greater precision and less effort than phenotypic and karyotypic analysis (for more details, see Chapter 8).

Somaclonal variations in micropropagated plants of strawberry for commercial cultivation are not desirable but these could be valuable in genetic improvement through breeding as desired variation in tissue-cultured plants can suitably be utilized in the breeding of superior varieties. In addition, under intensive observation and with suitable modification in the nutrient medium, it may be possible to create somaclonal variations in tissue cultures and further screen out regenerants resistant to biotic/abiotic stress (Jain, 2001). Certain changes such as variations in number and structure of chromosomes, amplification of genes, activation of transposable elements and alteration in DNA methylation have been associated with causes of somaclonal variation (Phillips et al., 1994) in tissue culture of strawberry, and these have also been produced through *in vitro* culture and selection. Usually, the genetic variation in tissue-cultured plants has been ascribed to regeneration from callus culture (Popescu et al., 1997) but phenotypic variations might be observed among plants regenerated from meristem culture (Sansavini et al., 1989). Several useful variations with respect to plant and fruit characteristics have been observed in tissue-cultured strawberry plants regenerated from leaf- and petiole-derived callus (Popescu et al., 1997). Thus it is clear that the occurrence of somaclonal variation is strongly influenced both by the genetic constitution of variety and the

Tissue Culture

type of explant, for example, a variant with modified colour (white) of fruit pulp has been induced from callus derived from petioles of strawberry cv. "Gorella" (Popescu et al., 1997). In addition, evidence of genetic/somaclonal variations such as reduced susceptibility to wilt-causing soilborne fungi (Battistini and Rosati, 1991; Toyoda et al., 1991), earliness, separation of calyx, susceptibility to mildews and ploidy level have also been observed in strawberry (Simon et al., 1987).

Sansavini et al. (1989) reported three types of somaclonal variations most commonly associated with micropropagated strawberry, viz., (1) chlorophyll-mutant white stripe, (2) chlorosis and (3) dwarfism. There are genetic differences in varieties with respect to their susceptibility to somaclonal variations that are passed on to runners but the degree of variation declines significantly over the growing season. The somaclonal variations are transmitted through sexual reproduction and the variations are more distinct when the mother plants are affected due to somaclonal variations. After selfing, S_1 offspring segregated, showing 26.7% incidence of white stripe, 60% dwarf and semi-dwarf and 66.7% of chlorosis. In crossing of affected and normal plants, white stripe affected only 15.4% and dwarfism 56.3% of F_1 seedlings (Sansavini et al., 1989). Variations such as leaf chlorosis, white streak and dwarfism have been observed in somaclonal plant population regenerated from anther culture of strawberry cv. "Pajoro" (Faedi et al., 1993). Variations among the regenerants of strawberry cultivars with respect to callusing, the rates of growth in cell suspension and, in iso-enzyme patterns of acid phosphatase, peroxidase and glutamate dehydrogenase (Damiano et al., 1995) or DNA content from callus cultures have been observed, which disappeared after transfer to greenhouse in most of the (four out of five) cultivars (Brandizzi et al., 2001). A significant change in DNA methylation status was observed in cryopreserved shoot tips of strawberry (Hao et al., 2002). In addition, for chromosome mapping of cultivated strawberry, standard laboratory protocols have been standardized for preparing linkage maps using single sequence repeat (SSR) markers. These include maceration for 25 min. to obtain sufficient number of somatic cells with 56 chromosomes, 45% acetic acid for reducing the extent of damage to chromosomes and obtaining clear chromosome images, 72 h incubation for detecting chromosomes having fluorescent signals, increased PCR cycles and an aluminium tape for maintaining the PCR temperatures (Tantivit et al., 2016).

7.2.2 VIRUS ELIMINATION

Strawberry is infected by more than 60 viruses and mycoplasma (Singh, 2011), which are generally transmitted to subsequent generations through asexual propagation (runners) and, therefore, it becomes necessary to replace the mother plant every year. It is important to note that the shoot apical meristem and adjacent tissues are generally free from viruses and the viral titre increases with the distance from shoot apical meristem. The reasons for this are not known properly but the absence of vascular bundles in the shoot apical meristem and comparatively slow rate of movement of viruses through the plasmodesmata is believed to be the contributing factors. Meristem culture is, therefore, often considered to be useful in recovery of virus-free plants from otherwise virus-infected plants. Some of the approaches for obtaining virus-free plants are as follows:

1. Thermotherapy
2. Meristem culture
3. Thermotherapy combined with meristem culture
4. Cryotherapy combined with meristem culture
5. Chemotherapy
6. Callus cultures

In thermotherapy, plants are exposed to 35–40°C (hot air or water) for a few minutes to several weeks under high humidity (85–95%). Thermotherapy is generally combined with meristem culture to obtain virus-free plants. The technique of hot air therapy to eliminate viruses originally was developed by Posnette (1953) refined further by Belkengren and Miller (1962), who demonstrated the practice of excising heat-treated meristems for placement onto culture media. In this technique, the potted plants are placed in a growth chamber at ambient temperature, and then the temperature is raised by a few degrees per day till it attains 38°C and grown further for six weeks (Lines et al., 2006). Shoot tips are removed after treatment and the meristems with two- or three-leaf primordia are removed and cultured on a nutrient medium. Mullin et al. (1974) grew tissue carrying strawberry mild yellow edge (SMYE) viruses for six weeks at 36°C in a growth chamber before excising 0.3–0.8 mm meristematic tips with leaf primordia, and was able to obtain 33–75% of the plants free from SMYE as validated through leaf insert graft indexing by indicator strawberry plants. In this way, the heat-treated strawberry propagated using apical meristem could be maintained under detectable graft transmissible disease free conditions for seven years in a greenhouse. Similarly, in cryotherapy, the plants are exposed to low temperature (5°C) for prolonged duration (four months) followed by shoot meristem culture. In chemotherapy, some chemicals such as virazole (ribavirin), cycloheximide, actinomycin D, which interfere with the multiplication of virus, are added into the nutrient medium which helps in eliminating viruses.

7.2.3 *In vitro* Selection for Biotic/Abiotic Stress Tolerance

In vitro selection is a useful tool in identifying plants resistant/tolerant to biotic/abiotic stresses such as phytotoxins produced by pathogens or herbicides, cold temperature, aluminium, manganese and salt toxicity (Chaleff, 1983). Usually, cells are subjected to a suitable selection pressure *in vitro* to recover any variant line that has developed resistance/tolerance to stress followed by regeneration of plants from the selected cell. This approach presumes that tolerance operating at the unorganized cellular level can act, to some degree of effectiveness, in the whole plant, and that the trait can be transferred to other plants if it has a genetic control.

In strawberry, research have been taken up to obtain plants resistant/tolerant to *Alternaria alternata* (Takahashi et al., 1992), *Botrytis cinerea* (Orlando et al., 1997), *Colletotrichum acutatum* (Damiano et al., 1997; Hammerschlag et al., 2006), *Fusarium oxysporum* (Toyoda et al., 1991), *Phytophthora cactorum* (Maas et al., 1993; Sowik et al., 2001), *P. fragariae* (Maas et al., 1993), *P. nicotianae* var. *parasitica* (Amimoto, 1992), *Rhizoctonia fragariae* (Orlando et al., 1997) and to *Verticillum dahliae* (Sowik et al., 2001, 2003). Hammerschlag et al. (2006) used an *in vitro* screening system to evaluate "Chandler", "Delmarvel", "Honeoye", "Latestar", "Pelican" and "Sweet Charlie" cultivars propagated *in vitro*, and shoots regenerated from leaf explants of these cultivars for resistance to anthracnose (*C. acutatum*) isolate Goff (highly virulent). Regenerants with resistance were genotype specific, and the highest levels of anthracnose resistance (2–6% leaf necrosis) were exhibited by regenerants from explants of cultivars "Pelican" and "Sweet Charlie". Aseptically sown strawberry seedlings can be clonally multiplied on medium containing a cytokinin and utilized for evaluation against resistance to anthracnose by spray inoculation with *C. gloeosporioides* (*Glomerella cingulata*) at a concentration of 1×10^4 spores/µl with reliable results obtained within approximately 120 d after seed sowing (Hirashima et al., 2015). Further, *in vitro* screening enables researchers to screen genotypes of strawberries for tolerance/susceptibility to iron content in the soil nutrient solutions in a relatively fast and economical way, which could further be utilized in breeding programmes (Torun et al., 2014).

Insufficient winter hardiness is one of the major constraints limiting strawberry production in cold climates. An *in vitro* screening technology was determined for cold resistant strawberry seedlings (Rugienius and Stanys, 2001). Strawberry plants were regenerated from an isolated embryo axis on growth regulator-free MS medium, and from rescued cotyledons on the medium with 1.0 mg/l BA + 0.5 mg/l NAA. The temperature range *in vitro*, at which genotypes differentiated according to cold resistance, was −8°C to 12°C. Differentiation of strawberry genotypes according to this character conformed to their differentiation *in vivo* with a strong correlation ($r = 0.93$), recorded between cold resistance under *in vitro* and *in vivo* conditions. Further, it was observed that for *in vitro* screening of strawberry seedlings, 120 mmol/l NaCl and 10 mmol/l $NaHCO_3$ could be considered as index

Tissue Culture 131

for tolerance to salt and alkali, respectively (Zhao et al., 2017). The ability of rooting of strawberry seedlings was restrained *in vitro* and further exhibited by decreased rate of rooting, number of roots and length of roots coupled with distinct symptoms on leaves due to injury by excessive salinity and alkalinity under NaCl and NaHCO$_3$, respectively. The cultivar Benihoppe was comparatively more tolerant to excess salinity (NaCl) and alkalinity (NaHCO$_3$) than the Sweet Charlie. The growth of strawberry seedlings was suppressed due to increased pH of the solution *in vitro* and, the stress due to combined toxicity of NaHCO$_3$ and high pH was more injurious than NaCl.

7.2.4 GENETIC TRANSFORMATION

With the recent advances in tissue culture techniques, the biotechnological approaches are being extensively utilized for creation of genetic variations in many important crop plants, including strawberry. The latest and most potent biotechnological approach is the introgression of specifically constructed gene assemblies through genetic transformation techniques collectively referred to as genetic engineering. There has been significant progress in genetic transformation of strawberry with the primary objective of using genes for pest and herbicide resistance, cold tolerance and con-trolled ripening of fruits. Transformation in diploid and octoploid strawberry species using different constructs has been well documented in various genotypes (Table 7.1). There are several reports on *Agrobacterium*-mediated genetic transformation of strawberry (Jelenkovic et al., 1986; James et al., 1990; Nehra et al., 1990a, 1990b). *Agrobacterium*-mediated transformation involves co-cultivation of the *Agrobacterium* strain with the explants on a nutrient medium containing BA or TDZ with an auxin (2,4-D, IAA) in an organogenic regeneration system (Schaart et al., 2002; Zhao et al., 2004; Oosumi et al., 2006) and subsequent selection of regenerants in medium with 25–75 mg/l kanamycin (Schaart et al., 2002; Chalavi et al., 2003) or 4.0 mg/l hygromycin (Oosumi et al., 2006). Cefotaxime, carbenicillin and/or ticarcillin are used to control *Agrobacterium* contamination after inoculation. The transfer of gene could be confirmed by PCR with target gene-specific primers, copy number through Southern blot analysis and target protein expression by ELISA or Western blotting. Cordero de Mesa et al. (2000, 2004) reported a microprojectile-mediated transformation protocol using leaf explants of strawberry cv. Chandler. Although a direct gene transfer method has been reported (Nyman and Wallin, 1992; Wang et al., 2004), but leaf disks or crown sections as explants for *Agrobacterium*-mediated transformation are generally employed (Jelenkovic et al., 1986; James et al., 1990; Nehra et al., 1990a, 1990b; El Mansouri et al., 1996; Haymes and Davis, 1998; Barcelo et al., 1998; Ricardo et al., 2003; Folta et al., 2006; Mezzetti and Constantini, 2006). But the use of *Agrobacterium* plasmids (Abdal- Aziz et al., 2006) often results in the integration of non-T-DNA sequences in ~66% (range 40–90%) of transgenic plants.

> *Agrobacterium* is a gram negative and soilborne bacteria, which causes crown gall (*A. tumefaciens*) and hairy root (*A. rhizogenes*) diseases in a number of dicotyledonous plants. *Agrobacterium tumefaciens* has *pTi* (tumor inducing plasmid) while *A. rhizogenes* has pRi (root-inducing plasmids). In nature, these infect plants through wounds, usually near crown of roots at the soil surface.

7.2.4.1 Mechanism of *Agrobacterium*-Mediated Gene Transfer

Agrobacterium genetically modifies the plant cells infected by it, as a result, the modified plant cells produce a copious amount of opines, which serve as food for the bacterium, whereas human beings utilize *Agrobacterium* to artificially bring about genetic modification of plants to serve the same purpose. On entry and establishment in the plants, the plant parts infected by *A. tumefaciens* pro-duce tumour-like growth, while *A. rhizogenes*-infected parts produce hairy root-like growth. Thus, the *Agrobacterium*-affected plant parts undergo cancerous or oncogenic alterations and are capable

TABLE 7.1

Reports on *Agrobacterium*-Mediated Genetic Transformations in Strawberry

Cultivar/genotype	Explant	*Agrobacterium* strain	Reporter gene(s)[a]	Target trait[b]	References
Rapella	Leaf, petiole	LBA 4404	*nptII*	–	James et al. (1990)
Redcoat	Leaf	M P90	*nptII, gus*	–	Nehra et al. (1990a, 1990b)
Chandler, Induka, Elista	Leaf	LBA 4404	*nptII, gus*	–	Barceló et al. 1998, Harpster et al. (1998), Gruchala et al. (2004)
Totem	Leaf	EHA 105	*nptII, gus*	–	Mathews et al. (1998)
Diploid Alpine accession FRA 197	Leaf, petiole	LBA 4404	*nptII, gus*	–	Haymes and Davis (1998)
Chandler	Leaf	LBA 4404, Biolistic	*nptII, gus*	–	Cordero de Mesa et al. (2000)
Gariguette, Polka, line no. 88312	Leaf	AGL0	*nptII, gus*	–	Schaart et al. (2002)
Chandler	Leaf	LBA 4404	*nptII, gus*	*njjs25*	Jimenez-Bermudez et al. (2002)
Pajaro	Leaf	LBA 4404	*nptII, gus*	–	Ricardo et al. (2003)
Joliette	Stipules	LBA 4404	*nptII*	*pcht28*	Chalavi et al. (2003)
Hecker, La Sans Rivale, diploid accessions (FRA 197, FRA198)	Leaf, petiole	EHA 105	*nptII, gus*	–	Zhao et al. (2004)
Toyonaka	Anther calli	Biolistic	*nptII*	LEA3	Wang et al. (2004)
Chambly	Shoot	GV 3101	*nptII*	*wcor410a*	Houde et al. (2004)
Tiogar	Leaf	LBA 4404	*nos, nptII*	APF	Khammuang et al. (2005)
Firework	Leaf	CBE 21	*nptII*	*thau II*	Schestibratov and Dolgov (2005)
Anther	Leaf	LBA 4404	*nptII*	*FagpS*	Park et al. (2006)
Diploid accession	Leaf	GV 3101, LBA 4404	Hygromycin, *gfp*	–	Oosumi et al. (2006)
Pájaro	Leaf	LBA 4404	*nptII*	*ch5B, gln2, ap24*	Vellicce et al. (2006)
Chandler	Leaf	LBA4404	GUS	*FaPG2 FaPG1*	Quesada et al. (2009)
F. vesca Accesion 551572	Leaf	–	*nptII*	*Hos-pathogen (F. oxysporum f. sp. Fragariae)* interaction	Pantazis et al. (2013)
Festival	Leaf	–	*nptII*	Fruit trait improvement	Ayub and Reis (2016)

[a] *gfp*: Green fluorescent protein, *gus*: glucuronidase, *nos*: nopaline synthase, *nptII*: neomycin phosphotransferase.

[b] *ap24*: Thaumatin-like protein gene from *Nicotiana tabacum*, *APF*: anti-freeze protein gene from Antarctic fish, *ch5B*: Chitinase gene from *Phaseolus vulgaris*, *FagpS*: antisense cDNA of ADP-glucose pyrophosphorylase (AGPase) small subunit, *gln2*: glucanase gene from *N. tabacum*, *LEA3*: late embryogenesis abundant gene from barley, *njjs25*: strawberry pectate lyase, *pcht28*: *Lycopersicon chilense* chitinase, *thau II*: thaumatin II, *wcor410a*: wheat dehydrin.

Tissue Culture

of growing in a growth regulator-free culture medium, unlike normal plant cells, which require exogenous application of auxin and/or cytokinins for normal growth in a culture medium. The plant cells affected by crowngall or hairy roots synthesize unique nitrogenous compounds, "opines", which are neither synthesized nor utilized by normal plant cells. The opines thus produced are utilized by *Agrobacterium* cells as a carbon and nitrogen source, the bacteria being usually present in the intercellular spaces of crowngalls. The crowngall cells continue to produce opines even when these are cultured *in vitro*. Furthermore, even the plants regenerated from these crowngalls produce opines. Different types of opines are produced depending upon the *Agrobacterium* plasmid. Usually, *A. tumefaciens* produces octopine or nopaline, whereas *A. rhizogenes* produces agropine or mannopine. An interesting fact is that the tumour-inducing plasmids (*pTi*) and root-inducing plasmids (*pRi*) carry genes responsible for biosynthesis of auxins, such as IAA, and cytokinins, which are largely responsible for regulation of cell division and cell growth. This is the prime reason for indefinite growth of crowngall cells on a growth regulator-free culture medium as there is in-built mechanism of biosysnthesis of growth regulators, and external application of growth regulators is not required. When *Agrobacterium* is subjected to exposure to a higher temperature (at least 28°C), they lose their virulence properties and become avirulent (non-pathogenic). The virulence-regulating mechanism of *Agrobacterium* lies in the plasmids and not in the chromosome complement.

> *Vectors are circular DNA molecules that have independent existence and replication such as plasmids and viruses. In genetic transformation of plants, mostly Agrobacterium-derived plasmids are used. The pTi and pTr plasmids have unique properties that they contain some genes within their T-DNA that have regulatory sequences recognized by plant cells, while the remaining genes have prokaryotic regulatory sequences. Therefore, the regulatory sequences recognized by plant cells are expressed only in plant cells while the sequences regulating prokaryotic functions are expressed only in bacterium. As a result, these plasmids transfer their T-DNA into the host plant genome. And due to this reason, the* Agrobacterum *is used as a natural genetic engineer.*

A rapid regeneration and transformation technique for germplasm accession, "Laboratory Festival No. 9" derived through self-pollination of a productive cultivar "Strawberry Festival" was standardized by Folta et al. (2006). Besides, an effective method for the production of transgenic plants from which selectable marker genes have been removed has also been reported (Schaart et al., 2004), which combines a chemically inducible recombinase activity and a bifunctional selection system that allows the production of marker-free transgenic strawberry plants. Similarly, transformation of "Toyonaka" calli obtained from anther cultures has been achieved by particle bombardment with plasmid pBY520 containing late embryogenesis abundant protein gene, *LEA3*, from barley (Wang et al., 2004). Of the bombarded calli, nearly 15% regenerants were obtained and selected on a medium containing 10.0 mg/l phosphinothricin (PPT), which confirmed the presence of the target gene in the strawberry genome, as had also been confirmed through dot blot analysis.

The transformation efficiency in *Agrobacterium*-mediated transformation system varies among different cultivars and it was reported to be 0.95% in cv. "Rapella" (James et al., 1990), 6.5% in cv. "Red Coat" (Nehra et al., 1990b) and 11% in cv. Firework (Schestibratov and Dolgov, 2005). Contrary to this, 20.7% transformation frequency was recorded in cv. Chandler by combining *A. tumefaciens* infection and biolistic bombardment techniques (Cordero de Mesa et al., 2000). A number of risks are associated with strawberry transformation that include escapes, formation of chimeric shoots (Mathews et al., 1998) and somaclonal variation. Oosumi et al. (2006) advocated the use of hygromycin instead of kanamycin as it suppresses the growth of untransformed cells and facilitates efficient transgenic shoot formation with minimum escapes and chimeras.

Genetic transformation has improved many traits in strawberry as is evident from several recent reports that cultivated strawberries may be engineered with specific pathogenesis-related (PR) genes that can minimize the severity of *Botrytis cinerae*-induced grey mould (Schestibratov and Dolgov, 2005; Vellicce et al., 2006). Genes conferring strawberry mild yellow edge virus-coat protein (SMYEV-CP; Finstad and Martin, 1995) and cowpea trypsin inhibitor (CPTI; Graham et al., 1995) have successfully been transferred into strawberry. Strawberry has also been transformed with a rice chitinase gene (*RCC2*) imparted by the fungus *Sphaerotheca humuli* and the resultant plants exhibited improved resistance to powdery mildew (Asao et al., 1997). The cowpea protease trypsin inhibitor gene (*CpTi*) was introduced into a strawberry cultivar and the CPTI-overexpressing plants exhibited significantly higher root mass than control (Graham et al., 2002). Enhanced resistance to *Verticillium daliae*-induced verticillum wilt was obtained in plants over-expressing *pct28* chitinase from *Lycopersicon chilense* when compared with non-transformed plants (Chalavi et al., 2003).

The CP4.EPSP synthase gene regulating resistance to herbicide glyphosate has been introduced into cv. Camarosa and on the basis of their tolerance to glyphosate, a total of 30 lines were selected (Morgan et al., 2002). Similarly, increased salt tolerance in transgenic "Toyonaka" strawberry was reported (Wang et al., 2004) with 18.6% and 42.8% wilting at 50 mmol/l and 100 mmol/l NaCl salt stress conditions in LEA3-transgenic shoots as compared to 62.2% and 96% wilting in normal plants, respectively.

Improvement in the selection procedure and the transfer of the wheat *Wcor410a* acidic dehydrin gene in strawberry, which prevents membrane injury and greatly improves freezing tolerance in leaves has been standardized (Houde et al., 2004), and genes regulating anti-freeze proteins have been transferred into strawberriy (Khammuang et al., 2005). Similarly, a cold-induced transcription factor (Orthologs of *CBF1*) important, in the cold acclimation response in *Arabidopsis thaliana*, has been cloned from strawberry (Owens et al., 2002) and the putative orthologs [*F.*×*ananassa CBF1* (*FaCBF1*)] being 48% amino acid identical to *CBF1* with mRNA levels upregulated in leaves following exposure to 4°C for a varying duration ranging from 15 min. to 24 h. However, mRNA of *FaCBF1* was not detected in pistils of strawberry following exposure to 4°C. *Agrobacterium*-mediated transformation of a *CaMV35S-CBF1* construct on cv. "Honeoye" crown discs was conducted and two resultant transgenic lines expressed the transgene *CBF1* at low levels, a cold-induced transcription factor with significantly greater freezing tolerance at −6.4°C than the wild-type "Honeoye" leaf discs (Owens et al., 2002).

To control the fruit softening, an antisense sequence of a strawberry pectate lyase gene has been successfully incorporated and transgenic lines with firm fruits were developed (Jimenez-Bermudez et al., 2002). Similar results were also reported in strawberry "Apel 14" and "Apel 23" lines, in which pectate lyase mRNA transcript level was reduced by 90% and 99%, respectively (Sesmero et al., 2007). But after insertion of two antisense divergent *endo*-β-(1,4)-glucanase (*Cel*1 and *Cel*2) genes it was, concluded that fruit firmness of strawberry is not affected, and that before ripening, the *cel2* gene appears to play an important role in fruit development (Palomer et al., 2006). The gene *FaGAST* encodes a small protein with 12 cysteine residues conserved in the C-terminal region that has diverse functions such as cell division, promotion or restriction of cell elongation under the control of the *CaMV35S* promoter. The ectopic expression of *FaGAST* in transgenic *F. vesca* causes both delayed growth of the plant and reduced fruit size, late flowering and low sensitivity to exogenous application of gibberellins (de la Fuente et al., 2006).

To modulate the soluble sugar content of strawberry fruits, transgenic plants incorporating an anti-sense cDNA of ADP-glucose pyrophosphorylase (AGPase) small subunit (*FagpS*) under the control of the strawberry fruit-dominant ascorbate peroxidase (APX) promoter (cv. "Anther") was regenerated (Park et al., 2006). Most transgenic fruits did not show significant differences with respect to weight and hardness but a decrease of 27–47% in starch content and an increase of 16–37% in total soluble solid content was recorded in transgenic plants compared to non-transformed control.

Tissue Culture

Similarly, hygromycin-resistant strawberry plants have been developed in temporary immersion bioreactors through *Agrobacterium*-mediated gene transfer and transgenes verified by TAIL-PCR (Hanhineva and Karenlampi, 2007).

REFERENCES

Abdal-Aziz, S.A., Pliego-Alfaro, F., Quesada, M.A. and Mercado, J.A. 2006. Evidence of frequent integration of non-T-DNA vector backbone sequences in transgenic strawberry plant. *Journal of Bioscience and Bioengineering*, 101: 508–510.

Amimoto, K. 1992. Selection in strawberry with resistance to *phytophthora* root rots for hydroponics. *Acta Horticulturae*, 319: 273–278.

Asao, H.G., Nishizawa, Y., Arai, S., Sato, T., Hirai, M., Yoshida, K., Shinmyo, A. and Hibi, T. 1997. Enhanced resistance against a fungal pathogen *Sphaerotheca fumuli* in transgenic strawberry expressing a rice chitinase gene. *Plant Biotechnology*, 14: 145–149.

Ayub, R.A. and Reis, L. 2016. Organogenesis inhibition of strawberry cultivar "Festival", by the kanamycin antibiotic. *Acta Horticulturae*, 1113: 147–150.

Barcelo, M., El-Mansouri, I., Mercado, J.A., Quesada, M.A. and Alfaro, F.P. 1998. Regeneration and transformation via *Agrobacterium tumefaciens* of the strawberry cultivar Chandler. *Plant Cell, Tissue and Organ Culture*, 54: 29–36.

Battistini, C. and Rosati, P. 1991. *In vitro* evaluation of somaclonal strawberry (*Fragaria × ananassa* "Brighton") variants for susceptibility to *Phytophthora cactorum*. In: A. Dale and J.J. Luby (eds.). *Strawberry into the 21st Century*. Timber Press, Portland, Oregon: pp. 121–123.

Belkengren, R.O. and Miller, P.W. 1962. Culture of apical meristems of *Fragaria vesca* strawberry plants as a method of excluding latent a virus. *Plant Disease Reporter*, 46: 119–121.

Boxus, P. 1974. The production of strawberry plants by in vitro micro-propagation. *Journal of Horticultural Sciences*, 49: 209–210.

Brandizzi, F., Forni, C., Frattarelli, A. and Damiano, C. 2001. Comparative analysis of DNA nuclear content by flow cytometry on strawberry plants propagated via runners and regenerated from meristem and callus cultures. *Plant Biosystems*, 135: 169–174.

Cameron, J.S., Hancock, J.F. and Nourse, T.M. 1985. The field performance of straw-berry nursery stock produced originally from runners or micro-propagation. *Advances in Strawberry Production*, 4: 56–58.

Cao, X. and Hammerschlag, F.A. 2000. Improved shoot organogenesis from leaf ex-plants of highbush blueberry. *HortScience*, 35: 945–947.

Chalavi, V., Tabaeizadeh, Z. and Thibodeau, P. 2003. Enhanced resistance to *Verticillium dahlia* in transgenic strawberry plants expressing a *Lycopersicon chilense* chitinase gene. *Journal of the American Society for Horticultural Science*, 128: 747–753.

Chaleff, R.S. 1983. Isolation of agronomically useful mutants from plant cell cultures. *Science*, 219: 676–682.

de Mesa, M.C., Jimenez-Bermudez, S., Pliego-Alfaro, F., Quesada, M.A. and Mercado, J.A. 2000. Agrobacterium cells as microprojectile coating: a novel approach to enhance stable transformation rates in strawberry. *Australian Journal of Plant Physiology*, 27: 1093–1100.

de Mesa, M.C., Santiago-Domenech, N., Pliego-Alfaro, F., Quesada, M.A. and Mercado, J.A. 2004. The *CaMV35S* promoter is highly active on floral organs and pollen of transgenic strawberry plants. *Plant Cell Reports*, 23: 32–38.

Damiano, C., Ascarelli, A., Frattarelli, A. and Lauri, P. 1995. Adventitious regeneration and genetic variability in strawberry. *Acta Horticulturae*, 392: 107–114.

Damiano, C., Monticelli, S., Frattarelli, A., Nicolini, S. and Corazza, L. 1997. Somaclonal variability and *in vitro* regeneration of strawberry. *Acta Horticulturae*, 447: 87–93.

Davey, M.R., Anthony, P., Power, J.B. and Lowe, K.C. 2005. Plant protoplasts: status and biotechnological perspectives. *Biotechnology Advances*, 23: 131–171.

de la Fuente, J.I., Amaya, I., Castillejo, C., Sanchez-Sevilla, J.F., Quesada, M.A., Botella, M.A. and Valpuesta, V. 2006. The strawberry gene FaGAST affects plant growth through inhibition of cell elongation. *Journal of Experimental Botany*, 57: 2401–2411.

Debergh, P.C. and Maene, L. 1981. A scheme for commercial propagation of ornamental plants by tissue culture. *Scientia Horticulturae*, 14: 335–345.

Debnath, S.C. 2006. Zeatin overcomes thidiazuron-induced inhibition of shoot elongation and promotes rooting in strawberry culture *in vitro*. *Journal of Horticultural Science and Biotechnology*, 81: 349–354.

Dhlamini, Z., Spillane, C., Moss, J. P., Ruane, J., Urquia, N. and Sonnino, A. 2005. *Status of Research and Application of Crop Biotechnologies in Developing Countries: Preliminary Assessment.* Food and Agriculture Organization of the United Nations, Viale delle Terme di Caracalla, 00100 Rome, Italy: p. 53.

Donnoli, R., Sunseri, F., Martelli, G. and Greco, I. 2001. Somatic embryogenesis, plant regeneration and genetic transformation in *Fragaria* spp. *Acta Horticulturae*, 560: 236–240.

El Mansouri, I., Mercado, J.A., Valpuesta, V., Lopez-Aranda, J.M., Pliego-Alfaro, F. and Quesada, M.A. 1996. Shoot regeneration and *Agrobacterium*–mediated transformation of *Fragaria vesca* L. *Plant Cell Reports*, 15: 642–646.

Faedi, W., Quarta, R., Persano, S., Aaoloni, F.M. and Damiano, C. 1993. Somaclonal variations in plants regenerated by anther culture of cv. *Pajaro. Acta Horticulturae*, 348: 427–429.

Finstad, K. and Martin, R.R. 1995. Transformation of strawberry for virus resistance. *Acta Horticulturae*, 385: 86–90.

Folta, K.M., Howard, L., Dhingra, A., Stewart, P.J. and Chandler, C.K. 2006. Characterization of LF9, an octoploid strawberry genotype selected for rapid regeneration and transformation. *Planta*, 224: 1058–1067.

Fong, R.A., Goldy, R. and Moxley, D.R. 1889. Temperature and photoperiod effects on isolation of protoplasts from strawberry mesophyll cells. Abstract of the 86th Annual Meeting of the American Society of Horticultural Science, 25: 569–571.

Galletta, G.J. and Maas, J.L. 1990. Strawberry genetics. *HortScience*, 25(8): 871–879.

George, E.F. 1993. *Plant Propagation by Tissue Culture*, Part 1 In Practice, Exegetics Ltd., Edington, UK: p. 574.

Germana, M.A. 2006. Doubled haploid production in fruit crops. *Plant Cell Tissue and Organ Culture*, 86: 131–146.

Glimelius, K., Djupsjobacka, M. and Fellner-Feldegg, H. 1986. Selection and enrichment of plant protoplast heterokaryons of Brassicaceae by flow sorting. *Plant Science*, 45: 133–141.

Graham, J. 2005. *Fragaria* Strawberry. In: *Biotechnology of Fruit and Nut Crops.* Biotechnology in Agriculture Series No. 29 (R. Litz; ed.), CAB International, Wallingford, UK: pp. 456–474.

Graham, J., Gordon, S.C., Smith, K., McNicol, R.J. and McNicol, J.M. 2002. The effect of the cowpea trypsin inhibitor in strawberry on damage by vine weevil under field conditions. *The Journal of Horticultural Science and Biotechnology*, 77: 33–40.

Graham, J., McNicol, R.J. and Greig, K. 1995. Towards genetic based insect resistance in strawberry using the cowpea trypsin inhibitor. *Annals of Applied Biology*, 127: 163–173.

Gruchala, A., Korbin, M. and Zurawicz, E. 2004. Conditions of transformation and regeneration of 'Induka' and 'Elista' strawberry plants. *Plant Cell Tissue Organ Culture*, 79: 153–160.

Hammerschlag, F., Garcés, S., Koch-Dean, M., Ray, S., Lewers, K., Maas, J. and Smith, B. 2006. *In vitro* response of strawberry cultivars and regenerants to *Colletotrichum acutatum. Plant Cell, Tissue and Organ Culture*, 84: 255–261.

Hanhineva, K.J. and Karenlampi, S. O. 2007. Production of transgenic strawberries by temporary immersion bioreactor system and verification by TAIL-PCR. *BMC Biotechnology*, 7: 11.

Hao, Y.J. and Deng, X.X. 2005. Cytological and molecular evaluation of strawberry plants recovered from in vitro conservation by slow-growth. *Journal of Horticultural Science and Biotechnology*, 80(5): 588–592.

Hao, Y.J., You, C.X. and Deng, X.X. 2002. Analysis of ploidy and the patterns of amplified fragment length polymorphism and methylation sensitive amplified polymorphism in strawberry plants recovered from cryopreservation. *Cryoletters*, 23: 37–46.

Harpster, M.H., Brummell, D.A. and Dunsmuir, P. 1998. Expression analysis of a ripening-specific, auxin-repressed endo-1,4-β-glucanase gene in strawberry (Fragaria × ananassa). *Plant Physiology*, 118: 1307–1316.

Hartmann, H.T. and Kester, D.E. 1989. *Plant Propagation: Principles and Practices (Fourth Edition).* Prentice Hall of India, New Delhi, India: pp. 10 & 239.

Haymes, K.M. and Davis, T.M. 1998. *Agrobacterium* mediated transformation of "Alpine" *Fragaria vesca*, and transmission of transgenes to R_1 progeny. *Plant Cell Reports*, 17: 279–283.

Hofer, M. 2011. Conservation strategy of genetic resources for strawberry in Germany. *Acta Horticulturae*, 908: 421–430.

Hofer, M. and Reed, B.M. 2011. Cryopreservation of strawberry genetic resources in Germany. *Acta Horticulturae*, 918: 139–14.

Houde, M., Dallaire, S., N'Dong, D. and Sarhan, F. 2004. Overexpression of the acidic dehydrin WCOR410 improves freezing tolerance in transgenic strawberry leaves. *Plant Biotechnology*, 2: 381–387.

Tissue Culture

Hummer, K.E. 2009. Global conservation of strawberries: a strategy is formed. *Acta Horticulturae*, 842: 577–580.

Infante, R. and Rosati, P. 1993. *Fragaria vesca* L. "Alpine" protoplast culture and regeneration. *Acta Horticulturae*, 348: 432–434.

Jain, S.M. 2001. Tissue culture-derived variation in crop improvement. *Euphytica*, 118: 153–166.

Jain, S.M. and Pehu, E. 1992. The prospects of tissue culture and genetic engineering in strawberry improvement. *Acta Agriculturae Scandinavica, Section B – Soil & Plant Science*, 42: 133–139.

James, D.J., Passey, A.J. and Barbara, D.J. 1990. Agrobacterium -mediated transformation of the cultivated strawberry (*Fragaria* × *Ananassa* Duch.) using dis-armed binary vectors. *Plant Science*, 69: 79–94.

Jelenkovic, G., Chin, C.K. and Billings, S. 1986. Transformation studies of *Fragaria* × *ananassa* by triplasmids of *Agrobacterium tumefaciens*. *HortScience*, 21: 695 (Abstract).

Jimenez-Bermudez, S., Redondo-Nevado, J., Munoz-Blanco, J., Caballero, J.L., Lopez-Aranda, J.M., Valpuesta, V., Pliego-Alfaro, F., Quesada, M.A. and Mercado, J.A. 2002. Manipulation of strawberry fruit softening by antisense expression of a pectate lyase gene. *Plant Physiology*, 128: 751–759.

Jungnickel, F. 1988. Strawberries (*Fragaria* spp. and Hybrids). In: Y.P.S. Bajaj (ed.). *Biotechnology in Agriculture and Forestry 6, Cops II Edition*. Springer-Verlag, Berlin, Germany: pp. 39–103.

Khammuang, S., Dheeranupattana, S., Hanmuangjai, P. and Wongroung, S. 2005. Agrobacterium-mediated transformation of modified antifreeze protein gene in strawberry. *Journal of Science and Technology*, 27: 693–703.

Larkin, P.J. and Scowcroft, W.R. 1981. Somaclonal variation – a novel source of variability from cell cultures for plant improvement. *Theoretical and Applied Genetics*, 60: 197–214.

Lines, R., Kelly, G., Milinkovic, M. and Rodoni, B. 2006. Runner certification and virus elimination in commercial strawberry cultivars in Australia. *Acta Horticulturae*, 708: 253–254.

Maas, J.L., Zhong, L. and Galletta, G.J. 1993. *In vitro* screening of strawberry plant and root cultures for resistance to *Phytophthora fragaria and P. cactorum*. *Acta Horticulturae*, 348: 496–499.

Mathews, H., Dewey, V., Wagner, W. and Bestwick, R.K. 1998. Molecular and cellular evidence of chimaeric tissues in primary transgenics and elimination of chimaerism through improved selection protocols. *Transgenic Research*, 7: 123–129.

Mezzetti, B. and Constantini, E. 2006. Strawberry (*Fragaria* × *ananassa*). *Methods in Molecular Biology*, 344: 287–95.

Mohamed, F.H., Omar, G.F. and Ismail, M.A. 2017. *In vitro* regeneration, proliferation and growth of strawberry under different light treatments. *Acta Horticulturae*, 1155: 361–368.

Mohan, R., Chui, E.A., Biasi, L.A. and Soccol, C.R. 2005. Alternative *in vitro* propagation: use of sugarcane bagasse as a low cost support material during rooting stage of strawberry cv Dover. *Acta Horticulturae*, 348: 496–499.

Moore, P.P., Robbins, J.A. and Sjulin, T.M. 1991. Field performance of "Olympus" strawberry subclones. *HortScience*, 26: 192–194.

Morgan, A., Baker, C.M., Chu, J.S.F., Lee, K., Crandall, B.A. and Jose, L. 2002. Production of herbicide tolerant strawberry through genetic engineering. *Acta Horticulturae*, 567: 113–115.

Mullin, R.H. and Schlegel, D.E. 1976. Cold storage maintenance of strawberry meristem plantlets. *HortScience*, 11: 100–101.

Mullin, R.H., Smith, S.H., Frazier, N.W., Schlegel, D.E. and McCall, S.R. 1974. Meristem culture of strawberries free of mild edge, pallidosis and mottle disease. *Phytopathology*, 64: 1425–1429.

Nehra, N.S., Chibbar, R.N., Kartha, K.K., Datla, R.S.S., Crosby, W.L. and Stushnoff, C. 1990a. Genetic transformation of strawberry by *Agrobacterium tumefaciens* using a leaf disk regeneration system. *Plant Cell Reports*, 9: 293–298.

Nehra, N.S., Chibbar, R.N., Kartha, K.K., Datla, R.S.S., Crosby, W.L. and Stushnoff, C. 1990b. *Agrobacterium*-mediated transformation of strawberry calli and recovery of transgenic plants. *Plant Cell Reports*, 9: 10–13.

Niemirowicz-Szczytt, K., Zakrzewska, Z., Malepszy, S. and Kubicki, B. 1983. Characters of plants obtained from *Fragaria* × *ananassa* in anther culture. *Acta Horticulturae*, 131: 231–237.

Nyman, M. and Wallin, A. 1988. Plant regeneration from strawberry (*Fragaria* × *ananassa*) mesophyll protoplasts. *Journal of Plant Physiology*, 133: 375–377.

Nyman, M. and Wallin, A. 1992. Improved culture technique for strawberry (*Fragaria* × *ananassa* Duch.) protoplasts and the determination of DNA content in protoplast derived plants. *Plant Cell Tissue and Organ Culture*, 30: 127–133.

Oosumi, T., Gruszewski, H.A., Blischak, L.A., Baxter, A.J., Wadl, P.A., Shuman, J.L., Veilleux, R.E. and Shulaev, V. 2006. High-efficiency transformation of the diploid strawberry (*Fragaria vesca*) for functional genomics. *Planta*, 223: 1219–1230.

Orlando, R., Magro, P. and Rugini, E. 1997. Pectic enzymes as a selective pressure tool for *in vitro* recovery of strawberry plants with fungal disease resistance. *Plant Cell Reports*, 16: 272–276.

Owen, H.R. and Miller, A.R. 1996. Haploid plant regeneration from anther cultures of three North American cultivars of strawberry (*Fragaria × ananassa* Duch.). *Plant Cell Reports*, 15: 905–909.

Owens, C.L., Thomashow, M.F., Hancock, J.F., Amy, F. and Iezzoni, A.F. 2002. CBF1Orthologs in sour cherry and strawberry and the heterologous expression of *cbf1* in strawberry. *Journal of the American Society for Horticultural Science*, 127: 489–494.

Palomer, X., Llop-Tous, I., Vendrell, M., Krens, F.A., Schaart, J.G., Boone, M.J., Van der Valk, H. and Salentijn, E.M.J. 2006. Antisense down-regulation of strawberry endo-β-(1,4)-glucanase genes does not prevent fruit softening during ripening. *Plant Science*, 171: 640–646.

Pantazis, C.J., Fisk, S., Mills, K, Flinn, B.S., Shulaev, V., Veilleux, R.E. and Dan, Y. 2013. Development of an efficient transformation method by *Agrobacterium tumefaciens* and high throughput spray assay to identify transgenic plants for woodland strawberry (*Fragaria vesca*) using *NPTII* selection. *Plant Cell Reports*, 32(3): 329–337.

Park, J.I., Lee, Y.K., Chung, W.I., Lee, I.H., Choi, J.H., Lee, W.M., Ezura, H., Lee, S.P. and Kim, I.J. 2006. Modification of sugar composition in strawberry fruit by antisense suppression of an ADP-glucose pyrophosphorylase. *Molecular Breeding*, 17: 269–279.

Phillips, R.L., Kaeppler, S.M. and Olhoft, P. 1994. Genetic instability of plant tissue cultures: breakdown of normal controls. *Proceedings of the National Academy of Science USA*, 91: 5222–5226.

Popescu, A.N., Isac, V.S., Coman, M.S. and Radulescu, M.S. 1997. Somaclonal variation in plants regenerated by organogenesis from callus culture of strawberry (*Fragaria × ananassa*). *Acta Horticulturae*, 439: 89–96.

Posnette, A.F. 1953. Heat inactivation of strawberry viruses. *Nature*, 171: 312.

Quesada, M.A., Blanco-Portales, R., Pose, S., García-Gago, J.A., Jimenez-Bermudez, S., Munoz-Serrano, A., Caballero, J.L., Pliego-Alfaro, F., Mercado, J.A. and Munoz-Blanco, J. 2009. Antisense down-regulation of the *FaPG1* gene reveals an unexpected central role for polygalacturonase in strawberry fruit softening. *Plant Physiology*, 150(2): 1022–1032.

Reinert, J. 1958. Morphogenese und ihre kontrolle an gewebekulturen aus ca-rotten *Naturwissenchaften*, 45: 344–345.

Ricardo, Y.G., Coll, Y., Castagnaro, A. and Diaz Ricci, J.C. 2003. Transformation of strawberry cultivar using a modified regeneration medium. *HortScience*, 38: 277–280.

Rosati, P., Devreux, M. and Laneri, U. 1975. Anther culture of strawberry. *HortScience*, 10: 119–120.

Rugienius, R. and Stanys, V. 2001. *In vitro* screening of strawberry plants for cold resistance. *Euphytica*, 122: 269–277.

Sansavini, S., Rosati, P., Gaggioli, D. and Toschi, M.F. 1989. Inheritance and stability of somaclonal variations in micropropagated strawberry. *Horticulturae*, 280: 375–384.

Schaart, J.G., Salentijn, E.M.J. and Krens, F.A. 2002. Tissue-specific expression of the β-glucuronidase reporter gene in transgenic strawberry (*Fragaria × ananassa*) plants. *Plant Cell Reports*, 21: 313–319.

Schaart, J.G., Krens, F.A., Pelgrom, K.T.B., Mendes, O. and Rouwendal, G.J.A. 2004. Effective production of marker-free transgenic strawberry plants using inducible site-specific recombination and a bifunctional selectable marker gene. *Plant Biotechnology Journal*, 2: 233–240.

Schestibratov, K.A. and Dolgov, S.V. 2005. Transgenic strawberry plants expressing a *thaumatin* II gene demonstrate enhanced resistance to *Botrytis cinerea*. *Scientia Horticulturae*, 106: 177–189.

Sesmero, R., Quesada, M.A. and Mercado, J.A. 2007. Antisense inhibition of pectate lyase gene expression in strawberry fruit: characteristics of fruits processed into jam. *Journal of Food Engineering*, 79: 194–199.

Simon, L., Racz, E., and Zatyko, J.M. 1987. Preliminary notes on somaclonal variations of strawberry. *Fruit Science Reports*, 14: 151–154.

Sowik, I., Bielenin, A. and Michalczuk, L. 2001. *In vitro* testing of strawberry resistance to *Verticillium dahlia* and *Phytophthora cactorum*. *Scientia Horticulturae*, 88: 31–40.

Sowik, I., Wawrzynczak, D. and Michalczuk, L. 2003. Ex vitro establishment and greenhouse performance of somaclonal variants of strawberry selected for resistance to *Verticillium dahliae*. *Acta Horticulturae*, 616: 497–500.

Steward, F.C., Mapes, M.O. and Meats, K. 1958. Growth and organized development of cultured cells, II. Organization in cultures grown from freely suspended cells. *American Journal of Botany*, 45: 704–708.

Tissue Culture

Swartz, H.J., Galletta, G.J. and Zimmerman, R.H. 1981. Field performance and phenotypic stability of tissue culture-propagated strawberries. *Journal of the American Society for Horticultural Science*, 106: 667–673.

Taji, A., Kumar, P.P. and Lakshmanan, P. 2002. *In Vitro Plant Breeding*. Food Products Press, New York, New York: p. 167.

Takahashi, H., Takatsugu, T. and Tsutomu, M. 1992. Resistant plants to *Alternaria alternate* strawberry pathotype selected from caliclones of strawberry cultivar Morioka-16 and their characteristics. *Journal of the Japanese Society for Horticultural Science*, 61: 323–329.

Tantivit, K., Isobe, S., Nathewet, P., Okuda, N. and Yanagi, T. 2016. The development of a primed *in situ* hybridization technique for chromosome labeling in cultivated strawberry (*Fragaria ananassa*). *Cytologia*, 81(4): 439–446.

Torun, A.A., Kacar, Y.A., Bicen, B., Erdem, N. and Serce, S. 2014. *In vitro* screening of octoploid *Fragaria chiloensis* and *Fragaria virginiana* genotypes against iron deficiency. *Turkish Journal of Agriculture and Forestry*, 38: 169–179.

Toyoda, H., Horikoshi, K., Yamano, Y. and Ouchi, S. 1991. Selection of Fusarium wilt disease resistance from regenerants derived from callus of strawberry. *Plant Cell Reports*, 10: 167–170.

Vasil, V. and Hilderbrandt A.C. 1965. Differentiation of tobacco plants from single, isolated cells in microcultures. *Science*, 150: 889–892.

Vellicce, G.R., Ricci, J.C.D., Hernández, L. and Castagnaro, A.P. 2006. Enhanced resistance to *Botrytis cinerea* mediated by the transgenic expression of the chitinase gene ch5B in strawberry. *Transgenic Research*, 15: 57–68.

Wallin, A. 1997. Somatic hybridization in *Fragaria*. *Acta Horticulturae*, 439: 63–66.

Wallin, A., Skjöldebrand, H. and Nyman, M. 1993. Protoplasts as tools in *Fragaria* breeding. *Acta Horticulturae*, 348: 414–421.

Wang, D., Wergin, W. P. and Zimmerman, R.H. 1984. Somatic embryogenesis and plant regeneration from immature embryos of strawberry. *HortScience*, 19: 71–72.

Wang, J., Ge, H., Peng, S., Zhang, H., Chen, P. and Xu, J. 2004. Transformation of strawberry (*Fragaria ananassa* Duch.) with late embryogenesis abundant protein gene. *Journal of Horticultural Science and Biotechnology*, 79: 735–738.

Zhao, X., Li, G., Li, L.J., Hu, P.P. and Zhou, H.C. 2017. Effects of NaCl and NaHCO$_3$ stress on the growth of *in vitro* culture seedlings *of Fragaria × ananassa* Duch. *Acta Horticulturae*, 1156: 883–888.

Zhao, Y., Liu, Q.Z. and Davis, R.E. 2004. Transgene expression in strawberries driven by a heterologous phloem-specific promoter. *Plant Cell Reports*, 23: 224–230.

Zimmerman, R. H. 1981. Micropropagation of fruit plants. *Acta Horticulturae*, 120: 217–222.

8 Markers and Genetic Mapping

Era Vaidya Malhotra and Madhvi Soni

CONTENTS

8.1 Markers 142
    8.1.1 Morphological Markers 142
    8.1.2 Biochemical Markers (Isozymes) 142
    8.1.3 Molecular Markers 143
        8.1.3.1 RFLP 143
        8.1.3.2 RAPD 143
        8.1.3.3 AFLP 145
        8.1.3.4 ISSR 145
        8.1.3.5 Microsatellite Markers (SSRs) 146
8.2 Genetic Mapping 147
    8.2.1 Linkage Mapping 147
    8.2.2 QTL Mapping 149
    8.2.3 Genome Mapping 151
8.3 Functional Genomics 151
References 153

Strawberries have experienced a series of polyploidizations and natural hybridizations during evolution. Crosspollination has led to the development of hybrids with significant variations in plant and fruit characteristics such as fruit size, firmness, composition, taste, aroma and resistance to biotic and abiotic stresses, thus used for cultivation and breeding for intended traits (Darrow, 1966; Hancock, 1999; Faedi et al., 2002; Ulrich et al., 2007). Nowadays, a large number of strawberry cultivars are grown worldwide, and new varieties are emerging rapidly (Nielsen and Lovell, 2000). Due to continued introduction of new strawberry cultivars to the market, it is important to prevent their misidentification and protect the breeders' rights (Garcia et al., 2002), thus requiring the development of reliable methods for identification and assessment of genetic diversity in *Fragaria* spp. (Degani et al., 2001). Earlier, the identification of cultivars was based on morphological features, however, due to strong influence of environmental factors and age of plants on morphology and close resemblance among cultivars, varietal assessment on the basis of only phenotypic traits does not give reliable information about the genetic makeup of an individual genotype.

Development of DNA-based markers has proved highly effective in filling the gaps of morphology-based markers. A molecular marker is defined as a DNA sequence with a known location on a chromosome and associated with a particular gene or trait. Molecular markers may or may not be related with phenotypic expression of a trait. Unlike phenotype-based markers, molecular markers are stable in nature, and independent of environmental factors, phenotypic stages, growth and developmental stages of organisms (Weising et al., 1995). Other advantages of molecular markers include their being less time consuming and highly informative. Molecular markers can easily differentiate among different species as well as within different strains of the same species. They detect variation on the basis of DNA polymorphism, that is, variations that occur due to alteration in the genomic DNA.

In the past few years, several PCR-based techniques for multi-locus analysis have been developed, including restriction fragment length polymorphisms (RFLPs), random amplified polymorphic

DNA (RAPDs), cleaved amplified polymorphic sequences (CAPSs), sequence characterized amplified regions (SCARs), amplified fragment length polymorphisms (AFLPs), inter simple sequence repeats (ISSRs) and simple sequence repeats (SSRs). In strawberry, DNA marker technologies have been widely used for identification of cultivars, analysis of diversity, true-to-type assessment, development of genetic linkage maps and mapping of quantitative trait loci (QTL). Each marker system has its own strengths and limitations; hence, the choice of marker is an important decision in practical application.

8.1 MARKERS

8.1.1 MORPHOLOGICAL MARKERS

Identification of strawberry on the basis of morphological characters involves recording variation in leaves, flowers, fruits, their colour, size, growth habit etc. (Dale, 1996). On the basis of morphology, plants can be clustered if they have similar quantitative and qualitative traits (Brown and Schoen, 1994). Garcia et al. (2002) used morphological markers to certify the identity of strawberry cultivars in Argentina. For filing a patent in the United States and Europe, in addition to isozyme markers, morphological markers are also added in plant patent descriptions (Nielsen and Lovell, 2000). Recently, 38 morphological characters have been used to evaluate 11 strawberry cultivars, to assess the important morphological differences among them (El-Denary et al., 2016).

However, the phenotype of a plant is a result of interaction between genotype and environment. Identification of cultivars only on the basis of morphological traits is difficult as their expression depends upon method of cultivation and environmental conditions, and also closely related cultivars express very minute phenotypic variations, which are difficult to differentiate (Degani et al., 2001), for example, morphology-based misclassification has been reported in the case of *Fragaria multicipita* (Jomantiene et al., 1998). During identification of strawberry cultivars from Argentina, no morphological variation was observed in three accessions of "Pajaro" but it was clearly observed using molecular markers (Garcia et al., 2002). Nielsen and Lovell (2000) used different sets of morphological markers such as length and breadth of leaf, leaf shape, petal length and spacing and fruit length, size and shape to identify strawberry cultivars. But in most cases of cultivar identification such as those dealing with infringement of breeders' rights, only the fruit and not the whole plant is available thus pointing to a need for additional forms of identification. Although there are disadvantages associated with identification of species only on the basis of morphological characters (traits) but combination of morphological markers with molecular markers gives more relevant information for germplasm characterization and identification. To study the relationships among the subspecies of *F. virginiana* and *F. chiloensis* from North America, the combination of morphological and RAPD markers have shown good success (Harrison et al., 1997).

8.1.2 BIOCHEMICAL MARKERS (ISOZYMES)

Isozymes are multiple forms of enzymes differing in amino acid sequence but catalyzing the same reaction. They were the first molecular markers to be developed. The use of isozymes as markers is based on the staining of proteins having same function, but differing in their movement in electric field; depending upon which, different forms can be easily distinguished. Isozymes have been extensively used as markers to study diversity and for identification of strawberry cultivars. Hancock and Bringhurst (1979) used isozymes to study diversity of 13 diploid (*F. vesca*) and 19 octoploid *Fragaria* populations from California using two enzymes, phosphoglucoisomerase (PGI) and peroxidase (PX). A high level of genetic variation was observed in both the species depending upon the site of collection. Similar studies were also taken up by Arulsekar et al. (1981) to differentiate between different genotypes of the cultivated strawberry using PGI, leucine amino peptidase (LAP) and phosphoglucomutase (PGM). These markers have since then been used for cultivar

Markers and Genetic Mapping

identification (Nehra et al., 1991) and linkage analysis (Williamson et al., 1995) in strawberry. Flores et al. (2000) investigated variations of somaclones from the strawberry cultivar "Vila Nova" through biochemical variation.

However, there are some restrictions in using isozymes for such studies, such as, only proteins encoded by coding regions of DNA can be used, hence reducing the number of available markers. Isoenzymatic characterization of strawberry clones using starch gel separation was also not found to be equally effective, and enabled the identification of 9 out of 23 clones (Greco, 1993). Isozyme patterns depend upon the age and environmental condition of the plant (Nehra et al., 1991). Variations in catalytic properties of the different isozymes expressed under different environmental conditions even within the same species has been observed, which illustrates the sensitivity of isozymes to the environment (Hancock and Bringhurst, 1979). Thongthieng and Smitamana (2003) failed to identify hybrid lines of strawberry progeny from alternate crosses of four parental lines at high (90–95%) similarity levels using four enzyme systems (malate dehydrogenase, malic enzyme, LAP and diaphorase). To overcome these problems, DNA-based fingerprinting using molecular markers has emerged as a powerful tool to study genetic diversity and for the identification of individual plants allowing the characterization of cultivars (Bhat et al., 1997).

8.1.3 MOLECULAR MARKERS

A molecular marker is a fragment of DNA associated with a specific location in the genome. In strawberry, molecular markers have been extensively used for cultivar identification and to study genetic relationships among them.

8.1.3.1 RFLP

RFLPs are co-dominant markers representing heritable changes in the length of fragments of genomic DNA. In RFLP, differential DNA fragment profile observed by southern hybridization of restriction enzyme digested DNA with specific probes results in detection of DNA polymorphism. Using 24 PCR–RFLP markers, Kunihisa et al. (2004) distinguished 65 Japanese strawberry cultivars and 96 progenies of a selfed "Sachinoka" strawberry line. The RFLP technique is time consuming and costly, as it requires elaborate laboratory techniques such as the development of a specific labelled probe, usage of radioisotopes, southern hybridization and autoradiography (Kesseli et al., 1994).

8.1.3.2 RAPD

RAPD markers are random markers present throughout the genome (Williams et al., 1990; Gidoni et al., 1994). No DNA sequence information is required before primer synthesis, and they can be employed across species using universal primers (Congiu et al., 2000). Due to the speed and efficiency of RAPD analysis, these markers have been continuously used in strawberries for cultivar identification and diversity analysis.

RAPD markers have proven to be good tools to study biodiversity at the genetic level. Hancock et al. (1994) examined genetic diversity among eight related strawberry cultivars and advanced breeding selections. Using RAPD markers, Graham et al. (1996) differentiated even closely related *Fragaria* genotypes. A high level of similarity, ranging from 62–89% was observed among the genotypes, indicating their closely related nature, although they resulted from four independent breeding programmes (Scotland, England, the United States and the Netherlands). RAPD analysis was done by Landry and coworkers (1997) using DNA from two micro-extractions, and detected diversity among 75 strawberry cultivars. By using RAPD markers as well as morphological parameters, Harrison et al. (1997) determined that *F. virginana* subsp. *platypetala* (western Cascade mountains and Olympic Peninsula) was more closely related to North American *F. chiloensis* than to *F. virginiana* subsp. *virginiana* (east of the Missouri river) and subsp. *glauca* (eastern Rocky Mountains and Black Hills). This relationship could not be determined on the basis of morphological parameters alone. On the basis of RAPD data analysis, Porebski and Catling (1998) separated North American plants (*F. chiloensis* ssp. *lucida* and ssp. *pacifica*) from the South American plants (*F. chiloensis* ssp. *chiloensis*) but could not separate the two North American subspecies. In another study, again using RAPD markers, Harrison et al. (2000) observed less difference among *F. virginiana* and *F. chiloensis* accessions and more variation among plants within *F. virginiana* and *F. chiloensis* populations collected from major provenances of North America and Chile.

Identification and verification of cultivars are essential to protect breeders' rights and prevent misidentification of cultivars. Hancock et al. (1994) identified 28 polymorphic bands using 10 RAPD primers, and obtained two–six unique amplification profiles with each RAPD primer in the pool of eight strawberry genotypes. Using four RAPD primers, Gidoni et al. (1994) observed 10 polymorphic cultivar-specific bands enabling differentiation of closely related varieties, and thus prevented unauthorized shipment of both plants and fruits of these varieties. Degani et al. (1998) distinguished 41 strawberry cultivars grown in the United States and Canada on the basis of 15 RAPD markers generated with seven primers. RAPD technique was used by Congiu et al. (2000) to settle a lawsuit involving unauthorized commercialization of a patented strawberry variety of high economic relevance ("Marmolada"®). The markers were able to distinguish 13 clones of the cultivar "Onebor" (Marmolada) from a group of 31 plants, and markers were reproducible enough to be accepted as evidence in court. Using six RAPD primers, Zebrowska and Tyrka (2003) identified 11 markers that were specific to six strawberry cultivars.

Although RAPD studies require no prior knowledge of the genome that is being analyzed, the major drawback of the method is that they are dominant markers, which means each primer is not able to distinguish heterozygous from homozygous individuals; plus their reproducibility is very low, especially between laboratories. These considerations reduce the utility of RAPDs for fingerprinting and cultivar identification.

To overcome the drawbacks of RAPD markers, modification of the RAPD technique in the form of two molecular markers was done, that is, CAPSs and SCARs.

The development of CAPS markers involves extensive sequencing and screening for restriction enzyme digested genomic locus combinations that yield polymorphic products. Kunihisa et al. (2003) developed nine CAPS markers for identification of 64 Japanese cultivars. Initially available nucleotide sequences were surveyed for restriction sites and primers were designed to amplify 400–650 bp regions around these sites. But these primers amplified multiple homologous alleles, making banding patterns difficult to interpret. To overcome this problem, the PCR products were sequenced and subjected to a cluster analysis to identify groups of similar sequences, and new primers were designed to target unique sequences. The resulting CAPS markers amplified from only one subgenome were confirmed by diploid inheritance ratios in a test population, showing their efficiency for cultivar identification.

SCAR markers are locus-specific PCR-based markers. Sequence specific oligonucleotide primers have been developed from sequences of cloned RAPD fragments, which are linked to a specific trait of interest (Paran and Michelmore, 1993; McDermott et al., 1994). Two SCAR markers specific for a gene conferring resistance to red stele root rot (*Rpf1*) have been developed for strawberry (Haymes et al., 2000). Initially, the SCAR primer set produced multiple bands in the resistant test progeny as well as in some susceptible progeny. Thus, a new set of SCARs were developed on the basis of sequence differences among the bands which gave amplification only in the resistant progeny. Similar studies were done by Rugienius et al. (2006), where RAPD markers, developed earlier for strawberry *Rpf1* gene were difficult to reproduce and were, therefore, converted to SCAR markers. A SCAR marker was constructed based upon RAPD, linked to the susceptibility allele of the *Rpf1* gene. Following the cloning and sequencing of this marker, they designed SCAR marker specific to this gene. By using this newly developed SCAR marker, identification of disease-resistant varieties was carried out based on the presence of the gene, while the other varieties lacked it. Differently from using RAPD primers, they observed that only one band clearly showed presence or absence of the *Rpf1* gene. Although these markers are highly polymorphic and reproducible in

Markers and Genetic Mapping

nature, but the major drawback of these markers is the laborious cloning and sequencing requirement for their development.

8.1.3.3 AFLP

AFLP is a PCR-based technique that includes specific amplification of restriction enzyme digested DNA. The DNA template is restricted, followed by ligation of oligonucleotide adapters and pre-amplification, using primers having only one selective nucleotide, reducing the number of DNA fragments generated and finally selective amplification of sets of restriction fragments. In AFLP studies, purity of DNA is a major concern as the complete technique depends on restriction and ligation, and contaminated or degraded DNA may result in incomplete digestion (Arnau et al., 2001; Perry et al., 1998). AFLP markers are highly reproducible and no prior information about the target DNA is required. It detects polymorphisms depending upon the presence or absence of restriction fragments.

Degani et al. (2001) studied the genetic relationships among 19 strawberry cultivars from United States and Canada based on pedigree, RAPD and AFLP markers. A total of 228 bands were obtained, out of which 35 were polymorphic and distinguished all the 19 strawberry cultivars and 9 were cultivar-specific bands. A higher level of correlation between pedigree data coefficients and RAPD was observed than with AFLP similarity coefficients which could be possible due to even distribution of the RAPD markers used across the strawberry genome. AFPL marker based on single cutting enzyme PstI was used by Tyrka et al. (2002) to differentiate 6 strawberry cultivars and 13 salinity tolerant clones. The level of polymorphisms obtained using the PstI–AFLP method was intermediate and a single primer produced six–20 polymorphic markers and eight–19 discrete profiles suggesting AFLP method to be more effective in generating genotype-specific bands. A comparative analysis was done by Zhang et al. (2006) using AFLP to analyze the genetic diversity of 107 strawberry samples from five countries, and observed high level of polymorphism (70.9%) in the populations according to their genealogical relationship. Han et al. (2013) also studied the genetic relationships among 96 cultivars from five countries using AFLP markers. The results showed that 93.5% bands were polymorphic. More genetic variation in European, American and Japanese cultivars was observed than in those from China, indicating narrow genetic base of Chinese germplasm. Using 10 AFLP markers, genetic diversity and relatedness was studied by Manbari et al. (2012) among 17 local strawberry cultivars from the Kurdistan province of northwest of Iran and 13 exotic cultivars. A high level of diversity was seen in local cultivars, while in exotic cultivars, narrow range of diversity was observed.

8.1.3.4 ISSR

ISSRs are small DNA fragments located between oppositely oriented microsatellite sequences (Zietkiewicz et al., 1994). Compared with RAPD markers, these markers are more reproducible and detect a high level of polymorphism. They cost less and are easier to use than AFLPs and no prior knowledge of flanking sequences is required. ISSR markers have proved effective to characterize 30 strawberry varieties. A total of 390 bands, 113 of which were polymorphic (30%), were generated using five primers. With only one primer, all the varieties were distinguished, including those with a common pedigree. Similar type of banding patterns was observed when DNA samples were taken from tissues (leaves, sepals, and rhizomes) of the same plant or from different clones of the same variety (Arnau et al., 2002). With ISSR markers, Carrasco et al. (2007) reported a lower genetic diversity among *F. chiloensis* than *F. patagonica* cultivars, while having high genetic diversity at the species level. In another report, Debnath et al. (2008) used 17 ISSR markers to evaluate the levels of genetic relatedness among 16 strawberry cultivars and 11 advanced selections of three agriculture and agri-food research centres in Canada, and detected a high level of polymorphism among strawberry genotypes. Hussein et al. (2008) used nine ISSR primers to study polymorphism within six strawberry varieties. Primers generated 102 amplified fragments, of which 46 were observed to be variety-specific. The estimates of the genetic similarity ranged from 45–83%.

Hua et al. (2013) detected a high level of genetic diversity among 34 strawberry accessions from six different countries, revealing the relationship among the varieties with the use of nine ISSR primers. They classified the samples into seven groups correlating with the geographical distribution of strawberry accessions.

8.1.3.5 Microsatellite Markers (SSRs)

Microsatellite or SSR markers are short nucleotide sequence repeats of, usually, equal to or less than six bases in length (Rafalski et al., 1996). These markers are co-dominant in nature, highly polymorphic and more discriminative than others (Yao et al., 2007). SSRs have been studied well for variability and diversity analysis in strawberry (Ashley et al., 2003; Hadonou et al., 2004) and to protect IP rights of strawberry breeders (Shimomura and Hirashima, 2006). More than 900 SSRs have been developed in *Fragaria* especially from *F. vesca*, diploid *F. viridis* (James et al., 2003; Sargent et al., 2003; Cipriani and Testolin, 2004; Hadonou et al., 2004; Monfort et al., 2006; Zorrilla-Fontanesi et al., 2010), *F. virginiana* and the domestic strawberry *Fragaria × ananassa* (octoploid) (Ashley et al., 2003; Spigler et al., 2010; Bassil et al., 2006; Gil-Ariza et al., 2006; Zorrilla-Fontanesi et al., 2010). Using SSR markers, Tyrka et al. (2002) reported a lower diversity and small genetic variation in 19 strawberry accessions, with 15 originating from Europe and four from Japan or Canada (Figure 8.1). Sixteen SSR primer pairs, developed for strawberry cultivar "Earliglow" were examined for their ability to distinguish commonly grown strawberry cultivars. Six primer pairs amplified consistently, and produced variable fragment sizes and one primer pair generated unique profiles for two accessions (Dangl et al., 2007). Govan et al. (2008) developed 10 SSR markers to fingerprint 60 octoploid accessions, including 56 *Fragaria × ananassa* cultivars and four wild octoploid *Fragaria* species. Later on, to fingerprint 26 cultivars grown in Florida and advanced breeding selections from the University of Florida, 9 of the 10 SSRs developed by Govan et al. (2008) were used by Brunnings et al. (2010) to distinguish the closely related cultivars. A reduced set of four SSRs from 91 primer pairs were identified by Njuguna (2010) to differentiate 22 *Fragaria* species, maintained at the Corvallis National Clonal Germplasm Repository. SSR set differentiated an impressive list of 187 accessions, spanning 22 *Fragaria* species.

Using *Fragaria* species derived SSRs, Njuguna et al. (2011) assessed genetic diversity among populations of diploids, *F. iinumae* Makino and *F. nipponica* Makino, and examined intra- and

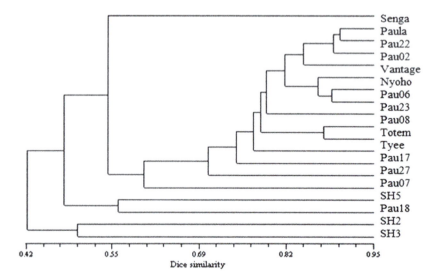

FIGURE 8.1 Dendrogram of 19 strawberry clones generated by UPGMA clustering of the similarity matrix. (From Tyrka et al., 2002 Simplified AFLP procedure as a tool for identification of strawberry cultivars and advanced breeding lines. *Euphytica*, 125: 273–280.)

interspecies relationships in overlapping populations. Yoon et al. (2012) used 18 SSR markers to analyze the genetic diversity of 59 accessions of cultivated strawberry from Korea, Germany, the United States, the UK and Japan. Average diversity statistic values of these five genetic groups showed that three genotypes from Germany were highly diverse, the second-highest heterozygosity was observed in the group from the United States, followed by the groups from Japan and Korea. The lowest gene diversity was identified in accessions from Japan. Using 20 SSR markers, Hosseini et al. (2013) reported genetic diversity in a collection of 13 exotic and 17 local strawberry cultivars collected from the Kurdistan province in the northwest of Iran, and observed a high level of polymorphism. They concluded that SSR markers could identify a high level of polymorphism in comparison to any other DNA marker. Using 11 SSR and six AFLP markers, Rugienius et al. (2015) characterized 44 strawberry genotypes and demonstrated the suitability of SSR and AFLP primer combinations for the genotyping of strawberry accessions and provided improved knowledge of genetic diversity of strawberry, developed in Lithuania. Cluster analysis, based on combined analysis of both marker systems evaluated, revealed two main groups consisting of 10 and 34 strawberry cultivars and clones, respectively.

The SSRs developed from expressed sequence tags (EST) are known as EST–SSRs. A number of studies have been carried out using EST–SSRs to study the genetic relationship and identification of strawberry cultivars (Dong et al., 2011; Gil-Ariza et al., 2006, 2009; Keniry et al., 2006). More than 59,800 *Fragaria* ESTs have been deposited in GenBank, serving an important source for SSR marker development, and approximately 7,000 unique ESTs from cultivated strawberry have now been published (Bombarely et al., 2010). Using EST–SSRs, Amaya (2009) analyzed genetic diversity in 92 selected strawberry cultivars with widely diverse origins and observed a reduction in diversity of these cultivars caused by breeding. Thirty-two SSR primer pairs were developed from *Fragaria × ananassa* "Strawberry Festival" ESTs that were highly polymorphic among 70 cultivars, with an average of 16 alleles per primer pair (Bassil et al., 2006).

SSR markers are very costly to develop but cross-species transferability of these markers can overcome this limitation. SSRs have shown high cross-transferability within *Fragaria* species (Gupta and Varshney, 2000). Several studies have shown a high level of transferability within the *Fragaria* genus (Ashley et al., 2003; Sargent et al., 2003; Lewers et al., 2005; Bassil et al., 2006; Davis et al., 2006; Monfort et al., 2006). The SSRs developed for *F. virginiana* gave amplification in *Fragaria × ananassa* and *F. chiloensis* (Ashley et al., 2003). Bassil et al. (2006) reported the cross-transferability of 32 SSR primer pairs developed from *Fragaria × ananassa* "Strawberry Festival" in *F. vesca*, *F. chiloensis* and *F. virginiana*. The marker gave 89% amplification in *F. vesca*, while it was 100% in *F. chiloensis* and *F. virginiana*. Thirty-one SSRs developed from *F. vesca* resulted in 77–100% amplification in other diploids and octoploid *Fragaria* species (Hadonou et al., 2004). Using 20 microsatellite primer pairs developed from *F. vesca*, 95% transferability was observed in *Fragaria × ananassa* (Cipriani and Testolin, 2004). In strawberry, studies on cross-transferability of SSR markers between octoploids and the diploids proved helpful in the studies of mapping and synteny studies (Rousseau-Gueutin et al., 2008).

8.2 GENETIC MAPPING

8.2.1 LINKAGE MAPPING

All molecular breeding efforts are aimed first at identifying linkage between DNA marker and important fruit traits. Linkage maps are utilized for identifying chromosomal regions containing genes associated with traits of interest. Linkage mapping is based on the principle of segregation of genes and markers in the progeny, via chromosome recombination (crossing over) during meiosis. Genes or markers located close together or tightly linked are transmitted together from parent to progeny more frequently than genes or markers placed distantly. Apart from propelling molecular breeding efforts, linkage maps come in handy for the dissection of complex traits and also in genome structure analysis. As linkage mapping helps in locating genes on their chromosomal positions through their association with molecular markers, they are important in efforts involving positional cloning of desired/known genes, mapping of QTL, and anchoring of contigs and scaffolds of physical maps.

The octoploid nature of cultivated strawberry sparked genomic and genetic research efforts in its diploid relatives and ancestors. Linkage analysis in *Fragaria* was first reported between an isozyme marker (SDH – shikimatedehydrogenase) and a morphological (yellow fruit colour) trait (1.1 cM*), using a cross between *F. vesca* cultivars "Yellow Wonder" and "Baron Solemacher" (Williamson et al., 1995). Another such report was published in the same year (Yu and Davis, 1995), using the same cultivars but a different isozyme locus (PGI-2 – phosphoglucoseisomerase 2) and morphological (non-runnering "*r*" locus) trait (18.9 cM). Thereafter, in 1997, the first *Fragaria* linkage map was constructed, again using the diploid *F. vesca* (Davis and Yu, 1997). This map was constructed in an F_2 population generated from a cross between "Baron Solemacher", a non-runnering European cultivar of *F. vesca* ssp. *Vesca*, and *F. vesca* ssp. *americana* W6, a wild runnering accession. The map was constructed using 75 RAPD markers, two previously described isozymes, the runnering locus, and an STS marker based on the alcohol dehydrogenase gene, and covered all the seven linkage groups with a map length of 445 cM.

The second diploid *Fragaria* map was based on the segregating F_2 population derived from an interspecific cross between *F. vesca* "815" and *F. bucharica* "601" (Sargent et al., 2004a). This map consisted of 78 markers, 68 out of which were SSRs, and it covered a map length of 448 cM. This was the first time that a SSR-based linkage map was constructed in strawberry. The map was constantly updated with more and more molecular markers (Sargent et al., 2004b, 2006a, 2007, 2008, 2011; Illa et al., 2011; Zorrilla-Fontanesi et al., 2011b). The map was enlarged to contain sequence tag sites derived from SSRs, gene-specific markers, RFLPs, SNPs and ESTs, covering all the seven linkage groups and spanning a genetic distance of 528.1 cM and is considered as the reference map for genus *Fragaria*.

A genetic map of octoploid strawberry (Lerceteau-Kohler et al., 2003) based on 515 AFLP markers was first constructed using a F_1 progeny of 119 individual plants of a two-way pseudo-test cross of "Capitola" ("CA75.121-101" × "Parker") and "CF1116" ["Pajaro" × ("Earliglow" × "Chandler")]. The markers mapped to 28 male and 30 female linkage groups, with the female map covering 1604 cM and the male map covering 1496 cM. This map was further extended and saturated by Rousseau-Gueutin et al. (2008). This extended map was based on 213 plants and consisted of the AFLP markers originally mapped, along with more AFLP markers, SCAR markers and SSRs. The map had 367 loci in 28 linkage groups in the female map with length of 2582 cM, and 440 markers, located in 26 linkage groups in the 2165 cM male map. The resulting map constituted the first comprehensive reference map of the octoploid strawberry.

Using this map, the researchers were able to identify four different *Fragaria* × *ananassa* linkage groups homoeologous to each of the seven *Fragaria* linkage groups in the *F. vesca* reference map. Another map, spanning 1541 cM was reported using 429 AFLPs in a population of 127 plants derived from a cross between "Tribute" × "Honeoye" (Weebadde et al., 2008). A linkage map was also constructed in *F. virginiana*. The resultant maternal and paternal maps had 33 and 32 linkage groups having 319 and 331 markers, respectively (Spigler et al., 2008, 2010).

One of the most comprehensive octoploid strawberry linkage maps was developed by Sargent and co-workers in 2012. It contained 490 transferrable SSR or gene-specific markers in 28 linkage groups and was based on a population of cross of "Redgauntlet" × "Hapil". This map covered 91% of *Fragaria* × *ananassa* genome with a length of 2140.3 cM and contained 549 mapped loci.

---

* Centimorgan (cM) was coined in honour of geneticist Thomas Hunt Morgan by his student Alfred Henry Sturtevant. A cM or map unit (m.u.) is a unit of genetic distance that represents a 1% probability of recombination in a single meiotic event. It depicts how likely two loci will segregate together during meiosis. For example if two genes are 1 cM apart, there is a 1% chance they will break apart during meiosis.

Markers and Genetic Mapping

Another octoploid map, based on "Holiday"×"Korona" population, containing 186 SSRs in 28 linkage groups with an estimated length of 2050 cM was constructed in 2014 (van Dijk et al., 2014). A group of researchers at the Research Institute of Horticulture in Skierniewice, Poland also developed a strawberry map using "Elsanta"×"Senga Sengana", mapping population. This map spanned 1450.4 cM in 30 linkage groups and consisted of 116 marker loci (Mohamed, 2014).

An integrated linkage map of *Fragaria×ananassa* was developed in 2013, using a total of 4474 SSR markers. Out of the total SSRs used, 3746 were *F. vesca* EST–SSRs, 603 were *Fragaria×ananassa* EST-derived and 125 were from *Fragaria×ananassa* transcriptome-derived. These markers were first mapped onto five parent-specific linkage maps, derived from three mapping populations: 02–19×"Sachinoka", "Kaorino"×"Akihime" and inbred lines "0212921"×"0212921", which were then assembled into an integrated linkage map. The constructed map was 2364.1 cM in length (Isobe et al., 2013).

Since all the linkage maps developed for strawberry were based on cross-transferable markers, it was possible to align maps of different mapping populations. One such study was carried out with a 241.6 cM linkage map of backcross progeny of a cross between *F. vesca* and *F. viridis*. When aligned to the reference map, it was observed that the marker order was conserved between the two maps (Nier et al., 2006). Conservation of marker order was also observed by Cipriani et al. (2006), between SSR markers mapped, in an intraspecific *F. vesca* population and the reference map. Generation of linkage maps in octoploid strawberry led to comparative studies between the diploid and octoploid maps. These studies revealed that the polyploid derivative did not have major chromosomal rearrangements when compared to the diploid ancestors (Rousseau-Gueutin et al., 2008; Sargent et al., 2009).

8.2.2 QTL MAPPING

QTL are chromosomal regions carrying genes governing a polygenic/quantitatively expressed trait. Mapping of QTL requires estimation and detection of linkage disequilibrium between the QTL and genetic markers. This requires both genotypic information and phenotypic values of the trait, measured for each individual of the mapping population. Identification of locations of important genes on the chromosomes is one of the major objectives of linkage map studies. A number of markers have also been linked to genes conferring particular traits in strawberry. Seven RAPD markers have been linked to the *Rpf1* locus conferring resistance to *Phytophthora fragariae* var. *Fragariae* (Haymes et al., 1997). SCAR markers linked to *Rca2*, a gene conferring resistance to *Colletotrichum acutatum* pathogenicity group-2 have been developed (Lerceteau-Kohler et al., 2005).

Fruit quality was one among the traits first selected for QTL analysis in strawberry. "Capitola" and "CF1116", genotypes that have many contrasting fruit traits were used to generate a mapping population of 213 plants. This population was used to map QTL governing 34 fruit quality traits (Lerceteau-Kohler et al., 2004). They observed the plants over a period of two years and detected 22 QTL, 8 on the female and 14 on the male map, responsible for 6.5–16% of the phenotypic variation for the observed traits. Using the same mapping as Lerceteau-Kohler, QTL governing *C. acutatum* and *Phytophthora cactorum* resistances were identified by Denoyes-Rothan et al. (2004). A total of 185 plants were screened for resistance and susceptibility reactions to *C. acutatum*, while the entire progeny was screened for reaction to *P. cactorum*. Five QTL each, for resistance to *C. acutatum*-pathogenicity group-1, with the effect ranging from 5.8–12.2% and 6.5–10.2% were identified, respectively.

Most of the strawberry varieties are short-day type, while few are everbearing or day-neutral. Day-neutrality is desired in strawberry to extend the harvest season. The everbearing trait has also been studied using molecular markers. Using a cross of "Honeoye"×"Tribute" for QTL analysis, Weebadde et al. (2008) studied flower blooming under long-day conditions. Based on multilocation trials, they detected eight QTL governing everbearing habit, with one of the loci responsible for nearly 36% of the phenotypic variation. They concluded that day-neutrality in plants was a polygenic character. In another report, an F_1 population from a cross of the same two plants as used by Weebadde and co-workers, was used to identify QTL for day-neutrality after multilocation testing of the plants (Castro et al., 2015). They identified a single marker locus, that is, ChFaM148-184 T, responsible for the total phenotypic variation for a number of weeks required for flowering at one location, and for 32.4% of the total phenotypic variation for runner production at another location. They were able to conclude that day-neutrality might be controlled either by a single major gene or a cluster of tightly linked genes.

The most comprehensive QTL analysis was carried out by Zorrilla-Fontanesi et al. (2011a). They analyzed 17 agronomic and fruit quality traits. This study holds importance, because they identified several candidate genes either underlying or within the confidence intervals of many QTL identified. *FaGaLUR*, a key gene in the d-galacturonic acid pathway in strawberry fruit was observed to be located within one of the linkage groups controlling L- ascorbic acid. Another gene, *FaMYOX* involved in myo-inositol biosynthesis was also located within a QTL. QTL for fruit firmness was co-located with an expansin gene *Fa-Exp2* in the same chromosome region. Similarly, a *SGR*-like gene involved in dismantling of photosynthetic chlorophyll–protein complexes during senescence was present in the same chromosome region as QTLs for yield and yield related traits. In total, 33 QTLs were identified for 14 of the traits studied. This report provided a number of useful markers which could be exploited in marker assisted selection (MAS) in breeding programmes, when developing populations for the respective traits. QTL for many other traits have also been identified in strawberry, some of which are enlisted in Table 8.1.

TABLE 8.1
QTLs Identified in Strawberry

Parents/Mapping Population	Traits Analyzed	No. of QTL Detected	Phenotypic Variation Covered	References
F. vesca, F. bucharica	6 flowering related traits	9	10.6–30.3%	Sargent et al. (2006)
F. ×ananassa cv. Capitola, *F. ×ananassa* breeding line CF1116	19 different fruit related traits	87	5–17%	Lerceteau-Kohler et al. (2012)
F. ×ananassa line 232, *F. ×ananassa* line 1392	17 agronomical and fruit quality traits	33	9.2–30.5%	Zorrilla-Fontanesi et al. (2011a)
F. ×ananassa cv. Redgauntlet, *F. ×ananassa* cv. Hapil	*Verticillium dahliae* resistance	11		Antanaviciute et al. (2015)
F. ×ananassa line 232, *F. ×ananassa* line 1392	48 volatile compounds	70		Zorrilla-Fontanesi et al. (2012)
F. ×ananassa cv. Capitola, *F. ×ananassa* breeding line CF1116	34 fruit quality traits	22	6.5–16.0%	Lerceteau-Kohler et al. (2004)
F. vesca, F. bucharica	Polyphenolic compounds	76		Urrutia et al. (2016)
F. ×ananassa cv Tochiotome, *F. ×annanasa* cv Itigotyukanbohonnou2gou (Nou2gou)	Strawberry anthracnose resistance	9		Iimura et al. (2013)
F. ×ananassa cv. Capitola, *F. ×ananassa* breeding line CF1116	Perpetual flowering and runnering	19		Gaston et al. (2013)

8.2.3 GENOME MAPPING

The allo-octoploid cultivated strawberry, *Fragaria × ananassa* (2n = 8x = 56), has a complex genetic architecture (Davis et al., 2007), with the genome size estimated to be around 708–720 Mb (1C = 708–720 Mb) based on flow-cytometric studies (Akiyama et al., 2001), where 1C is the DNA content of an unreplicated haploid nucleus. Cytological and genetic studies led to the prediction of three genome models for this fruit crop. Initially AAAABBCC genome model was predicted for the octoploid strawberry genome (Federova, 1946). Later another model, predicting the genome composition to be AAA'A'BBBB was given (Senanayake and Bringhurst, 1967), followed by a third model formula AAA'A'BBB'B' (Bringhurst, 1990). The model proposed by Bringhurst and co-workers is mostly accepted, and suggests the contribution of four distinct ancestral diploid genomes to form the genome of the octoploid species. Many studies using a variety of molecular and biochemical markers have also confirmed the disomic inheritance in octoploid strawberry, that is, the subgenomes acted as individual genomes and diploidized individually to result into the allo-octoploid genome, further supporting Bringhurst's genome model.

Fragaria × ananassa is genomically very complex due to the presence of eight sets of chromosomes, derived from four different ancestors. Thus, genomic efforts were first directed towards its simpler diploid ancestors. *F. vesca* (2n = 2x = 14), considered as one of the progenitors of cultivated strawberry has been selected as a genomic reference for the *Fragaria* genus (Shulaev et al., 2008). The reasons for selecting *F. vesca* as the reference genome included its short generation time, ease of vegetative and sexual propagation, ability to produce many seeds per plant, a short seed-to-seed cycle, a comparatively smaller genome size and ease of genetic transformation. The whole genome sequence of *F. vesca* ssp. *vesca* accession Hawaii 4 was deciphered in 2011 by an international collaborative effort, and was predicted to be ~240 Mb (Shulaev et al., 2011). The report highlighted the assembly of 209.8 Mb of the genome into 272 scaffolds and the prediction of 34,809 candidate genes. More than 6000 transposable elements were identified in the genome. When compared with *Arabidopsis*, orthologues and paralogues of a number of structural genes involved in biological processes such as flavour production, flowering and disease response were identified. This diploid progenitor species has been used for understanding the molecular genetics and genomics of the polyploid derivatives. Genome sequencing of polyploid crops is difficult as the homoeology of the subgenomes complicates the sequence assembly; while the high collinearity of the *F. ananassa* and *F. vesca* genomes further emphasized the exploitation of the diploid species to gain better insights into the more complex octoploid species (Rousseau-Gueutin et al., 2008).

Efforts to sequence the octoploid strawberry genome have been constantly made. One study was done to decipher its genome by comparing sequences of *Fragaria × ananassa* and four of its wild relatives, *F. iinumae*, *F. nipponica*, *F. nubicola* and *F. orientalis*. A virtual reference genome was generated by integrating genome sequences of homoeologous chromosomes. The assembled sequences of the wild species, and the heterozygous *Fragaria × ananassa* sequences were mapped onto the *Fragaria × ananassa* reference genome to elucidate the genome structure. Using the assembled sequences, 45,377 genes were predicted. The genome size of *Fragaria × ananassa* was estimated to be 692 Mb. This effort was carried out to understand the subgenomic structure of octoploid strawberry and was successful to a great extent (Hirakawa et al., 2014).

8.3 FUNCTIONAL GENOMICS

Functional genomics approaches are now being applied to strawberry to decipher roles of different genes, their encoded proteins and gene interactions. Different regulatory mechanisms are being elucidated at the transcriptional, translational and epigenetic levels. Functional genomics approaches are being used extensively for characterization of various fruit related traits (Table 8.2). The functions of many genes involved in fruit ripening, defence and fruit characteristics have been identified in strawberry.

TABLE 8.2
Genes Characterized in Strawberry

Gene	Species	Function	References
FaNES1	*F. ananassa*	Terpenoid production (linalool and nerolidol)	Aharoni et al. (2004)
FveLOM *FveWU*	*F. vesca*	Meristem regulator genes	Hollender et al. (2014)
Calchone synthase (*FaCHS*)	*Fragaria × ananassa*	Pigment formation, flavonoid biosynthesis	Hoffmann et al. (2006), Miyawaki et al. (2012),Ring et al. (2013)
Glycosyltransferase (*FaGT1*)	*Fragaria × ananassa* cv. Elsanta	Pigment formation, flavonoid biosynthesis	Griesser et al. (2008)
Eugenol synthase (*FaEGS*)	*Fragaria × ananassa*	Aroma, phenylpropene synthesis	Hoffmann et al. (2011)
Alcohol acyltransferase (*FaAAT2*)	*F. × ananassa* Duch. cv. Camarosa; *F. vesca; F. chiloensis; F. virginiana*	Aroma, fruit ester formation	Cumplido-Laso et al. (2012)
Sucrose transporter (*FaSUT1*)	*F. × ananassa*	Regulation of sucrose, ABA content and fruit ripening	Jia et al. (2013)
Rhamnogalacturonate lyase (*FaRGLyase1*)	*F. × ananassa*	Softening, pectin degradation	Molina-Hidalgo et al. (2013)
Cinnamyl alcohol dehydrogenase (*CAD*)	*Fragaria × ananassa* cv. Elsanta	Lignin biosynthesis, firmness	Yeh et al. (2014)
β–glucosidases (*FaBG3*)	*Fragaria × ananassa* cv. Albion	Defence to *Botrytis cinereal*	Li et al. (2013)
Mannose binding lectin (*FaMBL1*)	*Fragaria × ananassa* cv. Alba	Defence to *Colletotrichum acutatum*	Guidarelli et al. (2011)
Peroxidase (*POD27*)	*Fragaria × ananassa* cv. Elsanta	Lignin biosynthesis, firmness	Yeh et al. (2014)
Dihydroflavonol 4-reductase (*FaDFR*)	*F. × ananassa*	Pigment formation, flavonoid biosynthesis	Lin et al. (2013)
FaMADS9	*Fragaria × ananassa*	Fruit ripening	Seymour et al. (2011)
FaPE1	*Fragaria vesca*	Partial resistance to *Botrytis cinerea*	Osorio et al. (2011)

TIR (Toll/Interlukin I Receptor)-like resistance gene analogues (RGAs), which are resistance gene-like sequences and encode a nucleotide binding site (NBS) have been identified in strawberry. Apart from this, a number of resistance genes have been mapped and characterized, such as the *Rpf* gene for resistance to *Phytophthora fragariae*. Several ripening genes, for example pectatae lyase, dihydroflavonol 4-reductase, expansin, endopolygalacturonase and cystathionine γ-synthase have also been studied in strawberry fruits.

F. vesca has particularly gained importance as a reference for genomics studies in octoploid cultivated strawberry as well as the Rosaceae family. The use of *Fragaria* as a reference is mainly because of the similarity of coding regions of *Fragaria* genes to those of other Rosaceae species, and ease of transformation of Fragaria species. *Fragaria* species have the potential to be used for quickly testing the form and function of a gene, thus acting as a platform for further research in translational genomics (Folta and Davis, 2006).

With the dawn of the genomics era, rapid progress is being made in understanding the genetic basis of traits and their tagging with molecular markers. Future studies will be aimed at obtaining

Markers and Genetic Mapping

complete information on the gene networks governing fruit traits. There are many complex traits that are desirable to develop new strawberry varieties. Marker-based and genomics-assisted studies are the first steps to unravelling such complex traits. For developing better strawberry varieties, suitable to the current growing conditions, it is important to develop robust markers useful to breeders. More and more QTL need to be accurately identified, validated and positioned on the genome. Many research groups are developing genetic linkage maps for fruit quality attributes and resistance to different stresses. Reference maps for strawberry have also been established. These maps will be useful in identifying markers linked to traits of interest, and then further using them in MAS efforts in breeding programmes or in gene identification attempts.

The most important milestone in strawberry research will be the sequencing of the octoploid strawberry genome. The genome sequence will help in identifying key strawberry genes and their functions. An integrated approach to research is now being utilized to better understand the genetics of this crop. Technologies such as proteomics and metabolomics are being utilized to gain information on genetic control of different fruit quality characters. Strawberry omics research initiatives will lead to a better understanding of this crop and will assist in the creation of new cultivars, having stable production and better adaptation to the future environment.

REFERENCES

Aharoni, A., Giri, A.P., Verstappen, F.W.A., Bertea, C.M., Sevenier, R., Sun, Z., Jongsma, M.A., Schwab, W. and Bouwmeester, H.J. 2004. Gain and loss of fruit flavor compounds produced by wild and cultivated strawberry species. *The Plant Cell*, 16(11): 3110–3131.

Akiyama, Y., Yamamoto, Y., Ohmido, N., Oshima, M. and Fukui, K. 2001. Estimation of the nuclear DNA content of strawberries (*Fragaria* spp.) compared with *Arabidopsis thaliana* by using dual system flow cytometry. *Cytologia*, 66: 431–436.

Amaya, I. 2009. Impact of plant breeding on the genetic diversity of cultivated strawberry as revealed by expressed sequence tag-derived simple sequence repeat markers. *Journal of American Society of Horticulture Science*, 134: 337–347.

Antanaviciute, L., Surbanovski, N., Harrison, N., McLeary, K.J., Simpson, D.W., Wilson, F., Sargent, D.J. and Harrison, R.J. 2015. Mapping QTL associated with *Verticillium dahliae* resistance in the cultivated strawberry (*Fragaria* × *ananassa*). *Horticulture Research*, 2: 15009.

Arnau, G., Lallemand, J. and Bourgoin, M. 2001. Are AFLP markers the best alternative for cultivar identification? *Acta Horticulturae*, 546: 301–305.

Arnau, G., Lallemand, J. and Bourgoin, M. 2002. Fast and reliable strawberry cultivar identification using inter simple sequence repeat (ISSR) amplification. *Euphytica*, 129: 69–79.

Arulsekar, S., Bringhurst, R.S. and Voth, V. 1981. Inheritance of PGI and LAP isozymes in octoploid cultivated strawberries. *Journal of American Society of Horticulture Science*, 106: 679–683.

Ashley, M.V., Wilk, J.A., Styan, S.M.N., Craft, K.J., Jones, K.L., Feldheim, K.A., Lewers, K.S. and Ashman, T.L. 2003. High variability and disomic segregation of microsatellites in the octoploid *Fragaria virginiana* Mill. (Rosaceae). *Theoretical and Applied Genetics*, 107: 1201–1207.

Bassil, N. V., Gunn, M., Folta, K. M. and Lewers, K. S. 2006. Microsatellite markers for *Fragaria* from "Strawberry Festival" expressed sequence tags. *Molecular Ecology Notes*, 6: 473–476.

Bhat, K. V., Lakhanpaul, S., Chandel, K. P. S. and Jarret, R. L. 1997. Molecular markers for characterization and identification of genetic resources of perennial crops. Molecular genetic techniques for plant genetic resources. IPGRI report: pp. 107–117.

Bombarely, A., Merchante, C., Csukasi, F., Cruz-Rus, E., Caballero, J., Medina-Escobar, N., Blanco-Portales, R., Botella, M., Munoz-Blanco, J., Sanchez-Sevilla, J. and Valpuesta, V. 2010. Generation and analysis of ESTs from strawberry (*Fragaria* × *ananassa*) fruits and evaluation of their utility in genetic and molecular studies, *BMC Genomics*, 11: 503.

Bringhurst, R.S. 1990. Cytogenetics and evolution in American Fragaria. *HortScience*, 25: 879–881.

Brown, A.H.D. and Schoen, D.J. 1994. Optimal sampling strategies for core collections of plant genetic resources. In V. Loeschcke et al. (eds.). *Conservation Genetics*. Birkhuser Verlag, Basel, Switzerland: pp. 357–370.

Brunnings, A.M., Moyer, C., Peres, N. and Folta, K.M. 2010. Implementation of simple sequence repeat marker to genotype Florida strawberry varieties. *Euphytica*, 173: 63–75.

Carrasco, B., Garcés, M., Rojas, P., Saud, G., Herrera, R., Retamales, J.B. and Caligari, P.D.S. 2007. The Chilean strawberry [*Fragaria* chiloensis (L.) Duch.]: genetic diversity and structure. *Journal of American Society of Horticulture Science*, 132: 501–506.

Castro, P., Bushakra, J.M., Stewart, P., Weebadde, C.K., Wang, D., Hancock, J.F., Finn, J.J. and Lewers, K.S. 2015. Genetic mapping of day-neutrality in cultivated strawberry. *Molecular Breeding*, 35: 79.

Cipriani, G. and Testolin, R. 2004. Isolation and characterization of microsatellite loci in *Fragaria*. *Molecular Ecology Notes*, 4: 366–368.

Cipriani, G., Pinosa, F., Bonoli, M. and Faedi, W. 2006. A new set of microsatellite markers for *Fragaria* species and their application in linkage analysis. *Journal of Horticultural Science and Biotechnology*, 81: 668–675.

Congiu, L., Chicca, M., Cella, R., Rossi, R. and Bernacchia. G. 2000. The use of random amplified polymorphic DNA (RAPD) markers to identify strawberry varieties: a forensic application. *Molecular Ecology*, 9: 229–232.

Cumplido-Laso, G., Medina-Puche, L., Moyano, E., Hoffmann, T., Sinz, Q., Ring, L., Studart-Wittkowski, C., Caballero, J.L., Schwab, W., Muñoz-Blanco, J. and Blanco-Portales, R. 2012 The fruit ripening-related gene FaAAT2 encodes an acyl transferase involved in strawberry aroma biogenesis. *Journal of Experimental Botany*, 63: 4275–4290.

Dale, A. 1996. A key and vegetative descriptions of thirty-two common strawberry varieties grown in North America. *Advances in Strawberry Research*, 15: 1–12.

Dangl, G.S., Lee, E.W., Sim, S.T. and Golino, D.A. 2007. A new system for strawberry cultivar identification developed at Foundation Plant Services (FPS), University of California, Davis, using simple sequence repeat (SSR) primers. *Nass/Nasga Proceedings*: 118–121.

Darrow, G.M. 1966. *The Strawberry: History, Breeding, and Physiology*. 1st ed. Holt, Rinehart and Winston, New York, USA: p. 515.

Davis, T.M. and Yu, H. 1997. A linkage map of the diploid strawberry, *Fragaria vesca*. *Journal of Heredity*, 88: 215–221.

Davis, T.M., DiMeglio, L.M., Yang, R., Styan, S.M.N. and Lewers, K.S. 2006. Assessment of SSR marker transfer from the cultivated strawberry to diploid strawberry species: functionality, linkage group assignment, and use in diversity analysis. *Journal of American Society of Horticulture Science*, 131: 506–512.

Davis, T.M., Denoyes-Rothan, B. and Lerceteau-Kohler, E. 2007. Strawberry. In C. Kole (ed.). *Genome Mapping and Molecular Breeding in Plants IV: Fruits and Nuts*. Springer, Berlin, Germany: 189–206.

Debnath, S.C., Khanizadeh, S., Jamieson, A.R. and Kempler, C. 2008. Inter simple sequence repeat (ISSR) markers to assess genetic diversity and relatedness within strawberry genotypes. *Canadian Journal of Plant Sciences*, 88: 313–322.

Degani, C., Rowland, L.J., Hortynski, J. A. and Galletta, G.J. 1998. DNA fingerprinting of strawberry (*Fragaria* × *ananassa*) cultivars using randomly amplified polymorphic DNA (RAPD) markers. *Euphytica*, 102: 247–253.

Degani, C., Rowland, L.J., Saunders, J.A., Hokanson, S.C., Ogden, E.L., Golan-Goldhirsh, A. and Galletta, G.J. 2001. A comparison of genetic relationship measures in strawberry (Fragaria × *ananassa* Duch.) based on AFLPs, RAPDs, and pedigree data. *Euphytica*, 117: 1–12.

Denoyes-Rothan, B., Lerceteau-Kohler, E., Guerin, G., Bosseur, S., Bariac, J., Martin, E. and Roudeillac, P. 2004. QTL analysis for resistances to *Colletotrichum acutatum* and *Phytophthora cactorum* in octoploid strawberry (*Fragaria* × *ananassa*). *Acta Horticulturae*, 663: 147–151.

Dong, Q., Wang, X., Zhao, M., Song, C., Ge, A. and Wang, J. 2011. Development of EST-derived SSR markers and their application in strawberry genetic diversity analysis. *Scientia Agricultura Sinica*, 44: 3603–3612.

El-Denary, M.E., Ali, R.A.M. and Gomaa, S.E. 2016. Morphological markers for early selection in strawberry breeding programs. *Alexandria Science Exchange Journal*, 37(2): 104–114.

Faedi, W., Mourgues, F. and Rosati, C. 2002. Strawberry breeding and varieties: situation and perspectives. *Acta Horticulturae*, 567: 51–59.

Federova, N. J. 1946. Cross ability and phylogenetic relationships in the main European species of Fragaria. *Proceedings of the USSR Academy of Sciences*, 52: 545–547.

Flores, R., Rocha, B.G., Peters, J.A., Augustin. E. and Fortes, R.L.F. 2000. Isozymic analyses of somaclones from strawberry (*Fragaria* × *ananassa* Duch.) cv. *Vila Nova Ciencia Rural (Brazil)*, 30: 993–997.

Folta, K.M. and Davis, T.M. 2006. Strawberry genes and genomics. *Critical Reviews in Plant Sciences*, 25(5): 399–415.

Garcia, M.G., Ontivero, M., Ricci, J.C.D. and Castagnaro, A. 2002. Morphological traits and high resolution of RAPD markers for the identification of the main strawberry varieties cultivated in Argentina. *Plant Breeding*, 121: 76–80.

Gaston, A., Perrotte, J., Lerceteau-Kohler, E., Rousseau-Gueutin, M., Petit, A., Hernould, M., Rotan, C. and Denoyes, B. 2013. PFRU, a single dominant locus regulates the balance between sexual and asexual plant reproduction in cultivated strawberry. *Journal of Experimental Botany*, 64: 1837–1848.

Gidoni, D., Rom, M., Kunik, T., Zur, M., Izsak, E., Izhar, S. and Firon, N. 1994. Strawberry cultivar identification using Randomly Amplified Polymorphic DNA (RAPD) markers. *Plant Breeding*, 113: 339–342.

Gil-Ariza D.J., Amaya, I., Botella, M.A., Blanco, J.M., Caballero, J.L., Lopez-Aranda, J.M., Valpuesta, V. and Sanchez-Sevilla, J.F. 2006. EST-derived polymorphic microsatellites from cultivated strawberry (*Fragaria × ananassa*) are useful for diversity studies and varietal identification among *Fragaria* species. *Molecular Ecology Notes*, 6: 1195–1197.

Gil-Ariza, D.J., Amaya, I., Lopez-Aranda, J.M., Sanchez-Sevilla, J.F., Botella, M.A. and Valpuesta, V. 2009. Impact of plant breeding on the genetic diversity of cultivated strawberry as revealed by EST-derived SSR markers. *Journal of American Society of Horticulture Science*, 134: 337–347.

Govan, C.L., Simpson, D.W., Johnson, W., Tobutt, K.R. and Sargent, D.J. 2008. A reliable multiplexed microsatellite set for genotyping *Fragaria* and its use in a survey of 60 *F. × ananassa* cultivars. *Molecular Breeding*, 22: 649–661.

Graham, J., McNicol, R. and McNicol, J. 1996. A comparison of methods for the estimation of genetic diversity in strawberry cultivars. *Theoretical and Applied Genetics*, 93: 402–406.

Greco, I. 1993. Isoenzymic characterization of strawberry clones in South Italy. *Acta Horticulturae*, 345: 21–27.

Griesser, M., Hoffmann, T., Bellido, M.L., Rosati, C., Fink, B., Kurtzer, R., Aharoni, A., Munoz-Blanco, J. and Schwab, W. 2008. Redirection of flavonoid biosynthesis through the down-regulation of an anthocyanidin glucosyltransferase in ripening strawberry fruit. *Plant Physiology*, 146: 1528–1539.

Guidarelli, M., Carbone, F., Mourgues, F., Perrotta, G., Rosati, C., Bertolini, P., and Baraldi, E. (2011). *Colletotrichum acutatum* interactions with unripe and ripe strawberry fruits and differential responses at histological and transcriptional levels. *Plant Pathology*, 60: 685–697.

Gupta P.K. and Varshney R.K. 2000. The development and use of microsatellite markers for genetic analysis and plant breeding with emphasis on bread wheat. *Euphytica*, 113: 163–185.

Hadonou, A.M., Sargent, D., Wilson, F., James, C.M. and Simpson, D.W. 2004. Development of microsatellite markers in *Fragaria*, their use in genetic diversity analysis, and their potential for genetic linkage mapping. *Genome*, 47: 429–438.

Han, B.M., Zhao, M., Wang, J. and Yu H. 2013. Genetic relationships of strawberry cultivars by SSR analysis. *Journal of Plant Genetic Resources*, 14: 428–433.

Hancock, J.F. 1999. *Strawberries*. CABI Publishing, New York, USA: p. 237.

Hancock, J.F. and Bringhurst, R. S. 1979. Ecological differentiation in perennial octoploid species of *Fragaria*. *American Journal of Botany*, 66: 367–375.

Hancock, J.F., Callow, P.A. and Shaw, D.V. 1994. Randomly amplified polymorphic DNAs in the cultivated strawberry, *Fragaria × ananassa*. *Journal of American Society of Horticulture Science*, 119: 862–864.

Harrison, E.R., Luby, J.L., Furnier, G.R. and Hancock, J.F. 1997. Morphological and molecular variation among populations of octoploid *Fragaria virginiana* and *F. chiloensis* (Rosaceae) from North America. *American Journal of Botany*, 84: 612–620.

Harrison, E.R., Luby, J.L., Furnier, G.R. and Hancock, J.F. 2000. Differences in the apportionment of molecular and morphological variation in North American strawberry and the consequences for genetic resource management. *Genetic Resources and Crop Evolution*, 47: 647–657.

Haymes, K.M., Henken, B., Davis, T.M. and van de Weg, W.E. 1997. Identification of RAPD markers linked to a *Phytophthora fragariae* resistance gene (Rpf1) in the cultivated strawberry. *Theoretical and Applied Genetics*, 94: 1097–1101.

Haymes, K.M., Van de Weg, W.R., Arens, P., Maas, J.L., Vosman, B. and Den Nijs, A.P.M. 2000. Development of SCAR markers linked to a *Phytophthora fragariae* resistance gene and their assessment in European and North American strawberry genotypes. *Journal of American Society of Hortculture Science*, 125(3): 330–339.

Hirakawa, H., Shirasawa, K., Kosugi, S., Tashiro, K., Nakayama, S., Yamada, M., Kohara, M., et al. 2014. Dissection of the octoploid strawberry genome by deep sequencing of the genomes of fragaria species. *DNA Research*, 21: 169–181.

Hoffmann, T., Kalinowski, G. and Schwab, W. 2006. RNAi-induced silencing of gene expression in strawberry fruit (*Fragaria x ananassa*) by agroinfiltration: a rapid assay for gene function analysis. *Plant Journal*, 48: 818–826.

Hoffmann, T., Kurtzer, R., Skowranek, K., Kiessling, P., Fridman, E., Pichersky, E. and Schwab, W. 2011. Metabolic engineering in strawberry fruit uncovers a dormant biosynthetic pathway. *Metabolic Engineering*, 13: 527–531.

Hollender, C.A., Kang, C., Darwish, O., Geretz, A., Matthews, B.F., Slovin, J., Alkharouf, N. and Liu, Z. Floral transcriptomes in woodland strawberry uncover developing receptacle and anther gene networks. *Plant Physiology*, 165(3): 1062–1075.

Hosseini, S.A., Karami, E. and Talebi, R. 2013. Molecular variation of Iranian local and exotic strawberry (*Fragaria × ananassa* Duch.) varieties using SSR markers. *Biodiversity Journal*, 4(1): 243–252.

Hua, X.F., Zhong, F.L. and Lin, Y.Z. 2013. Genetic diversity analysis of strawberry accession with ISSR. *Fujian Journal of Agricultural Sciences*, 28: 232–236.

Hussein, T.S., Tawfik A.A. and Khalifa, M.A. 2008. Molecular identification and genetic relationships of six strawberry varieties using ISSR markers. *International Journal of Agriculture and Biology*, 10: 677–680.

Iimura, K., Tasaki, K., Nakazawa, Y. and Amagai, M. 2013. QTL analysis of strawberry anthracnose resistance. *Breeding Research*, 15(3): 90–97.

Illa, E., Sargent, D., Lopez Girona, E., Bushakra, J., Cestaro, A., Crowhurst, R., Pindo, M., et al. 2011. Comparative analysis of rosaceous genomes and the reconstruction of a putative ancestral genome for the family. *BMC Evolutionary Biology*, 11: 9.

Isobe, S. N., Hirakawa, H., Sato, S., Maeda, F., Ishikawa, M., Mori, T., Yamamoto, Y. et al. 2013. Construction of an integrated high density simple sequence repeat linkage map in cultivated strawberry (*Fragaria × ananassa*) and its applicability. *DNA Research*, 20(1): 79–92.

James, C.M, Wilson, F., Hadonou, A.M. and Tobutt, K.R. 2003. Isolation and characterization of polymorphic microsatellites in diploid strawberry (*Fragaria vesca* L.) for mapping, diversity studies and clone identification. *Molecular Ecology Notes*, 3: 171–173.

Jia, H., Wang, Y., Sun, M., Li, B., Han, Y., Zhao, Y., Li, X., Ding, N., Li, C., Ji, W. and Jia, W. 2013. Sucrose functions as a signal involved in the regulation of strawberry fruit development and ripening. *New Phytologist*, 198: 453–465.

Jomantiene, R., Davis, R.E., Dally, E.L., Maas J.L. and Postman, J.D. 1998. The distinctive morphology of *Fragaria multicipita* is due to phytoplasma. *HortScience*, 33(6): 1069–1072.

Keniry, A., Hopkins, C.J., Jewell, E., Morrison, B., Spangenberg, G.C., Edwards, D. and Batley, J. 2006. Identification and characterization of simple sequence repeat (SSR) markers from *Fragaria × ananassa* expressed sequences. *Molecular Ecology Notes*, 6: 319–322.

Kesseli, R.V., Paran, I. and Michelmore, R.W. 1994. Analysis of a detailed genetic linkage map of *Lactuca sativa* (lettuce) constructed from RFLP and RAPD markers. *Genetics*, 136: 1435–1446.

Kunihisa, M., Fukino, N. and Matsumoto, S. 2003. Development of cleavage amplified polymorphic sequence (CAPS) markers for identification of strawberry cultivars. *Euphytica*, 134: 209–215.

Kunihisa, M., Fukino, N. and Matsumoto, S. 2004. Development of PCR-RFLP marker on strawberry and the identification of cultivars and their progeny. *Acta Horticulturae*, 708: 517–521.

Landry, B.S., Rongqi, L. and Khanizadeh, S. 1997. A cladistic approach and RAPD markers to characterize 75 strawberry cultivars and breeding lines. *Advances in Strawberry Research*, 16: 28–33.

Lerceteau-Kohler, E., Guerin, G., Laigret, F. and Denoyes-Rothan, B. 2003. Characterization of mixed disomic and polysomic inheritance in the octoploid strawberry (*Fragaria × ananassa*) using AFLP mapping. *Theoretical and Applied Genetics*, 107: 619–628.

Lerceteau-Kohler, E., Moing, A., Guerin, G., Renaud, C., Courlit, S., Camy, D., Praud, K., Parisy, V., Bellec, F., Maucourt, M., Rolin, D., Roudeillac, P. and Denoyes-Rothan, B. 2004. QTL analysis for fruit quality traits in octoploid strawberry (*Fragaria × ananassa*). *Acta Horticulturae*, 663: 331–335.

Lerceteau-Kohler, E., Guerin, G. and Denoyes-Rothan, B. 2005. Identification of SCAR markers linked to *Rca2* anthracnose resistance gene and their assessment in strawberry germplasm. *Theoretical and Applied Genetics*, 111: 862–870.

Lerceteau-Kohler, E., Moing, A., Guerin, G., Renaud, C., Petit, A., Rothan, C. and Denoyes, B. 2012. Genetic dissection of fruit quality traits in the octoploid cultivated strawberry highlights the role of homoeo-QTL in their control. *Theoretical and Applied Genetics*, 124(6): 1059–1077.

Lewers, K.S., Styan, S.M.N., Hokanson, S.C. and Bassil, N.V. 2005. Strawberry GenBank-derived and genomic simple sequence repeat (SSR) markers and their utility with strawberry, blackberry, and red and black raspberry. *Journal of American Society of Horticulture Science*, 130: 102–115.

Li, Q., Ji, K., Sun, Y., Luo, H., Wang, H. and Leng, P. 2013. The role of FaBG$_3$ in fruit ripening and B. cinerea fungal infection of strawberry. *Plant Journal*, 76(1): 24–35.

Lin, X., Xiao, M., Luo, Y., Wang, J., and Wang, H. 2013. The effect of RNAi-induced silencing of FaDFR on anthocyanin metabolism in strawberry (*Fragaria × ananassa*) fruit. *Scientia Horticulturae*, 160: 123–128.

Markers and Genetic Mapping

Manbari, Y., Karami, E., Etminan, A. and Talebi, R. 2012. Characterization of genetic variation between local and exotic Iranian strawberry (*Fragaria × ananassa* Duch.) cultivars using amplified fragment length polymorphism (AFLP) markers. *International Journal of Biosciences*, 12(2): 159–167.

McDermott, J.M., Brandle, U., Dutly. F., Haemmerli. U.A., Keller, S., Muller, K.E. and Wolf, M.S. 1994. Genetic variation in powdery mildew of barley: development of RAPD, SCAR and VNTR markers. *Phytopathology*, 84: 1316–1321.

Miyawaki, K., Fukuoka, S., Kadomura, Y., Hamaoka, H., Mito, T., Ohuchi, H., Schwab, W. and Noji, S. 2012. Establishment of a novel system to elucidate the mechanisms underlying light-induced ripening of strawberry fruit with an Agrobacterium-mediated RNAi technique. *Plant Biotechnology*, 29: 271–277.

Mohamed, A.M.A. 2014. The genetic map of strawberry (*Fragaria × ananassa*) based on "Elsanta" × "Senga Sengana" mapping population. Ph.D. dissertation, Research Institute of Horticulture, Skierniewice, Poland: p. 108.

Molina-Hidalgo, F.J., Franco, A.R., Villatoro, C., Medina-Puche, L., Mercado, J.A., Hidalgo, M.A., Monfort, A., Caballero, J.L., Munoz-Blanco, J. and Blanco-Portales, R. 2013. The strawberry (Fragaria x ananassa) fruit specific rhamnogalacturonate lyase 1 (FaRGLyase1) gene encodes an enzyme involved in the degradation of cell walls middle lamellae. *Journal of Experimental Botany*, 64: 1471–1483.

Monfort, A., Vilanova, S., Davis, T.M. and Arus, P. 2006. A new set of polymorphic simple sequence repeat (SSR) markers from a wild strawberry (*Fragaria vesca*) are transferable to other diploid *Fragaria* species and to *Fragaria × ananassa*. *Molecular Ecology Notes*, 6: 197–200.

Mullis K.B. and Faloona F. 1987. Specific synthesis of DNA *in vitro* via polymerase chain reaction. *Methods in Enzymology*, 155: 350–355.

Nehra, N.S., Kartha, K.K. and Stushnoff, C. 1991. Isozymes as markers for identification of tissue culture and greenhouse-grown strawberry cultivars. *Canadian Journal of Plant Sciences*, 71: 1195–1201.

Nielsen, J.A. and Lovell, P.H. 2000. Value of morphological characters for cultivar identification in strawberry *(Fragaria × ananassa)*. *New Zealand Journal of Crop and Horticultural Science*, 28: 89–96.

Nier, S., Simpson, D.W., Tobutt, K.R. and Sargent, D.J. 2006. Construction of a genetic linkage map of an interspecific diploid *Fragaria* BC1 mapping population (*F. vesca* 815 × [*F. vesca* 815 × *F. viridis* 903]) and its comparison to the *Fragaria* reference map (FVxFN). *Journal of Horticultural Science and Biotechnology*, 81: 645–650.

Njuguna, W. 2010. A reduced molecular characterization set for *Fragaria* L. (strawberry), Chapter 2. In: Development and use of molecular tools in *Fragaria* L. PhD Dissertation, Oregon State University, Corvallis, Oregon, USA: pp. 66–127.

Njuguna, W., Hummer, K.E., Richards, C.M., Davis, T.M. and Bassil, N.V. 2011. Genetic diversity of diploid Japanese strawberry species based on microsatellite markers. *Genetic Resources and Crop Evolution*, 58: 1187–1198.

Osorio, S., Aureliano Bombarely, A., Giavalisco, P., Usadel, B., Stephens, C., Araguez, I., Medina-Escobar, N., Botella, M.A., Fernie, A.R. and Valpuesta, V. 2011. Demethylation of oligogalacturonides by FaPE1 in the fruits of the wild strawberry *Fragaria vesca* triggers metabolic and transcriptional changes associated with defence and development of the fruit. *Journal of Experimental Botany*, 62(8): 2855–2873.

Paran, I. and Michelmore, R.W. 1993. Development of reliable PCR-based markers linked to downy mildew resistance genes in lettuce. *Theoretical Applied Genetics*, 85: 985–999.

Perry, M.D., Davey, M.R., Power, J.B., Lowe, K.C., Bligh, H.F.J., Roach, P.S. and Jones, C. 1998. DNA isolation and AFLPTM genetic fingerprinting of shape *Theobroma cacao* (L.). *Plant Molecular Biology Reporter*, 11: 45–59.

Porebski, S. and Catling, P.M. 1998. RAPD analysis of the relationship of North and South American subspecies of *Fragaria* chiloensis. *Canadian Journal of Botany*, 76: 1812–1817.

Rafalski, J., Morgante, M., Powell, J. and Tinggey, S. 1996. Generating and using DNA markers in plants. In: Bruce Birren and Eric Lai (eds.). *Analysis of Non-Mammalian Genomes: A Practical Guide*. Academic Press, Boca Raton, Florida: pp. 75–134.

Ring, L., Yeh, S.Y., Hucherig, S., Hoffmann, T., Blanco-Portales, R., Fouche, M., Villatoro, C., Denoyes, B., Monfort, A., Caballero, J.L., Munoz-Blanco, J., Gershenson, J. and Schwab, W. 2013. Metabolic interaction between anthocyanin and lignin biosynthesis is associated with peroxidase FaPRX27 in strawberry fruit. *Plant Physiology*, 163: 43–60.

Rousseau-Gueutin, M., Lerceteau-Kohler, E., Barrot, L., Sargent, D.J., Monfort, A., Simpson, D., Arus, P., Guerin, G. and Denoyes-Rothan, B. 2008. Comparative genetic mapping between octoploid and diploid Fragaria species reveals a high level of collinearity between their genomes and the essentially disomic behavior of the cultivated octoploid strawberry. *Genetics*, 179: 2045–2060.

Rugienius, R., Siksnianas, T., Stanys, V., Gelvonauskienene, D. and Bedokas, V. 2006. Use of RAPD and SCAR markers for identification of genotypes carrying red stele (*Phytophtora fragariae*) resistance gene *Rpf1*. *Agronomy Research*, 4: 335–339.

Rugienius, R., Siksnianiene, J.B., Frercks, B., Staniene, G., Stepulaitienė, I., Haimi, P. and Stanys, V. 2015. Characterization of strawberry (*Fragaria × ananassa* Duch.) cultivars and hybrid clones using SSR and AFLP markers. *Zemdirbyste Agriculture*, 102(2): 177–184.

Sargent, D.J., Hadonou, M. and Simpson, D.W. 2003. Development and characterization of polymorphic microsatellite markers from *Fragaria viridis*, a wild diploid strawberry. *Molecular Ecology Notes*, 3: 550–552.

Sargent, D.J., Geibel, M., Hawkins, J.A., Wilkinson, M.J., Battey, N.H. and Simpson, D.W. 2004a. Quantitative and qualitative differences in morphological traits revealed between diploid *Fragaria* species. *Annals of Botany*, 94: 787–796.

Sargent, D.J., Davis, T.M., Tobutt, K.R., Wilkinson, M.J., Battey, N.H. and Simpson, D.W. 2004b. A genetic linkage map of microsatellite, gene specific and morphological markers in diploid *Fragaria*. *Theoretical and Applied Genetics*, 109: 1385–1391.

Sargent, D.J., Clarke, J., Simpson, D.W., Tobutt, K.R., Arus, P., Monfort, A., Vilanova, S., Denoyes-Rothan, B., Rousseau, M., Folta, K.M., Bassil, N.V. and Battey, N.H. 2006. An enhanced microsatellite map of diploid *Fragaria*. *Theoretical and Applied Genetics*, 112: 1349–1359.

Sargent, D.J., Rys, A., Nier, S., Simpson, D.W. and Tobutt, K.R. 2007. The development and mapping of functional markers in *Fragaria* and their transferability and potential for mapping in other genera. *Theoretical and Applied Genetics*, 114: 373–384.

Sargent, D.J., Cipriani, G., Vilanova, S., Gil-Ariza, D., Arus, P., Simpson, D.W., Tobutt, K.R. and Monfort, A. 2008. The development of a bin mapping population and the selective mapping of 103 markers in the diploid *Fragaria* reference map. *Genome*, 51: 120–127.

Sargent, D.J., Fernandez-Fernandez, F., Ruiz-Roja, J.J., Sutherland, B.G., Passey, A., Whitehouse, A.B. and Simpson, D.W. 2009. A genetic linkage map of cultivated strawberry (*Fragaria* x *ananassa*) and its comparison to the diploid *Fragaria* reference map. *Molecular Breeding*, 24: 293–300.

Sargent, D.J., Kuchta, P., Lopez Girona, E., Zhang, H., Davis, T.M., Celton, J.M., Marchese, A., Korbin, M., Folta, K.M., Shulaev, V. and Simpson, D.W. 2011. Simple sequence repeat marker development and mapping targeted to previously unmapped regions of the strawberry genome sequence. *Plant Genome*, 4: 165–177.

Sargent, D.J., Passey, T., Šurbanovski, N., Lopez Girona, E., Kuchta, P., Davik, J., Harrison, R., Passey, A., Whitehouse, A.B. and Simpson, D.W. 2012. A microsatellite linkage map for the cultivated strawberry (*Fragaria × ananassa*) suggests extensive regions of homozygosity in the genome that may have resulted from breeding and selection. *Theoretical and Applied Genetics*, 124: 1229–1240.

Senanayake, Y.D.A. and Bringhurst, R.S. 1967. Origin of the Fragaria polyploids – I: Cytological evidence. *American Journal of Botany*, 54: 221–228.

Seymour, G.B., Ryder, C.D., Cevik, V., Hammond, J.P., Popovich, A., King, G.J., Vrebalov, J., Giovannoni, J.J. and Manning, K. 2011. A *SEPALLATA* gene is involved in the development and ripening of strawberry (*Fragaria×ananassa* Duch.) fruit, a non-climacteric tissue. *Journal of Experimental Botany*, 62(3): 1179–1188.

Shimomura, K. and Hirashima, K. 2006. Development and characterization of simple sequence repeats (SSR) as markers to identify strawberry cultivars (*Fragaria × ananassa* Duch.). *Journal of Japanese Society of Horticulture Science*, 75: 399–402.

Shulaev, V., Korban, S.S., Sosinski, B., Abbott, A.G., Aldwinckle, H.S., Folta, K.M., Iezzoni, A., Main, D., Arus, P., Dandekar, A.M., Lewers, K., Brown, S.K., Davis, T.M., Gardiner, S.E., Potter, D. and Veilleux, R.E. 2008. Multiple models for Rosaceae genomics. *Plant Physiology*, 147: 985–1003.

Shulaev, V., Sargent, D.J., Crowhurst, R.N., Mockler, T.C., Folkerts, O., Delcher, A.L., Jaiswal, P. et al. 2011. The genome of woodland strawberry (*Fragaria vesca*). *Nature Genetics*, 43: 109–116.

Spigler, R.B., Lewers, K.S., Main, D.S. and Ashman, T.L. 2008. Genetic mapping of sex determination in a wild strawberry, *Fragaria virginiana*, reveals earliest form of sex chromosome. *Heredity*, 101: 507–517.

Spigler, R.B., Lewers, K.S., Johnson, A.L. and Ashman, T.L. 2010. Comparative mapping reveals autosomal origin of sex chromosome in octoploid *Fragaria virginiana*. *Journal of Heredity*, 101(1): 107–117.

Thongthieng, T. and Smitamana, P. 2003. Genetic relationship in strawberry cultivars and their progenies analyzed by Isozyme and RAPD. *Science ASIA*, 29: 1–5.

Tyrka, M., Dziadczyk, P. and Hortynski, J.A. 2002. Simplified AFLP procedure as a tool for identification of strawberry cultivars and advanced breeding lines. *Euphytica*, 125: 273–280.

Markers and Genetic Mapping

Ulrich, D., Komes, D., Olbricht, K. and Hoberg, E. 2007. Diversity of aroma patterns in wild and cultivated *Fragaria* accessions. *Genetic Resources and Crop Evolution*, 54: 1185–1196.

Urrutia, M., Schwab, W., Hoffmann, T. and Monfort, A. 2016. Genetic dissection of the (poly)phenol profile of diploid strawberry (*Fragaria vesca*) fruits using a NIL collection. *Plant Science*, 242: 151–168.

van Dijk, T., Pagliarani, G., Pikunova, A., Noordij, Y., Yilmaz-Temel, H., Meulenbroek, B., Visser, R.G.F. and van de Weg, E. 2014. Genomic rearrangements and signatures of breeding in the allooctoploid strawberry as revealed through an allele dose based SSR linkage map. *BMC Plant Biology*, 14: 55.

Weebadde, C.K., Wang, D., Finn, C.E., Lewers, K.S., Luby, J.J., Bushakra, J., Sjulin, T.M. and Hancock, J.F. 2008. Using a linkage mapping approach to identify QTL for day-neutrality in the octoploid strawberry. *Plant Breeding*, 127: 94–101.

Weising K., Nybom, H., Wolff K. and Meyer W. 1995. *DNA Fingerprinting in Plants and Fungi*. CRC Press Inc., Boca Raton, Florida, USA: p. 472.

Williams, J.G.K., Kubelik, A.R., Livak, K.J., Rafalski, J.A. and Tingey, S.V. 1990. DNA polymorphisms amplified by arbitrary primers are useful as genetic markers. *Nucleic Acids Research*, 18: 6531–6535.

Williamson, S.C., Yu, H. and Davis, T.M. 1995. Shikimate dehydrogenase allozymes: inheritance and close linkage to fruit color in diploid strawberry. *Journal of Heredity*, 86: 74–76.

Yao, M., Chen, L., Wang, X., Zhao, L. and Yang, Y. 2007. Genetic diversity and relationship of clonal tea cultivars in China revealed by ISSR markers. *Acta Agronomica Sinica*, 33: 598–604.

Yeh, S.Y., Huang, F.C., Hoffmann, T., Mayershofer, M. and Schwab, W. 2014. FaPOD27 functions in the metabolism of polyphenols in strawberry fruit (*Fragaria* sp.). *Frontiers in Plant Science*, 5: 1–18.

Yoon, M.Y., Moe, K.T., Kim, D.Y., Rho, R., Kim, S., Kim, K.T., Won, M.K., Chung, J.W. and Park, Y.J. 2012. Genetic diversity and population structure analysis of strawberry (*Fragaria* x *ananassa* Duch.) using SSR markers. *Electronic Journal of Biotechnology*, 15(2): 1–5.

Yu, H. and Davis, T.M. 1995. Genetic linkage between runnering and phosphoglucoisomerase allozymes, and systematic distortion of monogenic segregation ratios in diploid strawberry. *Journal of the American Society for Horticultural Science*, 120: 687–690.

Zebrowska, J.I. and Tyrka, M. 2003. The use of RAPD markers for strawberry identification and genetic diversity studies. *Journal of Food Agriculture and Environment*, 1(1): 115–117.

Zhang, Y., Feng, Z., Li, T., Dong, J., Wang, G., Zhang, K. and Han, Z. 2006. Genetic relationships of strawberry cultivars by AFLP analysis. *Acta Horticulturae Sinica*, 33: 1199–1202.

Zietkiewicz, E., Rafalski, A and Labuda, D. 1994. Genome fingerprinting by simple sequence repeat (SSR)-anchored polymerase chain reaction amplification. *Genomics*, 20: 176–183.

Zorrilla-Fontanesi, Y., Cabeza, A., Torres, A., Botella, M., Valpuesta, V., Monfort, A., Sanchez-Sevilla, J. and Amaya, I. 2010. Development and bin mapping of strawberry genic ssrs in diploid *Fragaria* and their transferability across the rosoideae subfamily. *Molecular Breeding*, 27: 137–156.

Zorrilla-Fontanesi, Y., Rambla, J., Cabeza, A., Medina, J.J., Sánchez-Sevilla, J.F., Valpuesta, V., Botella, M. A., Granell, A. and Amaya, I. 2012. Genetic analysis of strawberry fruit aroma and identification of o-methyltransferase FaOMT as the locus controlling natural variation in mesifurane content. Plant Physiology, 159: 851–870.

Zorrilla-Fontanesi, Y., Cabeza, A., Dominguez, P., Medina, J.J., Valpuesta, V., Denoyes-Rothan, B., Sanchez-Sevilla, J.F. and Amaya, I. 2011a. Quantitative trait loci and underlying candidate genes controlling agronomical and fruit quality traits in octoploid strawberry (*Fragaria* x *ananassa*). *Theoretical and Applied Genetics*, 123: 755–778.

Zorrilla-Fontanesi, Y., Cabeza, A., Torres, A.M., Botella, M.A., Valpuesta, V., Monfort, A., Sanchez-Sevilla, J.F. Amaya, I. 2011b. Development and bin mapping of strawberry genic-SSRs in diploid *Fragaria* and their transferability across the Rosoideae subfamily. *Molecular Breeding*, 27: 137–156.

9 Climatic Requirements

R. Yamdagni and A.D. Sharma

CONTENTS

9.1 Temperature 161
9.2 Photoperiod.. 163
9.3 Photoperiod and Temperature... 164
9.4 Light... 164
9.5 Ozone Exposure... 164
References.. 165

Most perennial fruit crops are versatile and have varying degrees of adaptation to different climatic parameters. Most of the fruit crops require specific climatic zones (tropical, subtropical or temperate) to grow, but strawberry has adapted for its cultivation in areas extending from the equator to the Arctic Circle. Climatic adaptability for optimum growth and fruit yield is the most important aspect in making the crop economically viable. Strawberry thrives best in a temperate climate. However, countries such as India where temperate areas are rainfed and have undulating topography and poor soil fertility, strawberries have shown remarkable adaptability for cultivation in subtropical areas and acceptability of the crop by growers as well as consumers due to higher yield and advanced availability of fruit (December–March) as compared to the fruits of temperate areas (April–June). The adaptation factors (Sherman and Beckman, 2003; Maltoni et al., 2006) affecting growth (such as plant growth, flower bud differentiation, flowering, fruit growth and fruit ripening), yield and fruit quality (such as flesh firmness, skin resistance, TSS, titratable acidity, skin colour, sugars and organic acids) are light (duration or photoperiod, intensity and quality), temperature (such as chilling and heat units), soil (such as type, and temperature), water tolerance (quality and quantity) and environment (viz., gaseous composition). The environmental factors influence plant responses through the production of endogenous hormones and biochemical constituents inside the plants (Hughes et al., 1969; Atkinson et al., 2005).

Strawberry is a short-day plant, which requires exposure to about 10 days of less than eight hours' sunshine for flower bud differentiation. In typical temperate areas (extreme cool areas), strawberry plants remain dormant during winter, and low temperatures are required to break their dormancy. With the advent of spring, the plants resume growth and begin flowering. The plants of varieties grown in subtropical climate start to grow after planting (October) and yield fruits from December–March. The role of climatic factors in strawberry growing is discussed in the following subheadings.

9.1 TEMPERATURE

A high growth rate of strawberries is maintained at day temperatures of 22–23°C. Root temperatures of 18–20°C tend to increase the growth of aerial parts of plants, while the top-root ratio increases with high root temperature (Shoemaker, 1954). The optimum day/night temperatures for leaf and petiole growth are 25/12°C, while for roots and fruits they are 18/12°C. At higher day and night temperatures (25/22 and 30/22°C), leaves, petioles and crowns have higher amounts of fructose and myoinositol, whereas at low temperatures (18/12 and 25/12°C), greater amounts of sucrose are accumulated. Plants grown at 25/12°C have the maximum starch and total carbohydrate contents. Cooler day/night temperatures (18/12°C) shift biomass from leaves to roots (Wang and Camp, 2000). The biosynthesis of volatile compounds in fruits of strawberry cv. Camarosa has been observed to be enhanced at 25/15°C day/night temperature, while maximum sucrose accumulation takes place at 25/15°C (Sanz et al., 2002). Plants grown at higher day/night temperatures (30/22°C) produce fruits with high phenolic content and antioxidant capacity (Wang and Zheng, 2001; Wang et al., 2003). The longest duration of fruit maturity, higher amino acids (asparagine, glutamic acid and alanine) and ascorbic acid contents have been observed at 23/10°C day/night temperatures, while at 23/30°C, shortest fruit maturity, low sugar-acid ratio and high levels of sugar and ellagic acid in strawberries have been recorded (Matsuzoe et al., 2006).

Higher temperatures affect the photosynthetic process through the modulation of enzyme activity as well as the electron transport chain (Sage and Kubien, 2007). High temperatures on the fruit surface due to prolonged exposure to sunlight hasten ripening and other associated events. The relationship between high temperature and the rate of ripening could be argued as a factor in reduced duration of crop cycle. Due to the relationship between yield and temperature, and between yield and solar radiation, climate change scenarios have been expected to result in reductions in duration of crop cycle (Palencia et al., 2013).

High temperature (40°C day/35°C night) has been observed more detrimental to photosynthesis and productivity than the moderate or low temperature (30/25 or 20/15°C). Net CO_2 assimilation rate (A) is markedly reduced at high temperature. Maintenance of higher A at higher temperature by an individual genotype indicates its tolerance to heat stress (Kadir et al., 2006). Exposure of strawberry plants to high temperature (42°C for four hours) tends to decrease most of the protein content, but a few new proteins are synthesized in response to heat stress called as heat shock proteins (HSP). The effect of heat stress on synthesis of HSP in strawberry plants varies with plant organ and cultivars (Ledesma et al., 2004). The studies on physiological changes in strawberry plants under heat stress indicate the significant changes in total protein and DNA contents by the type of heat stress (gradual heat stress or GHS and shock heat stress or SHT). The plants exposed to GHS exhibit significant increase in heat stress tolerance (HST) compared with the plants exposed to shock heat stress (SHS) (LT50 of 41.5 and 39°C, respectively). Consequently, gradual heat stress increases HST in strawberry leaves, which may be associated with the accumulation of several heat stable proteins in GHS plants (Gulen and Eris, 2003). In strawberry cv. Camarosa plants, the activity of peroxidase enzyme has been observed very high in response to high temperature treatment with a decrease in total protein content (Gulen and Eris, 2004). Heating treatment as a permeabilization method for plant cells is examined on the release of anthocyanin pigment from strawberry cells. The release of anthocyanin from viable cells probably depends on the functioning of the heat shock protein. After heating, the treatment with calcium has been observed to restore viability (Takeda et al., 2003). The effect of heat stress (root browning) in strawberry can be reduced by inoculating plants with vesicular arbuscular mycorrhizae (*Glomus mosseae* and *Glomus aggregatum*) with a greater number of leaves, leaf area, crown diameter and leaf and root dry weights (Matsubara et al., 2004).

A higher incidence of fruit deformation is the result of strawberry growing under higher temperature. As a general consideration, increasing air temperature/relative humidity seem to enhance the vulnerability of the strawberry to fruit malformation. The environmental factors and the fibrovascular system are involved in the onset of fruit deformity indicating the need for better control of growing conditions, particularly for susceptible cultivars such as "Capri" (Massetani et al., 2016). The susceptibility to fruit malformation changes in relation to genetic character of the cultivars (Carew et al., 2003), and can be magnified as a consequence of environmental and growing factors (Albregts and Howard, 1982; Lieten, 2002; Ariza et al., 2011b). The fruit deformity takes place due to loss of symmetry of the receptacle, which is often limited to the tip of the fruit (Ariza et al., 2011a). Poor pollen development and viability due to high temperature stress (30°C or higher) also tends to enhance the deformity in the strawberries (Song et al., 1999).

In cold climates, frost damage and winter injury are major constraints for production and maintenance of strawberry plantings (Mitra, 1991). Frost damage is of primary concern during spring as growers take advantage of cultural methods to produce an early crop (Ourecky and Reich, 1976). Yannai (1988) specified the level of temperatures for plant damage (–6°C), plant death (–12°C) and

Climatic Requirements 163

flower damage (–2°C and 40–50°C) for strawberries. There are four ways for damaging the reproductive parts of strawberry plants, viz., complete killing of whole primordium, variable damage within the flower truss, damage to the flower parts or injury to part of the vascular tissues of the peduncle. Varieties suitable to grow in subtropical areas are more prone to frost injury due to peak flowering during December–January.

The abnormalities of flowers during winter may also be associated with late lifting of the plants for planting. In most frost sensitive cultivars, a non-typical second season (August) fruit bearing phenomenon can be seen (Shokaeva, 2002). The winter hardiness of strawberries can be evaluated by testing differences in their ability to survive lengthy exposure to sublethal low temperatures or by applying conventional freezing tests to measure cold hardiness in different genotypes after field acclimatization (Linden et al., 2002). The stable or lower average leaf area, predominance of young leaves and higher amounts of chlorophyll and carotenoids during September and October than in early June are indicative of winter hardiness in strawberry plants (Rugienius and Brazaityte, 2001). In model plant systems, researchers have identified genes that play a role in tolerance to freezing by identifying proteins regulated in response to cold temperature. In two transgenic lines of strawberry cv. Honeoye, the temperature at which 50% electrolyte leakage occurred was –8.2 and –10.3°C, respectively. These values for tolerance to freezing were significantly greater than the value for the wild-type Honeoye leaf discs of –6.4°C (Owens et al., 2003). The observations on the influence of four combinations of day/night temperature (18/12, 25/12, 25/22 and 30/22°C) during growth have revealed significant changes in the contents of galactolipids (monogalactosyldiacylglycerol and digalactosyldiacylglycerol) and phospholipids (phosphatidylcholine, phosphatidylethanolamine, phosphatidylinositol, phosphatidylserine, and phosphatidylglycerol) and unsaturation of their fatty acids in the cool day/night growth temperature. Increasing temperatures during day/night has been observed to decrease MGDG/DGDG ratios. The shifts in saturation and composition of fatty acids observed with strawberry may be an adaptation response of plants to the temperature changes (Wang and Lin, 2006).

A small machine having propane gas cylinder with a ventilator, developed in Belgium, can protect 0.3–0.5 ha crop against frost. In a two-hectare field of Elsanta strawberry, the heated plot showed temperature of at least 2°C as compared to –2.7°C in non-heated plot, thus indicating less circulation of the warm air (Deckers and Schoofs, 2006) in the later one. In Russia, the shelter plants (oats, barley and mustard) are grown between the strawberry rows which fall (due to killing off the tops by first frost) and cover the strawberry plants at the end of the growing season like a mulch, and help to retain a protective snow cover throughout the winter with significant reduction of weed growth (Gurin, 2000)

9.2 PHOTOPERIOD

Strawberry cultivars are commonly classified on the basis of requirement of photoperiod for flower initiation. June-bearers are facultative short-day plants (Darrow, 1936), and initiate flowers during late summer and autumn in northern climates. Everbearers are long-day plants, which initiate flowers during long days under favourable temperatures (Darrow and Waldo, 1934), whereas day-neutral cultivars can initiate flowers independently irrespective of photoperiod (Bringhurst and Voth, 1980; Nicoll and Galletta, 1987).

Different clones of all the species are expected to react differently. At least some clones of the diploid *F. vesca* grow at all day lengths (Darrow and Waldo, 1934). *F. moschata* shows more response to long light periods than *F. vesca,* but less than cultivated varieties. The clones of *F. virginiana* express a greater response to longer periods in winter than do *F. moschata* and *F. vesca,* while those of *F. chiloensis* show the great variability, some grow well in the short days of winter and others showing response to longer light periods. Undoubtedly, a greater range of selection of *F. virginiana* clones, such as from Louisiana, Maryland and northern Canada, would have shown equal increase in range of response.

9.3 PHOTOPERIOD AND TEMPERATURE

Strawberry varieties have characteristic temperature-day length responses which determine their regional adaptation. Everbearing strawberry varieties are "long-day" plants and form flower buds in the long days of summer with more flower buds at 17-hour days than at 15 hours (Downs and Piringer, 1955) and relatively few at 11 hours or 13 hours. Ordinary varieties are "short-day" plants and form flower buds when the days are short and the temperatures are low during late summer and fall.

In strawberries, dormancy is induced during the autumn by a combination of day length and temperatures (Uleberg et al., 2016). Duration of flowering and the duration of flower induction period are very important traits with respect to adaptation of plants to different climatic conditions and enhanced fruit yields. Most strawberry genotypes are seasonal with respect to flowering (June-bearing) as short-day plants and they initiate flower buds under decreasing photoperiod during autumn, and consequently produce flowers and fruits during the next summer. Flowering often shows photoperiod–temperature interactions, and decreasing temperature may override the photoperiod in some genotypes (Heide, 1977; Verheul et al., 2007). The critical day length is about 14–15 hours and the optimal temperature for flower induction is about 15–18°C. In general, the strawberry plants should receive at least three weeks and preferably five weeks or more of short days for optimal flower induction (Uleberg et al., 2016).

Sonsteby and Heide (2007) suggested that in general, everbearing strawberry cultivars are qualitative (obligatory) long-day plants at high temperature (27°C), and quantitative long-day plants at intermediate (15–21°C) temperatures (long-day length enhances the flower induction process). These cultivars are day-neutral only at temperatures below 10°C. This photoperiodic effect is strongly temperature-dependent and increases with increasing temperature (Nishiyama and Kanahama, 2002; Nishiyama et al., 2003). The influence of light, derived from the extended photoperiod, clearly influences the flower induction, initiation and differentiation processes. The more advanced phenological stage suggests an advance and/or an acceleration of these processes allowing the production of more secondary inflorescences (Zucchi et al., 2016).

Runners are initiated only when the day length is 12 hours or longer with temperatures above 10°C. Downs and Piringer (1955) found runner production to increase with increasing day length up to 15 hours. Temperature of at least 22°C and a 15-hour day are effective for rapid runner production. Branching of crowns in ordinary varieties takes place most freely when the days are too short for runners and yet too long for flower buds.

9.4 LIGHT

Good-quality strawberry production essentially requires proper exposure of strawberry plants to sunlight, which facilitates advanced berry ripening due to higher absorption of solar heat energy. Besides light duration, light quality significantly influences the development of strawberry crop. Higher-irradiance LED is an effective method to improve strawberry yield (Hidaka et al., 2013). Continuous light with blue LEDs or white fluorescent lamps can shorten the nursery period by approximately 20 days (i.e., 25%; from 80 to 60 days), increasing fruit production efficiency over standard conditions of cultivation (a 16-hour light period) with fluorescent lamps (Yoshida et al., 2012). The supplementation of ambient light has been reported to improve the fruit quality of strawberry (Choi et al., 2015).

9.5 OZONE EXPOSURE

Ozone treatment/exposure affects growth and yield characteristics of strawberry plants in several ways; however, the extent of negative effects varies with the concentration, duration of exposure and tolerance of each variety. Reduced relative growth rate of leaf area, photosynthesis rate, initiation of inflorescence, production of root and shoot biomass, level of carbohydrate accumulation and leaf

Climatic Requirements

discolouration (red, yellow or brown) are the notable harmful effects of exposure of strawberries to ozone (Drogoudi and Ashmore, 2000; Manninen et al., 2003; Keutgen et al., 2005). Higher leaf Ca levels have shown good correlation with the tolerance of strawberry plants to ozone stress (Keutgen and Lenz, 2001).

REFERENCES

Albregts, E.E. and Howard, C.M. 1982. Effect of fertilizer rate on number of malformed strawberry fruit. *Proceedings of the Florida State Horticultural Society*, 95: 323–324.

Ariza, M.T., Soria, C., Medina, J.J., and Martinez-Ferri, E. 2011a. Fruit misshapen in strawberry cultivars (*Fragaria* x *ananassa*) is related to achenes functionality. *Annals of Applied Biology*, 158(1): 130–138.

Ariza, M.T., Soria, C., Medina, J.J. and Martinez-Ferri, E. 2011b. Incidence of misshapen fruits in strawberry plants grown under tunnel is affected by cultivar, planting date, pollination, and low temperatures. *HortScience*, 47: 1569–1573.

Atkinson, C.J., Nestby, R., Ford, Y.Y. and Dodds, P. A. A. 2005. Enhancing beneficial antioxidants in fruits: A plant physiological perspective. *Bio Factors*, 23(4): 229–234.

Bringhurst, R.S. and Voth, V. 1980. Six new strawberry varieties released. *California Agriculture*, 34: 12–15.

Carew, J.G., Morretini, M. and Battey, N.H. 2003. Misshapen fruit in strawberry. *Small Fruits Review*, 2(2): 37–50.

Choia, H.G., Moon, B.Y. and Kang, N.J. 2015. Effects of LED light on the production of strawberry during cultivation in a plastic greenhouse and in a growth chamber. *Scientia Horticulturae*, 189: 22–31.

Darrow, G.M. 1936. Interrelation of temperature and photoperiodism in the production of fruit-buds and runners in the strawberry. *Proceedings of American Society of Horticultural Science*, 34: 360–363.

Darrow, G.M. and Waldo, G.F. 1934. Responses of strawberry varieties and species to the duration of the daily light period. *USDA Technical Bulletin*: 453.

Deckers, T. and Schoofs, H. 2006. Frostguard: A new technique of frost control in small orchards? *Fruitteelt Nieuws*, 19(6): 26–30.

Downs, R.J. and Piringer, A.A. 1955. Differences in photoperiodic responses of everbearing and June-bearing strawberries. *Proceedings of the American Society for Horticultural Science*, 66: 234.

Drogoudi, P.D. and Ashmore, M.R. 2000. Does elevated ozone have differing effects in flowering and deblossomed strawberry? *New Phytologist*, 147(3): 561–569.

Gulen, H. and Eris, A. 2003. Some physiological changes in strawberry (*Fragaria* x *nanassa* "Camarosa") plants under heat stress. *Journal of Horticultural Science and Biotechnology*, 78(6): 894–898.

Gulen, H. and Eris, A. 2004. Effect of heat stress on peroxidase activity and total protein content in strawberry plants. *Plant Science*, 166(3): 739–744.

Gurin, A.G. 2000. Protective strips for commercial plantations of strawberry. *Sadovodstvo i Vinogradarstvo*, 1: 12–13.

Heide, O.M. 1977. Photoperiod and temperature interactions in growth and flowering of strawberry. *Physiologia Plantarum*, 40: 21–26.

Hidaka, K., Dan, K., Imamura, H., Miyoshi, Y., Takayama, T., Sameshima, K., Kitano, M. and Okimura, M. 2013. Effect of supplemental lighting from different light sources on growth and yield of strawberry. *Environmental Control Biology*, 51(1): 41–47.

Hughes, H.M., Duggan, J.B. and Banwell, M.G. 1969. Strawberry Bull. 95HMSO, 10, 6d, Min. Agric. Fish & Food, UK.

Kadir, S., Sidhu, G. and Al Khatib, K. 2006. Strawberry (*Fragaria* × *ananassa* Duch.) growth and productivity as affected by temperature. *HortScience*, 41(6): 1423–1430.

Keutgen, N. and Lenz, F. 2001. Responses of strawberry to long-term elevated atmospheric ozone concentrations-I. Leaf gas exchange, chlorophyll fluorescence, and macronutrient contents. *Gartenbauwissenschaft*, 66(1): 27–33.

Keutgen, A.J., Noga, G. and Pawelzik, E. 2005. Cultivar-specific impairment of strawberry growth, photosynthesis, carbohydrate and nitrogen accumulation by ozone. *Environmental and Experimental Botany*, 53(3): 271–280.

Ledesma, N.A., Kawabata, S. and Sugiyama, N. 2004. Effect of high temperature on protein expression in strawberry plants. *Biologia Plantarum*, 48(1):73–79.

Lieten, P. 2002. Boron deficiency of strawberries grown in substrate culture. *Acta Horticulturae*, 567: 451–454.

Linden, L., Palonen, P. and Hytonen, T. 2002. Evaluation of three methods to assess winter hardiness of strawberry genotypes. *Journal of Horticultural Science and Biotechnology*, 77(5): 580–588.

Maltoni, M.L., Magnani, S., Baruzzi, G. and Faedi, W. 2006. Studio sulla variabilita delle principali caratteristiche qualitative e nutrizionali della fragola. *Rivista di Frutticoltura e di ortofloricoltura*, 68(4): 56–60.

Manninen, S., Siivonen, N., Timonen, U. and Huttunen, S. 2003. Differences in ozone response between two Finnish wild strawberry populations. *Environmental and Experimental Botany*, 49(1): 29–39.

Matsubara, Y., Hirano, I., Sassa, D. and Koshikawa, K. 2004. Alleviation of high temperature stress in strawberry (*Fragaria ananassa*) plants infected with arbuscular mycorrhizal fungi. *Environment Control in Biology*, 42: 105–111.

Massetani, F., Palmieri, J. and Neri, D. 2016. Misshapen fruits in "Capri" strawberry are affected by temperature and fruit thinning. *Acta Horticulturae*, 1117: 373–379.

Matsubara, Y., Hirano, I., Sassa, D. and Koshikawa K. 2004. Increased tolerance to fusarium wilt in mycorrhizal strawberry plants raised by capillary watering methods. *Environmental Control in Biology*, 42: 185–191.

Matsuzoe, N., Kawanobu, S., Matsumoto, S., Kimura, H. and Zushi, K. 2006. Effect of night temperature on sugar, amino acid, ascorbic acid, anthocyanin and ellagic acid in strawberry (*Fragaria* x *ananassa* Duch.) fruit. *Journal-of-Science-and-High-Technology-in-Agriculture*, 18(2): 115–122.

Mitra, S. K. 1991. Strawberries. In: S.K. Mitra, D. S. Rathore and T. K. Bose (eds.). *Temperate Fruits*. Horticulture & Allied Pub., Calcutta, India: 594–596.

Nicoll, M.F. and Galletta, G.J. 1987. Variation in growth and flowering habits of June bearing and everbearing strawberries. *Journal of the American Society for Horticultural Science*, 112: 872–880.

Nishiyama, M. and Kanahama, K. 2002. Effects of temperature and photoperiod on flower bud initiation of day neutral and everbearing strawberries. *Acta Horticulturae*, 567: 253–255.

Nishiyama, M., Ohkawa, W. and Kanahama, K. 2003. Effect of photoperiod on the development of inflorescence in everbearing strawberry "Summerberry" plants grown at high temperature. *Tohoku Journal* of *Agricultural Research*, 53: 43–52.

Ourecky, D.K. and Reich, J.E. 1976. Frost tolerance in strawberry cultivars. *HortScience*, 11: 413–414.

Owens, C.L., Iezzoni, A.F. and Hancock, J.F. 2003. Enhancement of freezing tolerance of strawberry by heterologous expression of CBF1. *Acta Horticulturae*, 626: 93–100.

Palencia, P., Martinez, F., Medina, J.J. and Medina, J.L. 2013. Strawberry yield efficiency and its correlation with temperature and solar radiation. *Horticultura Brasileira*, 31: 93–99.

Rugienius, R. and Brazaityte, A. 2001. Investigations of leaf number, leaf area and pigments amount of different strawberry cultivars during vegetation. *Sodininkyste ir Darzininkyste*, 20 (3(1)): 154–163.

Sage, R.F. and Kubien, D. 2007. The temperature response of C_3 and C_4 photosynthesis. *Plant, Cell & Environment*, 30: 1086–1106.

Sanz, C., Perez, A. and Olias, R. 2002. Effects of temperature on flavor components in *in vitro* grown strawberry. *Acta Horticulturae*, 567(1): 365–368.

Sherman, W.B. and Beckman, T.G. 2003. The climatic adaptation in fruit crops. *Acta Horticulturae*, 622: 411–428.

Shoemaker, J.S. 1954. *Small Fruit Culture*. McGraw Hill Book Company, New York, New York.

Shokaeva, D.B. 2002. Effect of spring frosts on strawberry cultivars. *Sadovodstvo i Vinogradarstvo*, 5: 11–12.

Song, J., Nada, K. and Tachibana, S. 1999. Ameliorative effect of polyamines on the high temperature inhibition of in vitro pollen germination in tomato (*Lycopersicon esculentum* Mill.). *Scientia Horticulturae (Amsterdam)*, 80(3–4): 203–212.

Sonsteby, A. and Heide, O. M. 2007. Long-day control of flowering in everbearing strawberries. *Journal of Horticultural Sciences and Biotechnology*, 82(6): 875–884.

Takeda, T., Inomata, M., Matsuoka, H., Hikuma, M. and Furusaki, S. 2003. Release of anthocyanins from cultured cells with heating treatment. *Biochemical Engineering Journal*, 15(3): 205–210.

Uleberg, E., Sonsteby, A., Jaakola, L. and Martinussen, I. 2016. Climatic effects on production and quality of berries – a review from Norway. *Acta Horticulturae*, 1117: 259–262.

Verheul, M.J., Sonsteby, A. and Grimstad, S.O. 2007. Influences of day and night temperatures on flowering of *Fragaria* × *ananassa* Duch., cvs. Korona and Elsanta. *Scientia Horticulturae*, 112: 200–206.

Wang, S.Y. and Camp, M. J. 2000. Temperatures after bloom affect plant growth and fruit quality of strawberry. *Scientia Horticulturae*, 85(3): 183–199.

Wang, S.Y., Bunce, J.A. and Maas, J.L. 2003. Elevated carbon dioxide increases contents of antioxidant compounds in field-grown strawberries. Journal of Agricultural and *Food Chemistry*, 51(15): 4315–4320.

Wang, S.Y. and Lin, H. S. 2006. Effects of plant growth temperature on membrane lipids in strawberry (*Fragaria* x *ananassa* Duch.). *Scientia Horticulturae*, 108(1): 35–42.

Wang, S.Y. and Zheng, W. 2001. Effect of plant growth temperature on antioxidant capacity in strawberry. *Journal* of *Agricultural and Food Chemistry*, 49(10): 4977–4982.

Wang, S.Y., Zheng, W. and Maas, J.L. 2003. High plant growth temperatures increases antioxidant capacity in strawberry fruit. *Acta Horticulturae*, 626: 57–63.

Yannai, B. 1988. Report on Training Programme in Strawberry. Production Planning Cultivation Techniques and Post-harvest handling, Israel.

Yoshida, H., Hikosaka, S., Goto, E., Takasuna, H. and Kudou, T. 2012. Effects of light quality and light period on flowering of everbearing strawberry in a closed plant production system. *Acta Horticulturae*, 956: 107–112.

Zucchi, P., Martinatti, P. and Pergher, A. 2016. Photoperiod extension effect on nursery tray-plants of everbearing strawberry. *Acta Horticulturae*, 1117: 359–364.

10 Soil

P.K. Rai

CONTENTS

10.1 Site Selection 169
10.2 Soil Type 170
10.3 Irrigation 171
10.4 Soil Fertility 172
10.5 Growing Media 173
10.6 Soil Preparation 173
References 176

Soils are practically the universal medium of plant growth and fixed assets, so the economics of crop production largely depends on how well the soil is maintained. During the past few years, a great deal of attention has been paid to soil management practices that promote sustainable soil fertility and productivity (Doran and Jones, 1996; Pankhurst et al., 1997; Magdoff and Van Es, 2000). Human beings are dependent on soil and to a large extent, good soils are dependent upon human beings as they use the soils for varied purposes. Soil is the most important resource as fixed assets for strawberry production. It must be maintained at optimum fertility and productivity for beneficial strawberry production system. In this chapter, various aspects concerning selection and management of soils for profitable strawberry production are discussed.

10.1 SITE SELECTION

Success in strawberry production on a commercial scale depends primarily on selection of an appropriate site. The selected site should have uniform temperature, rainfall and drainage and proper protection from wind. A site having a gentle slope of about 2–4% with provision for drainage of excess water and cold air masses is considered optimum (Barney et al., 1992) for profitable strawberry cultivation. The site where the cold air at night drains to lower areas, the risk of damage to blossoms by frost is reduced (Henney, 1962). If the site lacks this type of microclimate, the grower must either create an ideal site or finally choose a site where there is ready access to a water supply. Irrigation is important for good plant establishment to maintain growth during dry periods, and is also used to prevent frost injury to strawberry flowers during spring. Areas that are prone to high-velocity hot or chilling winds should be avoided to protect the delicate plants from desiccation. Freezing temperature and frost during winter and spring are detrimental for strawberry plants and flowers, so low-lying pockets where cold air drains and accumulates should be avoided. Protection of plants and flowers from spring frost and hail storms is essential as spring and early summer are the prime times for plant growth and flowering. Strawberry plants need full sun for plant growth and fruiting but a harvestable crop can also be obtained with as little direct sunlight per day as six hours. Increased availability of sunlight improves yield and quality of strawberry fruits. Strawberry cultivation on steep slopes causes serious soil erosion and makes cultivation very expensive (Waldo et al., 1971). The orientation of selected sites should be northeast to northwest as these aspects are moderately warmer during winter. Strawberries on a northern slope ripen later (due to cold, more soil moisture and more soil depth) than those on a southern slope (due to more absorption of sun heat), but the yield on the northern slope may be much higher (Waldo et al., 1971; Childers et al., 1995).

169

As a rule, strawberries grow well under temperate climatic conditions but some cultivars are successfully grown in subtropical climatic conditions. Daylight period of 12 hours or less and moderate temperatures are important for flower bud formation. Each cultivar may have a different day length and temperature requirement. Another important point is proximity to market as fruits are very delicate and perishable in nature.

10.2 SOIL TYPE

Selecting a good site with suitable soil is the first step in the successful cultivation of strawberries (Childers et al., 1995) before planting them on a large scale.

The strawberries can be grown on several types of soils ranging from poor sandy to heavy clay – provided these have proper soil moisture holding capacity as well as optimum drainage and sufficient organic matter to support proper growth and fruiting of the crop. The best soil for strawberry production is deep, well-drained, sandy loam, sufficient in humus (more than 2% organic matter). Soil should also be porous and rich in humus and at least 60 cm deep. Strawberry roots are generally distributed within the top 15–40 cm soil depth and spread around 30–100 cm (Mann and Ball, 1927, Hansen, 1931). Based on dry weight determination and expressed as percentage of the total dry weight of the root system, 90% of the total root system was observed to be present in the top 15 cm of the soil and, of that, 73% was in the first 7.5 cm soil depth in England (Mann and Ball, 1927).

There is a definite cultivar adaptation to soils as the performance of different varieties varies with different categories of soils. Some grow better on heavier soils, while others perform well on light soils. Strawberries ripen somewhat earlier on sandy soils than on clay soils.

Water should not stagnate in the field, since most of its roots are confined to the top 15 cm of soil, so this layer need to be kept porous and enriched with humus. There should be no layer of lime underlying the top 15–20 cm soil profile, otherwise it causes burning of leaves. In drier areas, alkali soils must be avoided for the cultivation of strawberries.

On alkaline soils, particularly those derived from chalk or containing excessive free chalk, reduced iron absorption by roots may lead to chlorosis in leaves of strawberries (Hughes et al., 1969).

A field experiment was conducted on black and sandy soil of Poland to assess the effects of Ekosorb hydrogel (@ 3 and 6 g dm^{-3}) on the soil water content, floral damage due to frost and fruiting of strawberry cultivars Sengana, Dukat, Kent and Elsanta. Ekosorb application increased the number of flowers injured due to ground frost during spring which increased further with increasing rate of application of Ekosorb. Application of Ekosorb at both the rates increased availability of water in the soil, but this effect was higher in black soils than in sandy one. Regardless of the soil type, maximum fruit yield was obtained when Ekosorb at 3 g dm^{-3} was applied (Makoswska and Borowski, 2004).

Heavy clay soils that are usually poorly drained not only encourage the development of diseases but also hinder precise inter-cultural operations. Coarse-textured sandy soils are often infertile and prone to frequent soil moisture deficit, and require more frequent irrigation and greater attention to nutrient application practices. Plants established in low-lying muck or organic soils are more vulnerable to frost injury. Strawberries require well-drained soils. Beds for strawberry planting are artificially raised for sites with shallow soil (Barney et al., 1992).

Strawberries are less sensitive to soil pH and salts than raspberries or blueberries, and its production is also possible on slightly alkaline soils. But saline soils are not appropriate since strawberries have comparatively low tolerance to salts and prefer slightly acidic soils with pH ranging from 5.5–6.5. Good vigour has though been obtained on soils with a pH slightly higher than neutral (7.5); however, soil pH above 8 can adversely affect the availability of certain nutrients in strawberries, especially the iron levels. Leaf yellowing in strawberries is very common where soil pH is high. For each unit increase in salinity above the threshold value (1 mS/cm), yield decreases by 33%. Salinity (EC >2 mS/cm) exhibit symptoms such as scorching of leaves, and decreases the numbers of runners per plant, runner length, number of crowns per runner, number of leaves, fresh and dry weight of roots, and nutrients such as nitrogen, phosphorus and potassium; but EC >5 mS/cm is lethal to strawberry. The deleterious effect of high salinity (35 mM NaCl) in root zone is more striking than that of high pH (8.5), hence supplementary application of potassium can improve plant growth, fruit yield and quality of strawberries grown under saline conditions (Kaya et al., 2002; Khayyat et al., 2009). Salinity (10–16 mM l^{-1} NaCl) of irrigation water causes decrease in plant biomass and fruit yield and quality parameters such as titratable acidity and soluble solids content of strawberries (Kepenek and Koyuncu, 2002; Pirlak and Esitken, 2004). Camarosa has been observed to be comparatively more salt-tolerant than Chandler under saline conditions (Turhan and Eris, 2007). The use of green manures and acidifying fertilizers can alter the soil pH to some degree and help to improve plant growth, yield and quality of fruits. Thus, sandy loam to loamy soils with 5.7–6.5 pH are ideal for strawberry cultivation (Shoemaker, 1954; Anonymous, 1956) at a commercial scale.

Sites surrounded by natural bush stands may have native strawberry plants, which can harbour insect-pests, diseases and viruses and contribute as a source of initial inoculam for spread to nearby cultivated strawberry stands. To maintain good sanitation, it may be necessary to kill native strawberry stands within 400 m distance of commercial strawberry fields. It would be advisable to avoid planting strawberries in an area where strawberry, raspberry, alfalfa, tomatoes, potatoes, peppers or eggplant have been grown during the past four years. These crops are host to the pathogenic fungus *Verticillium*, causing root rot, which also attacks strawberries. Similarly, it is also advisable not to plant strawberries on recently ploughed grass or sod areas as it can lead to devastating weed problems, and damage by white grub, a common turf pest, which feeds upon strawberry roots. But, if a long-term crop-rotation that excludes the host crops is not feasible, it would be better to resort to soil fumigation. Inoculation of bio-fertilizers alone or in combination like *Azotobacter chroococcum*, *Azospirillum brasiliense* and *Pseudomonas striata* by root dipping and immediately planting, the runners have been observed to be helpful in enhancing the soil fertility and microbial activity in the soil thereby improved quality of fruits (Singh et al., 2010). Thus, integration of inorganic fertilizers with bio-fertilizers is beneficial for soil health, plant health, yield, and quality of fruits, and consequently gross as well as net returns.

10.3 IRRIGATION

Soils also influence the additional requirement of irrigation water for optimum growth and productivity of strawberries. Irrigation is essential for both consistent, high-yielding strawberry production and management of crops against damage due to frost. Irrigation water with significant salt content, combined with poor soil structure, may also cause buildup of undesirably high salinity levels. In low-rainfall areas, supplementary irrigation is essential to maximize production. An average of 30 cm of irrigation water is required over the growing season. Water applications may also be needed for management of damage to the crop possible due to spring frost and cooling the summer crop when temperatures are likely to reach very high. River or pond water is often used for irrigation. Well water may be suitable but should be tested to determine if the quality and quantity is satisfactory. Rolbiecki et al. (2004) tested four irrigation systems, viz., control (without irrigation), drip irrigation (half the water rate), drip irrigation (full water rate) and micro-sprinkler irrigation in an "Elsanta" strawberry field. Regardless of the irrigation regimes applied, all irrigated strawberry plants tended to increase the number and size of berries over control plants. Of the two irrigation methods tested, drip irrigation proved its superiority in respect of water use efficiency. The result obtained indicated the important role of irrigation in improving the yield of "Elsanta" strawberries, cultivated on loose sandy soils. In general, light or coarse-textured soil need more irrigation when compared with heavy or fine-textured soils (for more details see Chapter 14 on water management).

10.4 SOIL FERTILITY

High organic matter contents are highly desirable for successful strawberry production (Barney et al., 1992). Common agricultural practices, for example, soil tillage and organic amendment, may considerably influence the native communities of field earthworms. Abundance of earthworms and microbial biomass were studied in a strawberry field experiment (soil type: silty-clay) with a history of different crops (*Fragaria, Phleum pratense, Carum carvi, Secale cereale, Brassica rapa, Brassica napus, Amsinckia* spp., *Allium cepa* and *Fagopyrum esculentum*), and treatments with peat. After three years of perennial cropping of strawberry, the earthworm community consisted mainly of *Aporrectodea caliginosa* and *Lumbricus terrestris*. Soil peat amendment almost doubled the number of endogenic *A. caliginosa*, but had no effect on the anecic earthworm *L. terrestris*. The effect of cropping history on earthworms diminished after three years of strawberry cropping. Only the positive effects of *Carum carvi* on juvenile *Lumbricus* spp. was detectable following three years after finishing of cropping season. However, some crops had secondary effects on the earthworm distribution without significant influence on their numbers while they were grown, for example, more *A. caliginosa* were recorded from soil with a history of *Phleum pratense*. The effect of strawberry cropping was contradictory as six years of continuous strawberry cropping decreased the number of anecic earthworm (*L. terrestris*), but during the last three years, their proportion tended to increase from 6–40% with a concomitant big drop in the number of *A. caliginosa*. The abundance of *A. caliginosa* was associated with soil organic carbon (C), but not with soil microbial biomass (Kukkonen et al., 2006).

A field experiment with silt soil previously cultivated with oats was conducted to study the effects of peat amendment and crop production system on earthworms. All the crops (strawberry, timothy and caraway, and annual crops rye, turnip rape, buckwheat, onion and fiddle neck) were grown with or without soil amendment with peat. Soil organic carbon and microbial biomass was measured at 14 and 28 months. Peat increased the abundance of juvenile earthworm, *A. caliginosa*, by 74% in three growing seasons, without influencing the population of adult earthworms. The number of earthworms, *L. terrestris*, did not increase due to treatment with peat. Cultivation of caraway for three seasons favoured both *A. caliginosa* and *L. terrestris*. An equal abundance of *A. caliginosa* was also found in plots cultivated with turnip rape and fiddle neck. Total earthworm and especially the number of *A. caliginosa* was every less in plastic-mulched strawberry beds. This was mainly attributed to repeated use of the insecticide endosulfan. With the strawberry plots omitted, there was a significant correlation between soil microbial N measured at 14 months and number of juvenile *Aporrectodea* spp. and *Lumbricus* spp. measured at 28 months. The number of adult earthworms was not associated with either soil organic carbon or microbial biomass (Kukkonen et al., 2004). Twelve elements (macroelements and trace elements) were determined in strawberry fruits sampled from 41 commercial field plantations located in three provinces in Poland, and soil parameters were also evaluated. The result showed linear correlation between the copper and calcium contents of these fruit and the soils where they had been grown, but no such correlation was observed for lead, cadmium, zinc and magnesium. Increasing level of soil pH had a slightly negative impact on manganese, cadmium, zinc and chromium, but a positive impact on calcium, without affecting the contents of lead, copper and magnesium. There was also some slight impact of environmental factors on the content of these metals in strawberries (Szteke et al., 2006).

Worldwide, up to 20% of irrigated arable land in arid and semi-arid regions is salt affected. Climatic conditions enable continuous growing of several crops throughout the year, but increasing demand for irrigation water forces the growers to utilize water of poor quality. Salinity treatments (4, 6 and 8 dS/m) have shown the negative effects on total fresh berry yield, total number of fruits, fruit size, as well as on the number of runners and the length of longest runner. Furthermore, NaCl salinity stress accelerated leaf senescence and reduced the cropping season by 2–3 weeks (Ondrasek et al., 2006).

High NaCl concentrations affect plant growth and canopy gas exchange parameters. The treatment with 34 mM NaCl caused marked increase in the stomatal conductance (g_s) and transpiration

Soil

rate (E) of strawberry cv. Chandler. The leaf temperature also increased due to application of salt in Camarosa and Chandler strawberries. It is suggested that the reductions in g_s and E represent adaptive mechanisms to cope with excessive salt in cv. Camarosa. Strawberry cv. Camarosa was observed to be comparatively more salt-tolerant than Chandler under saline conditions (Turhan and Eris, 2007). Because of the need for developing fruit production in a balanced environment, rhizosphere studies help in determining precise doses of mineral fertilizers essential for proper vegetative growth and fruit production of economically important strawberry cultivars with specific nutrient requirements.

An experiment was conducted using rhizoboxes, and observation on changes in pH were recorded in the rhizosphere of strawberries depending on the dose and form of nitrogen applied (Paszt and Zurawicz, 2004). Addition of compost as a soil supplement has been reported to enhance the levels of ascorbic acid (AsA) and glutathione (GSH) and ratios of AsA/dehydroascorbic acid (DHAsA) and GSH/oxidized glutathione (GSSG) in the fruits of Allsar and Honeoye strawberries. The peroxyl radical, the superoxide radical, hydrogen peroxide, hydroxyl radical, and singlet oxygen absorbance capacity in strawberries increased significantly with increasing fertilizer dose and compost used. Fruits from plants grown in full-strength fertilizer with 50% soil + 50% or 100% compost, yielded fruits with maximum phenolics, flavonol, and anthocyanin content (Wang and Liin, 2003). Preplant applications of Composted Yard Waste (CYW) in field increased the soil pH, cation exchange capacity, organic matter content, and some nutrient element content in soil and, also influenced fruit nutrient elemental content while influence of fumigation and CYW's on fruit quality characteristics were not apparent (Funt and Scheerens, 2002).

10.5 GROWING MEDIA

Coir is a suitable media amendment for growing plants and may have beneficial effects on plant quality (Scagel, 2003). In a trial on "Elsanta" strawberry grown on peat bags, five ratios of NO_3–SO_4 were applied through nutrient solution from planting until end of harvest, viz., 12:0, 11:0.5, 10:1, 9:1.5 and 8:2 mmol/1 in the first trial and 13:0, 11:1, 9:2, 7:3 and 5:4 mmol/1 in two other trials. In all trials, withholding SO_4 significantly reduced fruit size and yield. For a short production cycle (winter planting), the ratios did not influence number, size and yield of fruits but for long production cycles (summer planting), the best plant performance and yield was obtained with NO_3–SO_4 ratios of 9:2 mmol/l. There was a significant reduction of total fruits, misshaped fruits and total yield with a NO_3–SO_4 ratio of 5:4 mmol l^{-1} (Lieten, 2004).

10.6 SOIL PREPARATION

Thorough preparation of soils is important. It is completely uneconomic to plant strawberries on sites infested with perennial weeds as it is impossible to eradicate these later without considerable damage to the plants (Hughes et al., 1969). If perennial weeds are present, adequate cultivation in fallow field or the use of suitable herbicides is essential to eliminate them before planting. The soil should be cultivated for one or two years before planting strawberries to make the field free from soil pests and other perennial weeds (Childers et al., 1995). Fumigation and solarization of soils are other alternative approaches to kill almost all types of hibernating insect-pests or pathogens in the soil or growing medium. Solarization of soil is a very effective and less expensive, nonchemical method for controlling soilborne diseases, insect-pests, nematodes and weeds. The efficacy of the thermal treatment for killing of pests and disease-causing agents is determined by the values of the maximum soil temperature and amount of heat accumulated, that is, duration × temperature (Rubin et al., 2007). In this practice, the moistened soils are mulched with transparent polyethylene, thereby increasing soil temperatures and killing pathogens. Soil fumigation also acts similar to the solarization. Fumigants are generally applied in liquid formulations, which in turn volatilize to form gases. Immediately after application of fumigant, soil is covered with polythene sheet (black) to prevent the chemical from escaping into the air. Higher soil temperature favours better volatilization and movement in the soil. At least two or three days are needed to fumigate the soil and thereafter the polythene cover is removed. Planting is done at three–four weeks after fumigation is over. For good fumigation, soil is ploughed 20–25 cm deep, and clods are broken during late August or early September for fall fumigation and, during late February–mid-March for spring fumigation. Methyl bromide was one of the most expensive fumigants of the registered materials for strawberries. Although soil fumigation with methyl bromide has ensured stability of strawberry production, its use has been discontinued because of its effect on stratospheric ozone. Methyl bromide (60 g/m^2) was a common soil fumigant for strawberry, but considering its hazardous impact on human health, its use has now been prohibited the world over.

The commercial use of fall fumigation of soil about once in five years is indispensable to control white grubs, wireworms, ants and weeds that harbour root aphids and many other pests. Several workers have suggested different chemicals such as 1,3.dicholopropene (18 g/m^2), metham-Na (108 g/m^2), dazomet (DAZ), metam sodium (MS), metam potassium (MK), dimethyl disulphide (DMDS), ethanedinitrile (EDN), propylene oxide (PO), sodium tetrathiocarbamate (ENZ), sodium azide (SEP), methyl iodide (MI) and furfural (FUR) as effective soil treating chemicals (Bartual et al., 2002; Cebolla et al., 2002). The likely short-term alternatives such as 1–3,dichloropropane, chloropicrin and methoam sodium (metam), although not ozone depleters, are potentially hazardous to the environment and human, if applied improperly.

In soil disinfection trials, 80% dicholoropropane and 20% methyl isocyanate showed herbicidal and nematicidal effects and stimulated plant growth of strawberry (Nikolova et al., 1976). Soil sterilization through a solar heating system is also very effective against many soilborne diseases. The use of 80% dicholoropropane and 20% methyl isocyanate or 98% dazomet granules @ 500 kg/ha effectively controlled weeds, *Pratylenchus penetrans* and other nematodes, and facilitated cultivation of strawberries in succession on the same land without the need for crop-rotation (Rebandel and Szczygie, 1991). Of the two fumigants, granular dazomet is recommended, being less toxic and easier to apply than dicholoropropane and 20% methyl isocyanate. As an chemical alternative to replace methyl bromide fumigation, Fennimore et al. (2003) attempted 1,3-dichloropropene, chloropicrin, and metam sodium, and applied emulsified formulations of these fumigants through the drip irrigation system as an alternative to the standard shank injection method of fumigation in strawberry production in California, and observed equal or better weed control than equivalent rates applied by shank injection, but the percentage of survival of weed seed at the edge of the bed was often higher in the drip-applied treatments than in the shank-applied treatments due to close proximity of the shank-injected fumigant to the edge of the bed. Application of mixture (61:35) of 1,3-dichloropropene and chloropicrin to the ridge under plastic mulch have been proved the best alternative to methyl bromide (Lopez et al., 2003; Baruzzi et al., 2005; Ajwa and Winterbottom, 2006).

As the strawberry roots are planted 12–15 cm deep which continues to grow downwards, it is desirable to correct any obvious deficiency in soil nutrients by applying half the recommended fertilizer dose of phosphorus and potash at the time of last ploughing and the rest as top dressing before planting (Hughes et al., 1969). If it is the practice to grow strawberries on the same land for two–three years, the organic matter content of the soil should be increased before setting each new planting. Grasses and meadow crops are probably the most effective means of increasing organic matter in soil (Waldo et al., 1971). In poor soils, the growers may need to plough two–three green manure crops such as oats, rye, wheat, barley, Sudan grass or peas, before planting of strawberries. *Brassica juncea*, is adaptable to many soil types and climates, and easy to manage in the field and therefore, can be used as a green manure crop. It can produce up to 138 t/ha dry matter, which may contain >1.6% N, and restricting the development of soil pathogens. Strawberry crop succeeding to *B. juncea* in the crop-rotation have shown no adverse reaction to glucosinolates, and have given results comparable to those following sterilization with methyl bromide (Lazzeri et al., 2003). The presence of high amounts of glucosinolates and of the enzyme myrosinase that catalyse their hydrolysis, in all Brassicaceae plant parts, linked to the high biodical activity of some glucosinolate enzymatic hydrolysis derivative products (mainly isothiocyanates and nitriles) have indicated the practical possibility of amending soil with these natural biodical compounds by the cultivation and green manure of selected species of this family. The glucosinolate-myrosinase (thioglucosidase) has been proved to be a good natural alternative for methyl bromide used for soil fumigation.

Water soluble formulations of alternative fumigants of methyl bromide such as 1–3, dichloropropane, chloropicrin and metom sodium can be applied through drip irrigation systems. In comparison to conventional shank methods of injection, application of soluble formulations though drip irrigation systems is economical and environment friendly, reduce exposure, and allow for simultaneous or sequential application of a combination of fumigants. The application of large amounts of organic matter in the form of farmyard manure may be beneficial in improving soil quality but in the presence of soil pathogens, it is likely to help them to proliferate. Solarization gives good control of both weeds and pathogens. It should be carried out during June–August, using nontransparent plastic film, on soil already banked and with underground irrigation devices in place. Green manuring with species producing biocidal glucosinolates, such as *B. juncea* and *Raphanus raphanistrum* is effective. However, in the case of *R. raphanistrum*, it must be combined with solarization to inhibit the germination of the manure crop. This technique has produced fruit with an average weight comparable to that produced by methyl bromide use. A combination of poisoned bait, *Bacillus* treatments and pheromone traps has controlled *Spodoptera littoralis*, the most damaging insect-pest in strawberries (Rosati, 2002).

Strawberry runners are a high-value cash crop that requires vigorous transplants free of pathogens. *Verticillium* wilt (caused by *Verticillium* sp.) and crown rot (caused by *Phytophthora cactorum*) are the main diseases of nursery which can effectively be managed by the use of chloropicrin, 1,3-dichloropropene and dazomet, which are well comparable with methyl bromide fumigation for control of strawberry diseases in nursery. Furthermore, 1,3-dichloropropene and methyl bromide applied at 50% rate under virtually impermeable film provided effective disease control in strawberry nurseries too (Cal et al., 2004). The effectiveness of *Trichoderma* strains for biocontrol of soilborne pathogens requires an improved understanding of the soil and root ecology of this fungus. Compost may be used as a substrate to establish and promote survival of *Trichoderma* in field soil (Leandro et al., 2007).

Chemical fumigants (metam sodium, dazomet, chloropicrin, chloropicrin +1,3 dichloropropene) and a steaming system exploiting the exothermic reaction between steam and CaO (Bioflash System™) tended to decrease the number of colonies of the fungus *Verticillium dahliae* in the soil, which corresponded to reduced incidence of *Verticillium* wilt (efficacy about 80%). The use of chemical fumigants had a positive impact on the size of the mother plants. The surface area covered by plants grown on the treated plots was 1.1–1.7 times larger than plants grown on non-fumigated control plots (Meszka and Malus, 2014).

After imposing ban on the use of methyl bromide fumigant since 2007, many countries in Europe are progressively restricting the use of other toxic fumigants such as dichloropropene, directing to adopt some alternative control techniques. Soil bio-solarization, a new technique combining soil bio-fumigation and soil solarization, has been observed to be a good option to cultivate the strawberry. In this technique, soil is bio-fumigated by amendment of chicken manure (25 t/ha), then solarized for 30 days by covering with a clear plastic mulch. This practice can increase the yield tremendously. This can also have higher yields than previously reported chemical fumigants namely 1,3-dichloropropene and chloropicrin. In addition, bio-solarization is about 20% cheaper than treatment with 1,3-dichloropropene and chloropicrin (Domínguez et al., 2014).

The presence of large amounts of sulfur compounds in the plant parts of *Allium* species has led to the suggestion that the residues of this plant family can be used for soil bio-fumigation. The active molecules in these *Allium* species are dimethyl disulfide (DMDS) and dipropyl disulfide (DPDS). The results have shown that onion byproducts and DMDS not only has a high level of bio-fumigant activity, but also stimulate vegetative growth. Incorporation of *Allium* byproducts into the soil helps to release DPDS, which persists up to one month afterwards. This treatment increases the strawberry productivity; a result that is comparable to those obtained using *Brassica*-based bio-fumigation. Hence, onion byproducts may have practical potential as new bio-fumigants and can be used as an alternative to methyl bromide (Arnault et al., 2013).

Anaerobic soil disinfestation (ASD), a biological alternative for soil fumigation developed in Japan (Shinmura, 2000; Momma, 2008) and the Netherlands (Blok et al., 2000; Lamers et al., 2010) has been proved quite effective in managing the soilborne pests. Overall, ASD treatment has been observed to be consistently effective at suppressing *Verticillium dahliae* and obtaining comparable yield with fumigant control in coastal California when 20 t/ha of rice bran (RB) was incorporated before planting and at least 75 mm of irrigation was applied in sandy loam to clay-loam soils. However, due to economic and high nitrogen application issues associated with use of 20 t/ha RB, there is interest in examining alternative carbon (C) sources used in ASD. Muramoto et al. (2014) conducted demonstration trials at four-local farms in which sugarcane molasses (Mol) 20 t/ha alone or in combination with RB (Mol 10 t/ha + RB 10 t/ha) were tested. Although Mol has advantages over RB in terms of ease of application and lower N content, the anaerobic condition created by molasses did not last long and split applications were needed. Further, cumulative marketable fruit yield from molasses @ 20 t/ha were as low as 70% of fumigated controls, whereas RB 20 t/ha and molasses @ 10 t/ha + RB @ 10 t/ha had similar yields as the control. Lack of effectiveness in molasses-based ASD may be related to uneven distribution of molasses across the bed profile and low soil temperatures at trial sites. Use of preplant fertilizer could increase the berry yield only ~5%, suggesting a possibility of reducing preplant fertilizer at ASD fields. Drip irrigation systems can serve as a vehicle to deliver water soluble formulations of fumigants to the target soil volume and may provide a more uniform distribution of chemicals in the soil than shank injection (Ajwa et al., 2002).

Lime supplies calcium and magnesium and also binds toxic aluminium. In humid regions, if the pH is 4.5–5.3, application of 1–2 tonnes of dolomite per acre (equivalent to 2.5–5 tonnes/ha) before one year of planting raises the pH to 5.5–6.5 (Childers et al., 1995). Where furrow irrigation is used, the plants are set on raised beds with furrows between them. The soil must be absolutely levelled so that water can run in either direction in the furrows between the beds (Waldo et al., 1971).

Bio-fumigation can be achieved by incorporating fresh plant material (green manure), seed meals (a by-product of seed crushing for oil), or dried plant material treated to preserve isothiocyanate activity (Lazzeri et al., 2004).

Strawberry plants grown continuously in the same plots and on the same site exhibited low strawberry plant density and fruit yield and high weed populations (Seigies and Pritts, 2006).

REFERENCES

Ajwa, A.H., Trout, T., Mueller, J., Wilhelm, S., Nelson, S.D., Soppe, R. and Shatley, D. 2002. Application of alternative fumigants through drip irrigation systems. *Phytopathology*, 92(12): 1349–1355.

Ajwa, H.A. and Winterbottom, C.Q. 2006. Alternatives to methyl bromide in Californian strawberry production. *Rivista di Fruticoltura e di Ortofloricoltura*, 68(4): 28–32.

Anonymous. 1956. *Wealth of India Vol. IV*. CSIR, New Delhi, India: pp. 57–60.

Arnault, I., Fleurance, C., Veyc, F., Fretayd, G.D. and Auger, J. 2013. Use of Alliaceae residues to control soil-borne pathogens. *Industrial Crops and Products*, 49: 265–272.

Barney, D.L., Davis, B.B. and Fellman, J.K. 1992. Strawberry production: Overview. The Alternative Agricultural Enterprises publication series was supported by a grant from the Northwest Area Foundation, St. Paul, Minnesota: p. 4.

Bartual, R., Cebolla, V., Bustos, J., Giner, A. and Lopez-Aranda, J.M. 2002. The Spanish project on alternatives to methyl bromide (2): The case of strawberry in the area of Valencia. *Acta Horticulturae*, 567: 431–434.

Baruzzi, G., Ceccaroni, S., Foschi, S., Lucchi, P., Maltoni, M.I., Mennone, C., Quinto, G. and Faedi, W. 2005. Use of 1, 3-D and chloropicrin as an alternative to methyl bromide on strawberry. *Rivista di Frutticoltura e di Ortofloricoltura*, 67(4): 42–48.

Blok, W.J., Lamers, J.G., Termorshuizen, A.J. and Bollen, G.J. 2000. Control of soil borne plant pathogens by incorporating fresh organic amendments followed by tarping. *Phytopathology*, 90: 253–259.

Cal, De A., Martinez-Treceno, A., Lopez-Aranda, J.M. and Melgarejo, P. 2004. Chemical alternatives to methyl bromide in Spanish strawberry nurseries. *Plant Disease*, 88: 210–214.

Cebolla, V., Bartual, R., Busto, J. and Giner, A. 2002. New chemicals as possible alternatives to methyl bromide in the area of Valencia. *Acta Horticulturae*, 567: 435–438.

Childers, N.F., Morris, J.N. and Sibbett, G.S. 1995. *Modern Fruit Science, Orchard and Small Fruit Culture*. Gainesville, Florida: Horticultural Publications.

Dominguez, P., Miranda, L., Soria, C., Santos, B., Chamorro, M., Romero, F., Daugovish, O., López-Aranda, J.M. and Medina, J.J. 2014. Soil biosolarization for sustainable strawberry production. *Agronomy for Sustainable Development*, 34: 821–829.

Doran, J.W. and Jones, A.J. 1996. *Methods for Assessing Soil Quality*. Special publication No. 49. Soil Science Society of America, Madison, Wisconsin.

Fennimore, S.A., Hasar, M.J. and Ajwa, H.A. 2003. Weed control in strawberry provided by shank-and-drip-applied methyl bromide alternative fumigants. *HortScience*, 38(1): 55–61.

Funt, R.C. and Scheerens, J.C. 2002. Effect of fumigation and composted yard waste o strawberry soil quality, fruit quality and fruit elemental content. Strawberry research – 2001, *Proceedings of the 5th North American Strawberry Conference*: pp. 116–123.

Gao, S.D., Qin, R.J., Ajwa, H.S. and Fennimore, S.T. 2014. Low permeability tarp to improve soil fumigation efficiency for strawberry production in California, USA. *Acta Horticulturae*, 1049: 707–714.

Henney, H.B. 1962. Bulletin on growing strawberries in E. Canada No. 1171: p. 3.

Hansen, H.C. 1931. Comparison of root and top development in varieties of strawberry. *American Journal of Botany*, 18: 658–673.

Hughes, H.M., Duggan, J.B. and Banwell, M.G. 1969. *Strawberry Bull*. 95, HMSO 10, 6d. Ministry of Agriculture Fishery and Food, UK.

Kaya, C. Higgs, D., Saltali, K. and Gezevel, O. 2002. Response of strawberry growth at high salinity and alkalinity to supplementary potassium. *Journal of Plant Nutrition*, 25(7): 1415–1427.

Khayyat, M.M., Reza, V., Soheila, R. and Salma, J. 2009. Potassium effect on ion leakage, water usage, fruit yield and biomass production by strawberry plants grown under NaCl stress. *Journal of Fruit and Ornamental Plant Research*, 17(1): 79–88.

Kepenek, K. and Koyuncu, F. 2002. Studies on the salt tolerance of some strawberry cultivars under glasshouse. *Acta Horticulturae*, 573: 297–304.

Kukkonen, S., Palojarvi, A., Rakkolainen, M. and Vestberg, M. 2004. Peat amendment and production of different crop plants affect earthworm populations in field soil. *Soil Biology and Biochemistry*, 36(3): 415–423.

Kukkonen, S., Palojarrvi, A., Rakkolainen, M. and Vestberg, M. 2006. Cropping history and peat amendment-induced change in strawberry field earthworm abundance and microbial biomass. *Soil Biology and Biochemistry*, 38(8): 2152–2161.

Lamers, J.G., Runia, W.T., Molendijk, L.P.G. and Bleeker, P.O. 2010. Perspectives of anaerobic soil disinfestation. *Acta Horticulturae*, 883: 277–283.

Lazzeri, L., Malaguti, L., Cinti, S. and Baruzzi, G. 2003. Biocidal plants for green manure in the rotation. *Colture Protette*, 32(1): 53–56.

Lazzeri, L., Leoni, O. and Manici, L.M. 2004. Biocidal plant dried pellets for biofumigation. *Industrial Crops and Products*, 20(1): 59–65.

Leandro, L.F.S., Guzman, T., Ferguson, L.M., Fernandez, G.E. and Louws, F.J. 2007. Population dynamics of Trichoderma in fumigated and compost-amended soil and on strawberry roots. *Applied Soil Ecology*, 35(1): 237–246.

Lieten, P. 2004. Nitrate-sulfate ratio for strawberries grown on peat bags. *Acta Horticulturae*, 649: 223–226.

Lopez-Aranda, J.M., Mediana, J.J., Miranda, L., Montes, F., Romero, F., Vega, J.M., Paez, J.I., et al. 2003. Alternatives to methyl bromide: Spanish trials. *Colture Protette*, 32(1): 59–65.

Magdoff, F. and Van Es, H. 2000. *Building Soils for Better Crops*, 2nd ed. Sustainable Agricultural Publications, University of Vermont, Burlington, Vermont.

Makoswska, M. and Borowski, E. 2004. The influence of the addition of Ekosorb to black soil and sandy soil on the content of water in soil, frost injury of flowers and on fruiting of strawberry. *Folia Horticulturae*, 16(1): 87–93.

Mann, C.E.T. and Ball, E. 1927. Studies in the root and shoot growth of the strawberries. *Journal of Pomology and Horticultural Science*, 5: 149–169.

Meszka, B. and Malus, E. 2014. Effects of soil disinfection on health status, growth and yield of strawberry stock plants. *Crop Protection*, 63: 113–119.

Momma, N. 2008. Biological soil disinfestation (BSD) of soilborne pathogens and its possible mechanisms. *Jarq Japan Agricultural Research Quarterly*, 42: 7–12.

Muramoto, J., Shennan, C., Baird, G., Zavatt, M., Bolda, M.P., Dara, S.K., Mazzola, M., et al. 2014. Optimizing anaerobic soil disinfestation for California strawberries. *Acta Horticulturae*, 1044: 215–220.

Nikolova, G., Ivanov, V., Mirkova, E. and Choleva, B. 1976. Disinfection of soil with di-trapex in greenhouse cultivation of strawberry. *Ovoshcharstvo*, 55: 31–35.

Ondrasek, G., Romic, D., Romic, M., Duralija, B. and Mustac, I. 2006. Strawberry growth and fruit yield in a saline environment. *Agriculturae Conspectus Scientificus (Poljoprivredna Znanstvena Smotra)*, 71(4): 155–158.

Pankhurst, C.E., Doube, B.M. and Gupta, V.V.S.R. 1997. Biological indicators of soil health. CAB International, Oxford, UK. 8(7): 1400–1402.

Paszt, L.S. and Zurawicz, E. 2004. The influence of nitrogen forms on root growth and pH changes in the rhizosphere of strawberry plants. *Acta Horticulturae*, 649: 217–221.

Pirlak, L. and Esitken, A. 2004. Salinity effects on growth, proline and ion accumulation in strawberry plants. *Acta Agriculture Scandinavica Section-B, Soil and Plant Science*, 54(3): 189–192.

Rebeandel, Z. and Szczygiel, A. 1991. Effect of soil sterilization and cultivation system on strawberry growth and cropping. *Pracer z Zakresu Nauk rolniczych*, 71: 101–112.

Rolbiecki, S., Rolbiecki, R., Rzekanowski, C. and Derkacz, M. 2004. Effect of different inozation regimes on smooth and yield of Elsanta strawberry planted on loose sandy soil. *Acta Horticulturae*, 646: 163–166.

Rosati, C. 2002. Non-chemical alternatives to methy brominde. *Rivista di Frutticoltura e di Ortofloricoltura*, 64(7/8): 33–37.

Rubin, B., Cohen, O. and Gamliel, A. 2007. Soil solarization- an environmentally-friendly alternative. *Technical Meeting on Non-Chemical Alternatives for Soil-Borne Pest Control*, 26–28 June 2007, Hungary: pp. 1–136.

Scagel, C.F. 2003. Growth and nutrient use of ericacepus plants grown in media amended with sphagnum moss peat or coir dust. *HortScience*, 38(1): 46–54.

Seigies, A.T. and Prints, M. 2006. Cover crop rotation alters soil microbiology and reduce replant disorders in strawberry. *HortScience*, 41(5): 1303–1308.

Shinmura, A. 2000. Causal agent and control of root rot of welsh onion. *PSJ Soilborne Disease Workshop Report* 20: pp. 133–143.

Shoemaker, J.S. 1954. *Small Fruit Culture*. McGraw Hill Book Company INC, New York, New York: p. 447.

Singh, S.R., Zargar, M.Y., Singh, U. and Ishaq, M. 2010. Influence of bio-inoculants and inorganic fertilizers on yield, nutrient balance, microbial dynamics and quality of strawberry (*Fragaria* x *ananassa*) under rainfed conditions of Kashmir valley. *Indian Journal of Agricultural Sciences*, 80(4): 275–281.

Szteke, B., Jedrzejczak, R. and Reczajska, W. 2006. Impact of some environmental factors on the content of metals in strawberries. *Journal of Elementology*, 11(2): 213–222.

Turhan, E. and Eris, A. 2007. Growth and stomatal behaviour of two strawberry cultivars under long-term salinity stress. *Turkish Journal of Agriculture and Forestry*, 31(1): 55–61.

Waldo, G., Bringhurst, R.S. and Voth, V. 1971. Farmers bull. *USDA, Washington State University v Plant Disease*, 88(2): 210–214.

Wang, S.Y. and Liin, H.S. 2003. Compost as a soil supplement increases the level of antioxidant compounds and oxygen radical absorbance capacity in strawberries. *Journal of Agricultural and Food Chemistry*, 51(23): 6844–6850.

11 Propagation

Nimisha Sharma, Laxuman Sharma, and B.P. Singh

CONTENTS

11.1 Methods of Propagation 179
    11.1.1 Sexual Propagation 179
    11.1.2 Vegetative Propagation 179
        11.1.2.1 Field Propagation of Runners 180
        11.1.2.2 Substrate Culture 182
        11.1.2.3 Factors Affecting Runner Production 183
        11.1.2.4 Micropropagation 184
References 188

The success of the strawberry industry at a commercial scale in different growing regions is largely dependent on the availability of good-quality planting material in a sufficient quantity and at reasonable prices. The propagation of recommended cultivars of strawberry is of special significance, due mainly to the involvement of high cost for frequent arrangement of the mother plants. Genetic unpredictability is a major reason for not propagating strawberry by seed. Conventionally, strawberry is propagated by runners to obtain true-to-type plants; however, because of some limitations such as limited rooting, seasonal and ecological limitations, requirement and availability of skilled workers, nursery infrastructure, prevalence of disease and pests etc. (Sakila et al., 2007), micropropagation has become a better alternative for large-scale multiplication of strawberries. Several researchers have suggested micropropagation as comparatively better technique than field regeneration for large-scale propagation of strawberries (Karhu and Hakala, 2002; Mahajan et al., 2001; Mohan et al., 2005).

11.1 METHODS OF PROPAGATION

11.1.1 SEXUAL PROPAGATION

Strawberry propagation through seeds is not a commercial practice as it leads to large variation in the progeny, and hence it is neither scientifically advisable nor feasible for commercial cultivation. But seed propagation is very common and limited for raising hybrid populations. Seeds have very poor rate of germination (<4.2%) due to the presence of certain inhibitory substances in the fruit (Hammami et al., 2005). Fresh seeds of strawberry require stratification for germination. The seeds may be stratified for 11 days at 1°C (Borgman, 1950). Seed treatments with 96% H_2SO_4 (Jonkers, 1985) or gibberellic acid (0.1 mg dm^{-3}) enhance the proliferation of axillary crown shoots or runners and concurrently decrease the callus growth along with the formation of roots and adventitious shoots (Litwinczuk et al., 2009).

11.1.2 VEGETATIVE PROPAGATION

Under normal practice, strawberry plants are commercially grown from the plantlets, which are multiplied at the runner nodes of the adult plants. Initially, the runner plants produce few roots, thereafter make fibrous root system. Once the runner plant is big enough, the stolon may be severed from mother plant, and runner is planted elsewhere. The runner production by a mother plant is variety

179

specific. So the potential of production of runners by the adult plants is one of the major parameters that must be taken into consideration in the propagation of individual strawberry cultivars.

The quality of a runner is determined by the size of crown as it is the most important organ for storage of carbohydrate in strawberry (Macias-Rodriguez et al., 2002). A crown diameter of 8 mm is ideal for selection as a strawberry transplant (Hochmuth et al., 2006b), as size and number of flowers have positive correlations with the crown diameter of daughter plant (Jemmali and Boxus, 1993; Mason, 1987), although the crown size varies with cultivars. Plants of >1 cm crown size perform better in all aspects of growth and fruiting, including number of crowns and leaves per plant, plant height, number of runners, area of terminal leaf, number of days to flower induction, number of flowers, number of fruits, average fruit weight and fruit size, so crown size can be set as a selection criterion for quality planting materials (Chercuitte et al., 1991). Larger crowns resulted in higher vigour and faster initial growth and earlier and higher fruit yield (Bish et al., 2002; Gimenez et al., 2009), due to early floral initiation and higher number of flowers per plant (Mason, 1987; Jemmali and Boxus, 1993). Plants having bigger crowns tend to have a well-developed root system, absorbing water and nutrients soon after planting efficiently, and resulting in rapid initial growth and early and higher total berry yield (Hochmuth et al., 2001; Durner et al., 2002; Takeda and Hokanson, 2003).

Exporters of strawberry runners have systematic multi-step Certification Programme to supply healthy and quality planting materials (Table 11.1).

11.1.2.1 Field Propagation of Runners

11.1.2.1.1 Fresh Bare-rooted Plants

The rate of runner production is influenced by photoperiod and temperature in association of inductive cycle (Zhang et al., 2000). The period during September–October is ideal for strawberry planting in subtropical and mid-temperate (at elevation <1500 m above mean sea level) zones, while in typical temperate areas, strawberry is planted during March–April for fresh bare-rooted runner production. In subtropical areas, deblossoming may be avoided from mother plants for runner production due to prolonged growing season (May–September) after fruit harvesting, but in temperate areas, total growing season is only five months (April–August), hence blossom removal is indispensable. In subtropical areas, high temperature (>40°C) following fruit harvesting adversely affects strawberry runner production; therefore, the use of green coloured agro-shade net (50–90% shading) may enhance runner production (Figure 11.1), although the level of shading is variety specific (Sharma et al., 2013). In India, strawberry runner production in young apple orchards is gaining popularity for earning extra income by interspace utilization in temperate orchards (Figure 11.2). Depending upon soil fertility, variety and other management conditions, runner production may vary from 600,000–1,200,000/ha under open field in temperate conditions.

TABLE 11.1

Steps in Certification of Propagation Programme of Strawberry Runners

Sl. No.	Category	Remark
i	Candidate plant (Foundation stock)	Maintained in quarantine greenhouses for one year. These are free from pests and diseases.
ii	Super Extra Elite (SEE)	Prebasic mother stock from candidate plants maintained in insect-proof screen houses in individual pots, filled with gamma irradiated substrate.
iii	Super Elite (SE)	Produced in aphid-proof greenhouses or tunnels in soil or soilless culture.
iv	Extra Elite (EE)	These are multiplied in the open field.
v	Elite (E)	Daughter plants from EE, and sold as fresh (bare-rooted, misted tips or plug plants) or dormant plants (frigo plants, waiting bed plants, tray plants etc.) to the growers.

Source: Lieten (2014).

FIGURE 11.1 Strawberry runner production under shade net (50% shading).

FIGURE 11.2 Strawberry runner production in young apple orchards.

11.1.2.1.2 Waiting Bed Plants

Waiting bed technique was developed during late-1960s in the Netherlands (Dijkstra, 1989), wherein freshly dug plants were maintained for 20–30 days temporarily; thereafter, the rooted plants are planted in crates with the application of slow release fertilizers through soil. These plants are (Lieten, 2000, 2014) relatively large (15–24 mm crown diameters) with more inflorescences (four–seven a plant) and fruits (40–65/plant). The grades of waiting bed plants are given in Table 11.2.

11.1.2.1.3 Cold-stored Plants (Frigo Plants)

Cold storage of strawberry runners was developed in the United States, to overcome difficulties associated with planting during severe winter temperatures (Hughes, 1966), wherein runners are stored (at −2 to 0°C for 4–6 weeks) in boxes holding 5000 plants/m^3 (Lieten, 2014). Temperature has the most significant influence on plant performance (Higgwe, 1986) compared with a storage period or preplant treatment with fungicide(s). Work done on runner plants of Redgauntlet and Cambridge Favorite cultivars of strawberry, showed that the plants stored at −1.1° C runnered freely during the year of planting, and grew erect in comparison with freshly lifted plants (Way, 1966). Winter dug strawberry runners can be stored in sealed or ventilated polyethylene bags at −2.2°C without any adverse effect on vigour in the field after planting. Pre-storage dusting of benomyl helps in minimizing the infection of mould during cold storage to planting of runners. The sealing of runners in polyethylene bags during storage is effective to improve the fruit yield after transplanting in field (Little et al., 1974).

The cold storage of runners is not only a good method to safely overwinter the strawberry plants and to compensate for the lack of natural chilling but it extends the strawberry production season

TABLE 11.2

Grades of Waiting Bed and Cold-Stored Plants

Grade	Crown Diameter (mm)	Number of Plants/Crate
Waiting bed plants		
A[++]	>22	80–100
A[+]	18–22	120–150
A	15–18	200–250
Cold-stored plants (frigo plants)		
A[++]	>20	150
A[+]	15–18	250
A	12–15	350
A-Standard[a]	9–12	600

Source: Lieten (2014).

[a] Often used as mother stock materials

from July until January (Lieten et al., 2005). Kulikov and Vysotskiy (2007) investigated the effects of long-term storage of strawberry plants in *in vitro* condition, and observed that storage of strawberry plants in refrigerated chambers was 3.2 times more economical than maintenance in field conditions. Nowadays, to keep bare-root strawberries alive just long enough to make it to their new homes, nurserymen use moist paper wrapping of the roots and put them in moisture-tight plastic bags, then shipping boxes, thereafter these are stored in freezers below freezing. The plants are then shipped to the destination. The grades of frigo plants are given in Table 11.2.

11.1.2.2 Substrate Culture

Substrate tip propagation for raising certified planting materials of strawberries has gained popularity. Mother plants of desired varieties of strawberry are grown in substrate at $1.5 \text{ m} \times 25–30 \text{ cm}$ plant spacing. Substrate culture can also be used in greenhouse to raise certified planting materials of short-day and day-neutral cultivars. Mother plants are grown in containers placed on gutters hanging 2–3 m from the ground and spaced at 1.25 m from each other (Lieten, 2014).

11.1.2.2.1 Plug Plants

Annual strawberry plantings are generally established by using either freshly dug or dormant, cold-stored "frigo" plants, as these are relatively cheap. Dormant plants dug during the late autumn season, and cold stored plants are used for planting during the spring or early summer. However, prolonged storage tends to decrease the viability of plants (Hokanson et al., 2004). Plug or tray plants offer an alternative approach used in annual hill and high-tunnel strawberry production systems. Plug plants are expensive as compared to fresh-dug or dormant plants but, typically, these plants offer an appropriate level of initial vigour for autumn planting. Plug plants are relatively easy to grow and can be produced even in small growing structures such as a greenhouse or cold frame. The cold-stored mother plants are planted during late spring/early summer to produce runner tips for multiplication. The use of containerized plug transplants improve plant survival in the field without use of any fumigants (Durner et al., 2002). Plug transplants are being used in Europe and North America. Plug plants are easier to transplant, giving high establishment with limited use of water (Hochmuth et al., 2006a,b; Takeda and Newell, 2006). The plug transplants production can be planned precisely to make them available for planting at the intended time (Fernandez and Ballington, 2003; Hochmuth et al., 2006a). The transplanting of plugs having an intact root system without any mechanical damage by the remaining attached substrate to the roots shows enhanced initial growth to advanced flowering and berry harvesting (Duval et al., 2006; Hochmuth et al., 2006a).

Propagation 183

Production of runner tips is the first step in producing strawberry plug plants. Runners can be produced from mother plants, either in a greenhouse or in open field conditions by growing them in the soil or in peat-filled growing bags placed on the soil or on benches. Profuse runner production requires high temperatures and long days, so open field production is restricted to summer only. In protected structures, the first runner tip is expected about 8–10 weeks after the establishment of mother plant (Durner et al., 2002). The number of runners produced per plant increases with the advancement in time and the vigour of mother plants. Greenhouses can provide comparatively ideal temperatures (above 24°C) during day time and help to augment photoperiods (about 16 hours) for initiation of the process of runner production. With supplemental heat and light, the activity of runner production can be continued throughout the year. Usually, the plants require to be irrigated once in every three or four hours, with each irrigation event lasting for three or four minutes at the time of peak runner production. Runner tips are harvested when root initials (little white or brown pegs) become visible (not >1.2 cm) on the runner tip. Sorting of tips according to size prevents larger plug plants from crowding out the smaller ones in the tray (Durner et al., 2002; Takeda and Newell, 2006). Fifty cell plug trays having about 7 inches^3 per cell have proved satisfactory for strawberry plug production. Runners have to be without any damage to trifoliate leaves, and runner tips must be planted immediately after harvest. The runners are usually planted or cooled to 0°C within 45 minutes. If necessary, The runner tips can be stored upto one week at 0–1.1° C with 95% RH (Durner et al., 2002).

While planting, the root pegs and anchor must be kept just below the soil surface, and the developing leaves and the crown be removed above the soil surface. The intermittent misting of the plants for 7–12 days is generally sufficient and thereafter, the plants should be hardened off in the protected structure for one–two weeks before their establishment in the field or high tunnel (Durner et al., 2002). Generally, a runner tip takes about one month to convert into a well-rooted plug plant. The plug plants are an excellent alternative method for autumn, which provides appropriate levels of vegetative vigour for successful spring harvests. Besides, plug plants can also be a valuable option for September plantings in subtropical areas, where bare-rooted plants do not survive due to high temperatures. Strawberry plants transplanted during September start flowering and fruiting from the last week of December, which is highly remunerative. The non-availability of plug plants at the optimum time for autumn planting is the major drawback of plug plant production (Rowley, 2010).

The cell size used for strawberry plug production in trays influences the output of plants significantly (Folta, 2004). Advanced strawberry fruit yield can be obtained by using plug transplants, even from trays with small cells of 26.5 or 50.0 cm^3 (Gimenez et al., 2009). A period of about 28–35 days is required for production of plug transplants (Durner et al., 2002), during which, the size of container influences the availability of water and mineral nutrients. Container size can influence the plant growth in soilless culture (Nesmith and Duval, 1998). The use of trays with large cell size facilitates the growth of transplants; however, it reduces the number of transplants per tray and increases the amount of substrate and the costs of package and shipping (Bish et al., 2002; Hochmuth et al., 2006a,b). On the other hand, the use of trays with small cells tends to reduce the cost but may affect the physiological quality of the plants (Durner et al., 2002).

11.1.2.2.2 *Tray Plants*

These are the main plant types grown on substrate for producing certified plant materials. Runner tips are taken from the mother plants and managed to initiate roots directly in trays filled with peat-moss mixture. Unrooted cuttings are maintained in 250cc (9 cm deep × 8 cm diameter), peat-filled multicell trays, arranged in a 9–16 cell in a 60–100 cm × 20 cm tray (Lieten, 2014).

11.1.2.3 Factors Affecting Runner Production

The growth of runner is a result of cell division and elongation in the internodes (Nishizawa and Hori, 1993). The runner plant production is very much specific to ecological conditions. Therefore, growers face difficulties in obtaining high-quality runner plants in sufficient quantity (Kaska et al., 1984). In most of the cases, the runner plants are propagated from their own stocks, which is the possible reason for poor berry yield and quality (Turkben et al., 1997). Some factors affecting output of runner plants are plant bio regulators (Franciasi et al., 1985; Wang, 1992), growing elevation (Pırlak et al., 2002) and management practices (Anna and Iapichino, 2002). The rooting media (peat: perlite + nutrients as first rooting medium; and 0.75:1:1:0.5:0.75 soil: peat: perlite: sand: straight cow manure + nutrients) and order of runner node have been observed to influence the runner production significantly, while layered in conical yellow pots, although the response is variety specific (Turkben, 2008).

Light is an important factor affecting runner production in strawberry. The short-day photoperiod tends to produce more crown than the runners. A day length of 15 hours and temperature >27°C have been observed to be the most effective conditions for profuse runner production in strawberry (Darrow, 1966; Hytonen et al., 2009). The long-day photoperiod increases the level of endogenous gibberellins, and encourages proliferation of buds in the plant (Taylor et al., 1994); this difference in peduncle length is attributed to greater number of epidermal cells which indicates that cell division, and internodal length is prolonged in long-day conditions (Nishizawa and Hori, 1993; Nishizawa, 1994).

The number and length of strawberry runners are not only dependent on the photoperiod but genetic nature of variety (Pipattanawong et al., 1996; Libek and Tamm, 2001). Photoperiod, cultivars and interaction effects of photoperiod with cultivars significantly influence the number of runners per plant, and increase the runner length and number of plantlets per runner (Hasan et al., 2011).

Mycorrhizal inoculation has been observed to produce a higher number (2.6 per mother plant) of runner formation and daughter plants (6.2 per mother plant) than the control plants (1.4 runners and 3.8 daughters per plant), as it improves the capacity of propagation of mother plants and better growth, and nitrogen and phosphorus uptake in daughter plants (Alarcon et al., 2000).

The effect of biostimulants on promotion of strawberry plant growth has been studied by many researchers (Roussos et al., 2009; Sarli et al., 2009; Abdel-Mawgoud et al., 2010). An animal protein hydrolysate, a biostimulant; has been observed to enhance biomass production of newly formed roots (Marfa et al., 2009). In the research conducted by Zurawicz et al. (2004), the treatment of mother plants with plant bioregulators improved the number of daughter plants by almost 300% as compared to untreated control mother plants. Sas-Paszt et al. (2008) also demonstrated significant influence of bioregulators on the number and size of strawberry runner plants. Aslantas and Guleryuz (2004) confirmed the beneficial influence of biostimulants on the quantity and quality of strawberry runners and daughter plants. However, Lisiecka et al. (2011) observed a different effect of the animal protein hydrolysate-based biostimulant on number, length and diameter of runners and number of strawberry daughter plants as well as their crown diameter, fresh weight and number of leaves. Results of several experiments have shown that the effect of biostimulators depends on dose and frequency of application as well as the genetic nature of strawberry cultivar (Sas-Paszt et al., 2008; Botta et al., 2009; Sarli et al., 2009).

11.1.2.4 Micropropagation

After the initial report on it in the late 1960s (Boxus, 1974, 1999), micropropagation has become an inseparable part of strawberry cultivation. Presently, millions of strawberry plants are multiplied annually by tissue culture, as it gives a definite answer to some of the serious problems of viruses and soilborne fungi. Interestingly, this technique has the potential to raise a large number of true-to-type planting materials in a very short span of time (Mohan et al., 2005; Mullin and Schlegel, 1976) in a limited space, which can be used for production or for germplasm conservation. Complete new plants can be derived from tissue either from preexisting buds through adventitious shoot regeneration or through the formation of somatic embryos. Plants raised by axillary branching is considered as the most reliable *in vitro* method for producing true-to-type planting materials, as they have unchanged genetic composition of the mother plant. It has also improved

Propagation 185

the capacity of these plants to produce runners for planting in the field (Lopez-Aranda et al., 1994; George, 1996). Tissue culture derived strawberry plants grow more vigorously and produce more crowns, runners and inflorescences with higher yields than traditionally propagated plants (Lopez-Aranda et al., 1994; Diengngan et al., 2016) due to increased vigour and axillary bud activity, possibly related to the forced proliferation through hormonally induced crown branching (Swartz et al., 1981). This is caused by the carry over effect of cytokinin in the shoot proliferation medium (Waithaka et al., 1980). Physiological influence of growth hormones as well as of higher leaf densities is mainly responsible for the increase in plant vigour of *in vitro* propagated plants (Swartz et al., 1981; Cameron et al., 1989; Lopez-Aranda et al., 1994; Zebrowska et al., 2003). *In vitro*-generated plants also develop more leaves, enhancing the light interception for photosynthesis earlier than conventionally propagated plants (Gantait et al., 2010); however, these plants cost four or five times more than plants raised through conventional propagation (George, 1996).

Most commonly used explant for strawberry micropropagation is the meristem from the runner tip (Sowik et al., 2001), leaf disk (Debnath, 2005) and petiole (Folta et al., 2006), which are placed on a medium containing higher levels of cytokinins without auxin to induce the axillary budding and preventing the callus formation. The cytokinins helps in overcoming apical dominance and accelerating the branching of lateral buds from the leaf axis. Further, additional shoots are raised by the growth of axillary bud (Debnath, 2003). Shoot proliferation in strawberry has successfully been achieved from single meristems (Boxus, 1974), meristem callus (Nishi and Oosawa, 1973) and from node culture (Bhatt and Dhar, 2000).

In addition, strawberry plants can also be raised *in vitro* from various plant parts, namely peduncles/peduncular base of the flower bud (Foucault and Letouze, 1987; Lis, 1993), stems (Graham et al., 1995), stipules and roots (Passey et al., 2003), stolons (Lis, 1993), runners (Liu and Sanford, 1988), mesophyll protoplasts (Nyman and Wallin, 1988), anther cultures (Owen and Miller, 1996) and immature embryos (Wang et al., 1984); however, it is essential to take explants from those plants and maintain them in controlled conditions to avoid contamination during *in vitro* propagation.

Nodal segments (2.5–3 cm) can be used as explant to produce micropropagated wild strawberry (Bhatt and Dhar, 2000), which need to be rinsed initially with Savlon (an antiseptic containing 3% antimicrobial agent centrimide + detergent) for 15 minutes. Further, explants are immersed in 80% ethanol for 30 seconds just before the surface sterilization in $HgCl_2$ (0.05%) along with few drops of Tween-20 for five minutes. Explants are cultured on MS medium supplemented with 4 μM BA and 0.1 μM α-NAA.

Borkowska (2001) used modified medium (Boxus, 1974, 1992), by decreasing the concentration of BA (2.2 μM) and IBA (0.5 μM) for strawberry cultures. The agar (0.6–0.8%, w/v) has been observed to be the most commonly used gelling agent for *in vitro* strawberry culture on semisolid medium. The agar/galactomannan (guar, Indian Gum Industries, Jodhpur, India) mixture (0.3/0.3 w/v) in MS medium is more effective to enhance the shoot proliferation than medium containing agar (0.6%, w/v) only (Lucyszyn et al., 2006).

Hormones, vitamins, amino acids, physical state and the source of power for the media play important roles in the development of strawberry plants under *in vitro* conditions. Moreover, cytokinins play a major role in regulation of shoots branching. Several researchers have shown the importance of N6-benzylamino-purine (BAP) for regeneration of strawberry (Adel and Sawy, 2007; Biswas et al., 2007; Sakila et al., 2007). Nowadays, cytokinins such as BAP and thidiazuron (TDZ) are used for shoot regeneration (Passey et al., 2003; Landi and Mezzetti, 2006). TDZ can be used either alone (Mohamed et al., 2007) or in combination with auxins such as IBA for shoot regeneration of strawberry (Yonghua et al., 2005; Landi and Mezzetti, 2006). The lower level of BAP (0.5 mg/l) is more effective for shoot proliferation in strawberry (Marcotrigiano et al., 1984). The best medium for regeneration of strawberry is the MS medium supplemented with BAP @ 1.5 mg/l and cytokinin KIN @ 0.5 mg/l, and the plantlets produce roots in the MS medium with addition of IBA @ 1 mg/l (Sakila et al., 2007). MS medium supplemented with 2 mM TDZ and 4 mM BAP has been observed optimum for shoot multiplication from the shoot tips (Haddadi and Aziz, 2010). Moradi et al. (2011) showed the propagation technique of *Fragaria* using nodal segments from *in vitro*-germinated plants. The optimum concentration of BAP for bud evocation was 0.5 mg/l alongwith KIN@ 0.2 mg/l. However, MS basal medium containing 40 or 80 mg/l adenine sulphate has been observed to be best for rooting (Mathur et al., 2008). Gibberellic acid (0.1 mg dm^{-3}) enhances the proliferation of axillary crown shoots or runners and concurrently decreases callus growth along with the formation of roots and adventitious shoots (Litwinczuk et al., 2009). Further, the type and concentration of amino acids have a significant influence on the somatic embryogenesis. Proline is the best source for embryo culture of strawberry plants (Gerdakaneh et al., 2012). Genotypes of strawberry also affect the induction of somatic embryos (Gerdakaneh et al., 2009). The available information from different studies has indicated that optimal concentrations of different amino acids and genotypes need to be optimized for further recommendation (Chukwuemeka et al., 2005; Homhuana et al., 2008; Han et al., 2009).

Adventitious shoot regeneration is influenced by various factors like genotype, culture medium (including growth regulators and their combinations), physical environment, explant development stage etc. (Table 11.3). TDZ is the most active cytokinin and widely used to induce shoot organogenesis of strawberries. It can be used alone (Debnath, 2005) or in combination (Passey et al., 2003)

TABLE 11.3

In vitro Adventitious Shoot Regeneration of Strawberries Using Different Explants and Plant Bio regulators (PBRs)

Varieties/Cultivar	Source of Explant	Medium Composition Other than MS Salts (mg/l)	Shoot Regeneration (PBR, mg/l)	Reference
Redcoat, Veestar, Bounty,Kent, Micmac, Glooscap, Honeoye, Hecker, Fern, Selva	Leaf disk	B-5 vitamins	2.3 BA + 1.8 IAA	Nehra et al. (1989)
Hiku, Jonsok	Leaf disk	39 Fe (III) Na-EDTA + 2000 KNO3 + 400–600 CH	3 BA + 0.1 IBA	Sorvari et al. (1993)
Chandler	Leaf disk	N30K macrosalts microsalts and vitamins	2 BA + 0.5 IBA	Barcelo et al. (1998)
Calypso, Pegasus, Bolero, Tango, Elsanta	Leaf disk, petiole, root, stipule	Only MS	1 TDZ + 0.2 2.4-D, 2 BA + 0.5 TDZ + 0.2 2,4-D, 1 TDZ + 0.2 NAA or 2 BA + 0.2 2,4-D	Passey et al. (2003)
Hecker, La Sans Rivale, diploid accessions (FRA197, FRA198)	Leaf, petiole	B5 vitamins	2.2 TDZ + 0.3 IBA	Zhao et al. (2004)
Bounty	Sepal, leaf disk, petiole half	BM-D	0.44–0.88 TDZ	Debnath (2005)
Toyonoka	Leaf disk	Only MS or with AgNO3	1.5 TDZ + 0.4 IBA	Qin et al. (2005a, b)
Bounty, Jonsok, Korona, Polka, Zephyr	Leaf	Only MS	2 TDZ + 0.5 IBA	Hanhineva et al. (2005)
LF9	Leaf, petiole, stolon	Only MS	0.11 BA + 0.011 2,4-D + 1 TDZ	Folta et al. (2006)

Media: BM-D = Debnath and McRae (2001), MS = Murashige and Skoog (1962), N30K = Margara (1984).

Propagation

with 2, 4-dichlorophenoxy-acetic acid (2,4-D) or IBA (Yonghua et al., 2005) for shoot regeneration from strawberry leaves. Moreover, sepals have been tested for shoot regeneration of *in vitro* strawberry cultures (Debnath, 2005). Shoot regeneration has been obtained from sepal, leaf and petiole explants by incorporating TDZ (2–4 µM) in the culture medium and a dark treatment for 14 days before incubating the explants in a 16-hour photoperiod.

TDZ is more effective than BAP in inducing shoot regeneration. Two weeks of dark treatment has shown to increase the shoot regeneration rate (90.09%). Addition of $AgNO_3$ to the medium can change the direction of cell differentiation, from the formation of adventitious shoots to that of somatic embryos (Xuemei et al., 2004). H_2O_2 has been found to be correlated with the morphogenetic process in strawberry callus by acting as a messenger in the formation of bud primordium (Tian et al., 2003).

11.1.2.4.1 Bioreactors for Micropropagation

A bioreactor is a self-contained device, where sterile environmental conditions are maintained. It utilizes a liquid nutrient or liquid/air inflow and outflow systems, fabricated for intensive culture and control over micro-environmental conditions such as aeration, agitation, dissolved oxygen etc. (Levin and Vasil, 1989; Paek et al., 2005). The use of large-scale suspension cultures and automation has the potential to resolve the manual handling of the various stages of micropropagation. Ultimately, it decreases the production cost significantly. Initially, two types of bioreactors; mechanically agitated bioreactors and pneumatically agitated and non-agitated bioreactors are utilized for optimal mixing of oxygen, nutrients and culture without severe shear stresses (Paek and Chakrabarty, 2003).

Suspension culture is advantageous for several plant species. However, it causes asphyxia and hyperhydricity that result in malformed plants and loss of material. The malformed plants are characterized by glossy hyperhydrous leaves with distorted anatomy. To solve these issues, the following two major solutions (Ziv et al., 2003) are advocated:

1. Growth retardants used for malformation control
2. Temporary immersion bioreactors (TIB) to control rapid proliferation

Further, inclusion of 0.1 mM L-α-aminooxy-β-phenylpropionic acid (or AOPP); a specific inhibitor of phenylalanine ammonialyase and the starting and core enzyme of the phenylpropanoid pathway, has been observed (Edahiro and Seki, 2006) to reduce cell clumping in cell suspension cultures of cell line FAR (*F. ananassa* R).

TIB bioreactors (RITA®, VITROPIC, Saint-Mathieu-de-Tréviers, France) containing liquid MS medium with 9 µM TDZ and 2.5 µM IBA, used for shoot regeneration from leaf explants of five strawberry cultivars (Hanhineva et al., 2005), have been proved to be well suited for shoot propagation and for subsequent subculture of the developing plantlets with renewal frequency of 70 ± 8 to $94 \pm 2\%$, as compared with semisolid medium (83 ± 5 to $92 \pm 3\%$). It is highly cost effective as it requires half the time for handling plant material than for cultivation on a semisolid medium (Debnath, 2008; Debnath and DaSilva, 2007).

11.1.2.4.2 Acclimatization

Strawberry plantlets can be well adapted and established by transplanting them into trays filled with peat:perlite (1:1 v/v), and placing in a transparent polypropylene cover with 90% RH for 21 days. Thereafter, the plants should be individually transferred to the pots with peat, sand, and perlite (7:2:1) for another 21 days (Medina et al., 2007). Transplanting of plantlets into substrate consisting of perlite + vermiculite + cocopeat (2:1:2 v/v/v) increases the plant survival (90%). Hoagland's solution is used for irrigation after 30 days of planting. The runners are ready to plant after 90 days of acclimatization (Haddadi and Aziz, 2010).

REFERENCES

Abdel-Mawgoud, A.M.R., Tantawy, A.S., El-Nemr, M.A. and Sassine, Y.N. 2010. Growth and yield responses of strawberry plants to chitosan application. *European Journal of Scientific Research*, 39(1): 161–168.

Adel, E.L. and Sawy, M. 2007. Somaclonal variation in micro-propagated strawberry detected at the molecular level. *International Journal of Agriculture and Biology*, 9: 72–725.

Alarcon, A., Ferrera-Cerrato, R., Gonzalez-Chavez, M.C. and Villegas-Monter, A. 2000. Arbuscular mycorrhizal fungi in runner dynamics and nutrition of strawberry plants cv. Fern obtained from *in vitro* culture. *Terra*, 18(3): 211–218.

Anna, F. D. and Iapichino, G. 2002. Effects of runner order on strawberry plug plant fruit production. *Acta Horticulturae*, 1(567): 301–303.

Aslantas, R. and Guleryuz, M. 2004. Influence of some organic biostimulants on runner production of strawberry. *Atatürk University Z.F.*, 35(1–2): 31–34.

Barcelo, M., El-Mansouri, I., Mercado, J.A., Quesada, M.A. and Alfaro, F.P. 1998. Regeneration and transformation via *Agrobacterium tumefaciens* of the strawberry cultivar Chandler. *Plant Cell, Tissue and Organ Culture*, 54: 29–36.

Bhatt, I.D. and Dhar, U. 2000. Micropropagation of Indian wild strawberry. *Plant Cell Tissue and Organ Culture*, 60: 83–88.

Bish, E.B., Cantliffe, D.J. and Chandler, C.K. 2002. Temperature conditioning and container size affect early season fruit yield of strawberry plug plants in a winter, annual hill production system. *HortScience*, 37: 762–764.

Biswas, M.K., Hossain, M., Ahmed, M.B., Roy, U.K., Karim, R., Razvy, M.A., Salahin, M. and Islam, R. 2007. Multiple shoots regeneration of strawberry under various colour illuminations. *American-Eurasian Journal of Scientific Research*, 2: 133–135.

Borgman, H.H. 1950. Propagation. In: R.M. Sharma and R. Yamdagni (eds.). *Modern Strawberry Cultivation*. Kalyani Publishers, Ludhiana, India: 58–62.

Borkowska, B. 2001. Morphological and physiological characteristics of micropropagated strawberry plants rooted *in vitro* or *ex vitro*. *Scientia Horticulturae*, 89: 195–206.

Botta, A., Marin, C., Pinol, R., Ruz, L., Badosa, E. and Montesinos, E. 2009. Study of the mode of action of Inicium®, a product developed specifically to overcome transplant stress in strawberry plants. *Acta Horticulture*, 842: 721–724.

Boxus, P. 1974. The production of strawberry plants by *in vitro* micro propagation. *Journal of Horticulture Science*, 49: 209–210.

Boxus, P. 1992. Mass production of strawberry and new alternative for some horticultural crops. In: K. Kurata and T. Kozai (eds.). *Transplant Production Systems*. Kluwer Academic Publishers, The Netherlands: 151–162.

Boxus, P. 1999. Micropropagation of strawberry via axillary shoot proliferation. In: R.D. Hall (ed.). *Plant Cell Culture Protocols. Methods in Molecular Biology. Part III. Plant Propagation In Vitro*. Humana Press Inc., Totowa, New Jersey , 111: 103–114.

Cameron, J.S., Hancock, J.F. and Flore, J.A. 1989. The influence of micropropagation on yield components, dry matter partitioning and gas exchange characteristics of strawberry. *Scientia Horticulturae*, 38: 61–67.

Chercuitte, L., Sullivan, Y.A., Desjardins, Y.D. and Bedard, R. 1991. Yield potential and vegetative growth of summer-planted strawberry. *Journal of American Society for Horticultural Sciences*, 116: 930–936.

Chukwuemeka, R.E., Akomeah, P. and Asemota, O. 2005. Somatic embryogenesis in date palm (*Phoenix dactylifera* L.) from apical meristem tissues from "zebia" and "loko" landraces. *African Journal of Biotechnology*, 4(3):244–246.

Darrow, G.M. 1966. *The Strawberry: History, Breeding and Physiology*. Holt, Rinehart and Winston, New York, New York.

Debnath, S.C. 2003. Micropropagation of small fruits. In: S.M. Jain and K. Ishi (eds.). *Micropropagation of Woody Trees and Fruits*. Kluwer Academic Publishers, Dordrecht, Germany: 465–506.

Debnath, S.C. 2005. Strawberry sepal: Another explant for thidiazuron-induced adventitious shoot regeneration. *In Vitro Cell*, 41: 671–676.

Debnath, S.C. 2008. Developing a scale-up system for the *in vitro* multiplication of thidiazuron-induced strawberry shoots using a bioreactor. *Canadian Journal of Plant Science*, 88: 737–746.

Debnath, S.C. and McRae, K.B. 2001. *In vitro* culture of lingonberry (*Vaccinium vitis-idaea* L.): The influence of cytokinins and media types on propagation. *Small Fruits Reviews*, 1: 3–19.

Debnath, S.C. and Da Silva, J.A.T. 2007. Strawberry culture *in vitro*: Applications in genetic transformation and biotechnology. *Fruit, Vegetable and Cereal Science and Biotechnology*, 1: 1–12.

Diengngan, S., Mahadevamma, M. and Srinivasa Murthy, B.N. 2016. Efficacy of *in vitro* propagation and crown sizes on the performance of strawberry (*Fragaria* × *ananassa* Duch) cv. Festival under field condition. *Journal of Agricultural Science and Technology*, 18: 255–264.

Dijkstra, J. 1989. The use of cold stored waiting-bed plants for a late harvest. *Acta Horticulturae*, 265: 207–214.

Durner, E.F., Poling, E.B. and Maas, J.L. 2002. Recent advances in strawberry plug transplant technology. *HortTechnology*, 12: 545–550.

Duval, J.R., Chandler, C.K. and Legard, D.E. 2006. *Effect of Mechanical Damage to Strawberry Transplants*. Document HS922, University of Florida, Gainesville Florida. http://www.edis.ifas.ufl.edu

Edahiro, J.I. and Seki, M. 2006. Phenylpropanoid metabolite supports cell aggregate formation in strawberry cell suspension culture. *Journal of Bioscience and Bioengineering*, 102: 8–13.

Fernandez, G.E. and Ballington, J.R. 2003. Double cropping of strawberries in an annual system using conditioned plug plants and high tunnels. *Acta Horticulturae*, 614: 547–555.

Folta, K.M. 2004. Green light stimulates early stem elongation, antagonizing light-mediated growth inhibition. *Plant Physiology*, 135: 1407–1416.

Folta, K.M., Dhingra, A., Howard, L., Stewart, P. and Chandler, C.K. 2006. Characterization of LF9, an octoploid strawberry genotype selected for rapid regeneration and transformation. *Planta*, 224: 1058–1067.

Foucault, C. and Letouze, R. 1987. In vitro regeneration de plantes de Fraisier apartir de fragmentes de petiole et de bourgeons floraux. *Biologia Plantarum*, 29: 409–414.

Franciasi, R., Salas, P., Yamashiro, E. and Duarte, O. 1985. Effect of gibberellic acid on runner formation in different strawberry cultivars. *Journal of the American Society for Horticultural Science*, 24: 127–129.

Gantait, S., Nirmal, M. and Prakash, K.D. 2010. Field performance and molecular evaluation of micropropagated strawberry. *Recent Research in Science and Technology*, 2: 12–16.

George, E. F. 1996. *Plant Propagation by Tissue Culture. Part 2. Exegetics*. Exegetics Ltd: Edington, UK.

Gerdakaneh, M., Mozafari, A. A., Khalighi, A. and Sioseh-Mardah, A. 2009. The effects of carbohydrate source and concentration on somatic embryogenesis of strawberry (*Fragaria* × *ananassa* Duch.). *American-Eurasian Journal of Agricultural & Environmental Science*, 6: 76–80.

Gerdakaneh, M., Mozafari, A. and Sioseh-Mardah, A. 2012. Comparative root colonisation of strawberry cultivars Camarosa and Festival by *Fusarium oxysporum* f. sp. *fragariae*. *Plant and Soil*, 358: 75–89.

Gimenez, G., Andriolo, J.L., Janish, D.J., Cocco, C. and Dal Picio, M. 2009. Cell size in trays for the production of strawberry plug transplants. *Pesquisa Agropecuaria Brasileira*, 44: 726–729.

Graham, J., McNicol, R.J. and Grieg, K. 1995. Towards genetic based insect resistance in strawberry using the cowpea trypsin inhibitor gene. *Annals of Applied Biology*, 127: 163–173.

Haddadi, F. and Aziz, M.A. 2010. Micropropagation of strawberry cv.Camarosa: Prolific shoot regeneration from *in vitro* shoot tips using thidiazuron with N6-benzylamino. *Horticulture Science*, 45(3): 453–456.

Hammami, I., Jellali, M., Ksontini, M. and Rejeb, M.N. 2005. Propagation of the strawberry tree through seed (*Arbutus unedo*). *International Journal of Agriculture & Biology*, 1560–8530/2005/07–3–457–459.

Han, G.Y., Wang, X.F., Zhang, G.Y. and Ma, Z.Y. 2009. Somatic embryogenesis and plant regeneration of recalcitrant cottons (*Gossypium hirsutum*). *African Journal of Biotechnology*, 8(3): 432–437.

Hanhineva, K., Kokko, H. and Karenlampi, S. 2005. Shoot regeneration from leaf explants of five strawberry (*Fragaria* × *ananassa*) cultivars in temporary immersion bioreactor system. *In vitro Cellular and Developmental Biology – Plant*, 41: 826–831.

Hasan, S.M.Z., Al-Madhagi, I., Ahmad, A. and Yusoff, W.A.B. 2011. Effect of photoperiod on propagation of strawberry (*Fragaria* × *ananassa* Duch.). *Journal of Horticulture and Forestry*, 3: 259–263.

Higgwe, T.E. 1986. Evaluation of preplant storage treatments and subsequent growth responses of strawberry plants after transplanting. M.Sc. Thesis. Kansas State University Manhattan, Kansas.

Hochmuth, G., Chandler, C., Stanley, C., Legard, D., Duval, J., Waldo, E., Cantliffe, D. and Bish, E. 2001. Containerized transplants for establishing strawberry crops in Florida. *HortScience*, 37: 443–446.

Hochmuth, G., Cantliffe, D., Chandler, C., Stanley, C., Bish, E., Waldo, E., Legard, D. and Duval, J. 2006a. Containerized strawberry transplants reduce establishment period water use and enhance early growth and flowering compared with bare root plants. *HortTechnology*, 16: 46–54.

Hochmuth, G., Cantliffe, D., Chandler, C., Stanley, C., Bish, E., Waldo, E., Legard, D. and Duval, J. 2006b. Fruiting responses and economics of containerized and bare root strawberry transplants established with different irrigation methods. *HortTechnology*, 16: 205–210.

Hokanson, S.C., Takeda, F., Enns, J.M. and Black, B.L. 2004. Influence of plant storage duration on strawberry runner tip viability and field performance. *HortScience*, 39: 1596–1600.

Homhuana, S., Kijwijana, B., Wangsomnuka, P., Bodhipadmab, K. and Leungc, D.W.M. 2008. Variation of plants derived from indirectsomatic embryogenesis in cotyledon explants of papaya. *Science Asia*, 34: 347–352.

Hughes, H.M. 1966. Runner storage and plant size in relation yield to of strawberries. *Efford Experimental Horticulture Station*, 16: 135–141.

Hytonen, T., Elomaa, P., Moritz, T. and Junttila, O. 2009. Gibberellin mediates daylength-controlled differentiation of vegetative meristems in strawberry (*Fragaria* × *ananassa* Duch). *BMC Plant Biology*, 9: 18.

Jemmali, A. and Boxus, P. 1993. Early estimation of strawberry floral intensity and its improvement under cold greenhouse. *Acta Horticulture*, 348: 357–360.

Jonkers, H. 1985. Accelerated flowering of strawberry seedlings. *Euphytica*, 1: 41–46.

Karhu, S. and Hakala, K. 2002. Micropropagated strawberries on the field. *Acta Horticulture*, 2: 182.

Kaska, N., Cınar, A. and Eti, S. 1984. Effects of runner plants, grown in Adana and Pozanti on earliness of strawberry, yield and quality. *Doga Journal TUBITAK D2*, 8(3): 259–264.

Kulikov, I. and Vysotskiy, V. 2007. Storage of strawberry plants *in vitro*, their stability and economical aspects. *Sodininky ir Darzininkyste*, 26: 226–234.

Landi, L. and Mezzetti, B. 2006. TDZ, auxin and genotype effects on leaf organogenesis in Fragaria. *Plant Cell Report*, 25: 281–288.

Levin, R. and Vasil, I.K. 1989. An integrated and automated tissue culture system for mass propagation of plants. *In Vitro Cellular & Developmental Biology – Plant*, 25: 21–27.

Libek, A. and Tamm, K. 2001. The influence of strawberry planting material on the formation of runner plants and yield. *Transactions of the Estonian Academic Agricultural Society*, 15: 23–26.

Lieten, F. 2000. La fragola in Belgio-Olanda. In: W. Faedi (ed.). *La Fragola verso il*. Camera di Commercio, Industria, Artigianato e Agricoltura, Verona, 1998: 83–94.

Lieten, P. 2014. The strawberry nursery industry in The Netherland: An update. *Acta Horticulturae*, 1049: 99–106.

Lieten, P., Evenhuisb, B. and Baruzzic, G. 2005. Cold storage of strawberry plants. *International Journal of Fruit Science*, 5: 75–82.

Lis, E.K. 1993. Strawberry plant regeneration by organogenesis from peduncle and stolon segments. *Acta Horticulturae*, 348: 435–438.

Lisiecka, J., Knaflewski, M., Spizewski, T., Frąszczak, B., Kałuzewicz, A. and Krzesinski, W. 2011. The effect of animal protein hydrolysate on quantity and quality of strawberry daughter plants cv. Elsanta. *Acta Scientiarum Polonorum Hortorum Cultus*, 10: 31–40.

Little, C.R., Kroon, K.H. and Proctor, R.G. 1974. Cool storage and post-storage treatment of strawberry runners for summer planting. *Australian Journal of Experimental Agriculture and Animal Husbandry*, 14: 118–121.

Litwińczuk, L., Okołotkiewicz, E. and Matyaszek, I. 2009. Development of *in vitro* shoot cultures of strawberry (*Fragaria* × *ananassa* Duch.) "Senga Sengana" and "Elsanta" under the influence of high doses of gibberellic acid. *Folia Horticulturae Ann*, 21(2): 43–52.

Liu, Z.R. and Sanford, J.C. 1988. Plant regeneration by organogenesis from straw-berry leaf and runner culture. *HortScience*, 23: 1056–1059.

Lopez-Aranda, J.M., Pliego-Alfaro, F., Lopez-Navidad, I. and Barcelo-Munoz, M. 1994. Micropropagation of strawberry (*Fragaria*×*ananassa* Duch.). Effect of mineral salts, benzyladenine levels and number of subcultures on the *in vitro* and field behaviour of the obtained microplants and the fruiting capacity of their progeny. *Journal of Horticulture Science*, 69: 625–637.

Lucyszyn, N., Quoirin, M., Ribas, L.L.F., Koehler, H.S. and Sierakowski, M.R. 2006. Micropropagationof "Durondeau" pear in modified-gelled medium. *In Vitro Cellular & Developmental Biology – Plant*, 42: 287–290.

Macias-Rodriguez, L., Quero, E. and Lopez, M.G. 2002. Carbohydrate differences in strawberry crowns and fruit (*Fragaria* × *ananassa*) during plant development. *Journal of Agriculture & Food Chemistry*, 50: 3317–3321.

Mahajan, R., Kaur, R., Sharma, A. and Sharma, D.R. 2001. Micropropagation of strawberry cultivar Chandler and Fern. *Crop Improvement*, 28: 19–25.

Marcotrigiano, M., Swartz, H.J., Gray, S.E., Tokaricky, D. and Popenoe, J. 1984. The effect of benzylaminopurine on the *in vitro* multiplication rate and subsequent field performance of tissue culture propagation strawberry plant. *Advances in Strawberry Production*, 3(Spring): 23–25.

Marfa, O., Caceres, R., Polo, J. and Rodenas, J. 2009. Animal protein hydrolysate as a biostimulant for transplanted strawberry plants subjected to cold stress. *Acta Horticulturae*, 842: 315–318.

Margara, J. 1984. *Bases de la multiplication végétative*. INRA, Paris, France: pp. 262.

Mason, D.T. 1987. Effect of initial plant size on the growth and cropping of the strawberry (*Fragaria*×*ananassa* Duch.). *Crop Research (Hort. Res.)*, 27: 31–47.

Mathur, A., Mathur, A.K., Verma, P., Yadav, S., Gupta, M.L. and Darokar, M.P. 2008. Biological hardening and genetic fidelity testing of micro-cloned progeny of *Chlorophytum borivilianum* Sant.et Fernand. *African Journal of Biotechnology*, 7: 1046–1053.

Medina, J.J., Clavero-Ramirez, I., Gonzalez-Benito, M.E., Galvez-Farfan, J., Lopez-Aranda, J.M. and Soria, C. 2007. Field performance characterization of strawberry (*Fragaria* x *ananassa* Duch.) plants derived from cryopreserved apices. *Scientia Horticulturae*, 113: 28–32.

Mohamed, F.H., Beltai, M.S., Ismail, M.A. and Omar, G.F. 2007. High frequency, direct shoot regeneration from greenhouse-derived leaf disks of six strawberry cultivars. *Pakistan Journal of Biology Science*, 10: 96–100.

Mohan, R., Chui, E.A., Biasi, L.A. and Soccol, C.R. 2005. Alternative *in vitro* propagation: Use of sugarcane bagasse as a low cost support material during rooting stage of strawberry cv. Dover. *The Journal Brazilian Archives of Biology and Technology*, 48: 37–42.

Moradi, K., Otroshy, M. and Azimi, M.R. 2011. Micropropagation of strawberry by multiple shoots regeneration tissue cultures. *Journal of Agricultural Technology*, 7: 1755–1763.

Mullin, R.H. and Schlegel, D.E. 1976. Cold storage maintenance of strawberry meristem plantlets. *HortScience*, 11: 100–101.

Murashige, T. and Skoog, F. 1962. A revised medium for rapid growth and bioassays with tobacco tissue cultures. *Physiologia Plantarum*, 15: 473–497.

Nehra, N.S., Stushnoff, C. and Kartha, K.K. 1989. Direct shoot regeneration from strawberry leaf disks. *Journal of the American Society of Horticultural Science*, 114: 1014–1018.

Nesmith, D.S. and Duval, J.R. 1998. The effect of container size. *HortTechnology*, 8: 495–498.

Nishi, S. and Oosawa, K. 1973. Mass propagation method of virus free strawberry plants through meristem callus. *Japan Agricultural Research Quarterly*, 7: 189–194.

Nishizawa, T. 1994. Effects of photoperiods on the length and number of epidermal cells in runners of strawberry plants. *Journal of the Japanese Society for Horticultural Science*, 63(2): 347–352.

Nishizawa, T. and Hori, Y. 1993. Elongation of strawberry runners in relation to length and number of cells. *Tohoku Journal of Agricultural Research*, 43(3–4): 87–93.

Nyman, M. and Wallin, A. 1988. Plant regeneration from strawberry (Fragaria× ananassa) mesophyll protoplasts. *Journal of Plant Physiology*, 133: 375–377.

Owen, H.R. and Miller, A.R. 1996. Aploid plant regeneration from anther cultures of three North American cultivars of strawberry (*Fragaria* × *ananassa* Duch.). *Plant Cell Reports*, 15: 905–909.

Paek, K.Y. and Chakrabarty, D. 2003. Micropropagation of woody plants using bioreactor. In: S.M. Jain and K. Ishii (eds.). *Micropropagation of woody trees and fruits*. Kluwer Academic Publishers, Dordrecht, Germany: 756–766.

Paek, K.Y., Chakrabarty, D. and Hahn, E.J. 2005. Application of bioreactor systems for large scale production of horticultural and medicinal plants. *Plant Cell Tissue and Organ Culture*, 81: 287–300.

Passey, A.J., Barrett, K.J. and James, D.J. 2003. Adventitious shoot regeneration from seven commercial strawberry cultivars (*Fragaria* x *ananassa* Duch.) using a range of explants types. *Plant Cell Report*, 21: 397–401.

Pipattanawong, N., Fujishige, N., Yamane, K., Ijiro, Y. and Ogata, R. 1996. Effects of growth regulators and fertilizer on runner production, flowering, and growth in day-neutral strawberries. *Japanese Journal of Agriculture*, 40(3): 101–105.

Pırlak, L., Guleryuz, M. and Bolat, I. 2002. The altitude affects the runner plant production and quality in strawberry cultivars. *Proceedings of the Fourth International Strawberry Symposium, Acta Horticulturae*, 1(567): 305–308.

Qin, Y., Zhang, S., Asghar, S., Zhang, L. X., Qin, Q., Chen, K. and Xu, C. 2005a. Rege-neration mechanism of Toyonoka strawberry under different color plasticfilms. *Plant Science*, 168: 1425–1431.

Qin, Y., Zhang, S., Zhang, L.X., Zhu, D. and Asghar, S. 2005b. Response of strawberrycv. Toyonoka *in vitro* to silver nitrate (AgNO$_3$). *HortScience*, 40: 747–751.

Roussos, P.A., Denaxa, N.K. and Damvakaris, T. 2009. Strawberry fruit quality attributes after application of plant growth stimulating compounds. *Scientia Horticulturae*, 119: 138–146.

Rowley, D. 2010. *Strawberry Plug Plant Production*. Extension.usu.edu. Horticulture/Fruit/2010–01pr.

Sakila, S., Ahmed, M.B., Roy, U.K., Biswas, M.K., Karim, R., Razvy, M.A., Hossain, R., Islam, R. and Hoque, A. 2007. Micropropagation of strawberry (*Fragaria* x *ananassa* Duch.) a newly introduced crop in Bangladesh. *American Eurasian Journal of Scientific Research*, 2: 151–154.

Sarli, G., De Lisi, A., Montesano, V. and Schiavione, D. 2009. Evaluation of biostimulating products on strawberry in Southern Italy. *Acta Horticulturae*, 842: 805–808.

Sas-Paszt, L., Zurawicz, E., Masny, A., Filipczak, J., Pluta, S., Lewandowski, M. and Basak, A. 2008. The use of biostimulators in small fruit growing. In: A. Sadowski (ed.). *Fruit Crops. Biostimulators in Modern Agriculture*. Monographs series. Wies Jutra, Warsaw, Poland: 76–90.

Sharma, R.M., Singh, A.K., Sharma, S., Masoodi, F.A. and Shankar U. 2013. Strawberry regeneration and assessment of runners quality in sub-tropical plains. *Journal of Applied Horticulture*, 15: 191–194.

Sorvari, S., Ulvinen, S., Hietaranta, T. and Hiirsalmi, H. 1993. Preculture mediumpromotes direct shoot regeneration from micropropagated strawberry leaf disks. *HortScience*, 28: 55–57.

Sowik, I., Bielenin, A. and Michalczuk, L. 2001. *In vitro* testing of strawberry resistance to *Verticillium dahliae* and *Phytophthoracactorum. Scientia Horticulture*, 88: 31–40.

Swartz, H.J., Galletta, G.J. and Zimmerman, R.H. 1981. Field performance and phenotypic stability of tissue culture propagated strawberries. *Journal of Science*, 106: 667–673.

Takeda, F. and Hokanson, S.C. 2003. Strawberry fruit and plug plant production in the greenhouse. *Acta Horticulturae*, 626: 283–285.

Takeda, F. and Newell, M. 2006. Effects of runner tip size and plugging date on autumn flowering in shortday strawberry (*Fragaria* x *ananassa* Duch.) cultivars. *International Journal of Fruit Science*, 6: 103–117.

Taylor, D.R., Blake, P.S. and Browning, G. 1994. Identification of gibberellins in leaf tissues of strawberry (*Fragaria × ananassa* Duch.) grown under different photoperiods. *Plant Growth Regulator*, 15: 235–240.

Tian, M., Gu, Q. and Muyuan, M. 2003. The involvement of hydrogen peroxide and antioxidant enzymes in the process of shoot organogenesis of strawberry callus. *Plant Science*, 165: 701–707.

Turkben, C. 2008. Propagation of strawberry plants in pots: Effect of runner order and rooting media. *Journal of Biological and Environmental Science*, 2(4): 1–4.

Turkben, C., Seniz, V. and Ozer, E. 1997. An investigation on strawberry production in Bursa. *Uludag University Faculty of Agriculture*, 11: 1–9.

Waithaka, K., Hildebrandt, A.C. and Dana, M.N. 1980. Hormonal control of strawberry axillary bud development *in vitro. Journal of the American Society for Horticultural Science*, 105: 428–430.

Wang, A.Y. 1992. Effects of GA_3 on strawberry propagation. *Horticultural Abstracts*, 62: 9884.

Wang, D., Wergin, W.P. and Zimmerman, R.H. 1984. Somatic embryogenesis and plant regeneration from immature embryos of strawberry. *HortScience*, 19: 71–72.

Way, D.W. 1966. The field performance of cold-stored strawberry runners in Southeast England. *Journal of Horticultural Sciences*, 41: 291–298.

Xuemei, W., Haoru, T., Guoqin, W. and Yan, L. 2004. Effects of different culture conditions on regeneration from leaves of strawberry (*Fragaria×ananassa* Duch.) "Toyonoka". *Acta Horticulturae Sinica*, 31(5): 657–659.

Yonghua, Q., Shanglong, Z., Asghar, S., Lingxiao, Z., Qiaoping, Q., Kunsong, C. and Changjie, X. 2005. Regeneration mechanism of Toyonoka strawberry under different color plastic films. *Journal of Peanut Science*, 168: 1425–1431.

Zebrowska, J.I., Czernas, J., Gawronski, J. and Hortynski, J.A. 2003. Suitability of strawberry (*Fragaria×ananassa* Duch.) Microplants to the field cultivation. *Journal of Food Agriculture and Environment*, 1(3–4): 190–193.

Zhang, X., Himelrick, D.G., Woods, F.M. and Ebel, R.C. 2000. Effect of temperature, photoperiod and pretreatment growing condition on floral induction in spring bearing strawberry. *HortScience*, 35(4): 556.

Zhao, Y., Liu, Q.Z. and Davis, R.E. 2004. Transgene expression in strawberries driven by a heterologous phloem-specific promoter. *Plant Cell Reports*, 23: 224–230.

Ziv, M., Chen, J. and Vishnevetsky, J. 2003. Propagation of plants in bioreactors: Prospects and limitations. *Acta Horticulturae*, 616: 85–93.

Zurawicz, E., Masny, A. and Basak, A. 2004. Productivity stimulation in strawberry by application of plant bioregulators. *Acta Horticulturae*, 653: 155–160.

12 Planting

Biswajit Das

CONTENTS

12.1 Planting Considerations 193
12.2 Plant Type 194
12.3 Planting Plan 196
    12.3.1 Planting System 197
        12.3.1.1 Hill Row 197
        12.3.1.2 Strawberry Plasticulture Planting 198
        12.3.1.3 Matted Row... 198
        12.3.1.4 Ribbon Row 199
        12.3.1.5 Spaced Row... 199
        12.3.1.6 Hedge Row System 199
        12.3.1.7 Barrel and Pyramid System 199
        12.3.1.8 Vertical System 199
        12.3.1.9 Container and Hydroponic Planting 200
12.4 Planting Time 201
12.5 Planting Density 203
References... 205

Strawberry runners being very delicate in nature require utmost care at the time of planting. Runners should be protected from any type of mechanical damage while they are separated from the mother plants, as well as at the time of processing for planting. Planting of strawberry in a systematic plan is very much essential for the success of strawberry culture. The planting plan should include different essential aspects such as selection of site, choice of cultivars, quality planting material, soil/soilless growing medium, bed preparation, planting design, planting density and aftercare. The quality of runners determines planting success irrespective of cultivars. Runners are separated from the mother plants and planted directly on the raised beds. Another method of runner collection is from cold storage. Performance of strawberry culture mainly depends on genotypic and environment interaction in different agro-climatic zones worldwide. The planting plan depends on type of climate (Das et al., 2012), such as tropical, subtropical, semi-arid, cold- or hot-arid, coastal, foothills, mid- and high-temperate hills, Mediterranean region, and prevailing weather conditions, namely autumn, severity of winter and summer temperature, winter and spring snow and frost, hailstorm, rainfall and availability of irrigation water. The growing environment, particularly whether it is open field or a type of protected structure, also determines the planting plans. More recently, various structures have been designed for soilless horizontal or vertical strawberry cultures.

12.1 PLANTING CONSIDERATIONS

Plants should be procured from certified nurseries that are free from viruses and any other diseases or insects. Runner collection from very old plantations is not advisable. Planting very early, that is before the plantlets have attained the right stage of growth for planting, in the wet or very dry season should be avoided. Transportation of planting material should be done with proper care so that runners do not desiccate or suffer any root or shoot injury. Plants should be kept under shade and moistened by water sprinkler. Roots are to be covered with soil or other medium, such as moss grass, to protect from desiccation. If necessary, plants can be stored at 1–3.5°C for two weeks in plastic bags, as well as by being kept under moist soil, and should be planted as soon as possible after removing from cold storage. At the time of storing in the soil (heel), a shallow trench is dug, which is deep enough for the roots, and runners are placed in a single layer against one side of the trench with crowns partially above the soil line. Roots are covered with soil and gently pressed. Water is applied at regular intervals. At the time of planting, holes are dug that are twice as wide and long as the plant and soil is added back into the hole. The fleshy portion of the plant between the top growth and the roots is known as the "crown". At the time of planting, the middle of the crown is placed at the soil level, so that the crown is not buried and roots are not exposed to avoid the death of the plants or delay in runner production. Establishment of newly planted runners by regenerating new roots and heeling, and initiation of normal growth activities of the roots that are damaged during transplanting can be greatly affected by water stress, especially in the new environment with higher light intensity and heat. An excellent planting system requires the old planting to be replaced every year. Runners raised in specialized nurseries are closely spaced with all growing conditions optimized under a controlled environment, such as shading, a microclimate with less light penetration and higher humidity. Planting of these runners in open field conditions with full sunlight, wind and less humidity and fluctuation in air temperatures may cause "transplanting shock" to those runners. Hardening of these plants for a few days before transplanting is essential. Old beds require proper cleaning in the early spring by removing old dead leaves and unrooted runners. These practices will eliminate any chance of diseases such as grey mould infection. "J" rooted plants are those plants that have roots longer than the holes, which are planted by forcing into a root bunch with a J-shaped pattern into the hole. However, it has been observed that J-rooting does not affect transplant performance under an annual hill system (Duval, 2005). The quality of the soilless (peat and cocopeat) plug plants raised under greenhouse conditions has been found to be very good (Treder et al., 2015).

12.2 PLANT TYPE

There are four sources of quality planting material of strawberry, namely plug plants, dormant runners, freshly dug runners and test tube (tissue-cultured) plants. "Plug/tray/balled plants" can be procured as such for direct planting or can be raised by planting runner tips in the plastic/polystyrene root trainer trays in batches from mid-July–last week of August (Lieten, 1998). Both fresh plugs or cold-stored/frozen plugs are used for planting. Development of the planting materials is influenced by climatic conditions. At the time of planting, clumps of soil-containing roots are removed from the plastic root trainer trays or cells containing the plant without any disturbances or damage to the soil clump, and planted as such in the holes. Plant establishment is easy and mortality is nil as they maintain their turgor pressure and retain many active roots. There is consistency in yield of quality fruits from season to season and they provide better option for mechanical transplanting (Poling and Maas, 1998). Plug plants are generally planted in late August or early September and also provide a wider option for planting dates as plants may be available early. Plug plants are produced in a controlled environment in greenhouses or poly tunnels, which take less time and are considered free from any pathogens or insects. Tray plants are waiting bed plants produced in greenhouses in peat beds, which are wrapped in plastic with their root balls and kept in cold storage. "Dormant or frigo plant" may be available with a large number of varieties and are less expensive. They are collected during dormancy and wrapped in plastic to store for overwintering at −1.5 to 2°C. However, such plants require deblossoming and trimming of extra roots and leaves. There may be difficulty in establishment of plants or mortality may occur in some plants if they are stored in the nursery for too long before planting. Frigo plants facilitate early planting. Large and well-developed plants are taken out from the cold store just one day before planting and smaller plants may be put out for rooting with rooted plants taken out six or 12 hours before

Planting

planting to avoid shock. Planting may be done at any time as they remain unchanged in cold store. These bare-rooted plants are suitable for manual planting and initially grow slowly. Care should be taken while transportation to protect these plants from desiccation/damage. Proper planting ensures early breaking of dormancy. Before planting, the roots are dipped in buckets containing mild warm water to rehydrate the plants and break dormancy. Frigo plants are sorted on the basis of crown diameter. Plants with crowns of more than 1.8 cm diameter are waiting bed plants and have more than four flower trusses, 1.5–1.8 cm diameter frigo plants are A+ and have more than two flower trusses. Plants that are of 1.3–1.5 cm diameter fall under A grade, and less than 1.3 cm diameter plants are graded as B. Large frigo plants are designated as waiting bed plants with a crown diameter greater than 1.8 cm. These plants can be planted at any time, and the harvest starts 60–70 days from planting, producing a high yield per plant. Moreover, the yield potential in later years is higher than with other types of plants due to quick branching of the crown. This grade of plants is stored at −2°C to break their dormancy and to maintain their optimal properties. Plants are transported in a refrigerated vehicle. Leafless plants must be planted soon after they have been taken out of cold storage.

"Fresh plants" are raised and uprooted from the runner beds just before planting during mid-October for warm areas and in September for foothill areas. "Multiple crown mother plants" are another source of fresh plants obtained after digging and separating crowns of the fresh mother plants. These plants provide the option for late planting. However, plugs and freshly dug bare-rooted plants are usually comparable in yield performance, but freshly dug plants may have an advantage given that they do not concentrate the ripening of the crop as much as plug plants in some spring seasons (Poling et al., 2005). Because of improved yield and flexibility in the date of planting, cold-stored plants are preferred over freshly rooted runners by growers worldwide (Roudeillac and Veschambre, 1987). Frigo plants of the cv. Mara Des Bois express a well-developed and denser root system due to accumulated food reserves, which tends to enhance the further development of the plant (Lutchoomun, 1999). Hence, cold-stored plantlets result in more vigorous plant growth, which can advance bearing and produce higher yield of better-quality fruits than fresh plantlets. Runner formation in "frigo plants" starts in conditions of long-day length and warm days during summer in temperate, tropical or subtropical areas. However, in the temperate climate, the plant's metabolic activities slow down during autumn (at less than 15°C), with carbohydrate reserves in the rhizomes and roots of the plantlets tending to accumulate, while leaf development is suppressed. Consequently, the plants become relatively dormant vegetatively and flowering is induced. In winter, when short day-length and low temperature (less than 5°C) conditions prevail, the vegetative and reproductive development of plants are stopped completely, which resumes in spring under warm temperature conditions, eventually leading to flowering and fruiting. Plant development is more vigorous in presence of multiple crowns, producing more leaves and taller and larger leaf canopy than fresh runners. Early and fast plant establishment is accompanied with the availability of stored assimilates, and induces the vigour of strawberry crop. Frigo (in the central northern areas) and fresh plants (in southern areas) are two types of plants used in Italy for annual strawberry plantations. Frigo and fresh plants are planted in the last week of July, but fresh plants showed very low adaptability in comparison to frigo for different genotypes tested (Capocasa et al., 2009). In northern California, "cutoff" or "tops off" plant are another type of plant available during late October from nurseries. These have accumulated considerable chilling in the nursery, and have been mowed before digging and harvest. Cutoffs are recommended only for the very mild winter areas in southeastern North Carolina, but not for areas further north or west. They may have some utility in coastal South Carolina and Georgia as well. Based on past research studies, fully dormant or "frigo" plants are not recommended for plasticulture of strawberry in southern parts (Poling et al., 2005).

"Tissue-cultured plants" are now getting widely popular for commercial production of high-quality planting materials, especially since they are virus and nematode free. In North Carolina, micropropagated plants of strawberry are raised as nuclear plants that are virus indexed, free from major diseases and insect-pests, and field tested for genetic fidelity. From the stock of nuclear plants, foundation and then registered plants are produced, which are ultimately used for certified plant production in the nurseries. In Australia, an extensive virus indexing programme is run by Crop Health Services and certificates are awarded to the nurseries producing strawberry runners (Lines et al., 2006). Tissue-cultured plants provide a large number of plants in a short span of time, fulfilling the huge demand for quality planting materials. Tissue-cultured plants, after being procured from the hardening/acclimatization units of the respective agro-climatic zones, are planted very cautiously in a protected environment in poly tunnels or polyhouses.

The size of the planting material also affects the establishment rate. Large runners/crowns with more leaves and long roots are damaged during transportation and require extra trimming before planting. Plant size may be decided on the basis of length of petiole, crown diameter, fresh weight of plant, leaf area and numbers as well as root numbers. Crown diameter may vary from 0.8–1.8 cm. Human (1999) planted one, two and three runners per hole having three crown diameters of >1.0, 0.5–1.0 and <0.5 cm, and it was observed that the overall yield grew with the increase in number of runners per hole irrespective of crown diameters. However, plants having >1.0 cm and 0.5–1.0 cm crown diameter gave better results. In another study, it was observed that the size of plantlets with one–three compound leaves of plug plants did not affect yield and berry size (Crawford et al., 2000). Hicklenton and Reekie (2000) dug first-, second-, and third-generation daughter plants at different dates and observed that third-generation plants showed reduced survival when dug at early dates. However, all the plant types when dug after November 7 survived fully. Bish et al. (2000) reported that strawberry plants showed variability in size and fruit yield because of developmental differences in vegetatively propagated daughter plants. Daughter plants with greater leaf development (two expanded leaves) had increased root growth during transplant propagation. Larger stolon stem diameter (4 cm) of daughter plants resulted in larger crown diameter and greater flower development after transplanting. A linear relationship has been observed between crown diameter at planting and fruit yield, with best results having 1.8 cm crown diameter at the time of planting (Bussell et al., 2003). Large or medium-sized plug plants gave maximum and early yield in the first season, but in the second season, higher yields were obtained only from the large plug plants (Hochmuth et al., 2006). Runners with more than four leaves were observed to be better for plant growth and quality fruit production. Among three grades of runner size, namely, small, medium and large, the average fruit weight was higher from plants propagated from large runners, when grown in a fine peat substrate (Kehoe et al., 2009). Large runners used later in the season resulted in lower fruit weight. In Sicily, early harvesting is possible by using fresh plants of varieties such as "Candonga Sabrosa", "Florida Fortuna", "Sabrina", and "Sant Andreas", in comparison to cold-stored plants used in the past, and such approach has expanded the harvest window from December to May, and provided fruits of higher quality (D'Anna et al., 2014 and Sabatino et al., 2017). Effect of plant type on plant growth and yield was studied by Nestby and Sonsteby (2017) in Norway, where they observed that bare-root and plug- plants when planted immediately or one day after delivery performed better in terms of establishment and yield parameters. Storage of bare-rooted plants at 2−4°C generally shows reduction in yield, but plug plants performed well when planted at different time intervals except, when planted very late.

12.3 PLANTING PLAN

A strawberry planting plan comprises a planting design or system, planting time and method of transplanting of suckers. Moreover, it all depends on the type of strawberry cultivars, June-bearing, everbearing or day-neutral, as well as the runner production efficiency of the cultivars. Another important factor is planting site, whether in open fields or protected structures. Commonly, raised beds are prepared for training systems. The strawberry culture system is designed so that there is ample space for the overground and underground parts of mother and

Planting

runner plants on the beds to restrict the hibernation of insect-pests, weed growth and disease infection and to facilitate intercultural operations for improved vegetative as well reproductive growth.

12.3.1 Planting System

Planting of strawberry basically consists of three systems: the hill system, matted row and spaced row. The matted row system and annual hill system are the main training systems commonly followed (Black et al., 2008). Apart from this, single hedge row, double hedge row, single hill row or double hill row etc. can also be followed (Sharma and Sharma, 2004).

12.3.1.1 Hill Row

This system of planting is suitable with limited space and for those cultivars that produce few runners accommodating more plants/unit area. Everbearing and day-neutral cultivars perform well under this system as they show less tendency to profuse runner production; however, June-bearing types can also be trained on this system with runner pinching at intervals. This system is typically maintained and produces satisfactory crop for three–six years and fruit production gets reduced during later years. Hill culture is used where runners are manually or mechanically removed during summer, and during fall or winter, plants do not regenerate runners. The growing conditions under this system favour only growth and development for mother plants. Runners are planted in single or double rows on raised beds with 15–25 cm height at a spacing of 30–35 cm in the rows and 20–35 cm between the rows under double row planting. The spacing can be adjusted depending upon the vigour of cultivars, for instance, cv. Sweet Charlie is benefited by comparatively closer spacing than more vigorous cv. Chandler. In the double row, planting may be done opposite row hill or staggered pattern (alternate) row hill. "Annual hill culture" is the training system wherein the strawberry plants are set out each year for fruit production, with the objective to produce large fruit with excellent quality on regular basis. This system is good for high tunnel production system as it gives early production of quality fruits. Two hill systems, 20 cm double and 20 cm triple hill systems were compared with 38 cm double matted row in annual or perennial production (Strik et al., 1997); more fruit yield was recorded in the 20 cm triple hill than the 38 cm double matted row in the first and second production seasons (Figure 12.1).

FIGURE 12.1 Double row on raised beds planting with mulching under open and protected environment.

12.3.1.2 Strawberry Plasticulture Planting

Strawberry plasticulture is technically an annual hill training system. Raised beds are prepared 25 cm high, 75–80 cm wide at top and 80–85 cm wide at base. Beds are slightly crowned and tops are given slope from the centre to the edges with a slope drop of 3.2 cm for better runoff of excess water from the plastic mulch. Mostly, 137–157 cm wide plastic films are used over these beds. Such "super beds" provide an ideal air-soil-water environment for vigorous strawberry root development. Well-drained sandy loam soils under raised beds favour downward root growth as deep as 60 cm; however, in clay soils, root growth is more horizontal. Raised beds in open fields or high tunnels are covered with black or white polyethylene mulch. Black polyethylene mulch helps in warming early in the spring time and continues to warm the soil until late in the fall. While, white polyethylene tends to keep the soils of beds slightly cooler in the hot summer time (Rowley et al., 2010b). Beds of June-bearing strawberries are covered with black polythene mulch for moisture conservation and restricting weed growth. Plants are removed before the hottest part of the summer, and planted after the heatwave is over. White polythene mulch is used for day-neutral cultivars when the objective is to produce more in the summer. However, for early plant establishment and late-season extension of production, black polyethylene may be used successfully. Another raised-bed system generally followed in small scale strawberry production system or experimental layout is $1–2 \text{ m}^2$ in dimension and 15–20 cm raised with plant and row spacing of 40–45 cm. In major parts of south India, planting is done on raised beds of $4 \text{ m} \times 3 \text{ m}$ or $4 \text{ m} \times 4 \text{ m}$ dimensions with 45 cm plant-to-plant spacing and 60–75 cm row-to-row distance. Black polyethylene or straw mulches are also used on these beds. Runners are not allowed as there is no extra space for runner production. If necessary, then runners are placed along the space available between two beds.

12.3.1.3 Matted Row

Strawberry cultivars with profuse runner-producing efficiency are better suited for this training system. This system is rather a perennial production system of June-bearing cultivars in cold regions. Everbearing cultivars that produce fruits in the same year are also trained on this system. Moreover, this system is most suitable in areas that experience freezing or fluctuating temperatures, causing crown injury during winter. In open field conditions, the matted row system is better, however, it is not suitable for high tunnel system. Plants are set at 20–60 cm apart in single row on 60–90 cm wide beds and 90–120 cm distance is maintained between the rows. The regenerated runners are allowed to root freely; however, plant row is kept 45 cm wide after removing extra runners or overcrowded leaves. The productive span of these plants is three years, and new beds are prepared thereafter. It is very productive, but not as efficient as other training system in terms of higher yield. Fruit quality deteriorates if rows get too crowded. Dead and diseased leaves and berries are removed regularly in the growing season. Flower beds are pinched off in the first year, especially in the case of June-bearing cultivars. This favours growth of healthy plants and regeneration of more runners. In the case of everbearing or day-neutral cultivars, flowers are pinched off only at the early stages of the growing period. This practice will encourage better production from the second year. Runners emerging from the mother plants are trained along the rows at 15–20 cm spacing and their base is slightly pressed into the soil for easy rooting. A density of five or six plants for each square foot is maintained. The advanced matted row system has been developed to share the advantages of annual hill row and traditional matted row systems (Black et al., 2002). Stevens et al. (2006) evaluated the performance of strawberry under three systems, viz., conventional matted row, advanced matted row and cold climate plasticulture, and observed that the conventional matted row gave the maximum yield (17.4 tonnes/ha), followed by advanced matted row (13.2 tonnes) and plasticulture (11.8 tonnes/ha) but the size of fruits was greater in the plasticulture system.

Renovation of the matted row system is essential to sustain vegetative and reproductive growth of plants each year. June-bearing cultivars trained on this system are renovated every year just after harvest to ensure good production. However, day-neutral or everbearing cultivars are not renovated annually. All the cuttings and, old and week plants are removed and only healthy plants are kept.

Mowing of the foliage and crown up to 2.5–5 cm is essential. Mowed crowns initiate new growth during mid-August. Narrowing of the rows to 15–20 cm is done for closer spacing, keeping 90–100 cm centres. If the strawberry field is maintained for the next year, the renovation process should be started as soon as possible after harvest. Renovation involves a series of steps aimed at invigorating plants after harvest and reducing pressure from insect-pests, diseases and weeds.

12.3.1.4 Ribbon Row

A ribbon row on 10–15 cm raised beds or flat beds of 90–100 cm row centres also gives higher yield. However, site selection and bed preparation are similar to those in a matted row. Plants are spaced much closer at 8–15 cm apart with 90–95 cm in rows accommodating 136,023–143,457 plants/ha. Plants are allowed to flower in the first season and any runner arising is immediately removed so that all the metabolites can be utilized for crown enlargement and branching for higher fruit production. Higher fruit yield starts from the second year due to high density and healthy crowns. Planting is done from late May to late July. Late planting does not encourage runnering. Manual planting is done due to very close spacing. After planting, alleys between the rows are mulched to suppress weeds. After harvesting season is over, plantings are renovated every year.

12.3.1.5 Spaced Row

This training system is more suited for cultivars with moderate runner-producing efficiency. Spacing is adjusted to 45–75 cm apart for mother plants in a row that is spaced at 90–120 cm. Only four daughter plants (runners) are set at 10 cm spacing surrounding the mother plant. All other runners are removed. Mother plants of day-neutral cultivars are spaced at 20–30 cm in a row with 75–90 cm spacing between rows. Flowers are removed for the first six weeks after planting.

12.3.1.6 Hedge Row System

This system is between the matted row and hill system, with a slight modification of the spaced row system. Mother plants are set at 60 cm apart in rows and rows are spaced at 90 cm. In the single hedge row system, only one runner is allowed along the row to set in either side of the mother plant at 20 cm apart from the mother plant. Sometimes another two sets of runners are allowed to grow on both sides of the mother plant at right angles to the row, that is, towards the edges of the row, and this system is called a double hedge row.

12.3.1.7 Barrel and Pyramid System

In home gardens where space is limited, these designs are preferred for decorative strawberry planting. The barrel system is prepared by making 2.5 cm holes in the side of a 95–190-litre wooden barrel at approximately 20 cm spacing. The barrel is filled with successive layers of soil mixture, and strawberry plants are carefully inserted through the holes from the outside by pressing into the soil. In the pyramid system, a square or round pyramid structure is constructed with wooden frames. The first largest or base frame is about 180–240 cm in width, and successive frames of 150–210, 120–180, 90–150, 30–60 cm wide are placed one above another so that a pyramidal shape is given to the structure. The margins of each flat frame are nailed with wooden planks of 20 cm height so that each stage of the pyramid can be filled with soil. Strawberry plants are planted on each stage at 30 cm apart. Runners are not allowed in either of these systems.

12.3.1.8 Vertical System

This planting system was described by Rowley (2010a, c) to maximize high tunnel growing area. The structure is constructed by fixing PVC rain gutters/channels (rain gutters are actually used to drain off rainwater from the roof by fitting them along the roof edges of any building), with approximate dimension of 3.1 m long, 10 cm wide and curved 8 cm deep, on two horizontally placed wooden stands. Gutters are fixed one above another at 23 cm apart on wooden support stands. The bottom gutter is placed approximately 10 cm above ground level. In a single-sided vertical system, seven gutters are fixed at each end on two wooden stands that are placed 3.1 m apart in a slanting position from top down. These two slanted wooden frames are fixed at the top end with two vertically positioned wooden stands (1.5 m tall). A gap of 90 cm is kept at the base between the vertical and slanting wooden stands at both ends. In a two-sided vertical system, two wooden stands are joined at the top and placed in a slanting position keeping a gap of 1.4 m at the base. Vertical height is kept to one metre. Similarly, another set is placed 2.5 m apart from the first one, so that five gutters (2.5 m long) can be fixed horizontally at both the sides. The bottom ends of all the wooden stands are inserted at least 30 cm deep into the soil firmly to provide the sufficient strength to the structure to support the weight of growing media and plants. The gutters are filled with growing media mixture such as peat, perlite and vermiculite or any other suitable growing medium. Plug plants are planted along the gutter row at 2.3 cm spacing.

12.3.1.9 Container and Hydroponic Planting

Day-neutral or everbearing cultivars are suitable for container planting as they produce few runners and can be grown even under shade. Containers such as hanging baskets, multiplant containers and strawberry pots are used for this system. A single plant is planted in each 30 cm wide and 20–25 cm deep container and exposed to sun for at least six–eight hours daily.

A vertical hydroponics growing system was developed by Noakes et al. (2006) in George, South Africa. The system was based on a moulded polystyrene pot and four plants were adjusted in each pot. A PVC pipe provided support and drainage. Each pipe supported nine stacked and lidded pots, so a total of 36 plantings were supported. Each column was connected to a drain at the base and a support wire at the top. Columns were spaced in rows at one metre apart and 0.7–one metre in the row, which gave plant densities of 36–50 plants/m². Plant were fertigated using button drippers with a four-way manifold. Arrow drippers at the ends of the spaghetti tube ensured equal flow to the four chambers, and were used to anchor the spaghetti tube above the coconut coir medium to prevent penetration by roots The Californian varieties performed well in this system and tended to be more vigorous. Another viable system, soilless growing systems, was examined by Lopez et al. (2006) as an alternative to achieve a more sustainable, efficient and environmentally friendly growing system for strawberry production. Open and closed systems with and without slow sand filtration for disinfestations of the reticulated nutrient solution were evaluated using coconut fibre as a substrate and the closed soilless system with or without slow sand filtration was observed to be viable.

Three commercially acceptable growing systems were evaluated in Belgrade (Milivojevic et al., 2009):

1. hydroponics system in bags filled with 15-litre substrate placed on a 1.1 m-high platform
2. hydroponics system in bags filled with 20-litre substrate placed in rows on floor of tunnel
3. hill system by using black plastic mulch in the form of double rows

Observations proved that hydroponics growing system in bags placed on a 1.1 m-high platform gave a positive influence on number of flowers (9.2), fruits (9/plant) and yield per plant (87.8 g). However, during the spring of the second year, maximum fruit weight (22.9 g) and yield (518 g/plant) were recorded in the hill growing system. The hanging bag hydroponic system, vertical growth system or frame-type structures such as aeroponics or specialized benching, which utilize vertical space, may be used successfully. Perlite, being relatively lightweight, is used as the growing medium. The plants hang suspended from a height, so in the greenhouse or protected structure, it is necessary to keep space for walkways to access and harvest the crop. Thus, maintenance and harvesting of fruit can be done from below the growing system. Moreover, this also helps to improve air circulation in the growing environment and ensure fast drainage in the root zone. Supply of nutrient solution to the perlite culture is essential and dripper application of N: 80 ppm, P: 50 ppm, K; 85 ppm, Ca: 95 to 100 ppm, Mg: 50 ppm, S: 56 ppm, Fe: 2.8 ppm, B: 0.6 ppm, Mn: 0.4 ppm, Cu: 0.1 ppm, Zn: 0.2 ppm, Mo: 0.03 ppm, pH: 6–6.2, EC: 1.4–1.6 mS/cm was observed to be beneficial.

12.4 PLANTING TIME

Planting of strawberry runners can be done in spring or fall. However, optimum planting dates are specific to agro-climatic zones, and dates should be determined to allow sufficient vegetative growth of the plants before initiation of flowering in the season. This ensures the development of more flower trusses, which in turn ensures higher production of quality fruits. June-bearing cultivars are generally planted as "plug plants" in the autumn. In cooler climates, planting should be completed in the first week of September, and in warmer climates, it may be done later also. Plugged or dormant runners of day-neutral cultivars are planted in the autumn or even in the late winter using dormant bare-rooted runners. The time also depends upon the time to fruit maturity of cultivars, for example, cv. Sweet Charlie is an early-season cultivar, Chandler is early-midseason and these two cultivars are suitable for the cold region of mid-South Carolina and Georgia (Poling et al., 2005). Meanwhile, cv. Camarosa is midseason and preferred in areas with mild winter growing (such as the coastal plains of North Carolina, South Carolina and Georgia). Optimum dates of plug planting for cv. Chandler varies with state and region in the United States and generally fresh-dug plants should be set three–five days earlier than plugs. Optimum dates of planting for cv. Sweet Charlie also vary with state and region. But, as a general rule, this variety must be transplanted at least seven days ahead of Chandler. Plug plants are usually preferred for cv. Sweet Charlie over freshly dug bare-rooted transplants. Field observations suggest that cv. Sweet Charlie may be more susceptible to Phythophthora cactorum (crown rot) than cvs. Chandler or Camarosa. Cultivar Camarosa is planted in Carolina at about the same time as Chandler, but most growers prefer to set this variety at least three days ahead of Chandler. Planting time also determines the ultimate plant size at the time of fruiting under particular climatic condition. Therefore, the optimum date of planting or week for any area should be chosen so that four branch crowns are formed by harvesting time over and above to the main crown. Four branch crowns are needed to produce desirable number of berries per plant (about 35–40). But very early planting will end up producing six or more branch crowns per plant, which can produce so many blossoms that fruit size will be reduced to the point where both harvesting and marketing the small berries will be a limitation. Plants that are transplanted late will have inadequate cropping potential due to production of fewer branch crowns. When runners are planted slightly later in an unseasonably warm fall, that will result in production of runners and if there is production of more than two to three runners per plant in an "average autumn", this is probably a good indication of planting too early.

In California, strawberries are grown as an annual crop and the plants are first grown in nursery, then transplanted into the fields. In the northern districts, plants are replaced each autumn at the end of harvest and lie dormant through the winter. In the Oxnard and Santa Maria region of California, planting during late July through September provides strawberries through the fall and winter. Thus staggering of planting and harvesting is practiced. Staggered planting of single or different cultivars with early, mid or late maturity on different dates helps to extend the harvest season. Planting in Ontario (Dale and Pritts, 1989) is done early in spring (first week of May), which gives peak production after 12–14 weeks (late August). Therefore, higher yields can be obtained later in the season by planting a portion of the field in June. Under Indian conditions, staggered planting from early September–mid-October can extend the harvesting season from January–April in the plains and February–July in the hills.

In temperate regions of India, cvs. Tioga, Torrey, Chandler, Senga Sengana, Blackmore, Camarosa and Oso Grande are preferred. In the plains of north India, cvs Gorella, Chandler, Pusa Early Dwarf and Sweet Charlie, and in south India, cvs Bangalore, Sujatha and Labella are grown. In tunnels, you can plant earlier even during February as the climate becomes warmer early. Under plasticulture, different times of planting for dormant crowns, plugs, fresh single crown and fresh multiple crowns are mid–late July, early September, early September and early–late September in the warmer regions, respectively. Whereas, in colder region, mid-June–mid-July, mid–late August and early September were found to be suitable for dormant crowns, plugs and fresh multiple crowns, respectively. In areas which experience extreme cold or severe spring frost, planting is done in the autumn, so that the plants can establish and put forth growth before the inclement weather starts. In northeastern Indian states, cvs. Festival, Ofra, Camarosa and Sweet Charlie are recommended and planting time is September–October.

Plants are stored in plug bags at 0°C and planted during the spring, that is, late March–April. Stored plants are trimmed off all old and dead leaves, and roots are dipped into the water for an hour, then planted immediately by setting the plants in such a way that crown of the plants remain even with the soil surface. Watering and starter solutions of fertilizers are applied to facilitate establishment. Bare-rooted plants supplied in bundles can also be planted in the spring. If runners are raised in the pot or poly bag, then it is easy to plant these during the autumn, September–November. If plug plants are transplanted too early, then they produce excessive runners and more branch crowns, whereas late planting results in poor plant establishment and crown formation. Planting on optimum dates is beneficial for balanced vegetative growth with four–six branch crowns at full maturity. In tunnels or protected conditions, plug plants are better; however, if bare-rooted plants are transplanted, then it should be done few weeks early as compared to plug plants. Planting during July–August requires adequate care for better plant establishment. Date of planting is very important within any planting season, influencing the vegetative growth, flowering, yield and harvesting of fruits as well as runner production. Winter planting under California conditions is done on November 1, which is the optimum time for all cultivars, and summer planting is on August 10 (cv. Tioga) and September 1, which is optimum for all varieties (Anonymous, 1982). Danek (1984/85) observed a significant positive effect of early summer planting on the number of runners in the following summer. Anna and Curatolo (1986) planted strawberry cv. Sequoina from June 15–September 15 at monthly intervals with a plant density of 6.6 plants/m². Establishment of plants was better due to planting during June and September, while earliness was recorded due to planting during June–July, but yield was higher in the first season due to planting on the later dates only. Faedi et al. (1986) planted cvs. Fevette and Sequoina on October 3, 15 and 27 and November 7 and observed that early transplanting resulted in higher yields in both the cultivars (300 q/ha and 253 q/ha, respectively) and harvesting was also earlier (December 15 and December 7). Schmitz and Lenz (1989) observed runner production and number of leaves to be decreased as planting date was delayed. Yield of Class-I fruits was highest when the date of planting was fixed on July 15 and decreased markedly due to planting on successive dates. The yield of cv. Tioga decreased drastically when planting was delayed after September 30 in the mid-hill conditions of the temperate zone of India (Badiyala and Bhutani, 1990). Poling (1991) observed September 23 as a suitable date of planting with respect to higher yield (34.1 t/ha) in a raised-bed system in comparison to a matted row.

Caruso and D'Anna (1995) recorded a declining trend in yield when planting was delayed after August 10. O'Dell and Williams (1999) recommended autumn planting of strawberry cultivars under plasticulture in colder regions of Virginia. It has also been recommended that July–mid-August is the optimum time for cv. Late Star and other eastern American cultivars in colder region. Early planting on May 14 or June 5 produced better growth of plants and yield, while planting delayed for two weeks deteriorated fruit size, fruit mass and chemical composition of fruits (Zmuda, 1999). Tripathi et al. (2000) observed August 25 as the best date of planting for cv. Chandler and September 10 planting gave significantly higher titrable acidity fruit. Late autumn-lifted fresh runners were rooted in plastic bags in December and field planted during March and cvs. Camarosa gave 266.67 g/plant, and Sweet Charlie gave 167.56 g/plant yield (Ilgin et al., 2002). Maximum fruit yield was observed by planting strawberry cvs. Dorit, Camarosa and Selva on July 1 in Turkey and planting during mid-July was observed to be ideal for optimum yield and quality of fruits. In the submountain tract of Western Ghats of Kerala, fruit maturity in cvs. Chandler, Labella and Sujatha was observed at 67.1, 88.4 and 96.9 days, respectively after planting due to planting in the last week of September (Kumar et al., 2002). Cultivars Camarosa, Chandler, Bish and Gaviota gave higher yield when planted on October 20 in South Carolina conditions (Hasell et al., 2006). Mid-September planting in the semi-arid region of the north Indian plains was observed to be successful in terms of plant growth and quality of fruits (Singh et al., 2007). In Sicily, the recommended time of planting is August–October, depending on method of cultivation and cultivar. Plug plants (tip plants) and fresh-dug plants from nurseries at higher elevation are generally preferred due to their early fruiting potential (Caracciolo et al., 2009). Four cultivars, namely Camarosa, Candonga, Tudla and Naiad, and a selection, "MT 99.163.22- Kilo" ("MT 91.143.5" × "Chandler") were planted on September 25 and October 10 in 2006, and it was observed that total yields were higher in plug plants established during late September as plants were able to achieve adequate growth before the cold winter season. Ahmad (2009) and Singh et al. (2010) reported mid-September to the first week of November as the ideal time for strawberry planting in the Kashmir valley. The subtropical climate in Australia and New Zealand is favourable for strawberry cultivation and planting time is March–April, whereas in cooler areas, the usual planting time is late winter or early spring. Planting time in South American countries such as Brazil starts from March in the northern part of the southeastern Brazil in areas such as Minas Gerais and Sao Paulo and continues up to May–June in areas such as Rio Grande do Sul (Antunes and Peres, 2013).

12.5 PLANTING DENSITY

Planting distance and density depend mainly on field condition, vigour of cultivar and planting system. Planting density is positively correlated with the yield of quality fruits (Hancock et al., 1984; Wilson and Dixon, 1988). However, increasing density decreases the number of inflorescences per shoot (Hancock et al., 1984) indicating a clear competition for light, water and nutrients for flower initiation. Oeydvin (1972) obtained maximum yield and early ripening of berries due to planting in double rows at a spacing of 15 cm apart and 30.6 cm between the pairs of rows. The optimum spacing for winter planting in California conditions has been observed to be 22 cm × 25 cm between plants and row, and 35 cm × 35 cm between plants and rows during summer planting (Anonymous, 1982). In both plantings, double rows with 122 cm bed width were compared with four rows with 162 cm bed width, which accommodated 107,593 and 46,112 plants/ha during winter and summer, respectively (Table 12.1).

The closer spacing results in in low yield especially for cvs. Douglas and Pajaro due to large plant size in the former case and fruiting in the month of May–June in the latter case. However, plants on wide beds can give 12–15 tons higher yield per hectare during winter planting and 5–7.5 tons higher yield per hectare during summer planting. The maximum number of fruits/bed was recorded by Pandey et al. (1980) at a spacing of 10 cm × 10 cm during two years, and maximum fruit yield (40.2 t/ha) in cv. Red Coat with a spacing of 10 cm × 40 cm, which was significantly more than at spacings of 15 cm × 15 cm, 20 cm × 20 cm, 20 cm × 30 cm and 20 cm × 40 cm. Among five different spacings, 30 cm × 15 cm, 30 cm × 22.5 cm, 30 cm × 30 cm, 30 cm × 45 cm and 30 cm × 60 cm, the closer the spacing, the higher the yield of good-quality fruits in the first year, whereas in the subsequent

TABLE 12.1

Strawberry Planting Time in Different Agro-Climatic Zones of India

Regions	Time
Temperate region: Jammu & Kashmir, Himachal Pradesh, Uttarakhand	September–first week of November and March–April
Subtropical region in north India: Punjab, Haryana, semi-arid areas	Mid-September–first week of November
Northeast India	September–October
Tropical areas in south India like Mahableshwar, Kerala	November–December and last week of September
Himalayan foot hills and plain area	Last week of September–first week of October

Source: Sharma and Sharma (2004); Kumar et al. (2008); Das et al. (2013).

years fruit quality was better in wider spacings (Badiyala and Joolka, 1983). Awasthi and Badiyala (1983) recorded significantly more fruit yield of cvs. Tioga and Torrey in the mid-hills of Himachal Pradesh at 30 cm×45 cm or 30 cm×60 cm plant spacing. Maximum fruit yield (3.84 kg/m^2) was recorded with cv. Riva on four-row beds of 90 cm width, while cv. Primella gave the maximum fruit yield (4.58 kg/m^2) on three-row beds of 70 cm width (Verwijs, 1984). More marketable fruit yields of cvs. Holiday, Earglow and Red Chief were recorded due to spacing at 45 cm with plant densities of 6.3, 7.6 and 7.8 plants/ft^2, respectively. In a plant density trial (Poling and Durner, 1986), with freshly dug plants set in mid-October with 33 cm between the double row and 15, 20, 25 and 30 cm between plants in each row, day-neutral cvs. Fern, Selva, Tribute and Tristar performed better under higher density. A staggered double row planting with plants set at 20 cm apart, offset 10 cm from the centre, with 1.2 m between-row centres, is a very efficient planting design. At plant densities of 50,000/ha, fruit yield was 30% more in double rows than in single rows (Dale and Pritts, 1989). In day-neutral cultivars raised as an annual crop, the plant densities can be increased even further to 12 cm spacing without sacrificing much on yield per plant. Hesketh et al. (1990) recorded maximum fruit yield at closer spacing with higher density of 133,912 plants/ha than the wider spacing of 29,193–78,155 plants/ha. The maximum fruit yield of cv. Tioga was recorded with 10 cm×10 cm plant spacing in mid-hill conditions in Himachal Pradesh (Badiyala and Bhutani, 1990). Cultivar Red Chief strawberry was planted with two cultural systems, runnerless and matted row and in both systems, the plots with narrowest between-row spacing (30 cm for runnerless rows and 45 cm for matted rows) had maximum fruit yield per unit area (Masiunas et al., 1991). Cultivar Late Star was spaced at 35cm apart in the row on 155cm beds adjusting 35,938 plants/ha (Zmuda, 1999). Maure and Umeda (1999) recommended planting during the last week of August for better fruit production for cvs. Chandler and Camarosa in low desert conditions in Arizona where the growing period is short. Similarly, summer planting dates are followed in the valley conditions of California, while autumn planting is followed in Florida and coastal California. Runners are planted in two rows spaced at 35–45 cm apart and a 60–68 cm wide bed covered with black mulch. For double rows per bed, plant spacing was 30–45 cm, which allowed 43,037 plants/ha on 152 cm bed centres and plants spaced at 30 cm apart in rows for cv. Chandler in Virginia (O'Dell and Williams, 1999). Tripathi et al. (2000) reported that runners planted on bed dimensions of 2 m×1 m wide and 12–15 cm raised beds with spacing of 20 cm×50 cm gave maximum fruit yield in mid-hills of Himachal Pradesh. Uselis (2000) compared seven planting schemes with plant densities of 15.6–94.3 thousand plants/ha. The greatest economics was obtained with 0.8 m×0.4 m (31.2 thousand plants/ha).

The suitable spacing under matted row is about 30 cm between plants as this usually favours that rows are runnered-in or matted by the end of the first season. In the case of vigorous cultivars, 40 cm spacing is considered to be appropriate. Day-neutral cultivars are shy runner producers and runners are generally removed to promote branch crowns. In such cases, runners are planted in double rows of 20 cm apart with 150 cm between rows accommodating 57,000 plants/ha. In plots of 2 m×2 m size with the spacing of 40 cm×30 cm, cvs. Chandler, Labella and Sujatha yielded 2.21, 0.98 and 0.48 kg fruit/plot, respectively, and Chandler was observed best in a submountain tract of Western Ghats of Kerala (Kumar et al., 2002). The effect of plant density of 11, 16 and 25 plants/m^2 was evaluated by Camacaro et al. (2004) and observed that plant growth and yield per plant increased with increase in spacing, but the maximum harvest index and yield per square metre were obtained at closer spacing. Plant spacing of 30 cm×30 cm or 40 cm×30 cm were observed to be better for cvs. Favete, Cortina, Evita, Selene, Eris and Madelein in Belgrade, Servia and Montenegro (Nenadovic et al., 2006). Poling et al. (2005) reported that cv. Chandler should not be planted any closer than 35 cm in-row plant spacing (for a double-row plant bed on 152 cm centre) accommodating 37,050 plants/ha. The normal spacing for Chandler followed in Carolina was 30 cm between the double rows of plants for standard beds (71–76 cm width on the top) with plastic mulch. In case of cv. Sweet Charlie, the planting plan was 30 cm in-row plant spacing (for a double-row plant bed on a 152 cm centre) accommodating 43,225 plants/ha. Whereas, cultivar Camarosa is generally planted at a 35 cm in-row plant spacing (for a double-row plant bed on a 155 cm centre) accommodating

Planting

37,050 plants/ha). Similarly, Milivojevic (2006) obtained better performance of strawberry cvs. Senga Sengana, Marmolaada and Elsanta with 30 cm × 30 cm and 40 cm × 40 cm spacing, whereas 15 cm × 30 cm and 20 cm × 30 cm spacing gave poor results. The maximum fruit yield was obtained from cv. Raurica in combination with white plastic mulch and tray plants in a one-row bed system where plants were spaced 20 cm apart and rows were 1.2 m apart, creating a planting density of 40,000 plants/ha (Duralija et al., 2006). Senga Sengana and Chandler strawberries were planted at spacing at 30 cm × 40 cm × 30 cm (single row), 30 cm × 50 cm × 60 cm (double row), 30 cm × 60 cm × 90 cm (triple row) accommodating 47,619, 60,606 and 66,666 plants/ha, respectively, and spacing 30 cm × 40 cm × 60 cm double row gave maximum yield with good fruit quality in the agro-climatic conditions of the Kashmir Valley, Jammu and Kashmir, India (Ahmad, 2009). Proper planting system along with spacing or plant density is optimized for maximization of production of quality fruits. Variations or modifications in the planting methods in matted row or hill row system may be done to adjust according to the site and agro-climatic conditions. Density also depends on the type of strawberry varieties: day-neutral varieties produce less runners, whereas June-bearing varieties are prolific runner producers. Accordingly, the spacing and plant density are adjusted (Janssen, 2005). In South American countries, cvs. Camarosa, Aromas, Camino Real, Cristal, Oso Grande, Albion, Festival etc. are planted at a density of 40,000–80,000 per/ha in open fields and up to 200,000 in soilless culture in Brazil, 55,000–60,000 in Chile and 35,000–40,000 in open fields and around 60,000 in greenhouses in Uruguay (Antunes and Peres, 2013). In north Indian plains areas such as Hisar, planting in September at 25 cm × 245 cm or 30 cm × 30 cm was found good in terms of plant growth and fruit yield (Bhatia et al., 2017).

Strawberry planting depends on agro-climatic zones and cultivars as well as growing environment (protected, open field or soilless culture). In general, different planting times are mid-to-late-autumn (autumn planting) with plants harvested in September–November; in summer planting, plants are harvested in January–February and stored for planting in April–June and even as late as August depending upon the location. Care should be taken that soil is properly solarized and treated with suitable disinfectant pesticides, beds are properly spaced and appropriate mulch material is used. Most preferable plating plan is hill row with double rows on 15–25 cm high raised beds at spacing of 25–30 cm plant to plant, 45 cm row to row and 90–120 cm between two hills (beds).

REFERENCES

Ahmad, M.F. 2009. Effect of planting density on growth and yield of strawberry. *Indian Journal of Horticulture*, 66(1): 132–134.

Anna, F.D. and Curatolo, G. 1986. Effect of dates of planting on the earliness and total yield of greenhouse strawberry. *Acta Horticulturae*, 176: 199–208.

Anonymous. 1982. The effect of bed type and plant spacing on strawberry plant performance. *Strawberry News Bulletin*, XXIV(6). California Strawberry Advisory Board.

Antunes, L.E.C.A. and Peres, N.A. 2013. Strawberry production in Brazil and South America. *International Journal of Fruit Science*, 13: 156–161.

Awasthi, R.P. and Badiyala, S.D. 1983. Spacing and performance studies in some strawberry cultivars. *Indian Journal of Horticulture*, 40(1): 29–34.

Badiyala, S.D. and Joolka, N.K. 1983. Effect of different spacing on the performance of strawberry cv. Tioga. *Haryana Journal of Horticultural Science*, 12(3/4): 165–167.

Badiyala, S.D. and Bhutani, V.P. 1990. Effect of planting dates and spacing on yield and quality of strawberry cv. Tioga. *South Indian Horticulture*, 38(60): 295–296.

Bhatia, S.K., Sharma, R. and Kumar R. 2017. Effect of different planting time and spacing on growth, yield and quality of strawberry (Fragaria×ananassa) cv. Ofra. *International Journal of Pure & Applied Bioscience*, 5(5): 207–211.

Bish, E.B., Cantliffe, D.J. and Chandler, C.K. 2000. Strawberry daughter plant size alters transplant growth and development. *Acta Horticulturae*, 533: 121–126.

Black, B.L., Enns, J.M. and Hokanson, S.C. 2002. Advancing the matted row strawberry production system. *Proceeding of the 5th North American Strawberry Conference*: January 14-16, 2001, Niagara Falls, Canada in association with NASGA; pp. 112–115.

Black, B., Pace, M. and Goodspeed, J. 2008. *Strawberries in the Garden*. Utah State University Cooperative Extension. http://extension.usu.edu/files/publications/publication/Horticulture_Fruit_2008-06pr.pdf.

Bussell, W.T., Ennis, I.L., Pringle, G.J., Perry, F.A. and Triggs, C.M. 2003. Relationship between crown size at planting, yield and growth in strawberries. A*gronomy New Zealand*, 32/33: 9–12.

Camacaro, P., Camacoaro, G.J., Hadley, P., Dennet, M.D., Battey, N.H. and Carew, J.G. 2004. Effect of plant density and initial crown size on growth, development and yield in strawberry cultivars Elsanta and Bolero. *Journal of Horticultural Science and Biotechnology*, 79(5): 739–746.

Capocasa, F., Bordi, M. and Mezzetti, B. 2009. Comparing frigo and fresh plants in non-fumigated and heavy soil: The response of 10 strawberry genotypes. *Acta Horticulturae*, 842: 129–134.

Caracciolo, G., Moncada, A., Prinzivalli, C. and D'Anna, F. 2009. Effects of planting dates on strawberry plug plant performance in Sicily. *Acta Horticulturae*, 842: 155–158.

Caruso, P. and D'Anna, F. 1995. Planting dates of strawberries using cold stored plants. *Culture Protette*, 24(4): 87–95.

Crawford, T.D., Himmelrick, D.G., Sibley, J.L. and Pitts, J.A. 2000. Effect of runner plantlet size on performance of strawberry plug plants. *Small Fruits Review*, 1: 15–21.

Dale, A. and Pritts, M. 1989. *Dayneutral Strawberries*. Factsheet. Ministry of Agriculture Food and Rural Affairs, Ontario, Canada.

D'Anna, F., Caracciolo, G., Alessandro, R. and Faedi, W. 2014. Effects of plant type on two strawberry cultivars in Sicily. *Acta Horticulturae*, 1049: 553–556.

Danek, J. 1984/1985. Effect of several cultural treatments on the number of strawberry sets produced during the summer. *Prace Instytyty Sadownictwa I Kwiaciarstwa. Ser A*, 25: 25–36.

Duralija, B., Cmelik, Z., Druzic Orlic, J. and Milicevic, T. 2006. The effect of planting system on the yield of strawberry grown out-of-season. *Acta Horticulturae*, 708: 89–92.

Das, B., Ahmed, N. and Attri, B.L. 2012. Performance of strawberry genotypes at high altitude temperate climate under different growing environments. *Progressive Horticulture*, 44(2): 242–247.

Das, B., Krishna, H., Ahmed, N., Attri, B.L. and Ranjan, J.K. 2013. Protected strawberry culture at high altitude temperate climate as influenced by planting time, mulching and soil moisture. *Indian Journal of Horticulture*, 70(4): 506–511.

Duval, J.R. 2005. "J" rooting does not affect the performance of Florida strawberry transplants. *Small Fruit Review*, 4(4): 3–5.

Faedi, W., Gufo, S. and Bernardini, D. 1986. Piantagioni autunnali di piante fresche di fragola in Sicilia. *Incontro Frutticolo S.O.I. su: "La coltura della fregola"*, 14: 42–44.

Hancock, J.F., Pritts, M.P. and Siefker, J.H. 1984. Yield components of strawberry maintained in ribbon and matted row. *Crop Research*, 24: 37–43.

Hasell, R.L., Phillips, T.L., Dufault, R.J., Hale, T.A. and Ballington, J.R. 2006. Fall transplanting date affects strawberry cultivar performance in South Carolina. *International Journal of Fruit Science*, 6(2): 73–85.

Hicklenton, P.R. and Reekie, J. 2000. Plant age, time of digging and carbohydrate content in relation to storage mortality and post storage vigour of strawberry plants. *Acta Horticulturae*, 513: 237–245.

Hesketh, J.L., Eaton, G.W. and Baumann, T.E. 1990. Strawberry plant spacing on raised beds. *Fruit Variety Journal*, 44: 12–17.

Hochmuth, G., Cantliff, D., Chandler, C., Stanley, C., Bish, E., Waldo, E., Legerd, D. and Duval, J. 2006. Fruiting responses and economics of containerized and bare rooted strawberry transplants establishment with different irrigation methods. *HortTechnology*, 16(2): 205–210.

Human, J.P. 1999. Effect of number of plant per plant hole and of runner plant crown diameter on strawberry yield and fruit mass. *South African Journal of Plant and Soil*, 16(4): 189–191.

Ilgin, M., Kaska, N. and Colak, A. 2002. Effect of spring planting of Camarosa and Sweet Charlie fresh runners on yield and quality. *Acta Horticulturae*, 567(2): 581–583.

Janssen, D. 2005. Growing *Strawberries*. University of Nebraska-Lincoln. Extension http://lancaster.unl.edu

Kehoe, E., Savini, G. and Neri, D. 2009. The effects of runner grade, harvest date and peat growing media on strawberry tray plant fruit production. *Acta Horticulturae*, 842: 699–702.

Kumar, T.P., Sajith Babu, D. and Aipe, K.C. 2002. Performance of strawberry varieties in Wayanad district of Kerala. *Journal of Tropical Agriculture*, 40(2002): 51–52.

Kumar, P.T., Suma, B., Bhaskar, J. and Satheesan, K.N. 2008. *Management of Horticultural Crops*. New India Publishing Agency, New Delhi, India: pp. 158–164.

Lieten, F. 1998. Recent advances in strawberry plug transplant technology. *Acta Horticulturae*, 513: 383–388.

Lines, R., Kelley, G., Milinkovic, M. and Rodoni, B. 2006. Runner certification and virus elimination in commercial strawberry cultivars in Australia. *Acta Horticulturae*, 708: 253–254.

Lopez-Medina, J., Peralbo, A. and Flores, F. 2006. Strawberry production in soilless systems with slow sand filtration. *Acta Horticulturae*, 708: 389–392.

Lutchoomun, B. 1999. *Influence of Fresh and Cold Stored Plantlets on Strawberry Yield*. AMAS 1999. Food and Agricultural Research Council, Réduit, Mauritius.

Masiunas, J.B., Weller, S.C., Hayden, R.A. and Janick, J. 1991. Effect of plant spacing on strawberry yield in two cultural systems. *Fruit Variety Journal*, 45: 146–151.

Maure, M.A. and Umeda, K. 1999. *Influence of Cultivar and Planting Date on Strawberry Growth and Development in the Low Desert*. University of Arizona College of Agriculture. *Vegetable Report*: pp. 1–5.

Milivojevic, J. 2006. The influence of planting distance on generative potential of strawberry cultivars. *Vocarstvo*, 40(2): 113–122.

Milivojevic, J., Nikolic, M. and Durovic, D. 2009. The influence of growing system on cropping potential of strawberry cultivar "Clery" grown in plastic tunnel. *Acta Horticulturae*, 842: 115–118.

Nenadovic, M.E., Milivojevic, J. and Urovic, D. 2006. The influence of planting distance on the fruit characteristic of newly introduced strawberry cultivars. *Vocarstvo*, 40(2): 123–132.

Nestby, R. and Sonsteby, A. 2017. Effect of plant type and delayed planting on growth and yield parameters of two short day strawberry (*Fragaria* x *ananassa* Duch.) cultivars in open field. *Journal of Berry Research*, 7(3): 179–194.

Noakes, L.G., Wilken, L. and Villiers, de S. 2006. High density, vertical hydroponics growing system for strawberries. *Acta Horticulturae*, 708: 365–370.

O'Dell, C. and Williams, J. 1999. *Hill System Plastic Mulch Strawberry Production Guide for Colder Areas*. Virginia Cooperative Extension Publications 438-018: pp. 1–41.

OEydvin, J. 1972. Good results with strawberry yields by increasing the plant number per acre. *Gartneryrke*, 62(16): 306–312.

Pandey, D., Phogat, K.P.S. and Shukla, S.N. 1980. A note on the effect of different spacing on growth and yield of strawberry cv. Red Coat. *Progressive Horticulture*, 12(3): 67–70.

Poling, E.B. 1991. The annual hill planting system for South West North Carolina. In: *The strawberry into the 21st century*, A. Dale and J. J. Luby (eds.). Timber Press, Portland, Oregon: pp. 258–263.

Poling, E.B. and Durner E.F. 1986. Annual strawberry hill culture system in South East North Carolina. *HortScience*, 21(2): 240–242.

Poling, E.B. and Maas, J.L. 1998. Strawberry plug transplant technology. *Acta Horticulturae*, 513: 393–402. www.smallfruits.org/assets/documents/ipm-guides/2005culturalguidepart1bs1.pdf

Poling, E.B., Krewer, G. and Smith, J.P. 2005. *Southeast Regional Strawberry Plasticulture Production Guide*: pp. 1–21.

Roudeillac, R. and Veschambre, D. 1987. *La Fraise*. CTIFL-CIREF, Paris, France: pp. 182–183.

Rowley, R.D. 2010. Season extension of strawberry and raspberry production using high tunnels. M.Sc. Thesis. Utah State University, Logan, Utah: pp. 1–190.

Rowley, D., Black, B. and Drost, D. 2010a. *High Tunnel Strawberry Production*. Utah State University Cooperative Extension: pp. 1–6.

Rowley, D., Black, B.L. and Drost, D. 2010b. Early-season extension using june-bearing "chandler" strawberry in high-elevation high tunnels. *Hortscience*, 45(10): 1464–1469.

Schmitz, F. and Lenz, F. 1989. Die Bedeutung des Pflanzzeitpunktes für Wachstum und Ertrag bei den Erdbeersorten "Elvira", "Tenira", "Korona" und "Bogota". *Erwerbsobstbau*, 31: 118–122.

Sharma, R.R. and Sharma, V.P. 2004. *The Strawberry*. ICAR, New Delhi, India.

Sabatino, L., Pasquale, C.D., Abou, D.F., Martinelli, F., Busconi, M., D'Anna, E., Panno, S., Iapichino, G. and D'Anna, F. 2017. Properties of new strawberry lines compared with well-known cultivars in winter planting system conditions. *Notulae Botanicae Horti Agrobotanici*, 45(1): 9–16.

Singh, R., Sharma, R.R. and Goyal, R.K. 2007. Interactive effect of planting time and mulching on Chandler strawberry (*Frageria* x *ananasa* Duch.). *Scientia Horticulturae*, 111(4): 344–351.

Singh, S.R., Zargar, M.Y., Singh, U. and Ishaq, M. 2010. Influence of bio-inoculants and inorganic fertilizers on yield, nutrient balance, microbial dynamics and quality of strawberry (*Fragaria* x *ananassa*) under rainfed conditions of Kashmir valley. Indian *Journal of Agricultural Sciences*, 80(4): 275–281.

Stevens, M.D., Black, B.L., Lea-Cox, J.D. and Hapeman, C.J. 2006. Sustainability of cold-climate strawberry production systems. *Acta Horticulturae*, 708: 69–72.

Strik, B., Stonerod, P., Bell, N. and Cahn, H. 1997. Alternative production systems in perennial and annual culture of June-bearing strawberry. *Acta Horticulturae*, 439: 433–437.

Treder, W., Tryngiel-Gac, A. and Klamkowski, K. 2015. Development of greenhouse soilless system for production of strawberry potted plantlets. *Horticultural Science* (Prague), 42(1): 29–36.

Tripathi, V.K., Dwivedi, M.P., Sharma, R.M. and Agrahari, P.R. 2000. Effect of planting date and spacing on yield and quality of Chandler strawberry (*Fragaria* x *ananasa* Duch.). *Haryana Journal of Horticultural Sciences*, 29(3/4): 185–186.

Uselis. 2000. Comparison of strawberry planting schemes. *Sodininkyst-ir-Darzuin—kyrte*, 19(2): 11–12.

Verwijs, A. 1984. Planting systems with early hothouse strawberries. *Groenten en Fruit*, 40(19): 34–35.

Wilson, F. and Dixon, G.R. 1988. Strawberry growth and yield related to plant density using matted row husbandry. *Journal of Horticultural Sciences*, 63(2): 240–242.

Zmuda, D.E. 1999. The effect of delayed planting date on growth and cropping of strawberry plants. *Acta Agrobotanica*, 52(1.2): 103–112.

13 Nutrition

Arti Sharma and Bindiya Sharma

CONTENTS

13.1 Plant Tissue Testing 210
13.2 Role of Nutrients 210
    13.2.1   Nitrogen 210
    13.2.2   Phosphorus 213
    13.2.3   Potassium 213
    13.2.4   Calcium 214
    13.2.5   Magnesium 214
    13.2.6   Zinc 214
    13.2.7   Boron 215
    13.2.8   Iron 215
    13.2.9   Copper 216
    13.2.10 Manganese 216
13.3 Nutrient Recommendations 216
13.4 Foliar Nutrition 217
13.5 Fertigation 218
13.6 Organic Production 219
References 222

Making provisions for the creation of an ideal environment for optimum growth of the root system in strawberries is an important step in obtaining an economically profitable planting. Since strawberries are a perennial and shallow-rooted crop, most root growth can be observed within the top 15 cm of soil profile. And because of their shallow root system, strawberries are very sensitive both to excess or deficit soil moisture conditions. Heavy demands for water are placed on the root system, especially in the short period during which berries are developing. The land to be planted with strawberries should be thoroughly cultivated and liberally manured with farmyard manure or compost. It is best to apply manures to the crop preceding the strawberries so that it may become well decomposed. In the absence or short supply of manure, commercial fertilizers should be applied in sufficient quantities, but these should be used in conjunction with green manure crops such as cowpeas (*Vigna unguiculata*), soybeans (*Glycine max*) and velvet beans (*Mucuna pruriens*).

As the strawberry may be cropped for one year only or for up to three years or more, application of manures and fertilizers varies according to the duration of time for which the plantation is to be retained (Hughes et al., 1969). Before planting, a sample of soil from the site should be analyzed and an estimate should be made for the requirement of lime, potash and phosphorus. If a dressing of 50–75 tonnes per hectare of farmyard manure is ploughed into the land and any soil deficiency of potassium or phosphorus is remedied before planting strawberry runners, application of additional fertilizers is usually not needed where cropping is needed to be retained for one year. A dressing of farmyard manure on fertile soils before planting of strawberries often provides all the needs of the crop over a few years. The availability of nutrients depends upon the extent of the roots, which is decided by the depth of soil and the available moisture in the soil.

209

13.1 PLANT TISSUE TESTING

The results of testing and analysis of plant tissues can determine if soil nutrients are being taken up by the strawberry plants in adequate amounts. Even when nutrients are in adequate supply, the ability of plants to utilize available nutrients can be influenced by several factors, such as soil temperature (high or low), root development, available soil moisture (deficit or excess) and the methods and time of fertilizer application. Results of testing and analysis of plant tissues can help growers to avoid mid-term nutrient deficiency problems. Occasionally, severe nutrient problems are encountered during the early crop growth period. The results of standard testing and tissue analysis may allow time for corrective measures to be taken in the same season, thereby reducing the potential loss of fruit yield.

Normal nutrient levels in strawberry tissues vary with plant growth stages such as flowering, fruiting and after fruiting. The values of nutrient content in plant tissues may also be influenced by the time of sampling, parts of the plant sampled, the age of the leaf sampled and the cultivars sampled, but the extent of deficiency levels is a reliable indicator for deciding whether nutrients must be added or not. Similar to soil sampling, the standard procedures for sampling of plant tissue for analysis are important. Some of important points to consider are:

- Care must be taken to ensure cleanliness in sampling and handling.
- Proper sampling equipment needs to be used.
- Specific plant parts should be taken as samples at the proper stage of growth.
- The representative sample should immediately be cleaned, washed and dried with a dry cloth or soft brush.
- Non-typical or non-representative plants and areas in the field should be avoided for sampling.
- Plant samples contaminated with fertilizers, pesticide residues or soil particles should be avoided.
- The samples should be collected in a clean brown paper bag, not a metal container.
- The youngest expanded mature leaf needs to be selected as a sample.
- About 25–30 leaves should be taken from a reasonable number of typical plants throughout the field.
- The sample should be dried at room temperature.

These samples should be subjected to tissue analysis. The critical concentration and deficiency/sufficiency ranges of nutrients at various stages of crop growth and development in strawberry are given in Table 13.1 and 13.2.

A nitrate test of petiole sap has been suggested as a decision-support tool that can be used to determine the nitrogen status of strawberry plants and thereby facilitate improvement in fertilizer application techniques (Raynal-Lacroix and Abarza, 2002).

13.2 ROLE OF NUTRIENTS

13.2.1 Nitrogen

Nitrogen is one of the most important elements in strawberry production. Nitrate nitrogen, the form most readily used by plants, is indicated in kg/ha in soil test results. Strawberries require annual applications of nitrogen at the right time and quantity. Improper timing and/or rates of nitrogen may lead to increase in winter injury, fruit softening and high incidence of disease (Nam et al., 2006), leaf burn injury (Jackson, 1972), low fruit yield (Brandstveit, 1979; Haynes and Goh, 1987) and fewer inflorescences and blooms per plant . Typical nitrogen fertilizers used on strawberries include urea (46% N), ammonium nitrate (34% N), potassium nitrate (13% N), and calcium nitrate (15% N).

TABLE 13.1

Leaf Nutrient Status of Strawberries (Dry Weight Basis)

Nutrient	Nutrient (Form)	Plant Part Tested	Unit	Tentative Critical Concentration	Range		
					Deficiency Symptoms Visible	Deficiency Symptoms not Visible	Sufficiency Range
Boron	B	Blade	ppm	25	18–22	35–200	
Calcium	Ca	Blade	%	0.3	0.08–0.2	0.4–2.7	0.77–1.48
Chlorine	Cl	Petiole	%	–	<0.07	0.07–0.41[a]	
Copper	Cu	Blade	ppm	3	<3	3–30	3–22.5
Iron	Fe	Blade	ppm	50	5–40	50–3000	58–114
Magnesium	Mg	Blade	%	0.2	0.03–0.1	0.3–0.7	0.25–0.75
Manganese	Mn	Blade	ppm	30	4–25	30–700	45–121
Molybdenum	Mo	Blade	ppm	0.5	0.12–0.4	0.5[a]	
Nitrogen	NO$_3$	Petiole	ppm	500	0.5	700–20,000	
Total-N		Blade	%	2.8	2–2.8	3.0[a]	2.07–3.04
Phosphorus	H$_2$PO$_4$	Petiole	%	0.07	0.02–0.07	0.1–0.5	0.2–0.38
Potassium	K	Petiole	%	1	0.1–0.4	1–6	
		Blade	%	1	0.1–0.5	1–6	1.84–2.21
Sodium	Na	Blade	%	–	<0.01	0.01–0.4[a]	
Sulfur	SO$_4$	Blade	ppm	100	25–80	100–500[a]	
Zinc	Zn	Blade	ppm	20	6–10	20–50[a]	15–33

Source: Modified from Ulrich et al., 1980, Strawberry deficiency symptoms: A visual and plant analysis guide to fertilizer. Publ. 4098. Division of Agricultural Sciences, University of California, Oakland, California, and Almaliotis et al., 2002, Leaf nutrient levels of strawberries (cv. Tudla) in relation to crop yield. ISHS. Acta Horticulture, 567: 447–449.

[a] Salt damage is indicated by values >0.5% for Cl and/or 0.1 for Na.

TABLE 13.2

Critical Levels and Deficiency Threshold Values (Percent Dry Weight) for Leaf Nutrient Contents in Strawberry

Nutrient and Stage of Sampling	Deficiency Threshold	Critical Level
N		
Flowering	2.5	3
Fruiting	2	2.6–3
After Fruiting	1.5	2
P		
Flowering	0.25	0.3
Fruiting	0.20	0.25–0.3
After Fruiting	0.15	0.2
K		
Flowering	1.0	2
Fruiting	1.0	1.5
After Fruiting	0.6	1

Source: Modified from Bould (1964).

Higher nitrogen rates are likely to increase the number of runners. This could be disadvantageous because too many runners can create high runner populations and excessive competition for carbohydrate. On the positive side, high nitrogen levels during summer may increase formation of branch crowns and benefit overall production. Guttridge (1960) observed the antagonistic effect of vegetative growth on reproductive growth, while Guttridge and Anderson (1973) reported that increase in plant growth may or may not be correlated with increased yield of fruits in strawberry.

Nitrogen-deficient plants have stunted growth with yellowish-green appearance, small leaves with upright stiff petioles, and bright orange-red margins of older leaves with subsequent browning and drying of margins. Low leaf nitrogen content (<3%) at the time of flower bud differentiation significantly delays the flowering in strawberry (Tanaka and Mizuta, 1976). The berry size of the strawberry cv. Korona has been observed to be depressed significantly with a lower dose of nitrogen (Nes and Hieltenes, 1992). Ram-Autar and Gaur (2003) observed the number of leaves per plant and plant height to increase with increasing rates of nitrogen application, whereas the number of fruits harvested per plant, total yield and ascorbic acid content and total acidity of fruits increased with increasing rates of nitrogen, up to 150 kg/ha and decreased thereafter. An increase in rate of nitrogen application may increase the protein and amino acid contents. In addition, the competitive ability of reproductive tissues to absorb nitrogen is higher than that of vegetative tissues, at the later growth stages (Guo et al., 2003). The characteristic aroma components of strawberry increase with increasing levels of nitrogen up to some extent but at higher application rates, it declines (Liu et al., 2004). Anthocyanin synthesis in strawberry fruits may be reduced by N deficiency (Yoshida et al., 2002).

Thus nitrogen plays a vital role in strawberry production by being involved in influencing plant growth (Breen and Martin, 1981; Singh et al., 1983; Haynes and Goh, 1987; Neuweiler, 1997; Singh et al., 2001; Rana and Chandel, 2003), flowering and fruit set (Breen and Martin, 1981; Yoshida et al., 1991; Yoshida et al., 2001), fruit yield (Keefer et al., 1978; Pandey and Mishra, 1983; Kohli et al., 1984; Haynes and Goh, 1987; Human and Kotze, 1990; Rana and Chandel, 2003), fruit quality (Saxena and Locascio, 1968; Haynes and Goh, 1987), nutrient accumulation (Breen and Martin, 1981; Archbold and Mackown, 1997; Bhat, 1999) and post-harvest quality (Noe et al., 1997).

The source and nature of nitrogen fertilizer may play a significant role in nutrient mineralization, absorption and leaching in small-fruit crops. The amount of released nitrogen varies markedly

Nutrition 213

due to weather and other factors (Fixen and West, 2002). The overall nitrogen use efficiency of a cropping system can be improved by achieving greater uptake efficiency from applied nitrogen inputs, and by reducing the loss of nitrogen both from soil organic and inorganic nitrogen pools (Cassman et al., 2002). Thus, it is important for the economic and environmental sustainability of the strawberry production system to consider the appropriate application and utilization of soil-applied nitrogen sources.

13.2.2 PHOSPHORUS

Phosphorus is an essential part of many sugar phosphates involved in photosynthesis, respiration and other metabolic processes; and it is also a part of nucleotides, as in RNA and DNA, and of the phospholipids present in the membrane (Salisbury and Ross, 1986). Phosphorus application increases plant growth (Dennison and Hall, 1956) and yield (Pandey and Mishra, 1983). Phosphorus-deficient plants are stunted and, in contrast to those lacking nitrogen, are often dark green in appearance. Sometimes accumulation of anthocyanin pigments can occur. The veins of older leaves turn bluish, while the leaf lamina looks bluish-purple in the phosphorus-deficient leaves of strawberries. If excess phosphorus is provided to the plants, root growth is often comparatively increased in relation to shoot growth. Phosphorus deficiency develops first on old leaves. Dry weight begins to get reduced when phosphorus concentration in plant dry matter is below 0.3% and fresh weight begins to get reduced when phosphorus concentration in plant sap is about 520 mg/kg (Jeong et al., 2001). Seolhyang strawberry has been observed to show restricted plant growth in phosphorus-deficient plants, with minimum values for leaf number, plant height, leaf length and width, petiole length and fresh and dry weights (Choi et al., 2013). The macro-element contents such as sodium, calcium and magnesium increase as phosphorus concentration increases but microelements (such as iron and manganese) behave *vice versa* in the dry matter of above ground tissues. When the phosphorus concentration in the soil solution is elevated, the pH and EC of root substrate decrease. Strawberry runners and their productivity are significantly related to the availability of soil phosphorus, soil water content and pH levels. High soil phosphorus nutrition at the pH level of 6.2 promotes strawberry nursery plant propagation (Hong et al., 2009).

Deficiency of both phosphorus and nitrogen reduce the rate of leaf production in strawberry. Decreased leaf areas are due to lower cell numbers and, to a lesser extent in severe deficiency, to smaller cells. The rate of cell production is markedly reduced due to phosphorus and nitrogen deficiency. The effect of phosphorus and nitrogen deficiency on potential fruit yield is confined to the inhibition of proliferation of branch crownd and the lack of available sites for flower initiation. There is no effect of nutrient level on the number of flowers initiated on each apical meristem. For increasing the berry yields, the manipulation of the status of nutrients should be aimed predominantly concerned with enhanced branch crown formation, which provides additional sites for flower initiation (Abbott, 1968).

13.2.3 POTASSIUM

Potassium is an activator of many enzymes that are essential for photosynthesis and respiration, and it also activates enzymes needed to form starch and proteins. The importance of potassium for strawberries has been observed in many experiments. On soils that are excessively deficient in potash and where farmyard manure has not been applied, the omission of potash from the nutrition schedule generally had serious effects on vigour, longevity and cropping (Hughes et al., 1969). Annual applications of potassium fertilizers are, therefore, needed, especially when adequate dressings of farmyard manure (FYM) in soil have not been done. Potassium deficiency leads to a great reduction in the plant growth and yield contributing factors. Being a major component of the strawberry fruit, potassium application has special significance as the size, colour and acidity of the fruit have a positive relation with potassium nutrition (Ricketson 1966; Behnamiyan and Masiha, 2002). But increasing soil potassium concentration sometimes decreases firmness of strawberry fruits (Dierend and Faby, 2002, 2003).

13.2.4 CALCIUM

Twisted and deformed tissues result from calcium deficiency, and the meristematic areas die early in the season. Calcium deficiency in strawberry is associated with the tipburn of young, unopened leaves, which subsequently results in deformed and crinkled mature leaves (Mason and Guttridge, 1974). Consequently, a small but continuous supply of calcium ions is very much essential for the normal development of fruit and to minimize leaf tipburn injury in strawberry plants (Chiu and Bould, 1976).

Albinism (a physiological disorder characterized by fruits turning white) in the Elsanta cultivar of strawberry has been observed to be caused by excessive fertilization with nitrogen and potassium, leading to over-vigorous growth and calcium deficiency and interfering with anthocyanin synthesis (Buchter, 1991). The post-harvest application of $CaSO_4$ significantly minimizes the incidence of fruit rot after harvest caused by the fungus *Colletotrichum acutatum* (Smith and Gupton, 1993). Similarly, post-harvest treatment by dip of fruits in 1.0% $CaCl_2$ solution is effective in management of post-harvest decay and in maintaining firmness and soluble solid contents without affecting the sensory qualities of the strawberry fruits (Garcia et al., 1996). There are many reports that the use of calcium fertilizers, during flowering and after flowering, in the form of foliar sprays has a beneficial effect on fruit quality. Fruit deficient in calcium loses firmness and marketable quality, and becomes vulnerable to bruising damage during transport (Markus and Morris, 1998; Lacroix and Carmentran, 2001). Besides, calcium deficiency impairs the permeability of cell membranes and can even lead to the disintegration of their structure (Marinos, 1962; Lara et al., 2004). It is advisable to go for preharvest foliar sprays of calcium to improve its content in berries.

Calcium sprays significantly increases the contents of calcium, nitrogen, zinc and manganese in leaves and fruits, but significantly decreases the potassium content of fruits in the first and second crops, whereas the boron content of leaves decreases significantly. The contents of phosphorus, magnesium, copper and iron have no significant correlation with the concentration of calcium spray (Chen et al., 2003) but supplementation of calcium in the nutrient solution increases the calcium contents in the roots and leaves, and also increases crop yield (Treder, 2004). CaO in combination with steam can be used for soil disinfection in strawberry production (Lenzi et al., 2004). Prohexadione-Ca, an inhibitor of gibberellin biosynthesis, when sprayed on strawberry plants decreases the number of autumn runners, and increases dry mass of the crown and roots, thereafter increasing the number of flowers in inflorescences of plants (Black, 2004; Reekie et al., 2005; Karhu and Hytonen, 2006).

13.2.5 MAGNESIUM

Besides its presence in chlorophyll, magnesium is an essential nutrient element as it combines with the energy system adenosine triphosphate (ATP), allowing ATP to function in many reactions. Further, magnesium activates (Salisbury and Ross, 1986) many enzymes involved in photosynthesis, respiration and formation of deoxyribonucleic acid (DNA) and ribonucleic acid (RNA). In strawberry, the requirement for magnesium is not very high but its deficiency sometimes occurs in poor soils, such as those overlying gravels or where excessive applications of potassic fertilizers cause imbalances between the levels of magnesium and potash (Hughes et al., 1969). The magnesium deficiency leads to turning the leaf margin into yellowish-green with subsequent downward and inward curling of leaves (Mitra, 1991). The main vein and the basal portion of leaf turn yellowish-orange, while the areas adjacent to the mid-rib remain green (Lineberry and Burkhart, 1943). Magnesium-deficient fruits ripen late with poor colour development (Bunemann and Gruppe, 1962). Preflowering spray of magnesium nitrate (3–5 g per liter water) can help to overcome magnesium deficiency.

13.2.6 ZINC

Zinc is involved in chlorophyll formation or prevents chlorophyll destruction so zinc-deficient plants exhibit pale green–yellow leaves.

Nutrition 215

Application of zince as $ZnSO_4$ (0.4%) increases the number of leaves, flowers, fruit set and yield, total soluble solids (TSS) and ascorbic acid contents, and the number of runners in Chandler strawberry (Chaturvedi et al., 2005). Spraying of strawberry plants with zinc tends to reduce the severity of infection of powdery mildew (*Spaerotheca macularis*) disease and helps increase fruit yield (Fouly, 2004).

Zinc concentration @ 30 µmol/l can induce zinc toxicity and iron deficiency symptoms in strawberries grown on peat bags, which results in reduced fruit yield with a higher proportion of malformed fruits (Lieten, 2003). Zinc toxicity reduces leaf area, plants' total dry matter production, TSS content in fruit, fruit size and leaf chlorophyll content (Casierra and Poveda, 2005).

13.2.7 BORON

Plants of short-day strawberry cv. Elsanta growing in peat and rock wool show leaf tipburn symptoms without application of boron. Boron-deficient plants produce very typical flowers with damaged pistils and typical fruit malformation, besides significant reduction in fruit weight, fruit number and total yield (Lieten, 2002). In the studies of Wojcik and Lewandowski (2003), sprays of boron increased the status of this micronutrient in fruit and leaf tissues of Elsanta strawberry.

Preharvest foliar application of calcium and boron is quite useful in reducing the incidence of disorders and getting higher marketable yield in the "Chandler" strawberry (Singh et al., 2007). Application of fertilizer formulations containing boron influences plant yield and chemical composition of strawberry cv. Fern (Esringu et al., 2011). Boron application has been observed to increase the phosphorus, potassium, manganese, zinc and copper contents in plants. Further, they (Esringu et al., 2011) concluded that addition of boron @ 5.5 kg/ha is sufficient to elevate soil boron levels to non-deficient levels.

13.2.8 IRON

Iron-deficient plants are characterized by development of pronounced interveinal chlorosis similar to that caused by magnesium deficiency, but it occurs first on the youngest leaves. Strawberry plants grown in sand culture showed dark green venation at 21 days after the trials were started. At 35 days, newly formed leaves showed slight interveinal chlorosis, while at 48 days, the chlorotic areas were bright yellow with dark green veins. Within the next seven days, the oldest leaves were observed to be necrotic marginally. Iron deficiency-induced chlorosis can greatly reduce fruit yields in strawberries grown in high pH, calcareous soils. The plants treated with iron sprays once a week produced higher fruit yield than the untreated control in strawberry cvs. Motto (13%), Chandler (30%) and Douglas (56%). Unsprayed plants of strawberry cv. Motto were observed to produce relatively more fruit yield but of inferior quality and had a shorter shelf-life than the unsprayed plants of other cultivars (Zaiter et al., 1993). Strawberry cultivars show different responses to various doses and types of application of sequestrene Fe-138. Some varieties respond negatively to the application of fertilizers containing iron, and the yield may decrease in comparison to no iron application (Turemifl et al., 1997). Lieten (2000) associated iron leaf values lower than 45 ppm on a dry matter basis with deficiency symptoms. A minimum concentration of 10 µmol Fe/l in the nutrient solution was considered satisfactory for growth and good fruit production of strawberry cv. Elsanta, grown on peat bags. There was A significant linear relationship between iron concentration and yield and quality (Almaliotis et al., 2002; Karp and Starast, 2002).

Foliar application of iron is an effective way to increase its concentrations in strawberry cultivars, because the concentration of iron in strawberry leaves increased continuously with repeated foliar applications of iron for its optimal concentration in leaves. $FeSO_4.7H_2O$, which is cheaper than other sources of Fe, can be used in foliar application; however, care should be taken about the negative impact of foliar iron application on leaf manganese and especially calcium concentration (Erdal et al., 2004) as the micronutrient mixtures contain impurities having manganese and calcium.

Strawberries have maximum iron accumulation in roots (13 times more than in the crown) followed by crowns and fruits (Gaweda and Ben, 2004). Besides, higher concentrations of ferrous sulfate have a toxic effect on the plant and retard the plant growth, yield and quality of strawberries (Chaturvedi et al., 2005). Feeding of iron in liquid form tends to increase the total NPK (Nitrogen, Phosphorous and Potassium) also in roots. A low level of iron restricts NO_3 concentration in fruits, which is most desired by the consumers (Taghavi et al., 2005).

13.2.9 COPPER

Copper is present in several enzymes or proteins involved in oxidation and reduction. Without copper, young leaves often become pale green with necrotic spots between the veins. The deficiency of copper in soils (<3 mg/kg dry weight) greatly depresses the yield of strawberry fruits (Van der Boon, 1961). $CuSO_4$ (5 g per liter water) can be used as a foliar feeding to address the copper deficiencies in strawberry.

13.2.10 MANGANESE

Manganese plays a structural role in the chloroplast membrane system, and one of its important roles is similar to that of chloride in the photosynthetic split of H_2O. Deficiency symptoms of manganese on strawberry plants can be characterized as interveinal chlorosis while the veins remain green.

Manganese deficiency symptoms appear with leaf manganese values <200 ppm (Lieten, 2004). Manganese deficiency significantly causes reduction of petiole length but has no significant effect on flower characteristics, fruit size or fruit set. Manganese toxicity causes reduction of dry matter production, leaf pigment and fruit quality in strawberries (Casierra and Poveda, 2005). Spraying strawberries with manganese significantly reduces the severity of infection of powdery mildew disease (*Spaerotheca macularis*) and helps in increasing fruit yield (Fouly, 2004).

13.3 NUTRIENT RECOMMENDATIONS

FYM has been reported to be best for strawberry cultivation (Bailey, 1963). A light dressing of bone meal (170 kg/ha) at the end of December can improve the size and flavour of the fruit (Anon, 1956). It is preferable to apply 50–75 t/ha FYM before planting strawberry runners. Deep litter from the broiler house, poultry manure and spent mushroom compost are often available, and useful, but as they contain more nitrogen, phosphorous and potash than does FYM, they should not be used at the same rate. Five–7.5 tonnes of deep litter or broiler house manure are equivalent to 50 t/ha of FYM. It is essential to apply bulky organic manures regularly every three or four years if an annual crop practice on the same site is to be followed (Huhges et al., 1969).

In greenhouse strawberry cultivation, the mixing of bark compost with volcanic ash oil has a positive impact on fruit yield, while it causes physiological disorders in the plants when mixed with volcanic heavy clay soil. Besides, growing strawberry in 75% white peat + 25% styromull (expanded polystyrene granules) and with broiler litter has been reported to be a good option for getting commercial yields of several strawberry cultivars (Bastelaere, 1987; Rubeiz et al., 1993).

Under most conditions, 454–680 kg fertilizer mixture containing 2–3% nitrogen and 6–8% each of phosphorous and potassium should be sufficient for strawberries. On fields where leguminous crops have been grown, a part of the nitrogen may be left out. It should be kept in mind that applications of commercial fertilizers in large quantities are profitable only when used on soils in good physical condition and well supplied with humus (Bailey, 1963). Calcium ammonium nitrate @ 90–180 kg/ha can be applied in a single application safely with respect to avoiding nitrate toxicity for the strawberries (Jackson, 1972). The use of 120 kg/ha of nitrogen as slow release fertilizer

Nutrition

applied two-thirds in autumn and one-third in spring has been observed to have best results in strawberry cv. Elista (Homann, 1973).

The best growth and maximum fruit yield of strawberry cv. Hokowase in long term cultivation can be obtained by maintaining <7–8 mg NO_3 N/100 g soil immediately after flower bud initiation. The basal and subsequent N treatments do not affect the total yield of Hokowase strawberry but greatly affect growth and time of cropping (Tanaka and Mizuta, 1976). In a study in Russia conducted on a derno-podzolic light loam soil supplied with 60 kg nitrogen, 120 kg P_2O_5 and 260 kg K_2O/ha along with simazine @ one kg/ha, increased the net profit by 20% in strawberry cultivation (Bagaev et al., 1976).

According to Maraes et al. (1977) significantly more yield of strawberry can be obtained with nitrogen at 50 and 100 kg/ha and P_2O_5 at 100 and 200 kg/ha. The addition of nitrogen increases fruit weight (12–13%) and fruit sugar contents, but the contents of acid and ascorbic acid decline (Panova et al., 1976). From the experiment on a series of manurial programmes, application of 10 tonnes FYM and 100 kg each of nitrogen and P_2O_5 and 500 kg ash/ha has been observed best for strawberries (Modoran and Hertea, 1977). Plants with high NPK and the highest ratio of phosphorus to potassium (in crown and root tissues) have been observed to be correlated with tolerance and resistance to cold, respectively (Zurawicz and Stuhnoff, 1977).

Increasing levels of nitrogen (3–12 meq/l N) and potassium (0.25–9.0 meq/l K) increase the growth and the number of leaves of strawberry plants but potassium at 4.0 meq/l is often detrimental when the crop is grown on sand (Martin and Rason, 1981). In Indian conditions, application of 75–100 kg nitrogen, 80–120 kg P_2O_5 and 50–70 kg K_2O/ha has been recommended for good fruiting in strawberry (Pandey and Mishra, 1983).

Optimum rates of boron ($Na_2B_2O_7.10H_2O$) and molybdenum ($Na_2MoO_2 H_2O$) can increase the production of strawberry fruits by 31% and 79%, respectively, and improve fruit quality with respect to vitamin C and sugar contents linearly (Cheng, 1994).

13.4 FOLIAR NUTRITION

The effect of application of fertilizers through foliar spray on fruit yield of strawberry depends, to a great extent, on the individual cultivar and mulching materials to be used (Karp and Starast, 2002). Foliar spraying with urea is not as common in strawberry as in many other perennial crops, but may be a good approach to improve plant nitrogen status if soil availability and/or root uptake of nitrogen are insufficient to meet short-term plant nitrogen demand. Problems with vigour and reserve content cause delay in the development or loss of strawberry plants. Spray application of urea (1%) increases the content of total soluble sugars in the crown. The reduced nitrogen content in root and stem increases with urea, but decreases with sucrose (Carrillo et al., 2005). Foliar spraying with urea represents a useful technique to increase strawberry nitrogen content in a short time and is an alternative or complement to soil nitrogen supply (Nestby and Tagliavini, 2005). Aton AZ, Ferticare 14/11/25 and Phosfik are good fertilizers for foliar feeding (Kahu and Klaas, 2005). Foliar feeding to strawberry plants with nitrogen improves fruit size (De La Rocha and Flores, 1955). Foliar application with K_2O and P_2O_5 at the time of flower bud formation increases the yield by 32.5% and with boron sprays by 18%. The application of NPK at this stage may reduce the yield by 15% but increases it by 21.3% when applied at floral initiation stage. Foliar application of phosphorus results in early flower bud differentiation in addition to maximum set of fruits and yield per plant, while the number of flowers have been reported to increase with nitrogen, phosphorus and potassium (Golikova, 1957).

The foliar application of urea along with full mineral nutrient not only enhances the photosynthetic activity of the plants, but also increases the leaf chlorophyll content and the fruit yield of strawberry. Four sprays of 0.5% N, 0.2% P_2O_5 and 0.5% K_2O applied between August and February have been observed to be highly effective in improving the fruit yield of strawberry (Singh et al., 1983). The application of iron (1–2%) with urea products and iron-containing substance (Fe EDDHA iron chelate) @ 1 kg Fe/ha, at weekly intervals from planting to harvesting, depending on the cultivars, are highly effective to increases the fruit yield by 13–56% (Tognoni et al., 1968).

Foliar application of Nutri-vit and Cal-max foliar fertilizers increases stem diameter, flower percentage, higher number of flowers, fruit yield and mean fruit weight in strawberries cvs. Premial and Elsanta (Chitu et al., 2002). Yalmaz et al. (2001) treated strawberry cultivars Tufts, Vista, Cruz and Brio with foliar applications of commercial foliar fertilizers (4% zinc, 4% iron, 3% manganese, 0.5% copper, 2% MgO, 1.5% boron, 0.05% molybdenum and 2.8% sulfur) at 500, 1000, 1500 or 2000 ppm to improve crop yield. Maximum fruit yield was obtained from the treatment due to 2000 ppm concentration, whereas minimum yield of fruits was obtained from control and 500 ppm treatments. Application of calcium fertilizers leads to increased calcium content in the fruits of strawberry cv. Jonsok (Karp and Starast, 2002).

Erdal et al. (2004) investigated the effects of foliar potassium, manganese and zinc concentrations in strawberry cultivars Addie, Dorit, Camarosa (iron-sensitive), Selva (iron-semi-sensitive) and Delmarvel (iron-resistant). Leaf iron and zinc concentrations increased with foliar applications of iron, while leaf phosphorus, magnesium and potassium concentrations were not affected, but calcium and manganese concentrations decreased, indicating that the leaf iron concentration of strawberry increased continuously with repeated foliar application of iron from both the sources. The use of $FeSO_4.7H_2O$ on leaf, iron concentrations was more effective than that of Fe- EDTA. Iron chelate is a soluble complex of iron, sodium and a chelating agent such as EDTA, EDDHA or other, used to make the iron soluble in water for the purposes of iron nutrition in plants.

Due to higher concentration of nutrients in the soil and unfavourable climatic conditions, sometimes foliar application of calcium nitrate does not show any effect on strawberry yield, but decreases the quantity of berries (Kukin et al., 2001) infected by grey mould (*Botrytis cinerea*) due to increased content of calcium in fruits (Karp and Starast, 2002). In sand culture studies on strawberries, the effect of application of foliar application of calcium nitrate to plants grown at higher salinity (NaCl, 35 m mol/l) has been reported to ameliorate the negative effects of salinity on plant growth, chlorophyll content and fruit yield. The increased membrane permeability caused by high NaCl is also reduced by $Ca(NO_3)_2$ sprays (Kaya et al., 2002). Sprays of calcium-containing Wuxal Aminocal can be recommended in strawberry culture, particularly in integrated fruit production, to improve the shelf-life of fruits and reduce the incidence of *B. cinerea* (Wojcik et al., 2006) causing grey mould.

Application of fertilizer preparations with calcium does not increase the yield and fruit mass of strawberries significantly, however, it significantly improves the firmness of the fruits, thus increasing the storability and transportability especially for cultivars with very delicate fruits (Bieniasz et al., 2012).

13.5 FERTIGATION

Fertigation has been observed to increase the fertilizer use efficiency (FUE) in irrigated crops. Yannai (1988) listed the following advantages of fertigation techniques:

1. Efficient application of fertilizers:
   a. Uniform distribution of fertilizers with irrigation water
   b. Avoiding fertilizer losses through volatilization from soil surface
   c. Deeper penetration and better distribution of fertilizers in soil solution
2. Labour savings
3. Crop damage and soil compaction due to use of machinery minimized
4. Flexibility in nutrient ratio
5. Transportation and storage losses minimized since liquid fertilizers are used in bulk
6. Ability to apply herbicides and other chemicals through the fertigation equipments

Nutrition 219

In an experiment conducted at Ellbridge Experimental Horticulture Station, Cornwall (Harnett, 1978) excellent strawberry crops were produced by growing plants in once used growing bags of sedge or sedge peat supplied with liquid feeding (150 kg nitrogen, 150 kg P_2O_5 and 300 kg $K_2O/10$ litres) via drip irrigation. Sande (1982) obtained consistently more fruit yield of strawberries with the plants grown by nutrient film technique. The runners of strawberry cv. Gorella were fertilized with N:P:K at 120 kg/ha each (Tosi, 1983) at planting and were compared with ferti-irrigation (2:1.5:3) to provide an equal dose of total nutrients. It was observed that the cumulative fruit yield of strawberries (autumn and spring) due to the later treatment (26,105 kg/ha and 326 g/plant) were relatively more than the former treatment (25,004 kg/ha and 312.5 g/plant). Accordingly, Tosi, 1986 recommended that optimum N:P:K ratios in the case of water and liquid fertilizers were 2:1.5: 3 and 2:1.5:4, respectively, for strawberry cv. Redgauntlet. In another study with cv. Confitura, the autumn application of 9% nitrogen, 12% phosphorus, 5.7% potassium and 9% magnesium of the total autumn/spring rate with the rest applied in the spring gave the maximum fruit yields during spring. Garate et al. (1991) recorded two times increased weight of roots of strawberry cv. Douglas plants with the use of 1 ppm nutrient solution of boron when compared with control.

Injected solution of fertilizers in combination with plastic mulching has positive effect on the quantity of the first and second grade fruits in strawberry cv. Senga Sengana (Karp et al., 2004) while Koszanski et al. (2005) recorded 28% increase in fruit yield of strawberry due to combined irrigation and fertilizer application. Plants of strawberry cv. Sweet Charlie has been observed to have a positive response to fertigation with different combination of $K-KCl:K-KNO_3$ particularly with the application of K–KCl at levels lower than 40% of the total potassium requirements. Therefore, KCl fertilizer, being cheap and highly soluble, can safely be used as source of potassium for strawberry plants (Ibrahim et al., 2004). Gutal et al. (2005) studied scheduling of nitrogen through the drip method of irrigation by applying four levels of recommended dose of nitrogen (50, 75, 100 and 125% RDF), and compared with the recommended dose (120:60:60 kg/ha) as band placement. The yields from 75% and 100% RDF were at par with each other, and both were superior to all other treatments. However, in terms of fertilizer savings (25%) and yield increase (15.6%), 75% RDF was the best treatment.

The risk of NO_3 leaching on a sandy soil may be reduced by using drip tapes with a 20–30 cm emitter spacing and irrigation events less than 900 l/100 m each (Simmonne et al., 2006). Critical concentrations of elements in soil and plants of strawberry can be determined by inducing deficiency symptoms by giving controlled concentrations in fertilizer solutions (Choi et al., 2001; Jeong et al., 2001).

The use of potassium @ 0.65 g/plant has been reported to be the most efficient for improving the number of fruits per plant and average fruit weight of strawberry cvs. Oso Grand and Summer. The increased concentrations of potassium (0.97–1.93 g/plant) reduces the productivity of these strawberry cultivars (Sousa et al., 2014). In fertigating strawberry crops, the use of the nutrient formulation should be based on the nutrient absorption characteristics for strawberry soil culture. This strategy would be sufficient for growth, and ensure environmentally friendly cultivation. Special attention should also be given to the nutrient conditions in soil. In case a high level of a certain mineral is present in soil, the amount of the mineral could be reduced from the nutrient formulation. In this way, imbalance of minerals in soil can be prevented (Yoon et al., 2014).

13.6 ORGANIC PRODUCTION

Production of strawberry through organic sources of input application is based on natural processes without the application of synthetic preparations, pesticides, artificial fertilizers, additives in processing and other matters, the presence of which could endanger the organically grown strawberries and requires a rotation of at least four years and should not follow potato or tomato crops. A green manure, preferably mixed grasses or *Brassica* spp., is desirable to help with weed and pathogen management. Resistant cultivars (Table 13.3) to pests, good soil structure and fertility, good

TABLE 13.3

List of Cultivars Resistant/Tolerant to Important Diseases and Useful for Organic Production System

Diseases	Resistant/Tolerant Cultivars
Leaf scorch	Alstar, Cavendish, Canoga, Earliglow, Guardian, Jewel, Redchief, Winona
Leaf spot	Cavendish, Canoga, Jewel, Lester, Ozark Beauty, Winona
Red stele	Albion, Cavendish, Clancy, Guardian, Idea, Mira, Northeaste, Tristar, Winona
Powdery mildew	Chandler, Clancy, Sparkle, Tribute, Tristar
Verticillium wilt	Albion, Clancy, L'Amour, Mesabi, Northeaster, Tristar
Anthracnose	L'Amour, Clancy

Source: Modified from Carroll et al., 2011, Production guide for organic strawberries. Cornell University, Cornell Cooperative Extension, New York State Department of Agriculture and Markets. New York State Integrated Pest Management Program: p. 45.

ploughing depth, drip/trickle irrigation and use of biological control agents are inseparable components of organic cultivation of strawberry. Weed control is a vital component against pathogens and pests. Effective preventive techniques include false seedbeds and burning of crop refuse before planting (Gengotti and Lucchi, 2000).

Saprophytic soilborne pathogens can be either actively suppressed by organic amendments or enhanced, depending on soil health conditions. Compost, FYM, peat-based substrate, humic acids, volcanic ash, vermicompost, poultry litter and animal litter can be used as organic amendments in growing strawberries (Zheng and Zhang, 2001; Yoshida et al., 2002; Tanaka et al., 2002; Pilanal and Kaplan, 2003; Arancon et al., 2003a, b; Bezhdugova, 2003; Bobev et al., 2004; Arancon et al., 2004; Preusch et al., 2004; Millner et al., 2004; Arancon et al., 2006).

Area of leaf, numbers of strawberry suckers, numbers of flowers, shoot weights and total marketable strawberry fruit yield increase significantly with the use of vermicompost compared with those for strawberries that received only inorganic fertilizers. The improvements in plant growth and fruit yields can be partially enhanced due to a large increase in microbial biomass in soil after application of vermicompost, leading to production of hormones or humates in the vermicomposts, which perform as plant growth regulators and are independent of nutrient supply (Arancon et al., 2003b). The plants of strawberry cv. Elsanta develop well with comparable fruit setting when grown either in peat amended with mineral fertilizer or peat mixed with compost, but show smaller, mechanically stronger and paler leaves with compost as substrate. The incidence of grey mould (caused by *B. cinerea*) is repressed with compost. Sporulation is inhibited in the immature fruits of plants grown in substrate with compost, whereas plants grown in substrates with mineral fertilizers are covered with sporulating *B. cinerea*. The index of leaf damage by powdery mildew (*Sphaerotheca macularis*) has been reported to be 30-fold higher in plants grown in substrates with mineral fertilizers than in those with compost. Higher sugar and glucose content, enhanced leaf thickness and increased hydrophobicity have been recorded in plants grown in substrate mixed with compost (Bobev et al., 2004)

Biofertilizers are important components in successful organic production of strawberries. Use of arbuscular mycorrhizal fungi (AMF) has shown significant potential for biocontrol of *Phytophthora* spp. Norman and Hooker (2000) established that microorganism-free exudates from roots colonized by AMF result in significantly less sporulation of *Phytophthora fragariae* than those from uncolonized plants. Among AMF, *Glomus versiforme* is the most effective. Inoculation with mycorrhiza has a positive effect on plant growth, fruit yield and quality of strawberry (Baruzzi et al., 2000; Qi et al., 2001; Gryndler et al., 2002; Vestberg et al., 2002; Sharma and Alok, 2004). In very high

Nutrition 221

soil phosphorus conditions, inoculation with AMF may be the only option for management of the symbiosis (Stewart et al., 2005).

Preininger et al. (2001) assessed a new approach for the biolistic gun, involving the bombardment of living bacteria into the target plant tissues. Applying the method of DNA transfer, living nitrogen-fixing bacteria and *Azotobacter vinelandii* were shot into the target inocules comprising regenerating callus cultures and callus from young leaves of strawberry to introduce nitrogen-fixing ability in the plants.

Rana and Chandel (2003) recorded maximum plant height, number of leaves per plant, leaf area, fruit size, fruit weight, TSS and total sugars in Chandler strawberry due to inoculation with *Azotobacter.* Wang et al. (2003) sprayed strawberry var. Fengxiang with two algal liquid fertilizers, which resulted in precocious fruiting and increased fruit production. Similarly, inoculation of strawberry cv. Sigiouao with *Azospirrillum brasilense* resulted in precocious fruiting and more fruit yield than the uninoculated control (El-Rewainy and Abd-Alla, 2005).

In the organic production system of strawberries, one of the most important factors is the right choice of cultivars. Application of nutrients at particular or specified rate is a cultivar-specific characteristic, which must be taken into account in evaluating the nutrient demands of strawberries as no fertilizers are added in organic production system (Daugaard, 2001). The higher incidence of pests and greater weed biomass in organic production systems than in conventional systems are constraints on enhanced fruit productivity/yield and consequently profitability. Due to its high productivity and low susceptibility to plant bugs and grey mould, Honeoye is one of the most profitable cultivars and appeared highly suitable for organic production management (Barth et al., 2002; Rhainds et al., 2002). Besides, some of the cultivars such as Madeleine, Patty, Onda (Daugaard and Linhard, 2000), Alba (Cozzolino, 2002), Polka and Bounty (Kahu and Klass, 2005) have also been observed to be suitable for cultivation in an organic production system.

Soils managed with organic farming practices generally have greater microbial activity. Consequently, the number of fungi- and/or bacteria-infecting native and pathogenic nematodes tend to increase in soils under organic system of crop management. A rhizobox experiment and a field study were conducted on "Elsanta" strawberry by Malusa et al. (2012) to monitor the structure of nematode communities and microbial populations in response to addition of five different organic manures and amendments, viz., dry manure, a seaweed extract, a compost extract, stillage from yeast production and a microbial consortium composed of mycorrhizal fungi, plant growth-promoting rhizobacteria (PGPRs) and *Trichoderma* spp., and their impact on plant growth of strawberries. Plant growth was observed to be enhanced by the products, and they tended to increase the total population of nematodes, by increasing the bacterial and fungal-feeding species, as well as the rhizosphere microbial populations. All products induced a decrease in the plant parasite nematode species. This resulted in reducing the ratio between the number of plant parasites and the other trophic nematode groups. The increase in bacterial and fungal-feeding nematodes induced by the microbial inoculums may affect the efficacy of this kind of products, and may benefit soil quality, and thereby plant health and resultant growth.

Many surveys indicate that consumers consider organic foods to be more beneficial to both human health and the environment and with better flavour than their conventionally integrated grown counterparts with respect to product quality (Valavanidis et al., 2009), while Hargreaves et al. (2008) concluded that this was not true for strawberries. However, consumer expectations and beliefs must be properly confirmed by more scientific studies. Roussosa et al. (2012) studied the effects of various crop management practices such as organic, conventional and integrated (combination of the both) on fruit quality of strawberries. Based on the statistical analysis, there were no significant differences between the three farming systems with respect to fruit quality. However, fruits produced from organic and integrated management practices were observed to contain more TSS. Fruit production was significantly greater under the integrated farming system, while both organic and integrated management systems resulted in production of fruits with more weight, which could be classified under the "Extra Large" category. They concluded that the integrated

TABLE 13.4

Organic Sources of Nutrition

Sl. No.	Sources of Nutrient	Nitrogen	Phosphorus	Potassium	Magnesium	Remarks
	Blood meal	13% N				
	Bone meal		15% P_2O_5			
	Sul-Po-Mag			22% K_2O	11% Mg	
	Soy meal	6% N	2% P	3% K_2O		(x 1.5)[a]
	Rock Phosphate		30% total P_2O_5			(x4)[b]
	Wood ash (dry, fine, grey)			5% K_2O		also raises pH
	Fish meal	9% N	6% P_2O_5			
	Alfalfa meal	2.5% N	2% P	2% K_2O		[c]
	Green sand or Granite dust			1% K_2O		(x 4)[d]
	Feather meal	15% N				(x 1.5)[a]
	Potassium sulfate					
	50% K_2O					

Source: Modified from Carroll et al., 2011, Production guide for organic strawberries. Cornell University, Cornell Cooperative Extension, New York State Department of Agriculture and Markets. New York State Integrated Pest Management Program: p. 45.

[a] Application rates for some materials are multiplied to adjust for their slow–very slow release rates.

[b] Application rates for some materials are multiplied to adjust for their slow–very slow release rates. Should be broadcast and incorporated before planting.

[c] Only non-GMO sources of alfalfa may be used.

[d] Application rates for some materials are multiplied to adjust for their slow–very slow release rates. Should be broadcast and incorporated before planting.

crop management system has resulted in more yield of high-quality fruits, that is, the organic crop management system yielded fruits with more health benefits, while the conventional crop management system resulted in less yield of fruits but with good quality. This is probably the result of improved soil properties/fertility (such as microbial activity, plant growth regulators produced by microorganisms etc.) due to addition of organic matter into the soil. Besides yield, fruit quality and susceptibility of plants to pests and diseases, other parameters to be considered for organic production of strawberries are heat tolerance (Archbold and Mackown, 1997), spring frost (Khanizadeh and DeEll, 2002) and cold hardiness during early development stage (Rugienius and Stanys, 2002). Some of the organic sources of nutrient supply for production of strawberries are given in Table 13.4.

REFERENCES

Abbott, A.J. 1968. Growth of the strawberry plant in relation to nitrogen and phosphorus nutrition. *Journal of Horticultural Science*, 43: 491–504.

Almaliotis, D., Velemis, D., Bladenopoulou, S. and Karapetsas, N. 2002. Leaf nutrient levels of strawberries (cv. Tudla) in relation to crop yield. ISHS. *Acta Horticulture*, 567: 447–449.

Anonymous, 1956. *The Wealth of India*, CSIR Pub. Vol. IV: pp. 57–60.

Archbold, D.D. and Clements, A.M. 2002. Identifying heat tolerant *Fragaria* accessions using chlorophyll fluorescence. *Acta Horticulturae*, 567(1): 341–344.

Arancon, N.Q., Lee, S., Edwards, C.A. and Atiyeh, R. 2003a. Effects of humic acids derived from cattle, food and paper-waste vermicomposts on growth of greenhouse plants. *Pedobiologia*, 47(5/6): 741–744.

Arancon, N.Q, Edwards, C.A, Bierman, P., Metzger, J.O., Lee, S. and Welch, C. 2003b. Effects of vermicomposts on growth and marketable fruits of field-grown tomatoes, peppers and strawberries. *Pedobiologia*, 47(5/6): 731–735.

Arancon, N.Q.C., Edwards, C.A., Bierman, P., Welch, C. and Metzger, J.D. 2004. Influences of vermicomposts on field strawberries: 1. Effects on growth and yields. *Bioresource Technology*, 93(2): 145–153.

Arancon, N. Q., Edwards, C. A., Lee, S. and Byrne R. 2006. Effects of humic acids from vermicomposts on plant growth. *European Journal of Soil Biology*, 42(1): S65–S69.

Barth, U., Spornberger, A., Steffek, R., Blumel, S., Altenburger, J. and Hausdorf, H. 2002. 10th International conference on cultivation technique and phytopathological problems in organic fruit growing and viticulture. *Proceedings of a conference, Weinberg Germany* 4–7 Feb. 2002: 212–216.

Baruzzi, G., Branzanti, B., Faedi, W., Genlili, M., Lucchi, P. and Neri, D. 2000. *Rivista di Frutticoltura e di Ortofloricoltura*, 62(12): 83–88.

Bezhdugova, M.T. 2003. *Sadovodstvo-i-Vinogradarstvo*, 5: 15–16.

Bhat, N.H. 1999. Response of strawberry cultivars to varied levels of organic manure. M.Sc. thesis, Sher-e-Kashmir Univ. of Agricultural Sciences and Technology of Kashmir, Shalimar, Srinagar.

Bieniasz, M., Malodobry, M. and Dziedzic, E. 2012. The effect of foliar fertilization with calcium on quality of strawberry cultivars "Luna" and "Zanta". *Acta Horticulturae*, 926: 457–461.

Bobev, S., Willekens, K., Goeminne, G., Desmet, T., Driessche, G. van, Claeyssens, M., Beeumen, J. van, Messens, E. and Maes, M. 2004. *Communications in Agricultural and Applied biological Sciences*, 69(4): 591–593.

Bould, C. 1964. Leaf analysis as guide to the nutrition of fruit crops. V. Sand culture N, P, K, and Mg experiments with strawberry (*Fragaria spp.*). *Journal of Science, Food and Agriculture*, 15: 474–487.

Brandstveit, T. 1979. Adverse effect of nitrogen on strawberries. *Landbruket*, 30: 55–67.

Breen, P.J. and Martin, L.W. 1981. Vegetative and reproductive growth responses of three strawberry cultivars to nitrogen. *Journal of the American Society for Horticultural Science*, 106: 266–272.

Carroll, J., Pritts, M. and Heidenreich, C. 2011. *Production guide for organic strawberries. Cornell University, Cornell Cooperative Extension*, New York State Department of Agriculture and Markets. New York State Integrated Pest Management Program: p. 45.

Cassman, K.G., Dobermann, A. and Walters, D.T. 2002 Agroecosystems, nitrogen-use efficiency, and nitrogen management. *AMBIO*, 31(2): 132–140.

Chiu, T.F. and Bould, C. 1976. Effect of calcium and potassium on Ca mobility, growth and nutritional disorders of strawberry plants (*Fragaria* sp.). *Journal of Horticultural Science*, 51: 525–531.

Choi, JongMyung, Jeong, Suckee, Cha, KiHyu, Chung, HaeJoon, Choi, JongSeung and Seo, Kwanseok. 2001. Korean Journal of Horticultural Science & Technology, 42(3): 142–146.

Cozzolino, E. 2002. Cultivation techniques and varieties for organic strawberries. *Informatore Agrario*, 58(27): 51–53.

Daugaard and Lindhard, H. 2000. Strawberry cultivars for organic production. *Gartenbauwissenschaft*, 65(5): 213–217.

Daugaard, H. 2001. Nutritional status of strawberry cultivars in organic production. *Journal of Plant Nutrition and Soil Science*, 24(9): 1337–1346.

De La Rocha, S. and Flores, C.D. 1955. Tests on nitrogen foliar nutrition. *Informe Mensual Estacion Experimental Agricola De La Molina*, 29: 5–9.

Dennison, R.A. and Hall, C.B. 1956. Influence of nitrogen, phosphorus, potash and lime on the growth and yield of strawberry. *Florida State Horticultural Society*, 69: 224–228.

Dierend, W. and Faby, R. 2002. Calcium and potassium nutrition of strawberries - Part I: Results of an inquiry. *Erwerbsobstbau*, 44(2): 40–48.

Dierend, W. and Faby, R. 2003. Calcium and potassium nutrition of strawberries - Part II: Potassium and calcium fertilization trials. *Erwerbsobstbau*, 45(2): 50–61.

El-Rewainy, H.M. and Abd-Alla, M.M.A. 2005. Response of strawberry cv. Siqiouao to inoculation with *Azospirillum brasilense* and foliar application of some micronutrients. *Journal of Agricultural Sciences*, 36(3): 1–11.

Erdal, I., Kepenet, K. and Kzlgoz, I. 2004. Effect of foliar iron applications at different growth stages on iron and some nutrient concentrations in strawberry cultivars. *Turkish Journal of Agriculture and Forestry*, 28(6): 421–427.

Fixen, P.E. and West, F.B. 2002. Nitrogen fertilizers: Meeting contemporary challenges. *Ambio*, 31: 169–176.

Fouly, H.M. 2004. Effect of some micronutrients and the antioxidant salicylic acid on suppressing the infection with strawberry powdery mildew disease. *Bulletin of the Faculty of Agriculture*, Cairo University, 55(3): 475–486.

Garate, A., Menzanares, M., Ramon, A.M. and Carpena, R.R. 1991. Boron requirements of strawberry (*Fragaria ananassa* L. cv. Douglas) grown in hydroponic culture. *Acta Horticulturae*, 287: 207–210.

Garcia, J.M., Herrera, S. and Morilla, A. 1996. Effects of post harvest dip of calcium chloride on strawberry. *Journal of Agricultural and Food Chemistry,* 44: 30–33.

Gaweda, M. and Ben, J. 2004. Zmianyzawartoscimikroelementow w roslinachtruskawki (*Fragaria x ananassa* Duch.) w zaleznosciodczasuuprawy. *Folia Universitatis Agriculturae Stetinensis, ser. Agricultura,* 240(96): 55–58.

Gengotti, S. and Lucchi, C. 2000. Useful advice for growing strawberries organically in Emilia-Romagna. *Frutticoltura Rivista di Fruit, coltura edi-Ortofloricoltura,* 62(12): 60–62.

Golikova, N.A. 1957. Influence of non-root feeding on the harvest of strawberries. *Udobrenie I Urozhai,* 2: 30–32.

Gryndler, M., vosatka, M. Hrselova, H., Catska, V., Chvatalova, I. and Jansa, J. 2002. Effect of dual inoculation with *Arbuscularmycorrhizal* fungi and bacteria on growth and mineral nutrition of strawberry. *Journal of Plant Nutrition and Soil Science,* 25(6): 1341–1358.

Guo, Y.Y., Jiang, Y.M. and Peng, F. 2003. Influence of different level of nitrogen on amino acid and protein in strawberry. *Journal of Fruit Science,* 20(6): 475–478.

Gutal, G.B., Barai, V.N., Mane, T.A., Purkar, J.K. and Bote, N.L. 2005. Scheduling of irrigation for strawberry through drip. *Journal of Maharashtra Agricultural Universities,* 30(2): 215–216.

Guttridge, C.G. 1960. Photoperiodic responses in *Fragaria. Bull Inst. Agron. Stat Rech. Gembloux,* 2: 941–948.

Guttridge, C.G. and Anderson, H.M. 1973. Relationship between plant size and fruitfulness in strawberry in Scotland. *Journal of Horticulture Research,* 13: 125–135.

Hargreaves, J.C., Adl, M.S., Warman, P.R. and Rupasinghe, H.P.V. 2008. The effects of organic and conventional nutrient amendments on strawberry cultivation: fruit yield and quality. *Journal of Science, Food and Agriculture,* 88: 2669–2675.

Haynes, R.J. and Goh, K.M. 1987. Effect of nitrogen and potassium applications on strawberry growth, yield and quality. *Communications in Soil Science and Plant Analysis,* 18(4): 457–471.

Hong, L., Truro, N.S. and Asiedu, S.K. 2009. Strawberry nursery plant propagation in relation to soil phosphorus and water variation. In *Proceedings of the International Plant Nutrition Colloquium XVI,* UC Davis, California.

Hughes, H.M., Duggan, J.B. and Banwell, M.G. 1969. Effect of potassium and magnesium levels on performance indicators of *strawberry* plants grown in hydroponic. Strawberry Bull. HMSO 10, 6d, Min. Agric. Fish Food, UK.

Human, C. and Kotze, W.A.G. 1990. Effect of nitrogen and potassium fertilization on strawberries in an annual hill culture system. Yield and fruit size. *Communications in Soil Science and Plant Analysis,* 21(9–10): 771–782.

Ibrahim, A., Abdel Latif, T., Gawish, S. and Elnagar, E. 2004. Response of strawberry plants to fertigation with different K-KCl/K-KNO$_3$ combination ratios. *Arab Universities journal of Agricultural Sciences,* 12(1): 469–480.

Jackson, D.C. 1972. Nitrate toxicity in strawberries. *Agrochemophysica,* 4: 2–45.

Jeong, S., Choi, J.M., Cha, K.H., Chung, H.J., Choi, J.S. and Seo, K. 2001. Deficiency symptom, growth characteristics, and nutrient uptake of "Nyoho" strawberry affected by controlled calcium concentrations in fertilizer solution. *Journal of Korean Society for Horticultural Science and Technology,* 42(3): 284–288.

Johanson, F. and Walker, R.B. 1963. Nutrient deficiencies and foliar composition of strawberries. *Proceeding of the American Society for Horticultural Science,* 83: 431–439.

Karhu, S.T. and Hytonen, T.P. 2006. Nursery plant production controlled by prohexadione-calcium and mechanical treatments in strawberry cv. "Honeoye". *Journal of Horticultural Science and Biotechnology,* 81(6): 931–942.

Karp, K. and Starast M. 2002. Effects of spring time foliar fertilization on strawberry yield in Estonia. *Acta Horticulturae,* 594: 501–505.

Kaya, C., Ak, B.E., Higgs, D. and Murillo-Amador, B. 2002. Influence of foliar-applied calcium nitrate on strawberry plants grown under salt-stressed conditions. *Australian Journal of Experimental Agriculture,* 42(5): 631–636.

Keefer, R.F., Hickman, C.E. and Adams, R.E. 1978. The response of strawberry yields to soil fumigation and nitrogen fertilization. *HortScience,* 13(1): 51–52.

Khanizadeh, S. and DeEll, J. 2002. Chlorophyll fluorescence: a new technique to screen for tolerance of strawberry flowers to spring frost. *Acta Horticulturae,* 567(1): 337–339.

Klaas, L. and Kahu, K. 2005. Influence of the foliar fertilization and moving of runners on the strawberry yield and quality of fruits. *Transactions of the Estonian Agricultural University Agronomy,* 220: 117–119.

Kohli, R.R., Iyenoar, B.R.V. and Reddu, Y.T.N. 1984. Growth, dry matter production and yield in Robusta banana as influenced by different levels of nitrogen. *Indian Journal of Horticulture*, 41(3–4): 194–198.

Koszanski, Z., Friedrich, S., Podsiado, C., Rumasz, Rudnicka, E. and Karczmarczyk, S. 2005. The influence of irrigation and mineral fertilization on morphology and anatomy, some physiological processes and yielding of strawberry. *Woda Srodowisko Obszary Wiejskie*, 5(2): 145–155.

Kukin, M., Karp, K. and Starast, M. 2001. The influence of foliar fertilization on the yield of strawberry cultivar Senga Sengana. *Transactions of the Estonian Agricultural University*, 212: 141–144.

Lacroix, C.R. and Carmentran, M. 2001. Fertilizers and the strawberry plant: yield and fruit quality. Fertilisation de fraisier, rendement et qualité des fruits. *Infos Ctifl*, 170: 41–44.

Lenzi, A., Lombardi P. and Tesi, R. 2004. Effects of steam and exothermic substances (KOH and CaO) on lettuce and strawberry production: two years of experimentation. *Advances in Horticultural Science*, 18(4): 155–160.

Lieten, F. 2000. Iron nutrition of strawberries grown in peat bags. *Small Fruits Review*, 1(2): 103–112.

Lieten, P. 2002. Boron deficiency of strawberries grown in substrate culture. *Acta Horticulturae*, 567(2): 451–454.

Lieten, F. 2003. Zinc nutrition of strawberries grown on peat bags. *Small Fruits Review*, 2(4): 63–72.

Lieten, P. 2004. Manganese nutrition of strawberries grown on peat. *Acta Horticulturae*, 649: 227–230.

Lineberry, R.A. and Burkhart, L.V. 1943. Nutrient deficiencies in the strawberry leaf and fruit. *Plant Physiology*, 18: 324–333.

Liu, S.Z., Jiang, Y.M. and Peng, F. 2004. Effects of different fertilizing methods on enhancing ammonium for strawberry grown in greenhouse. *Journal of Shandong Agricultural University*, 35(2): 183–186.

Malusa, E., Sas-Paszt, L., Trzcinski, P. and Gorska, A. 2012. Influences of different organic fertilizers and amendments on nematode trophic groups and soil microbial communities during strawberry growth. *Acta Horticulturae*, 933: 253–260.

Marinos, N.C. 1962. Studies on submicroscopical aspects of mineral deficiencies. I. Calcium deficiency in the shoot apex of barley. *American Journal of Botany*, 49: 834–849.

Markus, D. and Morris, J.R. 1998. Preharvest calcium applications have little effect on mineral distribution on ripe strawberry fruit. *HortScience*, 33(1): 64–66.

Mason, G.A. and Guttridge, C.G. 1974. The role of calcium, boron and some divalent ions in leaftip burn of strawberry. *Scientia Horticulturae*, 2: 299–308.

Millner, P.D., Ringer, C.E. and Maas, J.L. 2004. Suppression of strawberry root disease with animal manure composts. *Compost Science and Utilization*, 12(4): 298–307.

Mitra, S.K. 1991. The strawberry. In: *Temperate Fruits*, Horticulture and Allied Publishers, Calcutta, India: pp. 549–596.

Modoran, I. and Hertea, B.I. 1977. Determining the type and amount of organic and mineral fertilizers for strawberries. *British Exptl. Stat. Bistria. Masand Romandia*, 5: 303–316.

Nam, M.H., Jeong, S.K., Lee, Y.S., Choi, J.M. and Kim, H.G. 2006. Effects of nitrogen, phosphorus, potassium and calcium nutrition on strawberry anthracnose. *Plant Pathology*, 55: 246–249.

Nes, A. and Hjeltenes, A. 1992. The effects of covering and manuring on runner production, yield and berry size of the strawberry cultivar Korona. *Norsk*, 6(3): 195–203.

Nestby, R. and Tahliavini, M. 2005. Foliar uptake and partitioning of urea-N by strawberry plants as affected by timing of supply and plant N status. *Journal of Horticulture Science and Biotechnology*, 80(2): 272–275.

Neuweiler, R. 1997. Nitrogen fertilization in integrated outdoor strawberry production. *Acta Horticulturae*, 439: 747–751.

Norman, J.R. and Hooker, J.E. 2000. Sporulation of *Phytophthora fragariae* shows greater stimulation by exudates of non-mycorrhizal than by mycorrhizal strawberry roots. *Mycological Research*, 104(9): 1069–1073.

Pandey, R.M. and Mishra, K.K. 1983. *Fal Utpadan (Hindi)*. GBPUAT, Pantnagar, India.

Panova, Z.M., Kondakov, A.K. and Krainova, V.V. 1976. Fertilization and chemical composition of strawberry Trudy Taentr. *Inta. Skh*, 4: 10–12.

Pilanali, N. and Kaplan, M. 2003. Investigation of effects on nutrient uptake of humic acid applicant of different forms to strawberry plant. *Journal of Plant Nutrition and Soil Science,* 26(4): 835–843.

Preininger, E., Boka, K., Koranyi, P., Gyurjan, I and Zatyko, J. 2001. Introduction of atmospheric nitrogen fixing ability to plants: direct bacterium bombardment. *Acta Horticultrurae*, 560: 113–116.

Preusch, P.L., Takeda, F. and Tworkoski, T. 2004. N and P uptake by strawberry plants grown with composted poultry litter. *Scientia Horticulturae*, 102: 91–103.

Qi, G.H., Chen, G., Lu, G., Lanchun and Ding, P.H. 2001. Effects of *Arbuscular mycorrhizal* fungi on the yield and quality of strawberry grown in replanted soil. *International Journal of Fruit Science*, 18(6): 341–344.

Ram Autar and Gaur, G.S. 2003. Studies on the vegetative growth, yield and quality of strawberry (*Fragaria* x *ananassa* Duch.) as influenced by different levels of nitrogen. *Scientia Horticulturae*, 8: 71–74.

Rana, R.K. and Chandel, S. 2003. Effect of biofertilizers and nitrogen on growth, yield and fruit quality of strawberry. *Progressive Horticulture*, 35(1): 25–30.

Raynal-Lacroix, C. and Abarza, E. 2002. Reasoning fertilization of strawberry - the petiole nitrate testing, a tool. *Infos Ctifl,* 179: 44–48.

Reekie, J.Y., Hicklenton, P.R. and Struik, P.C. 2005. Prohexadione-calcium modifies growth and increases photosynthesis in strawberry nursery plants. *Canadian Journal of Plant Science*, 85(3): 671–677.

Ricketson, C.L. 1966. The relationships between certain berry characteristics of the strawberry and foliar concentrations of nitrogen, phosphorus, and potassium at harvest. *Proceedings of the XVII International Horticultural Congress*, 1: 418.

Rhainds, M., Kovach, J. and English-Loeb, G. 2002. Impact of strawberry cultivars and incidence of pests on yield and profitability of strawberries under conventional and organic management systems. *Biological Agriculture and Horticulture,* 19(4): 333–353.

Roussosa P.A., Triantafillidis, A. and Kepolas, E. 2012. Strawberry fruit production and quality under conventional, integrated and organic management. *Acta Horticulturae* (926): 541–546.

Rubeiz, I.G., Sabra, A.S. and Al-Assir, I.A. 1993. Broiler and layer poultry manures as nitrogen sources for "Douglas" strawberry in a tunnel production system. http://agris.fao.org/aos/records/US9446955

Rugienius, R. and Stanys, V. 2002. Formation of cold hardiness in strawberry during early plant development stage. *Acta Horticulturae*, 567(1): 33–336.

Salisbury, F.B. and Ross, C.W. 1986. *Plant Physiology*, CBS Publishers & Distributors 485, Shahdara, Delhi, India.

Saxena, G.K. and Locascio, S.J. 1968. Fruit quality of fresh strawberries as influenced by nitrogen and potassium nutrition. *Proceedings of the American Society for Horticultural Science*, 92: 354–362.

Sharma, M.P. and Alok, A. 2004. Influence of *Arbuscular mycorrhizal* fungi and phosphorus fertilization on the post-vitro growth and yield of micropropagated strawberry in an alfisol. *Canadian Journal of Botany*, 82(3): 322–328.

Simonne, E., Studstill, D. and Hochmuth, R.C. 2006. Understanding water movement in mulched beds on sandy soils: An approach to ecologically sound fertigation in vegetable production. *Acta Horticulturae*, 700: 173–178.

Singh, A.K., Singh, A.K., Pitam Chandra and Gupta, M.J. 2001. Effect of urea doses on growth and fruit yield of strawberry (*Fragaria* x *ananassa* Duch.) cultivated under greenhouse condition. *Progressive Horticulture*, 33(2): 194–198.

Singh, R.P., Srivastava, R.P. and Phogat, K.P.S. 1983. Effect of plant growth regulators on strawberry. *The Punjab Horticultural Journal*, 23: 64–68.

Smith, B.J. and Gupton, C.L. 1993. Calcium applications before harvest affects the severity of anthracnose fruit rot in greenhouse grown strawberries. *Acta Horticulturae*, 348: 477–482.

Sousa, G.G., Viana, T.V.A., Pereira, E.D., Albuquerque, A.H.P., Marinho, A.B. and Azevedo, B.M. 2014 Fertirriga ção potássica na cultura do morango no litoral Cearense. *Bragantia*, 73: 39–44.

Stewart, L.I., Hamel, C., Hogue, R. and Moutoglis, P. 2005. Response of strawberry to inoculation with *Arbuscular mycorrhizal* fungi under very high soil phosphorus conditions. *Mycorrhiza*, 15(8): 612–619.

Taghavi, J., Babalar, M., Ebadi, E., Ebrahimzadeh, H. and Asgari, M.A. 2005. The effects of different iron and boron concentrations on yield and mineral content in strawberry cv. Selva. *Iranian Journal of Agricultural Sciences*, 36(5): 1065–1073.

Tanaka, Y. and Mizuta, M. 1976. Nutrio-physiological studies on the strawberry "Hokowase" in the early forcing culture, 2: Influence of nitrogenous basal fertilizer supplied on the growth, yield and nutritional uptake. *Bulletin of Nara Agricultural Experimental Station*, 7: 31–37.

Tognoni, F., Alpi, A. and Sillari, B. 1968. Preliminary results of foliar fertilizing of strawberries. *Frutticoltura*, 30: 973–977.

Tosi, T. 1986. Some aspects of strawberry fertilizing. *Acta Horticulturae*, 176: 99–106.

Treder, W. 2004. The influence of calcium in nutrient solution on growth and fruiting of strawberries and its content in strawberry organs. *Folia Universitatis Agriculturae Stetinensis Agricultura*, 96: 197–202.

Turemifl, N., Ozguven, A.L., Paydas, S. and Idem, G. 1997. Effects of sequestrene Fe-138 as foliar and soil application on yield and earliness of some strawberry cultivars in the subtropics. *Acta Horticulturae*, 441.

Ulrich, A., Mostafa, M.A.E. and Allen, W.W. 1980. Strawberry deficiency symptoms: A visual and plant analysis guide to fertilizer. Publ. 4098. Division of Agricultural Sciences, University of California, Oakland, California.

Valavanidis, A., Vlachogianni, T., Psomas, A., Zovoili, A. and Siatis, V. 2009. Polyphenolic profile and antioxidant activity of five apple cultivars grown under organic and conventional agricultural practices. *International Journal of Food Science* and *Technology*, 44: 1167–1175.

Van De Sande, H.J.M. 1982. Trials with strawberry culture on substrate. *Groenten en Fruit,* 38: 32–33.

Van der Boon, J. 1961. De aardbei en het kopergehalte van zandgrond. *Fruitteelt*, 51: 1129.

Vestberg, M., Kukkonen, S. and Uasukainen, M. 2002. Occurrence and effectiveness of indigenous mycorrhiza of some strawberry fields in Finland. *Acta Horticulturae*, 567(2): 499–502.

Wang, Q., Shi, W.N. and Jiang, Y. 2003. Effect of sea algal liquid fertilizer on the growth and fruit quality of strawberry. *China Fruits*, 1: 27–28.

Wojcik, P. and Lewandowski, M. 2003. Effect of calcium and boron sprays on yield and quality "Elsanta" strawberry. *Journal of Plant Nutrition*, 26(3): 671–682.

Wojcik, P., Filipczak, J. and Alexander, A. 2006. Effect of foliar applications of Wuxal fertilizers on strawberry yield and fruit quality. *Acta Horticulturae*, 721: 207–211.

Yannai, B. 1988. Training programme on production, planning, cultivation techniques and post-harvest handling. Ministry of Agriculture, State of Israel.

Yoon, H.S., An, J.U., Hwang, Y.H., An, C.G., Chang, Y.H., Shon, G.M. and Rho, C.W. 2014. Improved fertilization strategy for strawberry fertigation culture. *Acta Hortticulturae*, 1049: 521–528.

Yoshida, Y., Goto, T, Hirai, M. and Masuda, M. 2002. Anthocyanin accumulation in strawberry fruits as affected by nitrogen nutrition. *Acta Horticulturae*, 567(1): 357–360.

Yoshida, Y., Ohi, M. and Fujimoto, K. 1991. Fruits malformation, size and yield in relation to nitrogen nutrition and nursery plants in large fruited strawberry (*Fragaria* × *annassa* Duch. cv. Ai-Berry). *Journal of Japanese Society for Horticultural Science*, 59(4): 727–735.

Zaiter, H.J., Saad, I. and Nimah, M. 1993. Yield of iron-sprayed and non-sprayed strawberry cultivars grown on high pH calcareous soil. *Communications in Soil Science Plant Analysis*, 24: 1421–1436.

Zheng, H.Q. and Zhang, Y.M. 2001. Effects of Shanghai Huiren concentrated organic fertilizer (SHCOF) on some vegetables. *Acta Agriculturae Shanghai*, 17(4): 65–68.

Zurawicz, E. and Stuhnoff, C. 1977. Influence of nutrition on cold tolerance of redcoat strawberries. *Journal of the American Society for Horticultural Science*, 102: 342.

14 Water Management

V. Pandey, R.M. Sharma, Dinesh Kumar,
S.D. Sharma, and S.K. Jena

CONTENTS

14.1 Role of Water in Plants 230
14.2 Classification of Soil Water/Moisture 231
14.3 Soil Moisture Constants 231
    14.3.1 Maximum Water Holding Capacity 231
    14.3.2 Field Capacity 231
    14.3.3 Wilting Point 231
    14.3.4 Available Soil Moisture 232
    14.3.5 Water Requirement 232
    14.3.6 Water Lost in Transpiration 232
    14.3.7 Evapotranspiration 232
14.4 Important Factors in Irrigation Scheduling 232
    14.4.1 Criteria for Scheduling Irrigation 233
    14.4.2 Depth of Irrigation 234
    14.4.3 Frequency of Irrigation 234
    14.4.4 Critical Time of Irrigation 234
    14.4.5 Stages of Irrigation 235
    14.4.6 Amount of Irrigation 235
    14.4.7 Time of Irrigation 236
    14.4.8 Varietal Response to Irrigation 236
    14.4.9 Effect of Growing Conditions 237
14.5 Irrigation Scheduling 238
    14.5.1 Soil Parameters 238
    14.5.2 Bed Capacity 240
    14.5.3 Plant Parameters 240
    14.5.4 Climatic Parameters 241
14.6 Water Requirement 241
14.7 Irrigation Systems 244
    14.7.1 Furrow Irrigation 244
    14.7.2 Pressurized or Micro-Irrigation Systems 244
        14.7.2.1 Sprinkler Irrigation 245
        14.7.2.2 Drip Irrigation 245
14.8 Decision Support Systems in Water Management 252
14.9 Biotic and Abiotic Stresses Associated with Water Management Practices 253
    14.9.1 Drought 253
    14.9.2 Flooding 254
    14.9.3 Salinity 254
    14.9.4 Weeds 254
    14.9.5 Pathogens 255

14.10	Water Saving Practices	255
	14.10.1 Mulching	255
	14.10.1.1 Organic Mulches	255
	14.10.1.2 Plastic Mulches	255
	14.10.1.3 Ideal Mulch Materials	256
	14.10.2 Anti-Transpirant	256
14.11	Indicators of Efficient Irrigation System	257
14.12	Methods for High Crop Water Use Efficiency	257
14.13	Water Quality	257

Summary 260
Terms used in this paper 260
Appendix: Conversion units from U.S. to SI and *vice versa* 261
References..261

Strawberries have long been recognized as an important item in the human diet, providing health benefits against a wide range of diseases, mainly due to their high content of certain bioactive compounds including ascorbates, anthocyanins, phenolic acids, carotenoids etc. (Gine-Bordonaba and Terry, 2011a; Manganaris et al., 2013). As most of the bioactive compounds within the plants' system are secondary metabolites, their synthesis can be triggered in response to biotic and abiotic stresses, such as ultraviolet (UV) radiations, drought, wounding and infections (Terry and Joyce, 2004; Terry et al., 2007; Jahangir et al., 2009). The biochemistry of strawberry fruits is influenced by preharvest treatments and cultivation practices (Terry et al., 2007a; Keutgen and Pawelzik, 2008; Crespo et al., 2010; Gine-Bordonaba and Terry, 2010), which ultimately form the basis of overall quality parameters including sensory attributes, such as taste and health-related biochemical compounds.

Strawberries return a greater profit by following water management practices than most other crops. Efficient water management practice is also useful in utilizing the applied nutrients in the soil, preventing abiotic stresses, especially frost damage of the blossoms (Mitra, 1991), hastening the growth of runners and increasing the size of berries and yields of late varieties (Heeney, 1962). Efficient irrigation management is not only of primary importance for the profitability and sustainability of field strawberry production, but it also influences the fruit yield, water use efficiency and diffusing pollution of ground and surface water. An efficient irrigation system is of great importance in areas where availability of water for crop production is a limiting factor. In the recent past, adoption of more efficient water management practices, such as subsurface drip irrigation (SDI) and use of plastic mulches, has greatly improved water use efficiency in commercial strawberry production. The duration of each irrigation event also varies in different soil textural classes. For lighter soils, the duration is generally less but the frequency is more, whereas for heavier soils, the duration is more but the frequency of irrigation is less.

14.1 ROLE OF WATER IN PLANTS

Water is the most important constituent of plant tissues and comprises 35–95% of different types of leaves at various stages of development, 60–90% of the roots and 70–90% of the most fleshy fruits. Fresh wood may contain 38–65% while dry seeds for storage contain 10–20%. In addition, the constituents of water, viz., hydrogen and oxygen, are nutrients, which, in association with carbon, form the major physiological biomass of plants. Water is most essential for all physiological and biochemical processes such as photosynthesis. Water acts as a medium in which all the nutrients, except carbon, are dissolved as solution and taken up in the plant system. Finally, water is the main constituent for transpiration and regulates plant/leaf temperature as well as cell turgidity. Like any other crop, strawberries have requirement for an optimum soil moisture regime and deviations from this optimum (excess or less soil moisture) cause adverse effects leading to poor growth, fruit yield and, in severe cases, quality.

Water Management 231

The ways in which the soil gets water are rain, irrigation, snowfall and dew, while it is lost through runoff, evaporation, drainage, infiltration, seepage, percolation and deep percolation from the soil or transpiration from plants' leaf surface.

14.2 CLASSIFICATION OF SOIL WATER/MOISTURE

The manner in which water is held in the soil and further translocated into the plant system forms the basis of its classification into physical or biological groups. In physical classification, water is classified into hygroscopic, capillary or gravitational water; while in biological classification, it could be classified into unavailable or desirably available water.

The fraction of water found in a layer around the soil particle in a non-liquid state, consequently immobile, being held with a tension of >31 atmospheres, is termed *hygroscopic water*. For all practical purposes, it is not available to plants. The available form of water, present in a liquid state and held between field capacity (0.33 atmospheres) and hygroscopic coefficient (31 atmospheres), is termed *capillary water*. When the tension reaches 15 atmospheres and above, the movement of water becomes very slow and at this stage, plants are deprived of their water intake from the soil. The availability of capillary water depends upon soil texture, structure and its humus content. In fine-textured soils with granular structure and a good amount of humus, the availability of capillary water is more. The remaining amount of water is present in soil macro-pores and is held at a tension of 0.33 atmospheres or less and is subjected to gravitational pull and usually has downward movement towards deep water table, passing freely and quickly from surface soil, and is termed *gravitational water*.

The water held between 31 and 15 atmospheres, below the wilting point (not available for normal growth of plants), is termed *unavailable water*. On the other hand, water found between the permanent wilting point and field capacity (held between 15 and 0.33 atmospheres) is termed *available water*. Further, superfluous or gravitational water is termed *free water*.

14.3 SOIL MOISTURE CONSTANTS

14.3.1 MAXIMUM WATER HOLDING CAPACITY

When all the pores and capillaries are filled with water, the soil is said to be at water saturation stage, and at this stage the soil is said to be at maximum water holding capacity. This condition minimizes the influence of gravity and results in a tension of <0.01 or even 0.001 of an atmosphere. For all practical purposes, it is not advisable to allow a cropped (except water logging tolerant or aquatic crops) soil to remain at this soil moisture level as so high a soil moisture saturation level does not allow aeration in soil, and depriving the plants of air is harmful and causes reduced growth and development, and ultimately yield. The maximum water holding capacity of a soil depends upon the amount of total pore spaces, being more in fine-textured soils and less in coarse-textured soils.

14.3.2 FIELD CAPACITY

When free water moves downward from a saturated soil as a result of gravitational force, it leads to a stage when the soil water is held at a tension of about 0.33 atmospheres or less. Or when the excess or surplus soil water is fully driven out of the horizon, then the soil at this stage is referred to be at field capacity, thus after the provision of adequate drainage, the percentage of water retained by soil against the gravitational movement is known as field capacity.

14.3.3 WILTING POINT

After the field capacity of the soil, there is further loss of soil moisture due to evaporation from the soil surface and absorption by plants growing in it. At this stage, the soil water is held with more tension and further absorption of soil water by the plant roots becomes more and more difficult against increasing tension. At this stage, the plants further try to adjust to diminished intake of water but soon after, this absorbed water supplements the water lost due to transpiration only and as a result, the plants start wilting temporarily, having reached what is known as the temporary wilting point, for example, sometimes, vigorous growing plants in the field show symptoms of wilting under intense sunlight at noon but recover in the evening. This condition arises due to an abnormally very high rate of transpiration and reverts back with reduced solar intensity. The amount of soil moisture at the permanent wilting point of plants is referred to as the *wilting coefficient* or the *critical moisture point.*

14.3.4 AVAILABLE SOIL MOISTURE

The amount of soil moisture found between the permanent wilting point and field capacity, or between a tension of 15 and 0.33 atmospheres, is referred to as available soil moisture. The availability of water increases as the tension reduces from 15 atmospheres and becomes less available with increase in tension.

14.3.5 WATER REQUIREMENT

Water management in strawberry production requires information about the water needs of the crop as well as the water holding characteristics of the soil. Excessive irrigation can leach crop nutrients from the root zone, while a soil moisture deficit can result in crop stress.

The amount of water required to produce a crop in a specific period, including evapotranspiration and other losses due to deep percolation, seepage and runoff and the water required to bring the soil to a requisite physical stage for plant growth (planting to maturity) is referred to as the water requirement of the crop.

14.3.6 WATER LOST IN TRANSPIRATION

It is necessary for all living plant cells to maintain turgidity and to facilitate rapid movement of nutrients and other compounds from one part to the other through transpiration. In addition, it is also essential to maintain the cell temperature by losing heat in the transpiration process. The transpiration ratio is a numerical value calculated by the amount of water transpired for every kilogram of dry matter produced.

14.3.7 EVAPOTRANSPIRATION

Practically, it is very difficult to differentiate and quantify the water lost in evaporation and transpiration separately; therefore, to designate the use of water by crops properly, evaporation and transpiration are combined into one term evapotranspiration. Further, the consumptive use includes evapotranspiration and the water used by the plants for their metabolic activities. Practically, evapotranspiration covers ≈99% of the total water requirement and the other 1% is used for metabolic activities, so evapotranspiration is generally considered to be the consumptive use of water.

14.4 IMPORTANT FACTORS IN IRRIGATION SCHEDULING

Relative water content, also known as relative turgidity is used for irrigation scheduling. Leaf thickness, stem diameter and fruit measurements are also indicators of water status in plants. Predawn leaf water potential is considered to be a good measure of plant water status for irrigation scheduling. Rate of transpiration from the leaf is also correlated with the plant's water status, since it is closely related to stomatal aperture and accordingly measurement of diffusive resistance or canopy temperature is also used for scheduling irrigation. Water use in relation to estimated evapotranspiration due to increase in evaporative leaf area increases in plants. Water use varies strongly between cultivars.

Leaf water retention and suction pressure are useful indices for evaluating plant adaptability to heat and drought. The magnitude of suction pressure reflects water availability to plants and can be used for determining irrigation dates. Weekly transpiration rates as determined by integrated heat pulse measurement are comparable with measurements by lysimeter for a well-watered strawberry plant. The physiological parameters of plants can be used to measure water stress and its effect on plant performance but its use is still limited due to several theoretical and practical difficulties as reviewed by Jones (2004). In strawberry production system, leaf temperature measurements with infrared thermometers were identified as a potential tool for irrigation management because they facilitated the detection of severe stress, which affects yield and biomass production (Penuelas et al., 1992). The physiological parameters like photosynthesis, stomatal conductance and leaf water potential measurements have also been successfully used in detection of water stress and understanding and differentiating the adaptation mechanism among strawberry cultivars especially for breeding drought-resistant cultivars (Blanke and Cooke, 2004; Klamkowsaki and Treder, 2008; Save et al., 1993). But the physiological parameters do not hold good as a criterion for irrigation scheduling or for assessing the appropriate volume of water to be applied (Jones, 2004).

14.4.1 CRITERIA FOR SCHEDULING IRRIGATION

Proper irrigation of strawberries is essential to maintaining a healthy and productive planting. Overirrigation slows root growth, increases iron chlorosis on alkaline soils and leaches nitrogen, sulfur and boron out of the root zone, leading to nutrient deficiencies. Excessive soil moisture also promotes root rot, particularly on heavy soils. Applying insufficient irrigation water results in drought stress. Drought stress during fruit development results in reduced fruit size and yield and poorer fruit quality. Soil properties should also be known and monitored for an effective irrigation scheduling program. The volume of readily available water to the crop depends on the soil water holding properties and the crop root zone. Water should be applied when no more than half the available water has been depleted.

Of the three criteria for scheduling irrigation in strawberry, partial root zone drying, deficit irrigation and full irrigation, from beginning of flowering to the end of fruit maturity in a field lysimeter under an automatic rainout shelter, it was observed that partial root zone drying had no advantage over deficit irrigation in terms of fruit yield and water use efficiency in strawberry cultivar "Honeoye". In full irrigation, the whole root zone was irrigated every second day to field capacity, viz., volumetric soil water content of 20%, while in deficit irrigation and partial root zone drying, 60% of full irrigation's water was given to either the whole or one-half of the root system, respectively, at each irrigation event. In the partial root zone drying treatment technique, the water application was shifted from one side to the other side of the plants when volumetric soil water content of the drying side had decreased up to 8–11%. Observations revealed that in comparison to full irrigation plants, leaf water potential was significantly lower in deficit irrigation and partial root zone drying plants in almost 30% cases of observations while stomatal conductance remained same across the three irrigation levels. Leaf area, fresh berry yield, individual berry fresh weight, berry water content and berry dry weight were significantly lower in deficit irrigation and partial root zone drying plants when compared to FI plants but the total number of fruits per plant remained *at par* in all three irrigation schedules. However, on the basis of berry fresh yield and berry dry weight, there was 40% saving in irrigation water by deficit irrigation and partial root zone drying schedules, which led to increased water use efficiency of irrigation (WUEI) by 28% in deficit irrigation and 50% in partial root zone drying. Similarly, partial root zone drying and deficit irrigation cannot be recommended as criteria for scheduling irrigation in strawberry as these decreased berry yield and yield components (Liu et al., 2007). Irrigation is generally needed to achieve acceptable yield and fruit quality (Serrano et al., 1992; Yuan et al., 2004) in strawberry production.

14.4.2 Depth of Irrigation

Irrigation is essential for strawberries, particularly in the first year of transplanting and rooting of runners. Therefore, maintaining an optimum soil moisture regime is essential for utilization of nutrients applied into the soil and prevention of frost damage of the blossoms and also during the fruit picking period. Strawberry is a shallow-rooted crop and almost entire root system is limited within the top 15–25 cm of the soil profile and most of the roots of strawberry are found within the top 15 cm of the soil profile (Bergeron, 2010).

The depth of irrigation varies at different growth stages. In the initial crop stage (after planting), most roots are concentrated within the top 10–12 cm and as such the irrigation water should not percolate below 12 cm soil profile. But at the full plant growth stage, the active root depth of strawberry extends approximately up to 25 cm of the soil profile. Accordingly, in an ideal irrigation design with efficient scheduling system, the water should not be allowed to percolate below 10–12 cm in the beginning and further below 25 cm soil profile at full growth. Soil moisture meters/probes are very useful in verifying the correct irrigation schedule and helpful in avoiding deep percolation of water (Espejo et al., 2009). Soil moisture meters placed in deeper soil profile (say 40 cm or 45 cm depth) than the effective root zone (~up to 25 cm deep soil profile) helps in monitoring the water in percolating to deeper soil profiles, which is not available to strawberry plants and further assists the operator in controlling the irrigation system. Fast percolation of water clearly indicates that the rate of water application needs to be slowed down to avoid loss of irrigation water. Under such situations, the number of emitters/drippers needs to be increased so that a larger proportion of the top soil profile is wetted, leading to maximum access of root volume to the moisture regime, facilitating maximum utilization of applied irrigation water.

14.4.3 Frequency of Irrigation

Although strawberry plants have been found capable of drawing water from deeper soil profiles but for optimum growth of plants the soil moisture tension should not exceed 1.0 atmosphere. In general, frequent irrigation with less water in each irrigation is a better strategy for improved efficiency in water management of this crop. Irrigation in newly planted beds facilitates early and improved rooting of runners. Maintenance of optimum soil moisture status is essential not only during the early development phase of fruits but also at the later stages during fruit maturity to facilitate berries attaining good size. In general, strawberries grown on sandy soils require more frequent irrigation than when grown on heavier soils.

Irrigation water for strawberries needs to be of excellent quality, as strawberries are very sensitive to salts, in particular chloride (but not sodium). Chloride will start to depress yield at very low levels. Ideally, the electrical conductivity of irrigation water needs to be below 0.75dS/m (750 µS/cm) or total dissolved solids (TDS) of 400 mg/l. Yield will drop by about 25% if water contains 650 mg/l TDS (1.20dS/m), and even more, if the water is saline.

Kirschbaum et al. (2003) have considered that irrigation frequency did not affect yield significantly. But in sandy soils, too low an irrigation frequency can lead to poor yields (Hoppula and Salo, 2007; Pires et al., 2006; Serrano et al., 1992). Lozano et al. (2016) have considered that high irrigation frequency (up to six irrigations per day during high evaporative demand) plays an important role in optimizing fruit yield in strawberry.

14.4.4 Critical Time of Irrigation

Blasse et al. (1984) stated that preblossom irrigation in strawberry was effective and the most important time for irrigation was during cropping and subsequent flower formation in September. But there is more accumulation of total soluble solids in fruits under non-irrigated plots in comparison with irrigated ones in fruits of cultivars Senga Sengana and Fratina. In an irrigation scheduling

Water Management

experiment based on Class A pan evaporimeter and four crop coefficient ratios, that is, 0.33, 0.66, 1.00 and 1.33 on cold-stored runners, it was observed that Aliso and Tuff cultivars planted in double rows at 30 cm × 35 cm spacing with a population of 50,000 plants/ha and mulched with black polyethylene sheet and irrigated for 15 days by sprinkler and labyrinth drippers (discharging 4 L/h, irrigation every 10 mm cumulative pan evaporation) under the mulch, Strabbioli (1988) observed increased individual fruit weight from 7.5 g/fruit to 8.5–10.4 g/fruit and fruit yield from 411–554 g/plant due to drip irrigation (at 1.00 crop coefficient) compared with unirrigated control (345 g/plant).

14.4.5 STAGES OF IRRIGATION

Usually, copious irrigation is provided through microsprinklers (Singh and Asrey, 2005) or furrows for about a month after planting for proper establishment of plants and drip/sprinkler irrigation with or without mulching is practised subsequently. Irrigation is usually not necessary during the fruiting period, especially with early and midseason varieties of strawberries. Excessive irrigation during this period leads to softness in fruits. If the weather is dry, irrigation of late varieties during the fruiting period prolongs production and increases the yield of strawberries (Heeney, 1962).

During the runner plant establishment stage, irrigations can be performed by sprinkling, divided into two–six times a day. The drip irrigation system is installed with drip emitters spaced at 30 cm and with a flow rate of 1.74 L/h. Two drip emitter tube lines are placed over each four plant-row bed, and each line is located between two rows of plants, except the central one. Until the beginning of the water level treatments, irrigations are performed whenever the soil water potential reaches −0.010 MPa at 10 cm depth (cru.cahe.wsu.edu).

14.4.6 AMOUNT OF IRRIGATION

In strawberry cultivation, direct evaporation losses from the top layer of soils are generally negligible as the planting is generally done on raised beds covered with impermeable polyethylene mulch (Martinez-Ferri et al., 2014). Strawberries require the application of a large amount of water because of its shallow rooting system, its high leaf area and the large water content of the fruit (Kruger et al., 1999; Klamkowski and Treder, 2006; Grant et al., 2010).

As a shallow-rooted plant (within 15–20 cm of the top soil), strawberry requires more frequent irrigation but less water at each irrigation. According to Heeney (1962), the strawberry plants need 19–31 mm of water per week during the growing season. The amount of water to be applied during each irrigation varies with the soil. In general, use 13–25 mm of water on sandy soil, 25–38 mm on loam and 38–50 mm on clay. In first year, the strawberry plants were found capable of drawing water from 46 cm depth (Harrold, 1955) and formed extensive and deep root systems (Rom and Dana, 1960). The optimum growth of strawberry plants occurs under soil moisture conditions where tension does not exceed one atmosphere. Thus frequent irrigations rather than fewer heavy ones are favoured for strawberries (Mortensen, 1934).

Gehrmann (1985) recorded a positive relationship between the amounts of water applied and fruit yield and quality, while reduction in the amount of water accelerated fruit maturity. Strabbioli (1988) recorded higher yield/plant and individual fruit weight of Aliso and Tuft strawberries when the irrigation was performed at 10 mm of crop cumulative evapotranspiration. In an experiment (Lozano et al., 2016) in Spain, application of irrigation water @ 8235 m^3/ha and 10,000 m^3/ha to Sabrina and Antilla cultivars of strawberry, fruit yield of 1027–1084 g/plant and 731–754 g/plant, respectively were obtained. Further, the land productivity varied from 74.7–78.9 t/ha and 48.7–50.3 t/ha in Sabrina and Antilla cultivars, respectively, while water productivity ranged from 16.5–18.3 kg/m^3 and from 13.8–14.3 kg/m^3 water applied in Sabrina and Antilla cultivars, respectively (Lozano et al., 2016).

Berry quality tends to decrease as the rate of irrigation increased. Particularly in flavoursome cultivars, high irrigation can be used without undue loss of fruit quality but in less flavoursome

236   Strawberries

cultivars irrigation rates need to be kept low (reducing yields) to maintain fruit quality. The water should be applied slow enough to avoid runoff. Excessive irrigation is, however, detrimental, which encourages growth of leaves and stolons at the expense of fruit and flower and also increases the incidence of botrytis rot.

14.4.7 Time of Irrigation

The susceptibility of strawberry plants to water stress (either shortage or excess soil moisture) is well known. The occurrence of guttation in the field is an indication of high plant and soil water availability. Young leaves are better indicators than the older leaves and guttation can be used for scheduling irrigation in strawberries (Glen and Takeda, 1989). Penuelas et al. (1992) concluded that leaf temperature, the difference between leaf and air temperature, and the derived indices such as accumulated stress degree days and crop water stress index are useful for the assessment of mild and very mild water stress in strawberries under protected cultivation.

In areas where the soil is sandy, frequent irrigation during summer is necessary, especially if the weather is warm and dry. The heavier soils generally do not need frequent irrigation, rather every two or three weeks during the rainless period in summer. If the strawberries are being planted during a dry spring or on a light sandy soil, the plants should be irrigated immediately after transplanting to get proper establishment of the plants. During the growing period, irrigation before the appearance of any sign of drought (Heeney, 1962) is ideal.

The ability of berries to swell after rain is affected by growing conditions at earlier stages of development, and that ample watering is necessary during the early development as well as in the final phase of fruit maturation, if the berries are to reach a good size. O'Neil (1983) stated that greater susceptibility of older leaves to water stress in the field during summer months appeared to be related to gradient in leaf osmotic potential within the plants and to an age dependency in the leaves to adjust osmotically when challenged to periodic water deficits.

14.4.8 Varietal Response to Irrigation

The water use efficiency and plant water relations in strawberry are variety specific, which is again controlled by leaf physiology and on the relative allocation of photosynthates to different plant parts, viz., fruits, leaves and roots. There appears to be a complex phenomenon involved in the relationship between yield and water use efficiency at the plant and leaf levels (Martinez-Ferri et al., 2016). The water requirement in different cultivars is governed by morphological and physiological features. In a study in the Huelva province of Spain, the negative effects of water shortage, in terms of total fruit production and water use efficiency, were more apparent in Sabrina and Camarosa as compared to Antilla, Benicia, Fortuna, Splendor or Santaclara varieties (Martinez-Ferri et al., 2016).

About 20% (7300 ha) of the total irrigated acreage (32,372.7 ha: CAP 2000) in Huelva province of Spain is planted under the putatively drought-resistant *F. chiloensis* (L.) Duch. × water stress-sensitive *F. virginiana* (L.) Duch. (Zhang and Archbold, 1993) hybrids developed by different breeding programmes oriented to achieve better agronomic traits.

The varieties with a deeper root system could exploit water from the lower soil profile; thus the cultivars with deep root system would be able to perform better and show greater water use efficiency even when water availability is reduced or deficit irrigation is given. The experimental results of a trial on water relations, growth and physiological response of seven strawberry cultivars showed (Martinez-Ferri et al., 2016) higher A/T (instantaneous water use efficiency) values in Camarosa and Sabrina cultivars in well-watered conditions, which indicated that these cultivars use water in a more efficient way. Similarly, the cultivars Camarosa and Sabrina also showed lower intercellular CO_2 concentration (C_i), which will reflect either a relatively low value of stomatal conductance (g_s), a relatively high photosynthetic capacity (A, amount and activity of photosynthetic machinery per unit leaf area) or a combination of these two (Farquhar et al., 1989).

Further, the strawberry cultivars with relatively higher instantaneous water use efficiency (A/T) could represent an agronomic advantage contributing to increased water productivity and the economic benefits of strawberry cultivation (Zhang and Archbold, 1993; Cattivelli et al., 2008).

Breeding for high A/T has been proposed as a target for improving water use efficiency in strawberry (Condon et al., 2004; Klamkowski and Treder, 2008).

Grant et al. (2010) have ascribed the low crop transpiration efficiency (TE_v) values with higher water use efficiency (WUE_c) but also with higher yields in different strawberry cultivars which suggests that selection for higher yield could indirectly lead to increased water use efficiency. On the other hand, Martinez-Ferri et al. (2016) could not observe higher leaf-level water use efficiency in more vigorous cultivars (i.e. Camarosa and Sabina). In fact, those cultivars with higher A/T and TE_v showed lower WUE_c. Thus it leads to the inference that the individual leaves are more efficient in water use, the bigger total leaf area per plant encompasses higher total plant transpiration and therefore, higher water consumption per unit weight of fruit produced leaving higher water footprint.

The results for Splendor, Fortuna and Benicia cultivars have indicated (Martinez-Ferri et al., 2016) that the higher WUE_c values do not necessarily translate into significantly higher yields but are related to higher values of harvest index (HI). Thus, the strawberry cultivars with higher HI or yield efficiency values are more efficient in the use of irrigated water with low water footprint. The cultivars with higher yield efficiency sometimes display certain degree of precocity, so planting these cultivars, besides representing substantial water saving during the period of more evaporative demand in the field, may enhance the economic benefit of strawberry cultivation by the early arrival of fruits into the market, thereby ensuring better prices to growers.

Thus, from agronomical point of view, it would be better to have a cultivar with high standing biomass or crop TE_v for assured and economically profitable productivity and high HI value to assure better crop WUE_c. This could be explained better by the following equation:

$$WUE_c = TE_v \times HI$$

Therefore, crop transpiration efficiency (TE_c) is closely related with HI and can appropriately be expressed by following equation:

$$TE_c = (TE_v \times HI) + TE_v$$

14.4.9 EFFECT OF GROWING CONDITIONS

For sustainable production system and minimization of adverse effects of drought on yield and quality of produce (Blatt, 1984; Martinez-Ferri et al., 2014), it is absolutely essential to understand the water relations of different strawberry cultivars at the physiological and agronomical level to determine the water use efficiency of applied irrigation.

With regard to open field cultivation, the mean total yield values per plant were always higher than 600 g plant^{-1} for the different water levels and soil mulches. In protected cultivation, irrigation levels of −0.010 and −0.035 MPa and the clear plastic mulch favoured vegetative growth, such as plant height, maximum horizontal dimension of plant, leaf area index and total marketable fruit yield and its components, such as mean number and weight of fruits per plant (Pires et al., 2006).

Protected strawberry cultivation, at irrigation levels of −0.010 and −0.035 Mpa with clear polyethylene mulch has been reported to favour vegetative growth and total marketable fruit yield. Therefore, total water depths in protected cultivation were 580, 496 and 474 mm, and in the open field 438, 380 and 336 mm, at irrigation levels of −0.010 MPa, −0.035 MPa and −0.070 MPa, respectively. Fruit yield per plant was high in the protected cultivation, ranging from 466.7–662.4 g plant^{-1}

238   Strawberries

(cru.cahe.wsu.edu), when compared with those achieved by Passos (1997) and Fernandes-Jr. et al. (2002), using the same cultivar in protected cultivation.

14.5 IRRIGATION SCHEDULING

For scheduling irrigation (when, how much and how) in strawberry, the time and quantity of water to be applied and the method of irrigation adopted are some of the important points to be considered. These involve the soil, plant and climatological parameters, in general. Common criteria for scheduling irrigation in strawberry production are based on climatic water balance (evapotranspiration), plant physiological properties, soil water status measurements or a combination of all these factors. Practically, irrigation scheduling includes how to irrigate, when or how frequently to irrigate and how much water to apply at each irrigation.

Irrigation should be scheduled on the basis of climatic water balance model and on tensiometer measurements. Results of the experiment demonstrate that irrigation influences yield and quality of strawberry (Kruger et al., 2002). Irrigation at 200 kPa resulted in best fruit yield. Fruit weight improved due to irrigation whereas fruit firmness decreased towards the end of the each harvest season and significantly reduced by irrigation.

14.5.1 SOIL PARAMETERS

The availability of soil moisture in the root zone provides reliable indication of irrigation requirement. The available pore spaces in the soil profile get immediately filled with water after irrigation or rain. A part of this water moves down after a period of free drainage and stops when the gravitational pull is balanced by the surface tension of a film of water around the soil particles and at this stage, the soil water content reaches a point called "Field Capacity". The roots obtain this water till it becomes too less with consequent rise in soil moisture tension. Later a stage comes when moisture is not available to roots and this stage is referred to as the "Permanent Wilting Point". Therefore, the soil moisture between field capacity and permanent wilting point is taken up by the plants and this quantity of water is known as available water. The available soil moisture varies depending on the texture and structure of soil, rooting depth and water extraction capability of the plant species (Table 14.1).

The appearance and feel of the soil at different profile depths give fairly good indications of its moisture status. The requirement for irrigation can be calculated by adding the moisture deficits of each profile and giving allowance for irrigation efficiency. The soil water status can be expressed more precisely by its content or energy status. Soil moisture content can be measured by drying samples at 105–110°C (gravimetric method) or by working out the ratio of volume of water to bulk volume of soil (volumetric method).

The component forces (F), which influence the energy status and contribute to total water potential are gravimetric (Z), matric (M) and osmotic (O) and the interrelationship among these factors is expressed by: $F = Z + M + O$.

TABLE 14.1

Field Capacity, Wilting Point and Available Moisture in Different Soil Texture Classes

Soil Type	Field Capacity (%)	Permanent Wilting Point (%)	Available Moisture (%)
Fine sand	3–5	1–3	2–4
Sandy loam	5–15	3–8	4–11
Silt loam	12–18	6–10	6–13
Clay loam	15–30	7–16	10–18
Clay	25–40	12–20	10–30

Water Management

Osmotic potential in most normal soils is negligible. Matric potential is usually measured by techniques such as pressure membrane apparatus and tensiometer. These methods of soil water status measurements have their own merits and limitations (Table 14.2).

The soil water status measurement methods either based on the soil water content or the soil matric potential (Ψ) have been thoroughly studied and well documented (Topp and Ferre, 2002; Young and Sisson, 2002) and have largely been used in irrigation scheduling of various crops. Further, the Ψ measurements have been preferred over soil water content measurements. The Ψ measurement method is easy to use in various field conditions (no calibration needed for soil types or salinity levels) and is directly related to the capacity of the soil to supply water at the rates required by the plants (Rekika et al., 2014). The soil water potential gradient is the force controlling the movement of water in soils, the data on Ψ also indicates the direction and magnitude of

TABLE 14.2

Relative Merits and Limitations of Different Soil Moisture Measurement Methods

Methods	Merits	Limitations
Gravimetric and Volumetric	Simple and accurate. Any range of measurement	Repeated measurement from same point impossible. Physically arduous and slow, particularly when sampling at greater depths
Atmometers	Inexpensive. Any range of measurement	Calibrated relation between soil moisture extraction by plant and evaporation data from atmometer is required.
Pressure Plate Apparatus	Provides simultaneous measurement of moisture content and water potential. Same soil sample can be used to derive moisture release curve.	Limited range of measurement between 200 and 0 joules/kg. Measures only matric potential. Extrapolation needed to obtain water potential values.
Pressure Membrane Apparatus	Provides simultaneous measurement of moisture content and water potential. Same soil sample can be used to derive moisture release curve.	Measurement range between −1000 and 200 joules/kg. Measures only matric potential. Measurement cumbersome.
Thermocouple Psychrometer	Very rapid and accurate. Wide range of measurement (−5000–0 joules/kg). Also measures plant water potential.	Very expensive. Repeated calibration needed. Precise temperature control required.
Tensiometer	Suitable for calculating soil profile permeability characteristics. Provides direct water potential values.	Limited range (−100–0 joules/kg) of measurement. Needs to be recharged when soil dries out. Preferential root growth of plants takes place near the porous cups. Intercultural operations hindered.
Electrical Resistance Meter	Relatively simple and rapid. Repeated measurements possible at a point.	Measurement range between −1500 and 0 joules/kg. Measurements are affected by soil salinity. Calibration required for each soil and repeatedly. Gypsum blocks are short lived. Very expensive.
Neutron Scattering	Measurement possible in any range. Repeated measurement possible at a point. Accurately measures changes in moisture content. Measurement not affected by salinity.	Measurements are affected by soil organic matter, Cl, Fe and B. Point measurement near soil surface not possible. Intercultural operations affected.
Dielectric Constant	Rapid measurement. Measurement not affected by salinity.	High temperature coefficient often requires associated temperature measurement.
Thermal conductivity Method	Wide range of measurement. Permits associated temperature measurement.	Sensitive to soil contact.

soil water fluxes and can be used in preventing leaching of applied irrigation water under the root zone (Kruger et al., 1999). Ψ–based data have successfully been used for irrigation scheduling and improving the yield of many agricultural crops (Pelletier et al., 2013; Periard et al., 2012; Shock and Wang, 2011).

The impacts of Ψ-based irrigation management on strawberry yield and water use efficiency has been studied by several workers and the results have indicated variable irrigation threshold of Ψ values for optimum results. In a sandy loam soil under a high tunnel, the fruit yield of strawberry was significantly maximum at –10 kPa Ψ in comparison to drier (–70 kPa) zones (Guimera et al., 1995; Penuelas et al., 1992). Significant decrease in fruit yield and quality of strawberry were observed due to irrigation scheduling at –30, –50 or –70 kPa Ψ with significantly maximum water use efficiency at an irrigation threshold of –50 kPa. In another field experiment, significantly maximum fruit yield of strawberry has been recorded by irrigation scheduling at –15 kPa and water use efficiency at –30 kPa (Hoppula and Salo, 2007) in sandy soil conditions. Both in open or protected cultivation, the irrigation regimes influence not only vegetative growth and fruit yield but also fruit quality. Contrary to open field conditions, deficit irrigation increases the sugar acid ratios, antioxidant capacity and total phenolic content of strawberries (Gine-Bordonaba and Terry, 2010; Terry et al., 2007).

But all the studies do not support the idea of scheduling irrigation in strawberry solely based on Ψ, as Kruger et al. (1999) have concluded (mainly because of economical and practical reasons), that evapotranspiration-based irrigation is more promising in improving fruit yield and water use efficiency in strawberry. In scheduling of irrigation in strawberry only on the basis of Ψ, installation of too many soil moisture meters/probes or tensiometers over a large field area could be a limiting factor, more so in undulating topography, for example, if Ψ data recorded from a field presented large spatial and temporal variability (place to place and time to time), which necessitated placement of several tensiometers for obtaining representative sample values over a large area (Hendrcikx and Wierenga, 1990; Hendrickx et al., 1994). In addition, the depth of installing the tensiometers, frequency of measurements of Ψ, frequency of irrigation, growing conditions such as open field, greenhouse or plastic tunnels and crop stages also need to be considered in scheduling irrigation. From fruit yield and other considerations, irrigating strawberries at –10 kPa to –15 kPa Ψ has been observed to be optimum (Guillaaume et al., 2015).

14.5.2 BED CAPACITY

This is a concept similar to field capacity but mostly adapted to drip-irrigated raised beds where the wetted bulb has a finite size similar to an infiltration from an infiltrometer (Reynolds and Elrick, 1990) and is not necessarily in hydraulic contact with the subsoil and water table. Drip-irrigated raised beds generally reach their own equilibrium at Ψ much higher than the expected in a standard flat soil (–33 kPa). In a dry soil profile, the water content generally approaches a saturation level after some time of irrigation or heavy rainfall. If there is rapid gravitational leaching, Ψ is reduced. After this stage, a transition period between gravitational (downward movement) and capillary water movement (lateral flow) normally occurs. Thereafter, for some time, further changes in Ψ due to water movement becomes very slow and almost negligible, which is mainly due to capillary water redistribution. And Ψ at the beginning of the capillary redistribution phase is referred to as the "bed capacity".

14.5.3 PLANT PARAMETERS

Wilting is the most important sign of water deficit and the growth of most of the plants is retarded at this stage. Changes in the colour of leaves, leaf angle, reduced growth etc. are other visual symptoms of water deficit in plants. For irrigation scheduling in strawberry, the measurement of plant water status is considered to be of most practical use. Moisture status in plant can be expressed

Water Management

either as content or its energy status, that is water potential. The water content in plant samples (per cent dry weight basis) can be measured by drying the samples at 60–70°C in digital hot air oven.

Water saturation deficit (or leaf relative water content) is calculated by:

$$\text{Water saturation deficit} = \frac{\text{Fully turgid weight-fresh weight}}{\text{Fully turgid weight-dry weight}} \times 100$$

Water saturation deficit (WSD) is also calculated by:

$$\text{WSD} = \frac{\text{Fresh weight-dry weight}}{\text{Fully turgid weight-dry weight}} \times 100$$

Water saturation deficit is also considered as the most practical approach in scheduling irrigation (Table 14.3).

14.5.4 CLIMATIC PARAMETERS

With the increase in age and ground coverage of strawberry plantations, the loss of water from soil surfaces by way of evaporation is progressively reduced but the loss of water from plant surfaces depends upon crop coefficient, a ratio between the evapotranspiration under a crop and the potential evapotranspiration or potential evaporation or simply pan evaporation. This evaporative demand mainly determines the amount and frequency of irrigation, keeping the incidental soil moisture conditions into account. The potential evapotranspiration is estimated by a USDA Class A open pan evaporimeter. The evapotranspiration for almost all practical purposes is equal to the consumptive use of water because the water used in metabolic processes is insignificant (≤1% of evapotranspiration). The cumulative pan evaporation data obtained from the observations on a USDA Class A open pan evaporimeter are practically used to schedule irrigation. Since the depth of irrigation water varies from soil to soil, the irrigation water: cumulative pan evaporation ratio is also used for scheduling irrigation in strawberries. The critical growth and development of the strawberry crop should always be taken into consideration while scheduling irrigation. In fully grown plantations, growth flushes, fruit setting and developmental stages are critical because water stress at this stage considerably reduces productivity, while in areas with water scarcity, a few irrigations at critical stages would result in maximum water use efficiency.

Several studies have indicated that evapotranspiration-based irrigation management is efficient for strawberry production (Hanson and Bendixon, 2004; Kruger et al., 1999; Yuan et al., 2004) but is being considered inefficient as it is neither able to account for rapid changes in climatic conditions nor the differences in water requirements of different cultivars (Gine-Bordonaba and Terry, 2010; Klamkowsaki and Treder, 2008; Kruger et al., 1999). The water management of strawberry based only on evapotranspiration cannot be used to assess if the applied irrigation water is lost due to percolation beneath the root zone (Simonne et al., 2012).

14.6 WATER REQUIREMENT

The crop water requirement of strawberry is assessed daily using the outdoor or under plastic cover climatic data recorded in a local weather station (Allen et al., 1998; Fernandez et al., 2010). These calculations are then converted into daily irrigation time which depends on the on-farm irrigation system and the irrigation efficiency (i.e., relationship of the volume of water required by the crop and the volume of water applied).

In fruit crops such as strawberry, the crop transpiration efficiency (TE_c) can be estimated as the sum of the crop water use efficiency (WUE_c, g fruit per unit volume (litre) of water transpired per

TABLE 14.3
Soil Moisture and Appearance Relationship

	Soil Moisture Deficit (cm/30 cm Soil Depth)										
	0.0 (FC)	0.5	1.0	1.5	2.0	2.5	3.0	3.5	4.0	4.5	5.0
Loamy sand	Leaves wet outline on hand when squeezed	Appears moist; makes a weak ball	Appears slightly moist; sticks together slightly	Very dry, loose; flows through fingers (PWP)							
Sandy loam	Leaves wet outline on hand; makes a short ribbon		Makes a hard ball	Makes a good ball	Makes a weak ball	Will not make a ball	PWP				
Loam	Leaves wet outline on hand; will ribbon out about one inch		Forms a plastic ball; slicks when rubbed	Forms a hard ball	Makes a good ball		Small clods crumble fairly easily	Small clods are hard (PWP[1])			
Clay Loam	Leaves slight moisture on hand when squeezed; ribbon out about two inches.		Will slick and ribbon easily	Will make a thick ribbon; may slick when rubbed	Makes a good ball		Will make ball, small clods will flatten out rather than crumble	Clods crumble			Clods are hard, cracked (PWP)

[1] PWP – permanent wilting point (Modified from Merriam, 1960, Field method of approximating soil moisture for irrigation. Transactions A.S.A.E., 3(1): 31–32.)

Water Management   **243**

plant) and the transpiration efficiency of standing biomass (TE_v, g plant dry weight per unit volume (litre) of water transpired):

$$TE_c = WUE_c + TE_v$$

Under well-watered conditions, the crop transpiration efficiency is directly related to the instantaneous water use efficiency (*A/T*, Condon et al., 2004). Accordingly, it can be presumed that the strawberry varieties with high *A/T* would have high crop transpiration efficiency.

The water requirement of strawberry can be estimated according to weekly crop evapotranspiration which are obtained by multiplying the reference evapotranspiration (Allen et al., 1998; Hargreaves and Samani, 1985) by the crop coefficient of strawberry tabulated by the Food and Agricultural Organization of the United Nations (Allen et al., 1998). The values for reference evapotranspiration can be obtained from a simple or automatic weather station installed nearby the commercial plantation of strawberry. In the absence of weather station, 10 years average values of reference evapotranspiration from the nearby and similar geographical locality can also be satisfactorily used for calculating water requirement and scheduling irrigation in strawberry.

The quantity of required irrigation water is calculated on the basis of water deficit, root zone depth, soil type and salt content in soil. Strawberries have a very shallow root system and the effective root zone commonly extends to about 30 cm depth but the maximum quantity of moisture is extracted from the top 10 cm soil profile. The quantity of water applied for normal growth and production of strawberries is called water requirement (WR) and is expressed by the following formula, where ET is evapotranspiration:

$$WR = ET + Wa + Ws + Wm \text{ or } Wi + Wr + Ws + Wg + Wd*$$

*Thus the water requirement consists of evapotranspiration plus non-avoidable losses (which could be) summarized as:

- runoff, seepage, deep percolation, weed growth etc. (Wa)
- water needs for special functions, such as leaching of salts, land preparation and planting (Ws)
- the quantity of water needed for metabolic activities (Wm)

Before applying irrigation water (Wi), the contribution of effective rainfall (Wr) and contribution from soil profile (Ws) through capillary rise of groundwater (Wg), from atmosphere in the form of dew, stem flow (Wd) etc., should, however, be deducted from this quantity.

The FAO has developed a computer model "CROPWAT", which facilitates the calculation of evapotranspiration using mean temperature, the minimum and maximum relative humidity, wind velocity at 2 m height and sunshine hours. Additionally, the altitude, the latitude and the longitude of the concerned locality are required. Reference crop evapotranspiration is evapotranspiration from a reference surface not short of water. FAO-Penman-Monteith method is considered the best method for determination of evapotranspiration, since it closely approximates gross evapotranspiration at the locations evaluated, is physically based and incorporates both physiological and aerodynamic parameters. This surface is a hypothetical gross reference crop with specific characteristics. The water requirement of strawberry depends upon its irrigation needs in addition to the factors associated with the losses and gains of water from the plantation. The optimum quantity of water required is that which results in maximum economic yield. The water requirement of strawberry varies at different stages of growth and development.

The efficiency of irrigation in strawberry plantations can be maximized by optimizing the efficiency in conveyance, application, distribution and storage of water in effective root zone and minimizing the losses. This provides maximum field water use efficiency, which is the ratio of yield to total amount of water used in the field. The objective of an efficient irrigation schedule is to achieve the highest water use efficiency. The water requirement of the strawberries is governed by cultivars, season and growing conditions.

14.7 IRRIGATION SYSTEMS

Irrigation in strawberries is applied through various systems, viz., surface, drip or trickle, sprinkler etc. depending upon the field conditions. Furrow irrigation can be used only where soil (or particularly the subsoil) is heavy and where the slopes are uniform. Sprinkler irrigation is however, valuable in areas where there may be salt or heat (>29°C) stress and need for blossom frost control (to −6.7°C) in spring by using irrigation water. Nowadays trickle or drip system is gaining popularity in a number of crops including strawberries. Reports have shown 15–20% increase in yields, better water supply during winter season and less rotting of strawberry fruits, which were hanging in sprinkler water (Childers et al., 1995).

While comparing furrow, sprinkler and drip irrigation methods, Tekinel et al. (1984) recorded maximum yield due to drip and minimum due to furrow irrigation. The amount of water used for drip irrigation was 38% less than that in the furrow and 20% less than that used in the sprinkler irrigation method. Albregts and Howard (1985) evaluated the continuous and intermittent sprinkler irrigation during two seasons; in first season at 3/17, 5/15, 10/20, 15/15 and in second season at 5/25, 5/15 and 5/10 minutes on/off, respectively. It was observed that leaf loss during the establishment period generally increased commensurately with the length of the "off" interval and decreased vice versa. The plant mortality was maximum with 3/17 and 5/25 minutes on/off treatment cycles. Hochmuth et al. (1993) observed that combinations of sprinklers and certain row cover treatments provided better fruit production than either of the treatments alone. Seong et al. (1993) recorded that a low mean percentage (14.3%) of grey mould disease affected fruits under drip irrigation compared with furrow irrigation (23.1%).

The type of soil, quantity of water available, initial cost of installation, operation, topography, crop duration and maintenance should be considered for choosing irrigation system for strawberries. On sloppy lands and in the hilly zones, contour basins and furrows, sprinkler and drip methods are suitable, whereas in plain lands rectangular plots, furrows, sprinkler, bubble and trickle methods are generally followed. For newly planted plants microsprinklers or a rain gun can be used profitably to increase the success of establishment, and to improve the vegetative growth of new plants.

14.7.1 Furrow Irrigation

In this system, the irrigation water from a head ditch or pipeline is allowed to flow in furrows and irrigate the soil surface in the planted area. Skill and care are required for uniform application of water from furrows because sometimes a portion of the target area remains unirrigated while a non-target area may get excess water. Generally, irrigation is applied in furrows between the rows and alleys are usually cultivated at two or three days after each irrigation. Care needs to be taken so that water does not wet the leaves as it is likely to aggravate foliar diseases.

14.7.2 Pressurized or Micro-Irrigation Systems

Micro-irrigation systems are helpful in applying small amounts of water at short time intervals. These are purposely designed in a typical manner to wet only the soil volume occupied by the root zone of plants. These systems maintain the wetted root zone of plants at or near an optimum moisture level. The micro-irrigation systems have a comparatively greater potential for precision irrigation than other conventional systems. These are meticulously controlled and are mostly automated in strawberry production systems (Phene and Howell, 1984; Meron et al., 1986; Dukes and Scollberg, 2004; Wanjura et al., 2004; Evett et al., 2006).

Research on micro-irrigation systems has been carried out primarily on horticultural crops that are planted at a wider spacing, such as in viticulture (Capraro et al., 2008 a, b) and many other perennial and woody orchard crops (Coates et al., 2004; Adhikari et al., 2008).

Use of sprinkler and drip/trickle irrigation techniques is more convenient to regulate water as well as to minimize the amount of water. On an average, the amount of water used for drip/trickle irrigation is approximately 30% less than what is required in furrows (Kamber et al., 1986). While comparing drip, sprinkler and furrow irrigation methods, Tekinel et al. (1984) observed maximum fruit yield due to drip irrigation and minimum in furrow irrigation.

The movement of irrigation water follows different patterns and is greatly influenced by soils of various textures. As a thumb rule, the downward movement of water is faster in sandy soils while in clayey soils the horizontal movement is comparatively more. In medium-textured soils, with balanced proportions of sand, silt and clay, the movement of water in horizontal and vertical directions from the ground surface of soil is more or less equal. Accordingly, the rate of water application in micro-irrigation (or say drip) system in strawberry plantation depends upon the soil textural class.

The root growth pattern in terms of depth and spread also influences the pattern of irrigation water application. The technical information about the irrigation requirements of strawberry and knowhow of the application of the calculated amount of water is governed by the complex water dynamics of the reference soil type. The optimum duration of the irrigation pulse and its frequency depend upon the crop, weather and soil conditions and need to be determined, which ultimately maximizes irrigation efficiency. Modelling techniques can be used for estimation of the optimum pulse duration with the help of numerical model Hydrus 2D/3D (Simunek et al., 2006).

14.7.2.1 Sprinkler Irrigation

In this system, lightweight perforated aluminium pipes are laid and irrigation water flows with force through these pipes for uniform irrigation. Nowadays, polyvinyl chloride (PVC) pipes with control valves and sections are used for the controlled flow of irrigation water. The system has further been improved to portable type of pipes by using suitable and quick coupling devices. Water use efficiency in sprinkler irrigation increases many times when compared with flood or furrow irrigation. Sprinkler systems can be useful under undulating topography with minimum soil erosion by adopting ridge and furrow systems. A sprinkler irrigation system is extremely useful for light-textured soils and undulating topography. Sprinklers are either fixed-head, rotating-head or perforated pipe types. Land grading and preparation are not essential for this system but the cost of installation is very high. Strawberries respond to sprinkler irrigation very well. The overhead sprinkler irrigation system has multiple uses such as pesticide and fungicide sprays and application of liquid fertilizers too.

Sprinkler irrigation reduces plant leaf temperature but increases the relative humidity and rapid air convection with increased CO_2 concentration in the plant microclimate. By sprinkler irrigation, frost injury to blossom can also be minimized or avoided. In Florida, where most strawberry cultivation is done in open field conditions, high-impact sprinkler irrigation (four or five gallons/minute per head) is commonly used primarily for transplant establishment and second for freeze protection of established plantations. For establishment of new plantations, the bare-root plants are planted into polyethylene-mulched beds and irrigated with sprinklers (~6000 gal/hour) for 8–10 hours per day for the initial 10 days after transplanting (Hochmuth et al., 1993).

This is primarily practised to reduce the temperature around the young crowns during the establishment and to provide enough soil moisture to newly emerged roots during the establishment period. But this is a very faulty method as it requires a large volume of water (14–20 inch/acre) and, further, the applied irrigation water goes waste as runoff and leaches down into the deep soil layers (Bish et al., 1997; Hochmuth et al., 2006).

14.7.2.2 Drip Irrigation

Drip/trickle irrigation is continuous or frequent application of water to a limited soil volume through drippers, emitters or extended small tubes. It is particularly important and useful in situations of water shortage, excessive soil permeability, steep land and scarce/costly labour. Clogging of drippers and high initial installation costs are, however, very serious limitations of the drip irrigation system.

The dimensions of the soil volume irrigated by each individual drippers and the water flow along the soil profile is determined by the reference soil characteristics (texture and hydraulic conductivity) and the discharge rate of the dripper. Water moves in the soil due to differences in water potential between different sections of soil.

Water potential is governed first by gravity, which forces water to move downward, and second by capillarity, which forces water to move in all directions, and ultimately water accumulates within soil pores. In coarse-textured sandy soils, the rate of water infiltration into the soil is fast but in fine-textured clay soils, the rate is slow due to very narrow passage between soil particles. The first drop of water that emerges from the drippers infiltrates immediately into the soil and as the irrigation continues, the subsequent drip water accumulates to form a puddle under the dripper. In fine-textured clay soils, the puddle formed on the soil surface is larger than in coarse-textured sandy soils. The slow penetration (infiltration) of water into the fine-textured clay soils in relation to the discharge rate of water from drippers causes accumulation of water on the soil surface. In general, downward movement tends to be slightly greater than lateral movement.

In spite of the fast infiltration of water due to relatively wide air spaces between the soil particles, the phenomenon of water accumulation also occurs in coarse-textured sandy soils in which the drip water moves mainly downward with limited expansion sideways. Under these conditions, the wetted zone will be typically carrot shaped. The discharge rate of drippers strongly determines the shape of the wetted volume of soil profile. Increased discharge rate increases the accumulation of water below the dripper resulting in extensive lateral spreading and less wetting downward. In a sandy soil, a dripper with a high discharge rate creates an onion-shaped wetted zone similar to the one attained in a clay soil, irrigated with a low-discharge-rate dripper, for example, in clay soil, a high discharge rate creates a puddle of about 1m diameter but the depth of wetting might be only 15 cm, which ultimately affects the root system, while the formation of a large puddle on the surface causes loss of water due to evaporation and the very purpose of increasing water use efficiency is defeated. The size of puddle under the dripper is independent of the duration of irrigation and is rather influenced practically by the type of soil and discharge rate of the dripper. When the soil profile is not uniform, and consists of layers of different texture and composition, the course of water movement is complicated and difficult to predict. Drip irrigation presents a number of advantages, contributing to avoiding the formation of a humid microclimate favourable to diseases (Howard et al., 1992; Madden et al., 1993; Tanaka et al., 2005; Tanaka, 2002). In addition, it allows automation, frequent irrigations, fertigation and obtaining high yields (McNiesh et al., 1985; Voth and Bringhurst, 1990; Serrano et al., 1992; Passos, 1997; Pires, 1998; Rolbiecki et al., 2004). Through the drip system, irrigation management is also critical, because strawberry is sensitive either to excess or deficit water supply (Gehrmann, 1985; McNiesh et al., 1985; Serrano et al., 1992; Pires, 1998; Krüger et al., 1999; Kirnak et al., 2001, 2003). In a trial, with less than 38% water used in drip irrigation than what was used in furrow and 20% less than what was used by sprinklers, the application of drip irrigation significantly increased the fruit size, yield and quality of strawberry cultivars Redgauntlet, Senga Sengana, Fratina and Gorella.

Drip irrigation at 100% of crop ET (1.0 "V" volume of water) has been found to give significantly higher fruit yield (57.07 q/ha) than surface irrigation (47.17 q/ha). The fruit yield improves further (78.63 q/ha) due to increased efficiency of drip irrigation when combined with black polyethylene mulch. Mulching with grasses increased the fruit size, sugar content, ratio of total soluble solids (TSS) to acidity and anthocyanin content of fruits when compared with black polyethylene mulch only. TSS and acidity of fruits remain higher in fruits harvested from rainfed, unmulched plots. Average seasonal requirement of water through drip irrigation (but without mulch) was calculated to be 475 mm, which decreased to 440–403 mm with grass and black polyethylene mulch, respectively. The corresponding values for surface irrigation were 696 mm (unmulched), 660 mm (+ grass mulch) and 608 mm (+ black polyethylene mulch). Water use efficiency under drip irrigation without mulch, drip + grass mulch and drip + black polyethylene mulch was 7.5, 8.8 and 11.4 kg/ha/mm, respectively, whereas, the respective values under surface irrigation were 4.7, 5.7 and 7.1 kg/ha/mm. Drip irrigation, besides saving 51% of irrigation water, resulted in about 19% higher fruit yield compared with the surface irrigation (Kumar et al., 2005).

Being a highly efficient (90–95%) water application system, drip irrigation is widely used on polyethylene-mulched strawberry, due to reduced water use and foliar diseases, ability to apply fertilizer with the irrigation, precise water distribution and the ability to electronically schedule irrigation on larger areas with relatively smaller pumps. The drip system can also be used as a subsurface drip system, placed at a depth of 60–90 cm (Locascio, 2005). In Brazil, protected cultivation, together with drip irrigation, was introduced in the beginning of the 1990s, as an attempt to control anthracnose in strawberry flowers and young fruits (Passos, 1997).

Withholding nutrition by solution during the flower initiation period and the onset of the differentiation period increased the flower number of the short-day strawberry cultivar Elsanta. The primary flower of the first inflorescence was more likely to develop phyllody if nutrient supply was limited during September. Withholding a complete nutrient solution until September or longer reduced crown diameter, length of inflorescences and petioles, number of leaves, flowers and yield significantly, but improved fruit weight and fruit set. Withholding application of nutrient solution after September reduced crown diameter and number of fruits significantly (Lieten, 2002).

14.7.2.2.1 *Causes of Variations in Uniformity of Water Application through Drip Irrigation System*

Although the potential efficiency of micro-irrigation (mainly drip) system is often referred to as more than 90% but practically the possibility of achieving such a higher level of uniformity in water application under commercial cultivation systems depends on meticulous management rather than the property of a system design (Smith et al., 2010.), for example, the drip irrigation application efficiency in sugar cane under commercial cultivation system varied from 30–90 percent (Shannon et al., 1996) primarily due to overirrigation and deep percolation.

Some of the factors causing variation in uniformity of performance of drip irrigation system in strawberry plantations could be:

1. differences in pressure in lateral pipes
2. clogging of emitters/drippers
3. inappropriate irrigation management zones in the plantation area

In the Huelva region of Spain, where strawberries are cultivated on a large scale, the most common emitters used in the irrigation system are one-year, low-density polyethylene tapes, which are comparatively cheap and easy to set up. But despite this cheapness and ease of setup, the polyethylene tapes are not a suitable system for pulse irrigation in strawberries planted in sandy soils as the irrigation water distribution uniformity was observed to drop dramatically to roughly 50% at the medium-end period of the season during evaluation 2011–12 and 2012–13 (Morillo et al., 2015). This was ascribed to the suction phenomenon that occurs when the system drains between each irrigation cycles (Garcia Morillo et al., 2012). It means that when one operation cycle of drip irrigation system is completed and the system is put to idle state, at that time the drippers/emitters placed at lower elevation in field drain out the excess water which causes the left over water in the drippers/emitters placed at higher elevations to move towards water draining points, which in turn sucks up muddy water and smaller soil particles from the irrigated field. Once all the water from the irrigation laterals is drained out through emitters/drippers or polyethylene tapes, then the soil particles along with other sediment impurities which remain stuck inside the drippers/drip lateral pipes, become the major source of blockage on drying. Even with only 5% of clogging, the water application efficiency and uniformity are drastically reduced in all the different types of emitters/drippers.

The situation is further aggravated in non-pressure-compensated drippers/emitters and under such situations, the flow rates differ greatly from the manufacturer's design range (Garcia Morillo et al., 2012; Gonalez Perea et al., 2014). When there is excess pressure in the lateral pipes and drippers/emitters, the flow rates are greatly increased from the manufacturer's design range, especially in non-pressure-compensated drippers/emitters.

Normally, the laterals have the inbuilt property of compensating the pressure variation up to 50 m lateral pipe length but in cases when the length of laterals is more than 50 m, the operating pressure is drastically reduced at the tail end of lateral pipes/plastic tapes causing reduced flow rate of water and variation in uniformity of irrigation water applied. It leads to differences in crop growth, fruit yield and quality due to deficit or excess soil moisture in the root zone of strawberry plants. It also causes variations in stages of fruit maturity and calls for extra labour for harvesting. Many times the managers of strawberry plantations do not verify the working pressure in the micro-irrigation systems and as such it remains unknown as to how much water is being applied or wasted. In addition, the farmers tend to use as much inputs as possible, such as water, fertilizers or pesticides, just to get increased productivity and ignore the economic principles of production or possible yield losses. An efficient irrigation system would be that which is able to supply the required quantity of water at the right time and place with a high efficiency. Hence, maximum use of tools and technologies should be made of in designing the irrigation system and the decision making process for enhanced irrigation and crop water use efficiency in a commercial strawberry production system.

14.7.2.2.2 Designing Precision (Drip) Irrigation System

Considering soil and local weather parameters, the designing of the precision irrigation system is a three-stage process:

1. Assessment of irrigation process

    In the first step, the irrigation process is assessed by a set of performance indicators to identify the possible inefficiencies in the irrigation system and its management. These denote the ratios that relate variables like irrigated area, the volume of irrigation water to be applied or the productivity. The total volume of irrigation water required during the crop growing period should also consider the amount of water essential for various crop growing processes like soil preparation, planting, irrigation for removal of salt from the soil (if essential) and to manage the ill effects of frosts etc.
2. Accurate irrigation scheduling

In the second step, accurate scheduling is done considering:

1. the accurate estimation of crop water needs
2. the design of the optimum irrigation pulse or interval
3. integration of technologies and designing a comprehensive precision irrigation system

The accurate estimation of crop water needs

The Relative Irrigation Supply (RIS) is calculated by:

$$RIS = \frac{WA_g}{10 \times (ET_c - P_e)}$$

where
WA_g = gross volume of water required/unit area (e.g., m^3/ha)
ET_c = crop evapotranspiration (mm/year)
P_e = effective precipitation (mm/year)

Water Management

The RIS confirms how irrigation matches the theoretical water requirements. Crop evapotranspiration (mm/year) is estimated as per Allen et al., 1998.

In areas where strawberries are planted in beds covered by black plastic and stay underneath polytunnels for most of the crop growth periods, or protected cultivation, natural precipitation is negligible so it does not affect soil water balance.

Crop evapotranspiration is estimated on the basis of the strawberry crop being grown outdoors or inside a greenhouse or polytunnels. Fernandez et al. (2010) have given a guideline for calculating reference crop evapotranspiration under plastic greenhouse.

Theoretically, RIS values below one indicate deficit irrigation and values above one show excess irrigation.

14.7.2.2.2.1 Calculating Reference Evapotranspiration and Crop Coefficient General principles and methodologies for calculation of reference crop evapotranspiration and crop coefficient have been given by Doorenboss and Pruitt (1977). The manual by Allen et al. (1998) describes some general values of the crop coefficient for scheduling irrigation in strawberries.

Since there are differences in cultivars, length of growing season or crop duration, weather conditions, soil properties, farming systems including intercultural operations and nutrient management practices, so water management strategy and irrigation scheduling in strawberry has a great variation. Moreover, major strawberry cultivation in the world is concentrated under protected cultivation in plastic/poly tunnels. Nevertheless, there is no standardized method for estimation of reference crop evapotranspiration and crop coefficient under poly tunnels for scheduling irrigation in strawberries. Most of the area under commercial strawberry cultivation in Europe and China are under energy-saving plastic tunnels, whereas in California and Florida, strawberries are cultivated mainly in the open field conditions.

Crop coefficients in strawberry for estimating water requirement and scheduling irrigation in strawberry as reported by various researchers show great variation, such as 0.45–0.7 during the summer production period (McNiesh et al., 1985), 0.8 in Florida when 60 cm-wide beds were nearly fully covered with plant canopy (Clark et al., 1996), whereas it was 0.7 in California when the maximum coverage of the crop was ~75% (Hanson and Bendixon, 2004) with crop evapotranspiration ranging from 310–396 mm. Jackson (1982) also reported crop coefficient values of 0.7 for final crop coverage of 75%. But Grattan et al. (1998) recorded crop evapotranspiration and estimated crop coefficient as nearly equal to the canopy cover when there was 100% coverage on the strawberry bed. And finally, Allen et al. (1998) have reported crop coefficient values for strawberry at 0.15 in the early season, 0.85 in the midseason and 0.7 in the late season.

More recently, Gavilan et al. (2014) have attempted to estimate the irrigation requirements of two commercial strawberry cultivars grown under poly tunnels in Huelva region of Spain. In this, the reference crop evapotranspiration (ET_0) was estimated under polytunnels using formula suggested by Fernandez et al. (2010) as detailed hereunder-

$$\text{If, DOY} < 220 \ ET_0 = \left(0.288 + 0.0019 \times \text{DOY}\right) \times R_{sin(i)}$$

$$\text{If, DOY} > 220 \ ET_0 = \left(1.339 - 0.0028 \times \text{DOY}\right) \times R_{sin(ii)}$$

where
   DOY = day of the year
   R_{sin} = daily solar radiation inside the polytunnels (mm/day)

Its accuracy has been reported to be similar to that of Makkink FAO24 and Penman-Monteith FAO56 Methods (Fernandez et al., 2010; Gavilan et al., 2015).

These equations are a practical way to determine reference crop evapotranspiration inside polytunnels.

However, these equations require greenhouse transmissivity (T) and outdoor solar radiation (R_{sout}) data ($R_{sin}=R_{sout}\times T$). The R_{sout} was estimated from data on forecast temperature supplied by the Spanish Meteorology National Agency using the Hargreaves equation (Hargreaves and Samani, 1982).

The transmissivity of polyethylene tunnels depends on the material used for coverage as well as the age and dirtiness of the polyethylene covers. The solar radiation is measured with help of pyranometers. Gavilan et al. (2015) have suggested an average value of 0.75 transmissivity to be suitable for average type of polyethylene tunnels.

The values of reference evapotranspiration under polytunnels are lower than outdoor values.

The design of the optimum irrigation pulse or interval

A well-designed drip irrigation system provides equal and desired volume of water to all the plants in a strawberry plantation with a higher degree of irrigation efficiency. Therefore, in a drip irrigation system, the selection of appropriate emitters/drippers is of utmost importance. Further, the network of irrigation layout should facilitate easy distribution of water over the field within the range of pressure provided from the service head. The service pressure should provide optimum pressure at the tail end for proper functioning of the drippers. The selection of the most suitable drippers is very essential for an efficient irrigation system as ultimately the drippers supply the right amount of water (flow and pulse duration) at the proper time to match the crop's water requirements (Martin et al., 2013)

- As far as possible, only pressure-compensated drippers should be used, especially in undulating terrain, so that even when the working pressure varies, the flow rates do not vary. Constant and regular flow with a right time of irrigation events improves the precision of the irrigation process and facilitates application of only the desired amount of water. Pressure-compensated drippers make management of the irrigation system easy for the managers as under these conditions there is no need either to check the flow of water nor the working pressure in the system and ultimately a well-designed system like this assures proper functioning of the irrigation system.
- As far as possible, anti-drain emitters should be selected, which ensures that the irrigation laterals are always full of water. Consequently, the irrigation system does not need a time-out for filling pipes (main, sub-main, sub-sub-mains and laterals) with water, especially in sandy soils where irrigation needs to be applied in short duration pulses. In addition, water filling the pipes and laterals does not allow the soil particles and other impurities to enter into the irrigation system, preventing the drippers/emitters from clogging.
- The use of anti-siphon drippers/emitters facilitates the prevention of accumulation of sediments and other materials, which cause clogging, which is the main reason for variation in irrigation water application uniformity and consequently irrigation efficiency (Burt et al., 1997).

In most cases, the irrigation process is accompanied with nutrient application, the practice commonly referred to as "fertigation". Under these conditions, when the pipes deliver a mixed solution of water and fertilizers and even sometimes pesticides, the risk of clogging of drippers/emitters or tapes is very high.

The emitter flow-pressure head equation can be calculated by following equation:

$$q = \alpha \times h^{x}$$

where
- q = flow rate (L/h),
- α = discharge coefficient (L/h/m)
- h = pressure head (m^x) and ^{x}=pressure head exponent

Water Management

The total gross irrigation depth can be calculated by following equation (Losada, 1995):

$$\frac{H_g}{H_r} = \frac{1 - C_d}{\text{IE}}$$

where

IE $=$ irrigation efficiency (ratio of H_n/H_g)
H_n $=$ the ratio of net irrigation requirement (mm)
H_g $=$ total gross applied depth (mm)
C_d $=$ deficit coefficient $= \dfrac{H_r - H_n}{H_r}$
H_r $=$ theoretical irrigation requirements

Here, the theoretical irrigation water requirement (H_r) is calculated as per Allen et al., 1998. IE is calculated as per the following equations:

$$\text{IE} = \frac{H_e - \left(\dfrac{1}{2}\right) \times \left[\left\{\left(H_e + \sqrt{3} \times \sigma_e\right)^2 - \left(H_r\right)^2\right\} / 2\sqrt{3} \times \sigma_e\right] + f \times H_r}{H_e}$$

where

f (fraction of the field without water deficit) $= \dfrac{H_e + \sqrt{3} \times \sigma_e - H_r}{2\sqrt{3} \times \sigma_e}$
H_r $=$ theoretical irrigation requirements

and

$$C_d = \frac{\left(1 - f\right) \times H_r - H_e + \left(1/2\right) \times \left\{\left(H_e + \sqrt{3} \times \sigma_e\right)^2 - H_r^2\right\}}{H_r}$$

Integration of technologies and designing comprehensive precision irrigation system

Although many tools and technologies that form the basis of a modern precision irrigation system are available, research and development are required before a truly precision irrigation system is tested and made available to end users in a particular geographical area (Smith et al., 2010). Accordingly, based on requisite data, in the third and the final step, a comprehensive irrigation system supported by latest automation technologies and computer software is integrated to achieve optimum system efficiency.

Finally, the irrigation system is designed on the basis of field, crop and weather data and needs to be supported by a locally installed weather station and soil moisture-sensing devices to provide real-time weather and soil moisture status. The real-time weather and soil moisture data help in making decisions on irrigation scheduling for maintaining the desired soil moisture in the root zone of plants at a commercial strawberry plantation.

In addition, a modern precision irrigation system should also have control devices such as a solenoid or electro-valves, pressure regulators, smart water meters with LED displays and irrigation/fertigation systems etc. that can be operated by a personal computer or an internet-enabled, hand-held small mobile cell device. A software-enabled and programmed computer device can provide daily/weekly information on optimum irrigation scheduling to commercial strawberry farm managers or small farmers. The drippers/emitters should be least sensitive to clogging, with important special features such as pressure-compensating, anti-drain or anti-siphon qualities. The pressure control and timing of irrigation should be considered in a holistic manner for increased water use efficiency.

The entire system can be controlled by well-defined software that finally provides the operation variable (irrigation time) for each day of the cropping season. It involves the installation of new electronic devices in a holistic and integrated manner that detects the current inefficiencies and designs and applies the best irrigation management strategies in the strawberry plantation.

For assessment of the sustainability of the strawberry production system in a given environment from the total irrigation water requirement point of view, the strawberry water footprint should also be considered. This indicator links the total water applied to the strawberry crop (SIWA, m^3/ha) and the fruit yield (kg/ha) as designated:

$$SWFA = \frac{SIWA}{yield}$$

where
SIWA = strawberry irrigation water applied
SWFA = strawberry water footprint applied

Thus, the SWFA is a measure of water use efficiency in the strawberry production process, wherein the smallest values indicate the best irrigation practices and represents the volume of water used per unit quantity of fruit produced.

14.8 DECISION SUPPORT SYSTEMS IN WATER MANAGEMENT

For a precision irrigation system, decision support systems with an irrigation planner, which calculates the water requirement daily, are very useful especially in large farm areas. These decision support systems consider the amount of water in the root zone of plants by taking into account the soil type, precipitation, irrigation, capillary water movement, percolation, root growth, evapotranspiration at different growth and developmental stages and the local weather parameters. Generally, it has to be calibrated daily but as the growth of the crop is governed by several factors, the theoretically assumed growth pattern is hardly achieved, which may lead to miscalculations in evapotranspiration and inaccurate irrigation scheduling.

The decision support systems (irrigation planners) calculate the soil moisture regime (water content) using a limited number of soil types but in actual cultivation practices, the soil types may differ from the one already set in the irrigation planner and under such circumstances miscalculations are logical to occur. Therefore, in most situations, the installation and use of tensiometers has been observed to be more satisfactory. Generally, a minimum of three tensiometers are necessary for decision support systems and further need to be checked regularly and maintained thoroughly.

In strawberry, fruit productivity under irrigation with a decision support system has been observed to increase at lower tensiometer values than the calibrated values (Evenhuis and Albas, 2002). Growing strawberries at field capacity (100 hPa tensiometer value) increased the productivity by one tonne per hectare over one irrigated at 200 hPa tensiometer value. Monitoring the soil moisture based on visual reading of tensiometers in a manually operated irrigation system is a time-consuming process and in areas where labour charges are higher, this adds to production costs. Under these circumstances, an automatic irrigation management system is of great help, not only in optimizing the amount of water used for irrigation but in reducing the labour cost and ultimately cost of production. Evaluation of manually operated (2.3 litre/m^2) and automatically controlled (1.3 litre/m^2) irrigation systems in strawberry cv. Clery with manually operated at 20 cbar applying irrigation two and three times per day (each 4.6 litre/m^2), respectively produced identical size fruits with at par fruit yield but there was saving of labour to the tune of 20 person hours per hectare, equivalent to US\$680/ha in addition to saving of water worth US\$1580/ha due to automation

Water Management 253

(Ancay et al., 2014). In all, it led to saving of 880 m³ of irrigation water per hectare per strawberry crop cycle, leading to reduced carbon footprint.

To minimize the loss of water due to percolation and to improve water use efficiency, it is always advisable to opt for frequent irrigation using a small amount of water rather than heavy irrigation at long intervals.

14.9 BIOTIC AND ABIOTIC STRESSES ASSOCIATED WITH WATER MANAGEMENT PRACTICES

14.9.1 DROUGHT

Soil moisture deficit leads to significant reduction in CO_2 assimilation rate (A), stomatal conductance (g_s) and transpiration rate (T), whereas instantaneous water use efficiency (A/T) increases with a positive correlation between gas exchange variables and stomatal conductance (g_s) which has strong positive correlation with leaf water content (Martinez-Ferri et al., 2016). In different strawberry cultivars, the yield and plant performance have been observed to be negatively affected under severe water shortage conditions (Serrano et al., 1992; Save et al., 1993; Liu et al., 2007; Grant et al., 2010). The decrease in plant growth and yield is closely related to reduced photosynthetic rates mainly due to a state of disturbed gas exchange, that is, diffusional but not from the non-diffusional limitations to CO_2 assimilation (Flexas and Medrano, 2002; Flexas et al., 2013). The water deficit may manifest in the reduction in number and or size of fruits (Liu et al., 2007).

In horticultural crops, deficit irrigation strategies with application of <100% crop evapotranspiration has been advocated (Grant et al., 2004; Liu et al., 2007; Jensen et al., 2009). Strawberry is very sensitive to water deficits (Voth, 1967; Hanson and Bendixon, 2004) and its sensitivity to water stress at different stages of growth and development strongly influences plant growth and yield (Kruger et al., 1999). Plant water status influences photosynthesis and consequently growth by regulating the metabolic and physiological processes, resulting in increased stomatal resistance (Iuchi, 1993). Due to deficit water stress conditions (Hsiao and Acevedo, 1974), the major physiological process, "stomatal opening", is affected, which is a very important activity as the stomatal openings are the main ways to pass water controlled by the plants during gaseous exchange.

The greater susceptibility of older leaves to water stress in the field during summer months is related to gradients in leaf osmotic potential within the plant and to an age dependency in the ability of leaves to adjust osmotically when challenged by periodic water deficits. In a trial with strawberry cv. Korona grown in sand and subject to 75%, 50% or 25% water stress (Gehrmann, 1985) as compared to daily water consumption (100% of PE), reduction in vegetative growth, fruit yield and quality but acceleration of fruit maturity was recorded due to imposition of irrigation stress.

Contrary to this, deficit irrigation in strawberry has been observed to improve the concentration of anthocyanins and antioxidants in fruits (Terry et al., 2007a) and such effects are cultivar dependent (Gine-Bordonaba and Terry, 2010), which many times leads to significant reduction in fruit size and yield (Blatt, 1984; Serrano et al., 1992; Kruger et al., 1999; Liu et al., 2007; Terry et al., 2007a).

Reduced preharvest irrigation has been found partially detrimental, while a small reduction in irrigation (75% pan evaporation replenishment; PER) had little or no effect but 50% or less of normal (100% PER) irrigation, although it did not reduce overall fruit yield but resulted in smaller fruits (Kirnak et al., 2003). Fruit yield, fruit size, leaf macronutrient composition and normal growth parameters were significantly reduced due to no irrigation and irrigation at less than 50% of normal (100% PER) even at two weeks before harvest. Individual fruit weight and water-soluble dry matter content in fruits significantly increased by irrigation at 50% or 25% PER (Kirnak et al., 2003).

The conventional deficit irrigation or partial root zone wetting with application of 60–65% of total crop evapotranspiration has been observed to cause significant reduction in berry yield due to low soil water potentials caused due to imposition of severe drought stress (Serrano et al., 1992; Save et al., 1993; Liu et al., 2007). Accordingly, the strategies of deficit irrigation (DI) with excessive water shortages may lead to significant economic losses for farmers and as such may not be advisable to growers as a technique for improving crop water use efficiency in strawberry. But the deficit irrigation strategy can advantageously be made use of to the extent it does not have any negative impact on yield and quality of strawberry fruits. With respect to their response to water stress, there are differences in strawberry cultivars at a vegetative growth and physiological level (Klamkowski and Treder, 2006 and Grant et al., 2010) and fruit quality (Bordonaba and Terry, 2010). Thus planting cultivars comparatively more tolerant to mild (low) soil moisture stress could be an important strategy to improve water use efficiency in commercial strawberry cultivation. The water relations and the relative tolerance of different strawberry cultivars to deficit soil moisture stress may vary from one region to the other and as such, a more efficient and/or relatively more tolerant strawberry cultivar could not only contribute to increased water productivity but also ascertain the environmental sustainability of strawberry cultivation in regions with short supply of irrigation water.

Water shortage of more than 30% of plant water requirements leads to significant losses of yield in most strawberry cultivars and as such deficit irrigation strategies below this threshold level (Martinez-Ferri et al., 2016) are not advisable in commercial strawberry cultivation. The Ψ range for which a compromise between optimal yield and water use efficiency is possible would be between -10 and -15 kPa, depending on the site. It is likely that an irrigation treatment of -10 Kpa would provide good results in most types of soils and under most climatic conditions (Letourneau et al., 2015).

Preharvest application of methyl jasmonate (0.01–0.1 mM) on strawberry plants brings in such metabolic changes in leaves (Gine-Bordonaba and Terry, 2016) by altering the stomatal opening which results in better transpiration control and enable it to withstand *in vitro* water stress (50 ml/day) in a better way (Wang, 1999) as compared to normal low water regime stressed plants (200 ml/day). This results due to its direct effects on restricted growth and hydraulic conductivity of roots (Staswick et al., 1992). In addition, the leaves in drought-stressed strawberry plants accumulate more carbohydrates and malic acid on the one hand while reduce respiration on the other which contributes to osmotic adjustment and enables the plants withstand deficit soil moisture stress conditions.

14.9.2 FLOODING

Similar to drought, the excess soil moisture conditions are also detrimental to plant growth and yield of strawberry (Voth, 1967). It encourages excessive vegetative growth and triggers production of stolon at the cost of flowering and fruiting. In fruiting plants, excess water increases the incidence of botrytis rot of fruits.

14.9.3 SALINITY

On the basis of percentage leaf damage and accumulation of salts in soil due to irrigation with saline water (ECi – 1.1, 1.6, 2.1, 2.5 and 2.9 ds/m), the Maraline cultivar has been observed to be the most tolerant to irrigation water salinity when compared with irrigation with river water (ECi – 0.5 ds/m). The Muir and Tudla cultivars were observed to be susceptible and showed severe salt injury by irrigation with saline water at 1.1 ds/m ECi. With increasing level of ECi of irrigation water, there was an increase in ECe in soil. In general, soil ECe was higher in Muir, followed by the Tundla cultivar (Kurunc and Cekc, 2005).

14.9.4 WEEDS

The combination of low rates of terbacil application (0.8–1.6 oz a.i./ha) at the three-leaf stage of strawberry followed by irrigation (10 cm) to remove the herbicide from foliage has been observed to be a safe option for growers to improve weed control and reduce costs involved in hand weeding in the planting year.

Water Management

14.9.5 PATHOGENS

Sometimes irrigation practices have been implicated in the transfer of pathogens from contaminated irrigation water to strawberry crops. Accordingly, outbreaks of foodborne illness have been attributed to these infected fruits, whereas consumer demand for fresh and minimally processed fruits has increased. Observation on the influence of irrigation methods (trickle irrigation with well or surface water and overhead irrigation with surface water) and water quality on the incidence of *Escherichia coli* on Ontario-grown strawberries in Canada during early, middle and late periods in the harvesting seasons of fruits revealed presence of *E. coli* in approximately half of the water samples tested (0.07–2.45 log cfu/100 ml) but fruits obtained from crops irrigated with surface water were free from *E. coli*. The fruits harvested from well or municipal water-irrigated fields had *E. coli* (1.17–2.64 log cfu/g). The revised guidelines for the year 2002 in terms of water quality permits maximum of 100 *E. coli*/100 ml and most of the water used for irrigation met the above standards. It was speculated that overhead irrigation with surface water does not seem to result in transfer of *E. coli* to strawberry fruits.

With irrigations at −10, −35 and −70 kPa, respectively, the number of irrigations were 188, 100 and 77.5, of which ~75% of the irrigations occurred at intervals of one day, from one–four days and from two–six days for −10, −35 and −70 kPa irrigation levels, respectively. The irrigation levels of −10 and −35 kPa and the clear plastic favoured total marketable fruit yield. The irrigation level at −70 kPa and the use of black plastic as soil mulch provided congenial conditions for greater incidence of soilborne diseases caused mainly by *Rhizoctonia* spp., *Colletotrichum fragariae*, *Verticillium dahliae*, *Fusarium* spp., *Pythium* spp. and *Phytophthora cactorum* late in the harvest season (Pires et al., 2007).

14.10 WATER SAVING PRACTICES

14.10.1 MULCHING

Webb (1973) suggested that if evaporative losses from ripe berries could be reduced, such as by clothing, tunnelling, wind breaks or by plant breeding for less permeable cell walls, increased yield might be expected. Mulching had great influence on the soil temperature (Gupta and Acharya, 1993). Transparent polyethylene raises the maximum soil temperature whereas, pine needles and grass mulches lowered it. All mulches are effective in maintaining higher soil moisture regime (Gupta and Acharya, 1993). Black polyethylene not only increases the total soil water content up to 60 cm soil depth but it increases nutrient uptake and enhances water use efficiency (121%) compared with unmulched plots (62%) combined with increased yield by 56%, bigger fruits and higher TSS.

Mulching in strawberry has been recognized as beneficial in minimizing low temperature (freeze or frost) injury during winter and suppressing early vegetative growth during spring season and thereby frost injury to fruit buds. But indirectly, mulching helps in reducing soil erosion and minimizing loss of soil moisture due to evaporation. In the beginning, however, plants should be exposed to a few frost spells (10–15) to make plants harden and grow slowly.

14.10.1.1 Organic Mulches

Mulching could be done by organic materials such as straw, grasses or similar other biodegradable farm wastes. Paddy straw has been observed to be the best option for use as mulch for strawberry in warmer location of northern India (Sharma and Sharma, 2003). However, the economic feasibility of using paddy straw needs to be considered in it being cattle feed across the length and breadth of the country. Organic mulches should also be free from weed or other undesirable seeds, which may help in increasing weed infestation in strawberry plantations.

14.10.1.2 Plastic Mulches

The utility of mulches in strawberry cultivation is governed by soil and climatic factors, cultivation conditions and cultivars (Passos, 1997). Sharma and Sharma (2003) observed better growth of strawberry cultivars such as Chandler, Douglas, Etna, Fern and Sweet Charlie due to mulching as compared with non-mulching. Mulching with black polyethylene sheets gives good weed control, reduces cost of cultivation on labourer by 25–30% and advances early cropping of cvs. Senga Sengana and Festival'naya by 4–8%, increases fruit yield by 25–34% with improved fruit quality and reduced disease incidence (Nikito-Chkina and Gusev, 1984). Row covers in combination with black polyethylene mulch (Sharma et al., 2008) significantly influenced plant physiology, growth and berry yield per plant (excluding albino fruits).

Black polyethylene sheet (50 μ) as mulch not only improved the fruit weight (12.77 g/fruit) but also fruit yield (2.758 kg/m^2) and proportion of early (0.759 kg/m2) and grade "A" (27.9%) fruits (Singh and Asrey, 2005). In general, mulching with black polyethylene sheets accelerated the beginning of flowering and ripening by about one–seven days and increased fruit yield by about 20% at the expense of larger number of peduncles and berries in northwest Russia (Plekhanova and Petrova, 2002); however, strong spring frost may damage flowers in early cultivars of strawberry, so care and meticulous planning need to be exercised in mulching with black polyethylene sheets in strawberry in extreme temperate and frost prone areas. Black (30 μm thick) or clear (50 μm in the greenhouse and 150 μm in the open field) low-density polyethylene film (LDPF) are used as soil mulch. Black plastic is the more frequently used soil mulch in the state of São Paulo, and clear plastic is being successfully used for several cultivars and locations in the world (Voth and Bringhurst, 1990).

However, there is a limitation to using polyethylene sheets for mulch as these need to be removed from the field and disposed of at the end of each fruiting season since it does not decompose in the soil. Waste paper mulches treated with appropriate fungicides and with or without a thin polyethylene layer can also be conveniently used as these have mulch properties but disintegrate easily on incorporation into the soil. Biodegradable paper mulches with a polyethylene coating on both sides or with a single polyethylene coating lasts satisfactorily for six or seven months depending upon weather conditions.

14.10.1.3 Ideal Mulch Materials

Ideal organic materials to be used for mulching must be loose and light in weight but sufficiently heavy enough so that they are not blown away by heavy winds. In India, Singh et al. (2007) reported that the strawberry cv. Chandler grown in the semi-arid conditions of northern plains and mulched with black polyethylene sheets showed significantly better growth, flowering (80.2 days), and fruited early (29.2 days), and produced larger fruits (12.6 g) and improved fruit yield (172.4 g/plant) but with slightly higher incidence of albinism (20.1%) and lower incidence of botrytis rot (7.3%) than those mulched either by clear polyethylene sheets or paddy straw. Black polyethylene mulch favoured better plant growth but enhanced the incidence of albinism in subtropical climate of north India when compared with paddy straw (Sharma and Sharma, 2003). In another trial in subtropical climate of New Delhi, India with 10 strawberry cultivars; improved vegetative growth in terms of crown height, crown spread, number of leaves and leaf area was recorded due to mulching when compared with no mulching. The vegetative growth improved significantly due to mulching with black polyethylene sheets and the improved plant growth was ascribed to better conservation of soil moisture, regulation in temperature and suppression of weeds (Sharma et al., 2004; Badiyala and Agrawal, 1981) but in general, the albinism incidence in fruits (Sharma et al., 2004) also increased due to black polyethylene mulching (38.6%) whereas, the albinism decreased significantly due to application of paddy straw mulch (22.5%) as compared even with unmulched control (36.6%) or white polyethylene mulch (32.7%).

14.10.2 Anti-Transpirant

The foliar spray application of kaolin (25 lb/acre) clay after six or eight days of sprinkler irrigation did not reduce plant establishment, canopy diameter or leaf greenness but facilitated water saving up to 20% and 40% of the total volume in strawberry. In addition, early and total marketable fruit

Water Management 257

yield were significantly influenced due to application of kaolin. The maximum strawberry fruit yield was obtained either due to six days of sprinkler irrigation followed by kaolin spray on seventh day (15.6 t/acre) or eight days of sprinkler irrigation followed by kaolin spray on ninth day (15.7 t/acre) and was equal to 10 days of sprinkler irrigation (Santos et al., 2012). It is presumed that the foliar spray of kaolin reduces leaf and crown temperature by reflecting solar radiation and decreases the evaporation around strawberry crowns and leaf transpiration, thus allows faster formation of new leaves and roots.

14.11 INDICATORS OF EFFICIENT IRRIGATION SYSTEM

Although the total fruit yield of strawberry may not differ significantly in different irrigation levels but increased percentage of Class "A" fruits, extra early and early crop maturity with standard fruit size and appropriate quality could be taken into consideration for making decisions on efficient irrigation system.

14.12 METHODS FOR HIGH CROP WATER USE EFFICIENCY

Higher crop water use efficiency can be achieved by:

1. enhancing the uptake efficiency of available water through the plant system, viz., by reducing evaporation and drainage losses
2. improving crop transpiration efficiency by acquiring more carbon per unit volume of water transpired, and
3. increasing the partitioning of the acquired biomass into the harvested product (fruit), viz., higher harvest index

Interestingly, all the three processes mentioned are not independent and their relative importance varies depending on water availability during the crop cycle (Condon et al., 2004). Martinez-Ferri et al. (2016) have concluded that, in horticultural crops such as strawberry, greater leaf-level water use efficiency is not always an agronomical advantage in terms of water use.

14.13 WATER QUALITY

With perceived threats of climate change and global warming causing increase in temperature and radiation leading to enhanced evapotranspiration of crop plants, the demand for irrigation water is likely to increase in future and therefore, an increase in quantity of water required for crop production can be easily expected by 2050 even in areas where otherwise sufficient water is presently available for irrigation (Fuehrer and Jasper, 2009).

The scarcity of water for crop production increases sectoral competition for its use for various purposes (Mesa-Jurado et al., 2012; Milano et al., 2012) and in these situations wastewater with proper treatment could be a satisfactory alternative to good quality irrigation water (Hamilton et al., 2007). As it is a rich source of several essential plant nutrients and can therefore judiciously be utilized for correcting nutrient deficiencies (Angelakis and Durham, 2008; Bahri, 1999; Bixio et al., 2006; Chu et al., 2004; Khurana and Singh, 2012) in many crop plants including strawberry. However, wastewater needs to be cautiously used with constant monitoring as these are laden with chemicals, biological contaminants and pollutants including heavy metals with very serious and negative environmental and health impacts (Fatta-Kassinos et al., 2011; Hamilton et al., 2006a; Muchuweti et al., 2006; Singh et al., 2004). Usage of wastewater along with sewage sludge may lead to contamination of soils with heavy metals and pollutants, which may find entry into food chain with great health consequences (Li et al., 2009; Rajaganapathy et al., 2011) and crops grown in such environments absorb and accumulate heavy metals (Mapanda et al., 2005; Muchuweti et al., 2006; Khan et al., 2008; Singh et al., 2010; Hashem et al., 2003) in the edible parts above the maximum permissible limits and may evoke public concerns of potential negative impacts on human health (Adam and Kizek, 2012; Skup-Jablonska et al., 2012). Although, the crop produced from advanced tertiary-treated wastewater have been found safe for consumption from consideration of heavy metal points of view (Christou et al., 2014b; Al-Lahham et al., 2007) but instead of using advanced treated wastewater, the reuse of untreated or insufficiently treated water accompanied with unscientific crop production practices may result into contamination of crop produce loaded with microbial and other anthropogenic pathogens, such as *Escherichia coli*, *Salmonella* spp., *Listeria* spp. or *Giardia* spp. (Peterson et al., 2001; Palese et al., 2009; Bernstein, 2011; Cirelli et al., 2012). Fortunately, the microbial risk assessment techniques developed to estimate the risk of using wastewater for restricted or unrestricted irrigation of crops and its effects on human health can appropriately be made use of in facilitating the establishment of acceptable limits of microbial contaminants in treated wastewater/effluents (Hamilton et al., 2006b; Mara et al., 2007; Mok et al., 2014). Although, irrigation with wastewater influences the yield and quality of crop plants as, depending on sewage origin and treatment technologies applied, it contains nutrients and heavy metals in considerable amounts. For example, irrigating tomato with advance-treated wastewater increased the size and weight, as well as decreased the firmness and weight loss of fruits (Al-Lahham et al., 2003), while in another study there was no effect on fruit weight, diameter and number of fruits in tomato due to irrigation with tertiary-treated wastewater (Christou et al., 2014). Although literature on using wastewater for irrigating strawberries is rare, low-quality industrial wastewater has been reported (Naaz and Pandey, 2010) to decrease the dry matter yield and photosynthetic pigments of lettuce plants but increase the activity of antioxidant enzymes and the contents of carotenoids, the antioxidant molecules.

To safeguard the consumers from possible health hazards by excreta-related pathogens and toxic elements due to reuse of wastewater in agriculture, well-established guidelines and criteria have been established (WHO, 2006; Brissaud, 2008; U.S. Environmental Protection Agency, 2012). The protection measures include wastewater treatment procedures (tertiary treatment and disinfection), wastewater application techniques (drip or subsurface irrigation) etc.

Irrigation in food crops with wastewater is, as such, prohibited in many countries while it is permitted with certain conditions and restrictions in some countries (U.S. Environmental Protection Agency, 2012), for example, reuse of wastewater for agricultural purposes is a well-established practice in Cyprus due to acute water shortages. It is estimated that currently 22 million cubic metres of tertiary-treated wastewater is being reused for irrigation and recharge of underground water bodies through a well-regulated comprehensive guideline and a code of good agricultural practices (Christou et al., 2014a) but crops that are consumed fresh, such as leafy vegetables, bulbs and tuber crops, should not be irrigated with wastewater but within the frames of these guidelines strawberry crops are allowed to be drip or surface irrigated as fruits are not allowed to directly come in contact with wastewater.

The results of an experiment (Christou et al., 2016) on strawberry irrigation with wastewater revealed that the wastewater did not significantly affect the fruits' marketability, taste and antioxidant capacity, or the heavy metal content, in comparison with controlled irrigation, irrespective of the irrigation techniques (drip, sprinkler or drip under plastic mulch) as the fruits heavy metal content was observed to be below the permissible levels set for fruit safety, whereas no microbial contamination (total coliform, *E. coli*, *Salmonella* spp., *Listeria* spp.) was observed in any irrigation water treatments. These results further highlighted the potential for the reuse of the advanced tertiary-treated effluent of good quality as a valid alternative for the irrigation of strawberry crops, even with sprinklers. However, further long-term studies and elaborate observations are needed for such a practice to be regulated. As such, a holistic approach needs to be followed while deciding on reuse of wastewater for irrigating strawberry crops (Table 14.4).

Water Management

TABLE 14.4

Fruit Productivity, Marketability and Taste Attributes of Strawberry Fruits as Influenced by Irrigation of Strawberry Plants with Treated Wastewater Effluent and Potable Water Applied through Different Techniques (Drip, Sprinkler or Drip Under Plastic Mulch)

Variable	Water Source and Irrigation Techniques*					
	PW-DR	WW-DR	PW=SPR	WW-SPR	PW-DRPM	WW-DRPM
Number of harvested fruits per plant	6.3	5.9	5.8	5.3	4.7	7.7
Marketable fruits' weight (g)	17.3±0.7a	15.1±0.5ab	17.4±0.6a	15.3±0.6ab	17.70.9a	13.50.5b
Unmarketable (deformed) fruits' weight (g)	12.2±1.0a	11.0±1.1a	11.8±1.1a	8.3±0.8b	10.2±1.0ab	9.3±0.06
Per cent marketable fruits	77.5±5.7	79.7±9.1	78.5±5.5	87.4±4.9	71.4±4.0	76.6±8.1
Per cent dry matter in fruits	7.16±0.01	7.19±0.12	7.09±0.05	6.92±0.41	7.68±0.27	7.24±0.09
pH of fruit juice	3.59±0.04	3.56±0.05	3.53±0.04	3.59±0.05	3.54±0.03	3.57±0.04
Total soluble solids in fruits (°Brix)	6.60±0.14	6.48±0.25	6.31±0.20	6.30±0.47	7.21±0.34	7.10±0.19
Titratable acidity (TA) (% citric acid)	0.88±0.04	0.85±0.04	0.83±0.06	0.80±0.04	0.90±0.05	0.90±0.05
TSS/acidity Ratio	7.50±0.21	7.63±0.31	7.60±0.59	7.87±0.18	8.01±0.44	7.88±0.93

Source: (Christou et al., 2016).

* Where, PW = Potable water, WW = Wastewater, DR = Drip, SPR = Sprinkler, DRPM = Drip under plastic mulch.

SUMMARY

Strawberries are water-intensive crops but are in great demand both for the fresh market and in the food industry. Despite its high cost and sales rate in the market, strawberry is a very profitable crop. For the long-term agronomic sustainability of strawberry cultivation system, it would be appropriate to have a cultivar with greater water use efficiency to minimize the gap between the yield potential and the actual yield in conditions of limited water availability (Condon et al., 2004 and Cattivelli et al., 2008). Thus strawberry cultivars with reduced crop area but competitive yield such as Fortuna, Benicia and Splendor, having comparatively less total transpiration with reduced water consumption, can ideally be chosen for commercial cultivation in conditions of low water availability.

With the objective of improving irrigation water use efficiency along with sustainability of water resources in a strawberry production system, the following two points should be considered in a holistic manner:

1. The chosen cultivars should have more water use efficient leaves (i.e. elevated instantaneous water use efficiency; A/T), which facilitate a high carbon assimilation rate and therefore, high biomass production and yield potential with low water consumption.
2. The cultivars chosen should have elevated HI, which facilitates increased partitioning of accumulated biomass towards reproduction at the expense of reduced leaf area.

The planting of varieties with competitive yields but reduced leaf area would decrease the total transpiration and thus can be considered as a method for minimizing irrigation needs. In such varieties, the reduced leaf photosynthetic area would be compensated by photosynthetically more efficient leaves.

For optimum benefits, irrigation should be a precision activity (Smith et al., 2010), which involves:

1. the accurate estimation of the crop water requirements
2. the precise application of the required amount of water at the right time using hydraulic elements with high volumetric efficiencies, which allows spatially uniform applications

The consumptive water use in crops depends upon several parameters such as crop type and variety, weather and soil parameters. A wide range of consumptive water use varying from 300 mm (Trout and Gartung, 2004) to 797 mm (Strand, 2008) have been reported in strawberry. Drip irrigation at 100% of crop ET (1.0 "V" volume of water) has been found to give significantly higher fruit yield (57.07 q/ha) when compared with surface irrigation (47.17 q/ha). The fruit yield improves further (78.63 q/ha) due to the increased efficiency of drip irrigation when combined with black polyethylene mulch. Mulching with grasses increased the fruit size, sugar content, TSS/acidity ratio and anthocyanin content of fruits when compared with black polyethylene mulch only.

A wide variation in crop yield of strawberries have been reported ranging from 179 g/plant (Rolbiecki et al., 2001) to 1.704 kg/plant (Larson and Shaw, 1996). Similarly, great variations in terms of land productivity ranging from 6 t/ha (Stevens et al., 2011) to as high as 83 t/ha (Serrano et al., 1992) have been reported. Even in terms of water productivity, there is wide variation ranging from 4 kg/m^3 (Kirschbaum et al., 2003) to as much as 14.67 kg/m^3 (Serrano et al., 1992) of water applied. Interestingly, great variations in amount of water applied have also been reported, ranging from 4071 m^3/ha to 15, 2014 m^3/ha (Garcia Morillo et al., 2015a). Strawberry cultivation is a highly profitable horticultural venture if all the water management principles discussed in this paper are followed.

TERMS USED IN THIS PAPER

1. Harvest Index, HI = total fruit yield per gram of vegetative biomass.
2. Transpiration efficiency, TE$_v$ = vegetative dry mass per liter of water transpired.

Water Management

3. Crop water use efficiency, WUE_c = total yield per liter of transpired water.
4. Yield efficiency, YE = a non-destructive estimation of the relative amount of biomass partitioned into reproductive and vegetative parts of the plants (i.e. HI).
5. Crop coefficient = It is a constant value and is a function of canopy coverage and stage of crop. It also denotes percentage crop coverage in relation to total area.

APPENDIX: CONVERSION UNITS FROM U.S. TO SI AND *VICE VERSA*

Sl. No.	To Convert U.S. to SI (multiply by)	U.S. Unit	SI Unit	To Convert SI to U.S. (multiply by)
1	0.4047	acre (s)	ha	2.4711
2	254.0000	acre-inch/acre	$m^3\,ha^{-1}$	0.0039
3	0.3048	ft	m	3.2808
4	3.7854	gal	L	0.2642
5	0.1242	gal/100ft	$L.m^{-1}$	8.0520
6	9.3540	gal/acre	$L.ha^{-1}$	0.1069
7	2.54	inch(es)	cm	0.3937
8	1.1209	lb/acre	$kg.ha^{-1}$	0.8922
9	1	mmho/cm	$dS.m^{-1}$	1
10	28.3495	oz	g	0.0353
11	2.2417	ton/acre	$Mg.ha^{-1}$	0.4461
12	°F–32÷1.8	°F	°C	1.8×°C +32

REFERENCES

Adam, V. and Kizek, R. 2012. Metal ions in cause, progression, treatment and diagnosis of genetic disorders, metabolic diseases and cancer. *Current Drug and Metabolism*, 13: 236.

Adhikari, D.D., Goorahoo, D, Cassel, F. and Zodolske, D. 2008. Smart irrigation as a method of Freeze Prevention: a proposed model. In: 2008 ASABE Annual International Meeting, Paper No. 085189, Providence, Rhode Island.

Albregts, E.C. and Howard, C.M. 1985. Effect of intermittent irrigation on establishment of strawberry. *Proceedings* of the *Soil and Crop Science Society of Florida*, 44: 197–199.

Al-Lahham, O., El Assi, N.M. and Fayyad, M. 2007. Translocation of heavy metals to tomato (*Solanum lycopersicon* L.) fruit irrigated with treated wastewater. *Scientia Horticulturae*, 113: 250–254.

Allen, R.G., Pereira, R.S., Raes, D. and Smith, M. 1998. Crop evapotranspiration: guidelines for computing crop water requirements. In: FAO Irrigation and Drainage, Paper No. 56, FAO, Rome, Italy.

Ancay, A., Vincent, M. and Barofio, C.A. 2014. Comparison of two irrigation systems in strawberry. Proceedings of 7th International Strawberry Symposium (Zhang, Y. and Maas, J. eds.) *Acta Horticulturae*, 1049: 529–533.

Angelakis, A.M. and Durham, B. 2008. Water recycling and reuse in EUREAU countries: trends and challenges. *Desalination*, 218: 3–12.

Badiyala, S.D. and Agrawal, G.C. 1981. A note on the effect of mulching on strawberry production. *Indian Journal of Agricultural Research*, 51: 832–834.

Bahri, A. 1999. Agricultural reuse of wastewater and global water management. *Water Science & Technology*, 40: 339–346.

Bergeron, D. 2010. Regie de Iirrigation goutte a goutte dans la production de fraises a jour neuters au Qubec. Fac. Des Sci. l'Agriculture l'Alimentation, department des Sols Genie Agroaliment. Universite Laval, Quebec, Canada.

Bernstein, N. 2011. Potential for contamination of crops by microbial human pathogen introduced into the soil by irrigation with treated effluent. *Israel Journal of Plant Sciences*, 59: 115–123.

Bish, E.B., Cantliffe, D.J., Hochmuth, G.J. and Chandler, C.K. 1997. Development of containerized strawberry transplants for Florida's winter production system. *Acta Horticulturae*, 439: 461–468.

Bixio, D., Thoeye, C., De Koning, J., Joksimovic, D., Savic, D., Wintgens, T. and Melin, T. 2006. Wastewater reuse in Europe. *Desalination*, 187: 89–101.

Blanke, M.M. and Cooke, D.T. 2004. Effects of flooding and drought on stomatal activity, transpiration, photosynthesis, water channel activity in strawberry stolons and leaves. *Plant Growth Regulation*, 42: 153–160.

Blasse, W., Grittner, I., Bischoff, S. and Eisemann, A. 1984. Strawberry irrigation in the Havelland fruit growing area. Horticulture (Berlin), 31: 183–185.

Blatt, C.R. 1984. Irrigation, mulch and double row planting related to fruit size and yield of "Bounty" strawberry. *HortScience*, 19: 826–827.

Bordonaba, J.G. and Terry, L.A. 2010. Manipulating the taste related composition of strawberry fruits (*Fragaria* × *ananassa*) from different cultivars using deficit irrigation. *Food Chemistry*, 122(4): 1020–1026.

Bordonaba, J.G. and Terry, L.A. 2016. Effect of deficit irrigation and methyl jasmonate application on the composition of strawberry (*Fragaria* × *ananasa*) fruit and leaves. *Scientia Horticulturae*, 199: 63–70.

Brissaud, F. 2008. Critaria for water recycling and reuse in the Mediterranean countries. *Desalination*, 218: 24–33.

Burt, C.M., Clemens, A.J., Strelkoff, T.S., Solomon, K.H., Bliesner, R.D., Howell, T.A. and Eisenhauer, D.E. 1997. Irrigation performance measures: efficiency and uniformity. *Journal of the Irrigation and Drainage Division, ASCE*, 123(6): 423–442.

Capraro, F., Patino, D., Tosetti, S. and Achugurensky, C. 2008a. Neutral Network-based precision agriculture. In: IEEE International Conference on Networking, Sensing and Control. April 6–8, Sanya, China: 357–362.

Capraro, F., Schugurensky, C., Vita, F., Tosetti, F., Lage, A. and Patino, D. 2008b. Intelligent irrigation in grapevines: a way to obtain to obtain different wine characteristics. In: Proceedings of the 17th World Congress of the International Federation of Automatic Control, Seoul, Korea.

Cattivelli, L., Rizza, F., Badeck, F.W., Mazzucotelli, E., Mastrngelo, E.M., Francia, E., Mare, C., Tondelli, A. and Stanca, A.M. 2008. Drought tolerance improvement in crop plants: an integrated view from breeding to genomics. *Field Crop Research*, 105: 1–14.

Childers, N.F., Morris, J.R. and Sibbett, G.S. 1995. *Modern Fruit Science Orchards and Small Fruits Cultivars*, Horticulture Publication, Gainesville, Florida: p. 22.

Christou, A., Maratheftis, G., Elia, M., Hapeshi, E., Michael, C. and Fatta-Kasinos, D. 2016. Effects of wastewaters applied with discrete irrigation techniques on strawberry plants' productivity and the safety, quality characteristics and antioxidant capacity of fruits. *Agricultural Water Management*, 173: 48–54.

Christou, A., Eliadou, E., Michael, C., Hapeshi, E. and Fatta-Kasinos, D. 2014a. Assessment of long-term wastewater irrigation impacts on the soil geochemeical properties and the bioaccumulation of heavy metals into agricultural products. *Environmental Monitoring and Assessment*, 186: 4857–4870.

Christou, A., Maratheftis, G., Eliadou, E., Michael, E., Hapeshi, E. and Fatta-Kassinos, D. 2014b. Impact assessment of the reuse of two discrete treated wastewaters for the irrigation of tomato crop on the soil geochemical properties: fruit safety and the crop productivity. *Agriculture, Ecosystem and Environment*, 192: 105–114.

Chu, J., Chen, J., Wang, C. and Fu, P. 2004. Wastewater reuse potential analysis: implications for China's water resources management. *Water Research*, 38: 2746–2756.

Cirelli, G.L., Consloi, S., Licciardello, F., Aiello, R., Giuffrida, F. and Leonardi, C. 2012. Treated municipal waste water reuse in vegetable production. *Agricultural Water Management*, 104: 163–170.

Clark, G.A., Albregts, E.E., Stanley, C.D., Smajstrla, A.G. and Zazueta, F.S. 1996. Water requirements and crop coefficients of drip irrigated strawberry plants. *Transactions, ASAE*, 39: 905–913.

Coates, R., Delwiche, M. and Brown, P. 2004. Design of a system for individual microsprinkler control. *Transactions, ASABE*, 49(6): 1963–1970.

Condon, A.G., Richards, R.A., Rebetzke, G.J., Farquhar, G.D. 2004. Breeding for high water-use efficiency. *Journal of Experimental Botany*, 55: 2447–2460.

Crespo, P., Gine-Bordonaba, J., Terry, L.A. and Carlen, C. 2010. Characterization of major taste and health related compounds of four strawberry genotypes grown at different Swiss production sites. *Food Chemistry*, 122: 16–24.

Doorenboss, J. and Pruitt, W.O. 1977. *Guidelines for Predicting Crop Water Requirements*. FAO Irrig. and Drain. Paper 24. FAO of The United Nations, Rome, Italy.

Dukes, M.D. and Scollberg, J.M. 2004. Automated subsurface drip irrigation based on soil moisture. In: ASAE, Paper No. 052188.

Espejo, A.J., Vanderlinden, K., Infante, M.J., Muriel, J.L. Septiembre, 2009. Empleo de sensores de humedad para caracterizar la dinámica de flujo de agua de riego en fresa. In: Vida RURAL. 15.

Water Management 263

Evenhuis, A. and Albas, J. 2002. Irrigation of strawberries by use of decision support systems. *Acta Horticulturae*, 567(2): 475–478.

Evett, S.R., Peters, R.T. and Howell, T.A. 2006. Controlling water use efficiency with irrigation automation: cases from drip and center pivot irrigation of corn and soybean. In: Southern Conservation System Conference, Amarillo, Texas.

Farquhar, G.D., Ehleringer, J.R. and Hubick, K.T. 1989. Carbon isotope discrimination and photosynthesis. *Annual Review of Plant Physiology and Plant Biology*, 40: 503–537.

Fatta-Kassinos, D., Kalavrouziotis, I.K., Koukoulakis, P.H. and Vasquez, M.I. 2011. The risks associated with wastewater reuse and xenobiotics in the agroecological environment. *Science* of the *Total Environment*, 409: 3555–3563.

Fernandez, M.D., Bonachela, S., Orgaz, F., Thompson, R., Lopez, J.C., Granados, M.R., Gallardo, M. and Fereres, E. 2010. Measurements and estimation of plastic green-house reference evapotranspiration in a Mediterranean climate. *Irrigation Science*, 28(6): 497–509.

Fernandes Jr, F., Furlani, P.R., Ribeiro, I.J.A. and Carvalho, C.R.L. 2002. Producao de frutos e estolhos do morangueiro em diferentes sistemas de cultivo em ambiente protegido. *Bragantia*, 61: 25–34.

Flexas, J. and Medrano, H. 2002. Drought inhibition of photosynthesis in C_3 plants: stomatal and non-stomatal limitations revisited. *Annals of Botany*, 89: 183–189.

Flexas, J., Ninemets, U. and Galle, A. 2013. Diffusional conductances to CO_2 as a target for increasing photosynthesis and photosynthetic water use efficiency. *Photosynthesis Research*, 117: 45–59.

Fuehrer, J. and Jasper, K. 2009. Bewasserungsbedurftigkeit in der Schweiz. Schlussbericht, Forschungsanstalt Agroscope Reckenholz-Tanikon (ART), 74 S.

Garcia Morillo, J.G., Martin, M., Camacho, E., Rodriguez Diaz, J.A. and Montesino, P. 2015. Toward precision irrigation for intensive strawberry cultivation. *Agricultural Water Management*, 151: 43–51.

Garcia Morillo, J., Montensinos, P., Rodriguez Diaz, J.A., Camacho, E. and Hess, T. 2012. Hacia la sostenibilidad del cultivo de la fresa: demanda real de riego y posibilidades de mejora. In: Actas del XXX Congreso nacional de riegos, D-1: 149–150.

Gavilan, P., Lozano, D. and Ruiz, N. 2014. El rego de la fresa en el entorno de Donana. Evapotranspiration, coefficient de cultivo y efficiencia del riego. In: XXXII Congreso Nacional de Riegos, Madrid 10 al 12 junio de 2014 (in Spanish).

Gavilan, P., Ruiz, N. and Lozano, D. 2015. Daily forecasting of reference and strawberry crop evapotranspiration in greenhouse in a Mediterranean climate based on solar radiation estimates. *Agricultural Water Management*, 159: 307–317.

Gehrmann, H. 1985. Growth, yield and fruit quality of strawberries as affected by water supply, 1st International Symposium on Water Relations in Fruit Crops, Pisa. *Acta Horticulturae*, 171: 463–469.

Gine-Bordonaba, J. and Terry, L.A. 2010. Manipulating the taste related composition of strawberry fruits (*Fragaria* × *ananassa*) from different cultivars using deficit irrigation. *Food Chemistry*, 122: 1020–1026.

Gine-Bordonaba, J. and Terry, L.A. 2011a. Strawberry. In: Terry, L.A. (Ed.), *Health-Promoting Properties of Fruit and Vegetables*. CABI, Wallingford, Oxfordshire, UK: pp. 291–320.

Gonalez Perea, R., Camacho, E., Mpntesinos, P. and Rodriguez Diaz, J.A. 2014. Critical points: interactions between on-farm irrigation systems and water distribution network. *Irrigation Science*, http.//dx.doi.org/10.1007/s00271-014-0428-2.

Glen, D.M. and Takeda, F. 1989. Guttation as a technique to evaluate the water status of strawberry. *HortScience*, 24: 599–601.

Grant, O.M., Johnson, A.W., Davies, M.J., James, C.M. and Simpson, D.W. 2010. Physiological and morphological diversity of cultivated strawberry (*Fragaria* × *ananassa*) in response to water deficit. *Environmental and Experimental Botany*, 68: 264–272.

Grant, G.M., Stoll, M. and Jones, H.C. 2004. Partial root zone drying does not affect fruit yields of raspberries. *Journal of Horticultural Science* and *Biotechnology*, 79: 125–130.

Grattan, S.R., Bowers, W., Dong, A., Snyder, R.L., Carrol, J.J. and George, W. 1998. New crop coefficients estimate water use of vegetables row crops. *California Agriculture*, 52: 16–21.

Guillaaume, L., Caron, J., Anderson, L. and Cormier, J. 2015. Matric-potential based irrigation management of field grown strawberry: effects on yield and water use efficiency. *Agricultural Water Management*, 161: 102–113.

Guimera, J., Marpha, O., Candela, L. and Serrano, L. 1995. Nitrate leaching and strawberry production under drip irrigation management. *Agriculture, Ecosystem and Environment*, 56: 121–135.

Gupta, R. and Acharya, C.L. 1993. Effect of mulch induced hydrothermal regime on root growth, water use efficiency, yield and quality of strawberry. *Journal of the Indian Society of Soil Science*, 41(1): 17–25.

Hamilton, A.J., Stagnitti, F., Premier, R., Boland, A.M. and Hale, G. 2006a. Quantitative microbial risk assessment models for consumption of raw vegetables irrigated with reclaimed water. *Applied Environmental Microbiology*, 72: 3284–3290.

Hamilton, A.J., Stagnitti, F., Premier, R. and Boland, A.M. 2006b. Is the risk of illness through consuming vegetables irrigated with reclaimed wastewater different for different population groups? *Water Science and Technology*, 54: 379–386.

Hamilton, A.J., Stagnitti, F., Xiong, X., Kreidl, S.L., Benke, K.K. and Maher, P. 2007. Wastewater irrigation: the state of play. *Vadose Zone Journal*, 6: 823–840.

Hanson, B.R. and Bendixon, W. 2004. Drip irrigation evaluated in Santa Maria valley strawberries. *California Agriculture*, 58: 48–53.

Hargreaves, G.H. and Samani, Z.A. 1982. Estimating potential evapotranspiration. *Journal of Irrigation and Drainage Engineering*, 108: 225–230.

Hargreaves, G.H. and Samani, Z.A. 1985. Reference crop evapotranspiration from temperature. *Applied Engineering in Agriculture,* 1: 96–99.

Harrold, L.L. 1955. Evapotranspiration rates for different crops. *Journal of American Society of Agricultural Engineers*, 36: 669–672.

Hashem, H.A., Hassanein, R.A., El-Deep, M.H. and Shouman, A.I. 2003. Irrigation with industrial wastewater activates antioxidant system and osmoprotectant accumulation in lettuce, turnip and tomato plants. *Ecotoxicology and Environmental Safety*, 95: 144–152.

Hendrickx, J.M.H., Neiber, J.L.H. and Siccama, P.D. 1994. Effect of tensiometer cup size on field soil water tension variability. *Soil Science Society of America Journal*, 58: 309–315.

Hendrcikx, J.M.H. and Wierenga, P.J. 1990. Variability of soil water tension in a trickle chilli pepper field. *Irrigation Science*, 11: 23–30.

Heeney, H.B. 1962. Bulletin of Canada Department of Agri. Publ., 1171.

Hochmuth, G., Cantliffe, D., Chandler, C., Stanley, C., Bish, E., Waldo, E., Legard, D. and Duval, J. 2006. Containerized transplants reduce establishment-period water use and enhance early growth and flowering compared with bare-root plants. *HortTechnology*, 16: 46–54.

Hochmuth, G.J., Locascio, S.J., Kostewicz, S.R. and Martin, F.J. 1993. Irrigation method and rowcover use for strawberry freeze protection. *Journal of the American Society for Horticultural Science*, 118: 575–579.

Hoppula, K.I. and Salo, T.J. 2007. Tensiometer based irrigation scheduling in perennial strawberry cultivation. *Irrigation Science*, 25: 401–409.

Howard, C.M., Maas, J.L., Chandler, C.K. and Albregts, E.E. 1992. Anthracnose of strawberry caused by *Colletotrichum* complex in Florida. *Plant Disease*, 76: 976–981.

Hsiao, T.C. and Acevedo, E. 1974. Plant responses to water deficits, water use efficiency and drought resistance. *Agricultural Meteorology*, 14: 59–84.

Iuchi, T. 1993. Crescimento da planta e do fruto de morangueiro (*Fragaria* x *ananassa* Duch.) em diferentes regimes hídricos. Vicosa: p. 187.

Jackson, A. 1982. Central Coast Crop Coefficients for Field and Vegetable Crops. California Department of Water Resources. Water Conservation Office, University of California (UC), Oakland, California.

Jahangir, M., Abdel-Farid, I.B., Kim, H.K., Choi, Y.H. and Verpoorte, R. 2009. Healthy and unhealthy plants: the effect of stress on the metabolism of Brassicaceae. *Environmental and Experimental Botany*, 67: 23–33.

Jensen, N.L., Jensen, C.R., Liu, F. and Petersen, K.K. 2009. Water relations and abscisic acid in pot grown plants under limited irrigation. *Journal of the American Society for Horticultural Science*, 134: 574–580.

Jones, H.G. 2004. Irrigation scheduling: advantages and pitfalls of plant based methods. *Journal of Experimental Botany*, 55: 2427–2436.

Kamber, R., Eylen, M. and Tok, A. 1986. Effect of trickle and furrow irrigation on strawberry yield. Research Report, Tarsus Research Institute for Village Affairs, pp. 1–39.

Keutgen, A.J. and Pawelzik, E. 2008. Quality and nutritional value of strawberry fruit under long term salt stress. *Food Chemistry*, 107: 1413–1420.

Khan, S., Cao, Q., Zheng, Y.M., Huang, Y.Z. and Zhu, Y.G. 2008. Health risks of heavy metals in contaminated soils and food crops irrigated with waste water in Beijing, China. *Environmental Pollution*, 152: 686–692.

Khurana, M.P.S. and Singh, P. 2012. Wastewater use in crop production: a review. *Resources and Environment*, 2: 116–131.

Kirnak, H., Kaya, C., Higgs, D. and Gercek, S.A. 2001. Long-term experiment to study the role of mulches in the physiology and macro-nutrition of strawberry grown under water stress. *Australian Journal of Experimental Agriculture*, 52: 937–943.

Kirnak, H., Kaya, C., Higgs, D., Bolat, I., Simsek, M. and Ikinci, A. 2003. Effects of pre-harvest drip irrigation scheduling on strawberry yield, quality and growth. *Australian Journal of Experimental Agriculture*, 43(1): 105–111.

Kirschbaum, D.S., Correa, M., Borguez, A.M., Larson, K.D. and Dejon, T.M. 2003. Water requirement and water use efficiency of fresh and waiting bed strawberry plants. *IV International Symposium on Irrigation of Horticultural Crops*, 664: 347–352.

Klamkowski, K. and Treder, W. 2006. Morphological and physiological response of strawberry plants to water stress. *Agricultural Scientific Journal*, 71: 159–165.

Klamkowski, K. and Treder, W. 2008. Response to drought stress of three strawberry cultivars grown under green house condition. *Journal of Fruit and Ornamental Plant Research*, 16: 179–188.

Kruger, E., Schmidt, G. and Bruchner, U. 1999. Scheduling strawberry irrigation based upon tensiometer measurement and a climatic water balance model. *Scientia Horticulturae (Amsterdem)*, 81(4): 409–424.

Kruger, E., Schmidt, G. and Rasim, S. 2002. Effect of irrigation on yield, fruit size and firmness of strawberry "Elsanta". *Acta Horticulturae*, 567: 471–474.

Kumar, S., Sharma, I.P. and Raina, J.N. 2005. Effect of levels and application methods of irrigation and mulch materials on strawberry production in North-West Himalayas. *Journal of the Indian Society of Soil Science*, 53(1): 60–65.

Kurunc, A. and Cekc, C. 2005. Response of three strawberry cultivars (*Fragaria* x *ananassa* Duch.) to different salinity levels in irrigation water. *Zahradnictvi Horticultural Science*, 32(2): 50–55.

Larson, K.D. and Shaw, D.V. 1996. Soil fumigation, fruit production and dry matter partitioning of field grown strawberry plants. *Journal of the American Society for Horticultural Science*, 121: 1137–1140.

Letourneau, G., Caron, J., Anderson, L. and Cormier, J. 2015. Matric potential based irrigation management of field grown strawberry: effects on yield and water use efficiency. *Agricultural Water Management*, 161: 102–113.

Li, M., Chen, L., Xiao, Y., Gan, L. Hu, Q.X. and An, S.Q. 2009. Effects of arbuscularmycorrhiza on absorption of nitrogen and phosphorus of *Spartina alterniflora* and *Phragmites australis*. Shengtai Xuebao. *Acta Ecologica Sinica*, 29: 3960–3969.

Lieten, P. 2002. The effect of nutrition prior to and during flower differentiation on phyllody and plant performance of short day strawberry Elsanta. *Acta Horticurae*, 567(1): 345–348.

Liu, F., Sevic, S., Jensen, C.R., Shahnazari, A., Jacobsen, S.E., Stikic, R. and Andersen, M.N. 2007. Water relations and yield of lysimeter grown strawberries under limited irrigation. *Scientia Horticulturae*, 111: 128–132.

Locascio, S.J. 2005. Management of irrigation for vegetables: past, present and future. *Hort Technology*, 15(3): 482–485.

Losada, A. 1995. *El Riego: Fundamentos Hidraulicos*. Mundi-Prensa, Cordoba, Spain.

Lozano, D., Ruiz, N. and Gavilan, P. 2016. Consumptive water use and irrigation performance of strawberries. *Agricultural Water Management*, 169: 44–51.

Madden, L.V., Wilson, L.L. and Ellis, M.A. 1993. Field spread of anthracnose fruit rot of strawberry in relation to ground cover and ambient weather conditions. *Plant Disease*, 77: 861–866.

Manganaris, G.A., Goulas, V., Vicentre, A.R. and Terry, L.A. 2013. Berry antioxidants: small fruits providing large benefits. *Journal of Science Food and Agriculture*, 94: 825–833.

Mapanda, F., Mangavayana, E.N., Nyamangara, J. and Giller, K.E. 2005. The effect of long-term irrigation using wastewater on heavy metal contents of soil under vegetables in Harare, Zimbabwe. *Agriculture, Ecosystems & Environment*, 107: 151–165.

Mara, D.D., Sleigh, P.A., Blumenthal, U.J. and Carr, R.M. 2007. Health risks in waste water irrigation: comparing estimates from quantitative microbial risk analyses and epidemiological studies. *Journal of Water and Health*, 5: 39–50.

Martinez-Ferri, E., Ariza, M.T., Dominguez, P., Medina, J.J., Miranda, L., Muriel, J.I., Montesinos, P., Rodriguez-Diaz, J.A. and Soria, C. 2014. Cropping strawberry for improving productivity and environmental sustainability. In: Malone, N. (Ed.), *Strawberries: Cultivation, Antioxidant Properties and Health Benefits*. Nova Science Publishers, New York: pp. 1–20.

Martinez-Ferri, E., Soria, C., Ariza, M.T., Medina, J.J., Miranda, L., Domiguez, P. and Muriel, J.L. 2016. Water relations, growth and physiological response of seven strawberry cultivars (*Fragaria* × *ananassa* Duch.). *Agricultural Water Management*, 164: 73–82.

Martin, M., Montesinos, P., Garcia, J., Rodriguez, J.A. and Camacho, E. 2013. Influencia de los emisores de riego en la sostenibilidad del uso del aqua en el entorno del parquet natural de Donana. *Actas del XXXI Congreso nacional de riegos*, C-1: 120–131.

McNiesh, C.M., Welch, N.C. and Nelson, R.D. 1985. Trickle irrigation requirements for strawberry in coastal California. *Journal of the American Society for Horticultural Science*, 110: 714–718.

Meron, M.R., Hallel, R., Shay, G. and Feuer, R. 1986. Soil-sensor actuated drip irrigation of cotton. In: *Proceedings International Conference on Evapotranspiration and Irrigation Scheduling*, November, San Antonio, Texas: pp. 886–892.

Merriam, J.L. 1960. Field method of approximating soil moisture for irrigation. *Transactions A.S.A.E.*, 3(1): 31–32.

Mesa-Jurado, M.A., Martin-Ortega, J., Ruto, E. and Berbel, J. 2012. The economic value of guaranteed water supply for irrigation under scarcity conditions. *Agricultural Water Management*, 113: 10–18.

Milano, M., Ruelland, D., Fernandez, S., Dezatter, A., Fabre, J. and Servat, E. 2012. Facing climatic and anthropogenic changes in the Mediterranean basin: what will be the medium-term impact on water stress? *Comptes Rendus Geoscience*, 344: 432–440.

Mitra, S.K. 1991. Strawberry. In: Bose, T.K., Rathore, D.S. and Mitra, S.K. (eds.), *Temperate Fruits*. Hort. Allied Pub., Calcutta, India: pp. 549–596.

Mok, H.F., Barker, S.F. and Hamilton, A.J. 2014. A propbabilistic quantitative microbial risk assessment of norovirus disease burden from wastewater irrigation of vegetables in Shepparton, Australia. *Water Research*, 54: 347–362.

Morillo, G., Martin, M., Camacho, E., Rodriguez Diaz, J.A. and Montensinos, P. 2015. Toward precision irrigation for intensive strawberry cultivation. *Agricultural Water Management*, 151: 43–51.

Mortensen, F. 1934. Runner plant production in southwest Texas. *Proceedings of the Society for Horticultural Science*, 32: 424–428.

Muchuweti, M., Birkett, J.W., Chinyanga, E., Zvauya, R., Scrimshaw, M.D. and Lester, J.N. 2006. Heavy metal content of vegetables irrigated with mixture of wastewater and sewage sludge in Zimbabwe: implications for human health. *Agriculture, Ecosystems and Environment*, 112: 41–48.

Naaz, S. and Pandey, S.N. 2010. Effects of industrial waste water on heavy metal accumulation: growth and biochemical response of lettuce (*Lactuca sativa* L.). *Journal of Environmental Biology*, 31: 273–276.

O' Neil, S.D. 1983. Osmotic adjustment and the development of freezing resistance in *Fragaria virginiana*. *Plant Physiology*, 72: 931–937.

Passos, F.A. 1997. Influência de alguns sistemas de cultivo na cultura do morango (*Fragaria* x *ananassa* Duch.). Piracicaba, pp. 106.

Palese, A.M., Pasquale, V., Celano, G., Figleuolo, G., Masi, S. and Xiloyannis, C. 2009. Irrigation of olive groves in southern Italy with treated municipal wastewater: effects on microbiological quality of soil and fruits. *Agriculture, Ecosystem and Environment*, 129: 43–51.

Pelletier, V., Gallichand, J. and Caron, J. 2013. Effect of soil water potential threshold for irrigation on cranberry yield and water productivity. *Transactions on American Society of Agricultural and Biological Engineers*, 56: 1325–1332.

Penuelas, J., Save, R., Marfa, O. and Serrano, L. 1992. Remotely measured canopy temperature of green house strawberries as indicator of water status and yield under mild and very mild water stress conditions. *Agricultural and Forest Meteorology*, 58: 63–77.

Periard, Y., Caron, J., Jutras, S., Lafond, A.J. and Houlliot, A. 2012. Irrigation management of romaine lettuce in histosols at two spatial scales: water, energy, leaching and yield impacts. *WIT Transactions on Ecology and Enviornment*, 168: 171–188.

Peterson, S.R., Ashbolt, N.J. and Sharma, A. 2001. Microbial risks from wastewater irrigation of salad crops: a screening level risks assessment. *Water Environment Research*, 73: 667–672.

Phene, C.J. and Howell, T.A. 1984. Soil-sensor control of high- frequency irrigation systems. *Transactions ASAE*, 27(2): 392–396.

Pires, R.C.M. 1998. Desenvolvimento e produtividade do morangueiro sob diferentes níveis de água e coberturas do solo. *Piracicaba*, pp. 116.

Pires, R.C.D.M., Folegatti, M.V., Passos, F.A., Arruda, F.B. and Sakai, E. 2006. Vegetative growth and yield of strawberry under irrigation and soil mulches for different cultivation environments. *Scientia Agricola*, 63(5): 417–425.

Pires, Regina Célia de Matos, Folegatti, Marcos Vinícius, Tanaka, Maria Aparecida de Souza, Francisco, Antonio Passos, Ambrosano, Glaucia Maria Bovi and Sakai, Emilio. 2007. Water levels and soil mulches in relation to strawberry diseases an yield in a greenhouse. Scientia Agricola *(Piracicaba, Braz.)*, 64(6): 575–581.

Plekhanova, M.N. and Petrova, M.N. 2002. Influence of black plastic soil mulching on productivity of strawberry Cultivars in Northwest Russia. *Proceedings of IV International Strawberry Symposium, Acta Horiculturae*, 567: 491–494.

Rajaganapathy, V., Xavier, F., Sreekumar, D. and Mandal, P.K. 2011. Heavy metal contamination in soil, water and fodder and their presence in livestock and products: A review. *Journal of Environmental Science and Technology*, 4(3): 234–249.

Rekika, D., Caron, J. Rancourt, G.T., Lafond, J.A., Gumiere, S.J., Jenni, S. and Gosselin, A. 2014. Optimal irrigation for onion and celery production and spinach seed germination in histosols: pp. 981–994. doi: 10.2134/agronj2013.0235.

Reynolds, W.D. and Elrick, D.E. 1990. Ponded infiltration from a single ring: I. Analysis of steady flow. *Soil Science Society of America Journal*, 54: 1233–1241.

Reynolds, W.D., Elrick, D.E. and Young, E.G. 2002. 3.4.2.2 Constant head soil core (tank) method. In: Dane, J.H., Topp, C.G. (eds.), *Methods of Soil Analysis. Part 4-Physical Methods*. Soil Science Society of America, Inc., Madison, Wisconsin: pp. 804–809.

Rolbiecki, S., Rolbiecki, R., Rzekanowski, C. and Derkacz, M. 2001. Effect of different irrigation regimes on growth and yield of "Elsanta" strawberries planted on loose sandy soil. *International Symposium on Irrigation and Water Relations in Grapevine and Fruit Trees*, 646: 163–166.

Rolbiecki, S., Rolbiecki, R., Rzekanowski, C. and Derkacz, M. 2004. Effect of different irrigation regimes on growth and yield of "Elsanta" strawberries planted on loose sandy soils. *Acta Horticulturae*, 646: 163–166.

Rom, R.C. and Dana, M.N. 1960. Strawberry root growth studies in fine sandy soil. *Journal of the American Society for Horticultural Science*, 75: 367–372.

Santos, B.M., Salame-Donoso, T.P. and Whidden, A.J. 2012. Reducing sprinkler irrigation volumes for strawberry transplant establishment in Florida. *HortTechnology*, 22(2): 224–227.

Save, R., Penuelas, J., Marpha, O. and Serrano, L. 1993. Changes in leaf osmotic and elastic potentials and canopy structure of strawberries under mild and very mild water stress. *HortScience*, 28: 925–927.

Seong, K.C., Cheong, S.R., Yu-I, C., Kim, K.Y. and Park, S.K. (1993). Effects of irrigation methods and covering materials on the occurrence of grey mould (*Botrytis cinerea*) in strawberry (*Fragaria* x *ananassa*) in protected culture. *RDA Journal of Agricultural Science Horticulture*, 35: 501–506.

Serrano, L., Carbonell, X., Save, R., Marpha, O. and Penuelas, J. 1992. Effects of irrigation regimes on the yield and water use of strawberry. *Irrigation Science*, 13: 45–48.

Shannon, E.L., McDoughal, A., Kelsey, K. and Hussey, B. 1996. Watercheck – a coordinated extension programme for improving irrigation efficiency in Australian cane farms. *Proceedings of the Australian Society of Sugar Cane Technology*, 18: 113–118.

Sharma, R.R., Sharma, V.P. and Pandey, S.N. 2004. Mulching influences plant growth and albinism disorder in strawberry under subtropical climate. *Acta Horticulturae*, 662: 187–191.

Sharma, R.R. and Sharma, V.P. 2003. Mulch type influences plant growth, albinism disorder and fruit quality in strawberry (*Fragaria × ananassa* Dusch.). *Fruits*, 58: 221–227.

Sharma, R.R., Singh, R., Singh, D. and Gupta, R.K. 2008. Influence of row cover and mulching interaction on leaf physiology, fruit yield and albinism incidence in Sweet Charlie strawberry (*Fragaria* x *ananassa* Duch.). *Fruits*, 63: 103–110.

Shock, C.C. and Wang, F.X. 2011. Soil water tension, a powerful measurement for productivity and stewardship. *HortScience*, 46: 178–185.

Simonne, E., Studstill, D., Hochmuth, R.C., Olczyk, T., Dukes, M., Munoz-Carpena, R. and Yuncong, C.L. 2012. *Drip Irrigation: The BMP Era – An Integrated Approach to Water and Fertilizer Management for Vegetables Grown with Plasticulture*. University of Florida IFAS Extension, Florida.

Singh, R. and Asrey, R. 2005. Growth, earliness and fruit yield of micro-irrigated strawberry as affected by planting time and mulching in semi-arid regions. *Indian Journal of Horticulture*, 62: 148–151.

Singh, R., Sharma, R.R. and Goyal, R.K. 2007. Intercative effects of planting time and mulching on "Chandler" strawberry (*Fragaria × ananassa* Duch.). *Scientia Horticulturae*, 111: 344–351.

Simunek, J., Jacques, D., van Genuchten, M.T. and Mallants, D. 2006. Multicomponent geochemical transport using the HYDRUS computer software packages. *Journal of the American Water Resources Association*, 42: 1537–1547.

Singh, A., Sharma, R.K., Agrawal, M. and Marshall, F.M. 2010. Health risk assessment of heavy metals via dietary intake of food stuffs from the wastewater irrigated site of a dry tropical area of India. *Food and Chemical Toxicology*, 48: 611–619.

Singh, K.P., Mohan, D., Sinha, S. and Dalwani, R. 2004. Impact assessment of treated/untreated wastewater toxicants treated by sewage treatment plants on health, agricultural, and environmental quality in the wastewater disposal area. *Chemosphere*, 55: 227–255.

Skup-Jablonska, M., Karakiewicz, B, Grochans, E., Jurczak, A., Nowak-Starz, G., Rotter, I. and Prokopowicz, A. 2012. Effects of blood lead and cadmium levels on the functioning of children with behavior disorders in the family environment. *The Annals of Agricultural and Environmental Medicine*, 19: 241–246.

Smith, R.J., Baillie, J.N., McCarthy, C.A., Raine, S.R. and Baillie, C.P. 2010. Review of Precision Irrigation Technologies and their application. National Centre of Engineering in Agriculture, University of Queensland, Toowoomba, NCEA Publication, 1003017/1.

Staswick, P.E., Su, W. and Howell, S.H. 1992. Methyl jasmonate inhibition of root growth and induction of a leaf protein are decreased in an *Arabidopsis thaliana* mutant. *Proceedings of the National Academy of Sciences USA*, 89: 6837–6840.

Stevens, M.D., Black, B.L., Lee-Cox, J.D. and Feuz, D. 2011. Horticultural and economic considerations in the sustainability of three cold-climate strawberry production system. *HortScience*, 46: 445–451.

Strabbioli, G. 1988. A study on strawberry water requirements. *Acta Horticulturae*, 228: 179–186.

Strand, L.L. 2008. Integrated pest management for strawberries, vol., 3351, UCANR Publication.

Tanaka, M.A.S. 2002. Controle de doenças causadas por fungos e bactéria em morangueiro. In: L. Zambolim, F.X.R. Vale, A.J.A. Monteiro and H. Costa (eds.). *Controle de Doencas de Plantas – Fruteiras*. UFV, Vicosa, Brazil: pp. 69–139.

Tanaka, M.A.S., Betti, J.A. and Kimati, H. 2005. Doencas do morangueiro – *Fragaria* x *ananassa* Duch. In: H. Kimati, L. Amorim, A. Bergamim Filho, J.A. Rezende and L.E.A Camargo (eds.). *Manual de Fitopatologia*. 3rd edn. Agronomica Ceres, Sao Paulo, Brazil, 2: pp. 556–571.

Tekinel, O., Kaska, N., Dinc, G. and Yurdakul, O. 1984. A research on comparison of furrow, sprinkler and drip irrigation methods upon early strawberry growth under Cukurow conditions (In Turkish). *Doga (Journal of Agriculture and Forestry), Series D, Bilim Dergisi*, 28(1): 48–55.

Terry, L., Chope, G.A. and Bardonaba, J.G. 2007. Effect of water deficit irrigation and inoculation with *Botrytis cinerea* on strawberry (*Fragaria* × *ananassa*) fruit quality. *Journal of Agricultural and Food Chemistry*, 55: 10812–10819.

Terry, L.A. and Joyce, D.C. 2004. Influence of growing condition and associated variable efficacy of acibenzolar in suppression of Botrytis cinerea on strawberry fruit. *Advanced Strawberry Research*, 23: 11–19.

Topp, C.G. and Ferre, T.P.A. 2002. Water content. In: J.H. Dane and C.G. Topp (eds.). *Methods of Soil Analysis, Part 4 – Physical Methods*. Soil Science Society of America, Inc., Madison, Wisconsin: pp. 417–546.

Trout, T.J. and Gartung, J. 2004. Irrigation water requirements for strawberries. *Acta Horticulturae*, 664: 665–671.

U.S. Environmental Protection Agency. 2012. Guidelines for water reuse. United States Environmental Protection Agency.

Voth, V. 1967. Grapes and berries. *Part II Strawberries. Irrigation of Agricultural Lands*, 37: 734–737.

Voth, V. and Bringhurst, R.S. 1990. Culture and physiological manipulation of California strawberries. *HortScience*, 25: 889–892.

Wang, S.Y. 1999. Methyl jasmonate reduces water stress in strawberry. *Journal of Plant Growth Regulation*, 18: 127–134.

Wanjura, D.F., Upchurch, D.R. and Mahan, J.R. 2004. Establishing differential irrigation levels using temperature-time thresholds. *Applied Engineering in Agriculture*, 20(2): 201–206.

Webb, R.A. 1973. A possible influence of pedicel dimensions on fruit size and yield in strawberry. *Scientia Horticulturae*, 1: 321–330.

WHO. 2006. WHO Guidelines for the safe use of wastewater, excreta and greywater. Wastewater Use in Agriculture, Vol. 2. World Health Organization, Geneva, Switzerland.

Young, M.H. and Sisson, J.B. 2002. Water potential. In: J.H. Dane and C.G. Topp (eds.). *Methods of Soil Analysis, Part 4 – Physical Methods*. Soil Science Society of America, Inc., Madison, Wisconsin: 547–670.

Yuan, B.Z., Sun, J. and Nishiyama, S. 2004. Effect of drip irrigation on strawberry growth and yield inside a plastic greenhouse. *Biosystems Engineering*, 87: 237–245.

Zhang, B. and Archbold, D.D. 1993. Water relations of a *Fragaria chiloensis* and a *F. virginiana* selection during and after water deficit stress. *Journal* of the *American Society* for *Horticultural Science*, 118: 274–279.

15 Weed Management

Manpreet Kour and A.S. Charak

CONTENTS

15.1 Components of an Integrated Weed Management Programme..273
    15.1.1 Weed Identification and Scouting 273
    15.1.2 Weed Biology..275
    15.1.3 Weed Threshold and Action Levels 276
    15.1.4 Management Methods 276
        15.1.4.1 Bed Renovation 276
        15.1.4.2 Soil Fumigation 277
        15.1.4.3 Thermal Controls 278
        15.1.4.4 Soil Solarization 278
        15.1.4.5 Cultural..278
        15.1.4.6 Mechanical Suppression of Weeds..279
        15.1.4.7 Biological 280
        15.1.4.8 Vinegar and Essential Oil Herbicides..280
        15.1.4.9 Chemical 281
15.2 Terms Used to Describe Herbicides 281
    15.2.1 Formulations 281
15.3 Herbicidal Injury..282
    15.3.1 Precautions..283
15.4 Summary of Chemical Weed Control Options..283
15.5 Problematic Weeds of Strawberry 285
    Management 285
    Management 285
    Management 286
References...286

Considering many other creeping plant species, strawberries are comparatively poor competitors against weeds for light, nutrients and moisture as these are shallow-rooted crops. Most weeds that invade strawberries fields are annuals. During plant establishment, little mallow (*Malva parviflora*), bur clover (*Medicago polymorpha*), sweet clover (*Melilotus officinalis*) and filaree (*Erodium cicutarium*) are common weeds because of the survival of their seeds after fumigation. After planting, grasses and broadleaf weeds with windblown seeds, including thistle and common groundsel, become problems. In certain sites, perennial weeds such as field bindweed and bermuda grass or yellow nut sedge may become serious, and their management is essential, especially in fields where the plantation is continued to a second cycle/year of production (Fennimore et al., 2010). Season long uncontrolled weed growth can reduce the strawberry productivity by 51%, and for every 100 g increase in weed biomass, fruit yield decreases by 6% (Prittis and Kelly, 2004). Moreover, in fields heavily infested by weeds, manual fruit harvesting becomes labour intensive, and it increases the cost of cultivation (Figure 15.1). Some of the commonly found weeds in strawberry field are illustrated from Figures 15.2 through 15.13. In the *pick your own berries* system of harvesting, weed-free fields are important for repeated sales and customer satisfaction. An efficient weed control programme should integrate the knowledge of how weeds enter the field to prevent infestation,

FIGURE 15.1 Strawberry field heavily infested with weeds.

FIGURE 15.2 *Gnaphalium uliginosum* (Marsh cupweed).

FIGURE 15.3 Carthamus oxyacantha.

FIGURE 15.4 *Melilotus indica*.

Weed Management 271

FIGURE 15.5 Anagallis arvensis.

FIGURE 15.6 Oxalis corniculata.

FIGURE 15.7 Parthenium hysterophorus.

FIGURE 15.8 Cannabis sativa.

FIGURE 15.9 *Cyperus* spp.

FIGURE 15.10 *Fumaria parviflora* (fineleaf fumitory/ Indian fumitory).

FIGURE 15.11 *Xanthium strumarium* (cocklebur).

FIGURE 15.12 *Conyza canadensis* (horseweed).

FIGURE 15.13 *Eleusine indica* (Indian goosegrass).

cultural control and chemical control practices. The primary goal of weed management should be to minimize weed competition for optimum yield.

Weeds are the cause of reduced yields due to their competing nature with the main crop for water, light and nutrients. Weeds also intercept sprays meant for crop protection, obstructing the applied material to crop foliage for specific purposes such as plant protection or nutrition. This way, weeds may promote development of disease by maintaining high humidity around the crop canopy, and some other weed species could be alternate hosts for pathogens and insect-pests. Timely intercultural operations, judicious use of herbicides and mulches and prevention are integral parts of a good weed management system. Among the factors that can influence weed control are the species present, their stage of growth, crop competition, soil characteristics and rainfall or irrigation. In addition, the strawberry weed management programme must be thought of as a continuous management effort and not as a seasonal operation. Often, repeated use of one successful control technique can lead to shifts in composition of the weed community. Therefore, an efficient and successful weed management programme does not rely on only one control method, but instead it must encompass a multitude of control measures, viz., integration of preventive, cultural, mechanical, biological and chemical control methods to achieve a sustainable production system that balances economic, health and environmental concerns.

15.1 COMPONENTS OF AN INTEGRATED WEED MANAGEMENT PROGRAMME

- Weed identification and scouting
- Weed biology
- Weed threshold and action levels
- Management methods
    1. Bed renovation
    2. Soil fumigation
    3. Thermal control
    4. Soil solarization
    5. Intercultural operations
    6. Mechanical suppression of weeds
    7. Biological materials
    8. Vinegar and essential oil herbicides
    9. Chemical applications
- Problematic weeds of strawberry

15.1.1 Weed Identification and Scouting

Weed identification and scouting is essential in the strawberry fields to make decisions on using appropriate control methods, for example, problematic annual weeds must be identified and controlled by using cultural methods such as solarization or summer fallowing before planting the crop. Also, patches of perennial weeds must be located and identified so that their propagules can be destroyed by herbicide spraying and cultivation well in advance of planting. Weeds not controlled are called *escapes*, and if the herbicide programme is not changed, in time the escapes will become the prevalent weeds in the field. For this reason, the identification of various species of weeds is required to differentiate between grasses, broadleaf and sedge species to choose the most effective herbicides for controlling weed species present in the strawberry ecosystem.

Scouting is the foundation of integrated pest management programmes as it provides awareness of the presence of weeds. Monitoring the weed population is essential to determine proper management practices. Weed data addresses the need of the crop and targets the weeds problems by reducing herbicide use or eliminating routine applications. The goal of monitoring is to locate and identify problematic weeds by walking in strawberry fields in a pattern. For the planting year the scouting should begin even before planting and should continue throughout the season, so that prevalent annual weeds can be identified to plan the control strategy. After planting of runners, the fields should be checked every two to three weeks throughout the season.

In the fruiting years, strawberry scouting should be done at least four times (spring, before renovation, late summer and in October) during the growing season (Anonymous, 2014). Scouts should watch for occurrence of new and invasive weeds and record the distribution of each species (Michael et al., 2006) as follows:

- General = found throughout the field
- Local = found in a small portion of the field
- Spotty = found in just a few places

Also the density of each species is recorded:

- One = scattered; just a few weeds
- Two = slight; one weed per 1.8 m of row
- Three = moderate; one weed per metre of row
- Four = severe; more than one weed per metre of row

After thorough monitoring, first attention should be given to low spots, wet areas and field margins, where new weed problems develop first. Many species fall into the general category and will be the primary targets of the weed control programme. Local distribution may indicate that the species has been recently introduced to the field, and eradication by preventing seed production may be possible. Sporadic distribution of common weeds, such as lamb's quarters (*Chenopodium album*), may indicate that herbicide-resistant biotypes have developed, in which case herbicides with alternative modes of action should be adopted (Michael et al., 2006). Strawberry crops cannot compete well with most weeds, hence maintenance of nearly weed-free conditions is essentially required for optimum production. An integrated approach combining hoeing, hand weeding and use of herbicides is an ideal strategy to maintain effective weed control. Cultivation, hoeing and hand weeding are most effective when weeds are small (<1.25 cm height). Herbicides need to be selected and applied at the appropriate time to control the most dominant species as it is more important than herbicide selection. A specially combined application of any of the two herbicides during the month of March providing broad spectrum activity during the period of spring growth eliminate much of the weed competition. Thereafter, a second application may extend weed control through the remainder of the growing season (Wehtje and Gilliam, 1991). In addition to scouting for weeds, any herbicide injury observed should also be noted and mapped to determine for future reference.

Weed Management

15.1.2 Weed Biology

Based on lifecycles, weeds within strawberry fields can be classified either as annuals (summer or winter), biennials or perennials (Table 15.1). Annual weeds can be classified as either summer or winter annuals. Summer annuals germinate in the spring and summer, complete the vegetative growth cycle, attain the reproductive stage and bear flowers, set seeds and dry out within the same growing season. A common characteristic of summer annuals is a rapid rate of growth and prolific seed production, which is detrimental to strawberry plants. Winter annuals germinate during late August–November, and overwinter as small plants called *rosettes*. In the following spring, winter annuals grow very quickly, bear flowers, seeds and later die. These weeds must be controlled during late summer and early fall before or soon after emergence.

TABLE 15.1

List of Common Weeds in the Strawberry Fields

Biology Group	Weeds	Scientific Name	Family
Annuals	Chickweed	*Stellaria media*	Caryophyllaceae
	Corn spurry	*Spergula arvensis*	Caryophyllaceae
	Cudweed	*Gnaphalium purpureum*	Compositae
	Field violet	*Viola arvensis*	Violaceae
	Grounsel	*Senecio vulgaris*	Asteraceae
	Hempnettle	*Galeopsis tetrahit*	Lamiaceae
	Lamb's quarters	*Chenopodium album*	Chenopodiaceae
	Mustards	*Brassica kaber*	Brassicaceae
	Nightshade	*Solanum nigrum*	Solanaceae
	Pineapple weed	*Matricaria discoidea*	Asteraceae
	Purselane	*Portulaca oleracea*	Portulacaceae
	Ragweed	*Ambrosia artemisiifolia*	Asteraceae
	Redroot pigweed	*Amaranthus retroflexus*	Amaranthaceae
	Scentless	*Tripleurospermum perforatum*	Asteraceae
	Chamomile	*Chamaemelum nobile*	Asteraceae
	Shepherd's purse	*Capsella bursa-pastoris*	Brassicaceae
	Smartweed	*Polygonum* spp.	Polygonaceae
Biennials	Wild carrot	*Daucus carota*	Apiaceae
	Evening primrose	*Oenothera biennis*	Onagraceae
	Common mullein	*Verbascum thapsus*	Scrophulariaceae
Perennials	Buttercup	*Ranunculus* spp.	Ranunculaceae
	Daisy	*Bellis perennis*	Asteraceae
	Dandelion	*Taraxacum officinale*	Asteraceae
	Horsetail	*Equisetum arvense*	Equisetaceae
	Plantain	*Plantago* spp.	Plantaginaceae
	Poison ivy	*Toxicodendron radicans*	Anacardiaceae
	Quack grass	*Elytrigia repens*	Poaceae
	Sheep sorrel	*Rumex acetosella*	Polygonaceae
	Stitchwort	*Stellaria holostea*	Caryophyllaceae
	Toadflax	*Linaria maroccana*	Scrophulariaceae
	Tufted vetch	*Vicia cracca*	Leguminosae
	Yellow nut sedge	*Cyperus esculentus*	Cyperaceae
	Yellow wood-sorrel	*Oxalis stricta*	Oxalidaceae
	Wild brambles	*Rubus fruticosus*	Rosaceae

Source: Hoover et al., 2016, Organic strawberry production in Minnesota. © Regents of the University of Minnesota.

Biennial weeds germinate during spring, complete the vegetative growth cycle during the fast growing season, overwinter as a rosette, bear flowers, produce seed and die during the second growing season. Although they do not usually cause serious problems in the strawberry fields, biennial weeds can be controlled effectively in the first growing year of their lifecycle.

Perennial weed species persist in the strawberry ecosystem year after year, and once established, their management becomes difficult. However, these can be eradicated from the already infested fields by avoided planting of strawberries for a few years. Perennials generally reproduce both sexually and asexually. In vegetative reproduction, new plants are regenerated from the growth of specialized reproductive structures such as rhizomes, tubers, stolons or roots. As a general principle, all perennial weeds must be controlled as soon as these emerge by suitable approaches such as hoeing or by using herbicides or combination of several suitable approaches.

15.1.3 WEED THRESHOLD AND ACTION LEVELS

The threshold level for weeds in strawberries plantations is based on the knowledge of the weed within the existing farming system. By understanding the biology, survival mechanisms and reproductive capabilities of the weeds, the long-term competitive effect of the weeds on the crop can be predicted, thereafter a decision should be taken to manage their population. From an economic perspective, weed control measures should be adopted when the cost of the crop damage is greater than the cost of the control measures.

15.1.4 MANAGEMENT METHODS

Weed management in a strawberry plantation is a year-round process. Accordingly, it should be started before planting of runners and continued through the lifecycle of the plantation. The most effective and economical weed management programmes in strawberries usually combine cultural, mechanical and chemical processes. Although several herbicides are available to control weeds in strawberries, they should be viewed as useful tools, not as substitutes for a good weed management programme. The continuous use of single control method (cultural, mechanical or chemical) may cause the emergence of a more tolerant or resistant weed species. Most common species tolerant to herbicides, for example, deep-rooted perennials such as common groundsel, can be controlled by integrating as many control measures as possible, such as crop rotation, hoeing, application of various herbicides with different modes of action and rotating herbicides from one season to the next.

15.1.4.1 Bed Renovation

Bed renovation is primarily the first step in the weed management programme. It should begin as soon as possible because early renovation allows more time for runner production, leading to larger crowns and more flower buds for the next year. Early renovation by tilling also improves weed management in many weeds before they produce seeds. For the bed renovation process, first the beds that are to be carried over for another year are identified and further, the beds having good production and normal plant stand with no major weed, insect or disease problems should be carried over for another year. Beds that do not meet these criteria should be ploughed down and rotated with other suitable crops for at least three years to reduce weed, insect and disease problems. Besides, the field should also be seeded with a suitable cover crop to increase soil organic matter content and reduce weed problems to improve the vigour and longevity of strawberry beds without the need for soil fumigation (Anonymous, 2015).

Bed renovation schemes are situation specific; however, once the harvest is completed, all beds should undergo the following steps (Anonymous, 2015):

- Management of broad-leaved weeds: For perennial broad-leaved weeds such as dandelion, shepherd's purse, daisy or golden rod and for a high population of annual broad-leaved

Weed Management 277

weeds such as lamb's quarters (*Chenopodium album*) and sorrel (*Rumex* spp.); 2,4-D amine (Formula 40®) is applied.
- Leaf mowing: Mow off the leaves of the strawberries about 3.75 cm above the crowns.
- Fertilizer application: Apply 45–67.5 kg N/ha. In addition, phosphorus and potassium applications should be made according to soil test recommendations.
- Plant thinning for the single matted-row system, strawberry plant rows should not be wider than 60 cm.
- Pre-emergent weed control: For annual weed control, terbacil (Sinbar® 80WP) @ 0.12–0.42 kg/ha is used.
- Subsoiling: Use of a subsoiling blade between the rows breaks up the compacted layers of soil and improves infiltration of water into soil.
- Irrigation- Strawberries will grow best if they receive 3.75 cm of water/week during the growing season.

15.1.4.2 Soil Fumigation

Fumigants are generally applied as liquid formulations, which become volatilized to form gases. Weed seeds are very difficult to kill with fumigation. Methyl bromide was once the standard soil fumigant because of its efficacy for killing weed seeds as well as other soilborne pests and diseases. However, due to problems with bromine in the atmosphere, its use is being phased out (Samtani et al., 2011; Whitaker et al., 2014). Chloropicrin or 1,3-dichloropropene (1,3-D)/chloropicrin followed five–seven days later by metam sodium or metam potassium has been reported to be a very effective fumigant for strawberries. Fumigation with 1,3-D plus chloropicrin mixture (Telone C35, Inline), chloropicrin, and metam sodium before bed preparation kills the seeds of most weeds and the reproductive structures of some perennials by respiration inhibition. Drip injection of fumigants such as 1,3-D plus chloropicrin mixture or chloropicrin often improves the weed control compared to shank fumigation. However, thorough wetting of the bed during fumigant injection ensures the proper weed control on the edges of the bed. Drip fumigation is used, when only, the bed is to be treated, and the space in between two beds has to be left without fumigation.

Soil fumigants kill the germinating seedlings and ungerminated seeds of weeds. However, it is essential that fumigants must be able to penetrate the seed coat and kill the seed embryo. Moist seeds are easy to kill, because the seed tissues swell with water and allow the fumigant to penetrate more effectively. Besides, moist seeds also have higher respiration rates, so are comparatively more susceptible to fumigants than dry seed. Proper irrigation before fumigation is one of the key factors for effective weed control with all fumigants. Soil temperature must be above 12°C for effective absorption of water by seeds. The seeds of bur clover, sweet clover, filaree, and little mallow weeds are difficult to kill because of their impermeable seed coats, limiting moisture and chemical penetration, consequently remaining dormant in the soil (Fennimore et al., 2010). Haar et al. (2003) found that chloropicrin followed by metam sodium reduced the percentage of viable seeds of *Sellaria media*, *Portulaca oleraceae* and *Polygonum aviculare*. However, Timothy et al. (2013) found that by flaming in strawberry fields, the weeding time is reduced to 10–12%.

For good fumigation results:

- Ideally, the site should be worked and fumigated in September.
- There should be no clods, large soil particles, or stones for soil fumigation.
- Soil temperature should be 15–21°C at 15 cm depth with soil moisture ranging from 50–75% of field capacity.
- Organic matter (cornstalks, grass, straw etc.) should be well decomposed.
- There should be a two-week–two-month period between fumigation and planting to avoid crop damage due to weedicide residue.

15.1.4.3 Thermal Controls

Thermal technologies (handheld flamers, mounted row crop flamers, infrared weeders, steamers, hot water and hot foam) are quite effective; however, timing is critical for effective thermal control of weeds. The younger the weed, the easier it is to desiccate. Sniauka and Pocius (2008) found that 1 mm weed blades of shepherd's purse (*Capsella bursa-pastoris*), common groundsel (*Senecio vulgaris*) and common chick weed (*Stellaria media*) heats (up to 70°C) comparatively 2.7 times faster than 2.8 mm thick blades. Grasses can be burnt back but the growing point usually puts forth new growth. Some of these devices may not fit in a particular system but others may be successful components of a well-planned weed control programme (forum.gon.com).

15.1.4.4 Soil Solarization

In summer, clear plastic sheets applied to preshaped beds several weeks before planting to solarize the soil and reduce the density of weed seeds and severity of soil borne diseases. Solarization is much more effective in areas where temperatures are consistently (30–45 days) hot enough in summer to produce soil temperatures of at least 50°C. Solarization can be even more effective if the residue of a cruciferous crop (especially broccoli or mustards) is incorporated into the soil just before the plastic is spread over the bed or following an application of metam sodium (375 l/ha) (Fennimore et al., 2010).

15.1.4.5 Cultural

Cultural controls are those good agricultural practices that minimize the growth of weeds, while optimizing crop growth. Many decisions and practices influence the effectiveness of cultural controls, as discussed in the following:

15.1.4.5.1 Site Selection

Since controlling perennial weeds in strawberries is difficult, select relatively clean fields with no history of perennial and hard-to-control weeds such as quack grass (*Elytrigia repens*), Canada thistle, field horsetail, chickweed (*S. media*), common groundsel (*S. vulgaris*), lady thumb and sow thistle. It is better to allow the first flush of annual weeds utilizing rainfall or irrigation to germinate before the beginning of tillage operations. Deep ploughing loosens the soil and facilitates dissipation of herbicides applied on preceding crops such as vegetables, ornamental bulbs and corn and thereby helps in minimization of risk of carryover of herbicide into the succeeding strawberries planted in the same field. A light field preparation with a disk harrow before planting of runners on the beds suppresses most of the germinated weeds and minimizes the competition of new plants with weeds.

As for as possible, the sites that have not been planted with crops like tomatoes, potatoes, peppers, eggplant, melons, okra, mint, brambles, chrysanthemums, roses or related crops for at least three–five years, should be selected to avoid the risk of residual effects herbicide in soil carried over from previous crops.

15.1.4.5.2 Sanitation and Prevention

Preventing new weeds from invading the farm and minimizing or eliminating seed production in the field reduces future weed problems. Prevention of weeds is the best practice by careful attention to cultural practices, such as cleaning equipment before moving it into the field, using clean straw mulch and preventing annual weeds from producing seeds. Proper field sanitation is also an inseparable part of a successful weed management programme in strawberry. As far as possible, field perimeters should be kept weed free, as they serve as initial source of weed seed to infest the field (cru.cahe.wsu.edu).

15.1.4.5.3 Crop Rotation

Weed infestations can be reduced by rotating with crops that have a different lifecycle, or grown with different set of cultural and chemical practices. Wheat, corn or vegetable row crops can be

Weed Management

grown while keeping the field weed free. Zavatta et al. (2014) reported weed suppression by rotating strawberry with broccoli. Dennis and Gail (2011) indicated lower weed biomass when strawberry was rotated with cover crops like *Sorghum bicolor, Panicum virgatum* or *Andropogon gerardii*. Cover crops may be grown and ploughed back before planting of strawberries.

15.1.4.5.4   Crop Competition

The use of certified strawberry planting material may cost more initially but proves cheaper in the long run. Medium-sized plants with large and healthy root systems are ideal to compete with weed populations. Management practices which stimulate healthy and vigorous strawberry plants reduce the losses from weeds. Some of these practices include timely bed renovation; selection of the proper amount, timing and placement of fertilizer; applying needs-based irrigation; using the right cultivars; maintaining optimum planting density and adopting standard pest management schedules.

15.1.4.5.5   Mulches

Mulching practice is an inseparable part of strawberry cultivation as it saves moisture, keeps berries clean, suppresses weeds, improves fruit yields and facilitates fruit picking. Wheat or pea vine straw can be used as a light mulch between the rows after hoeing or cultivation in the spring. This mulch keeps rain or sprinklers from packing the soil and saves the moisture. Several other materials, such as polyethylene sheets, sawdust, wood shavings, and well-rotten manure, lawn clippings, leaves and straw, are also suitable for mulching.

Opaque mulches are usually dark-coloured plastics (brown, black or green) that restrict light from penetrating the film effectively but blue plastic is not effective. Clear plastic is sometimes used in summer in warmer areas to solarize the soil, but in winter, it serves as a greenhouse and encourages both weed growth (if not mulched with black polyethylene) and strawberry plant growth. When using opaque mulches, these should be secured to the soil before transplanting and the strawberry plants placed in the soil after cutting a hole into the plastic at the desired spacing. Weed growth is greatly reduced with opaque mulches, but still there are chances of weeds that will need to be manually removed to grow in the hole made for the strawberry plant. Use the smallest possible hole to minimize weed growth around the strawberry plants and planting through slits in the mulch helps in minimizing weed growth (Fennimore et al., 2010).

15.1.4.6   Mechanical Suppression of Weeds

Mechanical suppression of weeds is an important component of weed management in a strawberry crop, particularly during the establishment year. Cultivation is most effective when emerging weeds are very small and observed between the rows. Cultivation at the weed's emergence can be shallow, preventing buried weed seed from being brought to the surface, where germination is then likely to occur (www.oardc.ohio-state.edu). A wiggle-hoe can be used to cultivate weeds between rows and between strawberry plants within rows. Mechanical methods of weed control include practices such as tillage, hand weeding, hoeing and mowing.

15.1.4.6.1   Tillage

Tillage can pull weeds from the soil, then bury them, cutting or weakening them by injuring the root or top growth. Annual and biennial weeds are generally easier to control with tillage than perennial weeds. Continuous destructions of the top growth through tillage help to manage the established perennial weeds by depleting their root reserves (www.gnb.ca). Cultivation (2.5 cm deep) should continue throughout the season as and when required. In general, shallow cultivation is preferred over deep cultivation. In new plantations, cultivation should be done at two–three weeks after planting with a root-tilling cultivator, followed by wiggle-hoe or spike-tooth harrow. Two months after planting, when strawberry runners start to grow, half disc-sweep equipped with blades or other suitable cultivators can be used to control weeds. In established strawberries, row width is set up with a rolling disc early in the spring and the first cultivation is made relatively deep (12–15 cm). Shallow cultivation may be performed with rotary tiller cultivators, wiggle-hoe, spike-tooth harrow, shield or spring tooth gangs or cultivator knives.

15.1.4.6.2 Hand Weeding

Hand weeding (pulling and hoeing) is necessary if weed-free strawberry fields are to be obtained. It is, however, time consuming and expensive. Hand pulling is one of the oldest and most effective methods against annual and biennial weeds or perennial seedlings. Removal of the entire established root system before flowering prevents seed production in perennial weeds. Hand pulling is easier after a rain or irrigation when the soil is wet. In the planting year, hoeing is an important part of any strawberry weed management programme, effectively controlling most annual weeds, many biennial weeds and seed-producing perennial weeds, but is only partially effective on established perennials. Well-established perennial weeds require continuous hoeing at one–two-week intervals during the growing season.

15.1.4.6.3 Mowing

Mowing is practised in the strawberry plantation system as a renovation process and not usually as a weed control method. It is used primarily as a means to prevent seed production or to restrict vegetative growth above the strawberry plants.

15.1.4.7 Biological

Many insects and animals are used as biological control (Anonymous, 2008). Some birds like geese prefer feeding on some serious weeds in strawberry ecosystem. Wherever infestations of chickweed, field horsetail and grasses are serious in a strawberry plantation, geese may be a great help. The population of geese should be just enough to keep down the weeds they like; otherwise, they eat buds and tender foliage. The number of geese needed varies with the amount of weeds available for grazing and the age of the geese (younger geese work better than grown up ones), and usually ranges from about 5–10 geese per hectare. As a rule, geese must be fed with a supplement in addition to the pasturage they get from weeding.

The corn gluten hydrolysate (a maize based biopesticide) used as a pre-emergence herbicide has potential in managing the weeds in matted-row strawberries. This product cannot duplicate the efficacy shown by synthetic herbicides, but can still reduce weed populations to densities below economic thresholds (Donald et al., 2001).

Martinez et al. (2006) recorded good control of many weed species, such as *Poa annua, Portulaca oleraceae, Lolium rigidum, M. parviflora, Medicago* sp., *Echinochloa crusgalli, Amaranthus retroflexus* and *Chenopodium album* by biofumigating the field with fresh hen droppings @ 3 kg/m². The sowing of cereal crops such as barley, oats, wheat or rye in the interrow spaces of strawberries can decrease the amount of weeds and increase its productivity. An increase in the yield of strawberry has been achieved due to rapeseed (8.18 t/ha), mustard (8.49 t/ha), oats (8.62 t/ha) and barley (8.66 t/ha) as intercrops when compared to sole strawberry plantation as control (8.07 t/ha) besides suppressing weeds (Gurin and Sukhochev, 2003, 2005).

15.1.4.8 Vinegar and Essential Oil Herbicides

The use of vinegar or acetic acid for weed control has been observed to be the least toxic choice of many home gardeners; however, its effectiveness varies with type of weeds sprayed and the concentration of acetic acid. Most commercially available vinegars have 5% acetic acid, which can be increased by distillation (up to 15%) and by other non-synthetic processes (up to 30%). Caution must be taken with formulations over 5%. Some commercial formulations of vinegar-based herbicides include lemon juice or citrus oil, which cause the desiccation of weeds due to degradation of the leaf's waxy cuticle layer. The thicker the cuticle layer on the weeds, the more frequent the applications or the more concentrated the solution should be (forum.gon.com).

Weed Management   281

15.1.4.9 Chemical

Herbicides are an important component of any integrated weed management programme in strawberries. The integrated use of herbicides in conjunction with good cultural practices and hand weeding is most effective in management of weeds in strawberry plantations. Several herbicides are available to manage the annual weeds in both new and established strawberry plantations, however, the herbicides that kill perennial weeds may also damage strawberry plants. Therefore, weeds that grow from underground vegetative parts (quackgrass, sheep sorrel, field horsetail, Canada thistle etc.) must be controlled before the planting of strawberry (cru.cahe.wsu.edu).

15.2 TERMS USED TO DESCRIBE HERBICIDES

- Preplant: A preplant herbicide is applied for management of weed foliage before main crops are planted, for example, in field preparation.
- Pre-emergence: A herbicide applied to the soil before weed (and/or crop) emergence. Pre-emergence herbicides may be selective or non-selective, depending upon the rate of application. These herbicides should be applied to weed-free areas because they have little or no activity when applied to emerged seedlings, for example, oxyflurofen, simazine and isoxaben are some broadleaf-active pre-emergence herbicides and Oryzalin, pendimethalin and prodiamine are used as grass-active pre-emergence herbicides (James et al., 2003).
- Post-emergence: Post-emergence herbicides are applied after weeds have emerged. They can be classified as having either selective or non-selective activity.
- Selective: A selective herbicide controls only certain species or only one type of weed. Grass-active herbicides with selective activity include products with the grass-active ingredients fluazifop-butyl, sethoxydim or clethodim. Selective herbicides control only certain species of weeds; for instance, Poast, Fusilade and Select control only grasses and can be safely applied over the foliage of strawberry.
- Non-selective: A herbicide that kills both crops and weed plants. These post-emergence herbicides should be limited to direct applications to avoid contact with strawberry crop (e.g., Roundup, Paraquat). The herbicides such as Gramoxone, Liberty and Roundup Ultra are non-selective herbicides and must be applied only to weeds or before planting of strawberry.
- Contact: A herbicide that kills only foliage or the plant parts that are contacted by the spray, with little or no movement in to non-contacted tissue (e.g., Gramoxone and Liberty).
- Systemic: A herbicide that is applied to an actively growing weed and absorbed by plant parts, moving throughout the root and shoot system, thus effectively controls perennial weeds with well-developed root systems (e.g., glyphosate).
- Residual: A herbicide that remains in the soil for one–several months, or more than a year, continuing to control weeds and potentially damage crops after its application in to the soil. Residual herbicides are usually applied pre-emergence to weed-free soil in the spring. Some herbicides, including those used on berries, may persist long enough to damage crops planted during one or more years after their last use. The best control is achieved when the application is not impeded by trash or actively growing ground cover.

15.2.1 FORMULATIONS

Herbicides are usually formulated as wettable powders (e.g., 50 WP), emulsifiable concentrates (e.g., 2 EC), aqueous suspensions (e.g., AS), or granules (e.g., 10 G). Granular formulations improve the ease of application, and minimize contact of herbicides with crop foliage.

15.3 HERBICIDAL INJURY

Herbicides are designed to selectively kill weeds and do not injure strawberry plants. However, no system is perfect. Herbicide drift or lift-off (co-distillation) can injure strawberry plants. These injuries are further aggravated due to the following:

- Weather conditions: Soon after the application of a pre-emergent soil-acting herbicide, the occurrence of exceptionally warm weather can lead to rapid growth and uptake of herbicide with subsequent crop damage. Cool and wet conditions that reduce the growth of the strawberry plants can enhance herbicide injury symptoms due to the inability of the strawberry to quickly recover from the herbicide injury. Stressed crops due to drought or other pests are more likely to be injured by herbicide residues in soils than healthy crops, whereas prolonged water stress can the make the waxy layer of leaves thicker and increase leaf hairiness, resulting in less herbicide penetration. Plants suffering from early frosts are more susceptible to foliage-applied herbicides. High wind speed increases plant susceptibility to foliage-applied herbicides. However, relative humidity helps to more rapid penetration into leaves and herbicide translocation within the plant.
- Soil texture: Strawberry injury resulting from the use of herbicides is common, particularly on light soils. Light, sandy soils require less herbicide than heavier soils for comparable levels of weed control. Most herbicide injuries can be traced to using too high a rate on light soils, incorrect timing of sprays, incorrectly calibrated sprayers, sensitive cultivars and weak plants growing under unfavourable conditions. Soil compaction may influence effects of herbicide damage. The soil compaction results in concentration of roots just below the top surface of the soil, and as a result excessive uptake of herbicide may occur.
- Plant type: Newly planted strawberries are especially sensitive to herbicides. Crop tolerance increases in late summer and autumn. This corresponds to the time when pre- emergence herbicides can be used to control many winter annual and perennial weeds. Conversely, healthy strawberry plants are most capable of tolerating recommended treatment rates. Some crop cultivars are more sensitive than others to particular herbicides. For example (Anonymous, 1992), "Kent" strawberries are comparatively more sensitive than Veestar to terbacil (Sinbar™).
- Organic matter: Strawberries growing on soils low in organic matter are especially prone to injury due to herbicidal toxicity. Accordingly, lower rates of herbicide need to be used on fields low in organic matter (less than 2%).
- Herbicide residues: Herbicide residues (Anonymous, 1992) on previously sprayed stubble and other trash can damage emerging crop seedlings upon contact (e.g., Gramoxone).
- Miscellaneous factors: Any other factor that injures the crop (e.g., other pests, winter injury, exposure of crowns and root systems due to erosion, improper mineral nutrition or wet spots in the field) can also make the crop more susceptible to injury to herbicide toxicity and to minimize the risk of crop injury, growers need to avoid the application of overdose of herbicides and could be advised to follow the instructions on the product labels. The growers should use herbicides in combination with hand weeding to obtain complete weed control. One need to be cautious about the uniform application of herbicides to get the desired results. Variations in the spray pattern, speed of the rig, worn nozzle tips etc. may change the application rate sufficiently to damage the crop or reduce weed control.

Samtani et al. (2011) observed a reduction in strawberry fruit yield due to preplant application of flumioxazin at 0.21 kg a.i/ha. Severe Sinbar-type (terbacil) injury also occurred when certain post-emergence grass herbicides and Sinbar were tank-mixed or applied excessively in close sequence. Herbicide injury has been observed to be greatest when terbacil is applied directly to the foliage rather than to the soil but is lowest when foliage is rinsed with water after herbicide application (Polter et al., 2004). Devrinol, Dacthal and Sinbar are wettable powders that do not dissolve in water. These herbicides form a suspension in water that can be maintained only by constant agitation in

Weed Management 283

the spray tank. Glyphosate injury to strawberry plants causes white leaf blades and veins on newly emerged leaves. Growth stops shortly after application or drift (Fisher and Huffman, 2010). The injury to strawberry is also caused by herbicides, such as Clopyralid, Glyphosate, Imazethapyr, Paraquat, S-Metolachlor, Simazine and Terbacil (Anonymous, 2009).

Therefore, several precautions need to be taken for a uniform and effective herbicide application.

15.3.1 PRECAUTIONS

- Screens in the line should be no more than 50 mesh to avoid clogging.
- Spray material should not be allowed to settle at the bottom of the tank and should constantly be stirred by mechanical or hydraulic stirrer.
- Wettable powders should not be used as herbicides in gear, roller, or impeller pumps as they are abrasive in nature.
- Wettable powders wear nozzles readily so stainless steel nozzles should be preferred over brass nozzles as they wear sooner.
- The spray rig must be properly designed and calibrated often (approximately every 20 hours of use with brass nozzles) so that the amount of spray material being applied per unit area is assured.

15.4 SUMMARY OF CHEMICAL WEED CONTROL OPTIONS

The results of weed management in strawberry are summarized (Anonymous, 2015) as follows:

DCPA (Dacthal®): It is a pre-emergent herbicide used during early spring, late autumn or after renovation, suitable for some annual broad-leaved weeds and grasses work best in lighter, warmer soils and can beneficially be used as an alternative to terbacil or napropamide under the conditions of high risk of injury to strawberry plants.

Napropamide (Devrinol®): It is a pre-emergent herbicide, suitable for annual grasses, volunteer grains and some broadleaf weeds, used just before mulching in the fall, is good for split application, for example, half the maximum rate of application after renovation or in late summer after the desired daughter plants have rooted, and a second half-rate application during dormancy gets activated by irrigation, rainfall or light cultivation within 24 hours of application. It is photosensitive and suitable for application in soils having <10% organic matter.

Terbacil (Sinbar®): It is a pre-emergent herbicide with some post-emergent activity. It is used at the time of renovation, that is, after mowing and tilling the beds, but before new growth begins, followed by late autumn application, which is injurious for runner production and effective in well-drained soils. To avoid development of terbacil-resistant weeds, do not use terbacil continuously. Rather, it should be rotated with other herbicides having a different mode of action, such as napropamide.

Sethoxydim (Poast®): It is a post-emergent herbicide used against actively growing grasses only and should not be used within six weeks of terbacil (Sinbar®) application to avoid leaf injury. It is suitable for use in combination with a crop oil concentrate but not with 2, 4-D. It should not be applied when rainfall is expected within one hour.

Clethodim (Arrow®, Prism®, Select®): It is a post-emergent herbicide used for control of actively growing grasses but not for broadleaf weeds. It should be used in combination with a crop oil concentrate. Use of Clethodim is restricted in bearing plantations and under conditions when the strawberry plants are under stress. If rain is expectED within one hour of application, then its use should be discouraged. Clethodim application @ 240 g a.i/ha can efficiently control *Agropyron repens* at the four–six-leaf stage as well as *P. annua* from the two–three-leaf stage until the end of tilling, whereas when applied @ 96 g a.i/ha, it can effectively control *E. crusgalli* (19Lisek, 2005).

Paraquat (Gramoxone Inteon®): It is a post-emergent and contact herbicide used to manage most annual weeds and many perennial weeds but is injurious to strawberries. Paraquat should not be applied within three weeks of harvest or more than thrice in one season.

Pelargonic acid (Scythe®): It is a post-emergent and contact herbicide for control of most annual weeds and suppression of many perennial weeds. Being injurious to strawberry plants, it should be applied with a sprayer between rows only and strawberry plants should be shielded to protect them from herbicide spray drift. Although this product has a relatively low toxicity without residual soil activity, it has a strong, unpleasant odour.

2,4-D amine (Formula 40, Amine 4): It is a post-emergent herbicide effective on most broadleaf perennial weeds only. It can be applied immediately after the harvest is over to manage broadleaf weeds. This material can also be used in late autumn or early spring to manage winter annuals and biennials. It is not compatible with sethoxydim (Poast®).

Flumloxazin (Chateau®): It is a pre-emergent herbicide used during dormancy to manage broadleaf weeds.

Pendimethalin (Prowl H20®): It is a pre-emergent herbicide used in bands of the rows because of its toxic nature to strawberry plants. Its waiting period after application is 35 days, so it should not be applied within 35 days of fruit harvest.

In addition to above described herbicides, the applications of clopyralid @ 200 g/ha (Figueroa and Doohan, 2006), sulfentrazone (Figueroa and Doohan, 2005) and metamitron @ 1.08 kg/ha (Lisek, 2005) have been recommended for weed management in strawberries, however, the combination of napropamide (4.5 kg/ha) and oxyfluorfen (0.57 kg/ha) has been proved effective to manage the weeds and resulting in increase of the strawberry yield by 20% (Gilreath and Santos, 2005) (Table 15.2).

TABLE 15.2
Recommended Herbicides for Strawberries

Weeds	Time of Application	Herbicide and Dose/ha
	Pre-emergence	
Annual grasses and some broadleaf weeds	Any time before or after transplanting and in established plants	Dacthal W-75 @ 9.08–13.6 kg
Annual grasses and some broadleaf weeds	After adequate number of runners have rooted, or in late autumn	Devrinol 50WP @ 4.53–9.05 kg followed by irrigation within two to three days
Broadleaf weeds and some grasses	After transplanting, but before rooting of runners	Sinbar@ 0.15–0.212 kg, followed by 1.25–2.5 cm irrigation or immediate rainfall
Winter annual weeds	Late summer to early autumn in young plantings. In established beds, before new growth begins and late autumn but before mulching.	Sinbar 0.15–0.212 kg, followed by 1.25–2.5 cm irrigation or immediate rainfall
	Post-emergence	
Grasses	When grasses are 5–20 cm tall (before seed head formation)	1. Fusilade DX @ 1.13 to 1.7 kg + 2.83 litre crop oil concentrate or 0.7 litre non-ionic surfactant 2. Poast @ 2.83 litre + 2.83 litre crop oil concentrate 3. In established plantings 2,4-D (F.40) @ 1.4–2.83 litres
Non-selective (kill all)	Before weed growth is 15 cm (Should not be sprayed on strawberry plants)	Gramoxone extra @ 2.13 litre in 189.25–567.75 litres water + non-ionic surfactant

Source: Michael et al., 2006, *Midwest Strawberry Production Guide.* Bulletin, 926.

Weed Management 285

Of the different herbicides such as EPTC (Eptam), fomesafen, halosulfuron and S-metolachlor applied on top of the bed, all except S-metolachlor @ 214 g a.i./ha, were observed to be safe to strawberry, whereas when all these herbicides were drip- applied, none of them damaged the strawberry plants or reduced its yields (Boyd and Reed, 2016). Further, the herbicides oxyflurofen and flumioxazin are safe to use and can be incorporated in strawberry production practices for satisfactory weed control (Samtani et al., 2012).

15.5 PROBLEMATIC WEEDS OF STRAWBERRY

Four problematic perennial weeds (yellow nut sedge, toad flax, oxalis and horsetail) in strawberries are difficult to manage with common weedicides such as glyphosate (Roundup), so innovative approaches are required to eradicate them from strawberry fields before planting. For managing these weeds, close attention needs to be given to their lifecycles (especially their propagation method such as, seed, rhizomes or spores) to manage them appropriately. Mowing of field edges and patches of these weeds and seeds can keep problems from getting worse and need to be managed through cultural methods, such as crop rotations before the planting of strawberries.

Yellow nut sedge (*Cyperus esculentus*) is an erect grass-like perennial that mainly reproduces by tubers that can persist by producing nutlets that grow at the end of rhizomes. It reproduces less often by seeds too when left unmanaged in the field.

Management: It is difficult to eradicate forever, but strawberry growers should try to start with clean strawberry fields through following these few steps:

- Crop rotation of strawberry with maize or vegetable should be done when dual II magnum (s-metolachlor) herbicide is used as a preplant application.
- Field should be ploughed before the formation of nutlets during July and dragging nutlets into uninfested fields should be avoided.
- Sinbar (terbacil) can suppress nut sedge if used at high rates.

Toadflax (*Linaria maroccana*) is a perennial weed from the family Scrophulariaceae in Kingdom Plantae and gets dispersed through extensive creeping root system that forms dense patches.

Management:

- Tillage suppresses this weed but dragging roots to clean fields should be avoided.
- Use of Dacthal on germinating seedlings is effective in its management.
- Try using Amitrol before cereals, corn or beans or after cereals.
- For established toadflax plants with 15 cm vegetative regrowth after summer fallow or tillage, glyphosate (Roundup) can effectively be applied for its management.

Oxalis *syn.* wood-sorrel (*Oxalis stricta*) produces more than 5000 seeds per plant (Guertin, 2003).

Management:

- Frequent tillage and moving of weed seedlings is useful.
- Integrated application of Dacthal and Sinbar (terbacil) may reduce the plant density of oxalis in the field to a manageable level.
- Application of herbicide goal (23.5% oxyfluorfen) before mulching may reduce oxalis density to satisfactory level.

Horsetail (*Equisetum arvense*) belongs to the family Equisetaceae of the Kingdom Plantae and mainly spreads from extensive rhizomes. Even short segments of broken rhizomes sprout easily.

Management:

- In strawberry plantations, horsetail can be suppressed by 2,4-D or Sinbar (terbacil).
- Crop rotation with corn and use of MCPA (2-methyl-4-chlorophenoxyacetic acid) or Amitrol (before or after the crop) is useful in its management.

Tricky weeds in strawberries:

- For field violet or pansy (*Viola* spp.), use of terbacil (Sinbar) and Dacthal on germinating seedlings and premulch application of Goal are effective.
- For dwarf snapdragon, using Dacthal in the spring before germination is beneficial.
- For management of groundsel (*S. vulgaris*), use of devrinol (technical name) is suggested. But, if groundsel is established in the field, application of Lontrel (clopyralid) at renovation is very effective for its management. Use of Gramoxone as chemical renovation can greatly reduce severity of groundsel in strawberry plantations.
- For vetch (*Vicia sativa*), oxeye daisy (*Leucanthemum vulgare*) and sheep *sorrel (Rumex acetosella),* applying Lontrel at renovation but not after mid-August is effective in their management.
- For bindweed (*Convolvulus* spp.), applying 2,4-D at renovation, or use of spot spray of 2% glyphosate when bindweed is in bloom are helpful in management of this weed.
- For quackgrass (*Elymus repens*), application of Venture (a.i. Fluazifop-P-butyl and S-isomer) or Poast Ultra (a.i. Sethoxydim) + Merge (a.i. Surfactant blend and solvent) in the spring is beneficial in its management. Or if quackgrass grows back after renovation, spot spray of 1% glyphosate at the three–six-leaf stage is effective in management of quackgrass.
- For milkweed (*Asclepias syriaca* L.), a wiper applicator with a 33% solution of glyphosate at flower bud can be very effective.
- For coltsfoot (*Cirsium arvense (*L.) Scop.*),* spot spray of 2% glyphosate after harvest but before mowing is beneficial.
- For Canada thistle (scientitific name), spot spray of 2% glyphosate at early flower bud, or a wiper application of 33% solution of glyphosate is effective in its management. 2,4-D or Lontrel at renovation could be helpful, especially if the thistles are at full leaf to flowering stage.

REFERENCES

Anonymous. 1992. *Herbicide Damage.* Department of Agriculture and Aquaculture. New Nouveau Brunswick, Canada.

Anonymous. 2008. US Forest Service, Pacific Southwest Research Station. "Biological Control: Insect Release Proposed To Control Exotic Strawberry Guava". *Science Daily*, 27 May 2008.

Anonymous. 2009. *Herbicide Injury Ontario Crop IPM.* Ontario Ministry of Agriculture, Food and Rural Affairs, Canada.

Anonymous. 2014. *Strawberry IPM Weed Management Guide.* http://www2.gnb.ca/content/dam/gnb/Departments/10/pdf/Agriculture/SmallFruits-Petitsfruits/StrawberryIPM.pdf. Accessed on 23 December 2016.

Anonymous. 2015. Renovation and weed management issue. Strawberry IPM Newsletter. No. 7. https://extension.umaine.edu/ipm/blog/tag/strawberry/accessed on 23-12-2016.

Boyd, N.S. and Reed, T. 2016. Strawberry tolerance to bed-top and drip-applied pre emergence herbicides. *Weed Technology*, 30: 492–498.

Dennis N.P. and Gail R.N. 2011. Rotation with cover crops suppresses weeds and increases plant density and yield of strawberry. *HortScience*, 46(10): 1363–1366.

Donald, L., Obrycki, J. and Gleason, M. 2001. Leopold Center Progress Reports. 10

Fennimore, S.A., Daugovish, O. and Smith, R.F. 2010 In: *IPM Pest Management Guidelines: Strawberry Statewide IPM Program, Agriculture and Natural Resources*, University of California.

Figueroa, R.A. and Doohan, D.J. 2006. Selectivity and efficacy of clopyralid on strawberry (*Fragaria* × *ananassa*). *Weed Technology*, 20(1): 101–103.

Figueroa, R.A., Doohan, D. and Cardina. 2005. Efficacy and selectivity of promising herbicides for common groundsel control in newly established strawberry. *HortTechnology*, 15(2): 261–266.

Fisher, P. and Huffman, L. 2010. *Herbicide Injury: Glyphosate (Roundup) Injury on Strawberries*. Ontario Ministry of Agriculture, Food and Rural Affairs, Canada.

Gilreath, J.P. and Santos, B.M. 2005. Weed management with oxyfluorfen and napropamide in mulched strawberry. *Weed Technology*, 19(2): 325–328.

Guertin, P. 2003. In: *USGS Weeds in the West Project Status of Introduced Plants in Southern Arizona Parks*. Factsheet for *Oxalis stricta*. http://sdrsnet.srnr.arizona.edu/data/sdes/ww/docs/oxalstri.pdf

Gurin, A.G. and Sukhochev, V.N. 2003. Biological method of controlling weedy vegetation in strawberry plantations. *Sadovodstvo i Vinogradarstvo*, 2: 9–10.

Gurin, A.G. and Sukhochev, V.N. 2005. Low cost methods for weed control in raspberry nurseries. *Sadovodstvo i Vinogradarstvo*, 2: 11.

Haar, M.J., Fennimore, S.A., Ajwa, H.A. and Winterbottom, C.Q. 2003. Chloropicrin effect on weed seed viability. *Crop Protection*, 22(1): 109–115.

Hoover, E., Rosen, C., Luby, J. and Burkness, S.W. 2016. Organic strawberry production in Minnesota. © 2016 Regents of the University of Minnesota. fruit.cfans.umn.edu

James, E.A., Charles, H.G. and Glenn, W. 2003. Weed control in field nurseries. *HortTechnology*, 13(1): 9–14.

Lisek, J. 2005. Herbicidal efficacy and selectivity chletodim for plants of strawberry. (Skutecznosc chwastobojcza i selektywnosc chletodymu dla roslin truskawki). *Progress in Plant Protection*, 45(2): 861–864.

Martinez, L., Castillo, N., Aguirre, I., Zamora, G.J.E., Avilla, C. and Medina, L.J. 2006. Effect of biofumigation on typical weeds of strawberry fields. *Acta Horticulturae*, 708: 193–196.

Michael, A.E., Richard, C.F., Shawn, W., Kathy, D., Elizabeth, W., Doohan, D., Welty, C., Williams, R.N. and Brown, M. 2006. *Midwest Strawberry Production Guide*. Bulletin, 926.

Polter, S.B., Doohan, D. and Scheerens, J.C. 2004. Tolerance of greenhouse-grown strawberries to terbacil as influenced by cultivar, plant growth stage, application rate, application site and simulated postapplication irrigation. *HortTechnology*, 14(2): 223–229.

Prittis, M.P. and Kelly, M.J. 2004. Weed competition in a mature matted row strawberry planting. *Weed HortScience*, 39(5): 1050–1052.

Samtani, J.B., Husein A.A., John B.W., Gregory T.B., Susanne K., Jonathan H. and Steven A.F. 2011. Evaluation of non-fumigant alternatives to methyl bromide for weed control and crop yield in California strawberries (*Fragaria ananassa* L.). *Crop Protection*, 30: 45–51.

Samtani, J.B., Weber, J.B. and Steven A.F. 2012. Tolerance of strawberry cultivars to oxyfluorfen and flumioxazin herbicides. *HortScience*, 47(7): 848–851.

Sniauka, P. and Pocius, A. 2008. Thermal weed control in strawberry. *Agronomy Research*, 6(Special issue): 359–366.

Timothy W.M., Carl R.L. and Brian G.M. 2013. Evaluation of organic amendments and flaming for weed control in matted-row strawberry. *HortScience*, 48(3): 304–310.

Wehtje, G. and Gilliam, C.H. 1991. Weed control in field-grown holly. *Environmental Horticulture*, 9: 29–32.

Whitaker, V.M., Price, J.F., Peres, N.A., MacRae, A.W., Santos, B.M., Noling, J.W. and Folta, K.M. 2014. Current strawberry research at the University of Florida. *Acta Horticulturae*, 1049: 161–166.

Zavatta, M., Shennan, C., Muramoto, J., Baird, G., Bolda, M.P., Koike, S.T. and Klonsky, K. 2014. Integrated rotation systems for soilborne disease, weed and fertility management in strawberry/vegetable production. *Acta Horticulturae*, 1044: 269–274.

16 Mulching

S.K. Singh, Prashant Kalal, and Pramod Kumar

CONTENTS

16.1 Organic Mulches 290
16.2 Inorganic Mulches 291
    16.2.1 Black Polyethylene Mulch 291
    16.2.2 White Polyethylene Mulch 292
    16.2.3 White, Coextruded White-on-Black/Silver Reflecting Mulch 293
    16.2.4 Red-Coloured Polyethylene Mulch 293
    16.2.5 Infrared-Transmitting Mulches 293
16.3 Colour effects 294
16.4 Effects of Mulching 294
    16.4.1 Growth and Earliness 294
    16.4.2 Soil Moisture Conservation 295
    16.4.3 Fruit Yield and Quality 295
    16.4.4 Plant Protection 296
    16.4.5 Weed Management 297
16.5 When to Mulch 297
16.6 Removal of Mulch 297
References 297

Strawberries are one of the most delicious and fragrantly sweet-flavoured fruits (Sharma and Sharma, 2004). It is believed that various types of straw that were used traditionally as mulch to suppress weeds in field and keep the berries (fruits) free from contamination of soil gave rise to the name "strawberry". However, some are of the opinion that the straw-like appearance of the runners led to its current English name (Pritts and Handley, 1998). However, according to Buczacki (1994), "the name strawberry does not seem to have anything to do with the practice of putting straw mulch around the plants".

Be that as it may, mulching is an important culturing practice in strawberry, which is known to influence plant growth, yield and fruit quality (Sharma and Singh, 1999). Mulching is commonly practised in strawberry cultivation to keep the fruit clean and protect it from its contact with the soil to avoid fruit rot. Mulches cover the soil surface and provide a microclimate favourable for plants growth and fruit yield. Different types of locally available straw are economically cheap, being the most popular mulches used for strawberry production, but synthetic mulches, for example, polyethylene films, are also used (Ochmian et al., 2007). Insulation is important during the summer as well as the winter. Loose and clean straw is an ideal insulator as it allows good airflow, and shields the soil/roots from the direct rays of the sun. This practice is also known for moderating the hydrothermal regime and increasing water use efficiency (Verma and Acharya, 1996). Natural mulches are derived from animal products and plant materials. Mulches reduce soil evaporation and augment yield through increasing water use efficiency (Adekalu et al., 2006). Organic and/or inorganic mulches cover soils, and form a physical barrier, thereby minimizing the loss of soil water through evaporation, controlling soil erosion and growth of weeds, improving soil structure and infiltration of water into the soil and protecting crops from contamination with native soil. The mechanism of increasing water use efficiency by synthetic mulch and organic mulch is different. The main factor preventing the evaporation by organic mulches is soil temperature (Bushnell and Welton, 1931). Natural mulches help in maintaining soil organic matter and tilth (Tindall et al., 1991), providing food and shelter for earthworms and other desirable soil biota (Doran, 1980).

The ideal mulch should have the following characteristics:

- Durable and comparatively inexpensive
- Easily available and degradable
- Free from contamination such as the seeds of weeds and inoculum of different pathogen
- Free from becoming host to harmful insect-pests and diseases
- Good incubator of heat
- Neither very hard nor too light in weight to be easily blown away by winds

Mulches are used as a soil covering, for a variety of reasons including moisture conservation, heat trapping, creation of pathways for preventing runoff and soil loss, weed prevention and control and protecting roots from fluctuating and extreme temperatures.

Among synthetic mulches, different types and colours have been observed to affect soil and air temperature differently and, accordingly, the growth of the candidate crop (Tarara, 2000). Research under different climatic conditions across the world has shown conflicting results for strawberries grown on different types or colours of plastic mulch films (Al Khatib et al., 2001; Johnson and Fennimore, 2005; Mohamed, 2002). Some researchers have concluded that strawberry plants are very sensitive to colour of mulch material, exhibiting differences in plant growth and yield (Johnson and Fennimore, 2005; Mohamed, 2002), while others have observed that black mulch remains the best option for strawberry production (Himelrick et al., 1983). Earlier literature revealed that the growth attributes of strawberry plantlets has direct impact on the microclimate around the plant by modifying the radiation budget of the surface and better moisture conservation, besides weed suppression (Hassan et al., 2000; Tarara, 2000). Plastic mulches stimulate growth of young plants by reducing evapotranspiration and restraining heat loss during cold nights (Lieten, 1991). Some mulches may have a positive effect on photosynthesis (Atkinson et al., 2006) and on young plant development and early production of quality fruits (Orzolek and Murphy, 1993). Mulch materials can be categorized as organic and inorganic.

16.1 ORGANIC MULCHES

Organic mulches can be made of several naturally occurring substances. Hardwood mulch and hardwood bark mulch are two of the most common. However, pine needles, leaves, or even grass clippings are organic materials that can be used as mulch. Organic mulches decompose over time. Usually organic mulches should be replaced every couple of years. Past studies documented that the use of a surface organic (straw) mulch resulted in protecting the crop from stronger and stormy water precipitation by reducing runoff and facilitating increased infiltration and decreased evapo-ration (Rao et al., 1998; Schertz and Kemper, 1998). Organic mulches such as straw-vetch provide environmental benefits including increased nitrogen, nutrient recycling, reduced soil erosion, weed emergence and water loss, addition of soil organic matter, lower soil temperature during the hot summer months; and after decomposition, these act as source of slow-release nutrients (Abdul Baki and Teasdale, 1993; Muhammad et al., 2009; Sarolia and Bhardwaj, 2012).

In the subtropical and tropical regions of the world, paddy straw is cheap and easily available, hence an ideal choice for use as mulch for strawberry in the warm localities. Paddy straw is a natu-rally available secondary product from the paddy field and easily degradable and eco-friendly. The incidence of albinism has been observed to be much less with application of paddy straw (22.5%), signifying its importance over other mulch materials under subtropical climatic conditions (Sharma and Sharma, 2003). Oat, hay, rye or wheat straws are also equally useful materials for mulch-ing as these are loose and mostly lightweight and do not smother the plants. This fact assumes greater significance in view of the environmental issues concerning burning of leftover crop refuse of paddy or wheat in the field in states such as Punjab, Haryana, Uttar Pradesh and Rajasthan in India (https://www.downtoearth.org.in/.../paddy-burning-ngt-orders-fine-imposition-on-erri) as the smoke emitted due to burning is considered to be one of the factors causing several health problems especially respiratory diseases in these regions.

But organic mulch may contain seeds of other crops and weeds, which may later interfere with main crop after germination. In addition, these may act as hosts for harmful insect-pests and pathogens. Besides, placing and removal of straw beneath the plant are sometimes considered as labour-intensive and difficult tasks.

16.2 INORGANIC MULCHES

Inorganic mulches include gravel, pebbles, polyethylene and landscape fabrics. These types of mulches do not attract insect-pests and pathogens but do not decompose, which can be both beneficial in certain situations but limiting factors in certain other situations. In this system, polyethylene sheets are laid down over raised beds and strawberry plants are planted through holes punched in the top of the sheets (Figure 16.1).

16.2.1 Black Polyethylene Mulch

Black polyethylene mulch considerably reduces severity of weeds from growing around strawberry plants (without having to cultivate the soil) and further help to conserve the soil moisture and keep fruit clean. Black polyethylene mulch, the predominant colour used in strawberry production is opaque and absorbs and radiates (Figure 16.2) solar energy. These mulches absorb most UV, visible and infrared wavelengths of incoming solar radiation, and reradiates absorbed energy in the form of thermal radiation or long-wavelength infrared radiation back into the environment. Much of the solar energy absorbed by black plastic mulch is lost into the atmosphere through radiation and forced convection (Lamont, 1999). Nowadays, most polyethylene mulch is based on linear low density polyethylene (LLDPE) because it is more economical to use. The application of black plastic film as mulch is becoming popular, and very good results have been achieved particularly in arid and semi-arid regions (Bhardwaj et al., 2011). Polyethylene mulches regulate the hydrothermal regime and protect delicate strawberry fruits from direct contact with the soil (Sharma, 2000). Currently, the use of black polythene sheets for mulching in strawberry is a common practice. Besides, polyethylene mulch in other colours is also being used to improve the yield and quality of strawberries. The mulching contributes to a considerable reduction in

FIGURE 16.1 Strawberry planting with polyethylene mulching.

FIGURE 16.2 Black polyethylene mulching in strawberry.

the evaporation of water from the soil surface, and consequently better water utilization by the plants (Treder, 2003). Mulch can also be used to alleviate the negative effects of a long-term drought. The incidence of albinism is also significantly influenced by mulching, reaching its maximum under black polyethylene mulch and its minimum when grown under paddy straw (Sharma et al., 2003). In addition, black polyethylene mulch has also been observed to promote nitrogen mineralization, which might also have favoured the incidence of albinism (Lieten and Marcelle, 1993a, b). The plants have better growth and produce fruits with better weight and total soluble solids when cultivated under black polyethylene mulch, but the fruits have higher incidence of albinism, a serious disorder in subtropical climatic conditions where the temperature is comparatively high during fruit maturity and ripening (Sharma and Sharma, 2003). Sharma et al. (2013) concluded that coloured plastic mulches (black, red and yellow) significantly increased growth attributes and extended the duration of flowering and fruiting, and improved the fruit set, yield and fruit quality of strawberry in protected cultivation. Himelrick (1982) observed that plants grown under black polyethylene mulch produced more runners and fruits than those grown on clear or white plastic mulches, and that total fruit mass was higher with black and clear plastic mulches than with bare soil, on the other hand Baumann et al. (1995) reported no difference in fruit yields due to use of green and black plastic mulches, but plants grown on black mulch produced larger berries than those grown on green or no plastic mulch.

Recycling in subsequent years, reduced weed growth, better water use efficiency, production of disease-free fruits and early harvesting are some advantages of black polyethylene mulch. However, it is expensive and non-degradable. Besides, its use brings a high risk of spring frost injury (Plekhanova and Petrova, 2002) and sun scalding in late strawberry production. In open cultivation, polyethylene mulch may cause increased incidence of berry rotting during winter rains.

16.2.2 WHITE POLYETHYLENE MULCH

Clear plastic mulches absorbs little solar radiation but transmit 85–95%, with relative transmission depending on the thickness and degree of opacity of the polyethylene (Figure 16.3). The undersurface of clear plastic mulches is usually covered with condensed water droplets. This water is transparent to incoming short-wave radiation but opaque to outgoing long wave infrared radiation; so much of the heat lost to the atmosphere from a bare soil by infrared radiation is retained by clear plastic mulches. Clear polyethylene mulches are used primarily to provide an even warmer soil environment, a mini-greenhouse effect that allows early production. Using clear plastic mulches requires the use of herbicide, soil fumigant or solarization to control weeds (Lamont, 1999). The clear plastic mulch allows for the greatest transmission of infrared radiation, which allows high transmission of visible light. High light transmission may allow for weed growth under the mulch, if soil sterilization and eradication of weed seeds and propagules are incomplete (Himelrick et al., 1992).

Mulching

FIGURE 16.3 White polyethylene mulching in strawberry.

16.2.3 WHITE, COEXTRUDED WHITE-ON-BLACK/SILVER REFLECTING MULCH

Reflective silver or aluminium mulches also help to minimize soil temperatures. They tend to repel aphids, which could serve as vectors for various viral diseases. These mulches can be used to establish a crop when soil temperatures are high. Depending upon the degree of opacity of white mulch, it may require the use of a fumigant or herbicide because of potential weed growth (Lamont, 1999). Clear polyethylene mulches elevate soil temperatures more than opaque mulches, while reflective and light-coloured mulches tend to keep soil temperatures at the minimum when compared with dark-coloured mulches. White and reflective mulches (aluminium foil and aluminium-painted plastic mulches) have the effect of changing the amount and quality of light also reflected up into the plant canopy (Ham et al., 1993). Reflective mulches have lower insect infestations and reduced insect-transmitted diseases (Greer and Dole, 2003). Increased yield on aluminium-painted mulch was attributed to increased photosynthetically active radiations being reflected (Porter and Etzel, 1982). The amount of reflected light can be altered with the use of plastic mulches placed between the planting beds (Ham et al., 1993). Gough (2001) recorded no difference in soil temperature at a depth of 5 cm in the beds covered with either black or silver mulch. Ham et al. (1993) examined the optical properties of several mulches, and determined that silver mulch would absorb less short and long wave radiation compared with black, but would also emit less long wave radiation, potentially making it a better insulator, trapping more soil heat compared with black mulch.

16.2.4 RED-COLOURED POLYETHYLENE MULCH

Red mulches are the really new coloured mulch materials to be experimented and investigated with. It has been observed to increase the soil temperatures and yield of strawberries.

New colours that are currently being investigated are red, yellow, blue, grey and orange, which have distinct optical characteristics, and thus reflect different radiation patterns into the canopy of a crop, thereby affecting photosynthesis and plant morphogenesis, and increase early yield. The colours of mulch material are also correlated with the behaviour of certain insects, for example, yellow and orange surfaces attract green peach aphids, hence must be avoided.

16.2.5 INFRARED-TRANSMITTING MULCHES

These mulches range from blue-green to brown and selectively absorb photosynthetically active radiation, while transmitting solar infrared radiation, and exhibit improved soil-heating characteristics, although generally not as well as clear plastic mulches.

16.3 COLOUR EFFECTS

The most popular plastic mulch worldwide is black, followed by white-on-black and clear mulches. Other colours that have been evaluated include blue, green, red, yellow, brown, white and silver (Brault et al., 2002; Ngouajio and Ernest, 2004). The optical properties of coloured mulches influence soil and air temperatures around the crop as well as affecting weed growth under the mulch. In India, strawberry growers are shifting from hay mulching to black plastic mulching to reduce the amount of weeds, increase the soil temperature during early spring, conserve soil moisture and prevent the soil from splashing on the fruits. Furthermore, black polyethylene sheets absorb solar energy and keep more heat underneath the plastic mulch. Since different colours are known to reflect different wavelengths of light, the question of whether other colours would increase plant growth and provide the same favourable conditions as the black plastic alone is in fact a matter that needs to be precisely investigated. Plants use light as a signal that enables them to compete with their surroundings. They do not know if the signal is a neighbouring plant, dead plants on the surface of the soil or even the colour of the soil. The plant recognizes far-red light as the signal. If the plant detects an abundance of far-red reflection then there must be other plants growing nearby. The phytochrome signals the plant to put more photosynthates into shoots instead of roots. Polyethylene with various pigment combinations is called coloured mulch. In coloured mulches, crops are expected to produce larger fruits of better quality. Since the mulches do reflect sunlight in its full spectral distribution, other benefits including enhanced sugar accumulation and better fruit size may be achieved. Coloured mulches mimic the reflective patterns of the green leaves of neighbouring plants. The plant responds to the increased ratio of far-red–red light as though it was reflected from the nearby plants, when, in fact, it is just the coloured mulch, because of which, the plants put in more energy into the shoots to outgrow other plants. In fact, it even initiates the plant into producing more and better-quality fruit. The comparative effectiveness of coloured polyethylene sheets is summarized in Table 16.1.

16.4 EFFECTS OF MULCHING

16.4.1 GROWTH AND EARLINESS

Earlier studies have documented the significant effects of organic/synthetic mulches on vegetative growth and flowering traits of strawberry plants (Lareau and Lamarre, 1990; Lieten and Baets, 1991). Plants mulched with black polyethylene sheets have been observed to significantly improve growth, and result in early flowering and fruiting than either with clear polyethylene or paddy straw

TABLE 16.1
Comparative Effectiveness of Colours of Plastic on Light and Weed Control

Mulch Colour	Soil Temp. (5–10 cm Depth)	Remarks
Black	Increases (3–5°F)	Most common, and does well in temperate climate with excellent weed suppression
Clear	Increases (6–14°F)	Clear plastic gives poor weed control. Best in cool regions and for all crops
White/Silver	Decreases (−2–0.7°F)	Reflection interferes with movement of aphids. Suited for tropical climate having excellent weed suppression
Infrared Transmitting (IRT)	Increases (5–8°F)	Selective light transmission with excellent weed suppression

Source: Angima (2009), Penn State Extension (2015), and Sanders (2001).

Mulching 295

(Singh and Asrey, 2005), mainly because of varying soil hydrothermal regimes under different mulches (Singh et al., 2007). Plants under red polythene mulch were first to come into flowering followed by yellow and black polythene mulches. Red polythene mulches enhanced the flowering by three weeks (Sharma et al., 2013). Red and yellow plastic mulches significantly extended the duration of flowering and fruiting and improved the fruit set. Shokouhian and Asghari (2015) reported the superiority of black polyethylene mulch, most effective for strawberry cv. Paros in Ardabil's climatic conditions.

16.4.2 SOIL MOISTURE CONSERVATION

Strawberry is a low surface-creeping herb and, therefore, mulching is considered the most important cultural practice for soil moisture conservation. Better moisture conservation and higher soil temperature with the use of black polyethylene mulch than other mulches have also been reported by Hassan et al. (2000) and Singh et al. (2006). Studies have revealed that the soil moisture content differed significantly, and appropriate mulching of top soil layer was quite useful in the field to accelerate growth and get higher fruit yield. The highest average weight-based soil moisture content of the active soil layer (0–40 cm) within three years (2006–2008) was 18% in the fields mulched with straw, while the lowest one (16.2%), was in the field without mulching; in the plot mulched with black polyethylene, it was 16.5%. The yield of two years (2007–08) obtained from the field mulched with the black polyethylene sheet was higher by 60% than that in the non-mulched field and by 56% in comparison to the yield in the plot mulched with straw (Taparauskiene and Miseckaite, 2014). Kumar and Dey (2011) evaluated the effects of different mulches (hay, black polyethylene) on water use efficiency of strawberry cv. Chandler under drip and surface irrigation, and concluded that soil moisture content was greater under black polyethylene mulch compared with hay and subsequently, increased by 2.8–12.8% under black polyethylene mulch.

16.4.3 FRUIT YIELD AND QUALITY

Yield and quality of plant produce are influenced by the light conditions that exist during plant growth and development. Black is the most widely used colour of polyethylene mulch (Blackhurst, 1962). Though red and yellow polyethylene mulches also increase the size, weight and yield of fruits with improved fruit chemical composition, but red plastic mulches were observed to be most effective for increasing yield and improving fruit quality of strawberries under protected cultivation (Sharma et al., 2013).

It is well documented that black polyethylene-mulched strawberry produces higher yields of better-quality fruits than unmulched controls (Kikas and Luik, 2000). It has been shown that mulches add growth regulatory effect of reflected wavelength combinations in the visible and far-red parts of the electromagnetic solar light spectrum. It is reported that fruit size of strawberries are larger when raised using red plastic mulches because the reflected far-red and red lights affect phytochrome-mediated allocation of photosynthates, and more is directed to developing fruits (Kasperbauer, 2000). Berries that ripen over the red light spectrum have been observed to grow 20% larger with higher sugar to organic acid ratios, along with emitting higher concentrations of favourable aroma compounds (Kasperbauer et al., 2001).

The effect of different coloured plastic mulches (black, red, and white) along with conventional practice was evaluated on yield and quality of strawberry cv. Camarosa. It was reported that mean fruit weight was significantly higher in red plastic mulches than in white mulches and control treatments. Fruit size significantly increased over red plastic mulch (Shiukhy et al., 2015). Singh et al. (2007) observed that black plastic mulch enhanced the weight and quality of strawberry fruit. Plants under black polyethylene mulch produced larger fruit and had higher yield per plant, mainly because of better plant growth owing to favourable hydrothermal regime of soil and a complete weed-free environment (Sharma and Sharma, 2003; Singh and Asrey, 2005; Singh et al., 2006).

Although, plants under clear polyethylene mulch also had better plant growth, but berry weight and fruit yield per plant were less than in plants mulched with paddy straw. Chopped wheat straw and black plastic were used as organic and inorganic mulches with comparison to non-mulched strawberry, and the average differences in berry weight between mulched and non-mulched plants were significant. The total number of berries per plant was 1.4 times bigger in the fields mulched with black polyethylene and lower in the fields mulched with straw than in the non-mulched control (Taparauskiene and Miseckaite, 2014). Earlier, Bunty Shylla and Sharma (2010) found that in cv. Chandler, yellow plastic mulch significantly increased the number of fruits, and were effective in early and higher total fruit yield when compared with black or silver-over-purple plastic mulches in the agro-climatic conditions of Solan, Himachal Pradesh, India. Un-mulched beds produced lower fruit yield of inferior quality. Yellow plastic mulches raised soil temperature by at least 2°C when compared with the unmulched bed.

Besides yield, mulches have significant effect on the fruit quality in terms of colour, shape, size and contents of sugars and organic acids (Rubeiz et al., 1991). Red and yellow plastic mulches have increased fruit yield, fruit size, weight and improved fruit chemical composition (total soluble solids, total sugars, reducing sugars, ascorbic acid and anthocyanin) of strawberry fruits under protected cultivation (Sharma et al., 2013). It is well documented that soil hydrothermal regimes and moisture conservation induced favourable conditions conducive for improved size and weight as the colour of mulches affect temperature below and above the mulch through the absorption, transmission and reflection of solar energy, which affects the micro-environment surrounding the plants (Lamont, 1999). Coloured mulches selectively absorb photosynthetically active radiation, while transmitting solar infrared radiation, thus red and yellow plastic mulches have the ability to increase the ratio of red:far-red wavelengths (R:FR) in the reflected light and resulted in increased photosynthetic activity of the plants, thereby accumulation of anthocyanin content in the fruits.

Strawberry is a high-value food crop, the aroma of which is important in consumer satisfaction. Twenty-three aroma compounds have been identified from fresh strawberry fruit. The greatest concentration of aroma compounds consisted of aliphatic esters, predominantly with the methyl and ethyl esters of butanoic and hexanoic acids. Concentration of fresh strawberry aroma compounds in fruits can be enhanced by growing the berries over a red plastic mulch that formulate to reflect more far-red (FR) and red light (R) and a higher FR:R photon ratio than standard black plastic mulch. It is reported that the concentration of esters produced by fruits grown over the red plastic mulch was >90% higher for Chandler and almost 50% higher for Sweet Charlie than for fruits grown over black plastic mulch (Loughrin and Kasperbauer, 2002). Miao et al. (2016) studied the effects of plastic films (red, yellow, green, blue, and white) on the content of anthocyanin in strawberry cv. Yueli. They observed that the red and yellow films led to a significant increase in total anthocyanin content (TAC) when compared with the white film, while the green and blue films caused a decrease in TAC. They also observed that there was improvement in flavonoids and upregulation of several related enzymes.

16.4.4 PLANT PROTECTION

Plant protection is a major problem in organic strawberry production. Grey mould (*Botrytis cinerea* Pers.) is the most serious disease, adversely affecting fruit yield in conventional and organic strawberry production. Mulching materials significantly affect grey mould incidence, which averaged 5.6% in plastic and 10.3% in green mulch. Buckwheat husk mulch causes the most severe losses compared with other mulches (Kivijarvi et al., 2002). Straw mulch provides a physical barrier between strawberry fruit and soilborne inoculums of pathogens such as *Phytophthora cactorum*; reduces splash dispersal of this organism and other pathogens (*Colletotrichum* spp.), and thus helps to keep the strawberry fruits healthy, clean and attractive. Laugale et al. (2004) reported that straw mulch laid between the rows is equally or more effective than fungicides for controlling leather rot (*P. cactorum*).

Mulching 297

16.4.5 WEED MANAGEMENT

Weed control is one of the greatest challenges faced by strawberry growers. Since strawberry plants are relatively slow growing and are poor competitors, weeds quickly invade and establish within bare areas. A good weed management programme is required to control weeds over the life of the strawberry planting through an integrated pest management strategy that integrates preventive, cultural, mechanical, biological and chemical control methods to have a sustainable production system that balances economic, health and environmental concerns. When combined with tillage techniques and herbicides, plastic mulches allow strawberry growers to maintain nearly weed-free fields. Black plastic mulch is not transparent to solar radiation, and, therefore, most effective to suppress the weed growth (Dittmar and MacRae, 2012). The practical use of a particular weed control method is related to the implementation costs. However, mulching the soil or the use of organic herbicides is several times more costly than the use of synthetic herbicides (Markuszewski and Kopytowski, 2008). Mulches obtained as cheap waste materials may be used locally (Rowley et al., 2011). A detailed discussion on weed management in strawberry has already been given in Chapter 15.

16.5 WHEN TO MULCH

In strawberry, mulching is usually done during the onset of winter to protect the plants from low-temperature stress and adverse frost. Based on prevailing temperature, the time of mulching is determined in the hills during winter. Mid-September plantings coupled with black polyethylene mulch favour better vegetative growth, with early and higher yield of better-quality fruits having a lesser incidence of botrytis rot in strawberry (Singh et al., 2007). Strawberry fruit production is greatly hampered by low-temperature stress during autumn, winter or spring in many temperate and subtropical regions of the world, which results in destruction of developing buds and flowers (Boyce and Read, 1983; Ourecky and Reich, 1976).

Too early winter mulching, therefore, may be as undesirable as too late. To harden the plants in the autumn when the growth has stopped, it is advisable to leave them exposed to temperatures up to a few degrees below freezing point for several days before mulching them. But the plants should be mulched, before they have been exposed to lower than −6.6°C. Well-hardened plants can often withstand −9.44°C. In general, the mulch should be applied after a week or more of near freezing temperature, and not after a few days of warm weather that may cause the plant to lose winter hardiness (Shoemaker, 1955).

16.6 REMOVAL OF MULCH

The time of removal of mulch is important in strawberry production. The best method to determine the proper timing for removal of mulch is the beginning of leaf growth under the mulch. It is generally ideal to remove the winter mulch in early spring when the incidence of adverse frost is over. If the mulching is removed very early, it may result in beginning of early flowering, and later the crop may face frost injury. The delayed removal of mulch tends to produce fruits smaller in size, and may enhance the incidence of several leaf diseases (Abbott and Gough, 1992). The time of retaining the mulch depends on cultivar, whether it is early, mid and late type.

REFERENCES

Abbott, J.D. and Gough, R.E. 1992. Comparison of winter mulches on several strawberry Cultivars. *Journal of Small Fruit and Viticulture*, 1: 51–58.

Abdul Baki, A.A. and Teasdale, J.R. 1993. A no tillage tomato production system using hairy vetch subterranean clover mulches. *HortScience*, 28: 106–108.

Adekalu, K.O., Okunade, D.A. and Osunbitan J.A. 2006. Compaction and mulching effects on soil loss and runoff from two southwestern Nigerian agricultural soils. *Geoderma*, 137: 226–230.

Al Khatib, B.M., Sleyman, A.S., Freiwat, M.M., Knio, K.M. and Rubeiz, I.G. 2001. Mulch type effect on strawberries grown in a mild winter climate. *Small Fruits Review*, 1: 51–61.

Angima, S. 2009. Season extension using mulches. Oregon State University Extension: Small Farms. http://smallfarms.oregonstate.edu/sfn/f09Season Mulches 4(3).

Atkinson, C.J., Dodds, P.A.A., Ford, Y.Y., Le Miere, J., Taylor, J.M., Blake, P.S. and Paul, N. 2006. Effects of cultivar, fruit number and reflected photosynthetically active radiation on *Fragaria* × *ananassa* productivity and fruit ellagic acid and ascorbic acid concentrations. *Annals of Botany*, 97: 429–441.

Baumann, T.E., Eaton, G.W., Machholz, A. and Spaner, D. 1995. Day-neutral strawberry production on raised beds in British Columbia. *Advances in Strawberry Production*, 14: 53–57.

Bhardwaj, R.L., Meena, C.B., Singh, N., Ojha, S.N. and Dadhich, S.K. 2011. *Annual Progress Report of KrishiVigyan Kendra, Sirohi*, MPUAT, Udaipur, India: pp. 45–46.

Blackhurst, H.T. 1962. Commercial use of black plastic mulch. In *Proceedings National Horticultural Plastics Conference*: p. 27.

Boyce, B.R. and Reed, R.A. 1983. Effects of bed height and mulch on strawberry crown temperatures and winter injury. *Advances in Strawberry Production*, 2: 12–14.

Brault, D., Stewart, K.A. and Jenni, S. 2002. Optical properties of paper and polyethylene mulches used for weed control in lettuce. *HortScience*, 37: 87–91.

Buczacki, Stefan. 1994. *Best Soft Fruit*. Hamlyn an Imprint of Reed Consumer Books Limited, London, UK.

Bunty, Shylla and Sharma, C.L. 2010. Evaluation of mulch colour for enhancing winter-strawberry production under polyhouse in mid-hills of Himachal Pradesh. *Journal of Horticultural Sciences* 5(1): 34–37.

Bushnell, J. and Welton, F.A. 1931. Some effects of straw mulch on yield of potatoes. *Journal of Agriculture Research*, 43: 837–845.

Dittmar, P.J. and MacRae, A.W. 2012. *The Florida Vegetable Handbook*. Weed management: chapter 5. University of Florida IFAS Extension. [Online: http://nfrec. ifas. ufl. edu/ Vegetable _Production/2012_VPG_main5.pdf]. Accessed 29 May 2014.

Doran, J.W. 1980. Microbial changes associated with residue management with reduced tillage. *Soil Science Society of America Journal*, 44: 518–524.

Gough, R.E. 2001. Colour of plastic mulch affects lateral root development but not root system architecture in pepper. *HortScience*, 36: 66–68.

Greer, L. and Dole, J.M. 2003. Aluminum foil, aluminum-painted, plastic and degradable mulches increase yields and decrease insect-vectored viral diseases of vegetables. *HortTechnology*, 13: 276–284.

Ham, J.M., Kluitenberg, G.J. and Lamont, W.J. 1993. Optical properties of plastic mulches affect the field temperature regime. *Journal of the American Society for Horticultural Science*, 228: 188–193.

Hassan, G.I., Godara, A.K., Kumar, J. and Huchche, A.D. 2000. Effect of different mulches on the yield and quality of "Oso Grande" strawberry (*Fragaria* × *ananassa*). *Indian Journal of Agricultural Sciences*, 70: 184–185.

Himelrick, D.G. 1982. Effect of polyethylene mulch colour on soil temperatures and strawberry plant response. *Advances in Strawberry Production*, 1: 15–16.

Himelrick, D.G., Dozler, W.A. and Akridge, J.R. 1983. Effect of mulch types in annual hill strawberry plantations. *Acta Horticulturae*, 348: 207–212.

Himelrick, D.G., Dozier Jr, W.A. and Akridge, J.R. 1992. Effect of mulch type in annual hill strawberry plasticulture systems. In: II *International Strawberry Symposium*, 348: 207–212.

Johnson, M.S. and Fennimore S.A. 2005. Weed and crop response to coloured plastic mulches in strawberry production. *HortScience*, 40: 1371–1375.

Kasperbauer, M.J. 2000. Strawberry yield over red versus black plastic mulch. *Crop Science*, 40: 171–174.

Kasperbauer, M.J., Loughrin, J.H. and Wang, S.Y. 2001. Light reflected from red mulch to ripening strawberries affects aroma, sugar and organic acid concentrations. *Photochemistry and Photobiology*, 74: 103–107.

Kikas, A. and Luik, A. 2000. The influence of different mulches on strawberry yield and beneficial entomofauna. In: *IV International Strawberry Symposium, Acta Horticulturae*, 567: 701–704.

Kivijarvi, P., Parikka, P. and Prokkola, S. 2002. Effects of biological sprays, mulching materials, and irrigation methods on grey mould in organic strawberry production. In: *XXVI International Horticultural Congress: Berry Crop Breeding, Production and Utilization for a New Century*, 626: 169–176.

Kumar, S. and Dey, P. 2011. Effects of different mulches and irrigation methods on root growth, nutrient uptake, water-use efficiency and yield of strawberry. *Scientia Horticulturae*, 127: 318–324.

Lamont, W.J. 1999. The use of different coloured mulches for yield and earliness. In: *Proceedings of the New England Vegetable and Berry Growers Conference and Trade Show*, Sturbridge, Massachusetts: pp. 299–302.

Lareau, M.J. and Lamarre, M. 1990. Effects of row cover and mulching in the production of day neutral strawberry cultivars. In: *Proceedings of the Third North American Strawberry Conference*, Houston, Texas.

Laugale, V., Bite, A. and Morocko, I. 2004. The effect of different organic mulches on strawberries. In: *V International Strawberry Symposium*, 708: 591–594.

Lieten, P. 1991. Multi-coloured crop production. *Grower*, 116: 9–10.

Lieten, F. and Beats, W. 1991. Research on strawberries. No future for coloured films. *Onderzoek-aardbei. Geentoekomstvoorgekleurdefolies. Groenteno Fruit, Vollegrondsgroenten*, 1: 10–11.

Lieten, F. and Marcelle, R.D. 1993a. Relationship between fruit mineral content and the albinism disorder in strawberry. *Annals of Applied Biology*, 123: 433–439.

Lieten, F. and Marcelle, R.D. 1993b. Relationship between fruit mineral content and the albinism disorder in Elsanta strawberry plants. *Acta Horticulturae*, 348: 294–298.

Loughrin, J.H. and Kasperbauer, M.J. 2002. Aroma of fresh strawberries is enhanced by ripening over red versus black mulch. *Journal of Agricultural and Food Chemistry*, 50: 161–165.

Markuszewski B. and Kopytowski J. 2008. Zachwaszczenieikosztyjegoregulowania w sadziejabłoniowym z produkcjaintegrowana. [Weed infestation and costs of their regulation in apple orchard with IPM]. *Zeszyty Naukowe Instytutu Sadownictwa i Kwiaciarstwa*, 16: 35–50.

Miao, L., Zhang, Y., Yang, X., Xiao, Z., Huiqin J., Zhang, Z., Wang, Y. and Jiang, G. 2016. Coloured light-quality selective plastic films affect anthocyanin content, enzyme activities, and the expression of flavonoid genes in strawberry (*Fragaria × ananassa*) fruit. *Food Chemistry*, 207: 93–100.

Mohamed, F.H. 2002. Effect of transplant defoliation and mulch colour on the performance of three strawberry cultivars grown under high tunnel. *Acta Horticulturae*, 567: 483–485.

Muhammad, A.P., Muhammad, I., Khuram, S. and Anwar-UL-Hassan. 2009. Effect of mulch on soil physical properties and NPK concentration in maize (*Zea mays*) shoots under two tillage system. *International Journal of Agriculture and Biology*, 11: 120–124.

Ngouajio, M. and Ernest, J. 2004. Light transmission through coloured polyethylene mulches affected weed population. *HortScience*, 39: 1302–1304.

Ochmian I., Grajkowski J. and Kirchhof R. 2007. Wpływ mulczowania gleby na ... skawki odmiany "Filon" Rocz. AR. Pozn. CCCLXXXIII, *OGRODN*. 41: 357–362.

Orzolek, M.D. and Murphy, J.H. 1993. The effect of coloured polyethylene mulch on the yield of squash and pepper. *Proceedings 24th of National Agricultural Plastics Congress*, Pennsylvania State University. University Park, PA; 24: 157–161.

Ourecky, D.K. and Reich, J.E. 1976. Frost tolerance in strawberry cultivars. *HortScience*, 11: 413–414.

Penn State Extension. 2015. Plastic mulches. Penn State Extension, College of Agricultural Sciences. http://extension.psu.edu/plants/plasticulture/technologies/plastic-mulches.

Plekhanova, M.N. and Petrova, M.N. 2002. Influence of black plastic soil mulching on productivity of strawberry cultivars in Northwest Russia. *Acta Horticulturae*, 567: 491–494.

Porter, W.C. and Etzel, W. 1982. Effects of aluminium-painted and black polyethylene mulches on bell pepper, *Capsicum annuum* L. *HortScience*, 17: 942–943.

Pritts, M. and Handley, D. 1998. *Strawberry Production Guide. Northeast, Midwest and Eastern Canada* (NRAES-88).

Rao, K.P.C., Cogle, A.L., Srinivason, S.T., Yule, D.F. and Smith, G.D. 1998. Effect of soil management practices on run off and infiltration processes of hardsetting Alfisol in semiarid tropics. In: L.S. Bhushan, I.P. Abral and M.S. Rama Mohan Rao (ed.) Soil and water conservation: Challenges and opportunities. *Proceeding of 8th ISCO Conference*, 1994. New Delhi, India. A.A. Balkema, Rotterdam, the Netherlands: pp. 1287–1294.

Rowley M.A., Ransom C.V., Reeve J.R. and Black B.L. 2011. Mulch and organic herbicide combinations for in-row orchard weed suppression. *International Journal of Fruit Science* 11 (4): 316–331.

Rubeiz., I.G. Naja Z.U. and Nimah, H.N. 1991. Enhancing late and early yield of greenhouse cucumber with plastic mulches. *Biological Agriculture and Horticulture*, 8: 67–70.

Sanders, D. 2001. Using plastic mulches and drip irrigation for home vegetable gardens. Horticulture information leaflet. North Carolina Extension Resources. http://content.ces.ncsu.edu/using-plasticmulches-and-drip-irrigation-for-vegetable-gardens

Sarolia, D.K. and Bhardwaj, R.L. 2012. Effect of mulching on crop production under rainfed condition: A review. *International Journal of Research in Chemistry and Environment*, 2: 8–20.

Schertz, D.L. and Kemper W.D. 1998. Crop-residue management system and their role in achieving a sustainable, productive agriculture. In L.S. Bhushan, I.P. Abral and M.S. Rama Mohan Rao (ed.). *Soil and Water Conservation: Challenges and Opportunities. Proceeding of 8th ISCO Conference*, 1994. New Delhi, India. A.A. Balkema, Rotterdam, the Netherlands: pp. 1255–1265.

Sharma, V.K. 2000. Effect of mulching and row spacing on growth and yield of strawberry. *Indian Journal of Horticulture*, 66: 271–273.

Sharma, R.R. and Sharma, V.P. 2003. Mulch type influences plant growth, albinism disorder and fruit quality in strawberry (*Fragaria × ananassa* Dusch.). *Fruits*, 58: 221–227.

Sharma, R.R. and Sharma, V.P. 2004. *The Strawberry*. ICAR, New Delhi, India,

Sharma, R.R. and Singh, S.K. 1999. Strawberry cultivation: A highly remunerative farming enterprise. *Agro India*, 3: 20–22.

Sharma, R.R., Sharma, V.P. and Pandey, S.N. 2003. Mulching influences plant growth and albinism disorder in strawberry under subtropical climate. In *VII International Symposium on Temperate Zone Fruits in the Tropics and Subtropics*, 662: 187–191.

Sharma N.C., Sharma, S.D. and Spehia, R.S. 2013. Effect of plastic mulch colour on growth, fruiting and fruit quality of strawberry under polyhouse cultivation. *International Journal of Bio-resource and Stress Management*, 4: 314–316.

Shiukhy, S., Raeini-Sarjaz, M. and Chalavi, V. 2015. Coloured plastic mulch microclimates affect strawberry fruit yield and quality. *International Journal of Biometeorology*, 59: 1061–1066.

Shoemaker, J. S. 1955. *Small Fruit Culture* (3rd edn). McGraw Hill Book Company, New York, New York.

Shokouhian, Ali Akbar and Asghari, Ali. 2015. Study the effect of mulch on yield of some strawberry cultivars in Ardabil condition. *International Conference on Agriculture, Ecology and Biological Engineering (AEBE-15)*, Antalya (Turkey), Sept. 7–8, 2015: pp. 1–5.

Singh, R. and Asrey, R. 2005. Growth, earliness and fruit yield of micro-irrigated strawberry as affected by planting time and mulching in semi-arid regions. *Indian Journal of Horticulture*, 62: 148–151.

Singh, R., Asrey, R. and Kumar, S. 2006. Effect of plastic tunnel and mulching on growth and yield of strawberry. *Indian Journal of Horticulture*, 63: 18–20.

Singh, R., Sharma, R.R. and Goyal, R.K. 2007. Interactive effects of planting time and mulching on "Chandler" strawberry (*Fragaria × ananassa* Duch.). *Scientia Horticulturae*, 111: 344–351.

Taparauskiene, L. and Miseckaite, O. 2014. Effect of mulch on soil moisture depletion and strawberry yield in sub-humid area. *Polish Journal of Environmental Studies*, 23: 475–482.

Tarara, J.M. 2000. Microclimate modification with plastic mulch. *HortScience*, 35: 169–180.

Tindall, J.A., Beverly, R.B. and Radcliffe, D.E. 1991. Mulch effect on soil properties and tomato growth using micro-irrigation. *Agronomy Journal*, 83: 1028–1034.

Treder W. 2003. Irrigation of strawberry plantations as a factor conditioning fruit quality. Ogólnopol. Conference on Strawberry pp. 88–92.

Verma, M.L. and Acharya, C.L. 1996. Waster stress indices of wheat in relation to soil water conservation practices and nitrogen. *Journal of Indian Society Soil Science*, 44: 368–375.

17 Flowering

Kiran Kour, Bikramjit Singh, and Tanjeet Singh Chahal

CONTENTS

17.1 Floral Morphology 301
17.2 Physiology of Flowering 303
    17.2.1 Auxins 303
    17.2.2 Gibberellins 303
    17.2.3 Cytokinins 304
    17.2.4 Primary Metabolites 304
    17.2.5 Nutrition 305
17.3 Factors Affecting Flowering 306
    17.3.1 Genetic Constitution of Varieties 306
        17.3.1.1 June-bearer/Short-Day Plants 306
        17.3.1.2 Everbearers/Long-Day Plants 307
        17.3.1.3 Day-Neutrals 308
        17.3.1.4 Remontant Strawberries...308
        17.3.1.5 Other Flowering Habits 309
    17.3.2 Light 309
    17.3.3 Temperature 310
    17.3.4 Photoperiod..310
    17.3.5 Light × Temperature 310
    17.3.6 Age of Plants..313
17.4 Molecular Insights into Flowering...313
References...315

Flowering is an important phase in the reproductive cycle and strawberry production. It is a very complex process and dependent on a series of growth events that establish and develop the plant body including potential flowering sites. In horticultural terms, the transition from vegetative to flowering marks a critical phase in production process, as the emergence of flower buds signals the anticipation of fruit set. Farmers in the past might have observed the process of flowering in strawberry for a long time, and resorted to manipulation of culturing practices as well as following calendars to anticipate, or perhaps they aided the induction of flowering to suit specific needs for optimized production.

17.1 FLORAL MORPHOLOGY

The principal thick stem of the strawberry inflorescence to the point of the first branch point is known as the peduncle. Stems leading from the peduncle branch point (node) up to the second subsequent branching points (nodes) are inflorescence branches. The stem supporting each individual flower, and later berry, is called a pedicel. The strawberry inflorescence is morphologically a dichasial cyme, but very variable in detailed structure (Guttridge, 1985). A dichasium is a determinate cluster of flowers that arise from a common peduncle by dichotomous branching just beneath the terminal flower, since this dichotomous branching or forking of the flower stem continues for two, three or four cycles behind each group of terminal flowers initiated. Thus the strawberry flower stalk is a type of compound dichasium. Darrow (1929) first indicated the complexity of strawberry inflorescence types and their subsequent effects on berry development in his pioneering study of strawberry flower counts and inflorescence habits in a number of hybrid and species clones.

Strawberry inflorescence peduncle and pedicels are continuations of the vegetative stem and its branches. At each node of the inflorescence, a bract replaces the subtending leaf, and the bud in the axil of each bract may develop into a branch of the inflorescence. The bract subtending the entire inflorescence is larger than subsequent bracts and may be almost the size of a true leaf, although variable in parts from one to three. A typical inflorescence axis has three internodes – two long internodes on each side of very short internodes, each subtended by a bract. The very short internode with two bracts close together makes it "appear that two opposite branches emerge from the same point and that there are three branches at each node just as there are three leaflets to each typical leaf" (Darrow, 1929). Besides, depending on cultivars, interspecific hybrids and a particular representative species under observation, there may be considerable variations in inflorescence types (Darrow, 1929).

In a complete and perfect flower, all four floral whorls and both male and female parts are present. The floral axis of the strawberry flower is swollen at the tip of the pedicel to form a receptacle. Into this swollen tip, hundreds of pistils are inserted spirally around the receptacle with the ovaries embedded in the receptacle, and connected to it by a vascular trace. The pistil styles and stigmas are pointed away from the receptacle. The aggregated pistils and the receptacle constitute the female portion of the flower. There are double rings of 20–35 stamens outside and surrounding the receptacle and pistils. Each stamen consists of a short stalk or filament to which the anther, the pollen-producing floral part, is attached. When the anther is mature, it splits along a longitudinal suture, and pollen grains exude from the opening. These airborne pollen grains are either carried away by wind or picked up by insects. Outside the rings of stamens, there are usually five separated white petals, which make up the corolla. When closed, the corolla offers protection to the reproductive parts of the flower. On opening, the corolla attracts animals with its colour, scent and sugary nectar secretions from glands at the base of the petals. The outer floral ring is the calyx, made up of one or two rings of five sepals. The principal function of the calyx is protection of flower in the bud stage, and some protection to fruit in early developmental stages.

Development of strawberry flowers normally proceeds from outside to inside, so that the floral part originates in the order: sepals, petals, stamens, pistils. The successively initiated flowers are progressively smaller and have fewer pistils; therefore, the fruits from late-initiated flowers are also smaller, so fruit size is said to be dependent on the number of pistils on a receptacle. The number of pistils varies from about 500 on a large primary flower to 50 or fewer on a very small quaternary or quinary flower. This view is a bit too simple, since pedicel size and hence water supply also decreases in successively initiated flowers. In addition to the position of inflorescence and the number of pistils, the completion of pollination and fertilization, internal hormonal levels, elasticity of receptacle and epidermis, number and enlargement potential of receptacle cells and nutrient and water supply are the other factors in ultimate fruit size.

Plants of some strawberry species are dioecious, including the parent species of the cultivated strawberry (*F. chiloensis*). They have male (staminate) or female (pistillate) flowers on separate plants. Under these circumstances, female plants must be interplanted with male or perfect-flowered plants to produce fruit. However, even in species where the sexes are borne on separate plants, some male plants bear few perfect flowers and produce a little fruit. Since the mid-19th century, practically all cultivars have originated through hybridization and human selection. These have been chosen with perfect flowers, hoping to effect self-fertilization. Occasionally, sterile or partially sterile plants are also found, which produce no fruit or no late fruit, even though the blossoms appear bisexual. Even among perfect-flowered cultivars, there are seasonal and geographical differences in flower and pollen productions, degree of fertilization and fruit set.

17.2 PHYSIOLOGY OF FLOWERING

Blanke (1993) studied the physiology of a bud by scanning electron microscopy of sepals of "Elsanta" strawberries. The epicuticular was composed of a crystalline layer in the form of wax ribbons lying upon a wax crust. Hairs and Trichomes were sparse. The stomatal frequencies were about $100/mm^2$ on the abaxial surface and $0.5–1/mm^2$ on the adaxial surface. The guard cells of the abaxial stomata opened to form an aperture measuring $3–4 \ \mu m \times 11–13 \ \mu m$. The sepals transpired @ 0.2 to 0.5 m mol $H_2O \ m^{-2} \ S^{-1}$ and showed a net photosynthesis rate of four to seven $\mu mol \ CO_2 \ m^{-1} \ S^{-2}$ under normal ecological conditions with a vapour pressure deficit of 1 to 1.3 kpa.

Flowering in strawberries is assumed to be controlled by promotive and inhibitive processes. Under flower-inducing conditions, certain recognizable biochemical or physiological changes occur such as the photoperiodic control of flowering (Thompson and Guttridge, 1960) through a flower inhibitor system produced in the leaves. Decline in the level of auxins after receiving required number of photo-inductive cycles (12–15) is a consequence of the changes from vegetative to floral states (Moore and Hough, 1962).

17.2.1 AUXINS

Hou and Huang (2005) observed that by using an anti-indole acetic acid (anti-IAA) monoclonal antibody and an anti-auxin-binding protein 1 (anti-ABPI), polyclonal antibody, IAA and ABPI are immunohistochemically localized in shoot apices during flower induction in strawberry. The IAA remains present in the shoot apices throughout the process of flower induction and gradually gets concentrated in the shoot apical meristem (SAM). The distribution of ABPI, and its dynamic changes are similar to those of IAA. In addition, the ABPI immune signal in SAM gradually increases, as the flower induction takes place. Whereas, external application of 1-N-napthylphtalamic acid (NPA) during induction of flowers prevents the accumulation of IAA in the SAM, and delays the process of flower bud differentiation. Besides, it leads to development of abnormal flowers. Thus an appropriate amount of IAA in the SAM and normal transport (polar) of auxin are essential for induction and differentiation of flowers in strawberries.

17.2.2 GIBBERELLINS

The gibberellins (GAs) are believed to be involved in regulating the flowering response in June-bearing varieties of strawberries (Guttridge, 1985; Taylor et al., 1994). All the identified derivatives of GAs are members of the early 13-hydroxylation biosynthetic pathway, and a similar GA spectrum has also been found in leaf petioles (Wiseman and Turnbull, 1999). GA is needed for runner initiation, and is one of the signals that mediate photoperiodic control of the fate of axillary buds, and thereby flowering, in strawberry. Remay et al. (2009) demonstrated that GA biosynthesis genes were downregulated, and that GA deactivation genes were upregulated upon floral initiation, indicating that a reduction in GA content may induce or enable flowering.

Inhibition of flowering can artificially be mimicked by application of GA (Guttridge and Thompson, 1963), as applications of inhibitors of GA biosynthesis increases crown formation, flowering and yield (Black, 2004; Hytonen et al., 2008). The inhibitors of GA biosynthesis, such as prohexadione-calcium (Pro-Ca), have the opposite effect, and can effectively inhibit runner formation and, at least under some conditions, enhance crown branching and increase flowering and yield under field conditions (Black, 2004; Hytonen et al., 2008). The levels of active GA_1 in bud tips are similarly reduced when plants are either moved from long-day to short-day conditions or are treated with Pro-Ca, compared with control plants in long-day conditions. The synthesis of GA in strawberry is blocked under short-day conditions (Hytonen et al., 2009; Sonsteby et al., 2009).

17.2.3 Cytokinins

Although cytokinins (CKs) are suspected to be involved in the flower initiation in strawberry and other species, the evidences for this is scarce. The composition of CK changes in strawberry crowns during flower initiation (Yamasaki and Yamashita, 1993). Zeatin has been observed to increase just before flower initiation, and thereafter decreases. Exogenous CK can increase inflorescence number in both short- and long-day strawberry cultivars, presumably by increasing branch crown formation (Weidman and Stang, 1983). The changes in free CK during flower induction plays important role in flower induction in strawberries (Eshghi and Tafazoli, 2007). Free CK in shoot tips and leaves of induced plants can increase from 3–12 days after the start of the short-day treatment (DASST), and thereafter decreases, while the bound CK level in shoot tips of induced plants decreased from 3 to 12 DASST. On the other hand, the free and bound CK in roots of induced plants decreased from 3–12 DASST and increased thereafter.

17.2.4 Primary Metabolites

The nutrient diversion theory (Sachs and Hackett, 1983) suggests that assimilates are the primary controlling factor in flowering, whereas other researchers assign a little subordinate role to carbohydrates (Bernier, 1984; Jones, 1990). An increase in carbohydrate levels in the apical bud is associated with the floral transition in photo periodically sensitive and cold-requiring plants (Bernier et al., 1993). The carbohydrate content of different parts of strawberry plants is supposed to have distinct roles in flowering. Concentration of soluble carbohydrate in different parts of June-bearing strawberry varieties may have a decisive role in flower bud induction. Of the various carbohydrates (sucrose, glucose, fructose and starch), sucrose is the most abundant soluble sugar available in all strawberry parts tested. Sucrose levels in shoot tips and leaves decreases at the beginning of the induction treatment, but soon increases to marked levels in shoot tips of induced plants and declines thereafter, whereas starch contents in shoot tips, leaves and roots of non-induced strawberry plants is comparatively higher than that of induced plants (Eshghi and Tafazoli, 2007).

Glucose content in the apical portion of strawberry shoots decreases after the plants are transferred to the inductive conditions and thereafter it increases gradually; almost corresponding to non-inductive plants. At the beginning of induction treatment, the sucrose content decreases but soon it increases to the level of the control plants whereas, fructose is higher in induced than control plants just before flower initiation (Hamano et al., 2003). The shoot tips (bud) of induced plants act as a strong "utilizing sink" and preferentially metabolized carbohydrates rather than storing them. It seems that non-structural carbohydrate contents in shoot tips, leaves and roots of strawberry may have an important role in flower bud differentiation (Eshghi and Tafazoli, 2007).

In short-day strawberry plants, polyamine conjugated spermidines are bound amines accumulated in the shoot tips during induction of flowering and before emergence of flowers. Variable associations of free amines and conjugated amines have been observed during induction of flowers and compared with the reproductive phase. During the period of flower development, phenylethylamine (an aromatic amine) has been observed to be the predominant amine representing 80–90% of the total free amine pool. Before fertilization, phenylethylamine conjugates are the predominant amides which are water-insoluble compounds. These substances are decreased drastically after fertilization. In plants grown continuously in long days, both free and bound polyamines (such as putrescine and spermidine) are low in vegetative shoot tips. Free amines (spermine and phenylethylamine), bound aromatic amines (phenylethylamine and 3-hydroxy, 4-methoxyphenylethylamine), conjugated spermidine and phenylethylamine do not appear. Male-sterile flowers are distinguished by their lack of conjugated spermidine and phenylethylamine and by decreased free phenylethylamine content. In normal (such as cv. Darsanga) and male-sterile (such as cv. Pandora) strawberry plants, α–DL- difluoromethylornithine (DFMO), a specific irreversible inhibitor of ornithine decarboxylase (ODC),

Flowering 305

causes inhibition of flowering and free as well as polyamine conjugates. When putrescine is added, polyamine titres and flowering are restored. A similar treatment with α –DL- difluoromethyarginine (DFMA), a specific, irreversible inhibitor of arginine decarboxylase (ADC), does not affect flowering and polyamine titres. These results suggest that ODC and polyamines are involved in regulating floral initiation in strawberries (Tarenghi and Martin-Tanguy, 1995).

17.2.5 NUTRITION

Besides light and temperature, other factors such as water supply and mineral nutrition can also influence the flowering process of strawberry. Growth inhibition caused by temporary stresses may trigger floral initiation under otherwise non-inductive conditions (Heide, 1977; Guttridge, 1985). Increasing the nutrient supply from a low level usually increases fruit yield, while a surplus, especially of nitrogen, can inhibit flower formation (Stadelbacker, 1963; Guttridge, 1985). Increase in nitrogen availability from a suboptimal level in the early stages of short-day flower induction significantly enhances and promotes flowering, while such an increase immediately before the commencement of short day may have the opposite effect. The time and rate of application of nitrogenous fertilizers during the flower bud induction period of nursery plants also influences yield of short-day strawberry varieties (Desmet et al., 2009; Yamasaki and Yano, 2009). Fertilizer application during the early part of a short day, night-chilling causes delayed flower bud induction when compared with delayed fertilizer application by 10 days (Yamasaki and Yano, 2009). Although, the lack of fertilizer nitrogen promotes flower bud induction, but once flower initiation starts, nitrogen is required for development of the flower parts (Yamasaki et al., 2002). In field-grown strawberries, adequate plant nitrogen status is important for maximizing growth, flower bud induction and flower bud differentiation. However, when granular fertilizer nitrogen is applied at renovation in perennial short-day strawberry systems, fertilizer use efficiency improves and plants produce more crowns (Strik et al., 2004), indicating the fact that number of crowns and leaves are correlated with the number of flowering sites (Strik and Proctor, 1988). Withholding nitrogen and phosphorus may not increase flowering (Abbott, 1968) but increasing nitrogen from a low nitrogen status increases flower development in strawberry (Latet et al., 2002), whereas application of high rates of nitrogen can inhibit flower bud induction and reduce yield (Darrow and Waldo, 1932).

More commonly, in perennial strawberry systems, nitrogen fertilizer has little effect on yield (Blatt, 1981; Strik et al., 2004), while good nutrition during the autumn period of flower bud induction is important in short-day strawberry (Long, 1939; Lieten, 2002). It has been observed that the application of fertilizers during the spring, when further flower bud induction and flower bud differentiation occurs, is not effective in increasing yield (Opstad and Sonsteby, 2008; Strik et al., 2004). When groups of short-day strawberry varieties with a low fertility regime are supplied with additional nitrogenous fertilizers for a period relative to a four-week short-day flower induction period, the extent and time of flowering are influenced by the time of application of nitrogenous fertilizers (Sonsteby et al., 2009). Application of nitrogenous fertilizers for three weeks, starting two weeks before the flower bud induction period, can delay flowering by one week. When plants are fertilized starting one week after the start of the short-day period, flowering can be advanced by eight days and the number of flowers can be increased by twofold without affecting the number of crowns in comparison to plants without additional nitrogen application. In strawberry, flower bud induction is thus limited when plants are at a high nitrogen status immediately before the short-day flower inductive period, whereas a high nitrogen status during the short-day induction period increases flower bud induction (Lieten, 2002; Sonsteby et al., 2009).

The time of irrigation can also have modifying effects on flowering in strawberry, for example, it has been shown that, shortly before the commencement of the natural short-day period for flower induction, generous irrigation during August significantly reduces flowering and thereby yield (Naumann, 1964).

Yamasaki et al. (2002) studied the involvement of carbon and nitrogen, allocation in the flowering of strawberry induced by low nitrogen using ^{13}C and ^{15}N tracer and observed that the allocation pattern of carbon and nitrogen in plants with flowers induced by low nitrogen was marked by:

1. low carbon content
2. recently fixed carbon primarily allocated to roots
3. recently absorbed nitrogen allocated comparatively more to crowns, young leaves and shoot apices than leaves

17.3 FACTORS AFFECTING FLOWERING

Induction of flowering in strawberry is generally influenced by genetic constitution of variety, climatic factors and management practices such as irrigation and nutrient management.

17.3.1 GENETIC CONSTITUTION OF VARIETIES

On the basis of flowering and fruiting habits, the strawberry genotypes have been classified into following categories:

1. the genotypes that produce only one flush of flowering and fruiting during the spring and are referred to as seasonal flowering genotypes
2. the genotypes which, in addition to spring flush, produce more or less continuous flowering and fruiting throughout the growing season (Heide et al., 2013) and are referred to as recurrent flowering or everbearing genotypes
3. the genotypes which are relatively insensitive to day lengths with respect to flowering and are referred to as the day-neutral genotypes (Galletta et al., 1981)

Everbearing cultivars are variously referred to as everbearers, perpetuals, rebloomers, remontants or cyclic flowering strawberries (Galletta and Bringhurst, 1990). The distinguishing features between everbearers and day-neutrals is that the older everbearers respond to long days while modern everbearers are day-neutral with respect to response to flowering (Durner et al., 1984).

Strawberry cultivars are mostly categorized as June-bearers, everbearers or day-neutrals as described in the following.

17.3.1.1 June-bearer/Short-Day Plants

The vast majority of strawberry cultivars are included in this group. June-bearers are classified as facultative short-day plants (Darrow, 1936), but flower bud formation in these varieties can occur under long-day conditions provided that the prevailing temperature remains below 16°C (Ito and Saito, 1962; Heide, 1977). June-bearers varieties produce only one flush of flowers per year (single cropping) in northern temperate climates. Several researchers have compared short and long-day lengths on the basis of natural light versus natural light supplemented with extension of day length by provision of artificial source of light, making it difficult to assign specific photoperiods. Although short-day photoperiods are needed for optimum flower induction, they may actually delay flower development. Studies on short-day cultivars have indicated that continuing short-day photoperiods may actually slow the development of previously initiated buds when compared with longer photoperiods (Moore and Hough, 1962; Durner and Poling, 1987). That is why strawberries have been considered to be short-day crops from the flower induction point of view and long-day plants from the flower development point of view (Salisbury and Ross, 1992).

17.3.1.2 Everbearers/Long-Day Plants

Everbearers are classified as long-day plants because they initiate flowers primarily in long-day (>12 hour) conditions (Darrow and Waldo, 1934). The cultivar "Oregon Everbearing" obtained in 1882 is considered to be the first North American everbearer cultivar, although Fletcher (1917) described an everbearing selection found in Ohio in 1852, which was most likely *F. virginiana*. The first commercially successful everbearer cultivar was probably "Pan-American", which is considered to a sport of the June-bearer "Bismarck" found in 1898 (Fletcher, 1917). Later on, the cultivar "Pan-American" turned out to be the source of the trait for a number of other successful everbearing cultivars such as "Progressive" and "Rockhill", the latter being the source of the trait in the proprietary everbearing cultivars developed by Driscoll Strawberry Associates (Darrow, 1966). Besides, the cultivar "Gem", a sport of "Champion" introduced in 1933, was also very popular for several decades as the source of the everbearing trait in North America. The European octoploid everbearers, which are also termed as *remontants* by the French, might have had their origin somewhat earlier. But unlike the American sources of the everbearing trait are largely based on mutations identified in cultivated plants, whereas the genetic source of the European everbearing cultivars, although claimed to be hybrids of large-fruited octoploids with the everbearing Alpine diploids, is not clear (Fletcher, 1917). It seems doubtful that this is truly the case because of the following two reasons:

1. It's difficult to construct such hybrids of *F. vesca* with octoploids (Mangelsdorf and East, 1927; Yarnell, 1931) and even more difficult to restore the adequate fertility.
2. The inheritance of the trait in the octoploid cultivars seems to be different than that in Alpine strawberries (Clark, 1937; Brown and Wareing, 1965).

Therefore, it is more likely that cross-pollination under natural conditions might have resulted in seedlings of pure *Fragaria × ananassa* pedigree, from which the original everbearer cultivars such as "Louis Gauthier", a double-cropping French cultivar (short-day genotype) with unusually permissive photoperiod or chilling requirements (Fletcher, 1917), were selected. But the cultivars such as "Mabille" and "Inepuisable", considered by some (Fletcher, 1917) to be pure Alpine strawberry, followed these lines, but none were commercially successful. Richardson (1914) suggested that the first true European octoploid everbearer was "Gloede's Seedling", introduced in 1866, and this may be the source for most or all European everbearers (Ahmadi et al., 1990). For the first time, breeding of strawberries for everbearing (long-day) trait was initiated by a French priest, Abbe Thivolet, and after more than a decade of breeding with both octoploids and diploids (Fletcher, 1917) a series of everbearers such as "St. Joseph" were introduced, which is also referred to as the "first true large-fruited everbearer cultivar" by Darrow (1966). Although a considerable improvement over everbearing cultivar "St. Joseph", Thivolet's later selection such as "St. Antoine de Padoue" was not productive, but these spurred the development of improved types later on. While everbearing plants derived from this source are in production even nowadays but this genotype has not contributed to American breeding programs. Although often referred to as "long-day" cultivars, these varieties do not have the characteristic behaviours associated with true long-day plants as neither do they appear to be regulated by the length of the dark period, nor do they respond identically to short days with a brief break of light during the night as they do to long photoperiods (Dennis et al., 1970; Durner et al., 1984), nor is flowering inhibited by short-day (Durner et al., 1984) conditions. Besides, these genotypes appear more dependent on the total amount of light received, as Dennis et al. (1970) observed no difference in flower formation in the everbearing cultivar "Geneva" grown either under a 12-hour photoperiod or under a 10-hour photoperiod with two hours break in light source during the night (Downs and Piringer, 1955; Piringer and Scott, 1964); a treatment that often inhibits flowering in short-day plants.

17.3.1.3 Day-Neutrals

The day-neutral plants are relatively insensitive in response to day length with respect to flowering. Observation on American, wild octoploid genotypes revealed that many plants of *F. virginiana* ssp. *Glauca* (syn. *Fragaria ovalis*) flowered during the summer and the fall, whereas most of the wild and commercial short-day cultivars (Darrow, 1966) did not flower at all. Evaluation of a large number of *F. virginiana* ssp. *Glauca* selections during the 1930s and 1940s at the United States Department of Agriculture (USDA) led to identification of a number of everbearing varieties including "Arapahoe" and "Ogallala" (Hildreth and Powers, 1941), which are under cultivation even in these days. Whether the trait expressed in these cultivars was the older everbearing trait or the day-neutral trait derived from *F. virginiana* ssp. *Glauca* is unknown, as sources of both are present in the pedigrees. The strong tendency of most remontant seedlings of "Arapahoe" and "Ogallala" cultivars for fruiting in their first year is distinct from other parents of everbearing cultivars (Ourecky and Slate, 1967), and more typical of day-neutral genotypes (Ahmadi and Bringhurst, 1991). Irrespective of the introduction of the day-neutral trait, the happening that has had the most significant impact came during the 1970s, when the University of California breeding programme, under Royce Bringhurst and Victor Voth, utilized a single selection of *F. virginiana* ssp. *glauca* introduced from the Wasatch Mountains in Utah for introgression of this trait into commercial strawberries (Bringhurst and Voth, 1984). Most strawberry cultivars planted in California during 1999 (Hancock, 1999) were derived from this source. These were distinct from the everbearing cultivars as the plants of these cultivars were truly insensitive to the length of the photoperiod received, and fruit at nearly the same rate over a broad range of photoperiods (Durner et al., 1984; Ahmadi and Bringhurst, 1991). Day-neutral cultivars had improved heat tolerance and longer harvest seasons when compared with the everbearing cultivars evolved earlier, but these were poor in runner production and as such considered as difficult-to-propagate (Durner et al., 1984) types. Although, day-neutral cultivars appear to differ from old everbearing genotypes, it has been difficult to distinguish between them, and in some cases the terms have been used interchangeably, leading to some confusion.

For the purposes of classification, "day-neutral" refers to the trait derived from *Fragaria virginiana* ssp. *Glauca,* while "everbearing" varieties are referred to those derived from "Pan-American", "Gloede's Seedling", or other old genetic resources, and the generic term "remontant" is being used to refer either genotypes where both are implied or the origin is not properly understood.

17.3.1.4 Remontant Strawberries

The first major exception to the common pattern of flowering was observed in the Alpine strawberry, *F. vesca* var. *semperflorens* Duch, sporadically observed in the European Alps mountain system. These include the group of strawberry genotypes originated due to mutation of the common diploid European woodland strawberry and were among the earliest genotypes to be cultivated. These were described as early as 1553 but could attain prominence by the mid-1700s only on introduction possibly from Turin, Italy into France and England. In 1811, a runnerless genotype of strawberry was observed by Labaute but this variety is of only minor commercial importance and only occasionally cultivated now (Darrow, 1966), although in the early-19th century, a white-fruited variety had been observed to be extensively cultivated in Quebec (Fletcher, 1917). Very little work has been done to characterize the physiology of flowering of these varieties, and since many workers have studied unnamed genotypes that are either seed-propagated or even collected directly from the wild (Chabot, 1978), it makes it difficult to consistently track the genotype-specific differences. Sironval and El Tannir-Lomba (1960) observed that their selection of *F. vesca* var. *semperflorens* required days longer than 12 hours to form flower buds, and that short-day treatments inhibited flowering in plants that had previously reached the flowering stage in long-day conditions. This instance appears to be an exception, not the rule (Darrow, 1966), as in other strawberry varieties, temperature has a significant effect, with moderately low temperatures favouring both reproductive development and biomass production (Chabot, 1978).

Flowering

17.3.1.5 Other Flowering Habits

A very specific but poorly characterized flowering habit referred to as the "amphiphotoperiodic" has also been observed in strawberry, as observed in the *F. chiloensis* selection CHI-24–1 (Yanagi et al., 2006), which behaves as a short-day plant under most conditions but long photoperiods inhibit flowering. However, under exceptional conditions, it has been observed to initiate flowering under continuous lighting as well. Although many varieties of strawberry do not flower altogether (Thompson and Guttridge, 1960), but a few cultivated short-day cultivars such as "Sparkle" (Collins and Barker, 1964) and "Sweet Charlie" (Stewart and Folta, 2010) have been observed to flower under such conditions as well and as such may fall into this category, but very little research has been done on this phenomenon. In recent years, there has been enhanced interest in identifying novel sources of remontancy and researchers have identified a number of wild accessions mostly among *F. virginiana* (Sakin et al., 1997; Hancock et al., 2002; Serce and Hancock, 2005a,b), which have responded with varying degrees of insensitivity to photoperiod or continual flowering. However, these germplasm resources have not yet been adequately characterized to determine whether they fall distinctly into either the day-neutral or everbearing categories, or whether they represent truly novel mechanisms. The studies on inheritance of various traits on flowering by Serce and Hancock (2005a) have revealed that there are multiple genetic mechanisms that are operating simultaneously.

17.3.2 LIGHT

It is now an established fact that the quality of light significantly influences the flowering of higher plants (Eskins, 1992; Izawa, 2005); therefore, the control of quality of light might facilitate earlier flowering, and thus shorten the cultivation cycle. However, the effects of quality of light on the flowering of everbearing strawberry varieties requires more clarification. Since everbearing strawberry varieties is a group of quantitative long-day plants, continuous exposure to light might promote earlier flowering compared with the exposure to standard 16-hour light period under open or protected cultivation. Therefore, it is presumed that the quality of light combined with prolonged duration of exposure to light might promote earlier flowering. However, the shortening of the nursery period is considered to inhibit the growth of seedlings and hence reduce fruit yield.

In a study, the effects of light quality on the flowering of everbearing strawberry under a 16-hour light versus continuous exposure to lighting illuminated by white fluorescent lamps, blue LEDs or red LEDs were investigated and it was observed that the flowering was significantly earlier in plants grown under:

1. blue LEDs compared with red LEDs, irrespective of light period
2. continuous lighting compared with the 16-hour light period, irrespective of light quality

The results showed that blue light advances the flowering of strawberry plants in everbearing varieties as compared to red light, while continuous exposure to light through blue LEDs advanced flowering and shortened the period of vegetative growth. In addition, the fruit yield of strawberry plants grown under continuous exposure to light through blue LEDs was higher than that of plants grown under standard conditions of cultivation, that is, a 16-hour light period by white fluorescent lamp (Yoshida et al., 2012). Yanagi et al. (2006) observed the effects of quality of light for continuous illumination at night on flower initiation of strawberry "CHI-24–1" plants grown under 24-hour daylight, consisting of neutral day length and continuous exposure to light at night by red and four types of far-red LEDs for 40 days during summer and autumn. Results revealed that >50% of the plants initiated flower buds under the incandescent and far-red fluorescent lamp, and the flower initiation was more distinct under the far-red LED light source whose peak wavelength was 735 mm.

It can be concluded that the shortening of nursery period by adjusting quality of light and period of light increases fruit production efficiency when compared with standard cultivation conditions

(Yoshida et al., 2012). The rate of flowering is influenced by greenhouse treatments, photoperiod and photo-inductive cycles, whereas runner production is influenced by photoperiod and temperature in association with photo-inductive cycle (Zhang et al., 2000).

17.3.3 TEMPERATURE

Similar to light, the temperature conditions also play important roles in the expression of flowering by conditioning the plant of strawberries for flowering before the beginning of the season. A certain amount of chilling, generally defined as the time between 0°C and 7°C, might be required to break the dormancy of flower buds and proceed with the normal cycle of development, though the extent of chilling requirement varies greatly in different cultivars of strawberries (Piringer and Scott, 1964; Durner et al., 1987; Darnell and Hancock, 1996). Increased chilling has been observed to be positively correlated with increased area of leaves, number of leaves, length of petiole and initiation of runners (Bringhurst et al., 1960; Piringer and Scott, 1964; Guttridge, 1969; Braun and Kender, 1985). Optimum chilling treatment promotes both vegetative and reproductive development (Darnell and Hancock, 1996). Optimum chilling treatment enhances vegetative growth, reduces induction of flowering and promotes flower bud differentiation in strawberries (Durner and Poling, 1987) but non-chilled plants, on the other hand, produce not only the fruits of smaller size (Harmann and Poling, 1997) but also lower quality (Bringhurst et al., 1960) than do the identical genotypes with adequate chilling. Since chilling is positively correlated with induction of flowering in strawberries, chilling treatment beyond the required amount is not only of no advantage but it can even markedly reduce the number of flowers, for example, there are substantial decreases in number of flowers in strawberry "Marshall" with increasing duration of chilling (Piringer and Scott, 1964). Although, the chilling treatments stimulate the development of previously initiated inflorescences (Tehranifar and Battey, 1997); but excessive chilling inhibits (Avigdori-Avidov et al., 1977; Tehranifar and Battey, 1997) and/or delays (Durner and Poling, 1987) induction of flower buds in strawberries. Thus, it can be concluded that excessive chilling delays the harvest and reduces the fruit yield of short-day varieties of strawberries.

17.3.4 PHOTOPERIOD

It is well-established scientific fact that the leaves are the primary recipients of photoperiodic signals, and that they are the sources of the putative stimulatory or inhibitory messengers that are transmitted to the apical meristematic regions (McDaniel, 1994; Thomas and Vince-Prue, 1997) and reflected as response to flowering. In seasonal flowering varieties of strawberries, there are evidences for the presence of both flower promoter(s) and one or more inhibitor(s) produced under short-day and long-day conditions, respectively, as has been observed from different studies (Guttridge, 1985; Durner and Poling, 1988). The presence of various flower-inhibiting and growth-promoting substances produced under long-day conditions can be transmitted between donor-receptor pairs of runner plants (Guttridge, 1959) through stolons. SF strawberries varieties are "negative short-day plant" which are induced to flower under short-day conditions as it releases the plant from long-day inhibition (Guttridge, 1985).

17.3.5 LIGHT × TEMPERATURE

Seasonal flowering cultivars of strawberries were first classified as facultative short-day plants, which require photoperiods of less than approximately 14 hours, or temperatures less than approximately 15°C to induce flowering (Darrow and Waldo, 1934). Temperatures as high as approximately 18°C are optimal for short-day induction of flowering in most seasonal flowering cultivars (Bradford et al., 2010; Opstad et al., 2011), while the flower-inducing effect of short days is markedly reduced at temperatures of ≤12°C and ≥22°C (Sonsteby and Heide, 2006; Verheul et al., 2006, 2007;

Opstad et al., 2011). In seasonal flowering or June-bearing varieties of the cultivated strawberries, the vegetative growth and flowering are regulated through the interaction effects of photoperiod and temperature (Darrow and Waldo, 1934; Guttridge, 1985; Taylor, 2002).

Generally, low night temperatures, to a large extent, can compensate for high day temperatures, and *vice versa* (Hartman, 1947; Verheul et al., 2006; Sonsteby and Heide, 2008), indicating no specific effect of dark-period and/or temperature on initiation of flowering in seasonal flowering varieties of strawberries. For the induction of flowering in seasonal flowering cultivars, the minimum number of short-day cycles required ranges from 7–14 under optimal conditions (Hartman, 1947; Ito and Saito, 1962; Jonkers, 1965; Guttridge, 1985), but can be influenced greatly due to genotypes, temperature and length of the photoperiod, and can be as many as 21 (Verheul et al., 2006) or more (Guttridge, 1985). Interruption of a long, inductive night by the so-called "night breaks" can effectively inhibit flowering in seasonal flowering cultivars under otherwise inductive conditions (Downs and Piringer, 1955; Ito and Saito, 1962; Jonkers, 1965).

For flower bud induction/initiation, the requirement of minimum number of photo-inductive cycles ranges from 7–24 (Hancock, 1999) with the number of required inductive cycles generally increasing with temperature (Ito and Saito, 1962).The interaction between photoperiod and temperature has been reviewed by several researchers (Darrow, 1936; Durner et al., 1984; Darnell, 2003; Darnell and Hancock, 1996; Durner and Poling, 1988; Guttridge, 1969, 1985; Heide, 1977; Strik, 1985; Taylor, 2002), suggesting that flower bud induction in short-day strawberry is insensitive to photoperiod at low (~10 to 15°C) and high (>25°C) temperatures as could be verified from the recent observations of Bradford et al. (2010), who observed insensitivity in short-day cultivar "Honeoye" to photoperiod at 14–20°C. In short-day conditions, the optimal temperature for flower bud induction is 15–18°C (Heide, 1977), while flower bud induction does not occur below 10°C or above 25°C (Durner and Poling, 1988; Heide, 1977; Sonsteby and Heide, 2006; Verheul et al., 2007). Flower bud development is optimum at higher (19–27°C) temperatures (Hartmann, 1947a; Le Miere et al., 1996) and the optimum night temperatures during short days also can vary among cultivars (Sonsteby and Heide, 2009).

In many short-day strawberry cultivars, flower bud induction occurs at low temperatures in long days (Heide, 1977; Ito and Saito, 1962; Verheul et al., 2007), with the temperature threshold varying greatly due to cultivars (Guttridge, 1985; Heide, 1977), as some cultivars in Nordic regions have no flower bud induction under long-day conditions at temperatures as low as 9°C (Sonsteby and Heide, 2006). In the coastal district of California near Watsonville, where the mean temperature is 17°C, the cultivars grown were able to flower and fruit through the long days of summer, while the same cultivars planted near Sacramento, experiencing a mean temperature of 23°C, behaved as strict short-day plants and produced only spring crop but under greenhouse conditions (Hartmann, 1947b) observations of the plants of four cultivars revealed that these flowered under long days at 17°C, but none flowered at 23°C (Hartmann, 1947b). Of these, three genotypes flowered freely under both temperatures under short-day conditions, while the fourth one, "Fairfax", failed to flower (even with 10-hour days) at higher temperatures. In addition, even the everbearing cultivars in long days are inhibited by high temperatures, as are day-neutral types, though in general the critical temperature is higher for day-neutral cultivars (Durner et al., 1984; Manakasem and Goodwin, 2001). Of the four Scandinavian short-day cultivars, two still flowered under 14-, 16- and even 24-hour photoperiods when grown at 12°C or 18°C temperatures, though flowering was substantially decreased under the longer photoperiods and these effects were greatly influenced due to the genetic constitution of different varieties (Heide, 1977).

The strawberry varieties behave variably at different locations, viz., in some cases the cultivars that are considered short-day at one location may behave as remontant at another location, for example, a British strawberry selection "Climax", behaved like a short-day when grown in the United States, but behaved as a two-crop variety in England, where temperatures are cooler and summer days are longer (Downs and Piringer, 1955). Similar to short-day photoperiods, which are essential for the development of flower buds but inhibit the development of flower trusses, low temperatures also influence induction of flowers in long photoperiods but may not be optimum for fruit production, because low temperatures slow down the development of the flower trusses (Darrow, 1966). The optimum temperatures for truss development has been observed to be about 19°C in the "Elsanta" cultivar (Le Miere et al., 1996) but temperatures that fluctuate diurnally are more effective in promotion of flowering than continuous temperatures, even at the same average temperature (Hartmann, 1947a) and these thermoperiodic rhythms have been observed to have similar effects in other plant species (Alvarenga and Valio, 1989) too, which facilitates more robust understanding of the effects of rhythms of the circadian clock (Salome and McClung, 2005).

Nishiyama and Kanahama (2000) observed the effects of variable day/night temperatures (20/15°C, 25/20°C and 30/25°C for 16 weeks) and photoperiods (8- and 24-hour) on the vegetative and reproductive growth of everbearing strawberry cv. Summerberry plants. Results revealed that flowering continued at 20/15°C and 25/20°C, under both photoperiods, whereas at 30/25°C, inflorescence formation was stopped in the eight-hour but promoted in the 24-hour photoperiods. Vegetative growth was most active at 20/15°C and seriously inhibited at 30/25°C in the 24-hour photoperiod.

The critical photoperiod for initiation of inflorescence and flower buds in an everbearing "Summerberry" variety of strawberry grown at high temperature (30/25°C) has been observed to be between 13 and 14 hours (Nishiyama et al., 2006). In some everbearing varieties, the initiation of flowering and formation of runners occurs irrespective of the day length (Guttridge, 1969), but in other varieties long days may be necessary for initiation of flowers (Darrow and Waldo, 1934). High temperature (>23°C) in long days enhances formation of runners in June-bearer and everbearer varieties, but the effects of temperature or photoperiod on the day-neutral varieties with respect to formation of runners or initiation of flowers is not known properly. Flowering and runner production are independent processes as in June-bearing cultivars, the flowers are initiated independently of the day length below a certain temperature level but above that level, long days or short days are needed for flower initiation. In the variety "Robinson", flowers were initiated at 8-, 12-, 16-, 20- and 24-hour day lengths at 9°C but at 17°C, a day length of 12 hours or less was necessary for flower initiation (Ito and Saito, 1962). Similarly, flower initiation in "Zefyr", "Jonsok" and "Glima" varieties at 18°C with day lengths of 10, 12, 14, 16 and 24 hour was successful but at 24°C, there was no flower initiation with day lengths of over 14 or 16 hours (Heide, 1977).

The flowering season of different strawberry cultivars in relation to inhibition by environmental factors has been summarized in Table 17.1 (Yanagi and Oda, 1993).

TABLE 17.1

Flowering Season and Environmental Factors That Inhibit Flower Initiation in Different Bearing Groups of Strawberries

Bearing Type	Reference Cultivars	Flowering Season	Environmental Factors Inhibiting Formation of Flowers
June-bearing	Fakuba and Hokowasa	(a) Spring and early summer	Winter chilling, long-day length in spring, high temperature in summer
		(b) Spring	Long-day length in spring
Everbearing	Ostara	(a) From spring to autumn	Winter chilling
	Rabunda	(b) Continuous	None
Intermediate	Aiko	(a) Spring and early summer	Winter chilling
	Kletter	(b) Continuous	None

Source: Modified from Yanagi and Oda, 1993. Effect of Photoperiod and chilling on floral formation of intermediate types between June and everbearing strawberries. Acta Horticulturae, 348: 339–346.

Flowering **313**

17.3.6 AGE OF PLANTS

More than two-months-old runners do not bear flowers, so runners developed late in the season (September) should be planted on runner beds instead of fruiting beds. Similarly, young micropropagated plants, while planted in fields, remain in the vegetative growth phase during the current planting season, and attain a floriferous phase for fruiting and yield in the subsequent season. The progenies of day-neutral strawberry varieties generally initiate flowers within four months of germination (Ahmadi et al., 1990) but older plants respond more robustly to both photoperiod and temperature (Ito and Saito, 1962), which could be, in part, a function of plant size, although a single intact leaf has been observed to be enough to perceive photoperiods desired for induction of flowering in strawberries (Hartmann, 1947a).

17.4 MOLECULAR INSIGHTS INTO FLOWERING

The ability to screen seedlings for flowering habits at a very early age could be of great help to breeders as these facilitate in saving the time and resources required for evaluation of otherwise useless seedlings lacking the desired trait. In a BC_1 population of 1049 plants derived from a cross combination of *F. vesca* ssp. *vesca* (*SFL/SFL*) × *F. vesca* ssp. *semperflorens* (*sfl/sfl*) Albani et al. (2004) successfully converted inter-simple sequence repeat (ISSR) markers to sequence characterized amplified region (SCAR) markers for identifying three DNA products associated with seasonal flowering of strawberries. Of these markers, two were observed to be linked with the seasonal flowering locus (SFL) at 1.7 and 3.0 cM, while the third one (SCAR2), was mapped to the same location as seasonal flowering locus. This represented the most tightly associated marker for flowering habit developed so far in *Fragaria* sp., but unfortunately the practical use of these markers in breeding has been limited because of the differences of *SFL* trait with the day-neutral and everbearing traits of the cultivated octoploid varieties of strawberries. Otherwise also, the commercial significance of *F. vesca* is very limited, which could be one of the reasons for insignificant headway in commercial exploitation of ISSR and SCAR markers in breeding and crop improvement programmes on *Fragaria vesca* strawberries. Through bulk segregant analysis of a small F_1 population, segregating in 1:1 proportion for remontancy; one random amplification of polymorphic DNA (RAPD) marker was identified (Kaczmarska and Hortynski, 2002) but the degree of its association and reliability for specific trait over a broad spectrum of progenies was not assessed. Five RAPD markers linked to the everbearing trait have been identified (Sugimoto et al., 2005), with linkages ranging from 11.8 to 24.3 cM, and these were considered to be weak. The weak linkages and difficulties in reproducibility of RAPD markers (Paran and Michelmore, 1993) have limited the practical use of these markers. Recently, the genomic sequences of *F vesca* genes that are integral parts of the photoperiodic flowering network in *Arabidopsis* and other systems have been inventoried and 30 kb genomic regions containing *PHYTOCHROME A (PHYA), CONSTANS (CO), SUPPRESSOR OF CONSTANS OVEREXPRESSION 1 (SOC), LEAFY (LFY),* including characterization of the polynucleotide repeat content physically associated with the gene has been identified and sequenced, which is expected to be useful in identifying allelic variations associated with traits for flowering. The results could further help in tracing these critical factors of photoperiod-regulated traits for flowering and facilitate understanding of associations between flowering behaviour and specific genes. These sequences may further facilitate the development and standardization of molecular tools to ascertain the role of these regulators in mechanisms of flowering in strawberries.

The understanding of the mechanism of photoperiod-regulated induction of flowers in strawberries depends upon the extent of unravelling of the molecular-genetic basis of the process. Studies on *Arabidopsis thaliana* (a facultative long-day annual species) has revealed involvement of several genes (Simpson et al., 1999) in regulation of flowering. The gene-regulated flowering in the diploid *F. vesca* strawberries is comparatively more clear-cut than in the octoploid *Fragaria × ananassa*. Further, *F. vesca* has been observed to be more amenable to the molecular-genetic approach for understanding the mechanisms of regulation of flowering. Researches at the University of Reading in the UK (Battey et al., 1998) and the University of New Hampshire (Davis, 1997) on *F. vesca* have resulted in locating a single dominant gene regulating seasonal flowering and identification of two molecular (RAPD) markers closely linked to this gene (Davis, 1997).

The perception of photoperiod begins with photoreceptors that are grouped into two broad categories, viz., phytochromes and cryptochromes. The phytochromes (phyA, B, C, D & E; conventionally referred to as *GENE1, PROTEIN1, mutant1*, holoreceptor1) respond to the red and far-red spectra of light. The cryptochromes (*cry1* and *cry2*) respond to the blue spectrum of light. A subset of these receptors entrain the complex feedback loop of the plant's central oscillator (Millar, 2003; Spalding and Folta, 2005), which in turn, guides a number of rhythmic outputs, including *CO* expression (Suarez-Lopez et al., 2001) and the base expression level of *CO* transcripts. The main regulation of the abundance of *CO* protein is post-translational as has been demonstrated with complementary biochemical, genetic and photo-physiological assays (Valverde et al., 2004). It has been observed that the red light sensor (*PHYTOCHROME B; phy B*) promotes the degradation of *CO* protein in response to red light signals whereas the far-red light sensor (PHYTOCHROME A; *phy A*) receptor counteracts the effects of *phy B* by stabilizing *CO* along with promotion of its nuclear partitioning. Further, the blue light, operating through *cry2*, imparts similar effects to *CO* as *phy A*, stabilizes the protein and promotes localization to the nucleus.

Different pathways (Amasino and Michaels, 2010; Higgins et al., 2010) integrate via factors referred to as the "floral integrator genes" including *FLOWERING LOCUS T (FT)* and *SUPPRESSOR OF CONSTANS1 (SOC1)*. After getting activated, these integrator genes in turn activate genes regulating activity of meristematic regions of flowers, such as *APETELA1 (AP1)*, *FRUITFULL (FUL)*, and *LEAFY (LFY)*, which transform the meristem from vegetative to generative state. In the photoperiodic pathway, *CONSTANS (CO)* measures seasonal changes in day length by integrating signals from the endogenous circadian clock and the external light signals perceived by phytochromes and cryptochromes (Yanovsky and Kay, 2002; Valverde et al., 2004). In long-day varieties of strawberries, the *CO* activates expression of *FT* in phloem companion cells in the leaves and, concurring with the definition of a *florigen*, the *FT* protein is transported in the phloem from the leaves to the apex where it initiates flowering. In addition to the floral integrator genes, floral suppressors such as *TFL1* are considered to be important regulators of flowering in plants such as in *Arabidopsis, TFL1* suppresses expression of *FT* targets such as *AP1, FUL*, and *LFY* (Hanano and Goto, 2011). The identification of *SFL* (Iwata et al., 2012) and the recent demonstration of its function (Koskela et al., 2012) in *F. vesca* is considered as a major advance in our understanding of the genetic mechanisms controlling flowering. In addition, the identification of the *SFL* gene in *F. vesca* as a homologue of the *Arabidopsis TERMINAL FLOWER1 (FvTFL1)* gene, and the recent demonstration of its function as a genetic switch (on/off) mechanism controlling the opposite photoperiodic responses of the seasonal flowering and evebearing genotypes of strawberry (Heide et al., 2013) could be considered as a major breakthrough in our overall understanding of the mechanism of regulation of flowering in plant systems in general and in strawberries in particular.

In a nutshell, flowering in strawberry is an important and complex process passing through a series of growth events that terminate in establishment and development of a plant body. Along these processes the potential flowering sites are also produced. The strawberry inflorescence is morphologically a "dichasial cyme". The transformation of the buds from vegetative to flower stem tip (flower bud initiation) depends on a series of complex control systems involving temperature and photoperiodic responses. Further, on the basis of photoperiodic responses, the strawberry cultivars are categorized as June-bearers (short-day), everbearers (long-day), day-neutrals or remontant types. Under flower-inducing conditions, changes in carbohydrates, phytohormones, nutrients and carbon:nitrogen ratio have a significant role in development of flowers in strawberry plants. The strawberries are very responsive to environmental stimuli, which may alter the patterns of their normal development. The principal influence of climatic factors identified on flowering of strawberries

Flowering

and fruit yield are photoperiodism, temperature, light, vernalization and plant age. Among these, photoperiod and temperature are the best studied factors. Photoperiodic control of flowering is mediated through a genetic pathway referred to as the photoperiodic pathway. In this pathway, the coincidence of photoperiodic signals perceived by photoreceptors, and internal gene expression during a specific phase determines flowering.

REFERENCES

Abbott, A.J. 1968. Growth of the strawberry plant in relation to nitrogen and phosphorus nutrition. *Journal of Horticultural Science*, 43: 491–504.

Ahmadi, H. and Bringhurst, R.S. 1991. Genetics of sex expression in *Fragaria* species. *American Journal of Botany*, 78: 504–514.

Ahmadi, H., Bringhurst, R.S. and Voth, V. 1990. Modes of inheritance of photoperiodism in *Fragaria*. *Journal of the American Society for Horticultural Science*, 115: 146–152.

Albani, M.C., Battey, N.H. and Wilkinson, M.J. 2004. The development of ISSR-derived SCAR markers around the Seasonal Flowering Locus (SFL) in *Fragaria vesca*. *Theoretical and Applied Genetics*, 109: 571–579.

Alvarenga, A.A. and Valio, I.F.M. 1989. Influence of temperature and photoperiod on flowering and tuberous root-formation of *Pachyrrhizus tuberosus*. *Annals of Botany*, 64: 411–414.

Amasino, R.M. and Michaels, S.D. 2010. The timing of flowering. *Plant Physiology*, 154, 516–520.

Avigdori-Avidov, H., Goldschmidt, E.E. and Kedar, N. 1977. Involvement of endogenous gibberellins in the chilling requirements of strawberry (*Fragaria* × *ananassa* Duch.). *Annals of Botany*, 41: 927–936.

Battey, N.H., Le Miere, P., Tehranifer, A., Cekic, C., Taylor, S., Shrives, K.J., Hadley, P., Greenl A.J., Darby, J. and Wilkinson, M.J. 1998. Genetic and environmental control of flowering in strawberry. In: *Genetic and Environmental Manipulation of Horticultural Crops* (Cockshull K.E., Gray, D., Seymour G.B., and Thomas, B., eds.), CAB International, New York: 111–131.

Bernier, G. 1984. The factors controlling floral evocation: an overview. In: D. Vince-Prue, B. Thomas and K.E. Cockshull (eds.). *Light and the Flowering Process*. Academic Press, New York: 277–292.

Bernier, G., Havelange, A., Hussa, C., Petitjean, A. and Lejeune, P. 1993. Physiological signals that induce flowering. *Plant Cell*, 5: 1147–1155.

Black, B.L. 2004. Prohexadione-calcium decreases fall runners and advances branch crown of "Chandler" strawberry in a cold-climate annual production system. *Journal of the American Society for Horticultural Science*, 129: 479–485.

Blanke, M.M. 1993. What do the flower buds of the strawberry look like under a microscope? *Commercial Horticulture*, 35: 9–12.

Blatt, D.R. 1981. Effect of nitrogen source, rate and time of application on fruit yields of the Micmac strawberry. *Communication in Soil Science and Plant Analysis*, 12: 511–518.

Bradford, E., Hancock, J.F. and Warner, R.M. 2010. Interactions of temperature and photoperiod determine expression of repeat flowering in strawberry. *Journal of the American Society for Horticultural Science*, 135: 102–107.

Braun, J.W. and Kender, W. J. 1985. Correlative bud inhibition and growth habit of the strawberry as influenced by application of gibberellic-acid, cytokinin, and chilling during short day length. *Journal of the American Society for Horticultural Science*, 110: 28–34.

Bringhurst, R.S. and Voth, V. 1984. Breeding octoploid strawberries. *Iowa State Journal of Research*, 58: 371–381.

Bringhurst, R.S., Voth, V. and Van Hook, D. 1960. Relationship of root starch content and chilling history to performance of California strawberries. *Proceedings of the American Society for Horticultural Science*, 75: 373–381.

Brown, T. and Wareing, P.F. 1965. Genetical control of everbearing habit and 3 other characters in varieties of *Fragaria vesca*. *Euphytica*, 14: 97–98.

Chabot, B.F. 1978. Environmental influences on photosynthesis and growth in *Fragaria vesca*. *New Phytologist*, 80: 87–98.

Clark, J.H. 1937. Inheritance of the so-called everbearing tendency in the strawberry. *Proceedings of the American Society for Horticultural Science*, 35: 67–70.

Collins, W.B. and Barker, W.G. 1964. A flowering response of strawberry to continuous light. *Canadian Journal of Botany*, 42: 1309–1311.

Darnell, R.L. 2003. Strawberry growth and development. In: N.F. Childers (ed.). *The Strawberry. A Book for Growers & Others*. Dr. Norman F. Childers Publications, Gainesville, Florida: pp. 3–10.

Darnell, R.L. and Hancock, J.F. 1996. Balancing vegetative and reproductive growth in strawberry). In: M.V. Pritts C.K.C. and T.E. Crocker (eds.). *Proceedings of the IV North American Strawberry Conference*: 144–150.

Darrow, G.M. 1929. Inflorescence types of strawberry varieties. *American Journal of Botany*, 16: 571–585.

Darrow, G.M. 1936. Interrelation of temperature and photoperiodism in the production of fruit-buds and runners in the strawberry. *Proceedings of the American Society for Horticultural Science*, 34: 360–363.

Darrow, G.M. 1966. *The Strawberry: History, Breeding and Physiology*. Holt, Rinehart and Winston, New York: p. 477.

Darrow, G.M. and Waldo, G.F. 1932. Effect of fertilizers on plant growth, yield and decay of strawberries in North Carolina. *Proceedings of the American Society for Horticultural Science*, 29: 318–324.

Darrow, G.M. and Waldo, G.F. 1934. Responses of strawberry varieties and species to the duration of the daily light period. *USDA Technical Bulletin*, 453: 32.

Davis, T.M. 1997. Mapping genes for day neutral flowering habit in strawberry. Plant and Animal Genome V, January 12–16, San Diego, California. Abstract W40.

Dennis, F.G., Lipecki, J. and Kiang, C.L. 1970. Effects of photoperiod and other factors upon flowering and runner development of three strawberry cultivars. *Journal of the American Society for Horticultural Science*, 95: 750–754.

Desmet, E.M., Verbraeken, L. and Baets, W. 2009. Optimisation of nitrogen fertilization prior to and during flowering process on performance of short day strawberry "Elsanta". *Acta Horticulturae*, 842: 675–678.

Downs, R.J. and Piringer, A.A. 1955. Differences in photoperiodic responses of everbearing and June-bearer strawberries. *Proceedings of the American Society for Horticultural Science*, 66: 234–236.

Durner, E.F. and Poling, E.B. 1987. Flower bud induction, initiation, differentiation and development in the Earliglow strawberry. *Scientia Horticulturae*, 31: 61–69.

Durner, E.F., Poling, E.B. and Albregts, E.A. 1987. Early season yield responses of selected strawberry cultivars to photoperiod and chilling in a Florida winter production system. *Journal of the American Society for Horticultural Science*, 112: 53–56.

Durner, E.F. and Poling, E.B. 1988. Strawberry developmental responses to photoperiod and temperature: a review. *Advances in Strawberry Production*, 7: 6–14.

Durner, E.F., Barden, J.A., Himelrick, D.G. and Poling, E.B. 1984. Photoperiod and temperature effects on flower and runner development in day-neutral, June-bearing and everbearing strawberries. *Journal of the American Society for Horticultural Science*, 109: 369–400.

Eshghi, E. and Tafazoli, F. 2007. Changes in mineral nutrition levels during floral transition in strawberry (*Fragaria x ananasa* Duch.). *International Journal of Agricultural Research*, 2(2): 180–184.

Eskins, K. 1992. Light-quality effects on *Arabidopsis* development. Red, blue and far-red regulation of flowering and morphology. *Plant Physiology*, 86: 439–444.

Fletcher, S. W. 1917. *The Strawberry in North America*. Macmillan & Co., New York.

Galletta, G.J. and Bringhurst, R.S. 1990. Strawberry management. In: G.J. Galletta and D.G. Himelrick (eds.). *Small Fruit Crop Management*. Prentice-Hall Inc., New York: 83–156.

Galletta, G.J., Draper, A.D. and Swartz, H.J. 1981. New everbearing strawberries. *HortScience*, 16: 726.

Guttridge, C.G. 1959. Further evidence for a growth-promoting and flower-inhibiting hormone in strawberry. *Annals of Botany*, 23: 612–621.

Guttridge, C.G. 1969. *Fragaria* In: L.T. Evans (ed.). *The Induction of Flowering*. Cornell University Press, Ithaca, New York: 247–267.

Guttridge, C.G. and Thompson, P.A. 1963. The effects of gibberellins on growth and flowering of *Fragaria* and *Duchesna. Journal of Experimental Botany*, 15: 631–646.

Guttridge, C.G. 1985. *Fragaria* x *ananassa*. In: A.H. Halevy (ed.). *CRC Handbook of Flowering*. CRC Press, Boca Raton, Florida: 16–33.

Hanano, S. and Goto, K. 2011. *Arabidopsis TERMINALFLOWER1* is involved in the regulation of flowering time and inflorescence development through transcriptional repression. *The Plant Cell*, 23: 3172–3184.

Hamano, M., Zamazaki, Y. and Miure, H. 2003. Sugar changes in the shoot apex of strawberry under flower bud inductive conditions. *Acta Horticulturae*, 626: 305–308.

Hancock, J.F. 1999. *Strawberries*. CABI Publisher, New York: p. 237.

Hancock, J.F., Luby, J.J., Dale, A., Callow, P.W., Serce, S. and El-Shiek, A. 2002. Utilizing wild *Fragaria virginiana* in strawberry cultivar development: Inheritance of photoperiod sensitivity, fruit size, gender, female fertility and disease resistance. *Euphytica*, 126: 177–184.

Flowering

317

Harmann, K.K. and Poling, E.B. 1997. The influence of runner order, night temperature, and chilling cycles on the earliness of "Selva" plug plant fruit production. *Acta Horticulturae*, 439: 597–603.

Hartmann, H.T. 1947a. Some effects of temperature and photoperiod on flower formation and runner production in the strawberry. *Plant Physiology*, 22: 407–420.

Hartmann, H.T. 1947b. The influence of temperature on the photoperiodic response of several strawberry cultivars grown under controlled environment conditions. *Proceedings of the American Society for Horticultural Science*, 50: 243–245.

Heide, O.M. 1977. Photoperiod and temperature interactions in growth and flowering of the strawberry. *Plant Physiology*, 40: 21–26.

Heide, O.M., Stavang, J.A. and Sonsteby, A. 2013. Physiology and genetics of flowering in cultivated and wild strawberries – a review. *Journal of Horticultural Science and Biotechnology*, 88(1): 1–18.

Higgins, J.A., Bailey, P.C. and Laurie, D.A. 2010. Comparative genomics of flowering time pathways using *Brachypodium distachyon* as a model for the temperate grasses. *PLoS ONE*, 5, 4, e 10065.

Hildreth, A.C. and Powers, C.L. 1941. The Rocky Mountain strawberry as a source of hardiness. *Proceedings of the American Society for Horticultural Science*, 38: 410–412.

Hou, Z.X. and Huang, W.D. 2005. Immunohistochemical location of IAA and ABPi in strawberry shoot apexes during flower induction. *Planta*, 222(4): 678–687.

Hummer, K.E. 2001. From the Andes to the Rockies: Native strawberry collection and utilization. *HortScience*, 36: 221–225.

Hytonen, T., Elomaa, P., Moritz., T. and Junttila, O. 2009. Gibberellin mediates daylength-controlled differentiation of vegetative meristems in strawberry (*Fragaria* x *ananassa* Duch). *BMC Plant Biology*, 9: 18.

Hytonen, T., Mouhu, K., Koivu, I., Elomaa, P. and Junttila, O. 2008. Prohexadione calcium enhances the cropping potential and yield of strawberry. *European Journal of Horticultural Science*, 73: 210–215.

Ito, H. and Saito, T. 1962. Studies on flower formation in strawberry plant I-Effects of temperature and photoperiod on the flower formation. *Tohoku Journal of Agricultural Research*, 13: 191–203.

Izawa, T. 2005. In: M. Wada, K. Shimazaki and M. Iino (eds.). *Light Sensing in Plants* Springer-Verlag, Tokyo, Japan, Berlin, Germany, Heidelberg, Germany, New York: p. 16.

Iwata, H., Gaston, A., Remay, A., Thouroude, T., Jeauffre, J., Kawamura, K., Oyant, L.H-S., Araki, T., Denoyes, B. and Foucher, F. 2012. The *TFL1* homologue *KSN* is a regulator of continuous flowering in rose and strawberry. *The Plant Journal*, 69: 116–125.

Jones, T.W.A. 1990. Use of a flowering mutant to investigate changes in carbohydrates during floral transition in red clover. *Journal of Experimental Botany*, 41: 1013–1019.

Jonkers, H. 1965. On the flower formation, the dormancy and the early forcing of strawberries. *Mededelingen van de Landbouwhooge school Wageningen*, 65(6): 1–59.

Kaczmarska, E. and Hortynski, J.A. 2002. The application of RAPD markers for genetic studies on *Fragaria* x *ananassa* Duch. (strawberry) with due consideration for permanent flowering. *Electronic Journal of Polish Agricultural Universities*, 5: 2.

Koskela, E.A., Mouhu, K., Albani, M.C., Kurokura, T., Rantanen, M., Sargent, D.J., Battey, N.H., Coupland, G., Elomaa, P. and Hytonen, T. 2012. Mutation in *TERMINALFLOWER1* reverses the photoperiodic requirement for flowering in the wild strawberry *Fragaria vesca*. *Plant Physiology*, 159: 1043–1054.

Latet, G., Meesters, P. and Bries, J. 2002. The influence of different nitrogen (N) sources on the yield and leaching in open field strawberry production. *Acta Horticulturae*, 567: 455–458.

Le Miere, P., Hadley, P., Darby, J. and Battey, N.H. 1996. The effect of temperature and photoperiod on the rate of flower initiation and the onset of dormancy in the strawberry (*Fragaria* × *ananassa* Duch.). *Journal of Horticultural Sciences*, 71: 361–371.

Lieten, P. 2002. The effect of nutrition prior to and during flower differentiation on phyllody and plant performance of short day strawberry Elsanta. *Acta Horticulturae*, 567: 345–348.

Long, J.H. 1939. The use of certain nutrient elements at the time of flower formation in the strawberry. *Proceedings of the American Society for Horticultural Science*, 37: 157–168.

Manakasem, Y. and Goodwin, P.B. 2001. Responses of day neutral and June-bearing strawberries to temperature and daylength. *Journal of Horticultural Science and Biotechnology*, 76: 629–635.

Mangelsdorf, A.J. and East, E.M. 1927. Studies on the genetics of *Fragaria*. *Genetics*, 12: 307–339.

McDaniel, C.N. 1994. Photoperiodic induction, evocation and floral initiation. In: *The Development of Flowers*. (Grayson, R.I. ed.). Oxford University Press, Oxford, UK: 25–43.

Millar A.J., 2003. Suite of photoreceptors entrains the plant circadian clock. *Journal of Biological Rhythms*, 18: 217–226.

Moore, J.N. and Hough, L.F. 1962. Relationships between auxin levels, time of floral induction and vegetative growth of the strawberry. *Proceedings of the American Society for Horticultural Science*, 81: 255–264.

Naumann, W.D. 1964. Bewasserungstermin und Blütenknospendifferenzierung bei Erdbeeren. *Gartenbauwissenschaft*, 29: 21–30.

Nishiyama, M. and Kanahama, K. 2000. Effects of temperature and photoperiod on the development of inflorescences in everbearing strawberry (*Fragaria* x *ananassa* Duch.). *Acta Horticulturae*, 514: 261–267.

Nishiyama, M., Ohkawa, W., Kanayama, Y. and Kanahama K., 2006. Critical photoperiod of flower bud initiation in ever bearing strawberry "Summerberry" plants grown at high temperatures. *Tohoku Journal of Agricultural Research*, 56: 1–8.

Opstad, N. and Sonsteby, A. 2008. Flowering and fruit development in strawberry in a field experiment with two fertilizer strategies. *Acta Agriculturae Scandinavica Section B: Soil and Plant Science*, 58: 297–304.

Opstad, N., Sonsteby, A., Myrheim, U. and Heide O.M. 2011. Seasonal timing of floral initiation in strawberry: Effects of cultivar and geographic location. *Scientia Horticulturae*, 129: 127–134.

Ourecky, D.K. and Slate, G.L. 1967. Behavior of the everbearing characteristics in strawberries. *Proceedings of the American Society for Horticultural Science*, 91: 236–248.

Paran, I. and Michelmore, R.W. 1993. Development of reliable PCR-based markers linked to downy mildew resistance genes in lettuce. *Theoretical and Applied Genetics*, 85: 985–993.

Piringer, A.A. and Scott, D.H. 1964. Interrelation of photoperiod, chilling, and flower-cluster and runner production by strawberries. *Proceedings of the American Society for Horticultural Science*, 84: 295–301.

Remay, A., Lalanne, D., Thouroude, T., LeCouvier, F., Oyant, Lh-S. and Foucher, F. 2009. A survey of flowering genes reveals the role of gibberellins in floral control in rose. *Theoretical and Applied Genetics*, 119: 767–781.

Richardson, C. 1914. A preliminary note on the genetics of *Fragaria*. *Journal of Genetics*, 3: 171–77.

Sachs, R.M. and Hackett, W.P. 1983. Source-sink relationships and flowering. In: W.J. Meudt (ed.). *Strategies of Plant Reproduction. Beltsville Symposium on Agricultural Research.* Allenheld, Osmun, Totowa, New Jersey: 263–272.

Sakin, M., Hancock, J.F. and Luby, J.J. 1997. Identifying new sources of genes that determine cyclic flowering in rocky mountain populations of *Fragaria virginiana* ssp *glauca* Staudt. *Journal of the American Society for Horticultural Science*, 122: 205–210.

Salisbury, F.B. and Ross, C.W. 1992. *Plant Physiology.* Wadsworth Inc, Belmont, California: pp. 3–26.

Salome, P.A. and McClung, C.R. 2005. What makes the Arabidopsis clock tick on time? A review on entrainment. *Plant Cell and Environment*, 28: 21–38.

Serce, S. and Hancock, J.F. 2005a. Inheritance of day-neutrality in octoploid species of *Fragaria*. *Journal of the American Society for Horticultural Science*, 130: 580–584.

Serce, S. and Hancock, J.F. 2005b. The temperature and photoperiod regulation of flowering and runnering in the strawberries, *Fragaria chiloensis, F. virginiana* x *ananassa*. *Scientia Horticulturae*, 103: 167–177.

Simpson, G.G., Gendall, A.R. and Dean, C. 1999. When to switch to flowering. *Annual Review of Cell and Developmental Biology*, 15: 519–550.

Sironval, C. and El Tannir-Lomba, J. 1960. Vitamin E and flowering of *Fragaria vesca* var. *Semperflorens* Duch. *Nature*, 185: 855–856.

Sonsteby, A. and Heide, O.M. 2006. Dormancy relations and flowering of the strawberry cultivars Korona and Elsanta as influenced by photoperiod and temperature. *Scientia Horticulturae*, 110: 57–67.

Sonsteby, A. and Heide, O.M. 2008. Temperature responses, flowering and fruit yield of the June-bearing strawberry cultivars Florence, Frida and Korona. *Scientia Horticulturae*, 119: 49–54.

Sonsteby, A. and Heide O.M., 2009. Temperature limitations for flowering in strawberry and raspberry. *Acta Horticulturae*, 838: 93–97.

Sonsteby, A., Opstad, N., Myrheim, U. and Heide, O.M. 2009. Interaction of short day and timing on nitrogen fertilization on growth and flowering of "Korona". *Science*, 303: 1003–1006.

Spalding, E.P. and Folta, K.M. 2005. Illuminating topics in plant photobiology. *Plant Cell and Environment*, 28: 39–53.

Stadelbacker, G.J. 1963. Why so much variation in strawberry fertilizer recommendations and practices. In: C.R Smith and N.F. Childers (eds.). *The Strawberry. Proceedings of the National Strawberry Conference.* Rutgers University, New Brunswick, New Jersey: 81–85.

Stewart, P.J. and Folta, K.M. 2010. A review of photoperiodic flowering research in strawberry (*Fragaria* spp.). *Critical Reviews in Plant Science*, 29: 1–13.

Strik, B.C. 1985. Flower bud initiation in strawberry cultivars. *Fruit Variety Journal*, 39: 5–9.

Strik, B., Buller, G. and Righetti, T. 2004. Influence of rate, timing and method of nitrogen fertilizer application on uptake and use of fertilizer nitrogen, growth, and yield of June-bearing strawberry. *Journal of the American Society for Horticultural Science*, 129: 165–174.

Strik, B.C. and Proctor J.T.A. 1988. The importance of growth during flower bud differentiation to maximizing yield in strawberry genotypes. *Fruit Variety Journal*, 42: 45–48.

Flowering

Suarez-Lopez, P., Wheatley, K., Robson, F., Onouchi, H., Valverde, F. and Coupl, G. 2001. Constant mediates between the circadian clock and the control of flowering in *Arabidopsis*. *Nature*, 410: 1116–1120.

Sudds, R.N. 1928. Fruit-bud formation in the strawberry. Pennsylvania Agricultural Expteriment Station, Bulletin, 230 (Annual Report of Director).

Sugimoto, T., Tamaki, K., Matsumoto, J., Yamamoto, Y., Shiwaku, K. and Watanabe, K. 2005. Detection of RAPD markers linked to the ever bearing gene in Japanese cultivated strawberry. *Plant Breeding*, 124: 498–501.

Tarenghi, E. and Martin, T.J. 1995. Polyamines, floral induction and floral development of strawberry (*Fragaria* x *ananassa* Duch). *Plant Growth Regulation*, 17: 157–165.

Taylor D.R., 2002. The physiology of flowering in strawberry. *Acta Horticulturae*, 567: 245–251.

Taylor, D.R., Blake, P.S. and Browning, G. 1994. Identification of gibberellins in leaf tissues of strawberry (*Fragaria* x *ananassa* Duch.) grown under different photoperiods. *Plant Growth Regulation*, 15: 235–240.

Taylor, D.R., Blake, P.S. and Crisp, C.M. 2000. Identification of gibberellins in leaf exudates of strawberry (*Fragaria* x *ananassa* Duch.). *Plant Growth Regulation*, 30: 221–223.

Tehranifar, A. and Battey N.H. 1997. Comparison of the effects of GA_3 and chilling on the Micmac strawberry. *Communication in Soil Science and Plant Analysis*, 12: 511–518.

Thomas, B. and Vince-Prue, D. 1984. Juvenility, photoperiodism, and vernalization. In: *Advanced Plant Physiology* (Wilkins M.B. ed.). Pitman, London, UK: 408–439.

Thomas, B. and Vince-Prue, D. 1997. *Photoperiodism in Plants*. 2nd Edition. Academic Press, London, UK: p. 428.

Thompson, P.A. and Guttridge C.G. 1960. The role of leaves as inhibitors of flower induction in strawberry. *Annals of Botany*, 24: 482–490.

Valverde, F., Mouradov, A., Soppe, W., Ravenscroft, D., Samach, A. and Coupl, G. 2004. Photoreceptor regulation of *CONSTANS* protein in photoperiodic flowering. *Science*, 303: 1003–1006.

Verheul, M.J., Sonsteby, A. and Grimstad S.O. 2006. Interaction of photoperiod, temperature, duration of short-day treatment and plant age on flowering of *Fragaria* x *ananassa* Duch. cv. *Korona*. *Scientia Horticulturae*, 107: 164–167.

Verheul, M.J., Sonsteby, A. and Grimstad, S.O. 2007. Influenceof day and night temperatures on flowering of *Fragaria* x *ananassa* Duch. cvs. Korona and Elsanta. *Scientia Horticulturae*, 112: 200–206.

Weidman, R.W. and Stang, E.J. 1983. Effects of gibberellic acid (GA_3), 6 benzyladenine (6-BA and promalin) (GA_{4+7}+ 6-BA) plant growth regulators on plant growth, branch crown and flower development Scott and Raritan strawberries. *Advances in Strawberry Production*, 2: 15–17.

Wiseman, N.J. and Turnbull, C.G.N. 1999. Endogenous gibberellin content does not correlate with photo-period-induced growth changes in strawberry petioles. *Australian Journal of Plant Physiology*, 26: 359–366.

Yanagi, Y.T. and Oda, O.Y, 1993. Effect of Photoperiod and chilling on floral formation of intermediate types between June and everbearing strawberries. *Acta Horticulturae*, 348: 339–346.

Yamasaki, A. and M. Yamashita. 1993. Changes in endogenous cytokinins during flower induction of strawberry. *Acta Horticulturae*, 345: 93–99.

Yamasaki, A. and Yano, T. 2009. Effect of supplemental application of fertilizers on flower bud initiation and development of strawberry – possible role of nitrogen. *Acta Horticulturae*, 842: 765–768.

Yamasaki, A., Yoneyama, T., Tanaka, F., Tanaka, K. and Nakashima, N. 2002. Tracer studies on the allocation of carbon and nitrogen during flower induction of strawberry plants as affected by the nitrogen level. *Acta Horticulturae*, 567: 349–352.

Yanagi, T., Yachi, T., Okuda, N. and Okomoto, K. 2006. Light quality of continuous illuminating at night to induce floral initiation of *Fragaria chiloensis* L.CHI-24-1. *Scientia Horticulturae*, 109(4): 309–314.

Yanovsky, M. J. and Kay, S. A. 2002. Molecular basis of seasonal time measurement in *Arabidopsis*. *Nature*, 419: 308–312.

Yarnell, S. H. 1931. Genetic and cytological studies on *Fragaria*. *Genetics*, 16: 422–454.

Yoshida, S., Hikosaka, S. and Goto, E. 2012. Effects of light quality and light period on flowering of everbearing strawberry in a closed plant production system. *Acta Hoticulturae*, 956: 107–111.

Zhang, X., Himelrick, D. G., Woods, F.M. and Ebel, R. C. 2000. Effect of temperature, photoperiod and pre-treatment growing condition on floral induction in spring bearing strawberry. *HortScience*, 35(4): 556.

18 Pollination

M.S. Khan, Poonam Srivastava, and R.M. Srivastava

CONTENTS

18.1 Flower Biology 321
18.2 Pollination Requirements.. 321
    18.2.1 Pollen Production, Stigma Receptivity and Nectar Production............................ 323
    18.2.2 Modes of Pollination and Pollinizers 324
    18.2.3 Pollinators.. 324
    18.2.4 Floral Activity of Honey Bees 327
18.3 Managed Pollination Practices.. 328
References.. 328

Pollination is an essential process in the lifecycle of most plants, and especially in higher plants such as fruit crops, as most of them require pollination to ensure proper fruit setting. Pollination is the basis of genetic reconstitution and creation of variabilities for making improvement and evolving new varieties. Pollination is the process of transfer of pollen grains from anther on to the stigma of same flower (self-pollination) or another flower of same plant species (cross-pollination). Self-pollinated plant species can produce fruits but their number and quality are often improved when cross-pollinated. Cross-pollination in crops has the potential to reduce loss and wastage of food so it could contribute to global food security (Klatt et al., 2014). To understand pollination biology in a fruit crop such as strawberry, it is essential to have knowledge of various pollination aspects such as floral biology, pollination requirement, effective pollination periods and the conditions, pollen germination, stigma receptivity, pollenizers, mode of pollination and pollinators.

18.1 FLOWER BIOLOGY

In strawberry, inflorescence is a modified stem and the flower cluster is a series of double-branches having one flower on each branch (Figure 18.1). The branching of inflorescence has one primary, two secondary, four tertiary, eight quaternary and next 16, if they develop, quinary. However, the types of inflorescence may vary with varieties or even in the same variety at different locations. The primary flower opens first and often produces the largest berry (Shoemaker, 1955).

The size of flowers in strawberries varies from 2.5–3.75 cm across; they are whitish and usually composed of 5–10 sepals, five oval petals, numerous styles and 20–35 stamens. The stamens are in multiples of five, usually arranged in three whorls, vary in size and are deep golden when containing viable pollen in abundance. In some cases, few flowers may also have poorly developed stamens. The number of pistils in a flower may vary greatly and ranges from 83–518 depending upon its position (Gardner, 1923). The primary flowers bear about 350 stigmas, secondary 260 and tertiary 180 (McGregor, 1976). The pistils are arranged in a regular spiral manner, each having an achene (commonly called the seed) at the base with an ovary. The ovary contains one ovule with a single-seed embryo.

18.2 POLLINATION REQUIREMENTS

Pollination has been demonstrated as the main limiting factor for strawberry development (Bigey et al., 2005). In earlier cultivars, pistillate flowers were common, requiring pollen from staminate

FIGURE 18.1 Strawberry flowers.

flowers for fruit setting (Darrow, 1927, 1937). The flowers in most of the present-day cultivars are hermaphrodite (perfect flowered) providing self-fertility, although a few varieties may have either pistillate flowers only, flowers with few stamens or flowers with stamens that fail to produce pollen and are thus practically self-sterile (Hughes, 1951; Shoemaker, 1955; Eaton and Smith, 1962; Hyams, 1962). For obtaining the maximum size and perfect shape of fruit, pollination of all of the pistils of a flower is necessary. In perfect-flowered cultivars, natural sterility is the primary cause for low fruit set in later-emerged flowers, but if the first flowers develop into nubbins (irregularly shaped berries) and yet later flowers produce good berries, the poor development could probably be ascribed to inadequate pollination (Darrow, 1966). Pollination results in fertilization of the seed embryo in the ovule that develops rapidly to form well-shaped fruit. Exclusion of insects from strawberry flowers gives decreased yield due to lesser fruit set and more malformed or misshapen fruits (Hughes, 1961, 1962; Moore, 1969; Nye and Anderson, 1974; Katayama, 1987; Dag et al., 1994). In protected cultivation, malformation of the earliest fruits in some varieties might be due to poor supply of viable pollen grains (Hughes et al., 1969). Owing to inadequate pollination, growth-promoting substances are not secreted because the fertilized seed in the achene is absent and the receptacle in its area fails to grow, leading to malformed and misshapen fruits (Robbins, 1931; Nitsch, 1952; Hughes, 1961, 1962). The percentage of malformed fruits in strawberry is increased due to insufficient pollination (Kronenberg, 1959; Kronenberg et al., 1959; Howitt et al., 1965; Carden and Emmett, 1973; Lieten, 1993; Blasse and Haufe, 1989; Vasquez et al., 2006; Kruistum et al., 2006). In the cv. Shasta grown in greenhouse conditions, no fruit set was recorded in caged plants while there was 77% fruit set when the plants were kept uncaged with air blown over them with a fan (Allen and Gaede, 1963). In another study (Petkov, 1965), only 31–39% of flowers could set fruits, with 60–65% of fruits being malformed when isolated from insects compared with 55–60% fruit set in exposed flowers. Free (1968) reported that plants of cv. Favourite when caged to exclude visits from pollinating insects had minimum (55%) fruit set as against 65.5% for the cages with bees. The weight of the berries (6.7 g, 8.3 g and 8.4 g) and percentage of malformed fruits (48.6, 20.7 and 15.4%) varied greatly in plots caged to exclude insects, caged with bees and open plots, respectively. Connor and Martin (1973) and Connor (1975) also recorded less fruit set (81%) in cloth-caged plots than in open plots (98%) in 11 cultivars. They also observed the impact of wind in increasing the development of achenes by an additional 14% over cloth-caged plots (53%). Shemetkova and Bachilo (1986) observed that cross-pollination by bees, in general, increased yields by 35–40% with improved berry quality. Irkaeva and Ankudinova (1994) recorded great variation ranging from 24–>75% cross-pollination in different strawberry plants.

Pollination

Dag et al. (1994) recorded 8% higher fruits and 30% higher yield in the variety Ofra in plants in open conditions than in covered plants. They also observed that the average percentage of malformed fruits varied from 13% in open to 32% in caged plots. Blasse (1981) established the necessity of insect-assisted pollination in cvs. Senga Sengana and Gauntlet and further recorded 15–50% more fruit yields in uncovered plants. In another study in Germany, varieties such as Redgauntlet, Tanira, Senga Sengana and Gorella produced 18% more fruit yield when the plants were pollinated in the open than when they were enclosed in cages (Blasse and Haufe, 1989). Bigey et al. (1997) reported that in cv. Gariguetle, under self-pollination, 7% of normal and 40% malformed flowers showed no fruit set but hand and open pollination greatly improved fruit set and the quality of fruits. Goodman and Oldroyd (1988) in Australia and Szklanowska (1995) in Poland, however, observed no effect on the total number of fruits that developed in the absence of insects but observed deformation and the average weight reduction was 37.8, 31.7, 29.3 and 21.9% in cvs. Kama, Dukat, Honeoye and Senga Sengana, respectively. In commercial strawberry crops in Colombia, Vasquez et al. (2006) introduced honey bees, *Apis mellifera*, colonies and obtained increased fruit yield (151.3 kg) with increased number of fruits per plant (61.1%) of improved quality compared with 74.5 kg (with no bees).

The pollination requirement of strawberry cultivars may also depend on the length of their stamen (Connor and Martin, 1973), as it has positive correlation with the percentage of achene development. The cultivars having flowers with shorter stamens as in "Surecrop" (2.5 mm) benefit most from insect pollination as compared with cultivars having flowers with lengthy stamens such as "Early Midway" (5.2 mm) that have the ability to auto-pollinate. The degree of self-compatibility in strawberry can also be predicted from the length of pollen grain and the pollen size index (as an indicator of pollen viability requires that inviable grains have a smaller or larger mean diameter than viable grains and that this difference is sufficiently large relative to natural size variation among viable grains). Genotypes producing long pollen grains with large pollen size index exhibit greater effects of autogamy (Zebrowska, 1998). Success in pollination in a crop may be influenced by farming system; for example, organic farming gives higher pollination success and better-developed fruits compared with conventional farms (Andersson et al., 2012). However, in organic farming, success in pollination of strawberries may further be influenced by leaf or flower-eating insects such as *Galerucella* spp. (Malm, 2013).

Various studies on the pollination requirement of strawberry plants, as discussed, suggest that although most modern cultivars have potentially self-pollinated and self-fertilized, the percentage of fruit set, fruit quality, marketable fruit yield increases and the percentage of malformed fruits decreases due to the presence of pollinating insects in the strawberry ecosystem.

18.2.1 POLLEN PRODUCTION, STIGMA RECEPTIVITY AND NECTAR PRODUCTION

The pollen production in different strawberry varieties may vary with areas and seasons. In some years, several varieties produce good anthers with abundant pollen, while there may be a little pollen production by these varieties in other years. A variety may behave differently in two localities, for example, the varieties Crescent and Sterling Castle are pistillate in the United States, but perfect flowered in England. Pollen viability, fertility and germination also differ among various cultivars affecting the fruit set and fruit size (Koyuncu et al., 2000; Kaczmarska, 2003). These differences in pollen viability, fertility and germination are attributed to the cultivars, use of plant growth regulators such as gibberellic acid GA_3, temperature and relative humidity (Leech et al., 2002; Ahn et al., 1988; Voyiatzis and Paraskevopoulou, 2002a, b; Ledesma and Sugiyama, 2005; Paydas et al., 2000a, b). The pollen are mature before the anthers dehisce, and are thrown onto some of the pistils as the anthers dehisce, resulting in self-pollination. However, the stigmas are receptive before pollen of the same flower is available, which encourages cross-pollination. The pollen can remain viable for several days but some flowers dry and shrink on the second day after opening (Connor, 1970). The stigma are receptive from seven–10 days after anthesis (Moore, 1964;

Darrow, 1966). The optimum time for pollination seems to be during the first one–four days after opening of flowers.

In the truss of the Ai-berry cultivar, the distal pistil of the first flower to open is small and immature, whereas the proximal pistils near the base of the receptacle are fully mature and receptive to pollen. The stigmas of the basal pistils become unreceptive at four days after anthesis, resulting in malformed fruits, while the distal pistils can remain non-receptive even at 10 days after anthesis (Yoshida et al., 1991).

In a strawberry flower, the nectar is secreted by a narrow ring of fleshy tissues in the receptacle between the stamens and stigmas. The sugar concentration in nectar varies from 26–30% (Shaw et al., 1954; Petkov, 1965). On an average, 0.6–0.8 mg nectar is produced by a single flower of strawberry (Petkov, 1965), which may increase to 1.7 mg/flower (Skowronek et al., 1985) in the most favourable situations.

18.2.2 MODES OF POLLINATION AND POLLINIZERS

Several researchers have recorded observations of various methods of pollination in strawberry. Pollination by honey bee may increase the yield by 22% over self- and wind pollination (Connor, 1970). Connor and Martin (1973) observed that achene development was 53% in self-pollination, 67% when wind aided and 91% in open pollination. Porter (1977) recorded a 33–44% increase in fruit yield of strawberries due to hand pollination but observed this method to be time consuming and uneconomic. However, Bigey et al. (1997) and Li (2006) suggested hand pollination in cvs. Gariguette, Toyonoka and Minbao for commercial cultivation. Vincent et al. (1990) reviewed the management of insect pollinators and pests in strawberry plantations. They reported that gravity could account for up to 80% of pollination in some strawberry varieties, with the wind contributing about 8%, but the heaviest fruits, with the maximum percentage well pollinated, are produced by a combination of gravity, wind and insect pollination. The bee pollination was superior to self-pollination (Carden and Emmett, 1973; Porter, 1977; Wilkaniec and Maciejewska, 1996; Lopez-Medina et al., 2006). Influence of the modes of pollination varies with cultivars. The greatest effects of self-pollination on yield components has been observed in cvs. B302 and Redgauntlet, while autogamy has the minimum effects in cv. Paula, showing greatest effects due to entomophily, whereas, the greatest effect of wind pollination has been observed in the Dukat cultivar (Zebrowska, 1998).

Planting of pollenizers adjacent to the main cultivar is useful in minimizing the number of malformed berries in an imperfect cultivar. The cultivars to be cross-pollinated should be in blocks eight–10 rows wide (Vinson, 1935) with a row of pollenizers to every three–five rows of an imperfect cultivar (Shoemaker, 1955). Free (1968) observed honey bees foraging on "Cambridge Favourite" strawberry plants, indicating that the greater the frequency of pollenizer rows, the more cross-pollinations occur (Free, 1993). Watters and Sturgeon (1992) evaluated some potential pollen donors for cv. Pandora.

18.2.3 POLLINATORS

As early as in the year 1917, Fletcher reported that 90% of the pollination of strawberries was performed by insects, mainly honey bees. Lounsberry (1930) and Pammel and King (1930) noticed that in Iowa, the bees worked on strawberry blossoms feverishly in certain conditions. However, in Russia, Shashkina (1950) observed flies as the principal visitors of strawberry flowers. The location of an apiary near a strawberry field increased the average production from 840 to 1225 lbs/acre (Muttoo, 1952). Andersonn (1969) compiled numerous references relating to strawberry pollination. Honey bees increase strawberry production in greenhouse and plastic greenhouses or tunnels (Mommers, 1961; Burd, 1970; Bonfante, 1970; Pinzauti, 1987; Katayama, 1987). Placement of honey bee colonies in strawberry under greenhouse and tunnels increases the yield and quality of the berries (Dag, 1994; Wilkaniec and Maciejewska, 1996). The pollination activity of honey

Pollination 325

bees (*A. mellifera*) and other insects was determined for eight cultivars of strawberry (Bagnara and Vincent, 1988) in Canada and it was estimated that insect pollination contributed 28% of the yield. In Japan, almost 85% of total strawberry production (200,000 tonnes/year) is from plastic green-houses where honey bees are used as pollinators (Sakai and Matsuka, 1988). The effect of honey bee pollination on fruit set, yield and quality of strawberry has also been reported by several other researchers (Abe, 1971, Japan; Canas and Gomez, 1993; Verma and Phogat, 1995, India; Hofmann and Haufe, 1995, Germany; Lee et al., 1998, Korea; Kakakhanian et al., 2010, Kurdistan Province). In strawberry, honey bees are considered the principal pollinators and the extent of their contribution in pollination differs under various situations and has been observed to vary from 25–54% (Pion et al., 1980, Canada); 58% (Singh, 1979, India); 77–98% (Skowronek et al., 1985, Poland); 35–42% (Shemetkova and Bachilo, 1986, Russia), 58.9% (Goodman and Oldroyd, 1988, Australia); 40–75% (Antonelli et al., 1988, the United States) and 56% (Oliveira et al., 1991, Canada). In Korea, about 98.9% of farmers use honey bee colonies compared with 0.4% farmers using bumble bees for pollination in strawberry (Yoon et al., 2011).

Numerous insects have been recorded to visit strawberry blossoms. Nye and Anderson (1974) listed 108 species of flower-visiting insects in the United States, whereas 32 species were recorded in Canada (Oliveira et al., 1991). Connor (1972) in the United States recorded populations of bees on strawberry fields and reported the share of different group of insects, such as honey bees at 25–32%; Andrenidae 14–20%; Halictidae 49–51%; Megachilidae up to 1%; *Ceratina* species 1–5% and *Bombus* species 1–2%. He further reported that more ground-nesting bees were present in fields surrounded by uncultivated land. From another field study in the United States, Nye and Anderson (1974) reported that bees and large dipterans, mostly drone flies (*Eristalis* spp.) were the most frequent visitors of strawberry blossoms. The most efficient pollinators were *A. mellifera, Osmia trevoris, Eristalis tenax, Eristalis brousii, Halictus rubicundus* and *Osmia panula*. In Korea, hover flies (Syrphidae) and solitary bees, especially Andrenidae, were common visitors to strawberry flowers (Woo et al., 1986). The Dipteran flies constituted 18–53%, wild bees (mainly Halictidae) 4–9% and the honey bees 40–75% of all insect visitors of strawberry flowers (Antonelli et al., 1988).

Chagnon et al. (1993) evaluated the relative efficiency of two groups of insects that pollinated strawberries in Canada and observed honey bees (*A. mellifera*) to be more efficient than indigenous bees (Andrenidae, Halictidae, and Megachilidae). Release of the syrphid fly, *Eristalis cerealis*, and honey bees, *A. mellifera*, was observed to result in production of 84–82.7% of strawberry fruits with normal shape, respectively, whereas it was 22.9% without the release of any pollinators (Kim et al., 1996).

Abrol (1989) and Partap et al. (2000) made observations on strawberry pollination by the Asian hive bee, *Apis cerana*. In Srinagar (J&K, India), the main insect visitors to strawberry flowers were honey bees (*A. cerana)* and *Lasioglossum* sp. The fruit set was 75% in open pollination as compared with 47% when insects were excluded, and malformed fruits were 11% and 23% in the respective conditions (Abrol, 1989). In two separate studies, Chang et al. (2001) in Korea and Chen and Hsieh (2001) in Taiwan compared pollination effects by the honey bees, *A. mellifera* and *A. cerana*. In an open strawberry field, the yield due to pollination with honey bees, *A. cerana*, was four–five-fold greater as than *A. mellifera*, indicating that *A. cerana* is the major pollinator of strawberry cv. Toyonoka in the Tahu area of Taiwan. When *A. mellifera* was introduced into the vinyl houses in Korea, the rate of deformed first fruits was zero in all three cvs. Bogyojosaeng, Suhong and Yeobong. However, when *A. cerana* was introduced, no deformed fruits were observed in Yeobong, but 2% and 45% fruits were deformed in cvs. Suhong and Bogyojosaeng, respectively (Chang et al., 2001). In a two-year study on five cultivars of strawberry (Dana, Etna, Belrubi, Chandler and Fern) at Kullu (Himachal Pradesh, India), Sharma et al. (2014) recorded superior-quality fruits with 9.85–45.8% increase in fruit yield in crops caged and pollinated with honey bees, *A. mellifera*, as compared with open pollinated crops.

Besides, honeybees, bumble bees (*Bombus terrestris*) have also been observed to be efficient pollinators for strawberry in Israel (Gundia, 1995) and Japan (Ikeda and Tadauchi, 1995). Eijnde (1992) observed no significant difference in average berry weight pollinated by bumble bees or honey bees in the Netherlands. Paydas et al. (2000b) obtained maximum fruit yield in Dorit, Douglas, Oso Grande, Camarosa and Chandler cultivars studied, the yields were maximum when crop was collectively pollinated with honey bee and bumble bees.

In a comparative studies of bumble bees (*Bombus* spp.) and the plant growth regulators (PGB; NAA) on yield and quality of "Camarosa" strawberry in plastic greenhouse in the Jordan valley, maximum fruit yield (719.7 g/plant) and fruit weight (21.2 g/fruit) was observed in plants subjected to bumble bees (Abu et al., 2004). Dimou et al. (2008) also obtained more well-shaped fruits when the strawberry plants were pollinated by bumble bees compared with a control in Greece. In Jordan (Zaitoun et al., 2006), pollination due to both honey bees and bumble bees improved the fruit quality in cv. Camarosa in plastic house by reducing the percentage of misshapen fruits to 6% and 13.3%, respectively, however, honey bees were most efficient in pollination where the fruit set was in excess of twice the control plots, and the bumble bees ranked second, causing 60% fruit formation over the control. They also obtained increased yield up to 29.6% due to release of bees in plastic houses and observed that size and length of strawberry berries were greater in plants pollinated by bumble bees compared with other treatments. Li et al. (2006) made detailed investigations into the behaviour and activity patterns of *Bombus lucorum* and *A. mellifera* for pollinating strawberry in greenhouses in China (Table 18.1).

In Spain, Ariza et al. (2012) observed that pollination reduced the incidence of misshapen fruit and increased the fruit yield of strawberry cv. Camarosa grown in tunnels. The bumble bee *B. terrestris* has also been observed to forage more on non-target wild flowers rather than the target crop of strawberry (Foulis and Goulson, 2014). Fu (2010) compared *B. lucorum, A. cerana cerana* and *A. mellifera ligustica* for their pollination efficiency in strawberry in China and observed that *A. cerana cerana* was the most efficient pollinator in greenhouses in low-temperature conditions.

In addition to pollination services, pollinators such as the bumble bees, *B. terrestris*, and honey bees, *A. mellifera*, can vector a microbial control agent to reduce the incidence of plant pathogen; the grey mould, *Botrytis cinerea*, in strawberries grown under greenhouse condition resulting in higher fruit yields (Mommaerts et al., 2011; Hokkanen et al., 2015).

Pinzauti (1992) suggested to explore the possibility of using the solitary bee, *Osmia carnuta*, as an effective pollinator for strawberry under greenhouse conditions in Italy while comparing pollination by another species of solitary bees (*Osmia rufa*) to wind with the use of a hairdryer. Wilkaniec et al. (1997) reported that the method of pollination did not influence the mean number

TABLE 18.1

Comparative Pollinating Behaviour of Bumble Bees, *Bombus Lucorum*, and Honey Bees, *Apis Mellifera*, in Strawberry

Behaviour and Activity Patterns	B. lucorum	A. mellifera
Time of flower visit	08.00–8.05 (AM) to 15.55–16.05(PM)	09.25–9.40 (AM) to 15.20–15.30 (PM)
Visiting temperature	12 to 13°C	>15°C
Individual diurnal activity time (seconds)	1.43 ± 4.48	180 ± 2.64
Foraging time (seconds)	105.71 ± 1.16	76.43 ± 3.83
Shuttling duration (seconds)	3.81 ± 0.42	6.0 ± 0.48
No. of visiting flowers/minute	8.44 ± 0.44	2.38 ± 0.15
Visit at young flowers/day	75%	31%
Mean row length per visit sequence	5 m	1.1 m

Source: Modified from Li et al., 2006, Strawberry pollination by Bombus lucorum and Apis mellifera in greenhouses. Acta Entomologica Sinica, 49(2): 342–348.

Pollination 327

of fruits/plant in six cultivars of strawberries under natural pollination grown in an unheated plastic tunnel. However, the percentage of malformed fruits was less due to pollination aided with bees, accounting for 3.7–5.6% of the total as compared with 25.3–34.6% for pollination by wind blowing and 29.2–38.7 for natural pollination. The commercial yield due to pollination with bees was 89% of the total yield compared with 67.4–74.4% for pollination by blowing of winds and 44.4–66.3% for natural pollination.

The Brazilian stingless bee (*Nannotrigona testaceicornis*) was introduced in to Japan from Brazil and used for pollination of strawberry. The flowers receiving four visits of bees produced well-formed fruits and it was estimated that 1200 bees/cage ($4.2 \times 8.1 \times 2.4$ m) are needed for economic fruit production (Maeta et al., 1992). Kakutani et al. (1993) introduced a colony of the honey bee, *A. mellifera*, to one compartment and four colonies of the stingless bee, *Trigona minangkabau*, to a similar compartment in a greenhouse containing strawberry plants. A third area had no bees. And it was concluded that the stingless bee as well as the honey bee could pollinate strawberry if more stingless bees were introduced. Another species of the stingless bee, *Tetragonisca angustula*, was an effective strawberry pollinator and it was shown that one colony of this species in a greenhouse with approximately 1350 strawberry plants was adequate to pollinate primary flowers of cv. Oso Grande in Brazil (Malagodi et al., 2004). Antunes et al. (2007) stated that in protected cultivation of strawberry cvs. such as Oso Grande, Tudla and Chandler, the presence of the stingless bees, *T. angustula* (Jatai bee) was indispensable for the production of marketable fruits. They recommended four beehives with 170 m² protected area. Roselino et al. (2009) investigated the success of two stingless bee species, viz., *Scaptotrigona depilis* and *Nannotrigona testaceicornis* in pollinating strawberries and concluded stingless bees to be more efficient pollinators of strawberry flowers cultivated in greenhouses producing fruits of better quality and greater commercial value.

18.2.4 FLORAL ACTIVITY OF HONEY BEES

Free (1968) observed the nectar and pollen collection behaviour of honey bees on strawberry. He stated that on nearly every flower visit, the nectar-collecting honey bees touched both stigmas and anthers. Some bees scrabbling for pollen also collected nectar. Connor (1970) recorded that the bees were normally most active from 10–15.00 hours at temperatures of 18–26°C in Michigan. In Korea, maximum activity of the honey bees (*A. mellifera* and *A. cerana*) was at 14.00 hours; however, the bees were active between 9.00–16.00 hours (Change et al., 2001). Goodman and Oldroyd (1988) reported that in Australia, only 26% of honey bees carried pollen whereas, in the United States, Antonelli et al. (1988) observed that 80% of honey bees carried pollen loads. In India, 60% of the honey bees, *A. cerana*, collected pollen (Singh, 1979). Chagnon et al. (1993) in Canada noted the foraging behaviour of *A. mellifera* and the indigenous bees (Andrenidae, Halictidae and Megachilidae), and observed that average-sized bees such as honey bees pivot at the top of the receptacle, and pollinated the apical stigmas, whereas, small bees circle on the stamens and around the receptacle, pollinating mainly the basal stigmas. In Hungary, honey bees gathered pollen predominantly (Soltesz et al., 2001).

A strawberry flower requires 15–20 bee visits to set fruit (Skrebtsova, 1957). Jaycox (1970) reported 11–15 bee visits to fertilize most of the ovules in each flower. Exposure of strawberry plants to insect pollination for a minimum of 15 days was necessary to obtain an acceptable fruit yield (Antonelli et al., 1988). However, Chagnon et al. (1989) reported that the pollination rate (number of fertilized ovules) had a relation to fruit weight and an adequate rate of pollination was achieved by four honey bee (*A. mellifera*) visits, lasting 40 seconds in total and pollination of all achenes was achieved in six visits in cv. Veestar.

The effectiveness of pollination of a strawberry flower depends on rates of pollination, that is, the number of fertilized ovules out of total ovules. Albano et al. (2009) observed that flowers visited by *A. mellifera* had 84% effectiveness, in contrast with only 53% in unpollinated flowers. Zapata et al. (2014) compared pollination in primary flowers of cv. Festival by the honey bee, *A. mellifera*, and the predator, *Chrysoperla carnea*, with self-pollination and open pollination. They recorded a significantly higher pollination rate (85.20 ± 2.41) and lower misshapen fruits rate (16.78 ± 1.20) due to pollination with honey bees, and suggested that use of *A. mellifera* could be an alternate to increase quality and production of fruits of strawberries cultivated in Irapuato, Mexico.

Honey bees may have preference for one cultivar over others or the cultivars may differ in their attractiveness to the honey bees and other pollinators. Strawberry cvs. Midway 1 and Midway 2 attract the most bees and flies, whereas cvs. Sunrise and Redchief are less attractive to pollinating insects. The former cultivars produce much pollen and the honey bees collected mainly pollen, whereas most honey bees on Sunrise collect nectar (Connor, 1972, 1975). In a greenhouse study in Belgium, Jacobs et al. (1987) observed more bee visits in cv. Karola than either Primella or Gorella. However, in an open field study, Sharma et al. (2014) did not find any preference of the pollinator species including *A. mellifera* for any of the five cultivars, Dana, Etna, Belrubi, Chandler and Fern. Studies conducted by Ceuppens et al. (2015) suggest that the pollination preference for cv. Sonata by the bumble bee, *B. terrestris* is due to being "less repellent" instead of "more attractive" than cv. Elsanta, with variety-specific flower emission lying at the basis.

18.3   MANAGED POLLINATION PRACTICES

Use and management of honey bees for strawberry pollination in greenhouses/Polyvinyl houses/ plastic houses and other cultivation environments are essentially required for quality berry production. As early as 1961, Mommers recommended the use of bees on strawberries in greenhouses. In Tochigi (Japan), the honey bees were first used in greenhouses during 1967 and Shimotori (1981) suggested the growers to hire four–six frames of colonies of bees from the beekeepers association during fruiting season. To minimize the mortality of bees in greenhouses and tunnels, Bonfante (1970) suggested that the hive should be kept outside to avoid loss of bees in the first few days of flight. Jaycox (1970) recommended use of one colony per acre whereas, Blasse (1981) recommended two beehives/ha. Blasse and Haufe (1989) also recommended the introduction of beehives into strawberry fields during flowering. Ahn et al. (1989) surveyed 175 greenhouses in four strawberry-growing areas of Korea, and reported the use of a three- or four-comb colony for pollination, and the hive was rotated between two or three plastic houses, with one or two days in each. More than half of the growers (56–90%) introduced honey bees before 10% flowering of the crop. Praagh (1983) also stressed on the importance of proper timing of introduction of hive into the plantation (when 10% of the flowers are open). Placement of the hives at different distances from the crop may influence fruit set, shape and size (Svensson et al., 1991). Connor (1970) observed fewer bees in the centre of large fields than around the edges that affected the achene development. For high fruit set, management of microclimate inside the beehives is necessary (Ohishi, 1999).

Use of synthetic pheromone, Nasonov and the commercial honey bee attractant, Bee Scent to attract honey bees to improve pollination and quality of strawberries has been suggested by Butts (1991) and Pardo et al. (1994). The use of Biovac (a self-propelled vacuum device designed to control insect pests of strawberry) may be passed before or after the activity period (08.00–18.00 h) of the pollinators to minimize the effect on pollinator populations (Chiasson et al., 1997). Recently, methods of artificial pollination and devices such as contact vibration devices and ultrasonic pollination devices have been developed for effective pollination in strawberries (Shimizu et al., 2015).

REFERENCES

Abe, Y. ed. 1971. *Strawberry Cultivation using Honey Bees as Pollinators*. Non-bun-kyo; Tokyo, Japan: pp. 235.

Abrol, D.P. 1989. Studies on ecology and behaviour of insect pollinators frequenting strawberry blossoms and their impact on yield and fruit quality. *Tropical-Ecology*, 30(1): 96–100.

Pollination 329

Abu-Rayyan, A., Nazer, I. and Qudeimat, E. 2004. Effects of some fruits setting induction means on yield and fruit quality of strawberry grown under plastic house in the Jordan Valley. *Dirasat-Agricultural-Sciences*, 31(3): 296–301.

Ahn, C.K., Choe, J.S., Um, Y.C., Cho, I.W., Yu, I.C. and Park, J.C. 1988. Effects of bee pollination and growth regulators treatment on preventing the malformation and accelerating the growth of strawberry fruit. *The Research Reports of the Rural Development Administration-Horticultural (Korea R.)*, 30: 3, 22–30.

Ahn, S.B., Kim, I., Cho, W.S. and Choi, K.M. 1989. Survey on the use of honeybee pollination of strawberries grown in plastic greenhouses. *Korean Journal of Apiculture*, 4(1): 1–8.

Albano, S., Salvado, E., Duarte, S., Mexia, A. and Borges, P.A.V. 2009. Pollination effectiveness of different strawberry floral visitors in Ribatejo, Portugal: selection of potential pollinators. Part 2. *Advances in Horticultural Science*, 23(4): 246–253.

Allen, W.W. and Gaede, S.E. 1963. Strawberry pollination. *Journal of Economic Entomology* 56: 823–825.

Andersonn, W. 1969. *The Strawberry, A World Bibliography*. Scarecrow Press Inc., Metuchen, New Jersey: p. 731.

Andersson, G.K.S., Rundlö, F.M. and Smith, H.G. 2012. Organic farming improves pollination success in strawberries. *PLoS ONE*, 7(2): e31599.

Antonelli, A.L., Mayer, D.F., Burgett, D.M. and Sjulin, T. 1988. Pollinating insects and strawberry yields in the Pacific Northwest. *American Bee Journal*, 128(9): 618–620.

Antunes, O.T., Calvete, E.O., Rocha, H.C., Nienow, A.A., Cecchetti, D., Riva, E. and Maran, R.E. 2007. Yield of strawberry cultivars pollinated by Jatai bees under protected environment. *Horticultura-Brasileira*, 25(1): 94–99.

Ariza, M.T., Soria, C., Medina-Mínguez, J.J. and Martínez-Ferri, E. 2012. Incidence of misshapen fruits in strawberry plants grown under tunnels is affected by cultivar, planting date, pollination, and low temperatures. *HortScience*, 47(11): 1569–1573.

Bagnara, D. and Vincent, C. 1988. The role of insect pollination and plant genotype in strawberry fruit set and fertility. *Journal of Horticultural Science*, 63(1): 69–75.

Bigey Schupp, J., Vaissiere, B. and Hennion, B. 1997. Pollination of strawberry: influence on yield and quality in early corps. *Infos Paris*, 130: 35–37.

Bigey, J., Vaissiere, B.E., Morison, N. and Longuesserre, J. 2005. Improving pollination of early strawberry crops. *International Journal of Fruit Science*, 5(2): 17–24.

Blasse, W. 1981. Strawberry production and bee introduction. *Gartenbau*, 28(11): 337–338.

Blasse, W. and Haufe, M. 1989. The influence of bees on yield and fruit quality in strawberry (*Fragaria* x *ananassa* Duch.). *Archiv fur Gartenbau*, 37(4): 235–245.

Bonfante, S. 1970. *Pollination of the Strawberry*. Atti 4 deg Convegno Nazionale Fragola, Cesena, Italy: pp. 291–296.

Burd, D.W. 1970. Tunnel strawberry pollination study: a step towards better fruit shape. *Grover*, 73: 1498–1499.

Butts, K.M. 1991. Bee attractants: improving strawberry quality? *Citrus and Vegetable Magazine*, 55(3): 16–17.

Canas Lloria, S. and Gomez Pajuelo, A. 1993. Honey bees in the pollination of crops. Hot-house crops. *Vida-Apicola*, 57: 38–43.

Carden, P.W. and Emmett, B.J. 1973. Artificially introduced blowflies show promise in strawberry pollination trials. *Grower*, 79(17): 1017–1019.

Ceuppens, B., Ameye, M., Van Langenhove, H., Roldan-Ruiz, I. and Smagghe, G. 2015. Characterization of volatiles in strawberry varieties "Elsanta" and "Sonata" and their effect on bumblebee flower visiting. *Arthropod-Plant Interactions*, 9(3: 281–287.

Chagnon, M., Gingras, J. and De Oliveira, D. 1989. Effect of honey bee (Hymenoptera: Apidae) visits on the pollination rate of strawberries. *Journal of Economic Entomology*, 82(5): 1350–1353.

Chagnon, M., Gingras, J. and De Oliveira, D. 1993. Complementary aspects of strawberry pollination by honey and indigenous bees (Hymenoptera). *Journal of Economic Entomology (USA)*, 86(2): 416–420.

Chang, Y.D., Yound, L.M. and Youngll, M. 2001. Pollination on strawberry in the vinyl house by Apis mellifera L. and A cerana Fab. *Acta Horticulturae*, 561: 257–262.

Chen, C.T. and Hsieh, F.K. 2001. Effect of honeybee pollination on the yield and fruit quality of strawberry variety "Toyonoka" (*Fragaria x ananassa* Duch.). *Plant Protection Bulletin Taipei*, 43(2): 117–127.

Chiasson, H., Vincent, C., Oliveira, D. and Richards, K.W. 1997. Effect of an insect vacuum device on strawberry pollinators. *Proceedings of the 7th International Symposium on Pollination*, Lethbride, Alberta, Canada, 23–28 June: pp. 373–377.

Connor, U. 1970. Studies of strawberry pollination in Michigan. *Report of the Ninth Pollination Conference*. Hot Springs, Arkansas : pp. 157–162.

Connor, U. 1972. *Components of Strawberry Pollination in Michigan.* : Doctoral dissertation, Ph.D. Thesis, Michigan State University, East Lansing, MI: vii + 84 pp.

Connor, U. 1975. The role of cultivar in insect pollination of strawberries. *Proceedings of the III International Symposium on Pollination*: pp. 149–154.

Connor, U. and Martin, E.C. 1973. Components of pollination of commercial strawberries in Michigan. *Horticulture Science*, 8(4): 304–306.

Dag, A. 1994. Use of bees for pollination of strawberries in greenhouses. *Hassadeh (Israel)*, 74(10): 1068–1077.

Dag, A., Dotan, S. and Abdul Razek, A. 1994. Honeybee pollination of strawberry in greenhouses. *Hassadeh*, 74(10): 1068–1070.

Darrow, G.M. 1927. Sterility and fertility in the strawberry. *Journal of Agriculture Research*, 34: 393–411.

Darrow, G.M. 1937. *Strawberry Improvement.* USDA Yearbook. Washington DC: pp. 445–495.

Darrow, G.M. 1966. *The Strawberry.* Holt, Rinehart and Winston, New York, Chicago, and San Francisco: p. 447.

Dimou, M., Taraza, S., Thrasyvoulou, A. and Vasilakakis, M. 2008. Effect of bumble be pollination on greenhouse strawberry production. *Journal of Apicultural Research*, 47(2): 99–101.

Eaton, G.W. and Smith, M.V. 1962. *Fruit Pollination.* Ontario Department of Agriculture No. 172. Ontario, Canada.

Eijnde, J-van-den. 1992. Pollination of greenhouse strawberries by bumblebees and honey bees. *Apidologie*, 23(4): 342–345.

Foulis, E.S. and Goulson, D. 2014. Commercial bumble bees on soft fruit farms collect pollen mainly from wildflowers rather than the target crops. *Journal of Apicultural Research*, 53(3): 404–407.

Free, J.B. 1968. The pollination of strawberries by honey bees. *Journal of Horticulture Science*, 43: 107–111.

Free, J.B. 1993. *Insect Pollination of Crops.* A. P. Harcourt Brace Jovanovich Publications, London, UK.

Fu, Y.A.N.G. 2010. Behavior of Bombus lucorum, Apis cerana cerana and Apis mellifera ligustica as Pollinator for Strawberry Grown in Greenhouse. *Journal of Anhui Agricultural Sciences*, 20: 91. http://en.cnki.com.cn/Article_en/CJFDTOTAL-AHNY201020091.htm

Gardner, V.R. 1923. Studies in the nutrition of the strawberry. *Missouri Agriculture Experiment Station Research*, 57: 31.

Goodman, R.D. and Oldroyd, B.P. 1988. Honeybee pollination of strawberries (*Fragaria* x *ananassa* Duchesne). *Australian Journal of Experimental Agriculture*, 28: 435–438.

Gundia, M. 1995. The use of bumble bees, *Bombus terrestris*, for pollination of greenhouse strawberry. *Hassadeh (Israel)*, 75(9): 49–50, 71–73.

Hofmann, S. and Haufe, M. 1995. Effect of pollination by honey bees on yield and fruit quality of strawberry (*Fragaria ananassa* Duch.). *Erwerbsobstbau (Germany)*, 37(5): 141–144.

Hokkanen, H.M., Menzler-Hokkanen, I. and Lahdenpera, M.L. 2015. Managing bees for delivering biological control agents and improved pollination in berry and fruit cultivation. *Sustainable Agriculture Research*, 4(3): 89–102.

Howitt, A.J., Pshea, A. and Carpenter, W.S. 1965. Causes of deformity in strawberries evaluated in a plant bug control study. *Michigan Agricultural Experiment Station Quarterly Bulletin*, 48(2): 161–166.

Hughes, H.M. 1951. *Fruit Cultivation for Amateurs.* Collingridge, London, UK.

Hughes, H.M. 1961. Preliminary studies on the insect pollination of soft fruits. *Experimental Horticulture*, 6: 44.

Hughes, H.M. 1962. Pollination studies. Progress report. M.A.A.F., N.A.A.S., Rep. Rept. Efford Exp. Hort. Stat. UK: p. 1961.

Hughes, H.M., Duggan, J.B. and Banwell, M.G. 1969. *Strawberry Bull.* 95, HMSO 10, 6d, Min. of Fish Food, UK.

Hyams, E. 1962. *Strawberry Growing Complete: A System of Procuring Fruit throughout the Year.* Faber and Faber, London, UK.

Ikeda, F. and Tadauchi, Y. 1995. Use of bumble bees as pollinators on fruit and vegetables. *Honeybee Science*, 16(2): 49–56.

Irkaeva, N.M. and Ankudinova, I.N. 1994. Evaluation of the outbreeding level of the strawberry *Fargaria vesca* L. using genetic markers. *Genetika (Russian Federation)*, 30(6): 816–822.

Jacobs, F.J., Houbaert, D. and Rycke, P.H. 1987. The pollinating activity of the honeybee (*Apis mellifera* L.) on some strawberry varieties (*Fragaria x ananassa* Duch.). *Apidologie*, 18(4): 345–348.

Jaycox, E.R. 1970. Pollination of strawberries. *American Bee Journal*, 110(5): 176–177.

Kaczmarska, E. 2003. Pollen viability and set of seeds in selected strawberry cultivars (*Fragaria x ananassa* Duch.) *Acta Scientiarum Polonorum Hortorum Cultus*, 2(1): 111–116.

Kakakhanian, A., Javaheri, D., Tahmasby, G., Bahmany, H., Kamangar, S. and Batobe, K. 2010. Pollination effect of honey bee and other insects on the quantitative and qualitative traits of strawberry in Kurdistan Province. http://agris.fao.org/agris-search/search.do?recordID=IR2012046019

Kakutani, T., Inoue, T., Tezuka, T. and Maeta, Y. 1993. Pollination of strawberry by the stingless bee, *Trigona minangkabau*, and the honey bee, *Apis mellifera:* an experimental studies of fertilization efficiency. *Researches* on *Population Ecology Kyoto*, 35(1): 95–111.

Katayama, E. 1987. Utilization of honey bees as pollinators for strawberries in plastic greenhouse in Tochigi prefecture. *Honeybee Science*, 8(4): 147–150.

Kim, I.S., Goh, H.G., Hong, K.J. and Lee, M.H. 1996. Increasing fruit fertilization by release of Eristalis cerealis in a vinyl house. *RDA Journal of Agricultural Science, Crop-Protection*, 38(1): 526–529.

Klatt, B.K., Klaus, F., Westphal, C. and Tscharntke, T. 2014. Enhancing crop shelf life with pollination. *Agriculture & Food Security*, 3(1): 1.

Koyuncu, F., Ylmaz, H. and Askn, M.A. 2000. An investigation on the pollen production capacities and germination rates of some strawberry cultivars. *Turkish Journal of Agriculture and Forestry*, 24(6): 699–703.

Kronenberg, H.G. 1959. Poor fruit setting in strawberries, I. causes of a poor fruit set in strawberries in general. *Euphytica*, 8: 47–57.

Kronenberg, H.G., Braak, J.P. and Zeilinga, A.E. 1959. Poor fruit setting in strawberries. II Malformed fruits in Jucunda. *Euphytica*, 8: 245–251.

Kruistum, G.V., Blom, G., Meurs, B. and Evenhuis, B. 2006. Malformation of strawberry fruit (*Fragaria* x *ananassa*) in glasshouse production in spring. *Acta Horticulture*, 708: 413–416.

Ledesma, N. and Sugiyama, N. 2005. Pollen quality and performance in strawberry plants exposed to high-temperature stress. *Journal of the American Society for Horticultural Science*, 130(3): 3451–347.

Lee, M.Y., Mah, Y.I., Park, I.K., Yoon, H.J., Chang, Y.D. and Kim, T.I. 1998. Effect of Apis mellifera on fruit production of strawberry cultivated in the vinylhouse. *Korean Journal of Apiculture*, 13(1): 21–26.

Leech, L., Simpson, D.W. and Whitehouse, A.B. 2002. Effect of temperature and relative humidity on pollen germination in four strawberry cultivars. *Acta Horticulture*, 567(1): 261–263.

Li, M. 2006. Some measures for improving strawberry fruit quality grown under shed. *China Fruits* (5): 67.

Li, J.L., Peng, W., Wu, J., An, J.D., Guo, Z.B., Tong, Y.M. and Huang, J.X. 2006. Strawberry pollination by *Bombus lucorum* and *Apis mellifera* in greenhouses. *Acta Entomologica Sinica*, 49(2): 342–348.

Lieten, F. 1993. Fruit misshaping of strawberries: over-visiting. *Fruitteelt*, 6: 27–29.

Lopez-Medina, J., Palacio Villegas, A. and Vazquez Ortiz, E. 2006. Misshaped fruit in strawberry, and agronomic evaluation. *Acta Horticulturae*, 708: 77–78.

Lounsberry, C.C. 1930. Visits of honey bees. In: L.H. Pammel and C.M. King (eds.). *Honey Plants of Iowa*. Iowa Geological Survey, 7: 1106–1109.

Malagodi, B., Katia, S. and Kleinert, A.M.P. 2004. Could Tetragonisca angustula Latreille (Apinae, Meliponini) be effective as strawberry pollinator in greenhouses? *Sustralian Journal of Agricultural Research*, 55(7): 771–773.

Malm, L. 2013. Effects of leaf beetle herbivory on pollination success and fruit development in woodland strawberry *Fragaria vesca*. Master's Thesis, Department of Ecology, Faculty of Natural Resources and Agricultural Sciences, Sweden.

McGregor, Samuel Emmet. 1976. Insect pollination of cultivated crop plants. http://www.beeculture.com/content/pollination_handbook/strawberry-1.html. Accessed 10 December 2009.

Meata, Y., Tezuka, T., Nadano, H. and Suzuki, K. 1992. Utilization of the Brazilian stingless bee, Nannotrigona testaceicornis, as a pollinator of strawberries. *Honeybee Science*, 13(2): 71–78.

Mommers, J. 1961. De bestuiving van aardbeien onder glas. *Meded. Dir, Tuinba*, 24: 353–355.

Mommaerts, V., Put, K. and Smagghe, G. 2011. Bombus terrestris as pollinator-and-vector to suppress *Botrytis cinerea* in greenhouse strawberry. *Pest Management Science*, 67(9): 1069–1075.

Moore, J.N. 1964. Duration of receptivity to pollination of flowers of the high-bush blueberry and the cultivated strawberry. In *Proceedings of American Society for Horticultural Science*, 85: 295–301.

Moore, J.N. 1969. Insect pollination of strawberries. *Journal of the American Society for Horticultural Science*, 94: 364, 367.

Muttoo, R.N. 1952. Bees produce more fruit. *Indian Bee Journal*, 1(7): 118–119.

Nitsch, J.P. 1952. Plant hormones in the development of fruits. *The Quarterly Review of Biology*, 27(1): 33–57.

Nye, W.P. and Anderson, J.L. 1974. Insect pollinators frequenting strawberry blossoms and the effect of honey bees on yield and fruit quality. *Journal of the American Society for Horticultural Science*, 99(1) 40–44.

Ohishi, T. 1999. Appropriate management of honeybee colonies for strawberry pollination. *Honeybee Science*, 20(1): 9–16.

Oliveira, D., Savoie, L., Vincent, C., Heemert, C. and Ruijter, A. 1991. Pollinators of cultivated strawberry in Quebec. *The Sixth International Symposium on Pollination*, Tilburg, the Netherlands, 27–31 August: pp. 420–424.

Pammel, L.H. and King, C.M. 1930. Cucurbitaceae, gourd family. *Bulletin of the Geological Survey Iowa*, 7: 650–662.

Pardo, P.R., Nates, P.G. and Nates, G. 1994. Increasing flower visits by *Apis mellifera* L. (Hymenoptera: Apidae) in corps by the use of synthetic Nasonov pheromone. *Revista Colombiana de Entomology*, 20(3): 187–192.

Partap, U., Matsuka, M., Verma, L.R., Wongsiri, S., Shrestha, K.K. and Partap, U. 2000. Pollination of strawberry by the Asian hive bee, *Apis cerana*. Asian bees and beekeeping: progress of research and development. *Proceedings of the Fourth Asian Apicultural Association International Conference*, Kathmandu, Nepal, 23–28 March: pp. 178–182.

Paydas, S., Eti, S., Kaftanoglu, O., Yasa, E. and Derin, K. 2000a. Effect of pollination of strawberries grown in plastic greenhouses by honey bees and bumblebees on the yield and quality of the fruits. *Acta Horticulturae*, 513: 443–451.

Paydas, S., Eti, S., Sevinc, S., Yasa, E., Derin, K., Kaska, N., Kaftanoglu, O., Plas, L.H.W. and Bockstaele, E. 2000b. Effects of different pollinators to the yield and quality of strawberries. Proceedings of the *XXV International Horticultural Congress*. Part 12 Application of biotechnology and molecular biology and breeding, general breeding, breeding and evaluation of temperate zone fruits for the tropics and subtropics, Brussels, Belgium, 2–7 August, 1998. *Acta Horticulturae*, 522: 209–215.

Petkov, V. 1965. Contribution of honey bees to the pollination of strawberries. *Gradinar Lozar Nauk*, 2: 421–431.

Pinzauti, M. 1987. Comportamento e ruolo delle api nell impollinazione della fragola. *Colture Protette*, 16: 65–68.

Pinzauti, M. 1992. Some observation on the bio-ethology, flight and foraging activity of Osmia cornuta Latr. (Hymenoptera: Megachilidae) during strawberry pollination. *Apicolturea (Italy)*, 8: 7–15.

Pion, S., Oliveria, D. and Paradis, R.O. 1980. Agents pollinisateurs et productivite du fraisier "Redcoat" *Fragaria x ananassa* Duch. *Phytoprotection*, 61: 72–78.

Porter, J. 1977. Strawberry yields increased by better pollination. *Horticulture Industry*, May: 375–376.

Praagh, J.P. 1983. Pollination of soft fruits. Effect of fungicides in the flight behaviour of bees in strawberry and raspberry crops. *Obstbau*, 5(5): 228–232.

Robbins, W.W. 1931. *The Botany of Crop Plants*. McGraw-Hill, New York, New York.

Roselino, A.C., Santos, S.B., Hrncir, M. and Bego, L.R. 2009. Differences between the quality of strawberries (*Fragaria x ananassa*) pollinated by the stingless bees *Scaptotrigona* all. *depilis* and *Nannotrigona testaceicornis*. *Genetics and Molecular Research*, 8(2): 539–545.

Sakai, T. and Matsuka, M. 1988. Honeybee pollination in Japan, with special reference to strawberry production. *Honeybee Science*, 9(3): 97–101.

Sharma, H.K., Gupta, J., Rana, B. and Rana, K. 2014. Insect pollination and relative value of honey bee pollination in strawberry, *Fragaria ananassa* Duch. *International Journal of Farm Sciences*, 4(2): 177–184.

Shashkina, L.M. 1950. Cross-Pollination of strawberries (*Fragaria grandiflora* Ehrh.). *Agrobiologiya*, 5: 45–47.

Shaw, F.R., Savos, M. and Shaw, W.M. 1954. Some observations on the collecting habit of bees. *American Bee Journal*, 94: 422.

Shemetkova, M.F. and Bachilo, A.I. 1986. Pollination of top and small fruit orchards with honey bees-obligatory means of agrotechniques. *Plodovodstvo*, 6: 139–143.

Shimizu, H., Hoshi, T., Nakamura, K. and Park, J.E. 2015. Development of a Non-contact Ultrasonic Pollination Device. *Environmental Control in Biology*, 53(2): 85–88.

Shimotori, K. 1981. Honey bees and the strawberry industry in Tochigi. *Honeybee Science*, 2(2): 57–60.

Shoemaker, J.S. 1955. *Small Fruit Culture*, 3rd edn. McGraw-Hill, London, UK.

Singh, Y. 1979. Pollination activity on strawberry at Jeolikote (District Nainital, India). *Indian Bee Journal*, 41: 17–119.

Skowronek, J., Jablonsi, B. and Szklanowska, K. 1985. Wplyw owadow zapylajacych na owocowanie 6 odmian truskawki (*Fragaria grandiflora* Ehrah). Pszczelnicze Zeszyty Naukowe, 29: 205–229.

Skrebtsova, N.D. 1957. Role of bees in pollination of strawberries. *Pchelovodstvo*, 34(7): 34–36.

Soltesz, M., Benedek, P., Nyeki, J., Szabo, Z. and Toth, T. 2001. Flower visiting activity of honey bees on fruit species blooming subsequently. *International Journal of Horticultural Science*, 7(1): 12–16.

Svensson, B., Heemert, C. and Ruijter, A. 1991. The importance of honeybee-pollination for the quality and quantity of strawberries (*Fragaria x ananassa* Duch.) in central Sweden. The Sixth International Symposium on Pollination, Tilburg, the Netherlands, 27–31 August: pp. 260–264.

Szklanowska, K. 1995. The role of insect-pollinator visits on fruiting in strawberry (*Fragara x ananassa* Duch.). Materialy ogolnopolskiej konderencji naukowe Nauka Praktyce Ogrodniczej z okazji XXV-lecia Wydzialu Orgodniczego Akademii Rolniczej w lublinie: pp. 907–910.

Vasquez, R., Ballesteros, H., Ortegon, Y. and Castro, U. 2006. A commercial strawberry crop (*Fragaria chiloensis*) being directly pollinated by bees (*Apis mellifera*). *Revista Corpoica*, 7(1): 50–53.

Verma, S.K. and Phogat, K.P.S. 1995. Effect of pollinators by honey bees on fruit setting and yield of strawberry (*Fragaria* spp.) under Himalayan valley conditions. *Progressive Horticulture*, 27(1–2): 104–108.

Vincent, C., Olivearia, D.D., Belanger, A., De-Oliveira, D.D., Bostanian, N.J., Wilson, L.T. and Dennehy, T.J. 1990. The management of insect pollinators and pests in Quebec strawberry plantations. In *Monitoring and Integrated Management of Arthropod Pests of Small Fruit Corps*, Bostanian, N. J.; Wilson, L. T.; Dennehy, T. J (eds.). Intercept Ltd, Andover, UK: pp. 177–192.

Vinson, R. 1935. Growing healthy strawberries. Cherries and soft fruits. Varieties and cultivation in 1935. Report of the conference held by the Royal Horticultural Society at the Greycoat Street Hall, July 16 and 17, 1935. pp. 24–9.

Voyiatzis, D.G. and Paraskevopoulou-Paroussi, G. 2002a. The effect of photoperiod and gibberellic acid on strawberry pollen germination and stamen growth. *Acta Horticulturae*, 567(1): 257–260.

Voyiatzis, D.G. and Paraskevopoulou-Paroussi, G. 2002b. Factors affecting the quality and in vitro germination capacity of strawberry pollen. *Journal of Horticultural Science and Biotechnology*, 77(2): 200–203.

Watters, B.S. and Sturgeon, S.R. 1992. Evaluation of some potential pollen donors for strawberry cv. Pandora. *Tests of Agrochemicals and Cultivars (United Kingdom)*, 13: 128–129.

Wilkaniec, Z. and Maciejewska, M. 1996. Utilization of honeybee (*Apis mellifera* L.) for the greenhouse cultivated strawberry. *Pszczelnicze Zeszyty Naukowe (Poland)*, 40(2): 227–234.

Wilkaniec, Z., Radajewska, B., Scheer, H.A.T., Lieten, F. and Dijkstra, J. 1997. Solitary bee *Osmia rufa* L. (Apoidea, Megachilidae) as pollinator of strawberry cultivated in an unheated plastic tunnel. *Proceedings of the Third International Strawberry Symposium*, Veldhoven, the Netherlands, 29 April–4 May, 1996. Vol. 1. *Acta Horticulturae*, 439: 489–493.

Woo, K.S., Choo, H.Y. and Choi, K.R. 1986. Studies on the ecology and utilization of pollinating insects. *Korean Journal of Apiculture*, 1(1): 54–61.

Yoon, H.J., Lee, K.Y., Kim, M.A., Park, I.G. and Choi, Y.C. 2011. Current status of insect pollinator use in strawberry crop in Korea. *Korean Journal of Apiculture*, 26(2). http://agris.fao.org/agris-search/search.do?recordID=KR2012004574

Yoshida, Y., Goto, T., Chiyo, T. and Fujime, Y. 1991. Changes in the anatomy and a receptivity of pistils after anthesis in strawberry. *Journal of the Japanese Society for Horticultural Science*, 60(2): 345–351.

Zapata, I.I., Villalobos, C.M.B., Araiza, M.D.S., Solís, E.S., Jaime, O.A.M. and Jones, R.W. 2014. Effect of pollination of strawberry by Apis mellifera L. and Chrysoperla carnea S. on quality of the fruits. *Nova Scientia*, 7(13): 85–100.

Zaitoun, S.T., Al-Ghzawi, A.A., Shannag, H.K., Rahman, A. and Al-Tawaha, M. 2006. Comparative study on the pollination of strawberry by bumble bees and honey bees under plastic house conditions in Jordan valley. *Journal of Food, Agriculture and Environment*, 4(2): 237–240.

Zebrowska, J. 1998. Influence of pollination modes on yield components in strawberry (*Fragaria x ananassa* Duch.). *Plant Breeding*, 117(3): 255–260.

19 Fruit Development

Sangita Yadav, Sandeep Kumar, Seema Sangwan, and Shiv K. Yadav

CONTENTS

19.1 Fruit Growth 336
19.2 Fruit Ripening 336
19.3 Changes Occurring during Fruit Development 337
    19.3.1 Cellular Changes 337
        19.3.1.1 Cell Wall Constituents 337
        19.3.1.2 Cellular Organelles 338
    19.3.2 Compositional Changes 338
        19.3.2.1 Enzymes 338
        19.3.2.2 Pigments 338
        19.3.2.3 Sugars 339
        19.3.2.4 Organic Acids 339
        19.3.2.5 Total Soluble Solids 339
        19.3.2.6 Phenolic Substances 339
        19.3.2.7 Amino Acids 340
        19.3.2.8 Volatile Constituents 340
    19.3.3 Physiological Changes 340
        19.3.3.1 Respiration Rate 340
        19.3.3.2 Ethylene Evolution Rate 341
19.4 Factors Affecting Fruit Development 341
    19.4.1 Climate 341
    19.4.2 Irrigation and Nutrition 341
    19.4.3 Number of Achenes 341
    19.4.4 Fruit Position on the Inflorescence 342
    19.4.5 Plant Bioregulators 342
References 343

Botanically, strawberry is an aggregate fruit, which develops by the simultaneous maturation of a number of separate ovaries, which occur on a common receptacle of a single flower. These develop into one-seeded fruits or achenes (a combination of seed and ovary tissues that originate at the base of each pistil) attached to the outer surface of the receptacle through vascular connection (Darrow, 1966; Takeda et al., 1990). In horticultural crops, the enlarged receptacle with achenes is considered the berry, but is often termed a "fruit" (Perkins-Veazie, 1995). Each berry can have 20–500 achenes depending on cultivar and growing conditions (Darrow, 1966). The edible part of the fruit consists of fleshy pith at the centre, next to a ring of vascular bundles branching out into the achenes – the seeds; then a fleshy cortex outside the ring and epidermis bearing a few hairs is connected to superficial achenes. The meristematic tissue is next to the epidermis without intercellular spaces (Avigdori-Avidov, 1986). In it, the cells continue to divide after the pith and cortex cells stop dividing. The vascular bundle leading to each achene supplies water and food from the stem through the central cylinder to the achene and surrounding parenchyma cells of the receptacle

(Antoszewski, 1973). The bundles are composed of longer and tougher cells compared with the flesh and seeds, and help to hold the berry together and make it firm.

The mature achene is composed of a hard and relatively thick pericarp, a thin testa, an endosperm consisting of one cell layer and a small embryo (Winston, 1902). The endosperm is a free-nuclear layer bounding the embryo sac unit upto two weeks after anthesis, and, thereafter it becomes cellular (Thompson, 1963). The stomata on ripe receptacle tissues appears to be protuberant and open, and may be involved in transpiration and respiration (Blanke, 2002).

19.1 FRUIT GROWTH

Fruits of the strawberry grow rapidly with full size being attained approximately 30 days after anthesis depending upon the growing conditions. Strawberry fruit growth may be sigmoid or double-sigmoid (Coombe, 1976; Rosati, 1993) depending upon the variety and part of fruit considered for the measurement of the degree of sigmoidicity. The Ozark Beauty cultivar has been observed to exhibit both sigmoid or double-sigmoid patterns of growth under different light conditions and sample frequencies (Mudge et al., 1981; Velutharnbi et al., 1985). The first peak is influenced by achene development and is associated with initial receptacle growth; the second peak is associated with the transport of carbohydrate assimilates into the receptacle, which enlarges rapidly (Strik and Proctor, 1988; Miura et al., 1990). Achene development occurs before the final increase in receptacle enlargement (Thompson, 1969). Upon maturation of the embryo, the rate of receptacle growth increases and initiates berry ripening. Growth of the achene is single-sigmoidal, and it achieves its full size before final fruit swell (Miura et al., 1990).

Receptacle growth after petal fall takes place initially due to a combination of cell division and cell expansion. Cell proliferation is completed at seven days and thereafter, growth takes place by means of an increase in cell volume amounting to about 100-fold during development (Knee et al., 1977). After fertilization, only 15–20% of the growth occurs due to cell division, while about 90% of the cortex growth after anthesis takes place due to cell enlargement (Havis, 1943), which is accompanied by major changes in the cell wall and subcellular structures. At petal fall, the receptacle cells have dense cell walls, small vacuoles, plastids with starch grains and abundant Golgi apparati and ribosomes. During development, the cell walls become swollen and increasingly diffuse, and the plastids lose their starch grains, eventually degenerating with the commencement of berry ripening (Knee et al., 1977).

19.2 FRUIT RIPENING

Fruit ripening is a genetically programmed stage of development overlapping with senescence (Watada et al., 1984). Strawberry has emerged as a suitable model plant for studying non-climacteric fruit ripening (Giovannoni, 2001, 2004). Although strawberry fruit is derived from the flower receptacle, it has the same development and ripening characteristics of true fruit, including the degradation of chlorophyll, the accumulation of anthocyanins, softening that is partially mediated by cell wall-hydrolysing enzymes, the metabolism of sugars and organic acids, and the production of flavouring compounds.

Strawberry is a non-climacteric fruit and it must be harvested at full maturity to allow the fruit to develop optimum quality in relation to flavour and colour. The main changes in fruit composition, which are usually associated with ripening, take place when the fruit is still attached to the mother plant. Consequently, strawberries should be harvested at the ready for consumption stage. This means there is a very short period of fruit at its best quality (Manning, 1996; Cordenunsi et al., 2003).

Ripeness, maturity, cultivar, irrigation and fertilization are major factors that can affect the quality of a product. Implicit in the use of single and multiple physical or chemical characteristics to determine optimum maturity is that changes in the selected parameter correlate with

Fruit Development **337**

the attainment of the general composition of quality characteristics of the product (Kays, 1991; Sturm et al., 2003).

In strawberry, the time taken for the berry to become fully ripe is closely related to the prevailing environmental temperature (Perkins-Veazie and Huber, 1987), and can vary from 20–60 days, probably reflecting an overall effect of temperature on metabolic rate as determined by enzyme activity. Strawberry ripeness is generally divided into four stages, viz., green, white, pink (or turning) and red. Fruits reach the white stage at about 21 days after anthesis and are fully red (ripe) within 30–40 days (Dennis, 1984). The entire ripening process is rapid, generally occurring within five–10 days following the white stage, depending on air temperature (Kano and Ashira, 1981; Perkins-Veazie and Huber, 1987).

Despite its non-climacteric nature (Perkins-Veazie, 1988), strawberry fruit undergoes distinct changes during ripening. Fruits continue to increase in size during the ripening process (Abeles and Takeda, 1990), which is marked by simultaneous changes in sensory attribute in normal fruit. However, "White Carter" (a mutant cultivar of *F. ananassa*) lacks red pigmentation (anthocyanin) even when fully ripe. This cultivar shows otherwise normal ripening behaviour with respect to rest of the sensory attributes. The strawberry fruits can attain full colour in storage even if detached at the white or pink stage (Smith and Heinze, 1958; Kalt et al., 1993) but being a non-climacteric fruit, changes in texture, sugars and acidity fail to develop properly.

19.3 CHANGES OCCURRING DURING FRUIT DEVELOPMENT

19.3.1 CELLULAR CHANGES

19.3.1.1 Cell Wall Constituents

The cell wall consists of pectic substances and cellulose as the main components in the strawberry fruits along with small amounts of hemicelluloses and noncellulosic polysaccharides (Jona and Foa, 1979). Proportions between protopectin and pectins are considered to indicate stage of maturity, softening and breakdown of cell walls of the fruit. A high proportion of free-soluble polyuronides exists in the receptacle at petal fall and in ripe fruit (Knee et al., 1977).

Strawberry fruits undergo extensive softening during ripening due to cell wall dissolution, caused by a coordinated action of cell wall proteins and enzymes including polygalacturonases (PGs). Villarreal et al. (2009) recognized two proteins that accumulated from the 25% red stage in the comparatively firm cultivars, such as Camarosa, and from the white stage in the slightly softer cultivars, such as Toyonoka. Woodward (1972) observed a very low level of water-soluble pectin at 14 days after petal fall, which increased to 90% of total pectin levels by 42 days (red-ripe stage) and resulted in the softening of fruits. Degradation of cellulose and hemicelluloses may also contribute to fruit softening (Abeles and Takeda, 1990). Huber (1984) observed that the proportion of total soluble polyuronides to increase from 30% in green fruits to 65% in ripe fruit. The increased hydration of the cell wall (Knee et al., 1977) during ripening is constant with increased solubility of polyuronides. Newly synthesized polyuronides added to the wall might affect its structural integrity. This could rise if the newly synthesized polyuronides were less firmly bound to the cell wall (Huber, 1984).

Ripening in strawberry is accompanied by an increase in the proportion of the neutral sugars associated with the soluble polyuronides, notably rhamnose and to a lesser degree arabinose and galactose (Huber, 1984). Increases in these sugars can be linked to the polygalacturonan backbone via the rhamnosyl moiety (McNeil et al., 1980), indicating that alterations may occur during ripening in the way in which carbohydrates are cross-linked with the cell wall.

Strawberry fruit ripening is related to pectin solubilisation and depolymerisation, although the magnitude of pectin depolymerisation is variety specific (Rosli et al., 2004). At harvest, firmness of half-red strawberries has been observed 20% higher than the firmness of full-red or dark-red strawberries (Forney et al., 1998; Menager et al., 2004). Menager et al. (2004) in a study using strawberries from the white to the dark-red stage, reported that firmness of the fruit decreased from white to half-red and then appeared to hold steady from three-quarters- to full- and dark-red.

19.3.1.2 Cellular Organelles

The cells of green fruit (one week after petal fall) contain dense cell walls, small vacuoles, starchy grains in plastids, dictyosomes, ribosomes and endoplasmic reticuli (Knee et al., 1977), having tubular proliferations of the tonoplast, which become extensive as berries ripen. These tubular proliferations may be involved in anthocyanin synthesis (Grisebach, 1982). At the white stage, cells are expanded and vacuolated, plastids have degenerated and most of the starch has disappeared. During ripening, the cells are connected only by small projections at the tips of the cells; these junctions are transverse by protoplasmic connections between the cells (Knee et al., 1977). In ripe fruit, cell walls are swollen and transversed by protoplasmic connections between adjacent cells, resulting in occlusion of the intercellular spaces by matrix material (Neal, 1965; Knee et al., 1977). Parenchymatous cortical cells continue to enlarge and become separated at the middle lamella as fruit ripen (Neal, 1965). Mitochondria appear to be normal in ripe fruit (Knee et al., 1977).

19.3.2 COMPOSITIONAL CHANGES

19.3.2.1 Enzymes

Little or no endo-or-exo-polygalacturonase activity has been observed in strawberry (Abeles and Takeda, 1990; Nogata et al., 1993) during fruit ripening. Nogata et al. (1993) isolated three forms of exo-polygalacturonase from fruits of strawberry but observed that their activity decreased from 0.313 unit/g fresh weight to 0.08 unit/g fresh weight as the fruits ripened, which was not sufficient to explain the rapid and complete softening of strawberry fruit.

Pectin methylesterase (EC 3.1.1.11) activity increases during ripening of fruits of strawberries (Barnes and Patchett, 1976) but is stoichiometrically insufficient to account for the large changes in water-soluble polyuronides (Huber, 1984). Endo-l,4-B-D-glucanase (cellulase) (EC 3.2.1.4) appears to be the most likely chemical compound observed so far in the enzymic softening of strawberry fruits, exhibiting a six fold increase in cellulase activity between the green and dark-red stages of ripening (Abeles and Takeda, 1990).

19.3.2.2 Pigments

The red colour in strawberry fruit is derived from the orange-brown anthocyanin pigments of pelargonidin glycosides (Bakker et al., 1994). Most strawberry cultivars have pelargonidin 3-glucoside as the predominant pigment, but some cultivars such as "Totem" have been observed to contain small amounts of cyanidin 3-glucoside, pelargonidin rutinoside and pelargonidin and cyanidin-3 glucosides acylated with succinic acid. Anthocyanin synthesis in strawberry fruits is believed to share a common pathway from the primary metabolic precursor phenylalanine. Phenylalanine ammonialyase (PAL) (EC4.3.1.5), the first enzyme in the pathway, has a pivotal role in directing synthesis towards secondary metabolites. The anthocyanin formation and expression in fruits of strawberries may be developmentally dependent on two regulatory enzymes, PAL and UDPGFT (uridine diphosphate glucose: flavonol O^3-D-glucotransferase) (EC 2.4.1.91) (Cheng and Breen, 1991). The accumulation of anthocyanin in ripening fruits depends on an increase in PAL activity, which could be used as biochemical marker of berry ripening (Given et al., 1988).

The biosynthesis of chlorophyll/carotenoids in green berries stops and that of the anthocyanin begins with the commencement of white stage (Woodward, 1972). Trace amounts of carotenoids are found in ripe fruit (Galler and MacKinney, 1965), originating from chloroplasts present in green berries (Gross, 1982). Chloroplasts disintegrate during ripening without becoming chromoplasts (Knee et al., 1977). Although anthocyanins are brought to be stored in the vacuoles of the epidermal and cortical cells (Woodward, 1972), at least part of anthocyanin synthesis, especially the glycosylation process, may be located in the cytosol rather than the vacuole (Runkova et al., 1972; Grisebach, 1982).

Fruit Development

19.3.2.3 Sugars

Sugars are one of the main soluble components in fruits of strawberries and provide energy for metabolic activities. In the ripe fruits of strawberries, glucose, fructose and sucrose account for >90% of the total sugars, with sorbitol, xylitol and xylose occurring in trace amounts (Makinen and Soderling, 1980). Approximately 80–90% part of the total sugars is contributed by the reducing sugars fructose and glucose in about a 1:1 ratio (Kader, 1991). Total sugars almost double, from about 2–5 mg per gram fresh weight between green and red-ripe stages (Woodward, 1972; Spayd and Morris, 1981).

During fruit development an increasing proportion of ^{14}C photoassimilates in leaves is exported to the receptacle with a decreasing proportion of this ^{14}C going to the fruits (Nishizawa and Hori, 1988). Sucrose is the main carbohydrate translocated to the fruit and fruit growth rates correlate with rates of sucrose uptake (Forney and Breen, 1985). Sucrose appears to be unloaded apoplastically and hydrolyzed in the free spaces before uptake (Forney and Breen, 1986) and this may account for the greater accumulation of glucose and fructose over sucrose in ripe fruit.

19.3.2.4 Organic Acids

Organic acids regulate pH of fruit, contribute to colour stability and inhibit enzyme activity. Similar to sugars, organic acids are important compounds contributing to flavour, and sugar/acid ratios are often used as an index of consumer acceptability and quality of fruits. In strawberries, the non-volatile organic acids are quantitatively most important in determining fruit acidity. Strawberry fruit may have 32–33 organic acids (Broderick, 1976; Mussinan and Walradt, 1975). The primary organic acid found in strawberry fruit at all stages of growth is citric acid, which represents 88% of the total organic acids in ripe berries (Green, 1971; Kim et al., 1993). Quinic and shikimic acids are found in very small amounts, and increase during maturation until the fruits are fully ripe, while both citric and malic acids decrease (Green, 1971; Sistrunk and Cash, 1973). In berries, total acidity expressed on a fresh weight basis increases modestly to maximum in mature green fruits before declining more rapidly in the later stages of ripening (Spayd and Morris, 1981). Moing et al. (2001) reported that the pH in strawberry fruits was close to 5 at 10 days after full bloom and decreased to roughly 3.7 at the turning stage, and subsequently remained unchanged until ripening. Acidity of fruits increases from the large green to white stages but decreases subsequently as the pink colour becomes visible (Woodward, 1972).

Organic acids are sequestered in the cell vacuole, and can be used in respiration or converted to sugars (Perkins-Veazie, 1995). Titratable acidity can be decreased by application of nitrogenous fertilizers or increased by preharvest shading of strawberry fruit (Saxena and Locascio, 1968; Osman and Dodd, 1992). Strawberry fruit is a rich source of ascorbic acid. The outer portion of the fruit is comparatively higher in ascorbic acid content than the inner tissues, and it increases as the fruit ripens (Ezell et al., 1947; Spayd and Morris, 1981).

19.3.2.5 Total Soluble Solids

The total soluble solids (TSS) of fruit are composed of sugars, acids and other substances dissolved in the cell sap (Perkins-Veazie, 1995). About 80–90% of the TSS content consists of sugar. The soluble contents in fruit increases steadily during development from 5% in small green fruit to 7.3% in overripe (dark-red) berries (Spayd and Morris, 1981). TSS content has been observed to be dependent on environmental conditions and intercultural operations during production than genetic inheritance (Shaw, 1990).

19.3.2.6 Phenolic Substances

Phenolic substances represent a diverse group of compounds including secondary plant metabolites such as polyphenols (tannins), proanthocyanidins (condensed tannins) and esters of hydroxybenzoic acids and hydroxycinnamic acids. Oxidation of phenolics by enzymes such as polyphenoloxidase and peroxidase causes enzymatic browning in strawberry fruits and processed products such as bruised fruits and thawed or pureed products.

Maturing strawberry fruits undergo dramatic changes in the levels of different types of phenolic compounds, which provide it with colour, flavour and resistance to pathogenic attack and environmental conditions such as exposure to UV rays.

Strawberry fruits contain polyphenols (tannins), chlorogenic acid, D-catechin and P-coumaric acid (Green, 1971; Wesche-Ebling and Montgomery, 1990). Total soluble phenols decrease (from 0.6–0.3%) as the strawberries grow from green berries and attain maturity/ripening as red berries (Spayd and Morris, 1981), probably due to the synthesis of anthocyanins. During early stages, non-tannin flavonoids and mainly condensed tannins accumulate to high levels and give strawberry an astringent flavour (Cheng and Breen, 1991). Later, when fruits start to ripen, other flavonoids such as anthocyanins (mainly pelargonidin glucoside), flavonols and cinnamoyl b-D-glucose accumulate to high levels (Latza et al., 1996; Manning, 1998; Moyano et al., 1998; Aharoni et al., 2000; Deng and Davis, 2001). Peroxidase and polyphenol oxidase activities also decrease by about 80% with fruit ripening. Polyphenol oxidase degrades fruit anthocyanin indirectly by reacting with D-catechin to form quinones that polymerize with anthocyanin pigments and form yellow-brown precipitates (Wesche-Ebling and Montgomery, 1990). Peroxidase may also degrade anthocyanins (Spayd and Morris, 1981).

19.3.2.7 Amino Acids

Amino acid metabolism generates liphatic and branched-chain alcohols (Kids, carbonyls and esters), some of which are used to generate volatile aromatic compounds (Perez et al., 1992). Perez et al. (1992) observed a declining trend of amino acids in Chandler strawberries as the fruits grew and attained ripening. Asparagine is the predominant amino acid in strawberry fruits and represents 50% of the total amino acid content at ripening (Perez et al., 1992). Other identified amino acids include aspartic and glutamic acid, a-alanine, proline, serine, valine and glutamine (Burroughs, 1960; Green, 1971; Perez et al., 1992).

19.3.2.8 Volatile Constituents

The quality of food in terms of flavour is a combination of the sensory impressions of taste detected by the tongue and aroma sensed by the nose (Manning, 1993). The fruits of strawberry accumulate (hydroxy) cinnamoyl glucose (Glc) esters, which may serve as the biogeneic precursors of diverse secondary metabolites such as the flavour constituents methyl cinnamate and ethyl cinnamate. Using quantitative real-time PCR analysis, Lunkenbein et al. (2006) reported that UDP-Glc:cinnamate glucosyl transferase (FAGT2) transcripts were accumulated to high levels during ripening of strawberry fruit. Further, using an antisense construct of FaGT2, they demonstrated that the enzyme only involved in formation of cinnamoyl Glc and p-coumaroyl Glc during ripening in comparison to *in vitro* studies where it glucosylates a number of aromatic acids. At least 35, possibly as many as 200, volatile compounds have been identified in fruit of strawberry (Mussinan and Walradt, 1975; Manning, 1993; Perez et al., 1993). Seven volatiles, viz., ethyl hexanoate, ethyl butanoate, methyl butanoate, methyl hexanoate, hexyl acetate, ethyl propionate and 3-hexenyl acetate are found in strawberry fruit and contribute to major fraction of aroma associated with ripe fruits of strawberries. Ethyl butanoate and ethyl hexanoate are the most prevalent volatiles in fruits of strawberry cv. Chandler (Perez et al., 1992). The volatiles such as 3-hexenyl acetate and hexyl acetate are responsible for the odour in green, immature berries; while ethyl esters are the major volatile components at all the stages of ripening and comprise 60% of the total volatile compounds observed in red berries. Isopropyl butanoate and propyl propanoate are detected only in fully ripe fruits. The relationship between the degree of maturity and the concentration of 2, 5-dimethyl-4-hydroxy-3(2H)-furanone and its derivatives has been investigated in seven strawberry varieties by Sanz et al. (1995).

19.3.3 Physiological Changes

19.3.3.1 Respiration Rate

Being non-climacteric, strawberry fruits fail to show sudden rise in respiratory rate during ripening; however, Ben-Arie and Faust (1980) reported a threefold increase in ATPase (EC 3.6.1.3) activity

Fruit Development 341

(on a fresh weight basis) in strawberry fruits between the green and dark-red stages, indicating that extra energy for transport process may be required during this phase of fruit development. Abeles and Takeda (1990) observed the rate of respiration in ripe strawberry fruits from 40–50 ml/kg/hour at 20°C. It was comparatively more (Hardenburg et al., 1986) when compared with both climacteric (tomato) and non-climacteric fruits (orange). Sas et al. (1992) observed increased rate of respiration in green, white and red berries by applying ethylene @ 10, 100 or 1000 µl/l, but respiration rates were similar at all the concentrations.

In ripe strawberry fruits, the respiration quotient (CO_2/O_2) is initially nearly one, then falls to 0.8 after a few days in storage (Li and Kader, 1989), indicating a shift from glucose to fatty acid as respiratory substrates (Wills et al., 1981).

19.3.3.2 Ethylene Evolution Rate

The strawberry fruits are generally accepted to be a non-climacteric fruit and, as such, would not be expected to express an increased synthesis of ethylene during ripening (Knee et al., 1977; Mitra, 1991). The endogenous production of ethylene in strawberry fruit is extremely low (15–80 pl/g fw/h), which declines from 60–30 pl/g fw/h, between green and white ripeness stages and increases to 80 pl/g fw/h at the dark-red stage (Abeles and Takeda, 1990).

19.4 FACTORS AFFECTING FRUIT DEVELOPMENT

19.4.1 CLIMATE

In addition to initiation of flower, the number of flowers per truss and branching structure of flower/fruit cyme, climate also influences fruit set and development in strawberries. Monselise (1986) reported promotary effects of high light intensities on berry ripening. The size of individual fruits is inversely proportional to the temperature during day hours, whereas, the temperatures during the night hours has no effect, which might be related to increased carbohydrate transport at lower temperature (Went, 1957). A long plant prechilling period has promotary effects on size of fruits (Monselise, 1986). The tip of the berry may become meristematic under some conditions, especially when growth is checked during periods of cool weather. The multiple tipped shapes of berries may be caused due to cold, dry weather in autumn, which affects flower buds (Darrow, 1966). Varieties grown under sunny days and cool nights have better flavour than those grown under cloudy, humid days and warm nights (Monselise, 1986).

19.4.2 IRRIGATION AND NUTRITION

Heavy irrigation from early summer until fruit harvesting increases mean berry size. Irrigation during development of the unripe fruit followed by a dry period increases the number of fruits in the next year. Heavy precipitation during July increases berry weight. The roles of nitrogen (Nes and Hjeltens, 1992), potassium (Ricketson, 1966), calcium (Chiu and Bould, 1976), magnesium (Bunernann and Gruppe, 1962), zinc (Jurgens, 1990) and iron (Zaiter et al., 1993) have been studied and are well understood/established in fruit development in a number of strawberry cultivars.

19.4.3 NUMBER OF ACHENES

Most of the growth of fruits in strawberries has been attributed to stimulation of assimilate transport (mobilization) by auxin secretion from achenes (Perkins-Veazie, 1995). The development of the fleshy receptacle is entirely dependent upon the presence of fertilized achenes, and if the achenes are removed, the receptacle immediately ceases to grow. Partial removal of achenes results in fruits of abnormal shape and size, as only the receptacle parts that are adjacent to fertilized achenes continue to grow (Nitsch, 1950). Webb et al. (1978) concluded that a minimum number of achenes were necessary for fruit development and that the maximum surface area of receptacle tissues influenced by achenes was limited to 0.165 cm^2. The spiral arrangement of achenes on the fruit surface may facilitate maximum diffusion of growth from achenes into receptacle tissues, thus accounting for the compensatory effects observed on number of achenes developed per unit receptacle.

19.4.4 FRUIT POSITION ON THE INFLORESCENCE

Fruit position on the inflorescence significantly influences the size of strawberry fruits. Primary fruits are larger than secondary or tertiary fruits. Sherman and Janic (1966) concluded that secondary and tertiary fruits attained 50% and 25% fresh weight, respectively, of the fresh weight of the primary fruit. Differences in fruit weight were due to the number of achenes (Abbott et al., 1970). Webb (1973) suggested that primary fruit had a shorter and more developed vascular system than secondary or tertiary fruit and were, therefore, more efficient in transport of assimilates. Overall, regulation of size of fruits is the result of the combined effects of number of achenes, position of fruits on the inflorescence, differential activity in hormone production and differential sensitivity of receptacle tissue to hormones (Moore et al., 1970).

19.4.5 PLANT BIOREGULATORS

It is well known that auxins are negative regulators of ripening of fruits (Given et al., 1988), as they have inhibitory effects on a number of cell wall-related genes, such as pectate lyase (Benıtez-Burraco et al., 2003), β-xylosidase (Martınez et al., 2004) and polygalacturonase (Villarreal et al., 2008). Nitsch (1950, 1955) reported that the growth of the strawberry receptacle is regulated by the achenes, that synthetic auxins can restore the growth of receptacles from which the achenes have been removed and that the achenes are a source of IAA. Thus the idea arose that the achenes (or seeds) are synthesizers and exporters of auxin to the fruit. Archbold and Dennis (1984) observed that a small quantity of free auxins appeared in the receptacles at 11 days post-anthesis and peaked in both receptacles and achenes just before the white stage which declined rapidly in the receptacle and gradually in the achenes as fruits turned red. A gradual decline in the supply of auxins from achenes at the later stages of growth has been implicated on the basis of ripening in strawberry fruits (Given et al., 1988). Application of the synthetic auxin (NAA) to de-achened large green or white fruits caused delay in development of colour but the patterns of translocation were similar to those of mature green fruit.

Gibberellins can act as fruit-ripening inhibitors (Martinez et al., 1996), as GA$_3$ treatment has been observed to reduce the anthocyanin levels as well as PG, mRNA and protein accumulation, although the inhibitory effect was not as marked as that provoked by auxins (Villarreal et al., 2009). The gibberellin GA$_3$ acts synergistically with NAAm in intact fruit cultured *in vitro* in promoting growth and ripening. Kinetin did not induce growth in tissue-cultured strawberries or when applied to intact berries (Lis and Antoszewski, 1979). Lis et al. (1978) detected peak of cytokinin- and gibberellin-like activity in fruits at seven days post-anthesis, which was more concentrated in the achenes than in the receptacles. After seven days post-anthesis, there was a sharp decline in the level of cytokinin in achenes as well as in receptacles, and remained at very low levels until the fruits were ripe. Gibberellin content was low in both achenes and receptacles and declined further as the fruit went through maturation. The effect of temperature upon achene development in relation to ripening was considered to be related to cytokinin activity in the achene, where it was mainly located (Kano and Asahira, 1979). It was further suggested that ripening in fruits grown at lower temperatures was delayed because high cytokinin-like activity was maintained in the achenes due to their later development.

Abscisic acid (ABA) accumulates in the achenes during ripening without increasing substantially in the receptacle. It has been suggested that the positive effect of ABA on anthocyanin accumulation, and colour development of fruits of strawberries are ethylene-mediated via enhanced

Fruit Development

PAL activity (Jiang and Joyce, 2003). Maximum concentrations of ABA in receptacle tissue were observed (Archbold and Dennis, 1984) at petal fall (0.9 µg/g dry weight).

Perkins-Veazie (1995) concluded that ethylene had little direct involvement in ripening of fruits of strawberries. Although exogenous application of ethylene (20 µl/l) can enhance the growth of botrytis rot on fruits and decrease firmness as much as 50% (El-Kazzaz et al., 1983). But Picon et al. (1993) did not find any effect of ethylene absorbed on pH, anthocyanin, TSS and acid contents of ripe and stored berries. Although strawberry ripening is not influenced by endogenous levels of ethylene, different ripening processes may be sensitized by ethylene and subsequently triggered by other hormones (Perkins-Veazie, 1995). In a study conducted by Villarreal et al. (2009), it was observed that polygalacturonase gene expression was upregulated by the exogenous application of the hormone, and that FaPG1 protein accumulation increased compared with control. However, when fruits were treated with an inhibitor of ethylene perception (1-MCP), anthocyanin, *PGs* expression and total PG activity were greatly reduced as compared to control.

REFERENCES

Abeles, F.B. and Takeda, F. 1990. Cellulase activity and ethylene in ripening strawberry and apple fruits. *Scientia Horticulturae*, 42: 269–275.

Abbott, A.J., Best, G.R. and Webb, R. A. 1970. The relation of achene number to berry weight in strawberry fruit. *Journal of Horticultural Science*, 45: 215–222.

Aharoni, A., Keizer, L.C., Bouwmeester, H.J., Sun, Z., Alvarez-Huerta, M., Verhoeven, H.A., Blaas, J., van Houwelingen, A.M., de Vos, C.H.R., van der Voet, H., Jansen, R.C., Guis, M., Mol, J., Davis, R.W., Schema, M., van Tunen, A.J. and O'Connel, A.P., 2000. Identification of the SAAT gene involved in strawberry flavor biogenesis by use of DNA microarrays. *Plant Cell*, 12: 647–662.

Antoszewski, R. 1973. Translocation of IAA in the strawberry plant in relation to the accumulation of nutrients in the fruit. *Acta Horticulturae*, 34: 85–88.

Archbold, D.D. and Dennis, F.G. 1984. Quantification of free ABA and free and conjugated IAA in strawberry achene and receptacle tissue during fruit development. *Journal of the American Society for Horticultural Science*, 109: 330–335.

Avigdori-Avidov, H. 1986. Strawberry. In: S.P. Monselise (ed.). *CRC Handbook of Fruit Set and Development*. CRC Press Inc., Boca Raton, Florida: pp. 419–448.

Bakker, J., Bridle, P. and Bellworthy, S.J. 1994. Strawberry juice colour: a study of the quantitative and qualitative pigment composition of juices from 39 genotypes. *Journal of Science, Food and Agriculture*, 64: 31–37.

Barnes, M.F. and Patchett, B.J. 1976. Cell wall degrading enzymes and the softening of senescent strawberry fruit. *Journal of Fruit Science*, 41: 1392–1395.

Ben-Arie, R. and Faust, M. 1980. ATPase in ripening strawberries. *Phytochemistry*, 19: 1631–1636.

Benitez-Burraco, A., Blanco-Portales, R., Redondo-Nevado, J., Bellido, M.L., Moyano, E., Caballero, J.L. and Munoz-Blanco, J. 2003. Cloning and characterization of tworipening-related strawberry (*Fragaria* x *ananassa* cv. Chandler) pectatelyasegenes. *Journal of Experimental Botany*, 54: 633–645.

Blanke, M.M. 2003. Photosynthesis of strawberry fruit. Proceedings of the Fourth Intrenational Strawberry Symposium, Tampere. *Acta Horticulturae*, 567: 373–376.

Broderick, 1976. Do- it –yourself strawberry. *International Flavours and Food Additives*, 6: 51–52.

Bunemann, V.G. and Gruppe, W. 1962. Buntersuchungenzurmineralischenernahrung von erdbeeren: iii. qualitatsmerkmale und minerals to ffgehalte der fruchte in abhangigkeit von der versorgungmitmakronahrstoffen. *Die Gartenbauwissenschaft Flight*, 27 (9): 166–182.

Burrough, L.F. 1960. The free amino acid content of certain British fruits. *Journal of Science Food and Agriculture*, 11: 14–18.

Cheng, G.W. and P. J. Breen. 1991. Activity of phenylalanine ammonia-lyase (PAL) and concentration of anthocyanins and phenolics in developing strawberry fruit. *Journal of the American Society for Horticultural Science*, 116: 865–869.

Chiu, T.F. and Bould, C. 1976. Effects of calcium and potassium on Ca mobility, growth and nutritional disorders of strawberry plants (*Fragaria* spp.). *Journal of Horticultural Science*, 51: 525–531.

Cordenunsi, B.R., Nascimento, J.R.O. and Lajolo, F.M. 2003. Physico-chemical changes related to quality of five strawberry fruit cultivars during cool-storage. *Food Chemistry*, 83: 167–173.

Coombe, B.G. 1976. The development of fleshy fruits. *Annual Review of Plant Physiology*, 27: 207–228.

Darrow, G.M. 1966. *The Strawberry: History, Breeding and Physiology.* Holt Rinehart & Winston, New York, New York.

Deng, C. and Davis, T.M. 2001. Molecular identification of the yellow fruit color (c) locus in diploid strawberry: a candidate gene approach. *Theoretical and Applied Genetics*, 103: 2–3.

Dennis, F.G. 1984. Fruit development. In: M.B. Tesar (ed.). *Physiological Basis of Crop Growth and Development.* American Society of Agronomy and Crop Science, Madison, Wisconsin.

El-Kazzaz, M.K., Sommer, N.F. and Fortlage, R.J. 1983. Effect of different atmospheres on postharvest decay and quality of fresh strawberries. *Phytopathology*, 73: 282–285.

Ezell, B.D., Darrow, G.M., Wilcox, M.S. and Scott, D.H. 1947. Ascorbic acid content of strawberries. *Food Research*, 12: 510–526.

Forney, C.F. and Breen, P.J. 1985. Growth of strawberry fruit and sugar uptake of fruit discs at different inflorescence position. *Scientia Horticulturae*, 27: 55–62.

Forney, C.F. and Breen, P.J. 1986. Sugar content and uptake of strawberry fruit. *Journal of the American Society for Horticultural Science*, 111: 241–247.

Forney, C.F., Kalt, W., McDonald, J.E. and Jordan, M.A. 1998. Changes in strawberry fruit quality during ripening and off the plant. *Acta Horticulturae*, 464: 506.

Galler, M. and Mackinney, G. 1965. The carotenoids of certain fruits (apple, pear, cherry, strawberry). *Journal Food Science*, 30: 393–395.

Giovannoni, J.J. 2001. Molecular regulation of fruit ripening. *Annual Review of Plant Physiology and Plant Molecular Biology*, 52: 725–749.

Giovannoni, J.J. 2004. Genetic regulation of fruit development and ripening. *The Plant Cell*, 16: 170–180.

Given, N.K., Venis, M.A. and Grierson, D. 1988. Phenylalanine ammonia-lyase activity and anthocyanin synthesis in ripening strawberry fruit. *Journal of Plant Physiology*, 133: 25–30.

Green, A. 1971. Soft fruit. In: A.C. Hulme (ed.). *Biochemistry of Fruits and their Products.* Vol. 2 (Hulme, A. C. Ed.) Academic Press, New York, New York: 375–409.

Grisebach, H. 1982. Biosynthesis of anthocyanins. In: P. Markasis (ed.). *Anthocyanin in Food Colours.* Academic Press, London, UK: 75–92.

Gross, J. 1982. Carotenoid changes in the juice of the ripening dancy tangerine (*Citrus reticulate*). *Food Science and Technology*, 15: 36–38.

Hardenburg, R.E., Watada, A.E. and Wang, C.Y. 1986. The commercial storage of fruits, vegetables, and florist and nursery stocks. USDA, Agr. Handbook 66 (revised) : 136

Havis, A.L. 1943. A developmental analysis of the strawberry fruit. *American Journal of Botany*, 30: 311–314.

Huber, D. J. 1984. Strawberry fruit softening: the potential roles of polyuronides and hemicelluloses. *Journal of Food Science*, 49: 1310–1315.

Jiang, Y. and Joyce, D.C. 2003. ABA effect on ethylene production, PAL activity, anthocyanin and phenolic contents of strawberry fruit. *Plant Growth Regulation*, 39: 171–174.

Jona, R and Fao, E. 1979. Histochemical survey of cell wall polysaccharides of selected fruits. *Scientia Horticulturae*, 10: 141–147.

Jurgens, J. 1990. Foliar fertilization in strawberry cultivation. *Erwerbsobstbau*, 32(4): 104–107.

Kader, A.A. 1991. Quality and its maintenance in relation to the postharvest physiology of strawberry. In: J.J. Luby and A. Dale (eds.). *The Strawberry into the 21st Century.* Timber Press, Portland, Oregon: 145–152.

Kalt, W., Prange, R.K. and Lidster, P.D. 1993. Postharvest color development of strawberries influence of maturity, temperature and light. *Canadian Journal of Plant Science*, 73: 541–548.

Kano, Y. and Asahira, T. 1979. Effect of endogenous cytokinins in strawberry fruits on their maturing. *Journal of the Japanese Society of Horticultural Science*, 47: 463–472.

Kano, Y. and Ashira, T. 1981. Roles of cytokinin and ABA in the maturity of strawberry fruits. *Journal of Japanese Society for Horticultural Science*, 50: 31–36.

Kim J.K., Moon, K.D. and Sohn, T.H. 1993. Effect of PE film thickness on MA (Modified Atmosphere) storage of strawberry. *Journal of the Korean Society of Food and Nutrition*, 22: 78–84.

Knee, M., Sargent, J. A. and Osborne, D. J. 1977. Cell wall metabolism in developing strawberry fruits. *Journal of Experimental Botany*, 28: 377–396.

Kays, S.J. 1991. *Postharvest Physiology of Perishable Plant Products.* AVI, New York, New York.

Latza, S., Gansser, D. and Berger, R.G. 1996. Identification and accumulation of 1-O-trans-cinnamoyl-beta-D-glucopyranose in developing strawberry fruit (*Fragaria ananassa* Duch. cv. Kent). *Journal of Agricultural and Food Chemistry*, 44: 1367–1370.

Li, C. and Kader, A.A. 1989. Residual effects of controlled atmospheres on postharvest physiology and quality of strawberries. *Journal of American Society of Horticultural Science*, 114: 629–634.

Lis, E.K., Borkowska, B. and Antoszewski, R. 1978. Growth regulators in strawberry fruit. *Fruit Science Reports*, 5 (3): 17–29.

Lis, E. K. and Antoszewski, R. 1979. *Photosynthesis and Plant Development* (Marcelle, R., Clijsters, H., van Poucke, M (eds.)). Dr W. Junk B.V. Publishers, The Hague: 263–270.

Lunkenbein, S., Bellido, M., Aharoni, A., Salentijn, E.M.J., Kaldenhoff, R., Coiner, H.A., Munoz-Blanco, J. and Schwab, W. 2006. Cinnamate metabolism in ripening fruit. Characterization of a UDP-glucose: cinnamate glucosyltransferase from strawberry. *Plant Physiology*, 140: 1047–1058.

Makinen, K. K. and Soderling, E. 1980. A quantitative study of mannitol, sorbitol, xylitol and xylose in wild berries and commercial fruits. *Journal of Food Science*, 45: 367–371.

Manning, K. 1993. Soft fruits. In: G.B. Seymour, J.E. Taylor and G.A. Tucker (eds.). *Biochemistry of Fruit Ripening.* Chapman & Hall, London, UK: 347–378.

Manning, K. 1996. Soft fruits. In: G.B. Seymour, J.E. Taylor and G.A. Tucker (eds.). *Biochemistry of Fruit Ripening.* Chapman & Hall, London, UK: 347–373.

Manning K. 1998. Isolation of a set of ripening-related genes from strawberry: their identification and possible relationship to fruit quality traits. *Planta*, 205: 622–631.

McNeil, M., Darvill, A.G. and Albersheim, P. 1980. Structure of plant cell wallsx. Rhamnogalacturonan I. A structurally complex pectic polysaccharide in the walls of suspension-cultured sycamore cells. *Plant Physiology*, 66: 1128–1134.

Martinez, G.A., Chaves, A.R. and Anon, M.C. 1996. Effect of exogenous application of gibberellic acid on colour change and phenylalanine ammonia-lyase, chlorophyllase and peroxidase activities during ripening of strawberry fruit (*Fragaria* x *ananassa* Duch.). *Journal of Plant Growth Regulation*, 15: 139–146.

Martinez, G.A., Chaves, A.R. and Civello, P.M. 2004. β-xylosidase activity and expression of a β-xylosidase gene during strawberry fruit ripening. *Plant Physiology and Biochemistry*, 42: 89–96.

Mitra, S.K. 1991. Strawberry. In: T.K. Bose, D.S. Rathore and S.K. Mitra (eds.). *Temperate Fruits.* Home-Allied Publishers, Calcutta, India: 549–596.

Miura, H., Imada, S. and Yabuuchi, S. 1990. Double sigmoid growth curve of strawberry fruit. *Journal of Japanese Society for Horticultural Science*, 59: 527–531.

Menager, I., Jost, M. and Aubert, C. 2004. Changes in physicochemical characteristics and volatile constituents of strawberry (cv. Cigaline) during maturation. *Journal of Agriculture and Food Chemistry*, 52: 1248–1254.

Monselise, S.P. 1986. *CRC Handbook of Fruit Set and Development.* CRC Press Inc. Boca Raton, Florida.

Mudge, K.W., Narayanan, K.R. and Poovaiah, B.W. 1981. Control of strawberry fruit set and development with auxins. *Journal of American Society of Horticultural Science*, 106: 80–84.

Moing, A., Renaud, C., Gaudillere, M., Raymond, P., Roudeillac, P. and Denoyes-Ruthan, B. 2001. Biochemical changes during fruit development of four strawberry cultivars. *Journal of the American Society for Horticultural Science*, 126: 394–403.

Moore, J.N., Brown, G.R. and Brown, E.D. 1970. Comparison of factors influencing fruit size in large-fruited and small-fruited clones of strawberry. *Journal of the American Society for Horticultural Science*, 95: 827–831.

Moyano, E., Portero-Robles, I., Medina Escobar, N., Valpuesta, V., Munoz-Blanco, J. and Caballero, J.L. 1998. A fruit-specific putative dihydroflavonol 4-reductase gene is differentially expressed in strawberry during the ripening process. *Plant Physiology*, 117: 711–716.

Mussinan, C.J. and Walradt, J.J. 1975. Organic acids from fresh California strawberries. *Journal of Agriculture and Food Chemistry*, 23: 482–484.

Neal, G.E. 1965. Changes occurring in the cell walls of strawberries during ripening. *Journal of the Sciences of Food and Agriculture*, 16: 604–611.

Nishizawa, T. and Hori, Y. 1988. Translocation and distribution of ^{14}C-photoassimilates in strawberry plants varying in developmental stages of the inflorescence. *Journal of the Japanese Society of Horticultural Science*, 57: 433–439.

Nitsch, J.P. 1950. Growth and morphogenesis of the strawberry as related to auxin. *American Journal of Botany*, 37: 211–215.

Nitsch, J.P. 1955. Free auxins and free tryptophane in the strawberry. *Plant Physiology*, 30: 33–39.

Nes, A. and Hjeltenes, A. 1992. The effects of covering and manuring on runner production, yield and berry size of the strawberry cultivar Korona. *Norsk*, 6: 195–203.

Nogata, Y., Ohta, H. and Voragen, A.G.J. 1993. Polygalacturonase in strawberry fruit. *Phytochemistry*, 34: 617–620.

Osman, A.B. and Dodd, P. B. 1992. Changes in some physical and chemical characteristics of strawberry (*Fragaria* × *ananassa* Duchesne) cv. Ostsra grown under different shading levels. *Acta Horticulturae*, 292: 195–207.

Perkins-Veazie, P. 1988. Development and use of an in vitro system to study the ripening physiology of strawberry fruit. PhD Thesis. University of Florida, Gainesville, Florida.

Perkins-Veazie, P. 1995. Growth and ripening of strawberry fruit. *Horticultural Reviews*, 17: 267–297.

Perkins-Veazie, P. and Huber, D. J. 1987. Growth and ripening of strawberry fruit under field conditions. *Proceeding of Florida State Horticultural Society*, 100: 89–93.

Picon, A., Martinez-Javega, J.M., Cuquerella, J., Del Rio, M.A. and Navarro, P. 1993. Effects of precooling, packaging film, modified atmosphere and ethylene absorber on the quality of refrigerated Chandler and Douglas strawberries. Food Chemistry, 48(2): 189–193.

Perez, A.G., Rios, J.J., Sanz, C. and Olias, J.M. 1992. Aroma components and free amino acids in strawberry variety Chandler during ripening. *Journal of Agricultural and Food Chemistry*, 40: 2232–2235.

Perez, A.G., Sanz, C. and Olias, J.M. 1993. Partial purification and some properties of alcohol acyltransferase from strawberry fruits. *Journal of Agricultural and Food Chemistry*, 41: 1462–1466.

Ricketson, C.L. 1966. The relationships between certain berry characteristics of the strawberry and foliar concentrations of nitrogen, phosphorus, and potassium at harvest. *Proceedings of the XVII International Horticultural Congress*, 1: 418.

Rosati, P. 1993. Recent trends in strawberry production and research: An overview. *Acta Horticulturae*, 348: 23–44.

Rosli, H.G., Civello, P.M. and Martinez, G.A. 2004. Changes in cell wall composition of three *Fragaria* x *ananassa* cultivars with different softening rate during ripening. *Plant Physiology and Biochemistry*, 42: 823–831.

Runkova, Liya V., Elwira, K. Lis, Tomaszewski, M. and. Antoszewski, R. 1972. Function of phenolic substances in the degradation system of indole-3-acetic acid in strawberries. *Biologia Plantarum*, 14: 71–78.

Sanz, C., Richardson, D.G. and Perez, A.G. 1995. 2, 5-Dimethyl-4hydroxy-3 (2H)-furanone and derivatives in strawberries during ripening. In: R.L. Rouseff and M.M. Leahy (eds.). *Fruit Flavor*. ACS Symposium Series 596, American Chemical Society, Washington, DC: 269–275.

Sas, L., Miszczak, A. and Plich, H. 1992. The influence of auxins, exogenous ethylene and light in the biosynthesis of ethylene and CO_2 production in strawberry fruits. *Fruit Science Report*, 19: 47–61.

Saxena, G.K. and Locascio, S.J. 1968. Fruit quality of fresh strawberries as influenced by nitrogen and potassium nutrition. *Journal of the American Society for Horticultural Science*, 92: 345–362.

Shaw, D.V. 1990. Response to selection and associated changes in genetic variance for soluble solids and titrable acids contents in strawberries. *Journal of the American Society for Horticultural Science*, 115: 839–843.

Sherman, W. B. and Janick, J. 1966. Greenhouse evaluation of fruit size and maturity in strawberry. *Proceeding of the American Society of Horticultural Science*, 89: 303–308.

Sistrunk, W.A. and Cash, J.N. 1973. Nonvolatile acids of strawberries. *Journal of Food Science*, 38: 807–809.

Smith, W.L. Jr and Heinze, P.H. 1958. Effect of color development at harvest on quality of post-harvest ripened strawberries. *Journal of the American Society for Horticultural Science*, 72: 207–211.

Spayd, S.E. and Morris, J.R. 1981. Changes in strawberry quality during maturation. *Ark Farm Res*, 30: 6–6.

Strik, C.B. and Proctor, J.T.A. 1988. Relationship between achene number, achene density, and berry fresh weight in strawberry. *Journal of the American Society for Horticultural Science*, 113: 620–623.

Strum, K., Koron, D. and Stampar, F. 2003. The composition of fruity of different strawberry varieties depending on maturity stage. *Food Chemistry*, 83: 417–422.

Takeda, F., Lightner, G.W. and Upchurch, B.L. 1990. A rapid method of determining carpel numbers in strawberry flowers. *HortScience*, 25: 230.

Thompson, P.A. 1963. The development of embryo, endosperm, and nucellus tissues in relation to receptacle growth in the strawberry. *Annals of Botany*, 27: 589–605.

Thompson, P. A. 1969. The effect of applied growth substances on development of the strawberry fruit: II. Interactions of auxins and gibberellins. *Journal of Experimental Botany*, 20: 629–647.

Veluthambi, K., Rhee, J. K., Mizrahi, Y. and Poovaiah, B. W. 1985. Correlation between lack of receptacle growth in response to auxin accumulation of a specific polypeptide in a strawberry (*Fragaria ananassa* Duch.) variant genotype. *Plant Cell Physiology*, 26: 317–324.

Villarreal, N.M., Rosli, H.G., Martonez, J.A. and Civello, P.M. 2008. Polygalacturonase activity and expression of related genes during ripening of strawberry cultivars with contrasting fruit firmness. *Postharvest Biology and Technology*, 47: 141–150.

Villarreal N.M., Martinez, G.A. and Civello, P.M. 2009. Influence of plant growth regulators on polygalacturonase expression in strawberry fruit. *Plant Science*, 176: 749–757.

Watada, A.E., Herner, R.C., Kader, A.A., Romani, R.J. and Staby, G.L. 1984. Terminology for the description of developmental stages of horticultural crops. *HortScience*, 19: 20–21.

Webb, R.A. 1973. A possible influence of pedicel dimensions on fruit size and yield in strawberry. *Scientia Horticulturae*, 1: 321–330.

Webb, R.A., Terblanche, J.H., Purves, Judith V. and Beech, M.G. 1978. Size factors in strawberry fruit. *Scientia Horticulturae*, 9: 347–356.

Went F.W., 1957. *The Experimental Control of Plant Growth*, Vol 17. Chronica Botanica Co., Waltham, Massachusetts and The Ronald Press Co., New York, New York: 129–138.

Wesche-Ebeling, P. and Montgomery, M.W. 1990. Strawberry polyphenoloxidase: Its role in anthocyanin degradation. *Journal of Food Science*, 55: 731–734.

Wills, R.B.H., Lee, T. H., Graham, D., McGlasson, W.B. and Hill, E.C. 1981. *An Introduction to the Postharvest Physiology and Handling of Fruits and Vegetables*. AVI, Westport, Connecticut.

Winston, A.L. 1902. The anatomy of edible berries. Conn. Exp. Station Report: 288–325.

Woodward J.R. 1972. Physical and chemical changes in developing strawberry fruits. *Journal of Science Food and Agriculture*, 23: 465–473.

Zaiter, H.J., Saad, I. and Nimah, M. 1993. Yield of iron-sprayed strawberry cultivars grown on high pH calcareous soil. *Communications in Soil Science and Plant Analysis*, 24: 1421–1436.

20 Use of Plant Bio-regulators

V.K. Tripathi, Sanjeev Kumar, and Vishal Dubey

CONTENTS

20.1 Plant Growth 350
20.2 Runner Production 351
20.3 Flowering 351
20.4 Earliness 352
20.5 Parthenocarpy 352
20.6 Yield 352
20.7 Fruit Quality 353
References 353

Plant hormones are naturally occurring substances that control growth and development. They are also known as growth substances. Plant hormones or phytohormones have been defined as "organic substances produced naturally in the higher plants, controlling growth or other physiological function at a site remote from its place of production and active in very minute amounts". Plant bioregulators or plant growth regulators are defined as "organic compounds other than nutrients which in small amount promote/inhibit or otherwise modify any physiological response in plants" (Prasad and Kumar, 1999). Hormones usually move within the plant from the site of production to the site of action. There are five classes of phytohormones, namely auxins, gibberellins, cytokinins, ethylene and abscisins.

Auxins are organic substances that at low concentrations (1–2 ppm) promote growth along the longitudinal axis when applied to shoots of the plant, or released within the plant system. The primary auxin in plants is indole-3-acetic acid (IAA). Naphthalene acetic acid (NAA), indole butyric acid (IBA), 2,4-dicholorophenxyacetic acid and 2,4,5-trichlorobenzoic acid (2,4,5-T) are some of the most popular auxins commonly used in strawberry production.

Gibberellins (GAs) were discovered by a Japanese plant physiologist, Kurosawa in 1926. They were first extracted from the culture of the fungus *Gibberella fujikuroi*. Developing fruits and seeds are the sites of GA biosynthesis. GAs are active in a variety of plant growth and development processes including dormancy, seed germination, stem elongation, flower initiation, flowers and fruit development.

The term cytokinin is a generic name for the substances that promote cell division and influence other growth regulatory functions of the plants. Zeatin is of natural occurrence and kinetin is synthetic cytokinin. The cytokinins play an essential role in the regulation of plant growth and differentiation, including the release of lateral buds from apical dominance and the delay of senescence.

The concept of endogenous inhibitors functioning as growth substances is quite old and these materials have been separated from higher plants for many years. Abscisic acid (ABA) is involved in many physiological and developmental processes throughout the lifecycle of plants. In the early phase of a plant's life, ABA regulates seed maturation and the maintenance of embryo dormancy. Later, at the onset of ontogenesis, it mediates several adaptation responses towards environment such as desiccation, cold as well as salt stress and acts as a negative growth regulator.

Ethylene is the only gaseous hormone that stimulates the process of ripening in fruits. It is biosynthesized essentially by all parts of higher plants, the rate varying with the type of tissue and its

350   Strawberries

stage of development. As the plant matures, ethylene influences sex determination and promotes fruit ripening. Finally, ethylene plays a role in senescing flowers and leaves.

Plant bioregulators are very effective in improving fruit production due to their direct influence on the quantitative as well as qualitative parameters of plant and fruit growth and production.

20.1 PLANT GROWTH

The use of GA_3 (30–200 ppm) has shown significant influence on plant growth, and proved to be an effective plant bioregulator for increasing plant height, spread, petiole length, leaf numbers, leaf area and number of crowns in many strawberry cultivars (Porlingis and Boynton, 1961; Singh and Kaul, 1969; Pathak and Singh, 1971; Agafonove and Solovei, 1975; Singh and Singh, 1978; Tafazoli and Shaybany, 1978; Lucchesi and Minani, 1980; Dwivedi, 1987; Anwar et al., 1990; Turemis and Kaska, 1997; Paroussi et al., 2002; Khokhar et al., 2004; Tripathi and Shukla, 2006; Kumar and Tripathi, 2009; Singh and Tripathi, 2010).

Pathak and Singh (1976) observed increased petiole length of the strawberries under GA treatments due to corresponding increase in epidermal and parenchyma cell length. They also observed that GA_3 application increased cell division and cell elongation in the subapical meristems of the shoots. The increase in the petiole length of strawberry plant following GA_3 treatment occurs due to increases in both the number and the length of cells in the epidermis (Guttridge and Thompson, 1959; El Shabasi et al., 2009). Induction of higher cell division due to cytokinins and a higher supply of assimilates mediated by application of benzyladenine (BA) might be the possible cause of enhanced petiole length, leaf area and number of leaves in strawberry (Dwivedi et al., 1999a,b; El-Shabasi et al., 2009). Different plant growth characters have been reported to have a highly significant positive relationship with strawberry yield (Hazarika et al., 2015). Besides increase in the root surface to volume ratio, hormonal levels and photosynthetic rate in plant take place due to the application of nitrogen-fixing bacteria along with BA (Singh and Singh, 2009). Enhanced carbohydrate content in foliage, roots and crowns of strawberry plants has been reported under the influence of GA_3 treatment, increasing the number of flowers and fruit yield (Ragab, 1996; El Shabasi et al., 2009). GA_3 application shifts the endogenous balance between promoters and inhibitors, favouring the process of fruit formation (Sharma and Sharma, 2006). Besides this, application of Ethrel (250 ppm) has also been reported to improve strawberry yield (El Shabasi et al., 2009). Singh and Phogat (1983) noticed an increase in number of leaves in Majestic cultivar of strawberry treated with 50 ppm GA_3 and a maximum number of crowns per plant with CCC at 1000 to 1500 ppm.

Dormant buds of strawberry contain relatively low level of endogenous GA_3 and enhanced level of ABA compared with non-dormant buds, so that exogenous application of GA_3 (50 ppm) to the lateral buds could achieve 97% break of their dormancy (Ahmed and Ragab, 2003).

Number of leaves and petiole length increased with 50 ppm BA, GA_3, BA+GA_3 when applied in Miyoshi, Enrai and Summer Berry cultivars of strawberry (Pipattanawong et al., 1996). Thakur et al. (1991) noticed that Activol (GA_3, BA, and GA_{4+7} mixtures) at 100 ppm and NAA at 20 ppm increased the crown height and vegetative growth in strawberry as compared to control, whereas, higher concentrations of CCC reduced crown height, leaf number and leaf area. Guo et al. (2006) reported that the net photosynthetic rate in leaves was promoted by the application of sodium bisulfite ($NaHSO_3$) and BA. They also recorded that the stomatal conductance and transpiration rates were significantly increased by BA. Zmuda et al. (2001) observed that when the leaves of cv. Senga Sengana were treated with 0.2% or 0.3% Betokson Super (2-naphthoxyacetic acid salt with triethanolamine), the fruits were smaller with reduced dry matter than the untreated ones. Different growth parameters, viz., plant height, plant spread, number of leaves per plant, petiole length and leaf area index were increased in cv. Sweet Charlie with increasing NAA level up to 35 ppm (Mir et al., 2004).

Chlormequat (1000 ppm), Daminozide (1000 ppm) and cycocel (1500 ppm) treatments increase the number of crowns per plant (Sach et al., 1972; Dwivedi et al., 1999). When CCC is applied at

Use of Plant Bio-regulators   351

1–2% to one-year-old strawberry plants at the start of runner development, it increases the number of crowns per plants (Agafonove et al., 1978).

Kumar et al. (1996) recorded maximum leaf area, number of leaves per plant and number of runners of Tioga strawberry due to application of triacontanol (5 ppm) during March before the emergence of flowers. Treatment of strawberry cv. Sophie with 300–400 mg L-tryptophane l^{-1} and 400 mg DL-tryptophanel^{-1} increased the plant height, stem and leaf blade thickness and leaf area index (Zhong et al., 2004).

Pollen germination was decreased from 42.5–5.8% by the application of jasmonic acid (JA) at 0.5 mM (Yildiz and Yilmaz, 2002). On the other hand, application of ACC (I-aminocyclopropane carboxylic acid) and ethephon (0.5 mM) increased pollen germination 55.6% and 60.7%, respectively.

20.2   RUNNER PRODUCTION

Non-availability of runners is an important limiting factor in sustainable strawberry production in several parts of the world, including India. Although it is a temperate crop, but strawberry has adapted well to mild subtropical areas particularly for fruit production. However, its regeneration is possible only in temperate areas. Hence, growers in the entire subtropical area are dependent on the supply of planting materials from temperate areas. PBRs have shown good effects on enhancing the efficiency of runner production of strawberry cultivars throughout the world. Use of GA_3 (50–100 ppm) has been very encouraging to enhance runner production in many cultivars such as Chandler, Cavalier, Olympus, Aliso and Tioga (Bruce, 1974; Tafazoli and Shaybany, 1978; Kalie et al., 1980; Anwar et al., 1990; Turemis and Kaska, 1997; Singh and Dhaliwal, 1982; Choma and Himelrick, 1984; Franciosoi et al., 1985; Singh and Tripathi, 2010).

Treatments of strawberry plants with 6-benzyl-aminopurine (BAP) are well known for its effects on runner growth (Waithaka and Dana, 1978) as the application of BAP (45–50 ppm) has been observed to enhance the regeneration capacity of Rabunda (Elizalde et al., 1979), Gorella and Chandler cultivars (Sharma et al., 2005). The application of this cytokinin could be of interest on varieties with low runner production behaviour or in areas with a short growing season.

Application of 50 ppm each of GA_3, BA, BA+GA_3 resulted in a two–three-times increase in the number of runners in Miyoshi and four–eight times in Enirai and Summer Berry cultivars (Pipattanawong et al., 1996). An increased number of daughter plants by almost 300% compared with the untreated control has been achieved in cultivars such as Selva, Mara des Bois and Vikat with the use of Gibrescol 10 mg (10% GA_3) and Promalin 3.6 SL (1.8% BA + 1.8% GA_{4+7}) at 0.2%.

Cycocel (0.6–2.4%) may also enhance runner production in strawberry (Agafonove and Solovei (1975). Pankov (1992) recorded maximum number of runners with 0.008% GA_3 and 1.2% CCC application, but lower and higher concentrations of GA_3 and CCC showed a tendency to decrease the number of runners per mother plant.

The strawberry cv. Nyoho suitable for greenhouse cultivation produces a large number of runners throughout the growing season, which was decreased with increasing concentration of paclobutrazol (from 0.01–0.2 mg *a.i.*/plant), whereas the number of lateral crowns and flower clusters were increased in the treated plants, and the negative effects of paclobutrazol application on growth of leaves could be minimized by transferring plants to untreated soils before flower bud induction (Nishizawa, 1993).

20.3   FLOWERING

Application of GAs generally results in early and profuse flowering in strawberry (Tavadze and Mazanashvilli, 1972; Ozguven and Kaska, 1993) but wide variations have been observed due to differences in varieties and planting times (Celestre and Pierandrei, 1971; Tafazoli and Vince, 1978). Enhanced flower production due to application of GA_3 (50–100 ppm) in strawberries has been reported by several researchers (Verzilow and Mikhteleva, 1975; Kalie et al., 1980; Tehranifar and Battey, 1997; Tripathi and Shukla, 2008; Kumar and Tripathi, 2009; Singh and Tripathi, 2010). Besides, GA_3 application @ 100 ppm extends the duration of flowering in the Chandler cultivar of strawberry (Tripathi and Shukla, 2006), but a higher concentration (200 ppm) of GA_3 may cause flower abnormality (Paroussi et al., 2002).

Application of CCC (1000–2000 ppm) enhances flowering in strawberry cultivars (Agafonove et al., 1978; Tripathi and Shukla, 2008), while NAA at 5–10 ppm (Thakur et al., 1991) or 25 ppm in strawberries (Stoyanov and Velchev, 1977) increases the number of flowers per plant.

20.4 EARLINESS

Application of GA_3 @ 75 ppm in strawberries may advance the flowering by one to two weeks and harvesting by one- to three-weeks (Pathak and Singh, 1971) but the response varies with the stage of application, e.g., application of GA_3 at about one month before the appearance of flower bud hastens flowering, whereas, at flower opening stage, it may hasten fruit maturation (Honda, 1972). The flowering and harvesting of strawberries may be advanced by GA_3 application, but in some cases, early flowering is highly susceptible to frost (Lopez et al., 1989; Galarza et al., 1989). Application of GA_3 reduces the time needed for inflorescence emergence, but the effect is variety specific (Paroussi et al., 2002; Singh and Tripathi, 2010).

Strawberry plants treated with CCC @ 1000 ppm may show advancement in flowering (Dwivedi, 1987), whereas application of BA @ 75–100 ppm advances the fruit ripening, if suitably supplemented with bee pollination (Ahn et al., 1988). Similarly, Yilmaz et al. (2003) also achieved the advanced fruit ripening by one week due to treatment of strawberry plants with jasmonic acid (0.25 to 1.0 mM).

20.5 PARTHENOCARPY

Earlier, the use of plant bioregulators was quite unsuccessful in production of the parthenocarpic fruits in a number of strawberry cultivars but nowadays some of the bioregulators such as GAs or auxins have been used successfully in inducing parthenocarpic fruit development in the strawberries. Beech (1983) recorded 100% parthenocarpic fruits in emasculated flowers of Red Gauntlet strawberries due to treatment with IAA, IBA (each at 1 ppm), NAA and NAAm (each at 0.05–0.10 ppm). However, some abnormalities, such as twisting of the pedicels and in some cases splitting of the receptacle tissue around the neck region of the fruit, were observed in treated plants. Although the application of NAA has beneficial effects (Archbold and Dennis, 1985) on receptacle growth, the fruits are comparatively smaller than those of the pollinated ones.

20.6 YIELD

The fruit yield improves due to application of GA_3 @ 50 to 100 ppm in different strawberry cultivars such as Gorella, Katrain Sweet, Senga Sengana, Chandler, Pajaro, Douglas and Benggata (Singh and Kaul, 1969; Celestre and Pierandrei, 1971; Pathak and Singh, 1971; Verzilow and Mikhteleva, 1975; Kalie et al., 1980; Singh and Phogat, 1983; Choma and Himelrick, 1984; Dwivedi, 1987; Kumar and Tripathi, 2009; Lopez et al., 1989; Anwar et al., 1990; Khokhar et al., 2004; Tripathi and Shukla, 2006; Tripathi and Shukla, 2008; Singh and Tripathi, 2010; Tripathi and Shukla, 2010), but there could be damage of the early emerging flowers by the GA_3 application at 25 ppm causing reduced yield over control (Ozguven and Kaska, 1990).

An increase (31.9%) in the yield of Senga Sengana cultivar has been reported with the application of Betokson Super (2-naphthoxy acetic acid salt with triethanolamine) at 0.2% (Masny et al., 2002). Strawberry cvs. Tufts and Cruz respond well to jasmonic acid (0.50 mM) in respect of fruit yield per plant (Yilmaz et al., 2003). The application of NAA @ 25–35 ppm has been proved to increase

Use of Plant Bio-regulators 353

the fruit number and yield of strawberries (Stoyanov and Velchev, 1977; Mir et al., 2004). Xiao et al. (1998) found that spraying of 100×10^{-6} m NAA+0.3% boric acid produced 27.5% higher yield of strawberries when compared with the untreated control.

Application of CCC at 1000–5000 ppm (Loveti and Vitagliano, 1970; Will, 1974; Benoit and Aerts, 1975; Agafonove et al., 1978; Soloved and Zakotin, 1979; Foda et al., 1979; Zakharova, 1979) or paclobutrazol @ 0.01–0.25 mg/plant increases the fruit yield of several cultivars strawberries (Nishizawa, 1993). Similarly, enhanced fruit yield due to treatment with Triacontanol at 5 ppm just before flower emergence (Kumar et al., 1996) or natural bamboo vinegar together with neutralized bamboo vinegar has been observed in strawberries (Bao et al., 2006).

20.7 FRUIT QUALITY

Application of GA_3 at 50–75 ppm increases the total sugars, reducing sugars, ascorbic acid, amino acid, total soluble solids (TSS) and anthocyanin contents in strawberries (Mikhteleva and Petrovskaya, 1975; Singh and Singh, 1978, 1979; Lopez et al., 1989; Ozguven and Kaska, 1990; Khokhar et al., 2004; Singh and Tripathi, 2010). Similarly, application of CCC at 1000–1500 ppm increases ascorbic acid, TSS and soluble protein contents but decreases acidity (Singh and Phogat, 1983; Dwivedi, 1987; Tripathi and Shukla, 2006; Tripathi and Shukla, 2007).

Preharvest (10 days before harvesting) spray with 50 or 100 ppm NAA has been observed most effective in improving the fruit quality of strawberries (Agrahari et al., 1998) at harvest and spray of 25 ppm GA_3 at 14 days before harvest has been found effective in not only extending the shelf-life up to nine-days but also reducing post-harvest losses (1.55%) due to decay and minimizing cumulative physiological loss in weight (7.26%) without adversely affecting the fruit quality of Chandler strawberry (Ram Asrey et al., 2004).

Activol (GA_3, BA and GA_{4+7} mixture) at 100 ppm, NAA at 20 ppm and CCC at 500 ppm increased the TSS and anthocyanin contents, whereas, 200 ppm Activol and 1000 ppm CCC reduced the TSS and anthocyanin contents in cv. "Zioga" (Thakur et al., 1991). Natural bamboo vinegar, together with neutralized bamboo vinegar, improves the storage life of the fruits, regulates the time of fruit maturation and improves the quality of fruits (Bao et al., 2006). Treatment with 300 ppm L-tryptophane or DL-tryptophane resulted in larger fruit size and, higher TSS and vitamin C contents in strawberry cvs. Sophia and Fengxiang when compared with untreated plants (Zhong et al., 2001). In strawberry cv. Sweet Charlie, the fruit quality components (TSS, total sugar, acidity, pH and vitamin C content) were improved by increasing level of NAA application up to 35 ppm (Mir et al., 2004).

The tunnel-grown strawberry cvs. Chandler, Douglas, Fern and Selva, treated at full bloom and 15 days later with mixtures of plant bioregulators, viz., GNB (GA_3 at 50, 100 and 200 ppm; β-NAA at 12.5, 25 and 50 ppm and 6-benzyl-aminopurine at 10, 20 and 40 ppm) and PG (phenothiol at 5 and 10 ppm; GA_3 at 2.5 and 5 ppm); but a higher dose of GNB induced depressive effect on the plants, which caused reduced fruit weight and yields, although the percentage of malformed fruit remained less than control. The application of GA_3 improved the earliness of the crops and slightly increased the fruit weight but did not reduce the percentage of malformed fruits (Galaza et al., 1993).

REFERENCES

Agafonove, N.V. and Solovei, E.P. 1975. Some characteristics of strawberry growth and productivity in relation to treatment with growth regulators. *Primenenie Eiziologocheski Aktivnykh Veshchestv Sadovodstve Moscoh, USSR*, 2: 12–17.

Agafonove, N.V., Solovei, E.P. and Blinovskii, I.K. 1978. Growth and fruiting of strawberry in relation to treatment with TUR and gibberellins. *Azvestiya Timiryazevak Oi Sel'sko Khozyaistvennoi Akademii*, 3: 150–162.

354   Strawberries

Agrahari, P.R., Thakur, K.S., Sharma, R.M., Singh, R.R. and Tripathi, V.K. 1998. Pre-harvest chemical manipulation in the storage quality of Chandler strawberry (*Fragaria* x *ananassa* Duch.). *New Agriculturist*, 9(1, 2): 7–11.

Ahmed, H.F.S. and Ragab, M.I. 2003. Hormone levels and protein patterns in dormant and non-dormant buds of strawberry and induction of bud break by gibberellic acid. *Egyptian Journal of Biology*, 5: 35–42.

Ahn, C.K., Choe, J.S., Um, Y.C., Cho, I.W., Yu, I.C. and Park, J.C. 1988. The effect of bee pollination and growth regulator treatment on preventing malformation and accelerating growth in strawberry fruits. *Research Report on Rural Development and Administration on Horticulture*, 30(3): 22–30.

Anwar, M., Imayatullah, H. and Hanana, A. 1990. Effect of different concentration of gibberellic acid on the growth and yield of strawberry. *Sarhad Journal of Agriculture*, 6(1): 57–59.

Archbold, D.D. and Dennis, F.G. 1985. Strawberry receptacle growth and endogenous IAA content as affected by growth regulator application and achene removal. *Journal for the American Society for Horticultural Science*, 110: 816–820.

Asrey, R., Jain, R.K. and Singh, R. 2004. Effect of pre-harvest chemical treatments on shelf life of Chandler strawberry (*Fragaria* x *ananassa*). *Indian Journal of Agricultural Science*, 74(9): 485–487.

Bao, B.F., Ma, J.Y., Zhang, Q.S., Ye, L.M. and Wang, P.W. 2006. Bamboo vinegar as potential plant growth regulator in stimulative effect (II) the field investigation. *Acta Agriculture Zhejiangensis*, 18(4): 268–272.

Beech, M.G. 1983. The induction of parthenocarpy in the strawberry cultivar Red Gauntlet by growth regulators. *Journal of Horticultural Science*, 58: 541–545.

Benoit, F. and Aerts, J. 1975. Growth control of forced strawberries with cycocel. *Fruitteelt*, 19: 44–47.

Bruce, H.B. 1974. The effect of gibberellic acid, blossom removal and planting date on strawberry runner plant production. *HortScience*, 9(1): 25–27.

Celestre, M.R. and Pierandrei, F. 1971. Effect of gibberellic acid on productivity and ripening date in three varieties of strawberry. *Annali dell'Instituto Sperimentale per La Frutticoltura*, 2: 61076.

Choma, M.E. and Himelrick, D.G. 1984. Response of day neutral, June bearing and everbearing strawberry cultivars to gibberellic acid and phthalimide treatments. *Scientia Horticulturae*, 22: 257–264.

Dwivedi, M.P. 1987. Effect of photoperiod and growth regulators on vegetative growth, flowering and yield of strawberry. Ph.D. Thesis, Dr. Y.S. Parmar University of Horticulture and Forestry, Nauni, Solan (H.P.), India.

Dwivedi, M.P., Negi, K.S., Jindal, K.K. and Rana, H.S. 1999a. Influence of photoperiods and bio-regulators on vegetative growth of strawberry under controlled conditions. *Advances in Horticulture and Forestry*, 7: 29–34.

Dwivedi, M.P., Negi, K.S., Jindal, K.K. and Rana, H.S. 1999b. Effect of bio-regulators on vegetative growth of strawberry. *Scientific Horticulturae*, 6: 79–84.

Elizalde, M.M., De, B., Guitman, M.R. and Biain-de-Elizalde, M.M. 1979. Vegetative propagation in everbearing strawberry as influenced by a morphactin, GA$_3$ and BA. *Journal of the American Society for Horticultural Science*, 104(2): 162–164.

El Shabasi, M.S.S., Ragab, M.E., El Oksh, I.I. and Osman, Y.M.M. 2009. Response of strawberry plants to some growth regulators. *Acta Horticulturae*, 842: 725–728.

Foda, S.A., Nassar, H.H. and Nansour, S.A. 1979. Effect of some growth regulator on runner production and yield of strawberry. *Horticulture*, 57: 119–125.

Franciosoi, R., Salas, P., Yamashio, E. and Duorte, O. 1985. Effect of gibberellic acid on runner formation and yield of strawberry. *Proceeding Tropical Region American Society for Horticulture Science*, 24: 127–129.

Galaza, S.L., Marota, J.V., Pascual, B.L. and Bono, M.S. 1993. Productive response of strawberry plants (*Fragaria* x *ananassa* Duch.) to different plant growth regulator mixes in Spain. *Acta Horticulturae*, 348: 307–314.

Galarza, S.L., Pasural, S.B., Alagarda, J. and Marota, J.V. 1989. The influence of winter gibberellic acid applications on earliness, productivity and other parameters of quality in strawberry cultivation (*Fragaria* x *ananassa* Duch.) on the Spanish Mediterranean Coast. *Acta Horticulturae*, 265: 217–222.

Guo, Y.P., Peng, Y., Lin, M.I., Guo, D.P., Hu, M.J., Shen, Y.K., Li, D.Y. and Zheng, S.J. 2006. Different pathways are involved in the enhancement of photosynthetic rate by sodium bisulfite and benzyladenine, a case study with strawberry (*Fragaria x ananassa* Duch.) plants. *Plant Growth Regulation*, 48(1): 65–72.

Guttridge, C.G. and Thompson, P.A. 1959. Effects of gibberellic acid on length and number of epidermal cells in petioles of strawberry. *Nature*, 183: 197–198.

Hazarika, T.K., Ralte, Z., Nautiyal, B.P. and Shukla, A.C. 2015. Influence of bio- fertilizers and bio-regulators on growth, yield and quality of strawberry (*Fragaria × ananassa*). *Indian Journal of Agricultural Sciences*, 85(9): 1201–1205.

Honda, F. 1972. The effect of gibberellic acid on growth and flowering of strawberry. *Bulletin of Horticultural Resesrach Station, Fukvoka, Japan*, 7: 45–47.

Kalie, M.B., Soetarto, I. and Usman, R.S. 1980. Observation on the effect of gibberellic acid on growth of strawberry var. Bengata. *Hortikultura*, 8: 45–51.

Khokhar, U.U., Prashad, J. and Sharma, M.K. 2004. Influence of growth regulator on growth, yield and quality of strawberry cv. Chandler. *Haryana Journal of Horticultural Science*, 33(3/4): 186–188.

Kumar, R. and Tripathi, V.K. 2009. Influence of NAA, GA_3 and boric acid on growth, yield and quality of strawberry cv. Chandler. *Progressive Horticulture*, 41(1): 113–115.

Kumar, J., Rana, S.S., Verma, H.S. and Jindal, K.K. 1996. Effect of various growth regulators on growth, yield and fruit quality of strawberry cv. Tioga. *Haryana Journal of Horticultural Science*, 25(4): 168–171.

Lopez, G.S., Pascual, B., Alagarda, J. and Maroto, J.V. 1989. The influence of winter gibberellic acid application on earliness, productivity and other parameter of quality in strawberry cultivation of the Spanish Mediterranean Coast. *Acta Horticulturae*, 265: 217–222.

Loveti, F. and Vitagliano, C. 1970. Further observations on the stimulatory effect of certain growth regulators on the productivity of strawberry plants. *Rivista de Ortoflorifrutticultura, Italy*, 54: 547–555.

Lucchesi, A.A. and Minani, K. 1980. The use of growth regulators in strawberries: Influence on growth cycle and final productivity. *Anais de Escota Superior de Agricultura Luiz de Quirz*, 37: 485–515.

Masny, A., Zurawicz, E. and Basak, A. 2002. Super influence of "Betokon" (BNOA) and other growth regulators on yield and quality in strawberry. *Acta Horticulturae*, 567(2): 467–470.

Mikhteleva, L.A. and Petrovskaya, B.T.P. 1975. The effect of gibberellin on metabolic content and localization of enzyme activity in the floral tissue of *Fragaria x ananassa*. *Bedrijbsontwkeling*, 6: 185–188.

Mir, M.M., Barche, S., Kirad, K.S. and Singh, D.B. 2004. Studies on the response of plant growth regulator on growth, yield and quality of strawberry (*Fragaria x ananassa* Duch.) cv. Sweet Charlie. *Plant Archives*, 4(2): 331–333.

Nishizawa, T. 1993. Growth and yield during the first year growing period as affected by paclobutrazol on greenhouse strawberry production. *Acta Horticulturae*, 329: 51–53.

Ozguven, A.I. and Kaska, N. 1990. The effect of GA_3 application on the yield and quality of strawberries grown in high tunnels. I. Summer planting. *Bache*, 19(1–2): 41–52.

Ozguven, A.I. and Kaska, N. 1993. The effect of GA on the precocity of strawberry crowns in Adana-Turkey. *Acta Horticulturae*, 345: 73–79.

Pankov, V.V. 1992. Effect of growth regulators on plant production of strawberry mother plant. *Scientia Horticulturae*, 52: 157–161.

Paroussi, G., Voyiatzis, D.G., Paroussis, G. and Drogoudi, P.D. 2002. Effect of GA_3 and photoperiod regime on growth and flowering in strawberry. *Acta Horticulturae*, 567(1): 273–276.

Pathak, R.K. and Singh, R. 1971. Effect of some external factors on the growth and fruiting of strawberry. II. Effect of GA, growth retardants and cloching on flowering and yield. *Progressive Horticulture*, 3(3): 53–63.

Pathak, R.K. and Singh, R. 1976. Effect of GA and photoperiod on yield and quality of strawberry. *Progressive Horticulture*, 8: 31–39.

Pipattanawong, N., Fujishige, N., Yamane, K., Ijiro, Y. and Ogata, R. 1996. Effect of growth regulators and fertilizer on runner production, flowering and growth in day neutral strawberries. *Japanese Society for Tropical Agriculture*, 40(3): 101–105.

Porlingis, I.C. and Boynton, D. 1961. Growth response of strawberry plants *Fragaria chiloensis* var. ananassa to gibberellic acid and to environmental conditions. *Proceeding for the American Society for Horticulture Science*, 78: 261–269.

Prasad, S. and Kumar, U. 1999. *Principles of Horticulture*. AGROBIOS (INDIA), Behind Nasrani Cinema, Chopasani Road, Jodhpur India: 367–401.

Ragab, M.E. 1996. Effect of GA_3 on number and some transplants characters of strawberry nurseries. Fourth Arabic conditions Conference. *Minia Society for Horticultural Science*, 78: 261–269.

Sach, M., Iszak, E. and Geisenberg, C. 1972. Effect of chlormequat and SADH on runner development and fruiting behavior of summer planted strawberry. *HortScience*, 7: 364–386.

Sharma, S.D. and Sharma, N.C. 2006. Studies on correlations between endomycorrhizal and Azotobater population with growth, yield and soil nutrient status of apple orchards in Himachal Pradesh. *Indian Journal of Horticulture*, 63: 379–382.

Sharma, R.M., Khajuria, A.K. and Kher, R. 2005. Chemical manipulation in the regeneration capacity of strawberry cultivarsunder Jammu plains. *Indian Journal of Horticulture*, 62(2): 190–192.

Singh, R. and Kaul, G.L. 1969. Effect of gibberellic acid on strawberry. I. Growth and fruiting. *Proceeding of International Symposium for Subtropical and Tropical Horticulture*: pp. 316–326.

Singh, H. and Singh, R. 1978. Effect of gibberellic acid and manuring on growth, flowering, fruiting and yield of strawberry (*Fragaria* spp.). *Indian Journal of Horticulture*, 35: 207–211.

Singh, H. and Singh, R. 1979. Effect of GA and manuring on the fruit quality of strawberry. *Punjab Horticulture Journal*, 19: 71–73.

Singh, K. and Dhaliwal, S.S. 1982. Effect of storage temperature and GA_3 treatments on runner production in strawberry. *Punjab Horticulture Journal*, 22(1–2): 87–92.

Singh, O.P. and Phogat, K.P.S. 1983. Effect of plant growth regulators on vegetative growth, yield and quality of strawberry (*Fragaria* spp.). *Progressive Horticulture*, 15(1–2): 64–68.

Singh, A. and Singh, J.N. 2009. Effect of bio-fertilizers and bioregulators on growth, yield and nutrient status of strawberry cv. Sweet Charlie. *Indian Journal of Horticulture*, 66: 220–224.

Singh, V.K. and Tripathi, V.K. 2010. Efficacy of GA3, boric acid and zinc sulphate on growth, flowering, yield and quality of strawberry cv. Chandler. *Progressive Agriculture*, 10(2): 345–348.

Soloved, E.P. and Zakotin, V.S. 1979. Application of TUR in strawberries. *Referativnyi Zhurnal*, 3: 55–58.

Stoyanov, S. and Velchev, V. 1977. Vegetative and reproductive behavior of plants of the strawberry cultivar Surprise des Halles treated with betanaphthyl acetic acid. *Ovoshcharstvo*, 56(5): 43–45.

Tafazoli, E. and Shaybany, B. 1978. Influence of nitrogen, deblossoming and growth regulator treatments on growth, flowering and runner production on the Gem everbearing strawberry. *Journal of the American Society for Horticultural Science*, 103(3): 372–374.

Tafazoli, E. and Vince, P.D. 1978. A comparison of the effects of long days and exogenous growth regulators on growth and flowering in strawberry (*Fragaria x ananassa* Duch.). *Journal of Horticulture Science*, 53: 255–259.

Tavadze, P.G. and Mazanashvilli, T.G. 1972. The effect of GA on the growth and yield of large fruited strawberries. *Vestrikgruz Botan Ova Akadeniya Nauka Gruz SSR*, 5: 13–15.

Tehranifar, A. and Battey, N.H. 1997. Comparison of the effects of GA_3 and chillings on vegetative vigour and fruit set in strawberry. *Acta Horticulturae*, 439: 627–631.

Thakur, A., Jindal, K.K. and Sud, A. 1991. Effect of growth substances on vegetative growth, yield and quality parameters in strawberry. *Indian Journal of Horticulture*, 48(4): 286–290.

Tripathi, V.K. and Shukla, P.K. 2006. Effect of plant bio-regulators on growth, yield and quality of strawberry cv. Chandler. *Journal of Asian Horticulture*, 2(4): 260–263.

Tripathi, V.K. and Shukla, P.K. 2007. Effect of plant bio-regulators, zinc sulphate and boric acid on yield and quality of strawberry cv. Chandler. *Journal of Asian Horticulture*, 4(1): 15–18.

Tripathi, V.K. and Shukla, P.K. 2008. Influence of plant bio-regulators and micronutrients on flowering and yield of strawberry cv. Chandler. *Annals of Horticulture*, 1(1): 45–48.

Tripathi, V.K. and Shukla, P.K. 2010. Influence of plant bio-regulators, boric acid and zinc sulphate on yield and fruit characters of strawberry cv. Chandler. *Progressive Horticulture*, 42(2): 186–188.

Turemis, N. and Kaska, N. 1997. Effect of gibberellic acid on the production and quality of strawberry runners. *Turkish Journal of Agriculture and Forestry*, 21(1): 41–47.

Verzilow, V.F. and Mikhteleva, L.A. 1975. The development and productivity of *Fragaria x ananassa* as affected by GA. *Skokhozyaistvennaya Bioligiya*, 10: 533–537.

Waithaka, K. and Dana, M.N. 1978. Effect of growth substances on strawberry growth. *Journal of the American Society for Horticultural Science*, 103(5): 627–628.

Will, H. 1974. The use of cycocel for earlier harvest of strawberries. *Erwerbsobstbau*, 16: 59–60.

Xiao, Y., Huang, J.C. and Li, H.B. 1998. Study on the effect of NAA and boron on the fruit growth and development of strawberry. *South China Fruits*, 27(4): 35–36.

Yildiz, K. and Yilmaz, H. 2002. Effect of jasmonic acid, ACC and ethephon on pollen germination in strawberry. *Plant Growth Regulation*, 38(2): 145–148.

Yilmaz, H., Yildiz, K. and Muradoglu, F. 2003. Effect of jasmonic acid on yield and quality of two strawberry cultivars. *Journal of the American Pomological Society*, 57(1): 32–35.

Zakharova, O.M. 1979. Application of growth regulators in strawberry propagation. *Referativnyi Zhurnal*, 7: 55–58.

Use of Plant Bio-regulators

Zhong, X.H., Shi, X.H. and Xiao, L.T. 2001. Effect of tryptophane on the fruit quality and production of strawberry. *China Fruits*, 2: 4–7.

Zhong, X.H., Shi, X.H., Ma, D.W., Huang, Y.F. and Dai, S.H. 2004. Study on the effect of tryptophan on the growth and fruit quality of strawberry cv. Sophie. *Journal of Fruit Science*, 21(6): 565–568.

Zmuda, E., Murawska, D. and Szember, E. 2001. Study on application of Betokson Super in growing of strawberry cv. Senga Sengana. *Zeszyty-Naukowe-Instytutu-Sadownictwa-i-Kwiaciarstwa w Skierniewicach*, 9: 185–191.

21 Special Cultural Practices

O.P. Awasthi and Sunil Kumar

CONTENTS

21.1 Pinching 359
21.2 Preconditioning 360
21.3 Defoliation 360
21.4 Plant Thinning.. 361
21.5 Deblossoming 361
21.6 Renovation (Renewing the Planting) 361
21.7 Intercropping.. 362
21.8 Ratooning 362
References... 362

Strawberry was introduced in the hilly regions of India by the British and later became a delicacy for tourists and gradually its cultivation was started in the plains of North India. The strong momentum of strawberry cultivation in general and in India in particular can be given further pushed through a few small but high-impact interventions, which could be collectively termed "special cultural practices". Being a high-input-requiring crop, some special cultural practices such as pinching and preconditioning have to be adopted. A brief special package of practices as presented here can help growers to obtain higher productivity of good-quality strawberry fruits with better economic returns.

21.1 PINCHING

The strawberry plant has very short internodes in the stem, known as the crown. Each node has one axillary bud, which develops into a stolon or branch crown or can remain dormant (Darrow, 1966). The uppermost axillary bud continues the vegetative extension of the crown, reflecting apical dominance over lower laterals. Whether meristems develop into stolon buds or branch crown buds is environment dependent (Guttridge, 1985). Many factors such as nutrition, photoperiod and temperature are the most important factors affecting flower induction in strawberries. Many researchers (Ito and Saito, 1962; Guttridge, 1985; Sonsteby and Heide, 2006) have opined that for short-day strawberry cultivars, vegetative and reproductive growth are opposite to each other and strongly regulated by photoperiod and temperature. Pinching refers to a process of removing soft growth from the growing tip of the plant to induce its natural tendency to branch. Pinching of shoot apices in young short-day strawberry plants tends to lead the lower axillary buds to develop into branch crowns even in long days and warm temperatures by releasing the buds from apical dominance (Neri et al., 2003).

Apical dominance, results from the production of auxin, a phytohormone, by terminal growing point and young leaves (Cline, 1997). Removing the terminal growing point and one or two of the uppermost leaves restricts the source of auxin and allows the development of dormant buds below the pinched portion to grow. Pinching is accomplished by using the thumb and forefinger or by cutting off the growth mechanically, using a knife, scissors or clippers. Pinching can also be done by plant growth regulators that limit the growth of terminal growing points and stimulate lateral branching. Although it delays the onset of flowering, it increases the number of blooms due to there

359

360 Strawberries

being more secondary branches than in non-pinched plants. Pinches are often made to control overall size and shape, to increase flower numbers and to create full, thick growth.

21.2 PRECONDITIONING

This is a forcing system for production of strawberries in winter and is more commonly practised in Japan (Mochizuki et al., 2009). For extended offseason production and to schedule harvests, short-day strawberry plants are preconditioned to increase the number of flower buds per plant, commonly called "waiting bed" plants. In this technique, plants are kept at temperatures above 5°C to avoid heavy dormancy and promote continued flower bud induction.

21.3 DEFOLIATION

Defoliation, defined as partial or complete removal of plant leaves, is a technique that has been used in several regions of the world for stimulating the photosynthesis and growth of young leaves, promoting better use of water and nutrients and also modifying the dry matter partitioning between source and sink compartments of plants (Chanishvili et al., 2005; Khan et al., 2007; Iqbal et al., 2012). However, elaborate studies are still required to evaluate whether defoliation would be a useful agronomic tool to avoid high temperatures that could reduce flowering in many varieties (Whitehouse et al., 2009). Plant leaf area is responsible for photosynthesis and its influence on the leaf-root ratio. The extent of defoliation treatments has been observed to significantly influence the flower initiation in strawberry due to differences in the inductive capacity of leaves of cultivars of differing groups of maturity and growth habit (Thompson and Guttridge, 1960). Since flower bud induction in strawberry is sensitive to thermo-photoperiod, and to some agronomical and nutritional factors, the timing of the defoliation is also important (Savini et al., 2005).

In the northern hemisphere, the June-bearing strawberries initiate flower buds from late summer–autumn in short-day conditions. The delay in defoliation time can cause changes in strawberry plant growth processes, reducing leaf area and plant size, which in turn has impact on formation of inflorescence and fruit (Casierra-Posada et al., 2012) in strawberry. Albregts et al. (1992) and Wildung (2000) suggested defoliation in the production of strawberry transplants to reduce pests and diseases, and to lower down the rate of transpiration during transport and planting. Reduced incidence of *Botrytis cinerea* through defoliation and increase in fruit size in the following cropping period have been reported by Daugaard et al. (2003).

Defoliation immediately after harvest in short-day strawberry cultivars without vernalization period has been observed to increase yield in the autumn, while delayed post-harvest defoliation reduces the yield (Zhang et al., 1992). Several studies on various aspects of post-harvest defoliation in strawberry including physiological changes (Casierra-Posada et al., 2012), the extent of defoliation (Albregts et al., 1992), cultivar response, cropping year and the developmental phase of the crop (Daugaard et al., 2003) have been made. Post-harvest defoliation has been observed to enhance the berry yield in strawberries (Nestby, 1985) but the yield could be affected tremendously by changing the extent of defoliation (Albregts et al., 1992; Casierra-Posada et al., 2012) and the phase of specific plant development response of strawberry plants to post-harvest defoliation depends on cultivar and cropping year (Daugaard et al., 2003; Whitehouse et al., 2009). The practice of burning as a means of defoliation should be avoided as it causes loss of useful biodiversity in the ecosystem including insects and microorganisms in the plantation, although it has been reported (Metspalu et al., 2000) to decrease the occurrence of spider mites (*Tetranycus urticae*). Flame defoliation can be used to remove the old strawberry foliage, but may damage the strawberry crown (Wildung, 2000).

Contrary to these benefits, changes in plant biomass production such as the number or weight of leaves due to defoliation may change the nutritional allocation patterns, which can directly affect the fruit yield and quality of berries (Correia et al., 2011). Decrease in the AAC content in strawberries through defoliation have been reported by Moor et al. (2004).

Special Cultural Practices   361

21.4 PLANT THINNING

The removal of plants or plant parts not only leads to compensatory development of vegetative organs, but also improves the formation and development of flowers and fruits (Hansen, 1996). The growth and yield recovery of everbearing strawberry, taking the thinning of reproductive organs such as crowns and clusters for summer season production in a cool region, was studied (Jong Nam et al., 2006), but no significant difference was observed in the number of fruits per flower cluster, although it reduced the harvesting interval, and due to increase in the number of crowns per plant, it enhanced the number and size of marketable fruits.

21.5 DEBLOSSOMING

Deblossoming or the removal of flower truss is sometimes a common practice to prevent the plant from fruiting, particularly in runner beds (Sharma and Yamdagni, 2000). A positive effect of the removal of flowers and runners on the growth and yield of strawberry plants has been observed (Daugaard, 1999; Paszko, 2002). Fruiting inhibits plant growth by depressing leaf size and area, and delays the emergence of leaves and runners. These inhibitions begin before or soon after flowering, not waiting for the fruiting period (Albregts, 1954). Removal of blossoms and young fruits from mother plants before completion of cell division not only increases the production of early saleable runners but also enhances the size of fruits (Liang et al., 2006). In perennial strawberry cultivation systems, deblossoming in the planting year has been observed (Daugaard, 1999; Lyu and Li, 2014) to increase fruit size, yield, and fruit quality in the second year. Deblossoming in cv. Everberry plants from spring to summer has been observed (Kumakura and Shishido, 1994) to have beneficial effect on yield and fruit characteristics of summer crop . If frigo plants are used in perennial cropping, removal of flowers in the first year is a common practice as it causes more vigorous vegetative growth and eventually, a higher yield in the succeeding year because of the sink effect (Rindom and Hansen, 1995).

As a good cultivation practice, the grower needs to maintain the optimum berry load on plants, besides maintaining a proper number of leaves and runners. If a strawberry plant produces many flowers, it is advisable to remove some young berries to achieve marketable-size berries, and to maintain the plant health in good state. The primary blossom can have five or six fruits per stem, while subsequent blossoms should not have more than three or four fruits per stem (Takei, 2010). Strawberry plants continuously produce flowers and berries during winter in subtropical conditions resulting in reduced berry size in late harvest. Liang et al. (2006) reported that fruit thinning significantly increased fruit weight and total sugar content but reduced the metabolic consumption of sucrose and hexose in strawberry.

21.6 RENOVATION (RENEWING THE PLANTING)

Renovation of strawberry plantings by selective removal of older plants is a practice to regulate plant density. After the first growing season, most strawberry plantings become overcrowded because of the perennial growth habit and prolific runner production causes reduction in fruit yield and quality. Diseases also increase when plantations become too dense. Normally twelve to sixteen plants per square meter need to be maintained in the spring for best yields. Zhang et al. (1992) observed 12–41% more yield due to renovation between two and four weeks after the last harvest in short-day strawberry cultivars when compared with nondefoliated plants. If renovation is delayed beyond mid-July, then leaves should not be removed because there will be insufficient time for emergence of a new set of leaves. The detailed schedule for renovation of strawberry is briefly summarized here:

- Within a week after last picking, strawberry leaves are mowed off with a rotary mower at about 2.5 cm above the crown. For management of weeds, 2,4-D alkanol amine salts may be applied between mowing and last picking. The weedicide should not be applied later in the growing season to discourage the misshaping of berries.

362 Strawberries

- The rows should be narrowed down with the help of a cultivator or roto-tiller and plants in the rows thinned by hand hoeing or harrowing across the rows depending on the planting system used. Rows may be narrowed to a 15–20 cm-wide strip with closer rows (90–100 cm), whereas a somewhat wider strip (25–30 cm) may be left for wider (more than 100 cm) row spacing.
- After renovation, nitrogen should be applied @ 62–65 kg/ha in heavy soils, while in light soils, 62–65 kg/ha nitrogen at renovation and an additional 30–32 kg/ha should be applied during late July.
- If essential, irrigation may be given to stimulate new growth and to activate the herbicidal action.

21.7 INTERCROPPING

Strawberry is sometimes planted as an intercrop in young orchards because of its quick growth and fruiting habit, which starts within 120 days after planting and continues up to three years, if managed properly. When the overstorey component crops are in their establishment phase, strawberry can be taken as an intercrop for supplementary income. However, it is advisable that the allelopathic effects of overstorey crops be kept in mind.

21.8 RATOONING

In subtropical conditions, where summer temperatures is very high (>40°C), there is mortality of mother plants. Further, the plants are severely affected by insects and eventually there is poor crop stand and low yield in the second crop. But strawberry is basically a perennial herb and therefore, it is possible to harvest fruits from same plantation for more than one season. However, for this, the management of plantation has to be excellent and scientific.

REFERENCES

Albregts, E.E. 1954. Blossom removal in the strawberry winter nursery. *Botanical Gazette*, 107: 352–361.

Albregts, E.E., Howard, C.M. and Chandler, C.K. 1992. Defoliation of strawberry transplants for fruit production in Florida. *HortScience*, 27(8): 889–891.

Casierra-Posada, F., Torres, I.D. and Riascos-Ortiz, D.H. 2012. Growth in partially defoliated strawberry plants cultivated in the tropical highlands. *Revista UDCA Actualidad and Divulgacion Científica*, 15(2): 349–355.

Chanishvili, S.S., Badridze, G.S., Barblishvili, T.F. and Dolidze, M.D. 2005. Defoliation, photosynthetic rates, and assimilate transport in grapevine plants. *Russian Journal of Plant Physiology*, 52(4): 448–453.

Cline, M. 1997. Concepts and terminology of apical dominance. *American Journal of Botany*, 84(8): 1064–1064.

Correia, P.J., Pestana, M., Martinez, F., Ribeiro, E., Gama, F., Saavedra, T. and Palencia, P. 2011. Relationships between strawberry fruit quality attributes and crop load. *Scientia Horticulturae*, 130(2): 398–403.

Darrow, G.M. 1966. *The Strawberry: History, Breeding and Physiology.* Holt, Rinehart and Winston, New York.

Daugaard, H. 1999. The effect of flower removal on the yield and vegetative growth of A+ frigo plants of strawberry (*Fragaria* x *ananassa* Duch). *Scientia Horticulturae*, 82(1): 153–157.

Daugaard, H., Sorensen, L. and Loschenkohl, B. 2003. Effect of plant spacing, nitrogen fertilisation, post-harvest defoliation and finger harrowing in the control of Botrytis cinerea Pers. in strawberry. *European Journal of Horticultural Science*, 68: 77–82.

Guttridge, C.G. 1985. Fragaria ananassa. In: A. Harvey (ed.). *CRC Handbook of Flowering.* Vol. 8. CRC Press, Boca Raton, Florida.

Hansen, P. 1996. Effects of plant and organ removals and light reduction on flower and fruit development in strawberry. *Acta Agriculturae Scandinavica B-Plant Soil Sciences*, 46(1): 74–78.

Iqbal, N., Masood, A. and Khan, N.A. 2012. Analyzing the significance of defoliation in growth, photosynthetic compensation and source-sink relations. *Photosynthetica*, 50(2): 161–170.

Special Cultural Practices

Ito, H. and Saito, T. 1962. Studies on the flower formation in the strawberry plants I. Effects of temperature and photoperiod on the flower formation. *Tohoku journal of Agricultural Research*, 13(3): 191–203.

Jong Nam, L., Eung Ho, L., Jun Gu, L., Su Jeong, K. and Yeoung Rok, Y. 2006. Growth and yield by controlled crowns and clusters of ever-bearing strawberry in highland. *Korean Journal of Horticultural Science and Technology*, 24(1): 26–31.

Khan, N.A., Singh, S., Nazar, R. and Lone, P.M. 2007. The source–sink relationship in mustard. *Asian and Australasian Journal of Plant Science and Biotechnology*, 1: 10–18.

Kumakura, H. and Shishido, Y. 1994. Effect of varietal difference and application of deblossoming treatment on yield and fruit quality of everbearing type strawberries (*Fragaria ananassa*) grown in the cooler region of Japan. *Bulletin of the National Research Institute of Vegetables, Ornamental Plants and Tea. Series A. (Japan).* 9:27–29

Liang, Y.R., Chen, J.W. and Qin, Q.P. 2006. The effects of fruit thinning on fruit weight, sugar metabolism and accumulation in fruit of strawberry (*Fragaria* x *ananassa* cv. Tochiotome) grown in the greenhouse. *Acta Agriculturae Zhejiangensis*, 18(4): 250.

Lyu, C.B. and Li, K.T. 2014. Flower thinning effects on annual plasticulture strawberry production in Chinese Taipei. *Acta Horticulturae*, 1049: 557–560.

Metspalu, L., Hiiesaar, K., Kuusik, A., Karp, K. and Starast, M. 2000. The occurrence of arthropods in strawberry plantation depending on the method of cultivation. In: *Proceedings of the International Conference "Fruit Production and Fruit Breeding":* Eesti Põllumajandusülikool, Tartu. Transactions of the Estonian Agricultural University, 207: 204–208.

Mochizuki, T., Yoshida, Y., Yanagi, T., Okimura, M., Yamasaki, A. and Takahashi, H. 2009. Forcing culture of strawberry in Japan – Production technology and cultivars. *Acta Horticulturae*, 842: 107–109.

Moor, U., Karp, K. and Põldma, P. 2004. Effect of mulching and fertilization on the quality of strawberries. *Agricultural and Food Science*, 13(3): 256–267.

Neri, D., Sugiyama, N., Iwama, T. and Akagi, H. 2003. Effect of apical pinching on the development of axillary buds in strawberry plants. *Journal of the Japanese Society for Horticultural Science*, 72(5): 389–392.

Nestby, R. 1985. Effect of planting date and defoliation on three strawberry cultivars. *Acta Agriculturae Scandinavica*, 35(2): 206–212.

Paszko, D. 2002. Effect of flower and runner removal on the yield and vegetative growth of two strawberry cultivars. *Annales Universitatis Mariae Curie-Sklodowska. Sectio EEE Horticultura*, 11: 19–27.

Rindom, A. and Hansen, P. 1995. Effect of fruit numbers and plant status on fruit size in strawberry. *Acta Agriculturae Scandinivica*, 45(2): 142–147.

Savini, G., Neri, D., Zucconi, F. and Sugiyama, N. 2005. Strawberry growth and flowering: An architectural model. *International Journal of Fruit Science*, 5(1): 29–50.

Sharma, R.M. and Yamdagni, R. 2000. *Modern Strawberry Cultivation*. Kalyani Publishers, Ludhiana, India: p. 172.

Sonsteby, A. and Heide, O.M. 2006. Dormancy relations and flowering of the strawberry cultivars Korona and Elsanta as influenced by photoperiod and temperature. *Scientia Horticulturae*, 110(1): 57–67.

Takei, M. 2010. The cultivation of strawberry in Japan. Farm Extension Officer, Nagano Prefecture, Japan The Japan Agricultural Exchange Council. Nikken Align Bldg. 8F, 5-27-14, Nishi-Kamata, Ota-ku, Tokyo 144-0051 Japan.

Thompson, P.A. and Guttridge, C.G. 1960. The role of leaves as inhibitors of flower induction in strawberry: With one figure in the text. *Annals of Botany*, 24(4): 482–490.

Whitehouse, A.B., Johnson, A.W. and Simpson, D.W. 2009. Manipulation of the production pattern of ever-bearing cultivars by defoliation treatments. In *VI International Strawberry Symposium*, 842: 773–776.

Wildung, D.K. 2000. Flame burning for weed control and renovation with strawberries. In: J. Cibrorowski (ed.). *Greenbook 2000 – Marketing Sustainable Agriculture*, pp. 91–95.

Zhang, L., Strik, B.C. and Martin, L.W. 1992. Date of renovation affects fall and summer fruiting of June-bearing strawberry. *Advances in Strawberry Research (USA)*, 11: 47–50.

22 Protected Cultivation

K.K. Pramanick and Poonam Kashyap

CONTENTS

22.1 Cultivars Suitable for Protected Cultivation 367
22.2 Advantages of Protected Cultivation.. 367
    22.2.1 Frost and Disease Control 367
    22.2.2 Plant Growth 368
    22.2.3 Advanced Berry Ripening.. 368
    22.2.4 Improvement in Yield and Quality of Fruits 368
    22.2.5 Runner Production 369
References.. 370

Throughout the world, strawberry cultivation is equally successful in climatic conditions that are characteristically temperate in the northern latitudes, subtropical in the plains or tropical at high altitude. In spite of attractive fruits and pleasing flavour, strawberry cultivation is still in its infancy in many countries, including India, primarily because of its perishable nature. Recently, some business houses have setup a number of pilot projects for large-scale strawberry production. Farmers in the subtropics in the vicinity of cities, such as New Delhi, Mumbai and Pune in India, have been profitably cultivating strawberries during winter months using plant material propagated in the cool climatic conditions such as temperate hilly regions (e.g., Himachal Pradesh in India). Strawberry growers in the hills, thus, have a double-pronged earning potential from strawberry cultivation (Pramanick et al., 2013), that is, runner and fruit production.

Protected strawberry cultivation practices can be defined as cropping techniques wherein the microclimate surrounding the plants is partially/fully controlled as per the requirement of the plant during the growing season. In early times, strawberry cultivation was initiated in open conditions, but with the introduction of plastic tunnels and greenhouses, its traditional objective was changed, and growers started to exploit these structures for extension of seasonal and offseason production worldwide, particularly in areas with mild winters. In the early 1950s, protected strawberry cultivation under plastic covers (polyvinyl chloride) was introduced in Japan. Thereafter, it became popular as a new cultivation system and the hectarage planted under protected cultivation and the quantity of production increased rapidly year by year during 1960s to 1970s. It has been estimated that more than 90% of strawberry fields are covered with plastic after making several modifications in growing structures, keeping in view the growing objectives (Oda et al., 2014). Now greenhouses are frequently used to hasten ripening, but they can also be used to produce autumn and winter crops. They are valuable in protection of crops against rains. The problem of high percentages of malformed, and small fruits frequently produced in strawberries can be resolved by improved pollination (Lopez-Galarza et al., 1997). In the UK, by combining glasshouses, plastic tunnels and field culture, strawberries are available almost year round (Takeda, 1999). In Europe, the production of strawberries for the fresh market under forced and protected conditions is increasing, which was initially aimed at enhancing the earliness of short-day varieties. Nowadays, the objective is to ensure year-round availability of fruits, forcing and preserving strawberry crops against adverse weather conditions (Neri et al., 2012).

At high altitude, harsh weather, such as freezing temperatures, frost and chilling wind during winter and hailstorm accompanied by low temperature, is the main hurdle for strawberry cultivation.

Another important factor is maintenance of optimum level of soil moisture at critical growth stages and fruiting of strawberry. Moreover, water shortage is becoming a serious problem in the hills during autumn, winter and spring due to change in precipitation/snowfall pattern over the years. However, given relatively high latitudes and weather conditions, which are sometimes extreme, the growing season is commonly constrained by temperature and available solar light. Consequently, protected cultivation in these region has increased significantly in recent years, including glasshouse growing for what is described as "offseason" production (Alsheikh et al., 2009; Daugaard, 2008).

Protected cultivation in strawberry plays a pivotal role in minimizing winter injury and plant mortality and increasing productivity. Covering the strawberry beds with low-height, clear plastic tunnels facilitated one-month-early cropping, prevented bed erosion and increased total fruit yields by 20% (Pramanick et al., 2000). Protected cultivation has been used for strawberry cultivation to protect plants from adversities of climatic conditions, and for better control of diseases (Passos, 1997; Goto and Tivelli, 1998; Andriolo et al., 2002; Fernandes et al., 2002).

Nowadays, protected and covered cultivation during winter is the most common style in Japan, having higher productivity (2–4.5 t/ha more) than open-field conditions. However, when productivity in this condition in Japan is compared with other producing countries or regions of the world, such as Brazil and California, where a long-term harvest is practised, it is not actually considered a large harvest, because of differences in sunlight during the course of production (Oda, 1988; Bertelson, 1995). Recently, these systems in plastic have been concentrated into two major types of overwinter cultures: that used by the June-bearer or the summer–autumn culture of the everbearer. The former system is important both economically and for quantity of fruit production (Oda et al., 2014).

During 1989–91, the technique of strawberry forcing with double film covering system proved very profitable. It is the only fruit crop that starts paying back within 100–120 days of planting. Protected cultivation has shown considerable levels of profitability in an economic analysis, despite heavy production costs (Asciuto et al., 2003). The greenhouse is generally covered by transparent or translucent material such as glass or plastic depending upon the availability and suitability of fabricating materials.

Besides greenhouses, plastic tunnels have been found highly effective for sustainable strawberry production. The introduction of polytunnels has revolutionized strawberry production. The growers have realized this as an opportunity to specialize in strawberry production, and to grow the strawberry, in place of less profitable, other agri-enterprises. Despite this clear trend, other farmers decided not to intensify their production methods and to produce for local markets. This divergence in farmer types intensified with the introduction of polytunnels for cultivation of strawberries in the late-1990s (Calleja et al., 2012). This dominant trend is believed to be driven by supermarket dominance in the horticulture sector (Hingley, 2005; Taylor and Fearne, 2009), as farmers are often left with no option other than to intensify production methods to maximize their output and with reduced cost of production.

High tunnels capture and retain heat, which protects the fruit and flowers from frost injury and helps to maintain the temperatures in the tunnel warm enough for regular and optimum growth of plants. High tunnels maintain optimal temperatures (21–27°C) for as much of the day as possible. Similarly, low tunnels/row covers are also very common used for protected strawberry cultivation. Low tunnels placed within the high tunnels are another tool for out-of-season strawberry production. Low tunnels are typically only 50 cm high and about 75 cm wide, just large enough to cover a single strawberry bed (Figure 22.1). Additionally, low tunnels can be used for winter-planted day-neutral varieties to enhance early plant growth and hasten berry growth and ripening. High tunnels generally provide 3–5°C higher temperature over outside conditions during the night. However, extreme caution should be taken with low tunnels as they have a tendency to get heated up very quickly and can lead to plant damage if not managed properly. It is better to uncover low tunnels and vent high tunnels during the daylight hours, thereafter re-cover low tunnels and close high tunnels during the night to retain warmth (Rowley et al., 2010). The use of agro-shade net has also been observed effective for runner production, particularly in subtropical climates (Sharma et al., 2013).

Protected Cultivation

FIGURE 22.1 Advanced strawberry production under low tunnel with polyethylene mulch.

22.1 CULTIVARS SUITABLE FOR PROTECTED CULTIVATION

Different varieties' adaptability to various growing conditions is a welcome characteristic for strawberry growing in protected cultivation. The country-specific list of strawberry cultivars recommended for protected cultivation is given in Table 22.1.

22.2 ADVANTAGES OF PROTECTED CULTIVATION

In protected cultivation, particular attention is given to maximizing good effects of CO_2 and fertigation on yield (Kawashima, 1991; Sung and Chen, 1991) and protection from the environmental adversities with appropriate plastic film (Rossi et al., 1989).

22.2.1 Frost and Disease Control

The size of a strawberry fruit varies with its position in the inflorescence, being largest from the primary flowers followed by secondary and tertiary flowers. In subtropical regions, primary flowers open during December and January, so they are highly prone to injury by frost, which may, in extreme situations, cause damage to the crop and reduction in income of up to 50% or more. This occurs often when several warm days are followed by frosty nights. Row covers and chemical protectants against frost such as "FrostGard" may help in protecting the blossoms from injury due to frost (Whitworth et al., 1993).

TABLE 22.1
List of Cultivars Suitable for Cultivation under Protected Conditions

Cultivar	Country	References
Rosella, Aliso, Adie, Honeoye, Gea, Queen Elisa and Irma	Italy	Lemaitre (1980); Faedi et al. (2004)
Brighton, Fern and Selva	Greece	Paraskevopoulou-Paroussi et al. (1991); Vlachonasios et al. (1995)
Lassen and Fresno	Argentina	Rodriguez (1981)
Camarosa and Blakemore	India	Das et al. (2012)
Camarosa, Dover, Oso Grande and Tudla	Brazil	Calvete et al. (2008)
Selva, Douglas, Brighton, Fern, Sweet Charlie, Belrubi, Seascape, Chandler, Shimla Delicious, Jutogh Special	India	Pramanick et al. (2000)

During winter (December–February), prevailing low temperatures minimize the rate of growth and development of strawberry plant, so low tunnels (50 cm height) of transparent polyethylene film (50 μ) can be installed with the help of GI wires or with locally available cheaper materials over raised beds to protect the strawberry plants from frost and chilling winds. These tunnels may be opened during daytime and closed during the night, which allows higher temperatures to build up in the top layer of soil (Pramanick et al., 2000), supporting root activity and facilitating plant growth.

For strawberry cultivation, soil disinfection by solar heating in closed plastic greenhouses is highly effective in controlling diseases in strawberry. Menzel et al. (2016) obtained 40% higher marketable fruit yields in plants in tunnels over yields of plants outdoors. This was mainly because fruits from the plants grown in tunnels had lower incidences of rain damage and/or grey mould. An anthracnose-susceptible cultivar may be grown successfully in a plastic house after covering with 80% shade nets (Miao et al., 2014)

22.2.2 Plant Growth

The use of row covers during autumn sustains carbohydrate metabolism longer and using row covers during spring with early removal of mulch advances the spring flush with an accelerated rate of carbohydrate metabolism in strawberry plants (Gast and Polland, 1989). Covered strawberry plants become photosynthetically independent earlier than uncovered plants as demonstrated by high nonstructural carbohydrate, glucose and fructose concentrations in covered rows. The covering prevents depletion of carbohydrate reserves in crowns and roots (Gast and Pollard, 1991). Protected cultivation at irrigation levels of –0.01 and –0.035 MPa with clear polyethylene mulch has been observed to favour vegetative growth of strawberry plant (Pires et al., 2006).

The plastic houses covered with 60% and 70% shade nets decrease temperatures by less than 1°C compared with the open conditions, while 80 and 90% shading nets can decrease it by 1.2°C and 1.3°C, respectively. Even in the cloudy dawn and early dawn, shading with 80% agro-shade nets maintained a light intensity of more than 10,000 lux and plantlets could still grow. The increasing shades tended to change the content of chlorophyll and MDA (malondialdehyde) and the activities of antioxidant enzymes superoxide dismutase and peroxidase, regularly and significantly to adapt to the low-light environment. Overall, the covering of the plastic house with 80% shading net has been observed to facilitate much higher growth of strawberry plantlets than open conditions (Miao et al., 2014).

22.2.3 Advanced Berry Ripening

Strawberries in protected conditions mature much earlier than those grown in open-field conditions. Covering the strawberry beds with polyvinyl/plastic tunnels, polypropylene fiber sheets, spun-bonded polypropylene, polyester, polyamide, polyethylene etc. has been shown to advance berry harvesting by 10 to 30 days (Pollard et al., 1989; Pramanick et al., 2000). It has been suggested that flower bud initiation can be induced even in long photoperiods if the light illuminating the strawberry crown lacks red and shorter wavelength light, whereas, excessive red light transmittance either with red LED lights or growing transplants in red photoselective shade net in August delays flower bud initiation (Takeda, 2012), thus affecting the time of flowering and fruiting.

22.2.4 Improvement in Yield and Quality of Fruits

In general, creation of a CO_2-enriched environment is a very useful technique in protected strawberry cultivation. Improved $^{14}CO_2$ fixation in strawberry plants has been observed by increasing the concentration of CO_2 in growth chambers (Desjardins et al., 1988). Enrichment of the environment with CO_2 (10,000 μl/l for 10 days) starting two weeks after initial bloom has resulted in enhanced leaf CO_2 exchange rate (CER) in rockwool cultured strawberry cv. Tzun Siang. Its enrichment throughout the fruiting period stimulated CER, reduced chlorophyll and leaf protein

Protected Cultivation

loss and enhanced fruit set and consequent fruit production (Sung and Chen, 1991). Enrichment of the growth environment with CO_2 (300–600 ppm) can enhance photosynthesis, starch synthesis and fruit yield and quality (by enhancing sugars and organic acids) in strawberry (Eckstein and Blanke, 2000). Increased CO_2 (300 and 600 μ mol/mol above ambient) tends to increase the ascorbic acid (AsA), glutathione (GSH) and ratios of AsA to dehydroascorbic acid (DHAsA) and GSH to oxidized glutathione (GSSG) and decrease in DHAsA in strawberry fruit. High anthocyanin and phenolic contents have also been observed in fruit of CO_2-treated plants. Growing strawberry plants in CO_2-enriched conditions significantly enhances fruit p-coumaroylglucose, dihydroflavonol, quercetin 3-glucoside, quercetin 3-glucuronide and kaempferol 3-glucoside contents, as well as cynidin 3-glucoside and pelargonidin 3-glucoside-succinate contents (Wang et al., 2003).

In the studies of Kawashima (1991), the fruit yield of strawberry was observed to be increased by 40–120% by increasing (up to 750 ppm) CO_2 concentration in the glasshouse, and setting the ventilation temperature at 28°C. The number of fruits and weight were generally increased by 20–30%. Yield increased with increasing CO_2 concentration in the range of 375–1000 ppm, but 750 ppm was considered to be a practical level.

The content of the anthocyanin cyanin 3-glucoside and the flavonols quercetin 3-glucuronide and kaempferol 3-glucoside has been observed to decrease in fruits grown in UV-blocking film when compared with open-field grown strawberries. By means of the UV-transparent film, the content of the flavonoids could be enhanced up to similar amounts as those in open-field-grown strawberries (Josuttis et al., 2010).

Changing the time of plugging and altering the light quality for production of plug plants creates an opportunity for double-cropping, for example, fruit production from short-day cultivars in autumn and early winter and again in the spring in the mid-Atlantic coast region in protected cultivation (Takeda, 2012). Simple reflective mulches can be used to support strawberry yield in protected cultivation when environmental light levels are low, and can reduce seasonal variation. Their use might also help to extend the harvesting season or expand the growing region into regions with lower irradiance (Grout and Greig, 2012).

22.2.5 RUNNER PRODUCTION

Production of runners in Sparkle strawberry has been observed to be enhanced when row-cover plastic films over wire hooks were placed over the crop during December indicating that runner production is better in protected structures (Austin, 1991). Normally, shade nettings are used to mitigate temporary heat-induced cessation in reproductive growth (Wagstaffe and Battey, 2008). Shading nets can also be used between April and July on runner beds to facilitate the runner production, particularly in the areas that experience high temperatures (up to 43°C) in the summer season (Table 22.2 and Figure 11.1).

The use of polyethylene in cultivation of strawberry on a commercial scale can play a pivotal role in minimizing winter injury and plant mortality, and increasing productivity. Covering the strawberry beds with low clear plastic tunnels induced one-month-early cropping, prevented bed erosion and increased total yields by 20%. Black polyethylene-mulched beds did not require any weeding. During summers at high elevation locations, the polyethylene sheets of the tunnels were replaced by plastic anti-hail nets or anti-bird nets, which resulted in advanced harvest, increased yield, and improved fruit quality (Pramanick et al., 2000). This possibility has also been reported in different agro-climatic conditions by Albregts and Chandler (1993). Mulching is an important component of the strawberry production system. Many synthetic and organic types of mulch are being used for strawberry cultivation in different parts of the country based on the climatic conditions and availability of raw materials. Himelrick (1982) showed that plants grown on black polyethylene produce more runners and fruits than with bare soil. Different workers have reported beneficial effects of organic mulches on strawberry production (Rebandel and Przysiccka, 1981; Badiyala and Aggarwal, 1981; Hassan et al., 2000; Lille et al., 2003).

TABLE 22.2
Effect of Shading Levels on Runner Production of Strawberries

| Parameter | Cultivar | Shading level (%) | | | | | LSD (P ≤ 0.05) | | |
		0	25	50	75	Mean	Cultivar (C)	Shading (S)	CxS
No. of runners/plant	Chandler	5.11	8.23	11.44	8.06	8.21			
	Oso Grande	6.02	12.33	16.33	11.21	11.47			
	Mean	5.57	10.28	13.89	9.64		1.67	2.31	2.86
No. of leaves/runner	Chandler	5.11	6.78	8.11	8.06	7.02			
	Oso Grande	5.73	8.21	8.65	8.13	7.68			
	Mean	5.42	7.50	8.38	8.10		NS	1.86	2.31
Crown diameter (mm)	Chandler	5.03	11.63	13.33	12.08	10.52			
	Oso Grande	6.66	11.83	13.65	11.45	10.90			
	Mean	5.85	11.73	13.49	11.77		NS	2.64	3.10

Source: Sharma et al., 2013, Strawberry regeneration and assessment of runner quality in subtropical plains. Journal of Applied Horticulture, 15: 191–193.

REFERENCES

Albregts, E.E. and Chandler, C.K. 1993. Effect of polyethylene mulch color on fruiting response of strawberry. *Proceedings – Soil Science Society of Florida*, 52: 40–43.

Alsheikh, M., Sween, R., Nes, A. and Gullord, M. 2009. Strawberry breeding in Norway: progress and future. *Acta Horticulturae*, 842: 499–501.

Andriolo, J.L., Bonini, J.V. and Boemo, M.P. 2002. Acumulacao de materiaseca e rendimento de frutos de morangueiro cultivado emsubstrato com different essolucoes nutritivas. *Horticultura Brasileira*, 20: 24–27.

Asciuto, A., Di Trapani, A.M. and Schimmenti, E. 2003. Protected cultivations of strawberry in Sicily: economic aspects of some recent product and process innovations. *Acta Horticulturae*, 614: 135–143.

Austin, M.E. 1991. Row covers for Sparkle strawberries. *HortScience*, 26: 603.

Badiyala, S.D. and Aggarwal, G.C. 1981. Note on the effect of mulches on strawberry production. *Indian Journal of Agricultural Science*, 51: 832–834.

Bertelson. 1995. The US Strawberry industry. USDA Eco. Res. Ser. Stati. Bull. No. 914.

Calleja, E.J., Ilbery, B. and Mills, P.R. 2012. Agricultural change and the rise of the British strawberry industry, 1920–2009. *Journal of Rural Studies*, 28: 603–611.

Calvete, E.O., Mariani, F., Wesp, C. de L., Nienow, A.A., Castilhos, T. and Cecchetti, D. 2008. Phenology, production and content of strawberry crops cultivars anthocyanins produced under protected environment. *Revista Brasileira de Fruticultura, Jaboticabal – SP*, 30: 396–401.

Das, B., Ahmed, N. and Attri, B.L. 2012. Performance of strawberry genotypes at high altitude temperate climate under different growing environments. *Progressive Horticulture*, 44(2): 242–247.

Daugaard, H. 2008. Table-top production of strawberries: performance of six strawberry cultivars. *Acta Agriculturae Scandanavica Section B – Soil and Plant Science*, 58: 261–266.

Desjardins, Y., Laforge, F., Lussier, C. and Gosselin, A. 1988. Effect of CO_2 enrichment and high photosynthetic photon flux on the development of autotrophy and growth of tissue-cultured strawberry, raspberry and asparagus plants. *Acta Horticulturae*, 230: 45–54.

Eckstein, K. and Blanke, M. 2000. Effects of enriched CO_2 and potassium nutrition on vegetative and reproductive growth, photosynthesis, and transpiration as well as fruit quality of strawberry. *European Journal of Horticultural Science*, 65(3): 129–133.

Faedi, W., Ballini, L., Baroni, G., Baruzzi, G., Lucchi, P. and Zenti, F. 2004. Queen Elisa and Irma, new strawberry cultivars for northern environments. *Informatore-Agrario*, 60: 45–48.

Fernandes, J.R.F., Furlani, P.R., Ribeiro, I.J.A. and Carvalho, C.R.L. 2002. Producao de frutos e estolhos do morangueiro em diferentes sistemas de cultivo em ambiente protegido. *Bragantia*, 61: 25–34.

Gast, K.L.B. and Pollard, J.E. 1989. Over wintering strawberry plants under row covers: effect on development of yield components. *Acta Horticulturae*, 265: 215–216.

Gast, K.L.B. and Pollard, J.E. 1991. Row covers and mulch alters seasonal strawberry carbon metabolism. *Acta Horticulturae*, 26: 421.

Goto, R. and Tivelli, S.B. 1998. *Production of Vegetables in Protected Environment: Subtropical Conditions*. Foundations Publishers of UNESP, Sao Paulo, Brazil: p. 319.

Grout, B.W.W. and Greig, M.J. 2012. Counteracting low light levels in protected strawberry cultivation using reflective mulches. *Acta Horticulturae*, 956: 489–492.

Hassan, G.I., Godara, A.K., Kumar, J. and Huchche, A.D. 2000. Effect of different mulches on the yield and quality of "Oso Grande" strawberry (*Fragaria* x *ananassa*). *Indian Journal of Agricultural Science*, 70: 184–185.

Himelrick, D.G. 1982. Effect of polyethylene mulch color on soil temperatures and strawberry plant response. *Advanced Strawberry Production*, 1: 15–16.

Hingley, M.K. 2005. Power imbalanced relationships: cases from UK fresh food supply. *International Journal of Retail & Distribution Management*, 33: 551–569.

Josuttis, M., Dietrich, H., Treutter, D., Will, F. and Linnemannstns, L. 2010. Solar UVB response of bioactives in strawberry (*Fragaria* x *ananassa* Duch.): a comparison of protected and open-field cultivation. *Journal of Agricultural and Food Chemistry*, 58: 12692–12702.

Kawashima, N. 1991. Studies on the CO_2 enrichment in a greenhouse. (3) Effect on growth of strawberry. Bulletin of the Agricultural Experiment Station of Nebraska, 22: 65–72.

Lemaitre, R. 1980. Outdoor strawberry varieties. *Fruit Belge*, 48 (389): 71–76.

Lille, T., Karp, K. and Varnik, R. 2003. Profitability of different technologies of strawberry cultivation. *Agronomy Research*, 1: 75–83.

Lopez-Galarza, S., Maroto, J.V., San Baautista, A. and Alagarda, J. 1997. Performance of waiting-bed strawberry plants with different number of crowns in winter plantings. *Acta Horticulturae*, 439: 439–448.

Menzel, C.M., Smith, L.A. and Moisander, J.A. 2016. Protected cropping of strawberry plants in subtropical Queensland. *Acta Horticulturae*, 1117: 273–278.

Miao, L.M., Zhang, Y.C., Yang, X.F. and Jiang, G.H. 2014. Effect of shading and rain-shelter on plantlet growth and antioxidant systems in strawberry (*Fragaria* × *ananassa*). *Acta Horticulturae*, 1049: 443–446.

Neri, D., Baruzzi, G., Massetani, F. and Faedi, W. 2012. Strawberry production in forced and protected culture in Europe as a response to climate change. *Canadian Journal of Plant Sciences*, 92: 1021–1036.

Oda, Y. 1988. Strawberry cultivation in subtropical and tropical highlands. Abstract of Soc. Tropical Agriculture Study 44–45 (In Japanese).

Oda, Y., Inoue, R., Saito, A., Sakurai, F. and Kawasaki, S. 2014. Protected overwinter strawberry cultural systems in Japan: origin, renovation, and consecutive improvements. *Acta Horticulturae*, 1049: 167–171.

Paraskevopoulou-Paroussi, G., Vassilakakis, M. and Dogras, C. 1991. Performance of five strawberry cultivars under plastic greenhouse or field conditions in northern Greece. *Acta Horticulturae*, 287: 273–280.

Passos, F.A. 1997. *Influecia de alguns sistemas de cultivo na cultura do morango (Fragaria x ananassa Duch.)*. Piracicaba: USP/ESALQ, pp. 106.

Pires, R.C. de M., Folegatti, M.V., Passos, F.A., Arruda, F.B. and Sakai, E. 2006. Vegetative growth and yield of strawberryunder irrigation and soil mulches for different cultivation environments. *Scientia Agricola (Piracicaba, Braz.)*, 63: 417–425.

Pollard, J.E., Gast, K.L.B. and Cundari, C.M. 1989. Overwinter row covers increase yield and earliness in strawberry. *Acta Horticulturae*, 265: 229–234.

Pramanick, K.K., Kishore, D.K. and Sharma, Y.P. 2000. Effect of polyethylene on the behaviour and yield of strawberry (*Fragaria* x *ananassa*). *Journal of Applied Horticulture*, 2: 130–131.

Pramanick, K.K., Kishore, D.K., Sharma, S.K., Das, B.K. and Murthy, B.N.S. 2013. Strawberry cultivation under diverse agro-climatic conditions of India. *International Journal of Fruit Science*, 13: 36–51.

Rebandel, Z. and Przysiccka, M. 1981. Evaluation of the suitability of some organic materials in strawberry cultivation. *Informator o Hadaniachprowadzonych w Zakladzie Sadownictwa Akadernii Rolniczej w Poznaniu*: pp. 113–118.

Rodriguez, J.P. 1981. The cultivation of strawberry. *Annals of the Rural Society Argentina*, 115: 20–23.

Rossi, F., Cristoferi, G., Bagioli, L. and Comporeale, S. 1989. Effect of plastic films on day neutral strawberry yields. *Acta Horticulturae*, 265: 235–242.

Rowley, D., Black, B. and Drost, D. 2010. *High Tunnel Strawberry Production*. Extension Bulletin, Utah State University Coperative Extension, Utah: p. 6.

Sharma, R.M., Singh, A.K., Sharma, S., Masoodi, F.A. and Uma, S. 2013. Strawberry regeneration and assessment of runner quality in subtropical plains. *Journal of Applied Horticulture*, 15: 191–193.

Sung, F.J.M. and Chen, J.J. 1991. Gas exchange rate and yield response of strawberry to carbon dioxide enrichment. *Scientia Horticulturae*, 48: 241–251.

Takeda, F. 1999. "Out-of-season" greenhouse strawberry production in soilless substrate. *Advances in Strawberry Research*, 18: 4–15.

Takeda, F. 2012. Newmethods for advancing or delaying anthesis in short-day strawberry. *Acta Horticulturae*, 926: 237–241.

Taylor, D.H. and Fearne, A. 2009. Demand management in fresh food value chains: a framework for analysis and improvement. *Supply Chain Management an International Journal*, 14: 379–392.

Vlachonasios, C., Vasalakakis, M., Dogras, C. and Mastrokokastas, M. 1995. Out of season glasshouse strawberry production in north Greece. *Acta Horticulturae*, 379: 305–312.

Wagstaffe, A. and Battey, N.H. 2008. Tunnel production of strawberry in the UK: a review. *Proc. 2007 N. Ames. Strawberry Symposium*, 14–16 Feb, 2007, Ventura, California: pp. 23–28.

Wang, S.Y., Bunce, J.A. and Maas, J.L. 2003. Elevated carbon dioxide increases contents of antioxidant compounds in field-grown strawberries. *Journal of Agricultural and Food Chemistry*, 51: 4315–4320.

Whitworth, J., Perkins-Veazie, P. and Collins, J.K. 1993. Fruiting and postharvest quality of three strawberry cultivars after frost protection treatments. *Acta Horticulutrae*, 348: 268–270.

23 Soilless Culture

Mahital Jamwal and Nirmal Sharma

CONTENTS

23.1 Advantages of Soilless Culture 375
    23.1.1 Enhanced Productivity 375
    23.1.2 Precise Use of Nutrient Solution.. 375
    23.1.3 Water Economy 375
    23.1.4 Reduced Labour Requirement... 375
    23.1.5 Sterilization Practices of Substrates 375
    23.1.6 Control of Root Environment.. 375
    23.1.7 Increased Crop Per Year 376
    23.1.8 Suitability in Problem Soil 376
23.2 Limitations of Soilless Culture 376
    23.2.1 High Initial Investment... 376
    23.2.2 Requirements of Diverse Technical Skill 376
    23.2.3 High Risk of Disease Infections... 376
23.3 Soilless Culture Systems... 376
23.4 Soilless Culture Techniques.. 378
    23.4.1 Ebb and Flow Technique/Flood and Drain Technique 378
    23.4.2 Deep Flow Technique 378
    24.4.3 Aeroponics System... 379
    23.4.4 Aerated Flow Technique 379
    23.4.5 Nutrient Film Technique 380
    23.4.6 Drip Irrigation Technique 380
    23.4.7 Root Mist Technique 381
    23.4.8 Fog Feed Technique 381
23.5 Suitable Cultivars.. 381
23.6 Substrates for Soilless Culture 381
23.7 Preparation of Solution for Nutrition.. 382
    23.7.1 Composition of Nutrient Solution.. 382
    23.7.2 Quality of Water 382
    23.7.3 Sources of Chemical Compounds or Fertilizers for Preparation of the
                Nutrient Solution 383
23.8 Factors Influencing Soilless Culture Systems... 383
    23.8.1 Composition of Nutrient Medium... 383
    23.8.2 PH of the Nutrient Solution.. 385
    23.8.3 EC of the Nutrient Solution.. 386
    23.8.4 Temperature of the Nutrient Solution... 387
    23.8.5 Concentration and Uptake of Nitrate (NO_3)... 389
    23.8.6 Reaction (pH) of the Nutrient Solution.. 389
    23.8.7 Anion Competition 390
    23.8.8 Salinity 390
    23.8.9 Root Aeration 390
    23.8.10 Substrate.. 390

23.9	Management of Nutrient Solution	390
	23.9.1 Regulation of pH	391
	23.9.2 Management of EC	391
	23.9.3 Temperature Control	392
	23.9.4 Oxygenation of the Nutrient Solution	392
References		393

Up to now, strawberry cultivation on a large scale for commercial purposes has been practised in soils in fields with conventional techniques of crop management, which had to address issues such as the environment, season, availability of land and manual labourers near large cities and phytosanitary standards. As a result, strawberry production in a soilless culture system has slowly become popular and nowadays it is being advocated across the strawberry-growing world. A well-organized soilless strawberry production system largely addresses these issues and helps minimize the problems associated with pests and disease management to a great extent. In traditional protected cultivation such as poly-tunnels, the cost of construction is very high and, therefore, high-value crops such as strawberries are an ideal choice for such typically specialized systems. Until recently, growers depended largely on methyl bromide to sterilize the soil before planting strawberries to minimize the problems associated with soilborne insect-pests and fungal diseases in the culture system. Bans and restricted use of agrochemicals/fumigants warranted a need to search for alternative methods for management of soilborne biotic stresses in new production systems, such as soilless culture referred to as "hydroponics". Worldwide, soilless crop production has increased significantly in recent years because of certain advantages, for example, environmental control, extended growing season, higher marketable yields, better-quality produce, efficient use of water and fertilizers and facilitated management of biotic stresses for enhanced yield and profit. Soilless strawberry production system provides an important alternative for an environmentally friendly, efficient and sustainable growing system for extended period or round the year production of strawberries. This way, the need for agrochemicals for sterilizing the soils for culture of strawberries can be minimized. Many commercial growers have preferred production of strawberries in "hydroponics" and slowly switched over to this system of cultivation. Soilless strawberry cultivation comparatively increases the produce quality and productivity of the crop resulting in enhanced incomes and competitive strawberry production system.

The potential of water in the culture of tomato at a commercial scale was conceptualized and promoted during 1929 (Gericke, 1937). In his model, he planted tomatoes in a layer of sand and supported the surface of the solution by netting and canvas, through which the roots passed into the nutrient solution. Originally, Gericke (1929) defined this method as "aquaculture"; however, since the term was in use for the culture of aquatic plants and animals (fish), other terms, viz., "water culture" and "solution culture" were introduced. Later, on the basis of the Greek words *hydro* (water) and *ponos* (labour), the term "hydroponics" was proposed by Setchell during 1937. In hydroponics or soilless culture, nutrients are supplied into the root system through a standardized solution and surplus solution is collected back, reconstituted and recycled. As per the recorded chronology, the first large-scale production in hydroponics was done by the United States Army during the Second World War (1939–1945), and a number of soilless production units were established in the Pacific Islands using volcanic substrates from the area with subsurface irrigation.

Nowadays, soilless technique is intensively used for production of a large number of horticultural crops and is successfully run by commercial enterprises in many developed countries. To advance and popularize the science of hydroponics further, the First International Society of Soilless Culture was established during 1955, when there was only one hectare of soilless cultivation in the world (Schwarz, 2001). In 1996, the world-over area in hydroponics was estimated to be about 12,000 ha, which increased to 25,000 ha during 2001 (Carruthers, 2002), of which about 90% was in 22 countries, and 66% of the area being distributed in four countries only: the Netherlands, Spain, Canada and France. There are no exact statistics available on area and production of crops in hydroponics, however, considering the growth trend in hydroponics, it is estimated that worldwide 50,000 ha

Soilless Culture 375

could be in soilless culture. Further, growth of strawberry in soilless culture in future depends on the development of easy to adopt and economically profitable and environmentally friendly production systems in soilless culture.

23.1 ADVANTAGES OF SOILLESS CULTURE

23.1.1 ENHANCED PRODUCTIVITY

Although higher yields of better-quality fruits are obtained in soilless culture of strawberries due to modified and improved growth conditions, especially nutrition and irrigation, but the fruit yield from soilless culture systems is generally not much superior to that of soil culture. Instead, soilborne constraints such as salinity and toxicities in soil culture can be well overcome in soilless culture. In addition, reduced cost of labour, higher yields and uniformity of produce are some of the other advantages. Besides, soilless culture in vertical sleeves offers opportunities for better use of the area in cultivation and facilitated crop management in comparison with the traditional strawberry culture in beds (Villagra et al., 2012).

23.1.2 PRECISE USE OF NUTRIENT SOLUTION

Supply of a regulated nutrient solution with the desired concentration at various stages of crop growth and its uniform application across the culture system are other important advantages of soilless culture. In addition, the concentrations of micronutrients such as manganese, boron, zinc, copper and lead can be regulated and maintained within safe limits and their toxicity can be avoided in the culture system. Other factors such as electrical conductivity (EC) and pH of the nutrient solution can be regulated according to the requirement of plants at different stages of growth and the environmental conditions.

23.1.3 WATER ECONOMY

The strawberries grown in protected conditions require large amounts of water because of the lack of rainfall. This demand is further aggravated if the production is taken up in regions with higher atmospheric aridity. In these conditions, the availability and judicious use of water governs the success or failure of the soilless strawberry production system. Therefore, the soilless strawberry production system has an advantage wherein the plant roots remain immersed in recyclable water with regulated supply of nutrient solution. In addition, the requirement for labourer and time for maintenance of irrigation system are largely saved in soilless culture.

23.1.4 REDUCED LABOUR REQUIREMENT

Less labour is needed in soilless culture because the system's functioning are controlled automatically.

23.1.5 STERILIZATION PRACTICES OF SUBSTRATES

Generally, there is no need for sterilization of substrate in soilless strawberry culture; however, in closed soilless culture systems there may be a requirement for a certain degree of sterilization of substrates, depending upon the system installed.

23.1.6 CONTROL OF ROOT ENVIRONMENT

In soilless culture systems, the temperature and oxygen supply in the root zone can be more easily controlled than in a soil culture.

23.1.7 Increased Crop Per Year

In a soilless strawberry culture system, the number of crop cycles per year increases because in such a protected production system, the time interval between successive plantations is reduced considerably.

23.1.8 Suitability In Problem Soil

The soilless system of strawberry culture provides an opportunity to produce commercial crop in the area prone with problem soils.

23.2 LIMITATIONS OF SOILLESS CULTURE

23.2.1 High Initial Investment

The soilless culture system involves higher capital investment for construction and maintenance than cultivation in soil. In addition, the extent of increased recurring investment on its maintenance depends on the system employed and the degree of automation integrated. All these expenses add to the overall cost of production.

23.2.2 Requirements of Diverse Technical Skill

Soilless culture is a highly specialized system and, therefore, for its successful operation the engaged manager should have expertise in crop cultivation, plant physiology, chemistry and knowledge of the control systems etc. Besides, crop growing in a restricted space requires a higher standard of management. Therefore, successful commercial soilless culture demands a better management and highly skilled staff, who have to be paid more. This also adds to cost of production. Many a times, availability of qualified and expert human labour is a limitation in itself.

23.2.3 High Risk of Disease Infections

There is a high risk of incidence of diseases in a closed soilless system compared with an open soilless system. The presence of fertilizers and high humidity creates an environment that stimulates growth of *Salmonella* and infection by *Verticillium* wilt. All these are limitations in soilless strawberry culture.

23.3 SOILLESS CULTURE SYSTEMS

Soilless culture is a broad term that involves different systems ranging from the simple, manual or semi-automatic to the refined and fully automatic ones. Not every system is suitable for all crops. Broadly, there are two different types of soilless culture:

- Water culture/solution culture: If the soilless culture uses only a nutrient solution, this system is categorized as water culture or solution culture.
- Substrate culture/aggregate culture: In this system, solid inorganic media such as sand, gravel, perlite and vermiculite or a solid organic medium such as peat moss is utilized for providing support to plants.

The major advantage of the solution culture system over a substrate culture is that a large volume of nutrient solution always remains in contact with the root system, providing adequate water and nutrient supply. The major disadvantage associated with solution culture is the difficulty of providing air supply and good support to the plant roots. If the nutrient solution is recycled, the system is called a closed soilless culture system, while if the nutrient solution is discarded after one use, it is called an open soilless culture system. Closed soilless culture systems sometimes create problems because of the changing composition of the nutrient solution with each use and cause the rapid spread of several diseases. Therefore, an open system is considered more reliable than closed systems. Pathogens and beneficial microorganisms recolonize the nutrient solution after disinfection. Therefore, it would be better to use methods stimulating microbiologically balanced growing environments by using more precise disinfection techniques such as slow sand filtration. This technique has been observed to get rid of pathogens without destroying the whole natural microflora (Van Os et al., 1997). Recirculation of the nutrient solution promotes environmental and economic advantages such as savings in water and nutrients. There are limitations in such system such as great risk of spreading pathogens through the nutrient solution and accumulation of organic compounds beyond limits, which could be phytotoxic. The propagules of pathogens, such as *Phytophthora cactorum*, can be removed by slow sand filtration and disease severity is consequently reduced (Martinez et al., 2010). To protect strawberry plants from *P. cactorum* or *Verticillium dahliae* grown in closed soilless culture system with slow sand filtration, Martinez et al. (2009) observed the presence of *Trichoderma* after inoculation in new and reused substrate, revealing the possibility to protect strawberry roots against *P. cactorum* by enriching the substrate with *Trichoderma.* However, *Trichoderma* was not effective against *V. dahliae.*

The accumulation of organic compounds has been reported in the nutrient solutions of closed soilless production systems (Jensen and Aoalsteinsson, 1992; Waechter-Kristensen et al., 1997). The primary contributors of these organic compounds in the nutrient solution are phenolic acids (group of organic acids) released by plant roots (McCalla and Haskins, 1964) and microorganisms as metabolites or as biotransformation products (Linch, 1976) in the rhizosphere, which are ultimately released in the nutrient solution. Modern disinfection techniques such as heat treatment, ozone treatment and UV radiation are high-tech and expensive. The main merit of slow sand filtration is that stabilizing microflora remain in the nutrient solution. In comparison with other filtration methods, the mechanical, chemical and biological properties may interact in slow sand filtration; however, the limitations associated with this method include the requirement for a large space and higher cost of cleaning of closed systems, which increases with enhanced turbidity of solution (Wohanka, 1992; Runia et al., 1996; Van Os, 1999). Slow sand filtration has been observed to control diseases (Rey et al., 2005). Some of the serious diseases, especially root rot, are less severe in a closed system than in open culture because of the higher amount of bacteria in the closed system (Tu et al., 1999). The populations of microorganisms vary significantly (Koohakan et al., 2004) with soilless culture systems used. The biological control of diseases in soilless culture system is more effective due to there being narrow diversity of microorganisms when compared with conventional culture methods on common soil (Paulitz, 1997). So conditions suitable for beneficial microorganisms can be established, relatively easily in soilless culture systems. The population of microorganisms in soilless systems may vary depending on the type of substrate used (Koohakan et al., 2004). The promotion of bacterial growth may be achieved by synthesis of plant hormones (directly) or antagonism and nutrient competition (indirectly) mechanisms, as the plants require colonization of the growth-promoting organism on the surface of their roots. Therefore, it is essential to enrich the rhizosphere with plant growth-promoting microbial strains during the early stages of plant growth and development. Seasonal changes also influence the type and magnitude of bacterial concentration in the rhizosphere and in the nutrient solution (Waechter-Kristensen et al., 1997). Microbial communities that established early in soilless growing systems become robust and resistant to perturbation (Calvo-Bado et al., 2006). Early introduction of *Pseudomonas chlororaphis* has been observed to be beneficial for the management of *Pythium aphanidermatum* in short-term experiments with soilless culture system (Chatterton et al., 2004). Bacteria and fungi compete for simple plant-derived substrates and develop antagonistic behaviour (Boer et al., 2005). Most of the bacteria colonize on the surface of plant roots (Campbell and Greaves, 1990) and can utilize a wide range of substances, such as carbon or nitrogen sources, and grow relatively faster than other microorganisms (Glick, 1995).

The most common soilless culture system, that is, the open system in wide use nowadays is drip irrigation. Soilless culture systems in water are hydroponics par excellence, in which the roots of the plants are in direct contact with the nutrient solution. In hydroponics systems with a substrate, the roots of plants grow and develop on inert substrates and the nutrient solution is usually applied through drip irrigation. Drip irrigation is the most widely used irrigation system for soilless culture of strawberries (Figure 23.1).

23.4 SOILLESS CULTURE TECHNIQUES

23.4.1 Ebb and Flow Technique/Flood and Drain Technique

In this technique, the nutrient solution is generally drained off three to four times in a day which facilitates the roots with aeration. This technique is good for regeneration of roots in cuttings, facilitating germination of seeds and young plants, which are fed with a simple method of flood and drain water.

23.4.2 Deep Flow Technique

This is also referred to as dynamic root floatation/raceway soilless culture technique. In this technique, plant roots are partially submerged in nutrient solution and the nutrient solution is circulated around the roots by a pump and drained with gravity. It is the most common commercially available technique and consists of polystyrene expanded plates floating on the nutrient solution and aerated intermittently with an air compressor. Polystyrene expanded plates provide mechanical support by holding a few plants and remain floating in liquid nutrient medium. Aeration is very important in this technique as the presence of dark roots is an indicator of poor oxygen supply, which limits water and nutrient uptake, and adversely affects the growth and development of plants (Figure 23.2).

Some limitations associated with this system include the higher cost of the containers, higher rate of water consumption, use of more fertilizer, need for frequent cleaning and replacement of polystyrene plates after each harvest and difficulty in management of pathogenic fungi such as *Pythium* in root infection. Nowadays, corrugated polyvinylchloride (PVC) roofing has become comparatively more popular than polystyrene plates as it needs less frequent replacements. However, this modification requires smaller containers (1–2 m^3), and does not require additional oxygen supply into the

FIGURE 23.1 Protected cultivation of strawberry with DIT.

Soilless Culture

FIGURE 23.2 Nutrient film technique for a soilless production system.

nutrient solution as the PVC roofing is supported on the edges of the container, creating a layer of air between the nutrient solution and the PVC roof, which facilitates continuous aeration of roots located between the two surfaces.

24.4.3 Aeroponics System

Aeroponics is an advance on the traditional hydroponics system, in which the roots are suspended in a tightly closed container instead of being submerged in nutrient solution provided with 16 mm diameter polyethylene hoses inserted inside at 60 cm distance from each other and operated with the help of nebulizers @ 30 litre flow per hour. The nebulization frequency of the aeroponic system is adjusted to keep the target roots moist and the leaves turgid. Round the clock, three minutes of irrigation every five minutes is enough to keep the plants turgid. Though not suitable for soilless strawberry production so far, the technique is adopted for production of mini-tubers in potatoes on the commercial scale (Otazu, 2010; Mateus-Rodriguez et al., 2012).

23.4.4 Aerated Flow Technique

This is a modified and improvised version of deep flow technique with provision for aeration of nutrient solution by special mechanisms. The Japanese "Kyowa technique" is similar to the aerated flow technique and is excellent for growing both leafy vegetables and fruit crops (Figure 23.3).

FIGURE 23.3 Nutrient solution reservoir and pump system.

23.4.5 NUTRIENT FILM TECHNIQUE

This technique was developed by Dr. Allen Cooper during 1965 at the Glasshouse Crops Research Institute, England (Resh, 1995). It is considered as the true soilless culture system in which the plant roots are maintained in direct contact with nutrient solution (Figures 23.2 and 23.4). This system is widely used for fast-growing and short-duration crops. In this system a thin film of nutrient solution flows through channels made of flexible sheets. The seedlings with little growing medium are placed at the centre of the sheet and both edges are drawn to the base of the plants and clipped together to prevent evaporation and to exclude light. The growing medium absorbs the nutrient solution for young plants, and when the plants grow, the roots extend to the channels. The maximum length of the channel is 5–10 m, placed on a slope so that the nutrient solution flows with gravity. At the lower end, the nutrient solution is collected and sent into the reservoir. The nutrient film technique is popular among farmers growing berry fruits as the plants of strawberry do not perform well when water accumulates around their roots. The optimum EC of nutrient solution is recommended (Luisi, 2002) to be 1.6–2 ms/cm, and if the crop is subjected to favourable winter temperatures, then on an average 8% higher yield than with conventional cultivation technique can be obtained, commanding higher prices.

23.4.6 DRIP IRRIGATION TECHNIQUE

In this technique, the plants are grown either on inert or organic substrates. The nutrient solution is cycled into the system and flows closely around the roots through drips or trickles to provide frequent nutrition (six or seven times) to the plants every day. Drip irrigation is the most widely used system in soilless culture worldwide, mainly with rock wool as a substrate (Donnan, 1998). The nutrient solution is supplied to each plant through emitters connected to spaghetti, which is inserted in black polyethylene drip hoses. Watering is done by applying small volumes of nutrient solution directly into the root zone at different time intervals to keep the substrate moist. This system is widely used for production of several fruit crops including strawberries. As rock wool is a relatively expensive substrate with his freight charges for its import, it is advisable to use alternative local media in countries where rock wool is not available. With the same principle as rock wool, natural substrates such as quarry sand or pumice are placed in polyethylene slabs, white outside and black inside. Generally, the size of growing bags is 100–120 cm × 25 cm.

The drip irrigation system is most commonly used in the column system of strawberry production as it facilitates accommodation of more plants per unit area, but is restricted to small plants. Plants growing in a column system must be provided with sufficient light; otherwise, they will be subjected to a low photosynthetic rate with reduced fruit yield. The system is mainly used for production of strawberries, but it is not very widely used among the hydroponics systems, as it demands somewhat enhanced initial investment on installation of system. The columns consist of expanded polystyrene pots of 3.5–4 litre capacity. The pots are stacked one above another through a PVC pipe

FIGURE 23.4 Strawberry production in an NFT system.

Soilless Culture

shaft of 0.5 in diameter. The appropriate height of a column is 1.7 m, with four pots per column. Up to 32 plants can grow in each column of stacked pots (four plants per pot, one in each corner). The pots contain a light substrate, such as pumice or perlite alone or mixed with sand, peat moss, rice hulls or coconut fibre. The nutrient solution is distributed by black polyethylene hoses that run over the columns. On each column, a drip is connected to a connector with four outlet, and 3 mm spaghetti in each outlet. The spaghettis are then anchored to the substrate of the first pot, with three on the substrate of the second pot. When the irrigation system is turned on, the nutrient solution enters at each spaghetti, wetting the substrate of the first and second pot. The drainage of the nutrient solution of the two first pots (the pots have holes in the bottom) wet the substrate of the other pots. It is preferable to have an open system, not closed, to prevent fungal disease in strawberry plants, such as *Botrytis* rot. The time of operation of drip irrigation system varies, depending on weather, age of plants and type of substrate, for example, strawberry plants during the full bloom and fruiting stage require eight irrigations per day, three–five minutes each, four times during morning and rest during afternoon. Over a period of six months, the fruit yield of strawberry may exceed 13 kg per column in soilless system of production.

23.4.7 ROOT MIST TECHNIQUE

In this technique, a mist of nutrient-rich solution is sprayed constantly onto the suspended roots of plants. This technique is also known as aeroponic, and good for initiating rooting of cuttings and for providing optimum oxygen levels into the root zone.

23.4.8 FOG FEED TECHNIQUE

This is similar to the root mist technique but the droplet size in this technique is comparatively smaller.

23.5 SUITABLE CULTIVARS

Although the principles of mineral nutrition in plants is the same for the soil as well as for the soilless culture, but different strawberry cultivars (e.g., Camarosa, Mrak and Selva) have been observed to respond differently to various substrates (Ameri et al., 2012). Camarosa had maximum leaf area, length of petiole, number of runners and total biomass, and Mrak had maximum fruit yield. In Italy, Moncada et al. (2008) also revealed differential responses of different strawberry cultivars grown in similar soilless culture conditions and observed Ventana and MT 99.20.1 (a selection) to have the good adaptability to soilless cultivation system. Miranda et al. (2014) obtained maximum fruit yield from cv. Festival followed by Oso Grande grown in closed hydroponic system. Paranjpe et al. (2003) recommended the Carmine, Treasure, Festival and Camarosa cultivars for protected cultivation, but these should be pollinated by wind, bees and other pollinators. For ensuring adequate pollination, closer planting at 18–36 cm × 40–60 cm is suggested.

23.6 SUBSTRATES FOR SOILLESS CULTURE

Substrate is the solid material used for mechanical support to the plants grown in a soilless culture system. The substrate facilitates comparatively better and stronger root growth than soil. The particle size and distribution are important properties of a substrate. There are different types of substrates, but no one is ideal as none has all the good properties essential for a substrate for soilless culture.

Some of the essential features of a good substrate are to be:

- chemically inert
- loose and not compact

- capable of high moisture retention
- easily available
- economically cheap

Some of the materials used as a substrate in soilless cultures include peat, rock wool or perlite, limestone-type substrates such as limestone and beach sand, cocopeat, coco-coir, cork thin waste mixed with rice hulls and ground reed canary grass. These could be used alone or in variable combinations to suit specific needs. Natural substances such as rock wool or perlite may also be used as substrate. Rock wool has over 90% pore space, so it has a very high water-retaining capacity. Limestone-type substrates (limestone and beach sand) are not suitable for growing plants as they alter the pH of the nutrient solution.

Cocopeat or perlite as a substrate has a high moisture-holding capacity at the upper layer and good air permeability of the lower layer, which provides good nourishment for well-developed roots of strawberries (Jun et al., 2006). As such, significant improvement in number of strawberry fruits has been observed due to application of cocopeat/perlite as substrate (Jun et al., 2006).

Kuisma et al. (2014) recommended the use of ground reed canary grass as a substitute for peat or coir in soilless culture of Elsanta strawberry. However, due to its low water holding capacity, irrigation regime has to be suitably adjusted. Martinez et al. (2013) observed higher microbial activity and pH and reduced fungal population in the substrate containing composted cork thin waste mixed with rice hulls (2:1 v:v). Significantly, maximum biomass and fruit yield, respectively, were observed in Camarosa and Mrak cultivars grown in soilless culture and adding vermicompost to substrates was effective in improving most growth and yield parameters (Ameri et al., 2012).

23.7 PREPARATION OF SOLUTION FOR NUTRITION

23.7.1 COMPOSITION OF NUTRIENT SOLUTION

Addition of nutrients for target plants in soilless culture systems depends upon the requirement for the specific plant nutrient at various stages of growth and development of plants. For this, the requirement for nutrients needs to be analysed, which describes their consumption. It can also be done by analyzing the total nutrient needs of plants, which are quantitatively adjusted according to the rate of growth and the amounts of water to be applied (Hansen, 1978). Thus, the composition and concentration of the nutrient solution depend on culture system, stage of crop development and the prevailing environmental conditions (Coic, 1973; Steiner, 1973). In soilless cultures, any ratio and concentration of ions can be given to the plants provided their precipitation limits for certain combinations are considered (Steiner, 1968). Thus, the selection of concentration of nutrients should be such that water and total ions are absorbed by the plant in the same proportion in which those are present in the solution. A nutrient formulation having 16.3 (13:3.3)- 2-7-3.5-1.8-1.8 me/l N $(NO3:NH4)$-P-K-Ca-Mg-S, respectively was observed to consume comparatively less of all mineral elements compared with other nutrient solutions in a soilless culture system (Yoon et al., 2014).

23.7.2 QUALITY OF WATER

In a soilless culture system, the quality of water refers to the concentrations of specific ions and phytotoxic substances relevant for nutrition of plants and the presence of organisms and other substances that can interfere with functioning and clog the irrigation systems (Tognoni et al., 1998). Regarding the presence of organisms both in water for preparing nutrient solution and in nutrient solution under recirculation, the quality of water can be regulated by heat treatment, UV radiation and membrane filtration. Although relatively cheap chemical treatments such as sodium hypochlorite, chlorine dioxide and copper silver ionization may partly address the problems associated with pathogens, this system has the limitation that there is possibility of introduction and accumulation

Soilless Culture 383

of other elements in closed systems (van Os, 2010). Therefore, it is necessary to carry out chemical analysis of water to be used in preparing the nutrient solution. Our knowledge on the kind and concentration of ions facilitates identification of the nutrients required in the nutrient solution and, therefore, it can accordingly be regulated at the time of preparation of original the nutrient solution. In addition, the chemical analysis of the water facilitates taking decisions on avoiding nutrients which are not needed in the nutrient solution. In specific conditions, the water taken from deep sea can also be used for preparing nutrient solutions for soilless culture system. It has been observed that plants treated with circulated water having an EC of 20 dS/m produced strawberry fruits of better quality with more soluble solids or increased yield by 30% in comparison to the control (Chadirin et al., 2007).

23.7.3 SOURCES OF CHEMICAL COMPOUNDS OR FERTILIZERS FOR PREPARATION OF THE NUTRIENT SOLUTION

A list of commonly used fertilizers and chemical compounds for preparation of nutrient mixtures for use in soilless crop production system and some of their useful characteristics of interest helpful in preparation of nutrient solutions for plant nutrition are summarized in Table 23.1.

23.8 FACTORS INFLUENCING SOILLESS CULTURE SYSTEMS

Composition of the nutrient solution, pH of solution, EC, temperature in the rhizosphere, concentrations of ammonium and nitrate salts, salinity and root aeration are some of the most important factors influencing soilless culture of strawberry.

23.8.1 COMPOSITION OF NUTRIENT MEDIUM

A nutrient medium in soilless production system is an aqueous solution containing mainly inorganic ions from soluble salts of nutrient elements essential for desired growth and development of plants. Essential element have a clear physiological role and their absence prevents the proper completion of the plant's lifecycle (Taiz and Zeiger, 1998). Of the total nutrient elements required, 16 are considered to be essential for normal plant growth and development. These are,

---

TABLE 23.1

Common Fertilizers Containing Macronutrients That are Generally Used for Preparation of Nutrient Solutions for Soilless Culture

Fertilizers	Chemical Formula	Nutrient content (%)	Solubility at 20°C (g/l)
Calcium nitrate	$Ca(NO_3)_2 5H_2O$	N: 15.5, Ca: 19	1290
Potassium nitrate	KNO_3	N: 13; K: 38	316
Magnesium nitrate	$Mg(NO_3)_2 6H_2O)$	N: 11, Mg: 9	760
Ammonium nitrate	NH_4NO_3	N: 35	1920
Monopotassium phosphate	KH_2PO_4	P: 23; K: 28	226
Mono ammonium phosphate	$NH_4H_2PO_4$	N: 12; P: 60	365
Potassium sulphate	K_2SO_4	K: 45; S: 18	111
Magnesium sulphate	$MgSO_4 7H_2O$	Mg: 10; S: 13	335
Ammonium sulphate	$(NH_4)_2SO_4$	N: 21; S: 24	754
Potassium chloride	KCL	K: 60; Cl: 48	330

Source: Modified after Libia et al., 2012, Nutrient solutions for hydroponic systems. In: Asao, T. (Ed.), *Hydroponics - A Standard Methodology for Plant Biological Researches.*

carbon (C), hydrogen (H), oxygen (O), nitrogen (N), phosphorus (P), potassium (K), Calcium (Ca), magnesium (Mg), sulphur (S), iron (Fe), copper (Cu), zinc (Zn), boron (B), molybdenum (Mo), manganese (Mn) and chlorine (Cl). Carbon and oxygen are taken by the plants from the atmosphere while hydrogen is supplemented from water. The rest of the 13 nutrient elements are supplemented through soil so these are to be taken up by the plants from the growth (nutrient) medium. Other elements such as sodium, silicon, vanadium, selenium, cobalt, aluminium and iodine are not essential elements but are considered beneficial because some of them can stimulate plant growth or can regulate the toxic effects of elements at higher concentration, or may replace essential nutrients in a less specific role (Trejo-Tellez et al., 2007). The concentration of different nutrient elements in the nutrient solution regulates its electrical conductivity and osmotic potential. Normally, the nutrient solution for soilless culture contains six essential nutrients, that is, nitrogen (N), phosphorus (P), sulfur (S), potash (K), calcium (Ca) and magnesium (Mg). On the basis of the concept of the mutual ratio of anions (NO_3^-, $H_2PO_4^-$ and SO_4^{2-}) and cations (K^+, Ca_2^+, Mg_2^+), Steiner (1968) postulated the principle of mutual ration between cations and anions. Improper ratio of cations and anions may affect performance of plants negatively and it is considered that the changes in the concentration of one ion must be accompanied by either a corresponding change in ions of the opposite charge and/or a complementary change in other ions of the same charge (Hewitt, 1966). When a particular nutrient solution is applied continuously, plants can take up ions at very low concentrations. Increased, the proportion of the nutrients in the growth medium is neither entirely used by plants nor their uptake has a positive impact on growth and development of plants including fruit production. The composition of the nutrient solution must reflect the uptake ratios of individual elements by crops as the demand and uptake for nutrients vary from species to species (Voogt, 2002) and one to another nutrient solution system. There are several formulations of nutrient solutions and some of the most commonly used formulations are summarized in Table 23.2.

As per estimates, the total quantity of nutrients removed varied from 78–91 kg N, 12–17 kg P, 92–125 kg K, 58–91 kg Ca, and 19–23 kg Mg per hectare respectively by the Idea and Marmolada cultivars of strawberry. Of these nutrient elements, calcium was mainly accumulated in leaves, magnesium was almost equally partitioned between leaves and fruit, while nitrogen, phosphorus and potassium were especially estimated to be in the fruit and roots. Further, only a small proportion of total plant nutrients was estimated in the crowns accounted at the end of fruiting season (Tagliavini et al., 2005). Observations on the effects of vesicular arbuscular mycorrhiza on the vegetative and

---

TABLE 23.2

Ranges of Concentration of Essential Mineral Elements

Nutrient	Hoagland and Arnon (1938)	Hewitt (1966)	Cooper (1979)	Steiner (1984)
N (mg/l)	210	168	60	168
P (mg/l)	31	41	60	31
K (mg/l)	234	156	300	273
Ca (mg/l)	160	160	170–185	180
Mg (mg/l)	34	36	50	48
S (mg/l)	64	48	68	336
Fe (mg/l)	2.5	2.8	12	2–4
Cu (mg/l)	0.02	0.064	0.1	0.02
Zn (mg/l)	0.05	0.065	0.1	0.02
Mn (mg/l)	0.5	0.54	2.0	0.62
B (mg/l)	0.5	0.54	0.3	0.44
Mo (mg/l)	0.01	0.04	0.2	–

Source: Hoagland and Arnon, 1938; Cooper, 1988; Steiner, 1984; Windsor and Schwarz, 1990.

Soilless Culture 385

reproductive characteristics of Camarosa and Maraline strawberry cultivars grown in soilless culture revealed that the fruit yield in the Maraline cultivar was increased significantly due to inoculation with mycorrhiza but there was no significant effect on other vegetative growth parameters. But phosphorus and mycorrhizal inoculations resulted in significant decrease in total soluble solids and pH in the fruits of both cultivars, hence it was concluded that mycorrhizal inoculation may cause minor biotic stress on the strawberry plant (Cekic and Yilmaz, 2011).

23.8.2 pH of the Nutrient Solution

The pH of the nutrient solution is the measure of the acidity or alkalinity, and it indicates the concentration of free H^+ and OH^- ions. In soils, a Trough diagram gives the effect of pH on status of availability of nutrients to the plants. Therefore, the variation in the pH of a nutrient solution influences its composition, elemental specificity and availability of different nutrient elements to plants grown in the medium. An important feature of the nutrient solutions is that these must contain ions in solution and in suitable chemical forms that can be absorbed by plants. Therefore, in soilless culture systems, the productivity of the target crop is closely related with regulation of pH and nutrient uptake from the solution (Marschner, 1995). Each nutrient element has differential responses to changes in pH of the nutrient solution, for example, in a nutrient solution, NH_3 forms a complex only with H^+ and at pH ranging from two and seven, it remains present entirely in the form of NH_4^+. When the pH of the nutrient solution increases up to or above seven, the concentration of NH_4^+ decreases with corresponding increase in the concentration of NH_3 (De Rijck and Schrevens, 1999). Nitrification has been observed to be significantly influenced (Tyson et al., 2007) by pH of nutrient medium. The increased levels of un-ionized ammonia at higher pH (8.5) levels reduces the uptake of plant nutrient.

The availability of phosphorus is strongly dependent on pH of the solution and the ideal pH for maximum availability of phosphorus to plants (De Rijck and Schrevens, 1997) ranges from five to six. On inert substrates, the availability of phosphorus to the plants is maximum when the pH of the solution is slightly acidic but decreases significantly in alkaline and highly acidic solutions (Dysko et al., 2008).

Potassium is present as a free ion in the nutrient solution at pH ranging between two and nine and only small amounts of K^+ form a soluble complex with SO_2^{4-} or is bound to Cl^- ions (De Rijck and Schrevens, 1998). Similarly, calcium and magnesium are also available to plants in a wide range of pH, however, the presence of other ions interferes with their availability due to the formation of compounds with different levels of solubility. Water naturally contains HCO_3^-, which is converted into CO_2 at >8.3 pH or into H_2CO_3 at <3.5 pH of the nutrient medium. Thus,, the Ca_2^+ and Mg^{2+} ions precipitate as carbonates at >8.3 pH and lead to clogging of the drippers or emitters of the irrigation/fertigation systems (Ayers and Westcot, 1987). In the increased pH condition of the nutrient solution, the concentration of HPO_4^{2-} ions predominates which gets precipitated with Ca_2^+ if the product of the concentration of these two ions exceeds 2.2 mol/m^3 (Steiner, 1984). Similarly, as the pH of nutrient solution increases from two to nine, the amount of SO_4^{2-}, forming soluble complexes with Mg_2^+ as $MgSO_4$ and with K^+ as KSO_4^- increases (De Rijck and Schrevens, 1999). Most of the micronutrients such as iron, copper, zinc, boron, and manganese become unavailable if the pH of the nutrient solution increases above 6.5 (Timmons et al., 2002; Tyson, 2007). The optimum pH of nutrient solution for satisfactory growth and development of most of the crops ranges between 5.5 and 6.5 and the availability of some of the nutrient elements such as Fe_2^+, Mn_2^+, PO_4^{3-}, Ca_2^+ and Mg_2^+ above seven pH may be restricted due to their precipitation into insoluble and unavailable salts (Resh, 2004). The pH and EC of nutrient solution and dry and fresh weight of fruits of strawberry cv. Seolhyang have been observed to be weakly correlated with changes in the concentration of magnesium in the nutrient solution (Regagba et al., 2014). In addition, the availability of nutrient elements in the form of ions also improves with increase in pH of the nutrient solution.

23.8.3 EC of the Nutrient Solution

The vegetative growth, reproductive development and production of plants are regulated by the concentration of ions in the growth (nutrient solution) medium (Steiner, 1961). The amount of ions of salts dissolved in the nutrient solution regulates the osmotic pressure, which is governed by the amount of solutes dissolved (Landowne, 2006) in the growth medium. The solutes reduce the free energy of water by diluting the water (Taiz and Zeiger, 1998). The terms solute or osmotic potential of a solution represents the effect of dissolved solutes on water potential and osmotic pressure and osmotic potential are used interchangeably (Sandoval et al., 2007).

EC estimates the osmotic pressure of the nutrient solution and is an index of salt concentration indicating the total amount of salts in a nutrient solution. Accordingly, EC of a nutrient solution could be considered as a good indicator of the amount of available ions in the rhizosphere solution (Nemali and Van Iersel, 2004). The osmotic pressure of a nutrient solution from EC can be estimated by the empirical relations Sandoval et al. (2007) as detailed below:

$$OP(atm) = 0.36 \times EC(in \ dS/m \ at \ 25°C)$$

$$OP(bar) = -0.36 \times EC(in \ dS/m \ at \ 25°C)$$

$$OP(MPa) = OP(bars) \times 0.1$$

The ions associated with EC are Ca^{2+}, Mg^{2+}, K^+, Na^+, H^+, NO_3^-, SO_4^{2-}, Cl^-, HCO_3^- and OH^- (Anonymous, 2001). The ratio of supply of micronutrients, such as iron, copper, zinc, manganese, boron, molybdenum and nickel, is comparatively much less than the macronutrients, so the effects of these nutrient elements on EC is negligible (Sonneveld and Voogt, 2009). Although the ideal EC is specific for each crop and dependent on environmental conditions, the required EC for a soilless culture system ranges from 1.5–2.5 ds/m. However, higher EC increases osmotic pressure and thereby restricts nutrient uptake, whereas lower EC may severely affect plant growth and yield (Samarakoon et al., 2006). A decrease in water uptake is strongly and linearly correlated with EC (Dalton et al., 1997). The threshold limit with respect to EC for different salinity groups using strawberry as sensitive crop is given in Table 23.3 (Jensen and Collins, 1985; Tanji, 1990).

The fruit quality such as soluble solids content, titratable acidity and dry matter content are improved due to increase (from 2–10 dS/m) in the EC level of the nutrient solution (Anna et al., 2003). It has been observed that the fruit quality in terms of sugar content in strawberries grown in soilless media was least influenced by the percentage of unmarketable fruits and fruit firmness increased with increase in electrical conductivity of the nutrient solution. This could be the reason behind use of deep-sea water for preparation of nutrient solution as it contains a high amount of Na^+, Mg_2^+, K^+ and Ca_2^+ (Chadirin et al., 2007).

TABLE 23.3
Threshold EC for Various Crop Salinity Groups

Salinity group	Threshold EC (dS/m)
Sensitive (Strawberry)	1.4
Moderately sensitive	3.0
Moderately tolerant	6.0
Tolerant	10.0

Source: Jensen and Collins (1985); Tanji (1990).

23.8.4 Temperature of the Nutrient Solution

The temperature of the nutrient solution influences the uptake of water and nutrients by the candidate crops in various ways. A low temperature of nutrient solutions decreases the uptake of water but increase the uptake of NO_3^- ions and encourage the production of thin white roots. The temperature of the nutrient solution also influences the photosynthetic apparatus. The effective quantum yield and the fraction of open Photo System II (PS II) have been observed to be comparatively greater in plants grown in a cold solution (Calatayud et al., 2008). In the long run, the solubility of oxygen is reduced while enzyme-mediated oxidation of phenolic compounds in root epidermal and cortex tissues are stimulated, but nutrient concentration in root xylem sap diminishes with reduced concentration of nitrogen, potassium and calcium compared with the nutrient solution (Falah et al., 2010) at higher temperatures. The reduction in requirement of nutrients is a consequence of reduction in a number of physiological processes, including respiration, which further influences plant growth. At $\geq 22°C$, oxygen demand is not covered by the nutrient solution as higher temperatures increase the diffusion of oxygen. At high temperatures of the nutrient solution, significant increase in vegetative growth than normal levels has been observed, which reduces fruiting and yield. The direct relationship of temperature with the amount of oxygen consumed by the plant and reverse relationship with dissolved oxygen in the nutrient solution are summarized in Table 23.4 (Libia et al., 2012).

The temperature in the rhizosphere solution greatly influence the uptake and transport of water in the stem (Ali et al., 1994). During morning hours, the temperature in the rhizosphere solution normally remains low which increases during the afternoon hours. Accordingly, the deep placement of plants where fluctuations in daily temperature remains modest can facilitate the plants to escape extremes of hot and cold temperatures in the rhizosphere. The influence of temperature on some physical properties of water (Kafkafi, 2001) are summarized in Table 23.5.

The rapid decline in rate of flow of sap due to cooling of the rhizosphere is a physical phenomenon influenced by the increase in viscosity and decline in permeability of water governed by the changes in the viscosity of root membranes (Kuiper, 1964) or closure of vascular bundles in the root (Johansson et al., 1998; Carvajal et al., 1999). Reduction in the size of stomatal openings alters the energy balance and temperature of the leaves (Hsiao, 1990) and the levels of cytokinins in xylem exudates increases slightly, though not significantly (Ali et al., 1996) due to increase in the rhizosphere temperature. Accordingly, the reduction in the rhizosphere temperature regulates the rate of supply of vital plant growth promoters such as cytokinins to

---

TABLE 23.4

Solubility of Oxygen in Pure Water at 760 mm Hg of Atmospheric Pressure as Influenced Due to Different Temperature Scale

Temperature (°C)	Solubility of Oxygen (mg/l of Pure Water)
10	11.29
15	10.08
20	9.09
25	8.26
30	7.56
35	6.95
40	6.41
45	5.93

Source: Modified from Libia et al. 2012, Nutrient solutions for hydroponic systems. In: Asao, T. (Ed.), *Hydroponics - A Standard Methodology for Plant Biological Researches.*

TABLE 23.5
Effect of Temperature on Physical Properties of Water

Temperature (°C)	Specific viscosity	Fluidity (l/poises)	Relative fluidity (%)
0	1.0000	55.80	56.0
8	0.7734	72.15	72.5
12	0.6899	80.89	81.3
16	0.6200	90.00	90.4
20	0.5608	99.50	100

Source: Modified from Kafkafi, 2001, Root zone parameters controlling plant growth in soilless culture. *Acta Horticulturae*, 554: 27–38.

the shoots rather than their concentration in the xylem sap. When the temperature of the atmosphere increases during the warm period of the day, there is enough synthesis of gibberellins, which subsequently remains more or less constant during cool nights. Decrease in gibberellin concentrations in the xylem exudates indicates that synthesis and transport of cytokinins and/or unloading of these plant growth promoters into the xylem vessels is adversely affected by low temperature in the rhizosphere (Milligan and Dale, 1988). Low root temperature during the day adversely affects the production and allocation of plant growth promoters such as cytokinins and gibberellins. Synthesis of cytokinins and to some extent gibberellins occurs in the roots (Van Staden and Davey, 1979), but is reduced due to nitrogen deficiency (Horgan and Wareing, 1980) but the series of events is not properly understood so far as what occurs first: reduction in nitrate uptake or reduction in biosynthesis of plant growth promoters in the roots. In a soilless culture system, efforts should be made to ensure that the temperature, atmospheric humidity, light and wind in the plant canopy do not fluctuate much throughout the period at 20°C at all temperatures in the rhizosphere solutions. The interrelationship of dynamics of water, light and temperature can be briefly elaborated further in the following:

1. A small increase in water flux starts at about five hours before beginning of light period.
2. The water flux increases immediately after light is switched on and reaches maximum at about five or six hours later.
3. The water flux starts declining at about five or six hours before the light is switched off.
4. The sudden reduction (from 20–12°C) in temperature of rhizosphere minimizes the water flux to nearly the same level as in the plants maintained constantly at 12°C.
5. In the roots maintained at 12°C (8°C during night and 16°C during the day), the transport of water is comparatively more during the day than plants that are maintained constantly at 12°C throughout the day and night.

Similarly, the patterns of opening of stomata (Talbot and Zeiger, 1996) has been observed to increase immediately after the light is switched on and reaches maximum about five or six hours later and, further the pattern of opening of stomata declines about five or six hours before the light is switched off.

The opening of stomata during the day is bell shaped (Talbot and Zeiger, 1996). Potassium chloride (KCl) acts as an osmotic solution and its flow into the guard cells during the morning hours regulates the opening of stomata. Potassium chloride is replaced by sucrose at about midday. Stomatal closure starts at five hours before dark and is followed by the decrease in level of sugars from the guard cells. These events are very important in soilless culture and should be verified for practical implementation. It might be possible to save energy, if the rooting media could be heated approximately one hour before sunrise, and can be kept low during the night.

Soilless Culture

23.8.5 Concentration and Uptake of Nitrate (NO_3)

During the entire season, observations on 15 m^3 nutrient solution circulating in a commercial greenhouse of 1000 m^2, revealed that when nitrate concentration is higher than 10 mM (140 ppm N), then its uptake is reduced and increases further with increase in seasonal temperature and radiation (Kafkafi et al., 1984). The uptake and upward transport of soluble potassium, and mainly the anions NO_3^- and $H_2PO_4^-$ increases with the increase in root temperature. In a recirculating nutrient solution system, plants grown in 8 litre sand medium, gave a healthy plant in solution, containing only nitrate (Ali et al., 1994). Addition of ammonium causes damage to the roots and induces magnesium deficiency (Kafkafi et al., 1971). When both ammonium and nitrate are present in the solution, ammonium uptake is preferred (Kafkafi et al., 1971). Sugar and oxygen are consumed during the process of nitrogen metabolism in the plant. The combined effects of rhizosphere temperatures and the ratios of ammonium and nitrate in the nutrient solution has revealed that when only nitrate is present in the rhizosphere solution, then the plants survive and roots remain alive at the rhizosphere temperature ranging from 10–32°C (Ganmore-Newmann and Kafkafi, 1983). The consumption of oxygen (3.5 mole) is comparatively more for metabolism of each mole of NH_4 nitrogen than the NO_3 nitrogen (1.5 mole). When ratio of ammonium in the rhizosphere solution is only 30% (2.1 mM) of the total N, the roots become black at high root temperatures but when total nitrogen is in the form of ammonium only, the roots die at 32°C. The site of metabolism of NH_4–N is roots and that of NO_3–N is partly in the roots but the larger part moves to the leaves, and is metabolized near the source of sugar production At any temperature, the sugar content in roots is lower when entire nitrogen is present only in NH_4 form in the rhizosphere solution as compared to that when entire nitrogen is present only in NO_3 form (Ganmore-Newmann and Kafkafi, 1985). With the increase of temperature in the root zone, the respiration increases and consumes sugar and at 32°C, no sugar remains for the metabolism of ammonium in the roots and; as a result, the ammonia (NH_3) produced in the cytoplasm kills the roots. The presence of ammonium in solution around the root zone is beneficial when the root temperatures are low and might be detrimental at high root temperatures (Ganmore-Newman and Kafkafi, 1980). The concentration of ammonium has great effect on different plants.

23.8.6 Reaction (pH) of the Nutrient Solution

The observations on the influence of NH_4 and NO_3 ions on the pH of the nutrient solution around the rhizosphere has revealed that the pH of nutrient solution around the rhizosphere decreases with increasing concentration of ammonium ions, while it increases with increased concentration of nitrate ions (Marschner and Romheld, 1983). This effect of the form of nitrogen ions is reflected by the pH of the rhizosphere nutrient solution at the point of entry into the roots. In the same plant, one side of the root system that is exposed to protons generated by NO_4 ions, while the other side that is taking up NO_3 ions, the pH of the rhizosphere nutrient solution around the roots increases.

In soilless culture, the pH influences the dissociation of phosphorus. At 5.2pH, nearly 100% of the total phosphorus in solution is present in the form of $H_2PO_4^-$ which proportionately decreases with increase in the pH and remains at around 50% of the total phosphorus in the solution when the pH reaches at 7.2. It should be kept in mind that mainly $H_2PO_4^-$ form of phosphorus is absorbed into the plants (Marschner, 1986). Increasing the ammonium concentration in the rhizosphere nutrient solution increases the phosphorus content in shoots. Root temperature also increases uptake and concentration of phosphorus (Ganmore-Newman and Kafkafi, 1980). The transport of nitrogen from bottom to the top is a function of root temperature. At low root temperature, nitrate accumulates in the roots but its transport to the top is decreased, whereas at high root temperatures, nitrate moves faster to the top and concentrates in the leaves (Ganmore-Newmann and Kafkafi, 1980; Ali et al., 1994). Therefore, ammonium containing solutions are safe at low root temperatures but might be detrimental when the root zone temperature increases.

23.8.7 Anion Competition

The major charge-balancing anion in the plant is chloride and its uptake is reduced with increase in concentrations of nitrate ions (Kafkafi et al., 1982) in the rhizosphere nutrient solution. At increased concentration of NO_3 ions in the solution, it accumulates in the leaves, causing a reduction in the concentration of chloride ions. The concentration of phosphorus in the plant is reduced with increasing concentration of NO_3 ions in the nutrient solution surrounding rhizosphere due to the effect of uptake of NO_3 ions on the pH of nutrient solution that increases the $H_2PO_4^-$ in the solution near the root.

23.8.8 Salinity

Nitrate can serve as an antidote to chloride (Cl^-) toxicity (Kafkafi et al., 1982). Increasing the concentration of NO_3 at very high chloride levels in the solution increases the growth, while the total EC in the solution increases from 1.6–3.2 dS/m (Bar et al., 1987; Xu et al., 2000). Observation on the effects of concentrations of sodium chloride (NaCl), calcium chloride ($CaCl_2$), calcium acetate ($Ca(AC)_2$) and calcium nitrate ($Ca(NO_3)_2$) on the growth and photosynthetic parameters of pot planted strawberry cv. Darselect in hydroponic culture revealed significant decrease in the biomass production, chlorophyll content, transpiration rate and stomatal conductance and these effects became more intense with an increase in the concentration of the quoted salts. The increase in the rate of net photosynthesis, transpiration and stomatal conductance decreased in the order of $Ca(AC)_2$, $Ca(NO_3)_2$, $CaCl_2$ and NaCl. The strawberry had reduced root–shoot ratio under NaCl stress and increased root–shoot ratio under calcium salt stresses when compared with control, which became intensified as the concentrations of the salts increased. The calcium salt stresses affected the strawberry more strongly than the NaCl stress, and the adverse effect was least due to NaCl stress and the most due to $Ca(AC)_2$ stress (Li et al., 2006).

23.8.9 Root Aeration

Root pressure during the night hours plays an important role in transport of calcium to the developing fruit and shortage of aeration in the root zone reduces root pressure and increases the frequency and severity of occurrence of calcium deficiency in flowers.

23.8.10 Substrate

Different substrates exert variable effects on strawberries in the soilless culture system and better management of the substrates facilitate induction of physicochemical properties favourable for the growth and development of the roots in strawberries. These materials either alone, such as rice husk (100%) or pumice (100%) or mixed in variable proportion such as rice husks + sand (75:25) or rice husks + pumice (50:50%) are generally used as substrates in soilless culture. Maximum (496.73 g/plant) fruit yield of strawberry has been recorded (Caso et al., 2009) due to use of rice husks (100%) as substrate which was reduced (300.95 g/plant) due to use of mixed substrate (75% rice husks + 25% sand). Results revealed that pumice could be an appropriate substrate for soilless culture of strawberries, but to avoid salinity problem due to frequent watering with the nutrient solution, it is essential to wash the substrate weekly with plain water (Caso et al., 2009).

23.9 MANAGEMENT OF NUTRIENT SOLUTION

Soilless culture of strawberries facilitates more precise control of the variable factors of production and creating increased production of improved quality fruits. This is particularly true for nutrient solution for soilless culture with special reference to temperature, pH, electrical conductivity and oxygen content etc. If these parameters are not controlled properly and timely, advantages can be translated into limitations.

23.9.1 REGULATION OF pH

In soilless culture system of strawberries, the pH of the nutrient solution regulates the availability and uptake of nutrient by plants. Therefore, adjustment of pH should be done day to day as the buffering capacity of soilless culture systems is comparatively less (Urrestarazu, 2004) than in open field cultivation systems. The changes in the pH of a nutrient solution with respect to the requisite balance of anions and cations depends on the differences in the magnitude of nutrient uptake by plants in the soilless culture systems. When the uptake of anions at higher concentrations is comparatively more than cations such as nitrate, then the plant excretes OH^- or HCO_3^- anions, to balance the electrical charges, which causes increase in the pH of the nutrient medium of the soilless culture system and this phenomenon is referred to as the physiological alkalinity (Marschner, 1995). Additional supply of NH_4 fertilizers as a source of nitrogen in the nutrient solution regulates the pH and thereby the availability of nutrient element is ensured. It has been observed (Breteler and Smith, 1974) that supply of NH_4 based fertilizers into the soilless nutrient medium decreases the pH of the solution even in the presence of NO_3 ions. In addition, Lorenzo et al. (2000) the addition of ammonium in a soilless nutrient solution having higher proportion of nitrate ions facilitated the enhanced uptake of total nitrogen during shoot elongation and increased concentration of phosphorus in the roots.

The proportion of total nitrogen added to the nutrient solution in the form of ammonium fertilizers is governed by the type of crops. The adjustment of pH of the nutrient solution by addition of chemicals such as acids to reduce the pH is widely used in the soilless culture systems. The pH of the nutrient solution in a soilless culture system is closely related to the concentration of bicarbonate (HCO_3^-) and carbonate (CO_3^{2-}) ions and; when some suitable acidic solution is applied into it, the CO_3^{2-}ion is transformed into HCO_3^-, and then $HCO_3^-$is converted into carbonic acid (H_2CO_3) because the carbonic acid is partially dissociated in H_2O and CO_2 (De Rijck and Schrevens, 1997). Therefore, the regulation of pH is normally done by using acids such as nitric, sulphuric or phosphoric acid, either individually or in varied combinations.

23.9.2 MANAGEMENT OF EC

EC is influenced by plants as they absorb nutrients and water from the nutrient solution surrounding rhizosphere. Therefore, concentration of some ions may decrease while the concentration of some other ions may increase in the nutrient solution at the same time both in close and open soilless culture systems. It has been observed (Lykas et al., 2001) that the concentration of iron decreases, but that of Ca^{2+}, Mg^{2+} and Cl^- increases, while that of K^+, Ca^{2+} and SO_4^{2-} remains nearly intermediate and does not reach critical levels in a soilless culture system. Increase in electrical conductivity by excess accumulation of ions such as, HCO_3, SO_4 and chlorine has been observed (Zekki et al., 1996) due to recirculation of nutrient solution in an open soilless culture system. The substrates can retain ions, which contribute to increased EC. The use of polyethylene or polypropylene sheets as mulch helps in reducing the consumption of water, increases the calculated water use efficiency, decreases the EC of the substrate (Ansorena, 1994) and is very useful in regulating the EC of the nutrient solution surrounding the rhizosphere (Farina et al., 2003).

Repeated use of nutrient solutions is a common practice for sustainable agricultural production systems (Andriolo et al., 2006), including the soilless culture of strawberries. Repeated use of nutrient solutions through management of EC is accomplished by adding a source of water complement to the drainage system to decrease EC and another complement nutrient solution to obtain the desired EC (Brun et al., 2001) level. A simple model for the changes in ion concentration and EC of a nutrient solution under circulation in a closed-loop soilless culture system has been developed on the basis of the balance equation for nutrient uptake by plants grown in soilless culture system

(Carmassi et al., 2003). The amount of water lost by way of crop evapotranspiration is compensated by refilling the nutrient reconstituting tank with a complete nutrient solution. Frequent monitoring of ions in nutrient solution of a soilless culture system is generally not essential (Bugbee, 2004) but the growth of a large number of harmful microorganisms in the nutrient solution makes it essential to carry out disinfection of nutrient solutions used in soilless culture systems to ensure crop sanitation and avoid the chances of spread of plant diseases (Calmina et al., 2008).

23.9.3 TEMPERATURE CONTROL

The temperature of the nutrient solution in a soilless culture system is directly related to the amount of oxygen consumed by the plants grown in it, but inversely related to the amount of oxygen dissolved in it. Since the temperature regulates the solubility of fertilizers and the capacity of uptake by plant roots, it is a very important factor, especially in extreme weather conditions. Every plant species has its specific minimum, optimum and maximum temperature requirement for growth and fruiting. Accordingly, suitable provision of heating or cooling for balancing the temperature of the nutrient solution is required in the soilless culture system of strawberries. It comprises an underground water pipe for energy saving, a large underground pipe filled with water for regulating the temperature of nutrient solution and a motor unit for circulating the nutrient solution under pressure between the cultivation bed and an underground pipe of water. This way, the temperature of water in the underground pipe (1.5 m below the soil surface) remains stable compared with that of outside greenhouses. During the circulation process, the nutrient solution passing through the smaller pipe gets heated through naturally warm water stored in the larger pipe during winter. On the other hand, the same principle works in the opposite direction during summer and the nutrient solution passing through the smaller pipe gets cooled by the cold water filled in the larger pipe during summer (Hidaka et al., 2008). There are different types of cooling systems, and observations have revealed that counter flow system with double pipes yielded best results in a soilless crop production system (Nam et al., 1996). Different varieties of strawberries respond variably to cooling of nutrient solutions: better productivity of strawberry cv. Sweet Charlie was recorded due to cooling of the nutrient solution to about $12°C$ by using a heating exchange device, while cooling of the nutrient solution affected the productivity of strawberry cv. Campinas (Villela et al., 2004).

Results of exploratory trials have indicated that water from the deeper layers of the sea can be used in the preparation of nutrient solutions as these are rich in nutrient element content. This way, the water from deeper layers of sea could be an alternative for nutrient-cooling systems in hot seasons. Cold deep-sea water pumped inside pipe could be beneficially utilized for reducing temperature of nutrient solution by heat exchange between the nutrient solution and the deep-sea water. After being used for cooling, deep-sea water containing abundant nutrients can be diluted into a standard nutrient solution and used for nutrition of desired plants. The deep-sea water is rich in nutrient elements and its use in irrigation might increase fruit quality (Chadirin et al., 2006).

23.9.4 OXYGENATION OF THE NUTRIENT SOLUTION

With increase in the temperature of nutrient solution the consumption of O_2 by plants also increases, which further increases the relative concentration of CO_2 in the root environment in the absence of adequate root aeration (Morard and Silvestre, 1996). The concentration of oxygen in the nutrient solution is governed by the demand of crop in soilless culture and it is more under enhanced photosynthesis stage (Papadopoulous et al., 1999). Reduction in the level of dissolved oxygen (3 or 4 mg/l) causes inhibition of growth of roots and leads to change in colour from whitish to brown, which is the first symptom of oxygen deficiency (Gislerod and Adams, 1983) in the nutrient solution. In long-term soilless culture, the organic matter content in the nutrient solution is usually increased, which promotes enhanced activity of microorganisms, leading to competition for oxygen in the root zone. But the roots being densely matted within the substrate in the soilless culture system alters the diffusion and supply of oxygen (Bonachela et al., 2010) in the nutrient solution. However, in special circumstances, the growers can resort to oxy-fertigation or oxygenation; an advanced method of oxygen enrichment through a pressurized system (Chun and Takakura, 1994) into the nutrient solution in soilless culture systems.

Soilless crop production is a versatile technology, appropriate for both countryside or backyard production systems to high-tech space stations. Soilless crop production technology can be an efficient technique for production of strawberries in extreme environmental conditions such as desert, hilly, temperate or arctic regions. In non-conventional areas, the soilless strawberry culture could facilitate production of locally grown high-value strawberries for the local population and for distant markets alike. There are some considerations for soilless culture such as its cost-competitiveness with the open field cultivation system. Issues such as artificial lighting, application of plastics and breeding of new cultivars with resistance to or tolerance of biotic and abiotic stresses will further help in increasing productivity and bringing down the cost of cultivation. There are bright prospects for the soilless culture of strawberries for improved production and distribution in non-conventional areas. Availability of credit facilities, public policies supporting soilless culture systems, related logistics, requisite support on various research and development issues and adequate human resource development. Besides economic benefits, the soilless culture of strawberries also facilitates conservation of water and energy, generation of employment, supporting nutritional security programmes and thereby improving quality of life of large masses. In the present circumstances, development and enhanced use of soilless culture system has potential for enhancing the economic well-being of stakeholders across the countries.

REFERENCES

Ali, I.A., Kafkafi, U., Sugimoto, Y. and Inanaga, S. 1994. Response of sand-grown tomato supplied with varying ratios of nitrate/ammonia to constant and variable root temperatures. *Journal of Plant Nutrition*, 17(11): 2001–2024.

Ali, I., Kafkafi, U., Yamaguchi, I., Sugimoto, Y. and Inanaga, S. 1996. Effect of low root temperature on sap flow rate, soluble carbohydrates, nitrate contents and on cytokinins and gibberellin levels in root xylem exudate of sand grown tomato. *Journal of Plant Nutrition*, 19: 619–634.

Ameri, A., Ali, T., Mahmoud, S. and Davarynejad, G.H. 2012. Effect of substrate and cultivar on growth characteristic of strawberry in soilless culture system. *African Journal of Biotechnology*, 11(56): 11960–11966.

Andriolo, J.L., Godoi, R.S., Cogo, C.M., Bortolotto, O.C., Luz, G.L. and Madaloz, J.C. 2006. Growth and development of lettuce plants at high NH_4^+: NO_3^- ratios in the nutrient solution. *Horticultura Brasileira*, 24(3): 352–355.

Anna, F.D., Incalcaterra, G., Moncada, A. and Miceli, A. 2003. Effects of different electrical conductivity levels on strawberry grown in soilless culture. Proceedings of International Symposium on Greenhouse Salinity. *Acta Horticulturae*, 609: 355–360.

Anonymous. 2001. *Soil Quality Test Kit Guide*. Natural Resources Conservation Service, USDA, Washington.

Ansorena, J.M. 1994. *Sustratos: Propiedades y caracterizacion*. Mundi-Prensa, Madrid, Spain: p. 172.

Ayers, C.J. and Westcot, D.W. 1987. *La Calidad del Agua en la Agricultura*. FAO. Serie Riego y Drenaje No. 29, Rome, Italy.

Bar, Y., Kafkafi, U. and Lahav, E. 1987. *Nitrate Nutrition as a Tool to Reduce Chloride Toxicity in Avocado*. South African Avocado Grower Association Yearbook. Vol. 10.

Boer, W., Folman, L.B., Summerbell, R.C. and Boddy, L. 2005. Living in a fungal world: Impact of fungi on soil bacterial niche development. *Microbiology Reviews*, 29: 795–811.

Bonachela, S., Acuña, R.A., Magan, J.J. and Malfa, O. 2010. Oxygen enrichment of nutrient solution of substrate grown vegetable crops under Mediterranean greenhouse conditions: Oxygen content dynamics and crop response. *Spanish Journal of Agricultural Research*, 8(4): 1231–1241.

Breteler, H. and Smith, A.L. 1974. Effect of ammonium nutrition on uptake and metabolism of nitrate in wheat. *Netherlands Journal of Agricultural Science*, 22(1): 73–81.

Brun, R., Settembrino, A., Couve, C. 2001. Recycling of nutrient solutions for Rose (*Rosa hybrida*) in soilless culture. *Acta Horticulturae*, 554: 183–191.

Bugbee, B. 2004. Nutrient management in re-circulating hydroponic culture. *Acta Horticulturae*, 648(1): 99–112.

Calatayud, A., Gorbe, E., Roca, D. and Martínez, P.F. 2008. Effect of two nutrient solution temperatures on nitrate uptake, nitrate reductase activity, NH_4^+ concentration and chlorophyll a fluorescence in rose plants. *Environmental and Experimental Botany*, 64(1): 65–74.

Calmina, G., Dennler, G., Belbahri, L., Wigger, A. and Lefort, F. 2008. Molecular identification of microbial communities in the recycled nutrient solution of a tomato glasshouse soil-less culture. *The Open Horticulture Journal*, 1(1): 7–14.

Calvo-Bado, L.A., Petch, G., Parsons, N.R., Morgan, J.A.W., Pettitt, T.R. and Whipps, J.M. 2006. Microbial community responses associated with the development of oomycete plant pathogens on tomato roots in soilless growing systems. *Journal of Applied Microbiology*, 100: 1194–1207.

Campbell, R. and Greaves, M.P. 1990. Anatomy and community structure of the rhizosphere. In: J.M. Lynch (ed.). *The Rhizosphere*. Wiley, England: 11–34.

Carmassi, G., Incrocci, L., Malorgio, M., Tognoni, F. and Pardossi, A. 2003. A simple model for salt accumulation in closed-loop hydroponics. *Acta Horticulturae*, 614(1): 149–154.

Carruthers, S. 2002. Hydroponics as an agricultural production system. *Practical Hydroponics and Greenhouse*, 63: 55–65.

Carvajal, M., Martinez, V. and Francisco, A.C. 1999. Physiological function of water channels as affected by salinity in roots of paprika pepper. *Physiologia Plantarum*, 105(1): 95–101.

Caso, C., Chang, M. and Rodríguez-Delfín, A. 2009. Effect of the growing media on the strawberry production in column system. Proceeding of International Symposium on Soilless Culture and Hydroponics. *Acta Horticulturae*, 843: 337–380.

Cekic, C. and Yilmaz, E. 2011. Effect of arbuscular mycorrhiza and different doses of phosphor on vegetative and generative components of strawberries applied with different phosphor doses in soilless culture. *African Journal of Agricultural Research*, 6(20): 4736–4739.

Chadirin, Y., Suhardiyano, H. and Matsuoka, T. 2006. Application of deep sea water for nutrient cooling system in hydroponic culture. *Proceedings of APAARI the International Symposium on Sustainable Agriculture in Asia, Bogor, Indonesia*, 18–21 September 2006: pp. 1–4.

Chadirin, Y., Matsuoka, T., Suhardiyanto, H. and Susila, A.D. 2007. Application of deep sea water (DSW) for nutrient supplement in hydroponics cultivation of tomato: Effect of supplemented DSW at Different EC Levels on Fruit Properties. *Bulletin Agronomy*, 35(2): 18–126.

Chatterton, S., Sutton, J.C. and Boland, G.J. 2004. Timing Pseudomonas chlororaphis applications to control Pythium aphanidermatum, Pythium dissotocum, and root rot in hydroponic peppers. *Biological Control*, 30: 360–373.

Chun, C. and Takakura, T. 1994. Rate of root respiration of lettuce under various dissolved oxygen concentrations in hydroponics. *Environment Control in Biology*, 32(2): 125–135.

Coic, Y. 1973. Les problèmes de composition et de concentration des solutions nutritives en culture sans sol. *Proceedings of IWOSC 1973, 3rd International Congress on Soilless Culture*, Sassari, Italy, 7–12 May 1973: pp. 158–164.

Cooper, A. 1988. The system. In: Grower Books (ed.): *The ABC of NFT: Nutrient Film Technique*. London, England: 3–123.

Cooper, A.J. 1979. *The ABC of NFT*, Growers Books, London, 181p.

Dalton, F.N., Maggio, A. and Piccinni, G. 1997. Effect of root temperature on plant response functions for tomato: Comparison of static and dynamic salinity stress indices. *Plant and Soil*, 192(2).

De Rijck, G. and Schrevens, E. 1997. pH influenced by the elemental composition of nutrient solutions. *Journal of Plant Nutrition*, 20(7–8): 911–923.

De Rijck, G. and Schrevens, E. 1998. Cationic speciation in nutrient solutions as a function of pH. *Journal of Plant Nutrition*, 21(5): 861–870.

De Rijck, G. and Schrevens, E. 1999. Anion speciation in nutrient solutions as a function of pH. *Journal of Plant Nutrition*, 22(2): 269–279.

Donnan, R. 1998. Hydroponics around the world. *Practical Hydroponics & Greenhouses*, 41: 18–25.

Dysko, J., Kaniszewski, S. and Kowalczyk, W. 2008. The effect of nutrient solution pH on phosphorus availability in soilless culture of tomato. *Journal of Elementology*, 13(2): 189–198.

Falah, M.A.F., Wajima, T., Yasutake, D., Sago, Y. and Kitano, M. 2010. Responses of root uptake to high temperature of tomato plants (*Lycopersicon esculentum* Mill.) in soil-less culture. *Journal of Agricultural Technology*, 6(3): 543–558.

Farina, E., Allera, C., Paterniani, T. and Palagi, M. 2003. Mulching as a technique to reduce salt accumulation in soilless culture. *Acta Horticulturae*, 609(1): 459–466.

Ganmore-Newmann, R. and Kafkafi, U. 1980. Root temperature and ammonium/nitrate ratio effect on tomato. II Nutrient uptake. *Agronomy Journal*, 72: 762–766.

Ganmore-Newmann, R. and Kafkafi, U. 1983. The effect of root temperature and NO_3^-/NH_4^+ ratio on strawberry plants. I. Growth, flowering and root development. *Agronomy Journal*, 75: 941–947.

Ganmore-Newmann, R. and Kafkafi, U. 1985. The effect of root temperature and $NO_3./NH_4^+$ ratio on strawberry plants. II. Nutrient uptake and plant composition. *Agronomy Journal*, 77: 835–840.

Gericke, W.F. 1929. Aquaculture, a means of crop production. *American Journal of Botany*, 16: 862.

Gericke, W.F. 1937. Hydroponics – crop production in liquid media. *Science*, 85: 177–178.

Gislerod, H.R. and Adams, P. 1983. Diurnal variations in the oxygen content and acid requirement of recirculanting nutrient solutions and in the uptake of water and potassium by cucumber and tomato plants. *Scientia Horticulturae,* 21: 311–321.

Glick, B.R. 1995. The enhancement of plant growth by free-living bacteria. *Canadian Journal Microbiology*, 41: 109–117.

Hansen, M. 1978. Plant specific nutrition and preparation of nutrient solutions. *Acta Horticulturae*, 82(1): 109–112.

Hewitt, E.J. 1996. *Sand and Water Culture Methods Used in Study of Plant Nutrition.* 2nd Edition. Commonwealth Agricultural Bureaux, Farnham Royal, Bucks, England. 547p.

Hidaka, K., Kitano, M., Sago, Y., Yasutake, D., Miyauchi, K., Affan, M.F.F., Ochi, M. and Imai, S. 2008. Energy saving temperature control of nutrient solution in soilless culture using an underground water pipe. *Acta Horticulturae*, 797(1): 185–191.

Hoagland, D.R. and Arnon, D.I. 1938. The water culture method for growing plants without soil. *California Agricultural Experiment Station Circulation*, 347: 32.

Horgan, J.M. and Wareing, P.F. 1980. Cytokinins and the growth responses of seedlings of *Betulapendula* Roth. and *Acer pseudoplatanus* L. to nitrogen and phosphorus deficiency. *Journal of Experimental Botany*, 31(121): 525–532.

Hsiao, T.C. 1990. Measurements of plant water status. In: B.A. Stewart and D.R. Nielsen (eds.). *Irrigation of Agricultural Crops.* ASA, CSSA and SSSA, Inc., Madison, Wisconsin: 243–279.

Jensen, M.H. and Collins, W.L. 1985. Hydroponic vegetable production. *Horticultural Reviews*, 7: 483–559.

Jensen, P. and Aoalsteinsson, S. 1992. Root dynamic in lettuce as affected by externally supplied phenolic acids in the nutrient solution. In: *Proceedings of the 8th International Congress on Soilless Culture*, Rustenburg, South Africa.

Johansson, I., Karlsson, M., Shukla, V.K., Chrispeels, M.J., Larsson, C. and Kjellbom, P. 1998. Water transport activity of the plasma membrane aquaporin PM28A is regulated by phosphorylation. *Plantcell*, 10(3): 451–459.

Jun, H., Hwang, J., Son, M., Lee, K. and Udagawa, Y. 2006. Effect of double layered substrate on the growth, yeild and fruit quality of strawberry in elevated hydroponic system. *Korean Journal of Horticulture Science and Technology*, 24(2): 157–161.

Kafkafi, U. 2001. Root zone parameters controlling plant growth in soilless culture. *Acta Horticulturae*, 554: 27–38.

Kafkafi, U., Walerstein, I. and Feigenbaum, S. 1971. Effect of potassium nitrate and ammonium nitrate on the growth, cation uptake and water requirement of tomato grown in sand soil culture. *Israel Journal of Agriculture*, 21: 13–30.

Kafkafi, U., Valoras, N. and Letey, J. 1982. Chloride interaction with nitrate and phosphate nutrition in tomato. *Journal of Plant Nutrition*, 5: 1369–1385.

Kafkafi, U., Dayan, E. and Akiri, B. 1984. Nitrate and phosphate uptake by tomato from nutrient solution in a commercial operation. In: *Proceedings, 6th International Congress on Soilless Culture*, Luntern. ISOSC, Wageningen, the Netherlands.

Koohakan, P., Ikeda, H., Jeanaksorn, T., Tojo, M., Kusakari, S.I., Okada, K. and Sato, S. 2004. Evaluation of the indigenous microorganisms in soilless culture: Occurrence and quantitative characteristics in the different growing systems. *Scientia Horticulturae*, 101: 179–188.

Kuiper, P.J.C. 1964. Water uptake of higher plants as affected by root temperature. *Mede. Landbowhogeschool Wageningen*, 64(4): 1–11.

Kuisma, E., Palonen, P. and Yli-Halla, M. 2014. Reed canary grass straw as a substrate in soilless cultivation of strawberry. *Scientia Horticulturae*, 178: 217–223.

Landowne, D. 2006. *Cell Physiology.* Miami, Florida: McGraw-Hill Medical Publishing Division.

Libia, I., Téllez, T., Fernando, C. and Merino, G. 2012. Nutrient solutions for hydroponic systems. In: Asao, T. (Ed.), *Hydroponics – A Standard Methodology for Plant Biological Researches.* http://www.intechopen.com/books/hydroponics-a-standardmethodology-for-plant-biological-researches/nutrient-solutions-for-hydroponic-systems

Linch, J.M. 1976. Products of soil micro-organisms in relation to plant growth. *CRC Critical Reviews in Microbiology*, 5: 67–107.

Li, Q.Y., Ge, H., Hu, S. and Wang, H.Y. 2006. Effect of sodium and calcium salt stresses on strawberry photosynthesis. *Acta Botanica Boreali Occidentalia Sinica*, 26(8): 1713–1717.

Lorenzo, H., Cid, M.C., Siverio, J.M. and Caballero, M. 2000. Influence of additional ammonium supply on some nutritional aspects in hydroponic rose plants. *The Journal of Agricultural Science*, 134(4): 421–425.

Luisi, P.L. 2002. Emergence in chemistry: Chemistry as the embodiment of emergence. *Foundations of Chemistry*, 4: 183–200.

Lykas, C.H., Giaglaras, P. and Kittas, C. 2001. Nutrient solution management re-circulating soilless culture of rose in mild winter climates. *Acta Horticulturae*, 559(1): 543–548.

Marschner, H. 1986. *Mineral Nutrition of Higher Plants*. London, UK.

Marschner, H. 1995. *Mineral Nutrition of Higher Plants*. Academic Press, New York, New York.

Marschner, H. and Romheld, V. 1983. *In vivo* measurements of root induced pH changes at the soil root interface: Effect of plant species and nitrogen source. *Zeitschrift für Pflanzenphysiologie*, 111: 241–251.

Martínez, F., Flores, F., Vázquez-Ortiz, E. and López-Medina, J. 2009. Persistence of *Trichodermaa sperellum* population in strawberry soilless culture growing systems. Proceedings of VIth International strawberry symposium. *Acta Horticulturae*, 842: 1003–1006.

Martínez, F., Castillo, S., Carmona, E. and Avilés, M. 2010. Dissemination of Phytophthora cactorum, cause of crown rot in strawberry, in open and closed soilless growing systems and the potential for control using slow sand filtration. *Scientia Horticulturae*, 125: 756–760.

Martínez, F., Castillo, S., Borrero, C., Pérez, S., Palencia, P. and Avilés, M. 2013. Effect of different soilless growing systems on the biological properties of growth media in strawberry. *Scientia Horticulturae*, 150: 59–64.

Mateus-Rodríguez, J., de Haan, S., Chuquillanqui, C., Barker, I. and Rodríguez-Delfín, A. 2012. Response of three potato cultivars grown in a novel aeroponics system for minituber seed production. *Acta Horticulturae*, 947: 361–367.

McCalla, T.M. and Haskins, F.A. 1964. Phytotoxic substances: from soil microorganisms and crop residues. *Bacteriological Review*, 28: 181–207.

Milligan, S.P. and Dale, J.E. 1988. The effect of root treatments on growth of the primary leaves of *Phaseolus vulgaris* L.: General features. *New Phytology*, 108: 27–35.

Miranda, F.R., Silva, V.B., Santos, F.S.R., Rossetti, A.G. and Silva, C.F.B. 2014. Production of strawberry cultivars in closed hydroponic systems and coconut fibre substrate. *Revista Ciencia Agronomica*, 45(40): 833–841.

Moncada, A., Miceli, A. and D'Anna, F. 2008. Evaluation of strawberry cultivars in soilless cultivation in Sicily. Proceedings of International Symposium on Greensys. *Acta Horticulturae*, 801: 1121–1128.

Morard, P. and Silvester, J. 1996. Plant injury due to oxygen deficiency in the root environment of soilless culture: A review. *Plant and Soil*, 184(2): 243–254.

Nam, S.W., Kim, M.K. and Son, J.E. 1996. Nutrient solution cooling and its effect on temperature of leaf lettuce in hydroponic system. *Acta Horticulturae*, 440(1): 227–232.

Nemali, K.S. and van Iersel, M.W. 2004. Light intensity and fertilizer concentration: I. Estimating optimal fertilizer concentration from water-use efficiency of wax begonia. *Hort Science*, 39(6): 1287–1292.

Otazu, V. 2010. *Manual on Quality Seed Potato Producing using Aeroponics*. International Potato Centre, Lima, Peru: p. 42.

Papadopoulous, A.P., Hao, X., Tu, J.C. and Zheng, J. 1999. Tomato production in open or closed rockwoolculture systems with NFT or rock wool nutrient feedings. *Acta Horticulturae*, 481(1): 89–96.

Paranjpe, A.V., Cantliffe, D.J., Lamb, E.M., Stoffella, P.J. and Powell, C. 2003. Winter strawberry production in greenhouses using soilless substrate: An alternative to methyl bromide soil fumigation. *Proceedings of the Florida State Horticultural Society*, 116: 98–105.

Paulitz, T.C. 1997. Biological control of root pathogens in soilless and hydroponic systems. *Horticultural Science*, 32: 193–196.

Regagba, Z., Choi, J.M., Latigui, A., Mederbal, K. and Latigui, A. 2014. Effect of various concentrations in nutrient solution on growth and nutrient uptake response of strawberry cv. Seolhyang grown in soilless culture. *Journal of Biological Sciences*, 14(4): 226–236.

Resh, H. 1995. *Hydroponic Food Production*. Woodbridge Press Publishing Company, Santa Barbara, California: p. 527.

Resh, H.M. 2004. *Hydroponic Food Production*. New Concept Press, Inc, Mahwah, New Jersey.

Rey, P., Déniel, F., Guillou, A. and Quillec, S. 2005. Management of bacteria to improve slow sand filtration efficiency in tomato soilless culture. *Acta Horticulturae*, 691: 349–355.

Runia, W.T., Michielsen, J.M.G.P., van Kuik, A.J. and Van Os, E.A. 1996. Elimination of root-infecting pathogens in recirculation water by slow sand filtration. In: *Proceedings of 9th International Jersey.* Congress on Soilless Cultures, St Helier, New Jersey: 395–407.

Samarakoon, U.C., Weerasinghe, P.A. and Weerakkody, A.P. 2006. Effect of electrical conductivity (EC) of the nutrient solution on nutrient uptake, growth and yield of leaf lettuce (*Lactuca sativa* L.) in stationary culture. *Tropical Agricultural Research*, 18(1): 13–21.

Sandoval, V.M., Sánchez, G.P. and Alcántar, G.G. 2007. Principios de la Hidroponía y del Fertirriego. In: G. Alcántar and L.I. Trejo-Téllez (eds.), *Nutrición de Cultivos*. Mundi-Prensa, México: 374–438.

Schwarz, M. 2001. Opening remarks. *Acta Horticulturae*, 554: 23–25.

Sonneveld, C. and Voogt, W. 2009. *Plant Nutrition of Greenhouse Crops*. Springer, New York, New York.

Steiner, A.A. 1961. A universal method for preparing nutrient solutions of a certain desired composition. *Plant and Soil*, 15(2): 134–154.

Steiner, A.A. 1968. Soilless culture. *Proceedings of the IPI 1968, 6th Colloquium of the International Potash Institute*, Florence, Italy: 324–341.

Steiner, A.A. 1973. The selective capacity of tomato plants for ions in a nutrient solution. *Proceedings of IWOSC 1973, 3rd International Congress on Soilless Culture*, Sassari, Italy, 7–12 May: 43–53.

Steiner, A.A. 1984. The universal nutrient solution. *Proceedings of IWOSC 1984, 6th International Congress on Soilless Culture*, Wageningen, the Netherlands, Apr 29–May 5 1984: 633–650.

Tagliavinia, M., Baldia, E., Lucchic, P., Antonellia, M., Sorrentia, G., Baruzzib, G. and Faedib, W. 2005. Dynamics of nutrients uptake by strawberry plants (*Fragaria* × *Ananassa* Dutch.) grown in soil and soilless culture. *European Journal of Agronomy*, 23: 15–25.

Taiz, L. and Zeiger, E. 1998. *Plant Physiology*. Sinauer Associates, Inc. Publishers, Sunderland, Massachusetts.

Talbot, L.D. and Zeiger, E. 1996. Guard cell osmo-regulation. *Plant Physiology*, 111: 1052–1057.

Tanji, K.K. 1990. *Agricultural Salinity Assessment and Management*. American Society of Civil Engineers, New York.

Timmons, M.B., Ebeling, J.M., Wheaton, F.W., Summerfelt, S.T. and Vinci, B.J. 2002. *Re-circulating Aquaculture Systems*. Cayuga Aqua Ventures, Ithaca, New York.

Tognoni, F., Pardossi, A. and Serra, G. 1998. Water pollution and the greenhouse environmental costs. *Acta Horticulturae*, 458(1): 385–394.

Trejo-Téllez, L.I., Gómez-Merino, F.C. and Alcántar, G.G. 2007. Elementos benéficos. In: G. Alcántar and L.I. Trejo-Téllez, L.I. (eds.). *Nutrición de Cultivos*. Mundi-Prensa, México: 50–91.

Tu, J.C., Papadopoulos, A.P., Hao, X. and Zheng, J. 1999. The relationship of *Pythium* root rot and rhizosphere microorganisms in a closed circulating and an open system in rockwool culture of tomato. *Acta Horticulturae*, 481: 577–583.

Tyson, R.V. 2007. Reconciling pH for ammonia biofiltration in a cucumber/tilapia aquaponics system using a perlite medium. *Journal of Plant Nutrition*, 30(6): 901–913.

Tyson, R.V., Simonne, E.H., Davis, M., Lamb, E.M., White, J.M. and Treadwell, D.D. 2007. Effect of nutrient solution, nitrate-nitrogen concentration, and pH on nitrification rate in perlite medium. *Journal of Plant Nutrition*, 30(6): 901–913.

Urrestarazu, M. 2004. *Tratado de cultivo sin suelo*. Mundi-Prensa, Madrid, Spain.

Van Os, E. A. 1999. Soilless growing systems in the Netherlands. *Practical Hydroponics and Greenhouse*, 45: 52–57.

Van Os, E.A., Bruins, M.A., Van Buuren, J., Van der Veer, D.J. and Willers, H. 1997. Physical and chemical measurements in slow sand filters to disinfect recirculating nutrient solutions. In: *Proceedings 9th Int. Congress on Soilless Culture*, Jersey: 313–327.

Van Os, E.A. 2010. Disease management in soilless culture systems. *Acta Horticulturae*, 883(1): 385–393.

Van Staden, J. and Davey, J.E. 1979. The synthesis, transport and metabolism of indigenous cytokinins. *Plant, Cell and Environment*, 2: 93–106.

Villagra, E.L., Minervini, M.G. and Brandán, E.Z. 2012. Determination of response in strawberry (*Fragaria* × *ananassa* Duch. cv. Camarosa) to different nutrition's in conventional agriculture on soil and soilless in vertical sleeves. Proceedings of IInd International Symposium on Soilless Culture and Hydroponics. *Acta Horticulturae*, 947: 173–178.

Villela, J., Luiz, V.E., Araujo, J.A.C. and Factor, T.L. 2004. Nutrient solution cooling evaluation for hydroponic cultivation of strawberry plant. *Engenharia Agrícola*, 24(2): 338–346.

Voogt, W. 2002. Potassium management of vegetables under intensive growth conditions. In: N.S. Pasricha and S.K. Bansal (eds.). *Potassium for Sustainable Crop Production*. International Potash Institute, Bern, Switzerland: 347–362.

Waechter-Kristensen, B., Sundin, P., Gertsson, U.E., Hultberg, M., Khalil, S., Berkelmann-Loehnertz, B. and Wohanka, W. 1997. Management of microbial factors in the rhizosphere and nutrient solution of hydroponically grown tomato. *Acta Horticulturae*, 450: 335–340.

Windsor, G. and Schwarz, M. 1990. *Soilless Culture for Horticultural Crop Production*. FAO, Rome Italy, Plant Production and Protection. Paper 101.

Wohanka, W. 1992. Slow sand filtration and UV-radiation: Low-cost techniques for disinfection of recirculating nutrient solution or surface water. In: *Proceedings of the 8th International Congress on Soilless Culture*, Rustenburg, South Africa: 497–511.

Xu, G., Magen, H., Tarchitzky, J. and Kafkafi, U. 2000. Advances in chloride nutrition of plants. *Advances in Agronomy*, 68: 97–150.

Yoon, H.S., An, J.U., Hwang, Y.H., An, Y.H., Chang, G.M.S. and Rho, C.W. 2014. Improved fertilization strategy for strawberry fertigation culture. Proceedings of 7th International Strawberry Symposium. *Acta Horticulturae*, 1049: 521–528.

Zekki, H., Gauthier, L. and Gosselin, A. 1996. Growth, productivity and mineral composition of hydroponically cultivated greenhouse tomatoes with or without nutrient solution recycling. *Journal of the American Society for Horticultural Science*, 121(6): 1082–1088.

24 Harvesting

Sushil Sharma and Kuldeep Singh

CONTENTS

24.1 Manual Harvesting 399
24.2 Mechanical Harvesting 400
    24.2.1 Night-time Picking Assistance 400
    24.2.2 Picking Machine 400
    24.2.3 Picking Robot 401
References 402

The ultimate objective of a strawberry plantation enterprise is optimized output. As the organization of the entire intercultural operations is both science and art, a timely judgement can make a great difference in this crop. Strawberries are high-paying crops but they also need special operations to make them an economically viable enterprise. Although strawberry fruits are tasty and nutritious, they are very soft and juicy with a highly perishable nature. In ordinary conditions, improper handling leads to spoilage of the crop and may cause great economic losses. Consequently, fruit harvesting is one of the most important operations in strawberry production, and any inefficiency in this operation may lead to reduce overall performance and minimize the good effects of other meticulous production practices. The strawberry fruit becomes rapidly overripe, and once picking gets delayed beyond a certain stage, it is almost impossible to continue, since rotting and overripe berries slow down the picking of the remainder of the crop, while berries picked immature do not have good flavour and appearance (Hughes et al., 1956). Strawberry fruit harvesting can either be done manually or it can be carried out mechanically.

24.1 MANUAL HARVESTING

Strawberry is ready for harvesting at four–six weeks after full blossoming. The time between first bloom and full bloom may be 10–12 days. A great increase in the number of ripe fruits occurs over the first four–six days of the harvest. Being non-climacteric fruits, strawberries are generally harvested when three-fourths of the fruit develops colour (www.almanac.com/plant/strawberries). The ripened red berries for fresh market and processing are picked every third and fourth day, respectively. Berries must be picked during the cool period of the day, by pinching and twisting, not by pulling. Berries are picked with the cap and stem attached to retain their firmness and quality. The stem is pinched off about 7–8 mm above the cap. Harvesting generally lasts up to three weeks. The caps from berries need to be removed, if harvested, for processing. Generally, 15–25 pickers are required to pick the fruit of 1ha of strawberry, depending on how fast the crop volume of berries ripen and pickers' experience. Presently, strawberries are harvested in small trays or baskets. Harvested berries should be kept in a cool place to avoid damage due to excessive heat in the open field.

The pick your own (PYO) method of harvesting has become a choice for growers as well as consumers, although small hectarages with high local demand may suffer from traffic jams, besides being problematic to growers too. Growers sell their produce by weight instead of volume with standard capacity containers to discourage unspecified harvesting containers being brought by the consumers. It is essential for a successful PYO harvesting method to take care about the

399

24.2 MECHANICAL HARVESTING

There are extensive labour requirements for strawberry production, and during the first year of commercial strawberry plantation establishment and production approximately 250 hours per hectare are required. The beginning of harvesting during the second year will require 1310.5 hours per hectare for hired pickers and approximately 325 hours per hectare for U-Pick operations (Kaiser and Ernst, 2014). Mechanical harvesting is a good option to get a good financial return in strawberry growing areas having labour scarcity. The cultivars producing a larger number of fruits on fewer long stems are more suitable for mechanical harvesting than those with a larger number of short stems carrying fewer fruits per stem. The basic functions of the mechanical harvester include stripping berries from the plants, separating them from leaves and other foreign materials and conveying them to transporting containers.

The berry harvesting technology is constantly evolving and advancing for commercial strawberry industry. In fact, one might consider the "boom" in technological advances to have continued unabated since the Industrial Revolution. The specialized equipment that is currently widespread automates many of the strawberry-specific cultivation tasks. Technologies allow for mounded rows with targeted irrigation and plastic mulch. Some of the recent strawberry harvesting interventions follow.

24.2.1 NIGHT-TIME PICKING ASSISTANCE

Night-time picking of high-moisture fruit allows growers to ship berries at the peak of freshness (Ohlemeier, 2014). In many parts of the world, such as Florida and California in the United States, strawberry growers harvest berries during the night to increase the productivity and quality, as availability of labour is a big concern during daytime. Therefore, arrangement of a lighting option is added to the conveyor belt of the strawberry harvest system (Donald, Ore.-based GK Machine Inc). The LED lighting should be of stadium quality. With this intense lighting, growers can begin picking as early as about 3 a.m. Lighting helps the crew to move fast, and growers can pick and grade the berries, and send the produce to the sale outlets before the day temperature rises. Picking during the night and very early hours brings better fruit quality so it works very well. The lights are powered via an inverter that runs through the alternator driven by the conveyor belt.

24.2.2 PICKING MACHINE

The harvesting season of berries generally lasts 30–35 days, and the physical strain caused by constant bending down to pick berries creates stress on workers. To find a way to relieve the worker from a tired back, a machine called the Picking Assistant has been manufactured (Zimmerman, 2011) by PBZ LLC (Paul B. Zimmerman Inc., a family owned company based in Lancaster County, Pennsylvania), which eases the physical strain on a worker, and protects him from the adverse weather as well. Both hands of the worker remain free, with a better view of what she/he is doing. The picker can stick with the task for longer without strain on the back, making the picking of strawberries a more efficient operation.

The machine consists of two high-torque, planetary gear motors powered by a 12-volt DC battery, giving it forward and reverse motion with adjustable height and width of operation. The battery and charger last up to 16 hours of picking. After harvesting, the picker keeps strawberries on trays and the machine can carry up to six of these trays. A tray provided on the back of the machine holds five trays, while the sixth tray remains in front of the picker. It can also be used for other types of operations, such as planting, weeding and row maintenance.

Harvesting 401

The worker lies on the machine with face down, her/his forehead, stomach and legs resting on adjustable pads. The worker's face remains free to view the row as the wheeled machine carries her/him over it. Simultaneously, the worker's hands are also free, and she/he can suitably pick whatever berries are observed. The acceleration of the machine can be controlled with a foot pedal. As the picker moves down the row for picking, a vinyl canopy protects him from sun and rain. Steering is manual. In a straight row, once the machine is pointed, it will stay in position, but if required it can be turned to any direction.

24.2.3 PICKING ROBOT

One of the latest advances in strawberry cultivation is the development of a strawberry picking robot, which not only performs the strawberry picking job, but also increases economic productivity with reduced labour costs by automating the most time-intensive aspect such as picking of fruits in strawberry cultivation. Automated harvesting robots were initially introduced in the 1980s but their software and computational ability have been improved with the passage of time. Thereafter, very good research (Bachche, 2015) on developing automatic robots was done in Japan (1999), Spain (2006) and China (2012) to upgrade the functioning of robots.

The idea for the specific strawberry picking robot is based on the driverless car (www.azorobotics.com). The DARPA (Defence Advanced Research Projects Agency) Grand Challenge (a competition for the development of driverless cars) prompted the formation of Robotic Harvesting, LLC by the California Strawberry Commission. It is based on the idea that if a car can navigate hundreds of kilometres of roads, why can a robot not navigate a few centimetres to pick a strawberry fruit? The robot grippers in strawberry should fulfil the special requirements (Bachche, 2015) such as adaptation to a variety of shapes, maximum adherence and minimal pressure, no berry damage, low maintenance, high reliability, low weight, be approved for contact with foodstuffs, low energy consumption, required positional precision for both gripping and releasing of the product, ease of cleaning and easy and fast ejection of the product (important for products of low weight).

Data collection, image finding and a supporting mobile platform are the important principles of robot functioning. The important components and their functioning of a strawberry harvesting robot are given in Table 24.1.

Limitations of picking robots are:

- difficulty in finding berries hidden under leaves and crown
- collection of rotten or diseased berries
- berry bruising due to grip of berries by picking arm because of its delicate nature
- being a new and advanced technology, being rarely cheaper

The harvested berries are picked in plastic punnets, which are placed in corrugated fibre trays. The punnet-filled trays are kept in a cool place to reduce water loss from the berries (Sharma and Yamdagni, 2000).

TABLE 24.1

Components of Strawberry Picking Robot

S. No.	Components	Functions
1	Stereo vision camera*	For 3D vision to locate the strawberry fruits
2	Software*	Analyzes and differentiates between ripe and unripe strawberry fruits
3	Arm	Analyzes berries with its three fingers clamping on the strawberry fruit
4	Fingers	Pluck berries and place the fruits onto a conveyor belt

* The combination of camera and advanced software improves the performance of picking robot.

REFERENCES

Bachche, S. 2015. Deliberation on design strategies of automatic harvesting systems: a survey. *Robotics*, 4: 194–222.

Childers, N.F., Morris, J.R. and Sibbett, G.S. 1995. *Modern Fruit Science Orchard and Small Fruit Culture*. Horticultural Publications, Gainesville, Florida: pp. 506–535.

Hughes, H.M., Duggan, J.B. and Banwell, M.G. 1956. Strawberry Bulletin 95, HMSO 10, 6D Ministry of Agriculture Fishery and Food, UK.

Kaiser, C. and Ernst, M. 2014. *Strawberries*. Centre for Crop Diversification Cooperative Extension Service University of Kentucky College of Agriculture, Food and Environment, Lexington, Kentucky.

Ohlemeier, D. 2014. Equipment sheds light on strawberry harvest at night. *The Packer*. http://www.thepacker.com/fruit-vegetable-news/Equipment-sheds-light-on-strawberry-harvest-at-night-23

Sharma, R.M. and Yamdagni, R. 2000. *Modern Strawberry Cultivation*. Kalyani Publishers, Ludhiana, India: p. 172.

Zimmerman, K. 2011. Picking Assistant's web page. Visit www.cropcareequipment.com and click on Vegetable Equipment.

25 Yield and Varietal Performance

Sanjeev Kumar, V.K. Tripathi, and Parshant Bakshi

The phenomenal increase in area and production of strawberry in the world has largely been due to the availability of high-yielding and pest- and disease-resistant cultivars. Varied quality, size, colour and flavour of fruits have further helped in its enhanced popularity among general population. Diversified uses of fruits, especially in processed products and confectionery, have helped in making strawberry a choice and most sought-after fruit in the market. All these factors put together contribute to the economic importance of strawberry both at local and international level, and as such, the yield and quality of fruits is a driving force in its economics *vis-à-vis* cultivation and marketing of strawberries. The performance of different cultivars with respect to yield and quality of fruits in different parts of world varies with soil, prevailing climatic conditions in the region and cultural management practices adopted. The significant effects of interaction of genotypes and environment on yield per plant, fruit weight (Pradeepkumar et al., 2002) and quality of fruits are well established. The impact of climatic conditions on strawberry productivity in future will depend on relative changes in temperature, precipitation and UV-B radiation, and the sensitivity of different varieties to all these environmental factors singly or in various combinations.

Strawberry production is influenced by the factors of climate change. There is a wide range of strawberry cultivars, which often show variability in their performance in different areas (Table 25.1). The effects of environmental factors on characteristic performance of strawberry have been reported by many researchers (Jouquand et al., 2008; Ara et al., 2009; Cocco et al., 2010). Due to the relationship between yield and temperature, and between yield and solar radiation, climate change scenarios have been observed to result in reduction in duration of crop cycle (Palencia et al., 2013). Temperature has been observed to have a strong and positive correlations with yield and fruit weight during May. On the other hand, during June, lower temperatures and humidity are needed for high productivity. During the autumn, lower temperature is suitable for inflorescence initiation while during late August, early September and November, high temperatures, water deficit and sunshine are beneficial (Hortynski et al., 1994) for fruiting.

The exposure of strawberry plants to ozone (78 ppb for two hours) can reduce leaf area, number of leaves, rubisco activity, chlorophyll content, plant biomass, root/shoot ratio and the distribution of carbohydrates. The reduced nitrogen concentration in leaves may be related with more distinct reduction in rubisco activity and chlorophyll content in older leaves of strawberry plants (Keutgen et al., 2005).

Ozone (O_3) fumigation (65 ppb for three months in closed chambers) inhibits growth and reduces fruit yield, but these reductions are variety specific. Ozone reduces the total soluble phenol content, especially agrimoniin, pedunculagin and GE-56. The reduction of gallic acid derivatives and ellagitannins in ozone-treated leaves may be explained by a shortage of their precursor substances, phosphoenolpyruvate and erythrose-4-phosphate, or by a consumption of soluble phenols in response to ozone stress (Keutgen and Lenz, 2001; Oertel et al., 2001). An earlier start to chlorophyll degradation and a tendency for reduced leaf numbers in the less ozone-tolerant cultivars indicate an acceleration of the leaf aging processes. The fact that calcium concentrations in the leaves of the more tolerant cultivars have been observed about twice as high as in susceptible cultivars favours the hypothesis that calcium plays a role in tolerance to ozone stress (Oertel et al., 2001).

The position of fruits on the inflorescence influences the decline of fruit size to a larger extent in small-fruited clones than in the large-fruited ones. The size of the fruit is controlled by the dimension of the receptacles and number of achenes. The stimulating effects of achenes are quite different

403

TABLE 25.1
Country-Specific Suitability of Strawberry Cultivars

Country	Cultivars	References
India	Chandler, Camarosa, Gorella, Sweet Charlie, Oso Grande etc.	Sharma et al. (2002), Sharma et al. (2007)
Italy	Alba, Kore, Cifrance Marjolaine, Queen Elisa, Idea, Nadina, Thethis, Darselect, Marmolada (Onebor), Kore, Camarosa, Idea, Miss, Onda, Pajaro, Seascape, Tudla (Milsei), Chandler, Don, Elsanta, Eris, Marascor, Miranda, Diamante, Granda, Madeleine (Civmad), Maya, Paros, Patty etc. Camarosa, Carlsbad, Clea, Tudla etc.	Testoni et al. (2006), Baruzzi et al. (2001), Faedi et al. (2004b) Funaro et al. (2000)
United States	Annapolis, Cavendish, Chambly, Governor Simcoe, Honeoye, Marmolada, Primetime, Seneca, Setter etc.	Kaps et al. (2003)
Estonia	Polka, Korona, Induka, Kent Bounty, Senga Sengana Edu, Siimoni 47, Jonsok, Senga Sengana, Senga Ducita, Hiku, Bounty, Elsanta, Gorilla, Manil, Lovril, Kouril, Bounty, Honeoye, Jonsok, Festivalnaja, Ostara, Pocahontas, Redgauntas, Venta	Kikas and Libek (2004) Karp et al. (2000) Libek and Kikas (1999), Libek (2002)
Poland	Honeoye, Elsanta, Kama, Kent, Vicoda, Pandora, Marmalada, Seal, Senga Sengana etc. Vikat, Kama, Elsanta, Senga Sengana	Maodobry and Bieniasz (2004), Dejwor and Radajewska (2001), Radajewska (2000), Gwozdecki (2000)
Finland	Ciloe, Vima, Reg Zinta etc. Kent, Inga, Lina, Emily	Hietaranta (2006) Hietaranta and Matala (2002)
Egypt	C9, C1, C2 and C6 (Camarosa-derived strains)	Okasha et al. (2003)
Iran	Queen Elisa, Ventana, Paros and Camarosa	Nankali et al. (2014)
Lithuania	Pegasus, Pandora, Tago and Senga Gigana, Oka, Senga Sengana	Uselis and Rasinskiene (2001)
France	Placartfre, Clery, Darlisette, Divine, Sonata, Matis, Daisy, Yamaska	Navatel et al. (2005)
China	Xingxioang, Fengxiang, Guinugan, American-6, Gorella, Allstar, Baijaaosheng, Dana, Maria	Jiang (2001)

in various genotypes, and the fruit weight per achene declines with the inferior blossom position. The larger leaves, a larger photosynthetic area, and thicker petioles and flower stalks are correlated with large berry size as their cells are larger (Hortynski et al., 1991). Plant morphology and anatomy have a close relationship with physiology, influencing plant growth and fruit yield of strawberry. Benihoppe, a high-yielding strawberry bears large leaves, supported by longer and upright petioles, which allow more solar radiation to penetrate the plant canopy (Mochizuki et al., 2014).

Salinity stress adversely influences the plant growth, fruit yield and quality of strawberries, although its magnitude varies with the tolerance of an individual cultivar, as some cultivars restrict the uptake of chlorine ions in leaves (Saied et al., 2005). Reduction of leaf area in strawberry is a common response in salt-stressed plants (Yilmaz and Kina, 2008). On a plant basis, smaller leaf areas could be a consequence of stress induced inhibition of cell division (Verslues and Zhu, 2007) and photosynthetic activity (Yilmaz and Kina, 2008), affecting the overall plant development (Keutgen and Pawelzik, 2009). However, reduction of leaf area in plants subjected to salinity could also be determined by an anticipated leaf abscission as a consequence of hormone-mediated premature senescence and/or ion toxicity (Munns and Tester, 2008; Turhan and Eris, 2005). Low stomatal density, and the consequent constitutively reduced transpirational flux, is a critical determinant of stress tolerance that may allow plants to adapt more effectively to salinity. The reduced transpiration rate of the cultivar Elsanta is functional in delaying the accumulation of chlorine ions in the

Yield and Varietal Performance

shoots (and their effects on leaf senescence) and helps in ameliorating the plant water status under stress. At low/moderate salinity, sodium ion exclusion systems may have also contributed to sustain salt-stress adaptation in the Elsanta cultivar (Orsinia et al., 2012).

Besides, growing systems have great influence on strawberry productivity. Growing strawberries in ethylvinylacetate-covered tunnels, heated with tepid waste water passing through polyethylene piping below the soil surface can increase winter harvests by 50–80% higher than those obtained from the same cultivars in an unheated tunnel, but the response depends upon the suitability of individual variety to grow under protected cultivation. Strawberry row covering with extruded polypropylene-polyamide mesh or spunbonded polypropylene during the autumn, winter and spring has been reported to increase the yield of Earligrow and Sparkle cultivars (Gast and Pollard, 1991). Application of plastic row cover in combination with barley straw mulch (10–12 cm thick) during autumn can provide the best protection in winters with below normal temperature and low snowfall, resulting in higher fruit yield as compared to open cultivation (Turner et al., 1992), mainly due to the higher temperatures (at crown level) under such protections (Moore et al., 1992).

The list of high-yielding cultivars, producing quality fruits are given in Tables 25.2 and 25.3.

Although, strawberry cultivation is basically taken up for fruit production, quality plant production also forms an integral part of strawberry plantations. Accordingly, the strawberry productivity could be assessed both in terms of runner and fruit production.

Yield is a polygenic character and greatly influenced by genetic constitution of varieties and environmental factors including crop management practices. It varies from 14.3 t/ha (cv. Marmalada) in Russia and 15.55 t/ha (cv. Chandler) to 30.9 t/ha (cv. Oso Grande) in India to as high as 43.83 t/ha (cv. Redgauntlet) in Lithuania. The land productivity has been observed to vary from 74.7–78.9 t/ha and 48.7–50.3 t/ha in Sabrina and Antilla cultivars, respectively on one hand, and the water productivity from 16.5–18.3 kg/m³ and from 13.8–14.3 kg/m³ water, applied in Sabrina and Antilla cultivars, respectively (Lozano et al., 2016). Similarly, drip irrigation at 100% of crop evapotranspiration (1.0 "V" volume of water) has been found to give significantly higher fruit yield (57.07 q/ha) when compared with surface irrigation (47.17 q/ha). The fruit yield improves further (78.63 q/ha) due to the increased efficiency of drip irrigation when combined with black polyethylene mulch (Kumar et al., 2005).

Interestingly, great variations in amount of water applied has also been reported ranging from 4071 m³/ha to 15,214 m³/ha (Garcia Morillo et al., 2015), which influence the productivity of strawberries. Similarly, great variations in terms of land productivity ranging from 6 t/ha (Stevens et al., 2011) to as high as 83 t/ha (Serrano et al., 1992) have been reported. Even in terms of water productivity, there is wide variation ranging from 4 kg/m³ (Kirschbaum et al., 2003) to as much as 14.67 kg/m³ of water used (Serrano et al., 1992).

TABLE 25.2

Some High-Yielding Strawberry Cultivars for a Few Specific Countries

Country	Variety	Yield	References
India	Chandler	15.55 t/ha	Nagre et al. (2005)
	Oso Grande	30.9 t/ha	Hassan et al. (2001)
Lithuania	Dukat, Senga Sengana, Zenit, Polka, Induka	11.2–13.4 t/ha	Rasinskiene and Uselis (2000)
	Saulene, Pandora, Senga Sengana,	11–17t/ha	Uselis (2005)
	Honeoye, Polka, Elkat, Dange	43.83 t/ha	Pranckietiene and Pranckietis (2001),
	Redgauntlet	(182 g/plant)	Austin (1991)
	Sparkle		
Poland	Honeoye	1.06 kg/plant	Maodobry and Bieniasz (2004)
Russia	Marmalada	14.3 t/ha	Prichko et al. (2005)
Serbia	Senga Sengana	1.62 kg/plant	Blagojevic (1999)

TABLE 25.3
Country-Specific Cultivars Identified for High-Quality Fruits and Specific Traits

Country	Variety	Trait	References
Italy	Camarosa	Sweetest fruit	Faedi et al. (2004a)
	Queen Elisa	Large berries, high acidity and ascorbic acid	Nuzzi et al. (2005)
	Irma, Queen Elisa, Demetra and Rubea	Good TSS:acid ratio, flavour, firm and colour and resistant to *Phytophthora cactorum, Colletotrichum acutatum, Sphaerotheca macularis* and *Alternaria alternate*	Faedi and Baruzzi (2004)
Poland	Elsanta	High vit. C	Skupien and Oszmianski (2004)
	Kent	High total sugars, polyphenols, total anthocyanin and ellagic acid	Mansy and Zurawicz (2004)
	Dukat	High TSS	
	Alice, Camarosa, Carisma, Onda Cigoulette, Cilady, Cireine, Darselect, Granda, and Evita Madeleine	High fruit quality	
India	Sea Scap	Large fruit size	Nagre et al. (2005)
	Chandler, Sweet Charlie and Florida-90	High vit. C	
United States	Mesabi	Frost tolerant	Robbins (2005)
	Camarosa	High furaneol (2, 5-dimethyl-4-hydroxy-3(2H)-furanon)	Loehndorf et al. (2000)
Australia	Flame and Kabarla	High titratable acidity	Chandler et al. (2003)
China	Meria	Large berry size Large fruits, glossy red colour with high sugars, acids and Vit. C	Zhou (2000)
	Tudla		Liu and Yin (2000)
Lithuania	Korona, Tenira, Polka and Venta	Very large berries	Rasinskiene and Uselis (2000)
Slovenia	Mohawk	High sucrose and TSS	Sturm et al. (2005)

REFERENCES

Ara, T., Haydar, A., Mahmud, H., Khalequzzaman, K.M. and Hossaini, M.M. 2009. Analysis of the different parameters for fruit yield and yield contributing characters in strawberry. *International Journal of Sustainable Crop Production*, 4(5): 15–18.

Austin, M.E. 1991. Rowcovers for "sparkle" strawberries. *HortScience*, 26: 603.

Baruzzi, G., Faedi, W. and Lucchi, P. 2001. Selection of strawberry varieties for cultivation in Italy. *Rivista-di-Frutticoltura e di Ortofloricoltura*, 63: 75–80.

Blagojevic, R. 1999. Biological properties of some strawberry cultivars in the conditions of Nis. *Jugoslovensko Vocarstvo*, 33(1/2): 17–25.

Chandler, C.K., Herrington, M. and Slade, A. 2003. Effect of harvest date on soluble solids and titratable acidity in fruits of strawberry grown in a winter, annual hill production system. *Acta Horticulturae*, 626: 345–346.

Cocco, C., Andriolo, J.L., Erpen, L., Cardoso, F.L. and Casagrande, G.S. 2010. Development and fruit yield of strawberry plants as affected by crown diameter and plantlet growing period. *Pesquisa Agropecuaria Brasileira*, 45: 730–736.

Dejwor, B.I. and Radajewska, B. 2001. The effect of plant age on mean berry weight of six strawberry cultivars. *Folia Horticulturae*, 13(2): 149–157.

Faedi, W. and Baruzzi, G. 2004. New strawberry cultivars from Italian breeding activity. *Acta Horticulturae*, 649: 81–84.

Faedi, W., Baruzzi, G., Capriolo, G., Milano, F.D., Stefano, A.D., Lucchi, P., Maltoni, M.L., et al. 2004a. Varietal studies for strawberry growing in southern Italy. *Rivista di Frutticoltura e di Ortofloricoltura*, 66(4): 20–27.

Faedi, W., Baruzzi, G., Maltoni, M.L., Manzecchi, L., Lucchi, P., Sbright, P. and Turci, P. 2004b. New strawberry varieties and selections tested in Romagna. *Rivista di Frutticoltura e di Ortofloricoltura*, 66: 30–38.

Funaro, M., Mercuri, F. and Spagnolo, G. 2000. Evaluation of strawberry varieties in Calabria. *Informatore Agrario*, 56(27): 41–45.

Garcia Morillo, J.G., Martin, M., Camacho, E., Rodriguez Diaz, J.A. and Montesino, P. 2015. Toward precision irrigation for intensive strawberry cultivation. *Agricultural Water Management*, 151: 43–51.

Gast, K.L.B. and Pollard, J.E. 1991. Rocovers and mulch alter seasonal strawberry carbon metabolism. *HortScience*, 26: 421.

Gwozdecki, J. 2000. Productive evaluation of new strawberry cultivars at Skierniewice. *Zeszyty Naukowe Instytutu Sadownictwa I Kwiaciarstwa w Skierniewicach*, 8: 249–254.

Hassan, G.I., Godara, A.K., Jitender, K. and Huchche, A.D. 2001. Evaluation of different strawberry (*Fragaria x ananassa* Duch.) cultivars under Haryana conditions. *Haryana Journal of Horticultural Sciences*, 30(1/2): 41–43.

Hietaranta, T. 2006. Strawberry cultivars for the Finnish fresh market. *Journal of the American Pomological Society*, 60(1): 20–27.

Hietaranta, T. and Matala, V. 2002. Preliminary varietal trial in Finland- identifying promising strawberry varieties for further testing. *Acta Horticulturae*, 567(1): 211–214.

Hortynski, J.A., Zebrowska, J., Gawronski, J. and Hulewicz, T. 1991. Factors influencing fruit size in the strawberry (*Fragaria ananassa* Duch.). *Euphytica*, 56(1): 67–74.

Hortynski, J.A., Liniewicz, K. and Hulewicz, T. 1994. Influence of some atmospheric factors affecting yield and single fruit weight in strawberry. *Journal of Horticultural Science*, 69: 89–95.

Jiang, Z.J. 2001. The types of strawberry and its cultural systems. *China Fruits*, 2: 47–48.

Jouquand, C., Chandler, C., Plotto, A. and Goodner, K. 2008. A sensory and chemical analysis of fresh strawberries over harvest dates and seasons reveals factors that affect eating quality. *Journal of the American Society for Horticultural Science*, 133: 859–867.

Kaps, M.L., Odneal, M.B. and Byers, P.L. 2003. Strawberry cultivar performance in Missouri, 1997–99. *Journal of the American Pomological Society*, 57(1): 19–25.

Karp, K., Starast, M. and Vernik, R. 2000. Yield and income of different strawberry varieties. *Transactions of the Estonian Agricultural University Agronomy*, 208: 56–59.

Keutgen, N. and Lenz, F. 2001. Responses of strawberry to long-term elevated atmospheric ozone concentrations-I. Leaf gas exchange, chlorophyll fluorescence, and macronutrient contents. *Gartenbauwissenschaft*, 66(1): 27–33.

Keutgen, A.J. and Pawelzik, E. 2009. Impacts of NaCl stress on plant growth and mineral nutrient assimilation in two cultivars of strawberry. *Environmental and Experimental Botany*, 65: 170–176.

Keutgen, A.J., Noga, G. and Pawelzik, E. 2005. Cultivar-specific impairment of strawberry growth, photosynthesis, carbohydrate and nitrogen accumulation by ozone. *Environmental and Experimental Botany*, 53(3): 271–280.

Kikas, A. and Libek, A. 2004. Evaluation of strawberry cultivars at the Polli Horticultural Institute (Estonia). *Acta Journal of the American Pomological Society*, 663(2): 915–918.

Kirschbaum, D.S., Correa, M., Borguez, A.M., Larson, K.D. and Dejon, T.M. 2003. Water requirement and water use efficiency of fresh and waiting bed strawberry plants. *IV International Symposium on Irrigation of Horticultural Crops*, 664: 347–352.

Kumar, S., Sharma, I.P. and Raina, J.N. 2005. Effect of levels and application methods of irrigation and mulch materials on strawberry production in North-West Himalayas. *Journal of the Indian Society of Soil Science*, 53(1): 60–65.

Libek, A. 2002. Evaluation of strawberry cultivars in Estonia. *Acta Journal of the American Pomological Society*, 567: 207–210.

Libek, A. and Kikas, A. 1999. Winter hardiness of strawberry cultivars at Polli. *Transactions of the Estonian Agricultural University, Agronomy*, 203: 102–105.

Liu, S.J. and Yin, K.L. 2000. Preliminary report on the performance of some strawberry varieties in Chongqing area. *South China Fruits*, 29(5): 43–44.

Loehndorf, J.R., Sims, C.A., Chandler, C.K. and Rouseff, R. 2000. Sensory properties and furanone content of strawberry clones grown in Florida. *Proceedings of the Florida State Horticultural Society*, 113: 272–276.

Lozano, D., Natividad, R. and Gavilan, P. 2016. Consumptive water use and irrigation performance of strawberries. *Agricultural Water Management*, 169: 44–51.

Maodobry, M. and Bieniasz, M. 2004. Yield structure of seven strawberry cultivars. *Folia Horticulturae*, 16(1): 79–85.

Masny, A. and Zurawicz, E. 2004. Field performance of selected strawberry genotypes collected at the Research Institute of Pomology and Floriculture (RIPF), Skierniewice, Poland. *Acta Horticulturae*, 649: 147–150.

Mochizuki, Y., Iwasaki Y., Fuke, M. and Ogiwara, I. 2014. Analysis of a high-yielding strawberry (*Fragaria × ananassa* Duch.) Cultivar "Benihoppe" with focus on root dry matter and activity. *Journal of the Japanese Society for Horticultural Sciences*, 83(2): 142–148.

Moore, P.P., Hummel, R.L., Bristow, P.R., Maas, J.L. and Galletta, G.J. 1992. Modern strawberry cultivation. *Acta Horticulturae*, 348: 253–256.

Munns, R. and Tester, M. 2008. Mechanisms of salinity tolerance. *Annual Review of Plant Biology*, 59: 651–681.

Nagre, P.K., Garad, B.V., Bulbule, A.V., Patil, V.S. and Mote, P.U. 2005. Varietal performance of strawberry under Igatpuri conditions of Western Ghat Zone. *Journal of Maharashtra Agricultural University*, 30(1): 120–122.

Nankali, A., Karami, F. and Sarseifi, M. 2014. Evaluation of new strawberry cultivars for Iran. *Acta Horticulturae*, 1049: 423–428.

Navatel, J., Baruzzi, L. and Bardet, A. 2005. Short-day strawberries: new varieties and production management. *Infos Ctifl*, 208: 13–17.

Nuzzi, M., Scalzo, R.L. and Tesoni, A. 2005. How to evaluate the "global quality" of two strawberry cultivars. *Informatore Agrario*, 61: 39–42.

Oertel, B., Keutgen, N. and Lenz, F. 2001. Responses of strawberry to long-term elevated atmospheric ozone concentrations II. Changes of soluble phenol contents in leaves. *Gartenbauwissenschaft*, 66(4): 164–171.

Okasha, K.A., Ragab, M.E., El-Domyati, F.M., Helal, R.M. and Mohamed, N.A. 2003. Production of new strawberry strains via anther culture 2-strains yield performance and fruits physical and chemical characteristics. *Annals of Agricultural Sciences Cario*, 48(1): 269–281.

Orsinia, F., Alnayefa, M., Boñab, S., Maggioc, A. and Gianquintoa, G. 2012. Low stomatal density and reduced transpiration facilitate strawberry adaptation to salinity. *Environmental and Experimental Botany*, 81: 1–10.

Palencia, P., Martinez, F., Medina, J.J. and Medina, J.L. 2013. Strawberry yield efficiency and its correlation with temperature and solar radiation. *Horticultura Brasileira*, 31: 93–99.

Pradeepkumar, T., Babu, D.S. and Aipe, K.C. 2002. Performance of strawberry varieties in Wayanad District of Kerala. *Journal of Tropical Agriculture*, 40(1/2): 51–52.

Pranckietiene, I. and Pranckietis, V. 2001. Evolution of strawberry cultivars by using conventional and organic growing technologies. *Sodininkyste ir Darzininkyste*, 20(3): 103–112.

Prichko, T.G., Chalaya, L.D. and Yakovenko, V.V. 2005. Studies of cultivars of strawberry in the South of Russia. *Sodovodstvo I Vinogradarstvo*, 1: 14–16.

Radajewska, B. 2000. Some biological characteristics and productive value of nine strawberry cultivars. *Folia Horticulturae*, 12(1): 13–18.

Rasinskiene, A. and Uselis, N. 2000. Estimation of biological and economic properties of 16 strawberry varieties. *Sodinikyste ir Darzininkyste*, 19(1): 53–68.

Robbins, J.A. 2005. Evaluation of selected raspberry and strawberry cultivars in Southern Idaho. *Horticulture Tech*, 15(4): 900–903.

Saied, A.S., Keutgen, A.J. and Noga, G. 2005. The influence of NaCl salinity on growth, yield and fruit quality of strawberry cvs. "Elsanta" and "Korona". *Scientia Horticulturae*, 103(3): 289–303.

Serrano, L., Carbonell, X., Save, R., Marpha, O. and Penuelas, J. 1992. Effects of irrigation regimes on the yield and water use of strawberry. *Irrigation Science*, 13: 45–48.

Sharma, R.M., Khajuria, A.K. and Kher, R. 2002. Evaluation of some strawberry (*Fragaria X ananassa* Duch) cultivars in Jammu plains. *Indian Journal of Plant Genetic Resources*, 15(1): 64–66.

Sharma, R.M., Singh, A.K., Sharma, S. and Masoodi, F.A. 2007. *Report on Training and Demonstration on Hi-tech Production and Postharvest Management of Strawberry under Jammu Plains.* Submitted to NHB, Government of India: p. 17.

Skupien, K. and Oszmianski, J. 2004. Comparison of six cultivars of strawberries (*Fragaria x ananassa* Duch.) grown in northwest Poland. *European Food Research and Technology*, 219(1): 66–70.

Stevens, M.D., Black, B.L., Lee-Cox, J.D. and Feuz, D. 2011. Horticultural and economic considerations in the sustainability of three cold-climate strawberry production system. *HortScience*, 46: 445–451.

Sturm, K., Koron, D. and Stamper, F. 2005. Evaluation of internal quality of strawberries, with regards to cultivation technology and degree of ripeness. *SAD Revija za Sadjarstvo Vinogradnistvo in Vinarstvo*, 16(6): 3–5.

Testoni, A., Lovati, F. and Nuzzi, M. 2006. Evaluation of post-harvest quality of strawberries in Italy. *Acta Horticulturae*, 708: 355–358.

Turhan, E. and Eris, A. 2005. Effects of sodium chloride applications and different growth media on ionic composition in strawberry plant. *Journal of Plant Nutrition*, 27: 1653–1665.

Turner, J.M., Stushnoff, C. and Tamino, K.K. 1992. Evaluation of row covers for overwintering of strawberries under prairie conditions. *Canadian Journal of Plant Science*, 72: 871–874.

Uselis, N. 2005. Evaluation of biological and economical properties of strawberry cultivars. *Sodininkyste ir Darzininkyste*, 24(4): 72–80.

Uselis, N. and Rasinskiene, A. 2001. Assessment of biological and economic properties of strawberry varieties. *Sodininkyste ir Darzininkyste*, 20(2): 18–31.

Verslues, P.E. and Zhu, J.K. 2007. New developments in abscisic acid perception and metabolism. *Current Opinion in Plant Biology*, 10: 447–452.

Yilmaz, H. and Kina, A. 2008. The influence of NaCl salinity on some vegetative and chemical changes of strawberry (*Fragaria* x *ananassa* L.). *African Journal of Biotechnology*, 7: 3299–3305.

Zhou, W.Z. 2000. Screening strawberry varieties for cold areas and cultural techniques. *China Fruits*, 3: 53–54.

26 Post-Harvest Handling and Storage

B.V.C. Mahajan and Alemwati Pongener

CONTENTS

26.1 Maturity and Harvesting 411
26.2 Precooling 412
26.3 Grading 413
26.4 Post-Harvest Treatments 413
    26.4.1 Role of Calcium (Pre- and Post-Harvest).. 413
    26.4.2 Waxing/Edible Coating 414
    26.4.3 Heat Treatment 416
    26.4.4 1-MCP.. 416
    26.4.5 Carbon Dioxide 416
    26.4.6 Oxygen 416
    26.4.7 Ozone 417
    26.4.8 Plant Bioregulators and Other Chemicals 417
    26.4.9 Ultrasound.. 417
    26.4.10 Ultraviolet (UV-C) Irradiation... 418
    26.4.11 Chlorine Dioxide 418
    26.4.12 Biocontrol... 419
    26.4.13 Titanium Dioxide 419
26.5 Packaging... 420
26.6 Storage Conditions.. 421
    26.6.1 Modified/Controlled Atmosphere Storage 422
    26.6.2 CO_2 Injury 423
26.7 Diseases.. 423
References.. 424

Strawberries are among the most popular and attractive fruits due to their high visual appeal and desirable flavour. As strawberry is a non-climacteric fruit, it is harvested essentially fully ripe. However, the fruit has a high rate of metabolism and deteriorates in a relatively short time, even without the presence of decay-causing pathogens. The fruits are very soft and very susceptible to mechanical injury, physiological disorders, fungal attack and water loss. The resultant short post-harvest life limits consumer access to and fresh market potential for this valuable fruit crop. The present chapter embodies appropriate post-harvest technologies for successful handling and storage of strawberry in light of available literature.

26.1 MATURITY AND HARVESTING

The harvest date in strawberries is generally determined on the basis of berry surface colour. All berries should be harvested at nearly fully ripe (more than three-quarters red in colour) as sugar content and eating quality do not improve after harvest. Appearance (colour, size, shape, and freedom from defects), firmness, flavour (soluble solids, titratable acidity, and flavour volatiles), and

412   Strawberries

nutritional value such as vitamin C are all important quality characteristics. A minimum 7% soluble solids and/or a maximum 0.8% titratable acidity in fruits are reliable indices of harvest maturity (Kader, 1999).

Strawberries are usually hand harvested and field packed. Berries are harvested with the calyx attached and must be held loosely in the hand to avoid bruising injury and discolouration. Harvest should be as frequent as needed to avoid overmature berries. Fruits should be sorted carefully, to discard any fruit with fungal infection, lesions or injuries. Considering the delicate and highly perishable nature of strawberries, harvesting, sorting, and packing operations should be undertaken simultaneously in the field. The strawberries must be handled with care and harvested fruits should not be dropped but rather placed gently into the container. Studies on fruit quality loss show that most damage occurs in the field during picking and packing. Packaging strawberries in the field has the advantage that berries are handled only once. Therefore, minimizing berry handling is critical to good quality maintenance and post-harvest life.

26.2 PRECOOLING

Good temperature management is the single most important factor in reducing strawberry deterioration and maximizing post-harvest life. The best way to delay spoilage is to quickly remove field heat and maintain the berries at a temperature close to 0°C. Deterioration of the ripe strawberry is enhanced by high fruit temperature, which hastens metabolic activities, decay development and internal breakdown. Failure to maintain produce at a low temperature during handling, storage and transportation results in loss of quality and marketable produce. Rapid removal of field heat within one hour of harvest is therefore essential. It is important to begin cooling within an hour of harvest to avoid loss of quality and reduction in amount of marketable fruit (Maxie et al., 1959). Also, prompt cooling of strawberries immediately after harvest delays the incidence and severity of decay (Nunes et al., 2005). Cooling delays of two, four, six and eight hours reduces marketability by 20%, 37%, 50% and 70%, respectively, after holding the fruit at 25°C. Therefore, the best way to maintain the fruit quality is rapid post-harvest cooling of the berries. Strawberry fruit quality is improved by precooling, through inhibition of decrease in ascorbic acid content, soluble sugar content and titratable acidity. An inefficient system increases the cooling time, thus increasing the operating cost and reducing the marketable produce and quality of the fruit. Therefore, selecting the right system/method of precooling is of paramount importance.

When pallets of berries are placed in large cold storage rooms, a long time is required to cool them. In spite of the fact that the fruits are rather small in volume and round in shape, unacceptably high weight loss occurs when this refrigeration system is used (Guemes et al., 1989). Forced-air cooling is the most common and fastest method to precool the strawberries, wherein cold air is forced to circulate rapidly through the containers, which are packed with the warm berries. The pallets having berries are maintained in such a way so that the cold air passes through the package openings and around the individual berries. The most common design consists of a tunnel, wherein the space is left between two rows of loaded pallets, and covering the top and one end of the tunnel with a tarp. The air is removed with the help of exhaust fans from the tunnel and a slightly negative air pressure is created. The rate and efficiency of cooling system depends on a number of factors, viz., the temperature difference between the fruit and the cold air, the air flow rate, the accessibility of the fruit to the cold air and the dimensions of the air channel.

There is potential for using hydro-cooling as a cooling method for strawberry. It is an effective way to cool strawberries to increase shelf life. According to Ferriera et al. (2006), better fruit quality is obtained through hydro-cooling than forced-air cooling. Recooling of the fruits after temperature abuse during shipment is beneficial in maintaining strawberry quality (Laurin et al., 2005). After cooling, strawberries can be cold stored and kept in good conditions for a maximum of only one week due to susceptibility to many varieties of rotting during and after storage. Berries should be protected from warming when they remain in the field after harvest. Due to their dark colour,

Post-Harvest Handling and Storage   413

strawberries under direct sun exposure absorb heat and get quickly warm to above air temperature. The amount of warming depends on the temperature difference between the berries and air, the duration of sun exposure, the amount of air flow (breezes) over the berries and the presence of moisture in the air and/or on the fruit surface. Berries held at 20°C have only one-quarter–one-half the life expectancy of those held at 0°C and market life is reduced to only a few hours if strawberry fruits are held $\approx$ 30°C.

26.3 GRADING

Fruits having uniform size and colour should be graded accordingly for packing. An admixture of large and small berries tends to look worse than a graded pack of uniform-sized berries. Slightly damaged fruits, with dry, slug or mechanical damage, may be graded out, and sold as second grade, which often meets a ready local market at the time of high prices when fruits are in short supply. As the strawberry is extremely perishable, each fruit should be examined by soundness. Even the smallest infection by *Botrytis cinerea* results in rotting of the whole lot. Berries marked or damaged by boxes or feet, slugs, beetles or birds should also be ruthlessly downgraded and discarded, or put in a separate punnet, and not be mixed with sound fruits. This is particularly important during wet weather when rotting occurs rapidly. The United States Department of Agriculture (USDA) has classified strawberry (Mitcham, 2016) into three different grades:

1. **U.S. No. 1 grade** – Calyx (cap) attached, firm, not overripe or undeveloped, of uniform size, and free of decay and damage. Each fruit must have at least three-quarters of its surface showing pink or red color.
2. **U.S. No. 2 grade** – Free from decay or serious damage and with at least half of each fruit showing pink or red color
3. **U.S. Combination grade** – Mixture of U.S. No. 1 and 2, with at least 80% meeting U.S. No. 1 grade.

26.4 POST-HARVEST TREATMENTS

26.4.1 ROLE OF CALCIUM (PRE- AND POST-HARVEST)

Post-harvest treatment with calcium chloride ($CaCl_2$) increases the percentage of bound calcium in the cell wall while preharvest sprays lead to greater vitamin C content, increase in total calcium, reduction in polygalacturunase and increase in pectin methylesterase activities (Scalon, 1999). Preservation of the cell wall and middle lamella structure by external application of calcium is related to higher uronic acid and sugar contents in the EDTA-soluble fraction, reflecting a slowdown of the process of dissolution of cell walls and thereby increased commercial life of treated fruits (Lara et al., 2004). $CaCl_2$-treated fruits retain the maximum level of total sugars, reducing sugars, non-reducing sugars, acidity and ascorbic acid and have a minimum percentage of rotting fruit and longer shelf life (Upadhayaya and Sanghavi, 2001; Asrey and Jain, 2005). Calcium gluconate (1%) applied in combination with chitosan coating (1.5%) was found to increase the firmness of strawberry with no sign of fungal decay for one week at 10°C or four days at 20°C (Hernandez-Munoz et al., 2006, 2008).

The major fruit rot problem in strawberry is caused by *B. cinerea*. Genetic differences in susceptibility to this fungus exist, but traits such as dry skin, resistance to bruising and firm flesh also have a positive effect on rot control. The beneficial effect of $CaCl_2$ sprays depends on the capacity of the sprayed plant to absorb and distribute calcium in the plant system, especially the fruits. Sprays increased the calcium content of the fruit, delayed post-harvest ripening and increased shelf life of fruits by delaying grey mould development, particularly in a variety with soft fruits (Cheour et al., 1990, 1991). Delay of strawberry ripening and mould development due to calcium treatment has been well demonstrated (Chéour et al., 1990; Eaves and Leefe, 1962). Foliar application of $CaCl_2$ on strawberry plants a few days before harvest increased fruit calcium content and influenced several post-harvest senescence changes involving free sugar, organic acid, anthocyanin contents, texture and electrical conductivity of fruits (Cheour et al., 1990).

Calcium application usually leads to an increase in apoplastic calcium concentration (Poovaiah, 1979) that may affect the structure and functions of cell walls and membranes and certain aspects of cell metabolism (Glenn et al., 1988) and delay leaf senescence (Poovaiah and Leopold, 1973) and fruit ripening (Tingwa and Young, 1974; Paliyath et al., 1984).

In Fern and Cardinal strawberry cultivars, pre-harvest calcium treatments (through soil or foliar application) reduced the decay in fruit held for two days at 4°C after foliar applications of $CaCl_2$. Anthocyanin, free sugar contents and tissue electrical conductivity increased, while titratable acidity and firmness decreased. In both cultivars, calcium treatment delayed ripening and grey mould development. Cheour et al. (1991) applied $CaCl_2$ on Kent and Glooscap cultivars repeatedly, three days, three and six days or three, six and nine days before harvest @ 0, 10 or 20 kg/ha. Calcium treatment influenced amounts of free sugars and organic acids, colour, texture and disease development during storage at 4°C. Calcium application had comparatively more effect on the fruits of the softer "Glooscap" cultivar, which contained relatively low levels of calcium at the time of treatment. The calcium content of the fruit appeared to depend mainly on the ability of the plant to accumulate and distribute calcium. Similarly, García et al. (1996) gave post-harvest dip treatment to strawberry with 1% $CaCl_2$ solution, which increased the fruit calcium content, controlled post-harvest decay and maintained firmness and soluble solid contents without affecting the sensory qualities of the fruits.

26.4.2 Waxing/Edible Coating

Consumer demand for high-quality foods has led to the development of technologies for preserving the quality of perishable products. Preservation of fresh produce through application of edible coatings has been achieved in several items of perishable produce. The edible coating significantly prevents weight loss and decay, maintains the firmness, delays alteration of color (redness) and total soluble solids content and improves the quality and storage features of the strawberries (Nadim et al., 2015; Trevino-Garza et al., 2015). Chitosan-edible active coatings (EACs) coated strawberries presented the best results in microbial growth assays. Sensory qualities (color, flavor, texture, and acceptance) improved and decay rate decreased (p< 0.05) in pectin-EAC, pullulan-EAC, and chitosan-EAC coated strawberries (Trevino-Garza et al., 2015). The mechanisms through which edible films preserve and extend the shelf life of fresh produce include decreased moisture loss, controlled gas exchange, and reduction in produce respiration rate. Strawberry fruits soaked in 1.5% soluble starch could be kept for five days at room temperature. Chitosan coating lowers the decay rate, respiration rate and released C_2H_4 evolution (Romanazzi et al., 2000). Chitosan coatings have been reported to limit fungal decay and delay the ripening of several commodities including strawberry. The direct antifungal effect of chitosan has been observed in *B. cinerea* and even preharvest treatment of strawberry plants has been found to reduce rot during post-harvest storage (Mazaro et al., 2008). Treatment of strawberries with chitosan and $CaCl_2$ can reduce fruit decay of all strawberry cultivars during cold storage (Chaiprasart et al., 2006). Additional reports on the use of chitosan to maintain quality and storage of strawberry fruits have been summarized in Table 26.1. These reconfirm the emerging importance of chitosan as an alternative for management of post-harvest diseases. A bilayer coating on strawberry fruits with wheat gluten and lipids has a significant effect on the retention of firmness, reduces weight loss and controls decay and maintains visual quality of the fruits (Tanada-Palmu and Grosso, 2005). Fruits coated with starch film and stored at ambient temperature showed the least weight loss, with better texture and longer shelf life (Henrique and Cereda, 1999). The use of mucilage coatings also leads to increased shelf life of strawberry fruits (Del-Valle et al., 2005).

TABLE 26.1

Improvement of Quality and Shelf Life of Strawberry Fruits with Post-Harvest Treatments Involving Chitosan

Chitosan Coating	Combination Treatment	Storage Conditions	Altered Physiology in Produce	References
Chitosan (1.5%)	Calcium gluconate (0.5%)	20°C	↓ Surface damage, delayed ripening, colour changes, fungal decay and loss of firmness	Hernandez-Munoz et al. (2006)
Chitosan (1–1.5%)	Calcium gluconate (0.5%)	10°C, 70 ± 5% RH	↓ respiration rate and ripening, and delayed changes in weight loss, firmness, and external colour.	Hernandez-Munoz et al. (2008)
Chitosan (1%)	–	0–1°C, 95–98% RH	↓ Gray mould, Rhizopus rot, and blue mould of strawberries	Romanazzi et al. (2013)
Chitosan (1%)	Carboxymethyl cellulose (1%) or hydroxypropylmethyl cellulose (1%)	11 ± 1°C, 70–75% RH	Delayed change in weight loss, decay, TSS, titratable acidity, and ascorbic acid. Maintained higher total phenolics and total anthocyanins, and inhibited cell wall-degrading enzyme activity	Gol et al. (2013)
Chitosan (0.5–1.5%)	–	5°C & 10°C	↓ fruit decay and extended shelf life by maintaining better fruit quality with higher fruit phenolics, antioxidant enzyme activity and oxygen radical absorbance capacity	Wang and Gao (2013)
Chitosan (1%)	Potassium silicate	2°C	Effective management of gray mould caused by *Botrytis cinerea*	Lopes et al. (2014)

Note: (↑: Increase; ↓: Decrease)

26.4.3 HEAT TREATMENT

Heat treatments have been applied in post-harvest technology of fruits for insect disinfestation, decay control, ripening delay and modification of fruit responses to other stresses. These changes are brought about probably through changes in gene expression and protein synthesis. Heat-treated strawberry fruits show lower weight loss and lower acidity and remain firm with minimum decay at 20°C (Vicente et al., 2002). Dipping fruit in water at 45°C for 15 minutes duration inhibited fruit decay, increased soluble solid content and decreased titratable acidity (Garcia et al., 1995), while hot water and hot air, both at 60°C for 20 seconds, reduced fruit decay (Jing et al., 2010). Heating fruit with hot air at 48°C for 30 minutes followed by treatment with hot water at 44°C for 20 minutes has been found to be effective in the control of *Rhizopus stolonifer* (Tu et al., 2006). Hot water treatment of strawberry fruits improves the resistance of fruits to fungal infection, while hot-air treatment preserved firmness (Lara et al., 2006). This maintenance of fruit firmness under heat treatment has been attributed to a decrease in the activity of cell wall-degrading enzymes and lower expression of expansion genes in strawberry fruits (Dotto et al., 2011).

26.4.4 1-MCP

1-Methylcyclopropene, a competitive inhibitor of ethylene action, binds to the ethylene receptor to regulate tissue responses to ethylene, and has been found to reduce respiration (Tian et al., 2000). Post-harvest treatment of strawberry with 1-MCP ($5-15$ nll^{-1}) extended shelf life by about 35% at 20°C and by 150% at 5°C (Ku et al., 1999). Use of 1-MCP @ $0.6-0.9$ µl/l inhibits the respiration of fruit and maintains high levels of superoxide dismutase activity (Li et al., 2006). Its use also helps to maintain strawberry firmness and colour (Jiang et al., 2001). Exposure to 1-MCP had a synergistic effect in combination with $CaCl_2$ and controlled atmosphere in slowing down fruit softening and deterioration (Aguayo et al., 2006). Despite its benefits in reducing fruit senescence, 1-MCP has been found to accelerate disease development in strawberry through lowering of phenolic compounds (Jiang et al., 2001) and its use has not been found to be cost-effective (Bower et al., 2003). In addition, the use of 1-MCP in higher concentration ($'500$ µl/l) has been found to have deleterious effects on strawberry as evidenced by accelerated loss of quality and decrease in post-harvest life (Ku et al., 1999).

26.4.5 CARBON DIOXIDE

Carbon dioxide (CO_2) shocks could be an alternative to conventional controlled atmosphere (CA) storage of strawberries (Tudela et al., 2003). Numerous reports on the beneficial effects of CO_2 application on post-harvest life of strawberries are available. Atmospheres containing $10-15\%$ CO_2 reduce the growth of *B. cinerea* and extend strawberry storage (Agar et al., 1990). Strawberry fruits treated with CO_2 (20kPa) during storage showed increased anthocyanin concentrations with less intense red flesh (Holcroft and Kader, 1999). Elevated CO_2 (10 to 30%) levels inside storage prevents fruit softening and slows down respiration rate, but levels beyond 30% cause development of off-flavour (Peng and Sutton, 1991). At the same concentration, CO_2 improves the flavour, maintains firmness and prolongs post-harvest life (Pelayo et al., 2003). Self-produced CO_2, from film-packed strawberries when retained within the package, has been found to delay fruit softening and colour development with significant reduction in decay (Vicente et al., 2003). Allende et al. (2007) also reported beneficial effects of super CO_2 application in maintaining the health-promoting compounds and shelf life of strawberry fruits.

26.4.6 OXYGEN

Strawberry is highly susceptible to attack by microbial organisms causing various types of diseases and fruit decays. The interest in use of a high oxygen atmosphere in post-harvest research has sprung

Post-Harvest Handling and Storage 417

from the knowledge that these conditions are detrimental to the growth of a few specific groups of bacteria and fungi. Treatment of strawberry with 100 kPa O_2 has been found to be the most effective in controlling decay during 14 days of storage (Wszelaki and Mitcham, 2000). Treatment with high oxygen maintains higher levels of superoxide dismutase, catalase and ascorbate peroxidase activities and vitamin C content in addition to delaying senescence and onset of decay of fruits. Strawberry fruits stored in high oxygen atmospheres ('40 kPa) showed higher antioxidant activity, total phenolics, less decay and longer post-harvest life than those stored in normal air (Ayala-Zavala et al., 2007).

26.4.7 OZONE

Ozone (O_3) has been used as an alternative to chlorine because of its better efficacy against a multitude of microorganisms. Besides, unlike chlorine, ozone leaves no toxic residue on the treated fruits. Consequently, numerous studies have already been focused on the bactericidal and fungicidal properties of O_3. Post-harvest treatment of strawberries with O_3 (0.35ppm) resulted in significant reduction in fruit decay for up to four days at 20°C (Perez et al., 1999). O_3 treatment has been found to reduce incidence of fruit decay, weight loss and fruit softening in naturally infected strawberry fruits (Nadas et al., 2003). O_3 treatment @ 0.28–0.70ppm on strawberry cv. Favetta effectively controlled fruit rot caused due to *B. cinerea* by inhibiting spore viability and development of mycelium (Chilosi et al., 2015). Combined application of ozone and high oxygen resulted in better control of fruit decay and keeping quality of strawberry (Wang et al., 2012).

26.4.8 PLANT BIOREGULATORS AND OTHER CHEMICALS

Nitric oxide fumigation @ 5–10 µl/l at ambient conditions helps in extending the post-harvest life of fruits by >50% (Wills et al., 2000) by inhibiting the ethylene production, respiration rate and the activity of ACC synthase (Zhu and Zhou, 2007). It is known that salicylic acid is the signal molecule in induction of systemic acquired resistance in plants, and it also exerts antifungal activity on some plants and harvested fruits. Salicylic acid @ 2 mmol/l is able to reduce post-harvest decay in strawberry and prevents post-harvest loss of fruits for 15 days (Babalar et al., 2007). Salicylic acid has also been found to enhance the biocontrol efficacy of antagonistic yeast, *Rhodotorula glutinis*, against post-harvest rot caused by *R. stolonifer* (Zhang et al., 2010). This has been attributed to the ability of salicylic acid to inhibit the spore germination of the fungus and increasing the activity of strawberry host defense enzymes and cell wall lytic enzymes. Methyl jasmonate (MeJA) is another naturally occurring plant growth regulator that plays an important role in promoting biosynthesis of secondary metabolites, inducing the expression of a set of defence genes, and activating resistance of the host against pathogens. The treatment of strawberries with MeJA (1 µmol/l) significantly inhibited the fruit decay caused due to *B. cinerea* during storage, probably by the induction of defence enzyme activities (Zhang et al., 2006). Apart from protection against pathogens and the subsequent reduction in post-harvest loss due to fruit decay, MeJA also enhances overall flavonoid and antioxidant content in several fruits such as blackberries, raspberries and strawberries. Flores et al. (2013) showed that MeJA stimulated production of phenolic compounds and anthocyanins, which promoted the antioxidant and anti-inflammatory capacity of strawberries.

It is possible to reduce the incidence of post-harvest grey mould on strawberries with a treatment of the vapours of allyl-isothiocyanate (AITC) @ 0.1 mg/l for four hours, opening a potential application of biofumigation in the post-harvest control of *B. cinerea* in strawberry (Ugolini et al., 2014).

26.4.9 ULTRASOUND

The use of ultrasound technology to maintain quality of fruit and vegetables has gathered momentum in recent years. Ultrasound not only improves microbial safety but also physical and chemical properties of the products. Strawberry cv. "Fengziang" treated with ultrasound (40 kHz) had significant reduction in the incidence of decay and number of microorganisms (Cao et al., 2010). Aday et al. (2013) reported that ultrasound treatment ranging from 30–60 W could improve the shelf life and quality of strawberry, while a power level as high as 90 W resulted in detrimental effects on strawberry quality. More recently Aday and Caner (2014) achieved remarkable shelf life (four weeks) in strawberry through combined application of ultrasound (30 W), O_3 (0.075 mg/l), and chlorine dioxide (6 mg/l). Another important aspect of usage of ultrasound technology for enhancing the shelf life of strawberry fruits is that it is not only safer, non-toxic and environmentally friendly, but also perceived as benign by public due to its widespread use in diagnostic imaging for human healthcare.

26.4.10 ULTRAVIOLET (UV-C) IRRADIATION

Ultraviolet irradiation has been used to maintain quality and reduce post-harvest losses in several fruits and vegetables. The dose as well as intensity of UV radiation significantly influences quality of produce through delayed ripening and senescence and reduced spoilage through its effect on genes associated with wound, defence and stress response. Physical treatments involving a low dose (4.6 kJm^{-2}) and heat treatment (45°C for three hours in a hot-air oven) have been assayed separately and in combination on strawberries and have been found to prolong the shelf life and maintain quality parameters (Pan et al., 2003). UV irradiation induces the synthesis of anthocyanins and improves quality in strawberry (Higashio et al., 2005). UV-C (4.35 kJ m^{-2}) treatment induces a variety of changes in the strawberry primarily via gene activation, resulting in firmer fruit with higher levels of bioactive molecules, as well as a stronger aroma. However, UV-C also induced accumulation of the putatively allergenic protein Fra a1 (Severo et al., 2015). UV-C (3.0 kJ/m^2) treatment may trigger hexose signaling transduction, regulate the biosynthesis and accumulation of anthocyanin, and finally contribute to red colour development in post-harvested strawberry fruit (Cai et al., 2015). Accumulation of antioxidants in strawberry has also been reported to be promoted through UV-C treatment (Erkan et al., 2008). The growth of *B. cinerea* is reduced by exposing the strawberries to UV-C with a dose @ 0.01 J cm^{-2} (Marquenie et al., 2002). The reports on successful use of UV-C in strawberry are summarized in Table 26.2.

Strawberry fruits irradiated with blue light at 40 µmol m^{-2} s^{-1} for 12 days at 5°C indicated that blue light treatment improved total anthocyanin content in strawberry fruit during storage. Meanwhile, the treatment increased the activities of glucose-6-phosphate, shikimatedehydrogenase, tyrosine ammonia-lyase, phenylalanine ammonia-lyase, cinnamate-4-hydroxylase, 4-coumarate/coenzyme-A ligase, dihydroflavonol-4-reductase, chalcone synthase, flavanone-3-beta-hydroxylase, anthocyanin synthase and UDP-glucose flavonoid-3-O-glycosyltranferase, which suggested that the enhancement of anthocyanin concentration by blue light might result from the activation of its related enzymes. Blue light might be proposed as a supplemental light source in the storage of strawberry fruit to improve its anthocyanin content (Feng et al., 2014).

26.4.11 CHLORINE DIOXIDE

The use of chlorine as a disinfectant in the food industry has been discouraged due to the generation of toxic byproducts. Besides, the use of aqueous sanitizers has the disadvantage of mould growth due to residual moisture. Chlorine dioxide (ClO_2) has been popularized by several researchers because it does not generate harmful bi-products and has higher biocidal efficacy than chlorine. Gaseous ClO_2 was found to be an effective sanitizer for killing Salmonella, *Escherichia coli*, and *Listeria monocytogenes*, as well as yeasts and moulds on fresh and fresh-cut produce (Sy et al., 2005). Treatment of strawberry with ClO_2 gas is an effective decontamination technique for improving the safety of strawberries while extending shelf life (Han et al., 2004). "Maehyang" strawberry treated with ClO_2 @ 50 mg/l in combination with UV-C (5 kJ m^{-2}) reduced the initial populations of total aerobic bacteria, yeast and moulds (Kim et al., 2010). The effectiveness of ClO_2 for maintaining the quality

TABLE 26.2
Effect of Ultrasound Treatment on Post-Harvest Life and Quality of Strawberry Fruits

UV-C	Variety	Combination Treatment	Storage Condition	Altered Physiology	References
1 kJ m^{-2}	Camarosa	Ozone (5000 mg l^{-1}) and MAP	2°C, 95% RH	↓ Fungal growth, total phenolic content and vitamin- C	Allende et al. (2007)
4.1 kJ m^{-2}	Aroma	–	20°C	Delayed fruit softening. ↓ Expression of genes of cell-wall-degrading enzymes viz. expansins, polygalacturonases, and endoglucanases	Pombo et al. (2009)
5 kJ m^{-2}	Maehyang	ClO$_2$ (50 mg l^{-1}) and fumaric acid (0.5%)	4 ± 1°C, 75 ± 5% RH	↓ population of total aerobic bacteria, yeast, and moulds. ↑ Sensory scores	Kim et al. (2010)
4.1 1 kJ m^{-2}	Toyonoka	–	20°C	↑ Expression and activity of phenylalanine ammonia-lyase and pathogenesis related (PR) proteins	Pombo et al. (2011)
5 kJ m^{-2}	Goha	ClO$_2$ (50ppm) and rice bran protein	4 ± 1°C	↓ population of total aerobic bacteria, yeast, and moulds. Better sensory score, maintained quality	Shin et al. (2012)
4 kJ m^{-2}	Camarosa	–	10°C	↓Decay, colour development, softening. Delayed ripening	Cote et al. (2013)

Note: (↑: Increase; ↓: Decrease).

of strawberry fruits during post-harvest storage has recently been reconfirmed by Shin et al. (2012). ClO$_2$, as a fogging disinfectant, significantly reduced the microbial population on the strawberry fruit surface, and reduced post-harvest diseases of strawberry (Vardar et al., 2012).

26.4.12 BIOCONTROL

Successful post-harvest handling and marketing of strawberry is limited by its short shelf life primarily due to its susceptibility to rot-causing pathogens, most importantly *B. cinerea* and *R. stolonifer.* Use of synthetic chemicals is still the most common means for controlling plant diseases including in strawberry. The use of microbial biocontrol agents has shown great potential as an alternative to chemical control of post-harvest decay of fruits and vegetables (Wisniewski and Wilson, 1992). Several researchers have reported the use of biocontrol agents for control of post-harvest decay in strawberry (Table 26.3).

26.4.13 TITANIUM DIOXIDE

Although strawberry is non-climacteric in nature, exposure to ethylene in storage may induce undesirable secondary ripening characteristics such as tissue softening. Photocatalytic oxidation in the presence of titanium dioxide (TiO$_2$) has been found to be a useful technique to reduce ethylene and CO$_2$ gases in storage rooms. Fruit ripening and senescence is delayed along with quality maintenance when strawberries are stored with TiO$_2$ photo catalyst in closed rooms (Nishizawa et al., 2006). Nano-packing of strawberry achieved through blending polyethylene with anatase TiO$_2$ (nano-powder) was able to maintain the sensory, physio-chemical and physiological quality of strawberry fruits (Yang et al., 2010).

420 Strawberries

TABLE 26.3

Control of Decay of Strawberry Fruits during Post-Harvest Handling and Storage Using Biocontrol Agents

Biocontrol Agent	Combination Treatment	Storage Condition	Effects	References
Candida pulcherima (1×10^3 CFU/ wound)	–	–	↓ lesion and conidiophore development and inhibited spore germination of *Botrytis cinerea*	Guinebretiere et al. (2000)
Cryptococcus laurintii (1×10^8 CFUml^{-1})	Hot water dips (55°C for 20 seconds)	20°C	↓ percentage of infected wounds of *Rhizopus stolonifer* as well as incidence of natural decay	Zhang et al. (2007a)
Rhodotorula glutinis (1×10^9 CFUml^{-1})	–	4°C for 20 days	↓ in incidence of grey mould by 95% and natural development of decay	Zhang et al. (2007a)
Rhodotorula glutinis (1×10^8 CFUml^{-1})	Salicylic acid (100 µg ml^{-1})	4°C & 20°C	↓ in disease incidence and lesion diameter of *Rhizopus* rot ↑ activity of strawberry host defense enzymes	Zhang et al. (2010)
Bacillus MFDÜ-2 (1×10^9 CFUml^{-1})	–	25°C	↓ disease incidence and prevented mycelial development of *Botrytis cinerea*	Donmez et al. (2011)
Sporidiobolus pararoseus strain YCXT3 (1×10^5 yeast cells/ml)	–	20°C	↓ disease incidence by 96–100% by inhibiting conidial germination and mycelial growth of *Botrytis cinerea*	Huang et al. (2012)
Rhodotorula mucilaginosa (4 mol/ml)	Phytic acid	4°C for 7 days	Effectively controlled natural spoilage of strawberry	Zhang et al. (2013)
Cryptococcus laurintii (log 8.ml/l)	Applied as pre-harvest spray at six days before harvest	20°C for 4 days; 4°C for 12 days	↓ in incidence of gray mould by 21% without adversely affecting the fruit quality	Wei et al. (2014)

Note: (↑: Increase; ↓: Decrease).

26.5 PACKAGING

The strawberries are packed in plastic punnets and are placed in the corrugated fibre trays. The punnet-filled trays should be kept in shade or shelter to reduce water loss from the berries. About 200–250 g fruits can be safely packed in a punnet. These packs provide proper ventilation as they have many perforations (10–12 holes/punnet). The pack should be properly labelled with name of variety and date of packing etc.

Modified atmosphere packaging (MAP) significantly prolongs the storage and shelf life of strawberry fruits by maintaining the fruit quality (Celikel et al., 2003). Refrigeration accompanied with MAP during transit or storage significantly prolongs the storage and shelf life of strawberry fruits. Results indicate that non-perforated film significantly increases shelf life due to an inherent modified atmosphere (Manleitner et al., 2002). A combination of 2.5% oxygen with 15% CO_2 has been

Post-Harvest Handling and Storage 421

found to be the optimum gas composition for strawberry MAP to prolong shelf life (Xiao et al., 2004). Low-density polyethylene (LDPE) has been found to be the best packaging film for maintaining strawberry quality (Artes et al., 2001) while a composite membrane of LDPE and polyvinyl chloride (PVC) has also been effective for strawberry packaging (Zhang et al., 2003).

Wrapping strawberries with semi-permeable film immediately after harvest minimizes the weight loss of the fruits and results in the formation of CO_2-enriched atmosphere (Barkai-Golan and Aharoni, 1983). Wrapping the fruits of strawberry varieties such as Fern, Brighton and Selva with plastic film reduced the weight loss significantly at 3°C or 6°C temperature, compared with uncovered fruits during storage (Paraskevopoulou-Paroussi et al., 1995).

El-Kazzaz et al. (1983) found that elevated CO_2 concentration and/or reduction of oxygen content of the atmosphere was very much effective to suppress the decay of strawberries not only in cold storage but also after their removal to ambient conditions. Further, fruits picked early in the season were less susceptible to fungal spoilage during storage at 2°C or shelf life at 15°C than fruits harvested late in the season (Browne et al., 1984). The pallet-packs stored in a modified atmosphere of 3%, 5% or 10% CO_2 (in air) did not consistently retard spoilage of fruits by *B. cinerea* or *Mucor pyriformis* during six days' storage at 2°C or during subsequent shelf life in air at 15°C. Storage at 10% CO_2 caused persistent off-flavours.

26.6 STORAGE CONDITIONS

Strawberry is a soft, easily perishable fruit characterized by a limited post-harvest life, mainly because of fungal decay (Barkai-Golan, 1981), loss of firmness, weight loss (Aharoni and Barkai-Golan, 1987), loss of brightness and colour darkening. To prolong the post-harvest life of strawberries, one must use the most suitable cultivars and design storage conditions in a way that the produce can be optimally protected.

Strawberries are highly perishable, as they have a relatively high rate of respiration (50–100 ml of CO_2 kg^{-1} h^{-1} at 20°C). Strawberries have a very low level of ethylene evolution (<0.1 ppm kg^{-1} h^{-1} at 20°C) and do not respond to exogenous application of ethylene with respect to stimulation of the ripening processes. However, removal of evolved ethylene from storage air is effective to suppress disease development in stored berries.

To keep berries at the field freshness stage, it is important to reduce fruit respiration rapidly after picking. This can be done by reducing the oxygen level (De La Plaza and Merodio, 1989). For long-distance shipping, partially coloured fruits are normally picked; pale fruits also occur in winter months. A post-harvest ripening process of 24–72 hours holds fruits in a ripening room at 28–30°C with 98% relative humidity (Rouddeillac and Bardet, 1989).

The strawberries can be stored for seven–10 days at 0°C and 90–95% relative humidity. The maximum freezing point is 0.8°C for strawberries, although berries with high soluble solids content are less prone to damage by freezing injury. Storing berries at recommended storage conditions even during short marketing periods is beneficial to quality retention. Low-temperature storage helps to retard the rates of respiration, softening, moisture loss, and decay development.

Strawberry fruits stored at a temperature higher than 0°C shows a higher content of aroma compounds and antioxidant capacity during the post-harvest period (Ayala-Zavala et al., 2004). This increase in antioxidant activity is probably due to an increase in antioxidant enzymes and other fruit antioxidants during the senescence stage (Hansawasdi et al., 2006). In terms of pulp firmness, titratable acidity and total soluble solids, the best results have been obtained when fruits were stored at 0°C, while the percentage of spoilage was less due to storage at −1.6°C (Brackmann et al., 2002).

Fruits of the strawberry cv. "Cambridge Favourite", harvested early in the session, could be stored up to six days in an ice-bank (about 0.5°C, 97–98% relative humidity) without any reduction in subsequent shelf life at 15°C. Such fruits, when stored for nine days, spoiled more rapidly on transfer to 15°C than those stored for a shorter period (Browne et al., 1984). It has been reported by Lidster et al. (1990) that fresh strawberry fruits can be stored maximum for 10 days at 0°C and

85–95% relative humidity. Sommers et al. (1973) reported that low temperature extended the market life of fruits by delaying the natural ageing process and by reducing the rate of development of post-harvest pathogens.

Strawberries are subject to rapid water loss, causing them to shrivel and deteriorate, as well as causing the calyx to wilt and/or dry out. These symptoms affect berry appearance before they affect eating quality. Water loss is governed by the vapour pressure deficit between the atmosphere and the product. The strawberry skin offers little protection to water vapour movement, and thus readily loses moisture to the surrounding air. Relative humidity of a storage room should be maintained at 90–95%, as strawberries will start to shrivel when stored below 90% relative humidity. However, excessive condensation of free water on the berries should be avoided.

Good sanitation of the storage and handling facilities is important to minimize contamination by pathogens causing decay of fruits. In addition, some moulds growing in storage rooms can impart off-flavors to the stored berries. Therefore, any decayed or contaminated berries and containers should be removed and disposed of promptly, and storage rooms should be periodically cleaned and sanitized.

26.6.1 Modified/Controlled Atmosphere Storage

Modified atmosphere (MA) storage generally refers to a sealed enclosure, in which the oxygen is lower and/or the carbon dioxide is higher than the concentrations found in fresh air. Whole pallet covers or consumer packages for containment of the MA are commonly used. MA with 15–20% CO_2 and 5–10% O_2 reduces the growth of *B. cinerea* and other organisms causing decay of strawberry fruits. In addition, this MA reduces the respiration and softening rates of berries, thereby extending the post-harvest life. However, exposure of strawberries to less than 2% O_2 and/or more than 25% CO_2 can cause off-flavours and brown discoloration of varying severity, depending on cultivar, duration of exposure and temperature.

A proper combination of CO_2 and oxygen can efficiently prolong the shelf life of strawberries by maintaining the quality parameters within acceptable values (Almenar et al., 2006) and they can be stored for up to one month (Kader, 2003). Controlled atmosphere (CA) storage is more effective than air storage in maintaining initial anthocyanin and soluble solids content (Nunes et al., 2002). Modified atmosphere prolongs the storage period of strawberries (Viskelis and Rubinskiene, 2006) and high CO_2 concentration is the main factor responsible for changes in alcohol acyltransferase activity and volatile composition of MA stored strawberries (Sanz et al., 2003).

Controlled atmospheres with reduced oxygen and/or elevated CO_2 concentrations have been successfully used in extending the post-harvest life and maintaining quality attributes of strawberry fruits include controlling post-harvest decay, reducing rate of respiration and ethylene production, retarding softening and increasing accumulation of certain volatiles (Smith 1957; El-Kazzaz et al., 1983).

Woodward and Topping (1972) in their experiment with strawberry cv. "Cambridge Favorite" reported that the output of CO_2 from fruits held at 4.5°C in air or at 1%, 2% or 5% O_2 after five days of storage fell to a minimum. Thereafter, the rate increased, more rapidly in air (in which rotting was more prevalent) than in 1% or 2% O_2. Fruits stored at 3°C in air or at 5%, 10%, 15% or 20% CO_2 remained in good condition for 10 days and all concentrations of CO_2 reduced rotting due to *Botrytis*. The alcohol content of the fruit increased with the length of storage and with higher concentrations of CO_2. However, 20% CO_2 caused severe injury after 30 days of storage. Long-term storage in oxygen at 1% or lower concentrations may lead to off-flavour and the use of higher CO_2 concentrations might be restricted to store the fruits for up to seven days in the absence of adequate refrigeration.

Lange et al. (1978) stored fruits of cv. "Senga Sengana" up to 12 days at 6°C in normal air, at 20% CO_2 in air, in 1% O_2 + 99% N_2 or under low atmospheric pressure (0.1 or 0.05 atm). The most efficient treatment for delaying rotting was 20% CO_2 in air, which extended the shelf life of fruits, but

Post-Harvest Handling and Storage

storage under reduced pressure showed no advantage over cold storage under normal atmosphere. The time limit for storage of "Senga Sengana" at 6°C was found to be eight days with the best treatment (20% CO_2). This treatment also had beneficial effect on fruit texture for up to eight days. Reduced atmosphere storage, however, was not beneficial to fruit quality. Smith (1992) observed that addition of CO_2 to the storage environment enhanced fruit firmness.

Ptocharaski (1982) recorded weight loss between 1.6% and 3.6% in fresh fruit of "Senga Sengana" stored for six days, and it was dependent upon the atmosphere. When the whole cold storage period was taken into consideration, the losses were amounted to about 0.6% per day for the control (normal atmosphere) and not less than 0.26% per day for any of the controlled atmosphere. The losses due to rotting after six days in storage were from 5% in the atmosphere containing 20% CO_2, 1% O_2 and 79% N_2, to about 15% for the normal atmosphere. With 0.5–3% O_2 or 5–15% CO_2, the respiration rate and incidence of decay of strawberry fruits were reduced and flesh colour and firmness maintained (El-Kazzaz et al., 1983; Li and Kader, 1989; Ke et al., 1991).

For keeping the strawberries at CO_2-enriched atmosphere (12–15% CO_2) it is essential to completely enclose pallet loads of berries in sealed plastic bags, pull a slight vacuum, then add CO_2 to the desired level within the bag and around the fruit. The accumulation of CO_2 by the respiring berries should be balanced with the permeability of bag, but puncturing the bags during handling needs to be avoided. It is essential that the berries should be thoroughly cooled before treatment, as the plastic pallet cover will impede further cooling and condensation can form when the fruit are not fully cooled.

26.6.2 CO_2 INJURY

Short term exposure to oxygen concentrations of 1% or lower and/or to CO_2 concentration of 20% or higher may have a potential for post-harvest insect disinfestations of strawberries (Ke et al., 1991; Ke and Kader, 1992). But extended exposure to these insecticidal atmospheres may result in accumulation of anaerobic volatiles, such as acetaldehyde, ethanol and ethylacetate (Li and Kader, 1989; Ke et al., 1993), which are correlated with development of off-flavour in the fruits. Hansen and Bohling (1981) reported that 20–30% CO_2 inhibited respiration and prevented decay of fruits for several days but 50% CO_2 resulted in development of off-flavour.

Negative aspects of high CO_2 concentrations are loss of flavour and an acidic taste, probably originating from the low oxygen level in the atmosphere, which induces fermentation processes. High alcohol levels have been found in the low oxygen and high CO_2-stored fruits, although the exposure of strawberries to this atmosphere for at least three days helped in maintaining quality (flesh firmness and color) during transport and following exposure to air (Li and Kader, 1989).

Ke et al. (1993) studied the changes in anaerobic products and enzymes in "Selva" strawberries in air, 20% CO_2 + 80% air, 50% CO_2 + 50% air and 0.25% O_2 + 99.75% N_2 at 5°C for one–nine days. The 20% CO_2 treatment slightly increased the concentrations of acetaldehyde, ethanol and ethyl acetate. Exposure to 50% CO_2 or 0.25% O_2 resulted in greater accumulation of the anaerobic volatiles, which was associated with reduction in fruit flavour. Activities of pyruvate decarboxylase (PDC) and alcohol dehydrogenase (ADH) were slightly increased by the elevated CO_2 and reduced oxygen atmospheres in the first five days. The 0.25% O_2 treatment caused a much greater increase in ADH activity after seven and nine days. Results from this research and previous studies indicated that induction of alcoholic fermentation in strawberries could be due to increased synthesis of PDC and ADH, activation of PDC and ADH by decrease in cytoplasmic pH and increased substrate concentrations of pyruvate, acetaldehyde and NADH.

26.7 DISEASES

Diseases are the greatest causes of post-harvest losses in strawberries. Prompt cooling, storage at the lowest safe temperature, preventing physical injury to the fruit and utilizing high CO_2 (10–15%)

are the best methods for disease control. In addition, care should be taken to keep diseased or wounded berries out of packages, as rot can spread from diseased to nearby healthy berries (for more details, please also see Chapter 29).

Grey mould rot, caused by *B. cinerea*, is the most common problem in strawberries. This disease can develop during storage if berries have been contaminated through harvest and handling wounds. The organism can also invade flower blossoms and remain dormant in the berry until ripening begins. Grey mould is most serious during rainy or foggy periods in the field. Surface mycelia from infected berries can directly penetrate adjacent healthy berries to produce a nest of rotting berries, which may continue to spread throughout the basket or flat. Avoiding mechanical injuries and good temperature management are effective measure to manage the disease.

Rhizopus rot (*R. stolonifer*) can also be a problem in strawberries. Cooling the fruits and keeping them below 5°C is very effective against this fungus, since it does not grow at these temperatures. Similarly to grey mould, *Rhizopus* rot can spread from infected berries to adjacent healthy ones, producing a nesting effect.

REFERENCES

Aday, M.S. and Caner, C. 2014. Individual and combined effects of ultrasound, ozone, and chlorine dioxide on strawberry storage life. *LWT – Food Science and Technology*, 57: 344–351.

Aday, M.S., Temizkan, R., Buyukcan, M.B., and Caner, C. 2013. An innovative technique for extending shelf life of strawberry: Ultrasound. *LWT – Food Science and Technology*, 52: 93–101.

Agar, T., García, J.M. and Streif, J. 1990. Effect of high CO_2 and low O_2 concentration on the growth of *Botrytis cinerea* at different temperatures. *Gartenbauwissenschaft*, 55: 219–222.

Aguayo, E., Jansasithorn, R. and Kader, A.A. 2006. Combined effects of 1-methylcyclopropene, calcium chloride dip, and/or atmospheric modification on quality changes in fresh-cut strawberries. *Postharvest Biology and Technology*, 40: 369–278.

Aharoni, T. and Barkai-Golan, R. 1987. Preharvest fungicide sprays and polyvinyl wraps to control *Botrytis* rot and the postharvest storage life of strawberries. *Journal of Horticultural Sciences*, 62: 177–181.

Allende, A., Marin, A., Buendia, B., Tomas-Barberian, F. and Gil, M.I. 2007. Impact of combined postharvest treatments (UV-C light, gaseous O_3, superatmospheric O_2 and high CO_2) on health promoting compounds and shelf life of strawberries. *Postharvest Biology and Technology*, 46: 201–211.

Almenar, E., Hernandez-Munoz, P., Lagaron, J.M., Catala, R. and Gavara, R. 2006. Controlled atmosphere storage of wild strawberry fruit (*Fragariavesca*L.). *Journal of Agricultural and Food Chemistry*, 54(1): 86–91.

Artés, F., Tudela, J.A., Marín, J.G., Villaescusa, R. and Artés, H.F. 2001. Modified atmosphere packaging of strawberry under selfadhesive films. *Acta Horticulturae*, 559: 805–809.

Asrey, R. and Jain, R.K. 2005. Effect of certain post-harvest treatments on shelf life of strawberry cv. Chandler. *ActaHorticulturae*, 696: 547–550.

Ayala-Zavala, J.F., Wang, S.Y., Wang, C.Y. and González-Aguilar, G.A. 2004. Effect of storage temperatures on antioxidant capacity and aroma compounds in strawberry fruit. *LWT – Food Science and Technology*, 37(7): 687–695.

Ayala-Zavala, J.F., Wang, S.Y., Wang, C.Y. and González-Aguilar, G.A. 2007. High oxygen treatment increases antioxidant capacity and postharvest life of strawberry fruit. *Food Technology and Biotechnology*, 45(2): 166–173.

Babalar, M., Asghari, M., Talaei, A. and Khosroshahi, A. 2007. Effect of pre- and postharvest salicylic acid treatment on ethylene production, fungal decay and overall quality of Selva strawberry fruit. *Food Chemistry*, 105: 449–453.

Barkai-Golan, R. and Aharoni, T. 1981. *The Volcanic Centre*. Israel Special Publications, Dagan, Israel. No. 194.

Barkai-Golan, R. and Aharoni, T. 1983. *Abstract of the 4th International Congress of Plant Pathol (Melbourne)*: p. 260.

Bower, J.H., Biasi, W.V. and Mitcham, E.J. 2003. Effects of ethylene and 1-MCP on the quality and storage life of strawberries. *Postharvest Biology and Technology*, 28(3): 417–423.

Brackmann, A., Frietas, S.T., Mello, A.M. and Neuwald, D.A. 2002. Effect of storage temperature on quality of "Oso Grande" strawberry. *Revista Brasileira de Agrociencia*, 8(1): 77–78.

Post-Harvest Handling and Storage

Browne, K.M., Geeson, G.D. and Dennis, C. 1984. The effects of harvest date and CO_2-enriched storage atmospheres on the storage and shelf-life of strawberries. *Journal of Horticultural Science*, 59: 197–204.

Cai, Y., Liyu, S., Wei, C., Xinguo, S. and Zhenfeng, Y. 2015. Effect of UV-C treatment on fruit quality and active oxygen metabolism of postharvest strawberry fruit. *Journal of Chinese Institute of Food Science and Technology*, 15(3): 128–136.

Cao, S., Hu, Z., Pang, B., Wang, H., Xie, H. and Wu, F. 2010. Effect of ultrasound treatment on fruit decay and quality maintenance in strawberry after harvest. *Food Control*, 21: 529–532.

Chaiprasart, P., Handsawasdi, C. and Pipattanawong, N. 2006. The effect of chitosan coating and calcium chloride treatment on postharvest qualities of strawberry fruit (*Fragaria × ananassa*). *Acta Horticulturae*, 708: 337–341.

Cheour, F., Willemot, C., Arul, J., Desjardins, Y., Makhlouf, J., Charest, D.M. and Gosselin, A. 1990. Foliar application of calcium chloride delays postharvest ripening of strawberry. *Journal of the American Society for Horticultural Science*, 115: 789–792.

Cheour, F., Willemot, C., Arul, J., Makhalouf, J., Charest, D.M. and Desjardins, Y. 1991. Postharvest response of two strawberry cultivars to foliar application of $CaCl_2$. *HortScience*, 26: 1186–1188.

Celikel, F.G., Kaynas, K. and Eronoglu, B. 2003. A study on modified atmosphere storage of strawberry. *Acta Horticulturae*, 628: 423–430.

Chilosi, G., Tagliavento, V. and Simonelli, R. 2015. Application of ozone gas at low doses in the cold storage of fruit and vegetables. *Acta Horticulturae*, 1071: 681–686.

Cote, S., Rodoni, L., Miceli, E., Concellón, A., Civello, P.M. and Vicente, A.R. 2013. Effect of radiation intensity on the outcome of postharvest UV-C treatments. *Postharvest Biology and Technology*, 83: 83–89.

De La Plaza, J.L. and Merodio, C. 1989. Effect of ethylene chemisorption on refrigerated strawberry fruit. *Acta Horticulturae*, 265: 427–434.

Del-Valle, V., Hernández-Muñoz, P., Guarda, A. and Galotto, M.J. 2005. Developement of a cactus-mucilage edible coating (*Opuntiaficusindica*) and its application to extend strawberry (*Fragariaananassa*) shelf-life. *Food Chemistry*, 91: 751–756.

Donmez, M.F., Esitken, A., Yildez, H. and Ercisli, S. 2011. Biocontrol of *Botrytis cinerea* on strawberry fruit by plant growth promoting bacteria. *Journal of Animal and Plant Sciences*, 21(4): 758–763.

Dotto, M.C., Pombo, M.A., Martinez, G.A. and Cibello, P.M. 2011. Heat treatments and expansin gene expression in strawberry fruit. *Scientia Horticulturae*, 130: 775–780.

Eaves, C.A. and Leefe, J.S. 1962. Note on the influence of foliar sprays of calcium on the firmness of strawberries. *Canadian Journal of Plant Sciences*, 42: 746–747.

El-Kazzaz, M.K., Sommers, N.F. and Fortlage, R.J. 1983. Effect of different atmospheres on postharvest decay and quality of fresh strawberries. *Phytopatholology*, 73: 282–285.

Erkan, M., Wang, S.Y. and Wang, C.Y. 2008. Effect of UV treatment on antioxidant capacity, antioxidant enzyme activity and decay in strawberry fruit. *Postharvest Biology and Technology*, 48: 163–171.

Feng, X., Shifeng, C., Liyu, S., Wei, C., Xinguo, S. and Zhenfeng, Y. 2014. Blue light irradiation affects anthocyanin content and enzyme activities involved in postharvest strawberry fruit. *Journal of Agricultural and Food Chemistry*, 62(20): 4778–4783.

Ferriera, M.D., Brecht, J.K., Sargent, S.S. and Chandler, C.K. 2006. Hydrocooling as an alternative to forced-air cooling for maintain fresh-market strawberry quality. *HortTechnology*, 16(4): 659–666.

Flores, G., Perez, C., Gil, C., Blanch, G.P. and Castillo, M.L.R. 2013. Methyl jasmonate treatment of strawberry fruits enhances antioxidant activity and the inhibition of nitrite production in LPS-stimulated Raw 264.7 cells. *Journal of Functional Foods*, 5: 1803–1809.

Garcia, J.M., Aguilera, C. and Albi, M.A. 1995. Postharvest heat treatment on Spanish strawberry (*Fragaria× ananassa* cv. Tudla). *Journal of Agricultural and Food Chemistry*, 43: 1489–1492.

Garcia, J.M., Herrera, S. and Morilla, A. 1996. Effects of postharvest dips in calcium chloride on strawberry. *Journal of Agricultural and Food Chemistry*, 44: 30–33.

Glenn, G.M., Reddy, A.S. and Poovaiah, B.W. 1988. Effect of calcium on cell wall structure, protein phosphorylation, and protein profile in senescing apples. *Plant Cell Physiology*, 29: 565–572.

Gol, N.B., Patel, P.R. and Ramana Rao, T.V. 2013. Improvement in quality and shelf-life of strawberries with edible coatings enriched with chitosan. *Postharvest Biology and Technology*, 85: 185–195.

Güemes, D.B., Pirovani, M.E. and Di Pentima, J.H. 1989. Heat transfer characteristics during air pre-cooling of strawberry. *International Journal of Refrigeration*, 12: 169–173.

Guinebretiere, M.H., Nguyen-The, C., Morrison, N., Reich, M. and Nicot, P. 2000. Isolation and characterization of antagonists for the biocontrol of the postharvest wound pathogen *Botrytis cinerea* on strawberry fruits. *Journal of Food Protection*, 63(3): 386–394.

Han, Y., Selby, T.L., Schultze, K.K., Nelson, P.E. and Linton, R.H. 2004. Decontamination of strawberries using batch and continuous chlorine dioxide gas treatments. *Journal of Food Protection*, 67(11): 2450–2455.

Hansawasdi, C., Rithiudom, S. and Chaiprasart, P. 2006. Quality and antioxidant activity changes during low-temperature storage of strawberry fruits. *Acta Horticulturae*, 708: 301–306.

Hansen, H. and Bohling, H. 1981. Studies on improving the quality of strawberries at normal temperatures using dry ice. *Besseres Obst*, 26: 165–167.

Henrique, C.M. and Cereda, M.P. 1999. Utilização de biofilms na conservação pós-colheita de morangos (*Fragaria ananassa* Duch). *Ciencia e Tecnologia de Alimentos*, 19(2): 231–233.

Hernández-Muñoz, P., Almenar, E., Ocio, M.J. and Gavara, R. 2006. Effect of calcium dips and chitosan coatings on postharvest life of strawberries (*Fragaria × ananassa*). *Postharvest Biology and Technology*, 39: 247–253.

Hernández-Muñoz, P., Almenar, E., Valle, V.D., Velez, D. and Gavara, R. 2008. Effect of chitosan coating combined with calcium treatment on strawberry (*Fragaria × ananassa*) quality during refrigerated storage. *Food Chemistry*, 110: 428–435.

Higashio, H., Hirokane, H., Sato, F., Tokuda, S. and Uragami, A. 2005. Effect of UV irradiation after harvest on the content of flavonoid in vegetables. *Acta Horticulturae*, 682: 1007–1012.

Holcroft, D.M. and Kader, A.A. 1999. Carbon dioxide-induced changes in color and anthocyanin synthesis of stored strawberry fruit. *HortScience*, 34(7): 1244–1248.

Huang, R., Che, H.J., Zhang, J., Yang, L., Jiang, D.H. and Li, G.Q. 2012. Evaluation of *Sporidiobolus pararoseus* strain YCXT3 as biocontrol agent of *Botrytis cinerea* on post-harvest strawberry fruits. *Biological Control*, 62(1): 53–63.

Jiang, Y., Joyce, D.C. and Terry, L.A. 2001. 1-methylcyclopropene treatment affects strawberry fruit decay. *Postharvest Biology and Technology*, 23(3): 227–232.

Jing, W., Tu, K., Shao, X.F., Su, Z.P., Zhao, Y., Wang, S. and Tang, J. 2010. Effect of postharvest short hot-water rinsing and brushing treatment on decay and quality of strawberry fruit. *Journal of Food Quality*, 33: 262–272.

Kader, A.A. 2003. A summary of CA requirements and recommendations for fruits other than apples and pears. *Acta Horticulturae*, 600: 737–740.

Kader, A.A. 1999. Fruit maturity, ripening, and quality relationships. *Acta Horticulturae*, 485: 203–208.

Ke, D. and Kader, A.A. 1992. Potential of controlled atmospheres for postharvest insect disinfestation of fruits and vegetables. *Postharvest News and Information*, 3: 31N–37N.

Ke, D., Goldstein, L., O'Mahony, M. and Kader, A.A. 1991. Effects of short-term exposure to low O_2 and high CO_2 atmospheres on quality attributes of strawberries. *Journal of Food Science*, 56: 50–54.

Ke, D., El Sheikh, J., Mateos, M. and Kader, A.A. 1993. Anaerobic metabolism of strawberries under elevated CO_2 and reduced O_2 atmospheres. *Acta Horticulturae*, 343: 93–99.

Kim, J.Y., Kim, H.J., Lim, G.O., Jang, S.A. and Song, K.B. 2010. The effects of aqueous chlorine dioxide or fumaric acid treatment combined with UV-C on postharvest quality of "Maehyang" strawberries. *Postharvest Biology and Technology*, 56: 254–256.

Ku, V.V.V., Wills, R.B.H. and Ben-Yehoshua, S. 1999. 1-Methylcyclopropene can differentially affect the postharvest life of strawberries exposed to ethylene. *HortScience*, 34(1): 119–120.

Lange, E., Plocharski, W. and Lenartowicz, W. 1978. Strawberries – Quality of fruits, their storage life and suitability for processing. Pt.1: The influence of controlled atmospheres and low pressure on the storage of strawberry fruits. *Fruit Science Report*, 5: 39–46.

Lara, I., Garcia, P. and Vendrell, M. 2004. Modifications in cell wall composition after cold storage of calcium treated strawberry (*Fragaria × ananassa* Duch.) fruit. *Postharvest Biology and Technology*, 34(3): 331–339.

Lara, I., Garcia, P. and Vendrell, M. 2006. Postharvest heat treatments modify cell wall composition of strawberry (*Fragaria × ananassa* Duch.). *Scientia Horticulturae*, 109(1): 48–53.

Laurin, E., Nunes, M.C.N. and Edmond, J.P. 2005. Re-cooling of strawberries after air shipment delays fruit senescence. *Acta Horticulturae*, 682(3): 1745–751.

Li, C. and Kader, A.A. 1989. Residual effects of controlled atmospheres on postharvest physiology of strawberries. *Journal of the American Society for Horticultural Science*, 114: 629–634.

Lidster, P.D., Hildebrand, P.D., Bérard, L.S. and Porritt, S.W. 1990. *Commercial Storage of Fruits and Vegetables*. Agriculture Canada Publication, 1532/E: pp. 31–32.

Li, Z., Wang, L., Gong, W. and Wang, Z. 2006. Effect of 1-MCP on senescence and quality of strawberry fruit. *Journal of Fruit Science*, 23(1): 125–128.

Lopes, U.P., Zambolim, L., Costa, H., Pereira, O.L. and Finger, F.L. 2014. Potassium silicate and chitosan application for gray mold management in strawberry during storage. *Crop Protection*, 63: 103–106.

Manleitner, S., Lippert, F. and Noga, G. 2002. Influence of packaging material on quality and shelf life of strawberries. *Acta Horticulturae*, 567: 771–773.

Marquenie, D., Schenk, A., Nicolai, B., Michiels, C., Soontjens, C. and Van Impe, J. 2002. Use of UV-C and heat treatment to reduce storage rot of strawberry. *Acta Horticulturae*, 567(2): 779–782.

Maxie, E.C., Mitchell, F.G. and Greathead, A. 1959. Studies on strawberry quality. *California Agriculture*, 13: 11–16.

Mazaro, S.M., Deschamps, C., Mio, L.L.M., Biase, L.A., Gouvea, L.A. and Sautter, C.K. 2008. Post harvest behavior of strawberry fruits after pre harvest treatment with chitosan and acibenzolar-S-methyl. *Revista Brasileira de Fruticultura*, 30: 185–195.

Mitcham, E.J. 2016. Strawberry. In: K. Gross, C.Y. Wang and M.E. Saltveit (eds.). *The Commercial Storage of Fruits, Vegetables, and Florist and Nursery Stocks*, 2nd edn. USDA Agriculture Handbook Number 66: pp. 559–561.

Nadas, A., Olmo, M. and Garcia, J.M. 2003. Growth of Botrytis cinerea and strawberry quality in ozone-enriched atmospheres. *Journal of Food Science*, 68: 1798–1802.

Nadim, Z., Ahmadi, E., Sarikhani, H. and Chayjan, R.A. 2015. Effect of methylcellulose-based edible coating on strawberry fruit's quality maintenance during storage. *Journal of Food Processing and Preservation*, 39(1): 80–90.

Nishizawa, T., Aikawa, T., Takahashi, M., Murayama, H. and Matsushima, U. 2006. Storage of horticultural products in closed rooms with TiO_2 photocatalyst: changes in room atmosphere and quality of fruits and cut flowers. *Acta Horticulturae*, 712(1): 261–268.

Nunes, M.C.N., Morais, A.M.M.B., Brecht, J.K. and Sargent, S.A. 2002. Fruit maturity and storage temperature influence the response of strawberries to controlled atmospheres. *Journal of the American Society for Horticultural Science*, 127(5): 836–842.

Nunes, M.C.N., Morais, A.M.M.B., Brecht, J.K., Sargent, S.A. and Bartz, J.A. 2005. Prompt cooling reduces incidence and severity of decay caused by *Botrytis cinerea* and *Rhizopusstolonifer* in strawberry. *HortTechnology*, 15: 153–156.

Paliyath, G., Poovaiah, B.W., Munske, G.R. and Magnuson, J.A. 1984. Membrane fluidity in senescing apples: Effects of temperature and calcium. *Plant Cell Physiology*, 25: 1083–1087.

Pan, J., Vicente, A.R., Martínez, G.A., Civello, P.M. and Chaves, A.R. 2003. Use of combined UV-C and heat treatments to improve postharvest life of strawberry. *ActaHorticulturae*, 628: 729–735.

Paraskevopoulou-Paroussi, G., Vassilakakis, M. and Dogras, C. 1995. Effects of temperature, duration of cold storage and packaging on postharvest quality of strawberry fruit. *Acta Horticulturae*, 379: 337–344.

Pelayo, C., Ebeler, S.E. and Kader, A.A. 2003. Postharvest life and flavor quality of three strawberry cultivars kept at 5°C in air or air + 20 kPa CO_2. *Postharvest Biology Technology*, 27: 171–183.

Peng, G. and Sutton, J.C. 1991. Evaluation of microorganism for bio-control of *Botrytis cinerea* in strawberry. *Canadian Journal of Plant Pathology*, 13: 247–257.

Perez, A.G., Sanz, C., Rios, J.J., Olías, R. and Olías, J.M. 1999. Effects of ozone treatment on postharvest strawberry quality. *Journal of Agriculutre and Food Chemistry*, 47(4): 1652–1656.

Ptocharaski, W. 1982. Strawberries-quality of fruits, their storage life and suitability for processing: Part III. Firmness and pectic substance changes of strawberries stored under normal and controlled atmosphere conditions. *Fruit Science Report*, 9: 111–122.

Pombo, M.A., Dotto, M.C., Martínez, G.A. and Civello, P.M. 2009. UV-C irradiation delays strawberry fruit softening and modifies the expression of genes involved in cell wall degradation. *Postharvest Biology Technology*, 51: 141–148.

Pombo, M.A., Rosli, H.G., Martínez, G.A. and Civello, P.M. 2011. UV-C treatment affects the expression and activity of defense genes in strawberry fruit (*Fragaria× ananassa* Duch.). *Postharvest Biology Technology*, 59: 94–102.

Poovaiah, B.W. 1979. Role of calcium in ripening and senescence. *Communications in Soil Science and Plant Analysis*, 10: 83–88.

Poovaiah, B.W. and Leopold, A.C. 1973. Deferral of leaf senescence with calcium. *Plant Physiology*, 52: 236–239.

Romanazzi, G., Nigro, F. and Ippolito, A. 2000. Effectiveness of pre and postharvest chitosan treatments on storage decay of strawberries. *Rivista di Fructticoltura e di ortofloricoltura*, 62(5): 71–75.

Romanazzi, G., Feliziani, E., Santini, M. and Landi, L. 2013. Effectiveness of postharvest treatment with chitosan and other resistance inducers in the control of storage decay of strawberry. *Postharvest Biology Technology*, 75: 24–27.

Rouddeillac, P. and Bardet, A. 1989. Fast post-ripening process for late fall strawberries. *Acta Horticulturae*, 265: 447–448.

Sanz, C., Olias, R. and Perez, A.G. 2003. Effect of modified atmosphere on alcohol acyltransferase activity and volatile composition of strawberry. *Acta Horticulturae*, 600: 563–566.

Scalon, S.P.Q. 1999. Effect of calcium chloride on strawberry. *Revista Brasileira de Fructicultura*, 21(2): 156–159.

Severo, J., Oliveira, I.R. de, Tiecher, A., Chaves, F.C., Rombaldi, C.V. 2015. Postharvest UV-C treatment increases bioactive, ester volatile compounds and a putative allergenic protein in strawberry. *LWT – Food Science and Technology*, 64(2): 685–692.

Shin, Y.-J., Song, H.-J. and Song, K.B. 2012. Effect of combined treatment of rice bran protein film packaging with aqueous chlorine dioxide washing and ultraviolet-C irradiation on the postharvest quality of "Goha" strawberries. *Journal of Food Engineering*, 113: 374–379.

Smith, R.B. 1992. Controlled atmosphere storage of "Redcoat" strawberry fruit. *Journal of the American Society for Horticultural Science*, 117(2): 260–264.

Sy, K.V., Murray, M.B., Harrison, M.D. and Beuchat, L.R. 2005. Evaluation of gaseous chlorine dioxide as a sanitizer for killing Salmonella, *Escherichia coli* O157:H7, *Listeria monocytogenes*, and yeasts and molds on fresh and fresh-cut produce. *Journal of Food Protection*, 68: 1176–1187.

Smith, W.H. 1957. The application of precooling and carbon dioxide treatment to the marketing of strawberries and raspberries. *Scientia Horticulturae*, 12: 147–153.

Sommers, N.F., Fortlage, R.J., Mitchell, F.G. and Maxie, E.C. 1973. Reduction of postharvest losses of strawberry fruits from gray mold. *Journal of the American Society for Horticultural Science*, 98: 285–288.

Tanada-Palmu, P.S. and Grosso, C.R.F. 2005. Effect of edible wheat gluten-based films and coatings on refrigerated strawberry (*Fragaria ananassa*) quality. *Postharvest Biology Technology*, 36: 199–208.

Tian, M.S., Prakash, S., Elgar, H.J., Young, H., Burmeister, D.M. and Ross, G.S. 2000. Responses of strawberry fruit 1-methylcyclopropene (1-MCP) and ethylene. *Plant Growth Regulation*, 32(1): 83–90.

Tingwa, P.O. and Young, R.E. 1974. The effect of calcium on the ripening of avocado (*Persea americana* Mill). *Journal of the American Society for Horticultural Science*, 99: 540–542.

Trevino-Garza, M.Z., Garcia, S., Flores-Gonzalez, M. del S. and Arevalo-Nino, K. 2015. Edible active coatings based on pectin, pullulan, and chitosan increase quality and shelf life of strawberries (Fragaria ananassa). *Journal of Food Science*, 80(8): M1823–M1830.

Tudela, J.A., Villaescusa, R., Artes, H.F. and Artes, F. 2003. High carbon dioxide during cold storage for keeping strawberry quality. *Acta Horticulturae*, 600: 201–204.

Tu, K., Chen, L. and Pan, X.J. 2006. Effects of different pre-storage heat treatments on the shelf quality and mold control of strawberry fruit. *Acta Horticulturae*, 712: 805–810.

Ugolini, L., Martini, C., Lazzeri, L., D'Avino, L. and Mari, M. 2014. Control of postharvest grey mould (Botrytis cinerea Per.: Fr.) on strawberries by glucosinolate-derived allyl-isothiocyanate treatments. *Postharvest Biology and Technology*, 90: 34–39.

Upadhayaya, A.K. and Sanghavi, K.U. 2001. Effect of various chemical and packaging methods on storage quality of strawberry *Fragariaananassa* cv. Chandler. *Advances in Plant Sciences*, 14(2): 343–349.

Vardar, C., Ilhan, K. and Karabulut, O.A. 2012. The application of various disinfectants by fogging for decreasing postharvest diseases of strawberry. *Postharvest Biology Technology*, 66: 30–34.

Vicente, A.R., Martínez, G.A., Civello, P.M. and Chaves, A.R. 2002. Quality of heat-treated strawberry fruit during refrigerated storage. *Postharvest Biology Technology*, 25: 59–71.

Vicente, A.R., Martínez, G.A., Chaves, A.R. and Civello, P.M. 2003. Influence of self-produced CO_2 on postharvest life of heat-treated strawberries. *Postharvest Biology Technology*, 27: 265–275.

Viskelis, P. and Rubinskiene, M. 2006. Effect of modified atmosphere on the quality of strawberry cv. Venta. *SodininkysteirDarzininkyste*, 25(1): 64–73.

Wang, S.Y. and Gao, H. 2013. Effect of chitosan-based edible coating on antioxidants, antioxidant enzyme system, and postharvest fruit quality of strawberries (*Fragaria* × *ananassa*Duch.). *LWT-Food Science and Technology*, 52: 71–79.

Wang, H., Li, Z., Gong, J. and Tang, J. 2012. Effects of ozone combined with high oxygen treatment on fresh-keeping in postharvest strawberry fruits. World Automation Congress, Puerta Vallarta, Mexico: pp. 1–4.

Wei, Y., Mao, S. and Tu, K. 2014. Effect of preharvest spraying *Cryptococcus laurentii* on postharvest decay and quality of strawberry. *Biological Control*, 73: 68–74.

Wisniewski, M.E. and Wilson, C.L. 1992. Biological control of postharvest diseases of fruits and vegetables: recent advances. *HortScience*, 27: 94–98.

Wills, R.B.H., Ku, V.V.V. and Leshem, Y.Y. 2000. Fumigation with nitric oxide to extend the postharvest life of strawberries. *Postharvest Biology Technology*, 18(1): 75–79.

Woodward, J.R. and Topping, A.J. 1972. The influence of controlled atmospheres on the respiration rates and storage behavior of strawberry fruits. *Journal of Horticulture Science*, 47: 547–555.

Wszelaki, A.L. and Mitcham, E.J. 2000. Effects of superatmospheric oxygen on strawberry fruit quality and decay. *Postharvest Biology Technology*, 20(2): 125–133.

Xiao, G., Zhang, M., Luo, G., Peng, J., Salokhe, V.M. and Guo, J. 2004. Effect of modified atmosphere packaging on the preservation of strawberry and the extension of its shelf-life. *International Agrophysics*, 18(2): 195–201.

Yang, F.M., Li, H.M., Li, F., Xin, Z.H., Zhao, L.Y., Zheng, Y.H. and Hu, Q.H. 2010. Effect of nano-packing on preservation quality of fresh strawberry (*Fragaria ananassa* Duch cv. Fengxiang) during storage at 4°C). *Journal of Food Science*, 75(3): C236–C240.

Zhang, M., Xiao, G., Peng, J. and Salokhe, V.M. 2003. Effects of modified atmosphere package on preservation of strawberries. *International Agrophysics*, 17(3): 143–148.

Zhang, F.S., Wang, X.Q., Ma, S.J., Cao, S.F., Li, N., Wang, X.X. and Zheng, Y.H. 2006. Effects of methyl jasmonate on postharvest decay in strawberry fruit and the possible mechanisms involved. *Acta Horticulturae*, 712: 693–698.

Zhang, H., Wang, L., Dong, Y., Jiang, S., Cao, J. and Meng, R. 2007a. Postharvest biocontrol of gray mold decay of strawberry with Rhodotorulaglutinis. *Biological Control*, 40: 287–292.

Zhang, H., Zheng, X., Wang, L., Li, S. and Liu, R. 2007b. Effect of yeast antagonist in combination with hot water dips on postharvest *Rhizopus* rot of strawberries. *Journal of Food Engineering*, 78: 281–287.

Zhang, H., Yang, Q., Lin, H., Ren, X., Zhao, L. and Hou, J. 2013. Phytic acid enhances biocontrol efficacy of *Rhodotorula mucilaginosa* against postharvest gray mold spoilage and natural spoilage of strawberries. *LWT-Food Science and Technology*, 52: 110–115.

Zhang, H., Ma, L., Turner, M., Xu, H., Zheng, X., Dong, Y. and Jiang, S. 2010. Salicylic acid enhances efficacy of *Rhodotorula glutinis* against postharvest *Rhizopus* rot of strawberries and the possible mechanisms involved. *Food Chemistry*, 122: 577–583.

Zhu, S. and Zhou, J. 2007. Effect of nitric oxide on ethylene production in strawberry fruit during storage. *Food Chemistry*, 100(4): 1517–1522.

27 Value Addition

Julie Dogra Bandral and Monika Sood

CONTENTS

27.1 Preprocessing Considerations 432
27.2 Processed Products... 432
    27.2.1   Strawberry Juice.. 433
        Preservation Tips 433
        Ingredients.. 433
    27.2.2   Squash/Syrup.. 433
        Preservation Tips 433
        Ingredients.. 433
    27.2.3   Jam 434
        Preservation Tips 434
        Ingredients.. 434
    27.2.4   Preserve... 434
        Preservation Tips 434
        Ingredients.. 435
    27.2.5   Chutney 436
        Preservation Tips 436
        Ingredients.. 436
    27.2.6   Sweet 'n' Sour Pickle.. 436
        Preservation Tips 436
        Ingredients.. 436
    27.2.7   Sauce 437
        Preservation Tips 437
        Ingredients.. 437
    27.2.8   Puree.. 438
        Preservation Tips 438
        Ingredients.. 438
    27.2.9   Frozen strawberries 438
        Preservation Tips 438
        Ingredients.. 438
    27.2.10 Strawberry wine 439
        Preservation Tips 439
        Ingredients.. 439
27.3 Effects of Processing on Quality..440
References..442

Strawberries are one of the most perishable fruit crops and are essentially harvested at the fully ripe stage. The high rate of metabolism destroys strawberries in a relatively short time even without presence of decay-causing pathogens. In general, strawberries have a maximum storage life ranging five to seven days at a temperature of 0°C and 95% relative humidity. They are usually harvested along with the attached calyxes and must be held loosely in the hand to avoid injury due to bruising. Good temperature management is the single most important factor in reducing deterioration of fruit quality and maximizing the post-harvest life of strawberry fruits. The best way to minimize post-harvest spoilage is quick removal of field heat and maintaining temperature of berries close to 0°C. Washing the strawberries first in a chilled chlorine bath or citric acid or tri-basic calcium solution followed by drying, packing and transportation in chilled conditions such as refrigerated/reefer vans also helps in enhancement of the shelf life of fruits. Since fruits are highly perishable and post-harvest losses are significant, processing the fruits into a range of value-added products could contribute to generating employment and enhance overall profitability and sustainability of strawberry enterprise. A brief account of processing for value addition in strawberry fruits is given in this chapter.

Strawberry jam is very popular but the fruits are also utilized for processing into food products such as juice, squash, wine, pickle, chutney, spread, yoghurt, ice cream and breakfast cereals (Haffner, 2002). In ice creams, fresh strawberries can be used as toppings or can be blended as small pieces into the semi-processed or finished ice cream recipe. The bulk of the strawberry fruits can be frozen. Partial removal of water (up to 50%) from fruits before freezing leads to some benefits such as the reduction of free water content, the depression of the freezing point, reduction in weight and volume of frozen fruit up to 50% and less drip loss during thawing, along with retention of natural taste and flavour (Rosa et al., 2006).

27.1 PREPROCESSING CONSIDERATIONS

Any defect in the quality of strawberries in terms of colour, firmness, ripeness, size, infection by mould or crushed/damaged fruits due to improper post-harvest handling has detrimental effects on the quality of the processed product. Therefore, care must be taken in procuring strawberry fruits with the best quality from processing point of view. Some of these considerations are:

- A relatively high rate of respiration (50–100 ml of CO_2 per kg per hour at 20°C) makes strawberry fruits very perishable, so they must be purchased a few days before use.
- The fruits produce very little ethylene (<0.1 ppm per kg per hour at 20°C) and do not respond to exogenous application of ethylene, being non-climacteric; however, removal of ethylene from storage air may reduce development and severity of storage diseases in the berries.
- Good temperature management is the single most important factor in reducing deterioration and maximizing post-harvest life of strawberry fruits.
- Fruits of strawberries that are firm, plump and free from mould with a shiny deep red look and attached green caps should be chosen for processing.
- Since strawberries do not ripen further after harvesting, the fruits with dull colour or having green or yellow patches are likely to be sour and of inferior quality, so they need to be avoided for harvesting to maintain standard quality.
- Medium-sized strawberries are often more flavourful than those that are excessively large.
- While buying strawberries that are prepackaged in a container, one should make sure that they are not packed too tightly, causing fruits being crushed or damaged.
- Before storing the strawberries in the refrigerator, mouldy or damaged strawberries should be removed to avoid further contamination in the lot.
- Being highly perishable, strawberries should not be washed until these are being taken for processing in to some value-added product.
- The caps and pedicel of fruits should not be removed until after a gentle wash of berries under cold running water and slight drying thereafter. This will prevent them from absorbing excess water, which can degrade the texture and flavour of strawberries.

27.2 PROCESSED PRODUCTS

The product-specific processing specifications have already been outlined with storage conditions for strawberries (www.oregonstrawberrycommission.org). Here, the product-specific systematic value addition procedures have been explained with flow sheets as follows.

Value Addition

27.2.1 Strawberry Juice

Preservation Tips

- As soon as the clear juice is obtained, it should be quickly heated to 60–80°C.
- Juices and squashes may be fortified with vitamins to enhance nutritive value to improve taste and texture. Usually vitamin C and β-carotene (in water dispersible form) are added @ 250–500 mg and seven–10 ppm, respectively.
- Juice-filled bottles are pasteurized for 25–30 minutes.
- Juices can be preserved by adding 66% sugar (Figure 27.1).

Ingredients

Juice : 1.0 litre
Sugar : 200 g

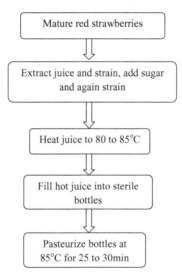

FIGURE 27.1 Flowsheet for juice preparation.

27.2.2 Squash/Syrup

Preservation Tips

- Squash and syrup contain about 30–65% sugar and this high concentration of sugar usually preserves them.
- Chemical preservatives such as sodium benzoate and potassium metabisulphite (KMS) can be added.
- Citric acid can be added to increase acidity, thereby enhancing the flavour and taste (Figure 27.2).

Ingredients

Strawberry juice : 1.2 litre
Sugar : 1.2 kg
Water : 600 ml
Citric acid : 37 g
Sodium benzoate : 2.0 gram (as preservative) (Essence and edible strawberry colour if desired)

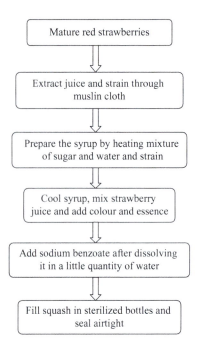

FIGURE 27.2 Flowsheet for syrup/squash preparation.

27.2.3 Jam

Preservation Tips

- Culled, ripe, unripe, undersized, deformed, unmarketable and damaged fruits of strawberries (cheaper) are used to prepare jam. The fruit pulp/extract is cooked with sugar, along with the correct proportions of acid and pectin for proper setting of jam.
- In case the pectin content of the fruit is low or negligible, commercial pectin @ 6 g/kg of sugar can be added.
- Jams generally do not spoil due to their high sugar content (more than 68.8%) and low pH (3–3.2) (Figure 27.3).

Ingredients

Strawberry pulp : 1.0 litre
Sugar : 1.0 kg
Pectin powder : 10 g
Citric acid : 5.0 g

27.2.4 Preserve

Preservation Tips

- Fruit is cooked in heavy syrup till it becomes tender, plump and transparent and is impregnated with syrup without shrivelling the fruit.
- Preserve should have at least 68% total solids and 55% of prepared fruit.
- By the principle of osmosis (water moves out of fruit and is replaced by sugar present in the syrup), the preserve is prepared and heated along with a high sugar concentration (Figure 27.4).

Value Addition

Ingredients

Whole red strawberries : 1.0 kg
Sugar : 1.0 kg
Water : 700 ml
Citric acid : 3.0 g

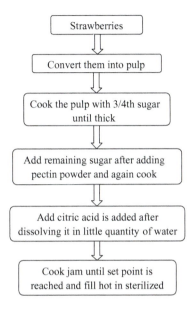

FIGURE 27.3 Flowsheet for jam preparation.

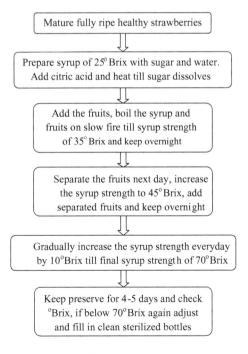

FIGURE 27.4 Flowsheet for preserve preparation.

27.2.5 Chutney

Preservation Tips

- Chutney should contain 40% minimum of fruit in the final product and total soluble solids of 50°Brix (FPO specifications, 1955). Culled, ripe, unripe, undersized, deformed, unmarketable and damaged strawberries, which are cheap, can be used to prepare chutney.
- Chutneys are delicious, palatable, hot and sweet, improve digestion and are good appetizers. Ginger, garlic, herbs and spices are added to impart flavour, vinegar to get the required acidity and sugar to enhance the flavour. All these together are responsible for the long shelf life of chutney (Figure 27.5).

Ingredients

Strawberries : 1.2 kg
Sugar : 800 g
Salt : 50 g
Onion, garlic, ginger : 50, 15 and 15 g, respectively (finely chopped)
Red chilli powder : 10 g
Black pepper, cinnamon, cardamom : 5.0 g each
Aniseed : 10 g
Vinegar : 100 ml

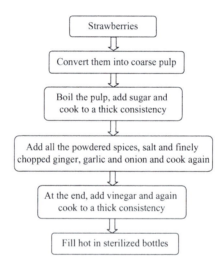

FIGURE 27.5 Flowsheet for chutney preparation.

27.2.6 Sweet 'n' Sour Pickle

Preservation Tips

By the principle of osmosis (water moves out of fruit and is replaced by sugar present in the syrup), the preserve is prepared by heating along with a high sugar concentration. Spices and oil act as preservative in case of sweet 'n' sour strawberry pickle (Figure 27.6).

Ingredients

Strawberries whole : 1.0 kg
Sugar : 900 g
Water : 300 ml
Citric acid : 5.0 g
Almonds, charmagaz, dates, raisins: 100 g each

Value Addition 437

Mustard oil : 200 ml
Spices : 1.0 g each of coarsely ground cumin seeds, cardamom (minor and major), aniseed, dalchini, black pepper
Salt : 20 g
Red chilli powder : 10 g

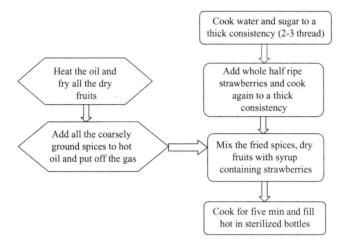

FIGURE 27.6 Flowsheet for sweet 'n' sour pickle preparation.

27.2.7 Sauce

Preservation Tips

- As per FPO specifications (1955), the sauce should contain 1.2% of acidity and 15°Brix total soluble solids.
- Thickening agents are added to prevent or retard sedimentation of solid particles in suspension in the sauce (Figure 27.7).

Ingredients

Strawberries ripe : 900 g
Granulated sugar : 75 g
Corn starch : 10 g
Citric acid : 2.5 g
Salt : One pinch

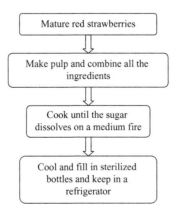

FIGURE 27.7 Flowsheet for sauce preparation.

27.2.8 Puree

Preservation Tips

- The minimum percentage of soluble solids in a medium puree free from salt should be 9°Brix and in heavy puree, it should be 12°Brix (Figure 27.8).

Ingredients

Strawberries ripe : 570 g
Granulated sugar : 50–65 g
Citric acid : Optional

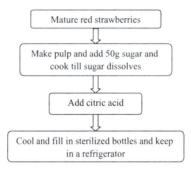

FIGURE 27.8 Flowsheet for puree preparation.

27.2.9 Frozen Strawberries

Preservation Tips

- They can be frozen with/without adding sugar. It is best to use them within six months if they are indeed sugar free. Otherwise, a light dusting of sugar before freezing will both help preserve their colour and prevent freezer burn (Figure 27.9).

Ingredients

Strawberries ripe : 570 g
Granulated sugar : 50–65 g

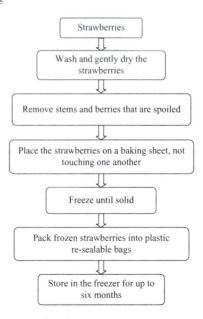

FIGURE 27.9 Flowsheet for frozen strawberries.

27.2.10 Strawberry wine

Preservation Tips

- In beverage preparation, fermentation is mediated through yeast, and in the process, it produces a range of products such as organic acids, alcohols, esters and sulphur compounds.
- High alcohol and acid content in the fermented wine keeps it stable and safe for prolonged storage (Figure 27.10).

Ingredients

Strawberries ripe : 570 g
Granulated sugar : 50–65 g

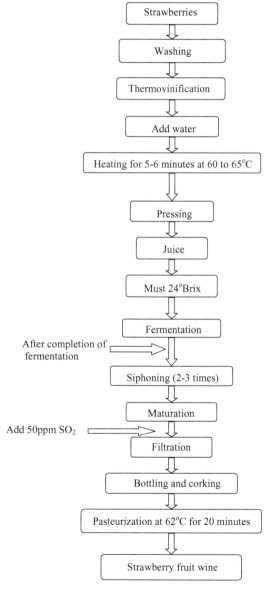

FIGURE 27.10 Flowsheet for strawberry wine. (Sharma et al., 2009, An overview on strawberry [*Fragaria x ananassa* (Weston) Duchesne ex Rozia] wine production technology, composition, maturation and quality evaluation. Natural Product Radiance, 8(4): 356–365.)

27.3 EFFECTS OF PROCESSING ON QUALITY

Washing the strawberries first in a chilled chlorine bath, slicing fruits, washing the sliced strawberries again in a second chilled chlorine bath, draining the sliced strawberries to remove moisture, washing the sliced strawberries in mixed citric acid and tribasic calcium solution, drying the sliced berries with dehumidified air, placing a selected weight of the strawberries into a number of gas impermeable containers and sealing the containers with breathable film extend the shelf life of the strawberries (US Patent 5616354).

Klopotek et al. (2005) processed strawberries into juice, nectar, wine and puree, and reported decreased ascorbic acid with the production time and processing steps, especially during heat treatment. Total anthocyanins decreased with high temperature and fermentation process, whereas, fermentation did not lead to heavy losses of total phenolics. Abers and Wrolstad (1979) reported about the factors responsible for the difference in colour and quality in commercially prepared preserves of Hood and Tioga varieties of strawberries and postulated that there is a major role of phenolics (leucoanthocyanin, flavanols etc.) in the deterioration of colour of preserves.

Although freeze-drying appears to be a great process, it does have some drawbacks. First of all, reconstitution can be an issue depending on the use of the freeze-dried fruit. The different concentrations of reconstitution media can alter the functional properties of the fruit (Mastrocola et al., 1997). This may be beneficial for producing formulated foods with a high added value. However, it may be detrimental when foods are reconstituted for consumption. The addition of sugar changes the calorie content of the berries as well as the amount of sugar and total carbohydrate present in the product (Singh, 2006).

Frozen processing has no impact on sodium, fibre, vitamin A or calcium content. It is interesting to note that frozen processing has impact on the Vitamin C content of the strawberry. Research findings suggest that there are changes in quality parameters such as anthocyanin content, total ascorbic acid, and total soluble sugars during cold storage (Cordenunsi et al., 2005), however, these changes are more closely linked to the differences in chemical composition of cultivars rather than the process of cold storage. It has also been observed that the loss of ascorbic acid occurs during initial 15 days of storage (Sahari et al., 2004; Skupien and Jakubowska, 2004). Furthermore, it is observed that the best storage temperatures are either −18°C or −24°C to preserve the characteristic quality of strawberries. In the study conducted by Ayala-Zavala et al. (2004), higher antioxidant capacity, total phenolics and anthocyanins were observed in strawberries stored at 10°C or 5°C in comparison to 0°C. This finding suggests that the antioxidant capability is reduced after the strawberries undergo frozen processing. Moraga et al. (2006) studied the changes in sugar composition (glucose, fructose and sucrose), citric acid, water and total soluble content as induced by partial dehydration and freezing–thawing processes in strawberries cv. Camarosa. Osmotic dehydration (OD) with 65°Brix sucrose solution, air drying (AD) at 45°C or combined treatments (OD–AD) were applied to reduce water content in strawberries to 70–85%. Fresh and dehydrated samples were frozen (−40°C, 24 hours) and stored at −18°C for 30 and 180 days. All samples processed by OD and OD–AD showed significant gain in sugar, and depending on the dehydration treatment, total or partial sucrose hydrolysis was observed. Dehydration treatments caused small losses of citric acid. During the freezing–thawing process, drip loss and enzymatic action also caused changes in sugar concentration. In another study, Paokkonen and Mattila (2006) observed that crushing before drying and low process temperatures improved sensory quality of dried strawberries and vacuum packaging improved storage stability.

Barbanti et al. (2004) while studying the effects of freezing (30 days) on the texture (fruit firmness and pulp cohesiveness) and on the drip loss (juice loss) of some strawberry cultivars observed that cultivars did not vary significantly in terms of juice and solid losses, however, firmness was reduced by approximately two-thirds after thawing frozen fruits, and the reduction in firmness was not correlated with the amount of drained juice. They suggested that the fruits of Maya and Paros remained firm and had lower juice loss after thawing and were identified as most suitable for industrial processing.

The strawberry varieties with less than 30% juice loss are suitable for jam and frozen food products. On the other hand, frozen strawberries with more than 60% juice loss are less desirable for processing (Kahnizadeh et al., 2005). El-Beltagy (2007) carried out thin layer solar enforced drying of strawberries (whole, halves, quarter and 3 mm discs) after pretreatment with the following solutions:

1. 1% sodium metabisulphite +1% citric acid
2. 1% ascorbic acid +1% citric acid
3. 1% citric acid and
4. 2% sodium metabisulphite

The required drying time were 28, 26, 20 and 24 hours for whole, halves, quarter and disc strawberries. Also, the halves, quarters and discs presoaked in 2% sodium metabisulphite had the highest pH and rehydration ratio. Similarly, Raghavan and Silveria (2001) and Ferrando and Spiess (2003a, b), while studying the shrinkage characteristics of strawberries osmotically dehydrated in combination with microwave drying, observed that shrinkage has a linear relationship with moisture ratio and the change in volume was bigger for the fruits osmotically dehydrated than for those non-osmotically dehydrated.

The pretreatments with calcium chloride, pectin methylesterase and crystallized sucrose increased the firmness and quality of strawberry jam after freezing (Suutarinen et al., 2000, 2002; Bugghenhout et al., 2005; Suutarinen and Autio, 2004). A moisture reduction of strawberry slices to about 60% before freezing improved the texture of strawberry fruit slices regardless of the dehydration method used (Cattaneo et al., 2002a, b). Ngo et al. (2007) used frozen and canned strawberries for preparation of jam and measured the changes in colour along with the following compositional determinations: total monomeric anthocyanins (ACN), total phenolic content (TPC) and per cent polymeric count. ACN in fresh strawberries ranged from 37.1–122.3 mg per 100 g of fresh fruit. Freezing resulted in apparent increase in ACN and transfer of 70.2% of the anthocyanins from the berries into the juice. Physical transfer of pigment to syrup also occurred with canning: there was approximately 70% loss in ACN, about 20% increase in polymeric colour and 23.5% decrease in TPC. Pronounced changes in colour and substantial losses of ACN and TPC in strawberry jams occurred during processing and nine weeks of storage. Storage of jam at 38°C over a period of nine weeks resulted in marked losses in TPC as compared to 21°C. Hilt et al. (2003) reported that the occurrence of phloridzin in strawberries was of particular interest for regulating the authenticity of strawberry products such as juices, jams and fruit preparations since phloridzin had so far been used for the detection of fraudulent admixtures.

Gao et al. (2004) reported that the rate of retention of ascorbic acid (AA) decreased in the pretreatment processes such as washing, heating and beating and there was no significant changes in dehydro-ascorbic acid (DHA). If the time of microwave treatment was under 80 seconds, it could achieve better softening effect and maintain the rate of retention of AA up to >90%. Addition of antioxidants in beating process could effectively control the loss of AA. Addition of 0.12% $(NaPO_3)_6$ in the process could make the rate of retention of AA come up to 97.18%. Addition of 0.03–0.04% tea polyphenolics could increase the rate of retention of DHA and, the effect of synthetic antioxidant was superior to that of natural antioxidant. According to Salem et al. (2006), reported that fermented milk can be prepared using *Bifidobacterium lactis* Bb-12 as starter culture where strawberry concentrate can be used up to 25% as a source of anthocyanins, vitamin C and minerals. According to Sharma et al. (2009) strawberry fruits are rich in several nutrients, especially phenolics with red colour due to anthocyanins, and, therefore, these hold promise for the production of quality red wine with medicinal properties. The presence of phenolic compounds imparts antioxidant properties and phenolic compounds can be modified by choosing appropriate method of wine production.

REFERENCES

Abers, J. E. and Wrolstad, R. E. 1979. Causative factors of colour deterioration in strawberry preserves during processing and storage. *Journal of Food Science*, 44(1): 75–81.

Ayala-Zavala, J.F., Wang, S.Y., Wang, C.Y. and Gonzalez-Aguilar, G.A. 2004. Effect of storage temperatures on antioxidant capacity and aroma compounds in strawberry fruit. *Lebensmittel Wissenschaft und Technologie*, 37(7): 687–695.

Barbanti, D., Parpinello, G.P. and Versari, A. 2004. Effects of freezing on the texture and on the drip loss of some strawberry cultivars. *Industrie Alimentari*, 43(435): 401–405.

Buggenhout, S.V., Maes, V., Messagie, I., Duvetter, T., Loey, A.V. and Hendrickx, M. 2005. Enzymatic firming of high-pressure frozen strawberry halves by means of exogenous pectin methylesterase. *Communications in Agricultural and Applied Biological Sciences* 70(2): 259–262.

Cattaneo, T.M.P., Maraboli, A., Leva, A.A. and Torreggiani, D. 2002a. Dehydrofreezing in the production of strawberry ingredients: influence on the quality characteristics of fruit yoghurt. *Acta Horticulturae*, 567(2): 791–794.

Cattaneo, T.M.P., Maraboli, A., Leva, A.A. and Torreggiani, D. 2002b. Improvement of strawberry ingredients using vacuum infusion: influence on the quality characteristics of strawberry yoghurt. *Acta Horticulturae*, 567(2): 787–790.

Cordenunsi, B.R., Genovese, M.I., Oliveira do Nascimento, J.R., Hassimotto, N.M.A., dos Santos, R.J. and Lajolo, F.M. 2005. Effect of temperature on the chemical composition and antioxidant activity of three strawberry cultivars. *Food Chemistry*, 91: 113–121.

El-Beltagy, A., Gamea, G.R. and Esse, A.H.A. 2007. Solar drying characteristics of strawberry. *Journal of Food Engineering*, 78(2): 456–464.

Ferrando, M and Spiess, W.E.L. 2003a. Mass transfer in strawberry tissue during osmotic treatment I: microstructural changes. *Journal of Food Science*, 68(4): 1347–1355.

Ferrando, M. and Spiess, W.E.L. 2003b. Mass transfer in strawberry tissue during osmotic treatment II: structure-function relationships. *Journal of Food Science*, 68(4): 1356–1364.

FPO Specifications. 1955. www.old.fssai.gov.in/Portals/0/Pdf/fpoact.pdf.

Gao, Y.J., Xiong, W.D., Xu, K.Y., Li, YR and Jiao, L.X. 2004. Study on the changes of AA and DHA during strawberry processing. *Scientia Agricultura Sinica*, 37(5): 773–775.

Haffner, K. 2002. Postharvest quality and processing of strawberries. *Acta Horticulturae*, 567(2): 715–722.

Hilt, P., Schieber, A., Yildirim, C., Arnold, G., Klaiber, I., Conrad, J., Beifuss, U. and Carle, R. 2003. Detection of phloridzin in strawberries (*Fragaria* x *ananassa* Duch.) by HPLC-PDA-MS/MS and NMR spectroscopy. *Journal of Agricultural and Food Chemistry*, 51(10): 2896–2899.

Kahnizadeh, S., Guathier, M., Rekika, D. and Deschenes, M. 2005. Evaluation of the drip loss of 30 cultivars and 9 advanced selections from agriculture and agri-food Canada National Strawberry Breeding Program. *Small Fruits Review*, 4(2): 35–40.

Klopotek, Y, Otto, Konrad and Bohm, Volker. 2005. Processing strawberries to different products alters contents of vitamin c, total phenolics, total anthocyanins, and antioxidant capacity. *Journal of Agricultural and Food Chemistry*, 53 (14), 5640–5646.

Mastrocola, D., Dalla Rosa, M. and Massini, R. 1997. Freeze-dried strawberries rehydrated in sugar solutions: mass transfers and characteristics of final products. *Food Research International*, 30(5): 359–364.

Moraga, G., Martinez, N.N. and Chiralt, A. 2006. Compositional changes of strawberry due to dehydration, cold storage and freezing-thawing processes. *Journal of Food Processing and Preservation*, 30(4): 458–474.

Ngo, T., Wrolstad, R.E. and Zhao, Y. 2007. Color quality of Oregon strawberries - impact of genotype, composition, and processing. *Journal of Food Science*, 72(1): C25–C32.

Paokkonen, K. and Mattila, M. 2006. Processing, packaging, and storage effects on quality of freeze-dried strawberries. *Journal of Food Science*, 56 (5): 1388–1392.

Raghvan, G.S.V. and Silveira, A.M. 2001. Shrinkage characteristics of strawberries osmotically dehydrated in combination with microwave drying. *Drying Technology*, 19(2): 405–414.

Rosa, M.D, Mastrocola, D. and Barbanti, D. 2006. http://www.actahort.org/book/265/265_.htm

Sahari, M.A., Boostani, F.M. and Hamidi, E.Z. 2004. Effects of low temperature on the ascorbic acid content and quality characteristics of frozen strawberry. *Food Chemistry*, 86(3): 357–363.

Salem, A.S., Gafour, W.A. and Eassawy, E.A.Y. 2006. Probiotic milk beverage fortified with antioxidants as functional ingredients. *Egyptian Journal of Dairy Science*, 34(1): 23–32.

Singh, 2006. http://www.calstatela.edu/faculty/hsingh2/articles/strawberry.research.pdf.

Value Addition

Skupien, K. and Jakubowska, B. 2004. Comparison of chemical composition of fresh and frozen fruit of selected strawberry cultivars. *Folia Universitatis Agriculturae Stetinensis Scientia Alimentaria* (3): 115–120.

Sharma, A., Joshi, V.K. and Abrol, G. 2009. An overview on strawberry [*Fragaria* x *ananassa* (Weston) Duchesne ex Rozia] wine production technology, composition, maturation and quality evaluation. *Natural Product Radiance*, 8(4): 356–365.

Suutarinen, M. and Autio, K. 2004. Improving the texture of frozen fruit: the case of berries. *Texture in Food*, Vol. 2 – Solid-Foods.: 388–409.

Suutarinen, J., Honkapaa, K., Heinio, R.L., Autio, K. and Mokkila, M. 2000. The effect of different prefreezing treatments on the structure of strawberries before and after jam making. *Lebensmittel Wissenschaft and Technologie*, 33(3): 188–201.

Suutarinen, J., Honkapaa, K., Heinio, R.L., Autio, K. and Mokkila, M. 2002. The effect of $CaCl_2$ and PME prefreezing treatment in a vacuum on the structure of strawberries. *Acta Horticulturae*, 567(2): 783–786.

United States Patent 5616354. http://www.freepatentsonline.com/5616354.html.

28 Physiological Disorders

A.K. Goswami, S.K. Singh, and Satyabrata Pradhan

CONTENTS

28.1 Malformed and Nubbins or Button Berries 445
28.2 Albinism...446
28.3 Phyllody 447
28.4 June Yellows...448
28.5 Sunscald 448
28.6 Low-Temperature Injury 448
28.7 Tip Burn 448
28.8 Uneven Ripening..449
28.9 Salts Injury 449
28.10 Bronzing...450
28.11 Dried Calyx 450
References..450

Disturbances in the normal metabolic activities in plant cells and resultant effects due to non-pathogenic factors such as extremes of macro- and microclimatic conditions, soil and water relations and imbalanced nutrient status in strawberry plants are termed physiological disorders. Adverse weather events, improperly planted and poorly maintained plantations and deficiency of various nutrients and interactions from fertilizers or pesticides are the possible causes of physiological disorders. Being a high-input-requiring, short-duration, herbaceous crop, strawberry is very easily affected by unfavourable weather conditions and improper nutrition, reflected in reduced yield of substandard fruits, causing great economic losses to growers. Some damages such as loss from lightning strikes or hailstorm cannot be corrected by human intervention, but others specially from mismanagement in standard cultural practices can be avoided by taking advanced corrective measures. In this chapter, attempts are made to holistically describe the occurrence of various physiological disorders, along with associated factors and mechanisms. In addition, based upon the latest scientific facts, suitable corrective measures have been suggested for various physiological disorders affecting successful strawberry production in the forthcoming sections.

28.1 MALFORMED AND NUBBINS OR BUTTON BERRIES

Malformed berries appear misshapen, look unattractive, and hence fetch a poor price in the market. Similarly, button berries or nubbins are misshapen, having small berries, with much lesser or negligible commercial value. The occurrence of fruit deformities is not only strongly dependent on the cultivar (i.e. genetic factor), but also promoted by unfavourable climatic factors, especially temperature (Ariza et al., 2009, 2011). As a general consideration, increasing air temperature alone or/and in combination with higher relative humidity seems to magnify the proportion of malformation of berries. The fruit deformity may derive from the loss of symmetry of the receptacle, which is often limited to the tip of the fruit (Ariza et al., 2011). A number of other factors contribute to the development of malformed berries, but important factors seem to be genotype, flower position within the inflorescence, viability of stamen and carpel, insufficient pollination, reduced development of stamens, improper growing environment and forced growth by heating (Carew et al., 2003;

Desmet et al., 2009). It is known that achenes are directly involved in secreting auxins to neighbourhood cells, which promote the thickening of the floral receptacle by increasing cell number and size (Nitsch, 1950; Mudge et al., 1981). Removal of only some of the achenes results in fruits of abnormal shape because only the parts of the receptacle adjacent to the remaining achenes continue to grow (Nitsch, 1950). Irrespective of the cultivars, misshapen fruits show a higher percentage of "small", non-functional achenes than well-developed fruits, indicating failure in achene development, which could be related to the functional integrity of reproductive structures (Ariza et al., 2011) including reduced biosynthesis and relocation of auxins, causing improper division and growth of cells causing deformed or misshapen fruits.

Low temperatures seem to decrease the rate of carpel maturation, especially in apical stigmas. Carpel fecundation leads to release of auxins, and perhaps this or other changes after fecundation can interfere and make immature carpels non-receptive. Besides carpel maturation, some other factors such as pollen grain germination and viability of ovarian cells also influence the severity of malformation of berries (Ariza et al., 2009).

The appearance of malformation is also related to high-temperature stress (30°C or higher), preventing pollen development and viability (Song et al., 1999). Similarly, low temperature restricts the activity of pollinators, so it is also an important factor contributing to the production of malformed berries. The environmental factors and the fibro-vascular systems are involved in the onset of fruit malformation indicating the need for better control of growing conditions (Massetani et al., 2016). Exclusion of insects from strawberry flowers gives decreased yield due to reduced fruit set and more malformed fruits (Katayama, 1987; Dag et al., 1994). The percentage of malformed fruits in strawberry is increased due to insufficient pollination, especially in protected cultivation (Hughes, 1962; Vasquez et al., 2006; Kruistum et al., 2006). Due to inadequate pollination, production of growth-promoting substances is greatly reduced on account of the absence of fertilized seeds in the achenes. As a consequence, the resultant receptacles fail to grow properly, causing malformation of berries (Robbins, 1931; Nitsch, 1952; Hughes, 1962). A sudden rise in temperature after flower bud initiation may reduce the rate of ovule fertilization and increase the proportion of unfertilized achenes, resulting in increased percentage of malformed fruits in first inflorescence. In addition, high temperature for long duration increases abortion of achenes during early embryogenesis due to low pistil potential, and finally induced fruit malformation in second inflorescence (Pipattanawong et al., 2009). Sharma and Singh (2010) observed a strong and an inverse relationship between total phenol content, activities of polyphenol oxidase and lipoxygenase (LOX) activity and the production of abnormal fruits in strawberry.

Similarly, some insects such as lygus bug are also associated with occurrence of malformed and button berries disorder (Garren, 1981). In spite of placement of beehives in strawberry plantations, the problem of frequent occurrence of malformed and button berries are observed and hence, these disorders appear to be due to disturbed physiological status in the plants and, most probably, improper nutrition might play a vital role in the occurrence of malformed fruits. Among different nutrient elements, the role of major nutrients such as nitrogen, potassium and calcium and minor elements such as boron appear to be important (Sharma and Sharma, 2004). Similarly, the correlation between LOX activity and, malformed and button berries have also been observed positive, indicating that excess of nitrogen and potassium, and deficiency of calcium and boron are related to the production of malformed and buttons or nubbins in strawberry (Sharma and Singh, 2008). Gibberellic acid has been reported to influence vegetative growth, flowering, fruiting and various disorders in strawberry and many other fruit crops (Thompson and Guttridge, 1959; Paroussi et al., 2002; Singh and Kaul, 1970).

28.2 ALBINISM

Strawberry suffers from various physiological disorders, but albinism is considered as the most serious problem, particularly under protected cultivation. Albino fruits grow normally but do not

Physiological Disorders 447

ripen properly (Lieten, 1999). Fruits suffering from albinism appear bloated, irregularly pink or totally white and sometimes swollen with pulp remaining pale. Affected fruits have poor flavour and tend to be acidic. Due to lack of firmness, albino fruits are susceptible to fruit rot during storage or transportation (Lieten and Marcelle, 1993; Sharma and Sharma, 2004). During the production season, periods of warm weather followed by overcast and foggy skies are particularly favourable for albino fruit. Hot weather during fruit ripening becomes a favourable condition for strawberry plants to exhibit albinism in the fruits. However, hot weather before flowering tends to reduce the incidence of albinism (Dale et al., 2017), as heat shock proteins are produced when the plants are subjected to heat before flowering, which could prevent the degradation of the enzymes involved in the biosynthesis of anthocyanins.

It has been observed that heavy fruit set causes poor leaf development to supply the required carbohydrate (Ulrich et al., 1980). Lower light intensities tend to increase the incidence of albinism, probably because of reduction in the supply of sugars being partitioned into the fruits (Lieten and Marcelle, 1993). Besides, albino berries occur primarily due to excess of nitrogen and potassium, and deficiency of calcium (Garren, 1981). Albino fruits are observed in fields of high fertility, especially with excess amounts of nitrogen supply, suggesting that overnutrition might be a cause of albinism. In addition, low light intensities increase the incidence of albinism possibly by decreasing the supply of sugars in fruits. This observation is consistent with the fact that more albino fruits occur during cloudy seasons. Less dry matter content in albino fruits than in normal fruits leads to infer that albino fruits become more hydrated, which may be due to increased competition for different nutrients between leaves and fruits during an excessive vegetative growth period, because luxuriant growth favours albinism in strawberry (Sharma and Sharma, 2003).

It appears that the proper balance needs to be maintained between sugar formation and its distribution for basic metabolism, leaf growth, root development and fruit production. In reality, however, sugar distribution may be disrupted at any time by environmental factors, such as insufficient sunlight, improper chilling, planting time, nutrition, irrigation, pests and diseases; as a consequence, leaves and fruits may either not develop properly or not in proper sequence, resulting in an increased incidence of albinism (Ulrich et al., 1980). Of the nutrients, potassium remains notably higher with lower calcium in albino fruits than the normal ones, which results in higher ratios for N:Ca and K:Ca, thereby giving some indications that albino fruit are physiologically more senescent on ripening than normal fruits, although they have poor colour development (Marcelle, 1984; Sharma and Sharma, 2003). Addition of silicon (as K_2SiO_3) to the nutrient solution may cause poor colour development in strawberry (Lieten et al., 2002). Production of albino fruits has also been observed to be correlated with mixoploidy nature of genotypes (Plante and Rousseau, 2001).

Selection of suitable variety, proper light conditions, management of insect-pests, nematodes and diseases and balanced nutrition are some corrective measures to avoid this problem. The flow of sugar to the fruit may also be stimulated by appropriate application of plant bioregulators (Ulrich et al., 1980).

28.3 PHYLLODY

In strawberries, phyllody includes abnormal development of floral parts into leafy structures. There are two causes of phylloidy, namely, infectious and non-infectious. Infectious phyllody occurs due to infection by phytoplasma (bacteria-like organisms) that are pathogenic to plants and are vectored by leafhoppers. Leafhoppers carry phytoplasma in their bodies and inject them while feeding on plants. Two diseases that cause phyllody are aster yellows and green petal. Strawberry plants infected with phytoplasma often continue to bear deformed fruit (Bolda and Koike, 2015). Non-infectious phyllody seems to be associated with the excess hours of supplemental chilling of the strawberry plants in storage.

28.4 JUNE YELLOWS

June yellows is a non-infectious genetic disorder affecting several cultivars of strawberry throughout the world. It is characterized by mild variegation to severe white streaking. The cause is not known, and many hypotheses have been suggested but there is no conclusive evidence to implicate a pathogenic agent, nutritional imbalance or physiological or genetical cause (Hughes and Beech, 1989). Affected plants do not recover. There is no known control for June yellows, however discouraging the runner production from affected mother plants helps in minimizing the occurrence of June yellows in subsequent plantations.

28.5 SUNSCALD

Sunscald is generally observed when there are sudden changes in temperature and light conditions. Symptoms appear on the upper side of fruit, near the calyx, just before ripening. The fruit becomes pale and mushy at first, eventually drying down to a distinct, firm, bleached lesion ranging from pink to off-grey. Berries growing on the south or west side of rows and those without adequate foliage canopy are most susceptible. Fruits exposed to sunlight are usually most affected, and the surface facing the sun shows greatest damage. Similarly, some varieties are more susceptible than others. High temperatures and overexposure of strawberry fruit to the sun are said to be the primary cause of sunscald (Polito et al., 2002). Symptoms are more common when hot sunny weather occurs after a prolonged period of cool cloudy weather. The spots of heat injury on the lower side of berries are also very common, when strawberries are retained in field late in the season with black polyethylene as mulch.

28.6 LOW-TEMPERATURE INJURY

Strawberry flowers are particularly susceptible to frost damage because the plant is low to the ground, and the blossoms open towards the sky during December–January, particularly in subtropical areas. Strawberry plants can tolerate temperatures as low as −7°C when dormant but actively growing, fruit-bearing plants are susceptible to low temperature (<15°C) and frost injury occurs when the temperature reaches −5°C or lower. Symptoms of low temperature injury include flowers with black centres that fail to produce fruit or/and misshapen fruits due to disrupted pollination during flowering. Additionally, freezing damage may result in loss of integrity of the tissue in existing fruit, marginal leaf burn/necrosis and in some cases crown damage. The tips and edges of new leaves may look water-soaked and then turn brown and dry. These leaves are misshapen because the damaged tissues cannot expand normally. Sometimes, the lower and upper leaf surfaces become separated from each other, giving the leaf a crinkled appearance. Frost injury usually occurs to primary flowers because these are the first to open. Primary flowers also produce the largest and most valuable fruit. Frost-damaged tissues are very susceptible to *Botrytis* grey mould disease too.

Strawberry plants develop symptoms of winter injury when ice crystals are formed in crown tissues disrupting cell membranes and functions. Applications of heavy doses of nitrogenous fertilizers in the autumn can increase the chances of winter injury to the plants because they don't harden off or acclimatize for winter. Raised beds can also increase the risk of winter injury because cold can penetrate the soil more deeply. Mulching with hay, straw, leaves or similar materials can reduce the probability of injury. A blanket of unpacked snow provides protection from winter injury (http://ucanr.edu/sites/sdim). Anticipating cold temperatures and preparing measures such as sprinkler irrigation, heaters, or plastic tunnels can prevent or minimize damage to plants in open field production systems.

28.7 TIP BURN

Expanded young leaves are cupped, puckered or distorted, with blunt tip. Tip burn occurs in folded emerging leaves. Globs of syrupy liquid may be found on blade midribs. Fruit develops a dense number of achenes (seeds) in patches or over the entire fruit; it may have a hard texture and acidic

Physiological Disorders   **449**

taste. The symptom of this disorder does not move from older to younger leaves like nitrogen deficiency. Fruits and younger leaves transpire less water or none at all, and these tissues are where the symptoms of calcium deficiency appear first. It is often confused with injuries from herbicides, pesticides etc.

Calcium deficiencies are related with this disorder but often it is caused by the plant's microclimatic conditions and is associated with periods of rapid growth or fluctuations in soil moisture. Calcium is taken up by the root tips, and once in the plant, it moves within the transpiration stream. It is transported to most actively transpiring plant parts such as the older leaves. Calcium does not easily move out of plant tissues. The incidence of tip burn is associated with foliar K:Ca and K:Mg ratios. Ratios above 3.4 for K:Mg and 1.77 for K:Ca represent a risk of more than 50% of tip burn incidence. The balance of potassium, calcium and magnesium in leaves of individual cultivars depends on transpiration rates, which is governed by temperature and humidity. Hence, the only way to prevent the occurrence of tip burn seems by way of bypassing the xylem transport, with foliar applications of calcium and eventually manganese. In the long-term, breeding for cultivars that accumulate large amounts of calcium and manganese in leaves could be a permanent solution for tip burn disorder in strawberry (Palencia et al., 2010).

28.8   UNEVEN RIPENING

This condition is thought to develop when high temperatures occur during fruit ripening, which affect the enzymes that cause normal pigmentation in ripe fruits. The strawberry fruits which are exposed to high temperatures during early stage of their development are not as susceptible as fruits exposed to cool temperatures early in the season. The variety Cavendish is especially susceptible to this condition, where white shoulders and blotches occur on otherwise ripe fruit. Although the affected fruits are ripe and sweet, but these fail to show uniform colour development.

28.9   SALTS INJURY

Strawberry plants are most sensitive to high concentrations of salts, because of shallow root systems. The concentration of salts may increase with high temperatures (high evapotranspiration losses). Reduced rate of vegetative growth and fruiting due to excess salt concentration in the root zone may occur before the characteristic marginal leaf "burn" is observed. Strawberry varieties with good early vigour may be slightly more tolerant to salt injury than slow-growing ones. Ability of salts to injure or kill strawberry plants depends on the composition of ions in salt and their concentration. Research results show that chlorides, especially calcium chloride, are most harmful to strawberry even at very low concentration (EC <1 dS/m). Sulphates are much safer, however, sodium sulphate can have negative effect on fruit production at EC <3 dS/m, while strawberry can tolerate EC up to 5 dS/m without significant yield losses when exposed to potassium sulphate.

The strawberry cv. Camarosa has been observed more tolerant than cv. Chandler against sodium chloride induced salinity stress (34 mM). The reductions in stomatal conductance and transpiration rate represent adaptive mechanisms to cope with excessive salt (Turhan and Eris, 2007). Treatments with sodium chloride generally reduces the leaf and root dry weight. Relative water content of leaves can be maintained despite increased salt concentrations, while loss of turgidity can be increased by sodium chloride. The treatment with sodium chloride generally increases sodium and chlorine contents in all plant parts. It can be concluded that "Camarosa" has the ability to regulate osmosis. "Chandler" also tolerates the salt injury at low salt concentrations (Turhan and Eris, 2009). In sodium chloride-treated plants, the amounts of sodium, chlorine, calcium and magnesium in the aerial plant parts are generally increased, while the amount of potassium and phosphorus decreases. The salt applications tend to increase the amount of sodium and chlorine but decreases the amount of potassium and magnesium in the roots, but they do not change the amount of calcium and phosphorus (Turhan and Eris, 2005). Drenching of potassium silicate (K_2O_3Si) at 1000 ppm in the nutrient solution of strawberry plants may be a routine strategy to maintain satisfactory growth and yield of strawberry under salinity prone growing conditions (Yaghubi et al., 2016).

28.10 BRONZING

Symptoms of bronzing consist of discolouration, desiccation and cracking of the fruit surface in response to injury due to heat or solar radiation. The damaged fruits are typically tan, brown or yellow-brown, hence it is termed "bronzing". The period of occurrence of bronzing coincides with high temperature, low relative humidity and high solar irradiance (Polito et al., 2002). The disorder occurs annually to some extent, but timing of its occurrence and severity varies from year to year. Bronzing results in serious economic losses due to production of significant proportion of unmarketable fruits (Larson and Sjulin, 2001).

28.11 DRIED CALYX

The strawberry dried calyx disorder (SDCD) was observed during 2005 in the United States of America and Spain. Initially newly opened and already formed flowers show slight darkening resembling to salt injury, and later progress towards the calyx causing burning, fruit discolouration and malformation. Strawberry cultivars such as Festival, Camino Real and Palomar are more severely affected than others. The exact cause is unknown, but reduction in fertilization rates during the days before an expected freeze and maintenance of regular irrigation particularly in susceptible cultivars appear to help to minimize the incidence of SDCD (Santos et al., 2009).

REFERENCES

Ariza, M.T., Lopez-Aranda, J.M., Cid-Atencia, M.T., Soria, C., Medina J.J. and Miranda, L. 2009. Effect of temperature on carpel maturation and misshapen fruit in strawberry. *Acta Horticulturae*, 842: 757–760.

Ariza, M.T., Soria, C., Medina, J.J. and Martinez-Ferri, E. 2011. Fruit misshapen in strawberry cultivars (*Fragaria × ananassa*) is related to achenes functionality. *Annals of Applied Biology*, 158: 130–138.

Bolda, M. and Koike, S. 2015. Phyllody on Strawberry. http://ucanr.edu/blogs/blogcore/postdetail.cfm?postnum=17588

Carew, J.G., Morretini, M. and Battey, N. H. 2003. Misshapen fruit in strawberry. *Small Fruits Review*, 2: 37–50.

Dag, A., Dotan, S. and Abdul Razek, A. 1994. Honeybee pollination of strawberry in greenhouses. *Hassadeh*, 74(10): 1068–1070.

Dale, A. , Pirgozliev, S. and Kermasha, S. 2017. Heat before flowering decreases albinism in strawberries. *Acta Horticulturae*, 1156: 453–456.

Desmet, E.M., Verbraeken, L. and Baets, W. 2009. Influence of heat spread system on malformation of "Elsanta" strawberries in spring. *Acta Horticulturae*, 842: 271–273.

Garren, R. 1981. Causes of misshapen strawberries. In: N. F. Childer (ed.). *The Strawberry*. Horticultural Publication, Gainesville, Florida: 326–328.

Hughes, H.M. 1962. Pollination studies. Progress Report M.A.A.F., N.A.A.S., Rept. Efford Exp. Hort. Stan.: p. 1961.

Hughes, J. D. and Beech, M.G. 1989. Strawberry june yellows. *Acta Horticulturae*, 265: 599–600.

Katayama, E. 1987. Utilization of honey bees as pollinators for strawberries in plastic greenhouse in Tochigi prefecture. *Honeybee Science*, 8(4): 147–150.

Kruistum, G.V., Blom, G., Meurs, B. and Evenhuis, B. 2006. Malformation of strawberry fruit (*Fragaria* x *ananassa*) in glasshouse production in spring. *Acta Horticulturae*, 708: 413–416.

Larson, K. D. and Sjulin, T. M. 2001. Influence of polyethylene bed mulch, drip irrigation rate, and intermittent overhead sprinkling on strawberry fruit bronzing. *HortScience*, 36: 443.

Lieten, F. and Marcelle, R.D. 1993. Relationships between fruit mineral content and the 'albinism' disorder in strawberry. *Annals of Applied Biology,* 123: 433–439.

Lieten, P. 1999. Albino berries. *Growers*, 132: 22–23.

Physiological Disorders

Lieten, P., Horvath, J. and Asard, H. 2002. Effect of silicon on albinism of strawberry. *Acta Horticulturae*, 567: 361–364.

Marcelle, R.D. 1984. Mineral analysis and storage properties of fruit. In: VIth *International Colloquium for the Optimization of Plant Nutrition*, Vol. 2. (Martin-Prevel, P. ed.) AIONP/GERDAT, Montpellier, France: 365–371.

Massetani, F. , Palmieri, J. and Neri, D. 2016. Misshapen fruits in "Capri" strawberry are affected by temperature and fruit thinning. *Acta Horticulturae*, 1117: 373–379.

Mudge, K.W., Narayanan, K.R. and Poovaiah, B.W. 1981. Control of strawberry fruit set and development with auxin. *Journal of the American Society for Horticultural Science*, 106: 80–84.

Nitsch J.P. 1950. Growth and morphogenesis of the strawberry as related to auxin. *American Journal of Botany*, 37: 211–215.

Nitsch, J.P. 1952. Plant hormones in the development of fruits. *The Quarterly Review of Biology*, 27(1): 33–57.

Palencia, P., Martinez, F., Ribeiro, E., Pestana, M., Gama, F., Saavedra, T., de varennes, A. and Correia, P. J. 2010. Relationship between tipburn and leaf mineral composition in strawberry. *Scientia Horticulturae*, 126: 242–246.

Paroussi, G., Voyiatzis, D.G., Paroussi, E. and Drogour, P.D. 2002. Effect of GA_3 and photoperiod regime on growth and flowering in strawberry. *Acta Horticulturae*, 567: 56–60.

Pipattanawong, R. , Yamane, K., Fujishige, N., Bang, S. and Yamaki, Y. 2009. Effects of high temperature on pollen quality, ovule fertilization and development of embryo and achene in 'Tochiotome' strawberry. *Journal of Japanese Society for Horticultural Science*, 78 (3): 300–306.

Plante, S. and Rousseau, H. 2001. Relationship between production of albino berries and mixoploidy in the strawberry (*Fragaria* x *ananassa*) cv. Kent. *Acta Horticulturae*, 560: 197–200.

Polito, V.S., Larson, K.D. and Pinney, K. 2002. Anatomical and histochemical factors associated with bronzing development in strawberry fruits. *Journal of the American Society for Horticultural Science,* 127: 355–357.

Robbins, W.W. 1931. *The Botany of Crop Plants*. McGraw-Hill, New York, New York.

Santos M.B., Chandler, C.K. and Whidden, A.J. 2009. Assessing the possible causes for the "strawberry dried calyx disorder" in Florida and Spain. *Acta Horticulturae*, 842: 829–832.

Sharma, R.R. and Sharma, V.P. 2003. Plant growth and albinism disorder in different strawberry cultivars under Delhi conditions. *Indian Journal of Horticulture,* 61: 92–93.

Sharma, V. P. and Sharma, R. R. 2004. *The Strawberry*. ICAR, New Delhi, India: p. 166.

Sharma, R.R. and Singh, Rajbir. 2008. Fruit nutrient content and lipoxygenase activity in relation to the production of malformed and button berries in strawberry (*Fragaria* × *ananassa* Duch.). *Scientia Horticulturae*, 119: 28–31.

Sharma, R.R. and Singh, Rajbir. 2010. Phenolic content pattern, polyphenol oxidase and lipoxygenase activity in relation to albinism, fruit malformation and nubbins production in strawberry (*Fragaria* × *ananassa* Duch). *Journal of Plant Biochemistry & Biotechnology*, 19: 67–72.

Singh, R. and Kaul, G.L. 1970. Effect of gibberellic acid on strawberry. I. Growth and fruiting. *Proceeding of the International Symposium on Horticulture*: 315–327.

Song, J, Nada, K. and Tachibana, S. 1999. Ameliorative effect of polyamines on the high temperature inhibition of *in vitro* germination in tomato *Lycopersicon esculentum* Mill. *Scientia Horticulturae*, 80: 203–212.

Thompson, P.A. and Guttridge, C.G. 1959. Effects of gibberellic acid on the initiation of flowers and runners in the strawberry. *Nature*, 184: 72–73.

Turhan, E. and Eris, A. 2005. Effects of sodium chloride applications and different growth media on ionic composition in strawberry plant. *Journal of Plant Nutrition*, 27: 1653–1665.

Turhan, E. and Eris, A. 2007. Growth and stomatal behaviour of two strawberry cultivars under long-term salinity stress. *Turkish Journal of Agriculture and Forestry*, 31: 55–61.

Turhan, E. and Eris, A. 2009. Changes of growth, amino acids, and ionic composition in strawberry plants under salt stress conditions. *Communications in Soil Science and Plant Analysis*, 40: 3308–3322.

Vasquez, R., Ballesteros, H., Ortegon, Y. and Castro, U. 2006. A commercial strawberry crop (*Fragaria chiloensis*) being directly pollinated by bees (*Apis mellifera*). *Revista Corpoica*, 7(1): 50–53.

Ulrich, A., Mostafa, M. A. E. and Allen, W. W. 1980. Strawberry deficiency: A visual and plant analysis guide to fertilisation. Pub. No. 4098. University of California, Department of Agricultural Science, Nerkeley, p. 58.

Yaghubi, K., Ghaderi, N., Vafaee, Y. and Javadi, T. 2016. Potassium silicate alleviates deleterious effects of salinity on two strawberry cultivars grown under soilless pot culture. *Scientia Horticulturae*, 213: 87–95.

29 Integrated Disease Management

M.K. Pandey, B.K. Pandey, and A.K. Tiwari

CONTENTS

29.1 Powdery Mildew 454
    29.1.1 Management 454
29.2 Anthracnose 456
    29.2.1 Disease Management 458
29.3 *Verticillium* Wilt 459
    29.3.1 Disease Management 460
29.4 Black Root Rot 461
    29.4.1 Disease Management 461
29.5 Red Stele 462
    29.5.1 Disease Management 462
29.6 Leather Rot 463
    29.6.1 Disease Management 463
29.7 Rhizopus Rot 464
    29.7.1 Disease Management 464
29.8 Grey Mould Rot 464
    29.8.1 Disease Management 465
29.9 *Fusarium* Wilt 468
    29.9.1 Disease Management 469
29.10 Leaf Spots 469
    29.10.1 Mycosphaerella Leaf Spot 469
    29.10.2 Disease Management 470
    29.10.3 *Pestaliopsis* Leaf Spot 471
    29.10.4 Yellow Leaf Spot 471
    29.10.5 Angular Leaf Spot 471
29.11 Leaf blight 472
29.12 Leaf Scorch 472
29.13 Viruses 473
References 474

Strawberries are one of the most delicate and soft fruits, and require very specific growing environments. Variations in growing conditions and field stresses favour the prevalence of a variety of fungi, bacteria, viruses and virus-like organisms causing diseases, which adversely affect growth, yield and fruit quality of strawberries. Being a crop having a very short duration of fruit development, management of diseases by using only synthetic chemicals is hazardous to human health and environment. This calls for a paradigm shift, favouring the integrated management of diseases to minimize risk with respect to food safety and environmental hazards as discussed in this chapter.

29.1 POWDERY MILDEW

Powdery mildew, the most important disease of strawberry, caused by the fungus *Sphaerotheca macularis* (Wall ex. Fr.) Jazc. (*S. macularis* f. sp. *fragariae*), is a major limitation for strawberry production in most strawberry-growing areas of the world (Hietaranta, 2006). The symptoms occur on all aerial parts, including the calyx and the fruits. Initially, the fungal infection starts from the underside of leaves and develops into white powdery growth. These leaves curl upward like a cup and show reddish blotches. In severe infections, leaves become burnt at the margins to a greater extent. White powdery mass also develops on the infected flower stalk, flower and fruits. Delayed colour development takes place in infected fruits (Wilhelm, 1961).

The pathogen *S. macularis* (Wall. ex. Fr.) Jacz f. sp. *fragariae* is an obligate pathogenic fungus and produces conidia from conidiophores and ascospores. The pathogen is air borne and overwinters through mycelium in old infected plants. It requires optimum temperature (16–18°C) and relative humidity (80–100%) for germination of conidia but maximum germination occurs at 20°C. Free water has a lethal effect on conidial germination and rain has a deleterious effect on spore dispersal (Peris, 1962; Jhooty and Mc Kean, 1965). Amsalem et al. (2006) also observed the best conidial germination and conidial germ tube length between 15°C and 25°C with relative humidity (RH) higher than 75%, but less than 98%. High light intensity reduced germination and hyphal growth. The viability of conidia on infected leaves has been observed at temperatures ranging from 15–35°C at 80–85% RH. Conidia survival declined over time, but a certain percentage of conidia remained active up to five months after incubation. The rate of conidial germination was significantly higher on young leaves than on older leaves. Cleistothecia as sexual fruiting bodies form in nature (Haward and Alberts, 1973), but the role of cleistothecia as a primary source of inoculum is not important because this stage is rarely formed. The shortest time from inoculation to appearance of the first disease symptoms is about four days, at 20°C and 30°C with RH above 75%. In growth chambers, appearance of symptoms requires temperatures of 10°C and 30°C, RH above 95% and radiation of 7000 lux (Blanco et al., 2004). On the basis of severity, disease classes have been standardized for fruit and leaf (Table 29.1).

29.1.1 MANAGEMENT

Use of resistant cultivars is the best method to manage the disease. In the cultivar "Yumenoka", the symptoms of powdery mildew are apparently less severe than in "Tochiotome" and "Akihime" (Ban and Yabe, 2005). Several cultivars that have shown resistance to susceptible reaction against powdery mildew pathogen are given in Table 29.2.

TABLE 29.1

Standards on Powdery Mildew Disease: Classes

Disease Class	Fruit or Leaf
0	None
1	Individual ill spot and *small ill spots less than 5 points
2	Lesion area <10%
3	10%< lesion area <30%
4	30%< lesion area <50%
5	Lesion area >50%

Source: Liang and Lin, 2014, Genetic study on resistance to powdery mildew in strawberry. Acta Horticulturae, 1049: 255–258.

* For leaf only

Integrated Disease Management

TABLE 29.2

Host Resistance

Response	Cultivars	Reference
Resistance	Milsei-Tudla	Brahm et al. (2005)
	Harmonie	Khanizadeh et al. (2005)
	Hongshimei	Gu et al. (2005)
	Solprins	Davik and Honne (2005)
	Mechyang	Kim et al. (2004)
	Brunswick	Jamieson et al. (2004b)
	Cabot	Jamieson et al. (2004a)
	Toyonoka, Akihime and Allstar	Zhang et al. (2004)
	Irma and Queen Elisa and Demetra and Rubea	Faedi and Baruzzi (2004)
	Arking and Real	Moore (1981)
Moderate resistance	Camarosa, Aromas, Oso Grande, Sweet Charlie and Campinas	Brahm et al. (2005)
	Evangeline	Jamieson et al. (2004c)
Moderate Susceptible	Marina	Lopez-Aranda et al. (2004)
	Dover	Brahm et al. (2005)
Susceptible	Burkley, Vila Nova and Santa Clara	Brahm et al. (2005)
	VimaReg, Zanta	Hietaranta (2006)
	FL 97-39	Rondon et al. (2004)
	Tamar	Linsley et al. (2006)

Biocontrol agents (*Ampelomyces quisqualis, Bacillus subtilis* and *Trichoderma harzianum*) are the promising tool for reducing the use of chemical pesticides against diseases. Biocontrol agents can manage the disease, but not as effectively as chemicals, and their level of activity is very much specific to the location, environment and time of application. When biocontrol agents are applied in combination with chemicals, a good reduction can be achieved, with maintaining the same control efficacy. Two sprays of augmenting native populations of the hyperparasite *Lecanicillium lecanii* (*Verticillium lecanii*) reduce the infection of powdery mildew partially (Miller et al., 2004). Invert emulsion formulation of *T. harzianum* is a promising treatment to prolong the post-harvest shelf life of fresh fruit from infection up to two months, and reduces the size of decay lesion (Batta, 2007). The Q-fect WP (*Ampelomyces quisqualis* 94013) is effective as a biofungicide for controlling powdery mildew (Lee et al., 2004). *Bacillus* TS02 is a biocontrol strain which has been isolated from soil. TS02 has been identified as a new strain of *Bacillus cereus* based on the homology of its 16SrRNA sequences to the reference strains registered in Genbank. TS02 has shown to form a monophyletic group with strains of the *B. cereus* complex group in the 16SrDNA sequence analysis, and share similarity values of 99.9% with the *B. cereus* strains in the sequence analysis. TS02 with 3×10^9 CFU/ml live bacteria can effectively control strawberry powdery mildew (Chen et al., 2014). A *Bacillus velezensis* antagonist, sprayed thrice at five-days intervals on powdery mildew-affected plants has successfully controlled the disease in strawberry with control efficacy of 88.7% (Shin et al., 2014).

Kanto et al. (2014) demonstrated the control of powdery mildew with UV-B radiation in soil culture and in drip fertigation. The lighting system was suspended two meter above the strawberry plants from the ceiling at intervals of five meter in a line in the middle of a vinyl house, and plants were radiated using UV-B fluorescent lamps daily between 9:00 and 15:00, with the light energy of ca. 2–7 kJ m^{-2} d^{-1}. The lighting system results in 100% control of powdery mildew. UV-B radiation induced transcription of strawberry genes associated with disease resistance, such as phenylalanine ammonia-lyase, chalcone synthase, chalconeisomerase, β-1,3-glucanase and osmotin-like protein. Furthermore, production of antifungal substances was induced in leaves. UV-B radiation also made strawberry to accumulate more anthocyanin in fruits than the control.

Spraying the plants with combination of salicylic acid (systemic acquired resistance-inducing) and mixture of micronutrients (iron, manganese and zinc) is three times more effective in reducing the disease than spraying with salicylic acid or micronutrients alone (Fouly, 2004). Similarly, application of "Luying", a mechanical oil emulsion (Liang et al., 2004), potassium silicate alone (Kanto et al., 2004) or in combination with calcium chloride (Palmer et al., 2006) and sodium bicarbonate (1%) with 0.5% polyoxyethylene sorbitanmonolaurate (Tween 20) have shown (Nam et al., 2003) effective control of powdery mildew in strawberry.

Many chemical fungicides (azoxystrobin, kresoxim-methyl, penconazole, hexaconazole, miclobutanil, tetraconazole, triadimenol, nuarimol, fenarimol, pyraclostrobin+nicobifen, mancozeb+famoxadon, pirifenox, pirazofos, polyoxin Al, sulphur, sulphur+pinolene, ultrafine oil, flusilazole, mycobutonil, tridemifon, fenarimol and buprimate etc.) have been reported effective in inhibition of germination and/or the rate of germ tube elongation. The chemicals that have no significant effect on germination are tetraconazole and triadimenol, but they can reduce germ tube elongation (Amsalem et al., 2004).

Sulphur-containing fungicides and benomyl are effective but sometimes phytotoxic, causing burning of leaves at above 27°C. Two to three sprays of buprimate or tridemifon at 15 days interval give good results at the nursery stage. Fungicides give effective control when they are applied after transplanting and just before flowering (Moore, 1981). Benzothiadiazole (BTH) can enhance the accumulation of soluble and cell-wall-bound phenolics in strawberry leaves which may then be involved in the BTH-induced resistance to powdery mildew infection (Hukkanen et al., 2007). Application of BTH (0.6–1g/litre) on young strawberry leaves provides effective mildew control and induces resistance in greenhouses. BTH and related bioactivators (viz. flavonols, quercetin and kaempferol) may be a realistic alternative for chemical control against powdery mildew in strawberries (Anttonen et al., 2003; Hua Xiu and Chen Xian, 2004). Kresoxim-methyl at a dilution of 3000 times has shown maximum control of powdery mildew in strawberry without phytotoxic symptoms in comparison to thiophanate-methyl, emulsifiable myclobutanil, sulfur and triadimefon. Another fungicide "Pancho" TF (cyflufenamid 3.4%+triflumizole 15%) @ 50 g/litre is highly effective against powdery mildew (Yokota, 2004). Signum (Hauke et al., 2004), flusilazole and myclobutanil (Zhang et al., 2004) are also effective fungicides for the control of powdery mildew.

29.2 ANTHRACNOSE

Strawberry anthracnose or strawberry black spot is a major disease throughout the world, infecting all parts: fruits, runners, crown and roots of strawberry (Freeman, 2008). Strawberry anthracnose is caused by several species of fungi in the genus *Colletotrichum*: *C. acutatum*, *C. fragariae*, and *C. gloeosporioides* (Smith and Black, 1993; Smith, 2013). It is difficult to raise strawberry plantlets during the main rainy season. The causal fungus, *C. fragariae*, infects all above-ground plant parts; however, the disease is most severe when the fungus infects the crown, causing crown rot, wilt, and death (Smith, 2013). Anthracnose disease of strawberries caused by *C. fragariae* Brooks was first reported from Florida (1926–29) by Brooks (1931). Later on, the disease was observed to spread in different countries. The first report of *C. acutatum* as the causal agent of anthracnose fruit rot of strawberry was documented in France during 1981 (Denoyes Rothan, 2006), Israel during 1995 (Koch-Dean et al., 2002), Norway during 1999 (Stensvand et al., 2001), Denmark during 2000 (Sundelin et al., 2005), Finland during 2000 (Parikka et al., 2006), the United States during 2000 (Turechek et al., 2002), Sweden during 2003 (Nilsson et al., 2005), China during 2004 (Dai et al., 2006), Malta during 2005 (Porta-Puglia and Mifsud, 2006), northwestern Tucuman, Argentina during 2000 (Ramallo et al., 2000) and in India, during 1974 by Singh (1974) from Bangalore. Monaco et al. (2000) reported *C. gloeosporioides* (*Glomerella cingulata*) as an anthracnose-causing agent in strawberry for the first time in northwestern Argentina.

Garrido et al. (2009) identified *C. acutatum* as the most important causal agent of strawberry black spot. The species has been subjected to a thorough taxonomic reassessment and is now suggested to be a species complex, comprising 31 species. *C. nymphaeae* (formerly known as *C. acutatum*, molecular group A2) is the most important agent of strawberry anthracnose (Damm et al., 2012). The pathogen can survive in infected strawberry debris at least over one winter or may be even for a longer time (Lilja et al., 2006). The pathogen is able to infect weeds and survive in weed debris for at least one year on the soil surface (Parikka and Lemmetty, 2009).

The pathogens can also survive in soil for at least nine months without host plants and make it difficult to raise the plantlets in the season of high temperature and humidity. In infected soils, plantlets become infected when soil containing spores is splashed onto crowns or stems by rain or irrigation water. In fumigated fields, the disease usually originates on infected nursery stock or from volunteer strawberry plants in adjacent fields.

C. acutatum J.H. Simmonds ex J.H. Simmonds causes anthracnose fruit rot and black leaf spots (Inada and Yamaguchi, 2006), whereas *C. gleosporiods* Penz and *C. fragariae* Brooks infect primarily the crown of the plants and cause crown rot and wilt. Symptoms may appear as sunken, dark brown and circular lesions on petioles, stolon (runners), which may then be girdled, resulting in wilting and death of the leaf or of the daughter plant beyond the lesion on the stolon. The fungus often spreads into the crowns of young plants, which it rots, and the plants then die in the nursery or after transplanting in the field.

On fruits, rot is initially restricted to only ripe fruits, and characterized by one or more light brown soaked lesions. Two or more lesions collapse and are occasionally covered by buff-coloured spores (Howard, 1972). On leaves, minute dull, violet-black or black spots appear and surrounded by the yellow region. In later stages, spots coalesce and form bigger spots causing defoliation. Other symptoms of anthracnose on plants are bud rot and flower blight.

The *C. acutatum* produces orange spore masses and develops orange colonies. The conidia become hyaline, straight and fusiform, and produce metallized, ellipsoid, ball- or pear-shaped appressoria at the end of germ tubes (Dai et al., 2006). *C. acutatum* is transported with plant material as latent infections and can also survive in soil and plant debris. The pathogen has been found to survive for over one winter in infected plant debris and soil (Parikka et al., 2006). In the survival test, after 20 months of storage, pathogens outdoor on soil surface (Lilja et al., 2006). Conidial germination was maximum at 25°C and minimum at 10°C. Production of secondary conidia was optimum at 25–30°C, and rarely at 10 and 35°C. The total number of appressoria on leaves by 132 hours after inoculation was maximum at 15°C and minimum at 35°C (Leandro et al., 2002). Survival and sporulation of *C. acutatum* on symptomless plants may have a significant role in the anthracnose disease cycle in strawberry.

C. fragariae mostly survives on infected or contaminated transplants, stolons in the soil, on weed hosts and possibly in plant debris in the soil. The fungus is able to survive in buried strawberry tissues for a month (Eastburm and Gubler, 1990). *Cassia obtusifolia*, a weed, can serve as a source of primary inoculum of *C. fragariae* (Howard and Albregts, 1973). Warm humid weather (>21°C) during flowering and fruiting favours the disease but extensive losses are expressed after rainy periods (Howard, 1972). The fungus produces fruiting body-like acervuli in lesions that contain numerous spores, that is, conidia borne in a mucilioginous matrix. These spores are splashed by irrigation or rain to nearby plants or carried by insects, animals and humans moving across the plants.

Plant phenolic compounds are strongly involved in the defence mechanism against pathogens by acting as soluble antimicrobial and deterrent compounds or making the cross-link with callose, proteins and polysaccharides in the cell walls, restricting the fungal penetration and the absorption of nutrients by the invading fungus (Ossipov et al., 2001). The most frequently identified phenolic groups in strawberries are hydroxybenzoic and hydroxycinnamic acids, hydrolysable tannins, flavonols, flavanols and anthocyanins (Aaby et al., 2012). *C. nymphaeae* infection has been observed to cause significant increase in total sugar content in mature fruits (1.1-fold) and decrease in total sugar content in semi-mature fruits (1.4-fold) and organic acid content at all stages of ripening (1.7-fold). Ellagic acid derivatives (1.9-fold), flavanols (1.5-fold) and flavonols (5.1-fold) significantly increase during all stages of ripeness after infection. The pattern of phenolic accumulation in strawberry runners can be altered by *C. nymphaeae* infection, in contrast to strawberry fruits, as most identified phenols decrease as a response to the infection (Weber et al., 2015).

29.2.1 DISEASE MANAGEMENT

Jiang et al. (2014) described a useful technology to raise strawberry plantlets in high temperature and humidity environments by using disease-free planting material, disinfecting soils with Cabrio and Mancozeb or chlorine dioxide and trichloroisocyanuric acid and cultivating under rain-shelter and shading cultivation. This measure can keep the plantlets free of rain. The most suitable sunshade net provides 80% shade to manage the disease.

Use of straw mulch (Madden et al., 1993) and use of planting stock obtained from disease-free nurseries may provide effective control of the disease. Growing nursery plants with as less fertilizer as possible, and removing all infected fruits at each harvest (Agrios, 2005) also reduce infection. Hot water treatment (for seven minutes at 35°C, followed by two or three minutes at 50°C) is successful in reducing the proportion of cuttings, infected with *C. acutatum* (Johnson et al., 2006). The infections by *C. acutatum* can be completely managed by the fumigation of soil and planting material with chloropicrine. Use of soil disinfectants such as chloropicrine and dazomet effectively controls the disease in the nursery bed (Matsuo et al., 1994). Use of overhead irrigation and plastic mulch in strawberry fields results in more severe anthracnose than when drip irrigation and straw mulch are used (Smith, 2006).

The progression of the pathogen *Colletotrichum* can be dependent upon reduction of the active defences of strawberry cultivars, and this is genotype and tissue dependent (Casado Diaz et al., 2006). Cultivars such as Sequoia, Florida Balle, Doner (Delp, 1981) and FL 79–1126 are resistant to *Colletotrichum* (Chandler et al., 1988).

Chandler, Elsanta and Valeta are more susceptible than Dover and Seascape. Honeoye and Pandora have useful level of resistance, even against the most aggressive isolates of *C. acutatum* (Simpson et al., 2006). The cultivars such as Sweet Charlie, Carmine, and Earlibrite are resistant against anthracnose fruit rot; while cv. Festival is intermediate in susceptibility; and cvs. Camarosa and Treasure are highly susceptible. The symptoms of anthracnose are apparently less severe in cv. Yumenoka as compared to cvs. Tochiotome and Akihime (Ban and Yabe, 2005). The cv. Treasure is more resistant to *C. gloeosporioides* caused by either species than any other cultivar (MacKenzie et al., 2006). The Sweet Charlie cultivar is field-immune to anthracnose (*C. acutatum*) fruit rot disease (Chandler, 2005). Ovation strawberry is moderately resistant to anthracnose fruit rots (Lewers et al., 2004) x Dover. Strawberry Parental Line No. 2 selected from a cross between "83118-41" (Florida 693 x Hinomine) is resistant to anthracnose, and this line can be used in breeding for evolving strawberry varieties resistant to anthracnose (Okimura et al., 2004). Shuman and Ballington (2003) reported that the susceptible cvs. Camarosa and Chandler had largest rates of increase in diameter of lesion, as compared with cvs. Sweet Charlie and NCR 95-08, which had much lower rate, and cv. Pelican had the lowest rate of anthracnose fruit rot. Gimenez and Ballington (2002) have classified the resistance in strawberry runners of Chandler (very susceptible), FL 87-210 (tolerant), MS/US 541 (very resistant), *F. chiloensis* NC 92-01 (resistant), NCH 87-10 (tolerant-susceptible) and NCC 89-39 (susceptible) against *C. acutatum*.

A few isolates of fungus *Trichoderma* are known for their ability to control plant pathogens. *T. harzianum* isolate T-39 as the commercial biological control product TRICHODEX is effective in controlling anthracnose (*C. acutatum*) in strawberry in controlled and greenhouse conditions. A combined application of *T. harzianum* (T-39) and *T. longibrachiatum* (T-166), or *T. atroviride* (T-161) at 0.4% is as effective as the chemical fungicide fenhexamide (Freeman et al., 2004). *C. fragariae* (F7), a virulent strain, triggers a plant defensive response (Salazar et al., 2007), elicited by one or more diffusible(s) compound(s) produced against virulent strain *C. acutatum* (M11).

Mycorrhizal fungal species (*Gigaspora margarita, Glomus fasciculatum, Glomus mosseae* and *Glomus aggregatum*) are highly effective to manage the anthracnose disease in strawberry cv. Nohime. After 30 days of inoculation with *C. gloeosporioides*, incidence and severity of browned crown and roots became lower in arbuscular mycorrhizal treated plots than non treated plots. *G. aggregatum* gave maximum reduction (100%) in disease incidence followed by *G. mosseae* (33%) compared with non-arbuscular mycorrhizal plot under capillary watering conditions, and the incidence varied depending on the arbuscular mycorrhizal fungal species. Tolerance to anthracnose occurred in arbuscular mycorrhizal fungus-infected strawberry plants, and no close relation appeared between the tolerance and phosphorus concentration in strawberry plants (Li et al., 2006).

Azoxystrobin + fixade and fluazinam + prochloraz are the efficient fungicides against anthracnose (Domingues et al., 2001). Use of benomyl with captan also gave good results (Cheah and Soterosi, 1984). Ergosterol biosynthesis inhibitors (EBI), a systemic fungicide can reduce infection of *C. fragariae* efficiently (Smith and Black, 1993). Among the commercial fungicides, cyprodinil + fludioxonil, fenhexamid, kresoxim-methyl (kresoxim), fosetyl-Al (fosetyl), vinclozolin, iprodione, benomyl and *Trichoderma* spp., cyprodinil + fludioxonil and fenhexamid have been ranked first in efficacy (Bascon et al., 2003) against anthracnose disease.

Treatment of runner cuttings infected with *C. acuatum* by immersing for seven minutes at 35°C, followed by two or three minutes at 50°C has been observed successful in reducing the disease (Johnson et al., 2006). Prochloraz and difenoconazole are the most effective fungicides for control of black flower (*C. acutatum*) disease in strawberry (Domingues et al., 2001).

Octave Reg. (462 g/kg prochloraz as the $MnCl_2$ complex) is highly effective in reducing the incidence of field symptoms of crown and stolon rot (*C. gloeosporioides*) in runner production from symptomless petioles (Hutton, 2006) of strawberries. Fluazinam 50 WF (1.5–2.0 litres/ha) is significantly reduced infection caused by *Gloeosporium* spp. in controlling anthracnose. Flowering is the best time to initiate fungicide treatment while *B. subtilis* (50 WP) is not effective as fungicide (Aguilar et al., 1995).

29.3 *VERTICILLIUM* WILT

Verticillium wilt (*Verticillium* spp.) or vascular wilt occurs on strawberries worldwide but it is more important in temperate zones of the world. In the United States and England, this disease has gained much importance due to its resultant heavy losses (Wilhelm and Paulus, 1980; Simpson et al., 1994). In England, both *Verticillium dahliae* Kleb. and *Verticillium albo – atrum* have been reported as pathogen responsible for wilt (Keyworth and Benett, 1951) but in the United States, only *V. albo – atrum* is reported as a pathogen (Wilhelm, 1952) causing this disease.

The symptoms of *Verticillium* wilt are almost identical to those of *Fusarium* wilt but *Verticillium* induces wilt at lower temperatures than *Fusarium*. *Verticillium* wilt causes upper leaves to drop and other leaves have irregular chlorotic patches that become necrotic. Older plants infected with *Verticillium* remain stunted with brownish discoloration of their vascular tissues. *Verticillium* infection may result in defoliation, gradual wilting and death of successive branches, or abrupt collapse and death of the entire plant. From India, Mitra (1993) reported that the symptoms of *Verticillium* wilt in strawberry are characterized by browning of wood at the base of crown and wilting tends to decrease the runner production, causing reddening of petiole and stolon late in the season. This disease causes total collapse of strawberry plants during peak of the first year's growth.

Both *V. dahliae* and *V. albo-atrum* produce short-lived conidia. *V. albo-atrum* produces dark, thick-walled mycelium but not microsclerotia. Fungus overwinters in the soil as mycelium or as microsclerotia, and can survive up to 15 years. *V. albo-atrum* grows best at 20–25°C, whereas *V. dahliae* prefers slightly higher temperatures (25–28°C), and sometimes more common in warmer regions (Agrios, 2005) (Figure 29.1).

FIGURE 29.1 *Verticillium* wilt affected plant.

29.3.1 Disease Management

Wilt disease is difficult to control but adoption of long-term crop rotation and use of disease-free planting materials, soil disinfactants and growing resistant cultivars are good alternatives to combat the disease (Ercole, 1976; Agrios, 2005). The thermal inactivation via soil solerization is useful for the control of *Verticillium* wilt in regions with high summer temperature and low rainfall. The effect of solarization on pathogenic microorganisms in the soil increases as the thickness of the polyethylene sheet decreases and vice versa (Shalaby and Mohamed, 2005). Use of black polyethylene mulch with low dose of nitrogenous fertilizers also helps to manage the disease. *Verticillium* wilt is very destructive in light textured, sandy soils. Growing resistant cultivar is a best method to manage the soil borne disease. Darselect, Clery and Queen Elisa are less susceptible than Elsanta (Spornberger et al., 2006), while Alba, Alice, Daroyal, Record and Salsa are resistant against *Verticillium* wilt. The cv. Flamenco was resistant to *V. dahliae*, Medina is less susceptible to *Verticillium* sp. than cv. Camarosa (Lopez-Aranda et al., 2005). Cultivars such as Calypso, Tang and Eros (Simpson and Blanke, 1996) and Pegasus (Simpson et al., 1994) are also reported to be resistant to wilt disease. Somaclonal variants of strawberry cvs. Elsanta and "K 40" with enhanced resistance to *V. dahliae* have been identified (Sowik et al., 2004).

Best control of soilborne diseases has been achieved by the use of fumigation with chloropicrine (8 kg/ha) than land fumigated with ethyle dibromide (Wilhelm and Paulus, 1980). In addition to chloropicrine, dozomat was also found effective against *V. dahliae* infection (Harris, 1989). The preplanting treatment of soil with dozomat and chloropicrine followed by covering the land with polyethylene sheet or nylon laminate is highly effective to manage the disease (Harris, 1989). Total fumigation by chloropropene followed by strip fumigation with methyle bromide also controls the disease effectively (Meesters et al., 1996). Soil drenching with fungicides, that is, benomyl and thiophanate M @ 0.1% in the holes, and soil solerization during hot months followed by fumigation with chloropicrine reduces the infection of *Verticillium* in strawberry (Hartz et al., 1993). Duniway (2005) observed that among the chemical alternatives to methyl bromide tested, 1,3-dichloropropene (1,3-D) mixed with chloropicrin (Telone C35TM) can be as effective as methyle bromide in controlling soil borne pathogens. Sequential applications of Pic and metam sodium (metam) improve *Verticillium* wilt control. Methyl iodide (iodomethane) gives disease control and increases yield similar to methyl bromide when mixed with chloropicrin.

Integrated Disease Management 461

Dimethyl disulfide fungicide at 0.08 kg/m^2 is effective against *V. dahliae* as a new potential chemical alternative to methyl bromide in soil disinfestations (Fritsch, 2005). A significant decrease in the number of micro sclerotia in the soil has been observed with incorporation of Vitasso intercrop (Steffek et al., 2006). Vitasso, brown mustard is used as a green manure. It acts as a biofumigant and thereby controls pests, diseases and weeds, and improves the soil's organic matter and nitrogen content.

29.4 BLACK ROOT ROT

Black root rot is a serious and common problem of strawberries. The term "black root rot" is the general name for several root disorders that produce similar symptoms. The disorders are generally referred to as root rot complex disease of strawberry, which is reported to be incited by a number of soil-inhabiting fungi, viz., *Laptosphaeria coniothyruim* (Fckl). Sacc., *Fusarium ortocereras* App. and Wollenw, *Pezizella lytheri* Shear and Dodge, *Ramullaria* species, *Corticium vagum* Berk and Cut., *Packybasidium cadium* Sacc., *Pythium* species Trow, *Pythium irrigulare* Buis, *Pyrenochaeta* spp., *Cylindrocarpon destrctans*, *Ceratobasidium* species (Bhardwaj and Sharma, 1999), *Rhizoctonia fragariae* and nematodes such as *Pratylenchus penetrans* (Particka and Hancock, 2005). Besides the fungal pathogens and nematodes, winter injury, phytotoxicity due to imbalanced application of fertilizers, soil compaction, drought and excess salt, water or improper soil pH are also involved in this disease complex. The interaction of lesion nematodes (*P. penetrans*) also increases *R. fragariae* root infection, and the severity of black root rot (LaMondia and Cowles, 2005).

The affected plants show an uneven patchy appearance due to dwarfing and leading to death. The root system is much smaller than normal. The main roots exhibit brown areas, and small, fibrous rootlets are killed. Lesions extend and the entire root is blackened. Leaves remain small with short petioles and turn yellowish and old leaves wither. Berries, if formed, remain small or wither before ripening. Soilborne fungi such as *R. fragariae, Pythium* spp. and nematode, *P. penetrans*, cause the death of feeder roots and the degradation of structural roots (Particka and Hancock, 2005). The disease is more prevalent in high clay content and waterlogged soils. Asad Uz Zaman et al. (2015) recommended *T. harzianum* isolate STA7 + Provex-2000 (100 ppm) + mustard oil cake to manage the black root rot (*R. solani* Kuhn) of strawberry.

29.4.1 DISEASE MANAGEMENT

For soilborne diseases, integrated management is the best method to manage the disease below the economic losses. Avoiding continuous cultivation of strawberries on the same land, disease-free planting material and provision of drainage in cultivated area are some of the precautionary measures for reduced disease incidence. Soil Solarization is very efficient in controlling strawberry root rot (Shalaby and Mohamed, 2005). Soil solarization with transparent polyethylene mulch (25 μm) for 40 days during June–July has been found effective for the control of root rot caused by *Sclerotium rolfsii,* which completely inhibited the germination of sclerotia of the fungus (Harender and Sharma, 2005). In solarized plots, root dip of strawberry runners in carbendazim 12% + mancozeb 63% (0.2%) or thiophanate-methyl (0.1%) fungicides has been found effective to control the black root rot (Harender and Sharma, 2005). Soil treatment with chemicals can reduce the population of undesirable nematodes and fungi in the soil. Another effective method is soil fumigation for management of this complex disease. Raised bed soil solarization (RBSS) (for seven weeks alone or with chicken manure amendment (10 t/ha), metam sodium (50 ml/m^2) after two weeks) of RBSS, and methyl bromide (50 g/m^2) significantly reduces soilborne diseases (*Rhizoctonia* spp. and *Phytophthora cactorum*) (Benlioglu et al., 2005). Dimethyl disulfide fungicide at 0.08 kg/m^2 is effective against *R. solani* (Fritsch, 2005), *R. fragariae, Pythium* and nematode, *P. penetrans* (Particka and Hancock, 2005) as a new chemical potential alternative to methyl bromide in soil disinfestations. Khosla and Kumar (2005) reported that combination of thiram and *Trichoderma*

462   Strawberries

viride talc formulation (0.4 + 0.5%) is very effective against root rot. They observed Dana and Etna cultivars to have some field resistance to root rot caused by *S. rolfsii*. The cultivars Bounty, Cabot and Cavendish are reported to be tolerant to the pathogens that cause black root rot (Particka and Hancock, 2005).

29.5 RED STELE

Red stele, also called red core, red root rot, brown stele and black stele, is one of the most destructive diseases of strawberry, wherever it is cultivated commercially in the world which is caused by the soilborne fungus *Phytophthora fragariae* Hickman. The disease first of all was noticed in Scotland during 1920 (Alcock et al., 1930) but the identity of the causal agent, that is, *P. fragariae*, was established in 1940 by Hickman. In Lithuania, *Phytophthora fragariae* was first identified on the roots of strawberry cv. Polka in 2000 (Jovaisiene, 2004). Worldwide 15 physiological races of *P. fragariae* have been reported (Mass, 1984). In India, the valleys of Solan, Sirmour, Kangra, Mandi and Kullu (Himachal Pradesh) are the major strawberry areas attacked by fungal pathogens causing this disease (Gupta and Bhardwaj, 2005).

The fungus attacks young roots and causes reddish discolouration of the stele, which in the advanced stage of infection, produces rat-tail like symptoms on lateral roots of the affected plants. The pathogenic fungus produces non-papillate terminal sporangia, oogonia, terminal or lateral, antheridia, amphigynous or paramours and spherical oospores. High soil moisture (>75%) and low temperature (15°C) are the most congenial factors for the development and spread of the pathogen (Bhardwaj and Gupta, 2005). The affected plants show stunting, wilting, drying and neither produce runners nor bear fruits. As the rot progresses, lateral roots are destroyed giving the main roots a characteristic rat-tail appearance (Bhardwaj and Gupta, 2005). Infected plants remain stunted and lose their shiny green luster. Younger leaves often have a metallic bluish-green cast. The size of fruits formed on infected plants remains very small with poor quality; sometimes fruits are not borne. Older leaves turn prematurely yellow or red and with the first hot and dry weather of early summer, diseased plants wilt rapidly and die. Diseased plants have very few new roots compared to healthy plants that have thick, bushy white roots with many secondary feeder roots. Infected strawberry roots usually appear grey, while the new roots of a healthy plant are yellowish-white. The inside or central portion of the root of wilted plant stele becomes pink to brick red or brownish red hence the disease is referred to as red stele disease, while stele of normal plants is yellowish-white. This red colour may show near the tip of the root or it may extend to entire root length. The red stele disease is best seen at the time of fruiting.

Heavy rainfall, frequent irrigation and waterlogged conditions helps in libration of the zoospores. Oospores survive in soil and primarily cause root infection. Germinated oospores produce sporangia and sporangia, germinate indirectly to produce many zoospores. The zoospore germination on the host roots, and oospore development in the host tissues occurrs at 15–20°C. Root exudates attract the zoospores which encyst around root tips and produces germtube to penetrate epidermis (Anderson and Colby, 1942). Fungus reaches the phloem and pericycle within three days of infection (George and Milholland, 1986, 1987).

29.5.1 DISEASE MANAGEMENT

For the management of the disease, various practices such as use of pathogen-free propagating material, chemical control, resistant cultivars and use of biological control agents have been reported to be the effective means to control this disease.

The disease can be controlled by use of resistant varieties but due to pressure of physiological races, it is not easy to maintain resistance. Rugienius et al. (2006) has identified *P. fragariae* resistance gene *rpf1* in the interspecific hybrids. Strawberry cultivars such as Ovation (Lewers et al., 2004), Evangeline and Brunswick (Jamieson et al., 2004b, c), Loke Delmarvel, Mohawk, Primetime and Erose (Igarashi et al., 1994; Galletta et al., 1995a, b; Simpson and Blanke, 1996) have been reported to grow in *Phytophthora*-affected areas.

The bacterial antagonists (*Pseudomonas fluorescens* T1R2K4, *P. fluorescens* IISR 8, *P. fluorescens* 81 and *B. subtilis*) inhibit the test fungus significantly, however, maximum inhibition has been observed with *P. fluorescens* IISR 8 and *P. fluorescens* T1R2K4 followed by *P. fluorescence* 81 and *B. subtilis*. In pot culture experiments, maximum disease reduction with all the antagonists has been achieved when bacterial antagonists is applied eight days before pathogen inoculation against *P. fragariae* (Gupta et al., 2005). In greenhouse and field experiments, root dip treatment with bacterial antagonists, *Raoultella terrigena* (G-584), *Bacillus amyloliquefaciens* (G-V1) and *P. fluorescens* (2R1-7) exhibit good control of fungal diseases, which is nearly comparable with the chemical fungicide aliette (Jayamani and Zeller, 2004). The disease can be controlled by use of disease-free planting materials taken from disease-free areas. Acidic soils should be avoided for the planting of strawberries. Soil fumigation with chloropicrine significantly reduces the inoculum potential (Joffens, 1957) and drenching the plants with systemic fungicides such as metalaxyl helps effectively manage the red stele disease (Mcintyre and Walton, 1981).

29.6 LEATHER ROT

Phytophthora cactorum causes crown rot of the strawberry rhizome and leather rot of strawberry fruit (Figure 29.2) (Eikemo and Stensv, 2015). In the future, *Phytophthora* diseases will most likely benefit from climate change in northern areas due to higher temperatures and increase in rainfall, which are predicted in the IPCC panel report (Pachauri and Reisinger, 2007). *Phytophthora citricola* can also cause crown rot. Of the major fungal pathogens attacking strawberry, *Phytophthora cactorum* causing leather rot of strawberry is the most damaging one in the valley areas of Solan, Sirmour, Kangra, Mandi and Kullu in Himachal Pradesh, India. Based on pathogenicity tests, *P. cactorum* was pathogenic to fruits only (Gupta and Bhardwaj, 2005). Symptoms occur on mature as well as immature berries. Moisture and temperature play important roles in the development of leather rot disease (Eikemo et al., 2004).

29.6.1 DISEASE MANAGEMENT

The use of polyethylene sheet and rice straw mulch reduces fruit rot incidence compared to the control (Sharma et al., 2005). Solarization with 50 μm low-density polyethylene sheet can reduce the soil *P. cactorum* population by 100% (Porras et al., 2007). Different species of *Trichoderma* fungus, namely *Trichoderma hamatum*, *Trichoderma harzianum*, *Trichoderma viride* and *Trichoderma virens* and bacterial antagonists, namely *P. fluorescens* strains and *Bacillus* sp. are effective against

FIGURE 29.2 Leather rot.

leather rot of strawberry (Sharma et al., 2005). Applications of *Trichoderma* spp. reduces soil populations of *P. cactorum* and reduces leather rot disease incidence (Porras et al., 2007). The combination of solarization and application of *Trichoderma* spp. (10^8 conidia/m^2) is very effective against *P. cactorum*, and considered to be the future alternatives to traditional chemicals for disease control (Porras et al., 2007; Porras et al., 2009). Raised bed solarization (for seven weeks) alone or with chicken manure amendment (10 t/ha), metam sodium (50 ml/m^2) after two weeks and methyl bromide (50 g/m^2) significantly reduces the leather rot (Benlioglu et al., 2005).

Foliar spray with fosetyl-aluminium (fosetyl) and metalaxyl can provide 82–96% control of leather rot (Sharma et al., 2005) disease. Foliar sprays of azoxystrobin, pyraclostrobin and potassium phosphate (ProPhyt), applied weekly from bloom through harvest, and mefenoxam (Ridomil Gold) applied twice as soil drench at early plant growth and at fruit set stage have shown to significantly reduce the leather rot disease incidence. Mefenoxam and potassium phosphite both strobilurin fungicides provide excellent control of strawberry leather rot (Rebollar-Alviter et al., 2005).

29.7 RHIZOPUS ROT

The fungi such as *B. cinerea* and *R. stolonifer* are two major pathogens affecting strawberries during storage (Reddy et al., 2000). This rot is a fungal disease caused by *R. stolonifer* (Ehrenb. Fr.), Vuill. and the infected fruits turn light brown which become soft, watery and exudate juice even at slight pressure. The whole fluffy mycelial growth with black sporangia are formed on severely infected fruits during transportation and storage. The pathogen is a wound parasite and can be prevented by pre-cooling of fruits before shipment. Tee tree oil (TTO) vapor has exhibited a higher activity against spore germination and mycelial growth of *Botrytis cinerea* and *R. stolonifer* under *in vitro* conditions. TTO vapors at 0.9 g/l significantly reduces artificially inoculated grey mold and soft rot *in vivo*, and treated strawberries maintain a fresher quality than untreated strawberries during storage. In addition, this treatment also enhances the resistance of strawberries against *B. cinerea*, which causes a higher hydrogen peroxide level and activities of superoxide dismutase, phenylalanine ammonia-lyase, peroxidase and β-1,3-glucanase during the first period of incubation. TTO can reduce fruit decay, possibly by inhibiting pathogen growth directly and inducing disease resistance indirectly, proving to be an alternative means of managing disease in strawberries (Shao et al., 2013).

29.7.1 DISEASE MANAGEMENT

The combination of *Cryptococcus laurentii* and hot water dips for a short period can be an alternative to chemicals for the control of post-harvest *Rhizopus* rot in strawberries (Zhang et al., 2007).

29.8 GREY MOULD ROT

Grey mold is a major storage disease of strawberry that considerably reduces yields and quality, limiting its international commercialization. In New Zealand, strawberry industry experiences losses due to fungal fruit rot diseases that can cost up to $4.4 million per annum or 20% of the crop value (Timudo et al., 2005).

Grey mould rot disease is caused by *B. cinerea*. It appears as fruit rot under humid conditions as the fungus produces a noticeable grey mould fruiting layer on the affected tissues (Agrios, 2005). Under high humidity conditions, white cottony mycelial growth may be formed (Bhardwaj and Sharma, 1999). Overwintered green leaves often turn necrotic during incubation, and frequently develop high numbers of conidiophores in coming crop season. Mummified fruits and weeds are generally of minor importance because they are present in less numbers. Sclerotia are found on the laminae and petioles of dead leaves, stolons and inflorescence (including pedicels), but appear most frequently on the latter.

Strawberry fruit decay caused by fungal infection usually results in considerable losses during post-harvest storage; thus discerning the decay and infection type in the early stage is necessary and helpful for reducing the losses. Electronic nose (E-nose) and gas chromatography–mass spectrometry (GC–MS) is able to detect early diagnosis of fungal disease, in addition to an accurate classification of the type of pathogenic fungi in the fruits during post-harvest storage (Pan et al., 2014).

Antifungal activity is found in all tissue types (viz. pith, cortex, epidermis) of green stage I of fruit but not present in white and red stage fruit. Antifungal compounds (e.g. phenolics) in strawberry fruit are inhibitory to *B. cinerea* (Terry et al., 2004).

Chandler et al. (2006) reported low incidence of *Botrytis* fruit rot disease in "Camarosa" and "Carmine" strawberries, which was two to three times higher in "Sweet Charlie" and "Earlibrite". "Honeoye" is moderately resistant to grey mould (Ludvikova and Paprstein, 2003). Legard et al. (2002a) reported "Sweet Charlie" to be highly susceptible and "Rosa Linda" the least. The cv. Jonsok has low resistance to *B. cinerea* (Laugale and Morocko, 2000). Some cultivars, such as Kent and Honeoye, are only slightly affected (Abanowska and Bielenin, 2002) while cv. Kentaro was observed to be less susceptible to *Botrytis* rot than Kita-ekubo (Kawagishi et al., 2001). The reaction of some strawberry cultivars against grey mould during storage is given in Table 29.3.

29.8.1 DISEASE MANAGEMENT

Optimization of cultural methods in susceptible strawberry cultivars has greater importance in reducing the losses due to *Botrytis* rot (Daugaard et al., 2003). Removal of infected leaves and debris from the field and storage rooms by providing conditions for proper aeration and quick drying of plants and plant products is a good preventive measure (Legard et al., 2002a,b; Agrios, 2005). Proper drainage and aeration in fields also help to reduce the incidence of grey mould (Daubeny and Pepin, 1977; Mass, 1978; Barritt, 1980). Polyethylene tunnels effectively reduce the risk of infection on strawberry flowers, resulting in decrease in fruit rot due to less precipitation under the polyethylene tunnels compared to the open field. Prevention or shortening of the leaf wetness period may also reduce the infection risk to the crop (Evenhuis and Wanten, 2006). Spacing significantly affects the incidence of fruit rot. Plants with wider spacing (46 cm) show less fruit rot (Legard et al., 2002) than narrow spacings (23, 30 and 38 cm). Mid-September planting favours lesser incidence of botrytis rot than other plantings of "Chandler" strawberry (Singh et al., 2007).

Hot water (44–48°C for 20–30 minutes) immersion of freshly harvested strawberries is an effective control of post-harvest decay of strawberry after one day of cold storage (1°C) and during three days of shelf life (18°C) conditions (Tu et al., 2006). Heat treatment at 45°C or 47°C delays *B. cinerea* proliferation, giving best retention of firmness and initial quality (Chira et al., 2002).

TABLE 29.3

Reaction of Some Strawberry Cultivars to Grey Mould

Reaction	Cultivars	References
Resistant	Harmonie	Khanizadeh et al. (2005)
Moderate resistant	Camarosa, Rosalinda and Capitola	Saber et al. (2003)
	Evangeline	Jamieson et al. (2004c)
Moderate Susceptible	Selva, Sea Scape and Chandler	Saber et al. (2003)
	Brunswick'	Jamieson et al. (2004b)
	Kent and Honeyoe	Abanowska and Bielenin (2002)
Susceptible	Kelia, Dorite, Tamar and Sweet Charlie	Saber et al. (2003)
	Senga Sengana, Syriusz, Marmolada and Dana	Abanowska and Bielenin (2002)
	Kita-ekubo	Konno et al. (2001)

Mulching materials (black plastic, green mulch, straw, buckwheat husks, flax fibre matting and wood chips) significantly influence grey mould incidence (Prokkola et al., 2003). Mulching with black polyethylene (50 μm) has a lower incidence of *Botrytis* rot (Singh et al., 2007) than those mulched either with clear polyethylene (50 μm) or paddy straw (10 cm thickness).

The *Trichoderma* species of fungi have been identified as potential biological control agents for more than 70 years due to the ability of several strains to antagonize other filamentous fungi, including many plant pathogenic species. The commercial biological control product TRICHODEX, has been reported as effective in controlling grey mould in strawberry (Freeman et al., 2001, 2002), with efficacy as good as the chemical fungicide fenhexamide (Freeman et al., 2004). The *Trichoderma* spp. protein solutions reduce the development of *B. cinerea* on strawberry leaves and promote the antifungal activity of *Trichoderma* spp. through and inactivating the action of the grey mould pathogen (EL–Zayat et al., 1993; Zhang et al., 2007). Treatment with the invert emulsion formulation of *T. harzianum* protects fruit from infection by the primary post-harvest pathogens of the fruit, and reduces the diameters of decay lesion thus prolonging the post-harvest shelf life of fresh strawberry fruits (Batta, 2007).

T. harzianum is more effective in controlling grey mould when vectored by honey bees (Maccagnani et al., 1999; Shafir et al., 2006) and flowering time as the best period to apply this antagonist (Boff et al., 2002). Bumble bees (*Bombus impatiens*)- and honey bees (*Apis mellifera*)- delivered *T. harzianum* reduces *B. cinerea*-infected strawberries equivalent to the fungicide treatments, while sprayed *T. harzianum* is less effective against *B. cinerea*-infected fruit (Kovach et al., 2000). *Trichoderma* inoculum when mixed with "Triwaks" dispenser and delivered to strawberry flowers by bees is highly effective in control of grey mould on strawberry fruit than season long chemical treatment (fludioxonil + cyprodinil and pyrimethanil), and can be integrated in an IPM scheme in this crop (Bilu et al., 2004). Biological control of *B. cinerea* is also carried out with a *Gliocladium catenulatum* product spread by honeybees (Parikka and Tuovinen, 2014).

The application of antagonistic bacteria such as *Pseudomonas lurida, P. rhizosphaerae, P. parafulva,* and *Bacillus megaterium*; along with natural plant volatiles such as aliphatic aldehydes hexanal, (E)-2-hexenal, (Z)-3-hexenal and (E)-2-nonenal; has been observed to enhance the efficacy of biocontrol agents against *B. cinerea in vitro* dual culture. In addition, (E)-2-nonenal has strong inhibitory effect on different strains of plant pathogenic fungus *B. cinerea* than on the epiphytic bacteria (Abanda et al., 2006). The fungus *Clonostachys rosea* (syn. *Gliocladium roseum*) is a potential biocontrol agent for management of *Botrytis* blight (Nobre et al., 2005) and the bacterial culture of *B. subtilis* has been reported to inhibit growth of hypha of *B. cinerea* (Chen et al., 2004). In greenhouses, the efficacy of the yeast antagonist (*Metschnikowia fructicola*) is equally effective as chemical control (fenhexamid) against preharvest fruit rot, however, the yeast can suppress post-harvest incidence of fruit rot significantly better than fenhexamid (Karabulut et al., 2004a).

Bacikol (based on *Bacillus thuringiensis*) has shown good results in controlling the grey mould rot of strawberry (Kandybin, 2003). The fungus, *Aureobasidium pullulans* var *pullulans* has been identified as a potential biological agent for control of grey mould on strawberry fruit, however, its efficacy has greater antifungal activity on green rather than on ripe fruits (Adikaram et al., 2002). Further, the fungus, *Pichia guilermondii* (yeast), inhibits the conidial germination of *B. cinerea*, whereas, *Bacillus mycoides* (bacterium) causes breakage and destruction of conidia, when both biological control agents are applied in a mixture. The fungus, *P. guillermondii* competes with *B. cinerea* for glucose, sucrose, adenine, histidine and folic acid. On the other hand, the bacterium, *Bacillus mycoides*, secretes volatile and non-volatile inhibitory compounds and activates the defence systems of the host (Guetsky et al., 2001).

Uses of three biological control agents such as BCAs (*Bacillus amyloliquefaciens* FZB42, *Aureobasidium pullulans* DSM 14940 and DSM 14941 and *Beauveria bassiana* ATCC 74040) have been found to improve control of *B. cinerea* (Sylla et al., 2015). Treatment with liquid formulation of *B. velezensis* antagonist isolate thrice per week for three weeks in the field resulted in higher control efficacy (50.2%) of grey mold when compared with a few registered biopesticides (Shin et al., 2014).

Integrated Disease Management

Plants of strawberry (cv. Pajaro) transformed with defence-related gene ch5B (encodes for a chitinase from *Phaseolus vulgaris*) using *Agrobacterium tumefaciens* can be used to introduce transgene-mediated resistance to grey mould in strawberry (Vellicce et al., 2006). A thaumatin II (thau II) cDNA (thaumatin II protein) driven by the CaMV 35S promoter introduced into strawberry via *Agrobacterium*-mediated transformation has shown a significantly higher level of resistance against grey mould when compared to the control plants (Schestibratov and Dolgov, 2005).

Fruits sprayed with calcium or boron and calcium imparted resistance to botrytis rot during storage (Wojcik and Lewandowski, 2003). Preharvest calcium (calcium sulphate) spray did not show any sign of grey mould during storage for more than 10 days at 10°C and 90±5% RH (Naradisorn et al., 2006). Foliar spray of inorganic salts such as monobasic potassium phosphate (10 g/litre) on fruiting plants has significantly reduced the incidence and severity of fruit rot (El-Shami et al., 2004). Sodium bicarbonate or Kaligreen (potassium bicarbonate) have good control of strawberry fruit rots, while potassium and calcium chloride are moderately effective salts, but ammonium bicarbonate is the least effective one. The antioxidants salicyclic acid and ascorbic acid can decrease the incidence and severity of fruit rots. Oxalic acid is moderately effective in this respect (Saber et al., 2003). A natural volatile compounds methyl salicylate has not only the antifungal activity but also the disease resistance-inducing activity as a fumigant, which causes reduction in grey mould incidence in strawberry fruits (Kim and Choi, 2002).

Strawberry flowers are more susceptible to *B. cinerea* than green fruit, and stamens are the principal infection court. Fungicide applications should focus on peak bloom periods to minimize fungicide use and optimize control of preharvest *Botrytis* fruit rot. During these periods, applications should be made at close intervals (weekly) to minimize losses due to *Botrytis* (Mertely et al., 2002). Regular spray of fungicides such as captan or thiram and benomyl or thiophanate methyle has been found effective against grey mould in strawberry (Horn, 1961; Borecka et al., 1973; Gourley, 1974). The rotting of fruits due to grey mould can be reduced by fumigation with acetaldehyde vapour (Prasad and Stadelbacher, 1974). Fenhexamid (750 mg l^{-1}) has been reported to arrest the infection of *B. cinerea* (Sallato et al., 2007).

Rhodotorula glutinis significantly reduces the natural development of decay of fruit following storage at 20°C for three days or 4°C for five days followed by 20°C for three days (Zhang et al., 2007). The use of low-rate applications of captan during the early season and fenhexamid during the second peak bloom period has been observed effective for controlling *Botrytis* fruit rot (Legard et al., 2005). Protective preharvest spray of three fungicides that is, Benlate [benomyl], prochloraz and vinclozolin can be given to reduce the level of grey mould but vinclozolin (2g/l) was most effective (El-Neshawy et al., 2003).

Signum (6.7% pyraclostrobin + 26.7% boscalid) has given a very good control of botrytis fruit rot (Hauke et al., 2004). Procymidone is highly effective against the pathogen even during seasons of high grey mould pressure. Switch 62.5 WG (cyprodinil+fludioxonil), applied only twice during the blossom period after the occurrence of infection (based on prediction analysis) has been reported very effective (Meszka, et al., 2004), and its efficacy is higher or equal to the standard product Sumilex 500 SC (procymidone), while BAS 516 GAF (strobilurin fungicide) is comparatively less effective than procymidone (Meszka and Bielenin, 2002). Procymidone, used as a standard fungicide, is the most active substance against preharvest rots (Pennisi et al., 2002).

Applications of iprodione during second peak blooming period reduces fruit rot (Legard et al., 2002a,b). In integrated crops, applications of cyprodinil + fludioxonil, fenexamide, mepanipirim, procymidone or pyrimethanil all give satisfactory control (Gengotti et al., 2002). The efficacy of dichlofluanid is similar to that of procymidone, which is considered to be the best fungicide against grey mould (Meszka et al., 2000). Chitosan and benzothiadiazole are the most effective alternative treatments, and preharvest spraying with these alternative chemical compounds can complement the use of conventional fungicides in the control of post-harvest decay of strawberry fruit, especially when disease pressure is low (Feliziani et al., 2015).

Controlled atmosphere storage technology to extend the shelf life of wild strawberry "Reina de los Valles" fruits stored at 3°C for three weeks in 10% CO_2/11% O_2 combination inhibited the development of *B. cinerea* (Almenar et al., 2006). Ethanol at 50% (v/v) significantly reduces the severity of decay after storage for three days at 1°C followed by two days at 24°C. Post-harvest hot water dips at 55 and 60°C for 30 seconds significantly reduce the severity of decay (Karabulut et al., 2004b). Ultraviolet light and heat treatment are other alternative techniques to the use of chemicals to reduce the development of the spoilage fungus *B. cinerea* on strawberry, during storage. Fungal growth on strawberries can be significantly retarded using UV-C at 0.05 J/cm^2 and higher doses (Terry et al., 2004). Ozone-enriched cold storage (for three days at 2°C in air with or without 1.5 µl ozone/litre and then transferred to room temperature) of naturally infected fruits reduces severity of decay due to *B. cinerea* (Nadas et al., 2003). Garcia et al. (2002) observed no spoilage, if the food product is maintained in an atmosphere containing >20% CO_2. Prompt cooling is important for minimizing post-harvest decay of strawberries but temperature management alone may not sufficiently controls post-harvest decay when the disease pressure is high (Nunes et al., 2005). Fruits treated with the combination of heat (63°C for 12 seconds), biocontrol (yeast, *Pichia guilliermondii* Wickerham), and controlled atmospheres (CA; 5°C under air or 15 kPa CO_2 for five and 14 days followed by two days at 20°C to simulate market conditions) had significantly less decay than those in alone but the combination treatments did not provide better control than the current commercially used treatment of 15 kPa CO_2 (Wszelaki and Mitcham, 2003).

Based on weather conditions and inoculums, different forecasting models have been developed that help to signal these situations to us, and suggest the right time to initiate the management of the disease. Models including both weather variables and inoculum gave the best predictions but models using weather variables only gave more accurate predictions than models using inoculum only. In a weather-based model, two important weather variables that affect daily incidence of flowers infection are day-time vapour pressure deficit and night-time temperature (Xu et al., 2000).

The forecast model BOTMAN is relatively simple, based on meteorological data, which allows effective, ecologically and economically more acceptable control, based on integrated chemical and biological measures (Milicevic et al., 2006). Currently, disease control for grey mould is based on preventive spraying every five–seven days during flowering and harvest. Replacing preventive spraying with applications based on infection warnings helps to reduce the amount of sprays. Of the three infection models (BOTEM, BoWaS, DSS-Italy) towards the control of *B. cinerea*, Laer et al. (2005) observed that BOTEM was the best model, and reduced 17–60% fungicide use and an efficacy between 66% and 93%. Berrie et al. (2000) suggested that mathematical model (BOTEM), can be incorporated into a management system to assist in decisions on fungicide sprays to control *Botrytis* fruit rot in main-season strawberry crops.

29.9 *FUSARIUM* WILT

Fusarium wilt, the most serious soilborne disease caused by the pathogen *Fusarium oxysporum* f. sp. *fragariae* (Fof), exists worldwide in strawberries. It was first found in Australia in 1962 (Winks and Williams, 1965), thereafter in Japan in 1969 (Okamoto et al., 1970), Korea in 1982 (Kim et al., 1982), China in 2005 (Huang et al., 2005; Zhao et al., 2009), the United States in 2006 (Koike et al., 2009), Spain in 2007 (Arroyo et al., 2009), and Serbia in 2013 (Stankovic et al., 2014). Consistent with the narrow host ranges of almost all forms of *F. oxysporum*, strawberry is the only known host to Fof (Kodama, 1974). Fusarium wilt in strawberry occurs after the plants are well established and begin to flower or produce fruit, at which time, the older leaves wilt, turn grey-green and begin to dry up (Figure 29.3). Poor and stunted plant growth, initial wilting of older leaves only, discoloration of internal crown tissues and collapsing and death of plants are the important field symptoms of *Fusarium* infection. As disease progresses, virtually all of the foliage collapse, and dry up with the exception of the central, youngest leaves (Koike and Gordon, 2015).

FIGURE 29.3 *Fusarium* wilt.

29.9.1 DISEASE MANAGEMENT

The ideal approach for management of *Fusarium* wilt is through the use of resistant strawberry cultivars such as Festival, Aromas, Camino Real, Portola and San Andreas (Fang et al., 2011; Gordon, 2012, 2013). Wild clones of *F. chiloensis*, collected originally from California, also exhibited resistance to *Fusarium* spp. and therefore could provide another source of gene for resistance breeding (Davalos Gonzalez et al., 2006).

Preplant treatment of infested soil with fumigants remains a useful management tool. Besides, solarization, anaerobic soil disinfestations, crop rotation with non-hosts, planting of *Brassicaceae* crops and pathogen-free transplants and sanitation of equipments are some effective means to reduce the risk of damage from *Fusarium* wilt (Koike and Gordon, 2015). Continuous planting of strawberry facilitates population buildup of *Fusarium* and other soilborne pathogens in the soil. Therefore, successive plantings of strawberry crop on the same piece of land, year after year, should be avoided. Broccoli may be particularly useful as a rotation crop due to its status as a non-host to *Fusarium* and the suppressive effect of broccoli crop residues (Njoroge et al., 2011; Subbarao et al., 1999). Anaerobic soil disinfestation involves incorporating an organic carbon source (such as wheat or rice bran, molasses, or ethanol) into the soil, covering the field or beds with plastic tarpaulines, and subsequently irrigating the site so as to maintain field capacity and create anaerobic soil conditions (Momma et al., 2013; Shennan et al., 2014) to control wilt disease. Control of *Fusarium* wilt causing soilborne disease can be tried by soil drenching after addition of rice and wheat bran (1-2 t/10 acre) (Shin et al., 2014).

29.10 LEAF SPOTS

29.10.1 MYCOSPHAERELLA LEAF SPOT

The disease *Mycosphaerella* leaf spot (Figure 29.4) was first reported from Niglar, in the Bhowali hills of Uttar Pradesh, India during October, 1952 (Bose, 1970). During the initial stages, 2–6 mm diameter, circular, purple scattered spots appear on the upper surfaces of young leaves, appearing most frequently on the blades of leaflets, but also on petioles, fruits and fruit stems. These leaf spots later on enlarge, then turn white, and are surrounded by dark purple margins, rendering a bird's eye effect. On lower surfaces of leaves, prominent veins touching the spots become reddish-purple and the entire leaf may die in the advanced stages. The lesions on fruit stalks are elongated as well as

FIGURE 29.4 *Mycosphaerella* leaf spot.

circular. Infection is not common on berries however, the pulp of infected berries becomes discoloured and rendering the fruits unmarketable under high disease pressure (Fulton, 1958).

Younger leaves are more susceptible to infection. The optimal temperature for infection is 25°C. The number of lesions developed, for all leaf wetness durations, increases gradually from 5–25°C and decreased sharply from 25–30°C but at 35°C, lesion development is arrested. For most optimum temperature ranges, a minimum of 12 hours of leaf wetness is necessary for infection (Carisse et al., 2000).

Mycosphaerella leaf spot is a fungal disease caused by *Mycosphaerella fragariae* Tul. (= *Ramularia tulasnii* Sacc, imperfect stage). The ascomata are usually very small, mostly emerging in the host tissues, commonly in leaves. The single septate hyaline ascospores usually formed on the upper surface of the leaves, sometimes formed on petioles and calyx also. On dead leaves, the sclerotia may also produce conidia. Sporulation starts at 5°C, increases up to 20°C with increasing temperature, and decreases thereafter. At temperatures of 10–25°C, sporulation increases over time from 24–96 hours, and stabilized thereafter (Carisse and Peyrachon et al., 1999).

29.10.2 Disease Management

The cultural practices such as planting density (single or double row system) and removal of dead and infected leaves help in significantly decreasing leaf spot incidence (Schmid et al., 2004). Planting the strawberry runners in well-drained soils, keeping out the weeds and maintaining proper spacing can also help in reduced incidence of this disease. Excessive use of nitrogenous fertilizers needs to be avoided (Harnendo and Casada, 1976). During sprouting (in spring) and bud emergence, the spray of micronutrients such as manganese, copper and boron can reduce the infection, and enhance the yield (Dorozhkin and Grisanovich, 1972). This disease is chemically managed by spraying of Bordeaux mixture, zeneb, ferbam, captan, cuprex and benomyl (Stubbs, 1956). The use of benomyl and thiophanate – M as dips to nursery plants is effective in controlling early infection (Paulus et al., 1978). Varieties such as Combridge (Festic and Delkic, 1977), Takane Kurumal 103, Horida's Wonder, Hukuda, Benzuru (Kim et al., 1978), Arking (Moore, 1981), Premier, Covaliar, Dilpasand, Albitron (Sindhan and Roy, 1981), Tarda Vicoda, Marmalada (Zurawicz and Daminikowski, 1995), Joliette (Khanizadeh et al., 1996), Brunswick (Jamieson et al., 2004b), Cabot (Jamieson et al., 2004a), Saint-Pierre (Khanizadeh et al., 2002a), Saint Laurent d'Orleans (Khanizadeh et al., 2005a), Rosalyne (Khanizadeh et al., 2002b) and AC-LAcadie (Khanizadeh et al., 2002c) have been reported resistant, while Evangeline is reported as moderately resistant to leaf spot disease

Integrated Disease Management 471

(Jamieson et al., 2004b). Dange, Elkat and Pegasus are relatively tolerant to leaf spot disease (Uselis et al., 2006).

Leaf age has significant effect on the susceptibility of strawberry to infection by the leaf spot pathogen *M. fragariae* (Tul.) Lindau. Young, rapidly expanding, one–four-weeks-old leaves are highly susceptible to infection and a large number of characteristic red/brown lesions are formed. Leaf buds and leaves older than 12 weeks at the time of inoculation are resistant to infection (Harrison and Stewart, 1998). Soil solarization and *T. harzianum* are good options to systemically induce resistance to foliar diseases in strawberries, which may be either due to direct effect on the plant or an indirect one, via stimulation of beneficial microorganisms in the rhizosphere (Levy et al., 2015).

29.10.3 *PESTALIOPSIS* LEAF SPOT

A common leaf spot disease of strawberry caused by the fungus *Pestaliopsis disseminate* (Theum) Stey was recorded first time by Singh et al. (1975) from Bangalore (India) on the Pusa Early Dwarf cultivar. It is prevalent in all the strawberry-growing areas of India.

The disease symptoms are characterized by circular, dark brown to chocolate-coloured spots surrounded by reddish-brown or yellowish margins. In advanced stages, the spots may coalesce and show a patchy appearance. The infection normally starts from the leaf margin and extends towards the midrib, which covers the whole leaf. The infected areas become brittle and get detached from the healthy ones. Severely infected leaves become defoliated. Dot-like fruiting bodies (acervuli) of the pathogenic fungus appears on the necrotic areas of the infected leaf. The acervuli are amphiginous, scattered, distinct sometime gregarious, globose to centricular, conic or pyramidal, puntiform, erumpent rupturing the epidermis or diplacing it completely, exposing the black conidial mass. Short conidiophores emerge from the basal cells, slightly broader at the base with tapering end. Conidia elliptical or clavate, fusiform, tapering at the base, straight or slightly constricted at septa, five-celled, measuring $18.5 \times 26 \times 5.5$ µm (Singh et al., 1975).

29.10.4 YELLOW LEAF SPOT

Yellow leaf spot is caused by fungus *Dendrophoma obscurans* (Ell. and Ev.) Anderson, and prevalent in every strawberry grown region. In India, it has been reported from hills of Uttar Pradesh (now in Uttarakhand) and Himachal Pradesh (Bose, 1970). It has been found in moderate to severe form from an altitude of 1,000 to 2,400 meters above sea level. The symptoms appear with onset of the rains and can be observed first on leaves causing maximum damage during July to September. During initial stage of infection minute, circular, radish to purple spots appear on the upper surface of leaves and calyces. The spots rapidly increase in size with yellow or yellow brown centre surrounded by a purple margin. The spots may coalesce and form irregular or oblong patches, covering more than half of the surface of leaflets. The large spots can be distinctly divided in three layer; outer purple, inner light brown and centre dark brown zones, often extend towards the edge of lamina, which become "V" or fan shaped. The affected areas with disease may dry up and become puckered or crumpled. Sometimes, shot holes are formed. Pycnidia found in necrotic area of spot, which are deeply embedded in host tissue with a distinct neck extending through the epidermis, and act as a source of primary infection (Anderson, 1956; Bose, 1970).

29.10.5 ANGULAR LEAF SPOT

This is an important bacterial disease of strawberry caused by *Xanthomonas fragariae* and reported from Carolina, Ethiopia, Argentina and Australia (McGechan and Fahy, 1976; Alipi et al., 1989; Navarthan et al., 1991; Ritchie et al., 1993; Braun and Hildebr, 2013). Initially, water-soaked spots appear on the upper surfaces of leaves and later on become dark brown and angular. In severe conditions, the affected plants may die. Disease can effectively be controlled by spraying streptomycine or streptocyclene, starting from the appearance of symptoms (Alipi et al., 1989). The cv. Earliglow is resistant to angular leaf spot (Hartung et al., 2003).

Bacterial angular leaf spot is transmitted to production fields almost exclusively through infected nursery stock. None of the methods can completely eliminate the pathogen *X. fragariae* from planting stock but integration of practices such as removing or trimming of remnant leaves and petiole tissue from nursery-trimmed plants, dippin plants in 10% chlorine bleach solution, use of UV-C radiation, surface-sterilization and heat treatment are the strategic practices to manage the disease (Turechek et al., 2013).

29.11 LEAF BLIGHT

This disease was first reported by Lele and Pathak (1965) in India. The disease is caused by *Rhizoctonia bataticola* (Taub). Bull. [=*Macrophomina phaseolina* (Tassi.) Goid] and *Dendrophoma obscurans* (*Phomopsis obscurans*).

The initial symptoms of this disease are found on all plant parts during September–October. More or less circular to spreading spots with ash grey centre and purplish dark margins are found on leaves. In advanced stage, these spots become oval to irregular in shape and lesions start spreading from margin of the leaves towards the centre. Infected runners and stalks turn dark brown or black with irregular lesions. Due to infection in plants, new growth is very much suppressed. The poor development in new roots and shredded appearance in older ones at their distal portion is found, but no irregular rot is seen. The cv. AC-L Acadie has been observed to be resistant to leaf blight (Khanizadeh et al., 2002c).

29.12 LEAF SCORCH

This disease was first reported by Wolf (1924) worldwide, and in India, this was recorded by Bose (1970) from the Kumaon hills of the then Uttar Pradesh (now Uttarakhand). This disease is caused by the fungus *Marssonina fragariae* (Sacc.) Kleb [(*Diplocarpon carliana*) (E & E) Wolf]. Leaf scorch disease mostly occurs in the months of April or May, and symptoms appear on leaves, petioles, peduncles and runners (Figure 29.5). Irregular, oval or angular and purple-coloured symptoms surrounded by a reddish or pinkish halo are found on leaves. Elongated, reddish-brown or purple streaks develop on petioles and peduncles. Similar symptoms appear on runners, which become discoloured and girdled (Bose, 1970). The fungus overwinters in the form of mycelia and produces two types of fruiting bodies (apothecia and acervuli) in early spring. Primary infections are caused

FIGURE 29.5 Leaf scorch.

Integrated Disease Management

by ascospores and conidiospores from the acervuli. During further vegetation, the fungus forms only acervuli (which appear only during the anamorph stage), and secondary infections are caused only by conidiospores. The first symptoms (red spots) appear only a few days after the emergence of the first spores (Milicevic et al., 2002).

Leaf scorch reduces photosynthetically active leaf area and dry matter accumulation. The vegetative characters most affected are leaf area and number, crown number and dry weight and root dry weight. A strong relationship has been established between maximum disease severity during late summer and autumn (which includes the time of flower bud initiation) and yield in the following season (Mutisya et al., 2005). In the impact study of strawberry leaf scorch, "physiological lesion" size has been estimated to be 1.6 and 2.1 for greenhouse-grown and field-grown plants, respectively, meaning that the total area of leaflet tissue affected by leaf scorch is 1.6–2.1 times larger than the leaflet area showing visual symptoms of the disease. The larger physiological lesion found in field-grown plants may be due to field-associated factors causing stress or variability in cultivars (Turechek et al., 2007).

Disease can be effectively managed by the spraying of carbendazim (Khanizadeh et al., 2005b). Fungicide formulations such as Folicur M 50 WP (tebuconazole+tolylfluanid) and Kidan SC (iprodione) are most effective, while Euparen M WP (tolylfluanid) and Quadris KS (azoxystrobin) are moderately effective against leaf scorch disease of strawberries (Milicevic et al., 2002).

Strawberry cv. Saint Laurent d'Orleans, has resistance to leaf scorch caused by *Diplocarpon earlianum* (Khanizadeh et al., 2005). Of the three cultivars, the incidence of leaf scorch disease was maximum in strawberry "Kent", followed by "Jewel" and minimum in Blomidon (Mutisya et al., 2005). Further, the strawberry cvs. Evangeline (Jamieson et al., 2004), Brunswick (Jamieson et al., 2004b), Cabot (Jamieson et al., 2004a), Saint-Pierre (Khanizadeh et al., 2002a), Roseberry, Rosalyne (Khanizadeh et al., 2002b) and AC-L'Acadie (Khanizadeh et al., 2002c) have also been observed to be resistant to leaf scorch.

29.13 VIRUSES

Strawberry crinkle virus (SCV) and strawberry mild yellow edge virus (SMYEV) are transmitted obligately by the aphid *Chaetosiphon fragaefolii* (Cockerell), which also occasionally infests the genus *Potentilla*, related to *Fragaria*. Several wild strawberry species have been observed to be hosts of these viruses (Yohalem et al., 2009). Strawberry plants showing viral symptoms include stunting, mild chlorosis and reddening; occasionally wrinkled, curled, and deformed leaves that may exhibit mottling; and chlorotic spots, forming a putative virus complex (Silva-Rosales et al., 2013). Besides strawberry mottle virus (SMoV), strawberry latent ringspot virus (SLRSV), strawberry necrotic shock virus (SNSV), SMYEV; *Fragaria chiloensis latent virus* (FClLV), *strawberry pallidosis associated virus* (SPaV), *Fragaria chiloensis cryptic virus* (FClCV) are some of the serious viruses of strawberries. As part of a US AID-MERC-funded project, "disease-indexing and mass propagation of superior strawberry cultivars" Ragab et al. (2009) made efforts to evaluate the virus status of strawberries in Egypt.

SCV is a common aphid (*Chaetosiphon fragaefolii* and *C. jacobii*)-transmitted virus, infecting strawberry plants (Richardson et al., 1972; Vaughan et al., 1933), and reducing fruit yield and quality due to stunted growth and distorted leaves (Perotto et al., 2014). The natural host range of SCV is restricted to members of the genus *Fragaria*. Schoen et al. (2001) studied the complete nucleotide and determined the nucleotides of SCV. The sequence revealed seven open reading frames (ORFs) on the viral complementary sequence. Four putative proteins of SCV have been postulated by comparison of the similarities to other rhabdovirus proteins: nucleocapsid (N), matrix protein (M), glycoprotein (G), and polymerase (L). Nucleotide sequences and deduced amino acid sequences of N, M, G and L of SCV have exhibited significant homology to the corresponding cytorhabdoviruses, Northern cereal mosaic virus and lettuce necrotic yellows virus. A series of two ORFs (genes P2 and P3) have been observed between N and M gene sequences, and one small ORF (P6) was present

REFERENCES

Aaby, K., Mazur, S., Nes, A. and Skrede, G. 2012. Phenolic compounds in strawberry (*Fragaria x ananassa* Duch.) fruits: composition in 27 cultivars and changes during ripening. *Food Chemistry*, 132: 86–97.

Abanda, N., Kpwatt, D., Krimm, U., Schreiber, L. and Schwab, W. 2006. Dual antagonism of aldehydes and epiphytic bacteria from strawberry leaf surfaces against the pathogenic fungus *Botrytis cinerea in vitro*. *Bio Control*, 51(3): 279–291.

Abanowska, B.H. and Bielenin, A. 2002. Infestation of strawberry cultivars with some pests and diseases in Poland. *Acta Horticulturae*, 567(2): 705–708.

Adikaram, N.K.B, Joyce, D.C. and Terry, L.A. 2002. Biocontrol activity and induced resistance as a possible mode of action for *Aureobasidium pullulans* against grey mould of strawberry fruit. *Australasian Plant Pathology*, 31(3): 223–229.

Agrios, G.N. 2005. *Plant Pathology* (fifth edition). Elsevier Academic Press, UK, pp. 922.

Aguilar Calvo, C., Mendoza Zamora, C. and Ponce Gonzalez, F. 1995. Chemical control of anthracnose (*Gloeosporium* sp.) and strawberry leaf spot (*Ramularia tulasnei* Sacc.). *Revista Chapingo Serie Protección Vegetal*, 2(1): 71–76.

Alcock N.L., Howells, D.V. and Foister, E.C. 1930. Strawberry disease in Lankashire. *Scottish Journal of Agriculture*, 13: 242–251.

Alipi, A.M., Ronco, B.L. and Carrainza, M.R. 1989. Angular leaf spot of strawberry, a new disease in Argentina. Comparative control with antibiotics and fungicides. *Advances Horticultural Science*, 1: 3–6.

Almenar, E., Hernandez Munoz, P., Lagaron, J.M., Catala, R. and Gavara, R. 2006. Controlled atmosphere storage of wild strawberry (*Fragaria vesca* L.) fruit. *Journal of Agricultural and Food Chemistry*, 54(1): 86–91.

Amsalem, L., Zasso, R., Pertot, I., Freeman, S., Sztjenberg, A. and Elad, Y. 2004. Efficacy of control agents on powdery mildew: a comparison between two populations. *Bulletin-OILB/SROP*, 27(8): 309–313.

Amsalem, L., Freeman, S., Rav-David, D., Nitzani, Y., Sztejnberg, A., Pertot, I. and Elad, Y. 2006. Effect of climatic factors on powdery mildew caused by *Sphaerotheca macularis* f. sp. *fragariae* on strawberry. *The European Journal of Plant Pathology*, 114(3): 283–292.

Anderson, H.W. 1956. *Diseases of Fruit Crops*. McGraw Hill Book Company Inc., New York, New York, Toronto, Canada, London, UK.

Anderson, H.W. and Colby, A.S. 1942. Red stele of strawberry on Path finder and Aberdeen. *Plant Disease Report*, 26: 291–292.

Anttonen, M., Hukkanen, A., Tiilikkala, K. and Karjalainen, R. 2003. Benzothiadiazole induces defence responses in berry crops. *Acta Horticulturae*, 626: 177–182.

Arroyo, F.T., Llergo, Y., Aguado, A. and Romero, F. 2009. First report of *Fusarium wilt* caused by *Fusariumoxysporum* on strawberry in Spain. *Plant Disease*, 93: 323.

Asad Uz Zaman, M., Bhuiyan, M.R., Khan, M.A.I., Bhuiyan, M.K.A. and Latif, M.A. 2015. Integrated options for the management of black root rot of strawberry caused by *Rhizoctonia solani* Kuhn. *Plant Biology and Pathology*, 338: 112–120.

Ban, Y. and Yabe, K. 2005. A new strawberry variety "Yumenoka". *Research-Bulletin of the Aichiken Agricultural Research Center*, 37: 17–22.

Barritt, B.H. 1980. Resistance of strawberry clones to Botrytis fruit rot. *Journal of American Society of Horticultural Science*, 105: 160–164.

Bascon, J., Gonzalez, L., Paez, J.I., Vega, J.M. and Montes, F. 2003. Effects of preharvest treatments with fungicides on strawberry yield and postharvest fruit rots. *Boletin de Sanidad Vegetal Plagas*, 29(3): 441–452.

Batta, Y.A. 2007. Control of postharvest diseases of fruit with an invert emulsion formulation of *Trichoderma harzianum* Rifai. *Postharvest Biology and Technology*, 43(1): 143–150.

Benlioglu, S., Boz, O., Yldz, A., Kaskavalc, G. and Benlioglu, K. 2005. Alternative soil solarization treatments for the control of soil-borne diseases and weeds of strawberry in the Western Anatolia of Turkey. *Journal of Phytopathology*, 153(7/8): 423–430.

Berrie, A.M., Harris, D.C., Xu Xiang Ming and Burgess, C.M. 2000. A system for managing botrytis and powdery mildew of strawberry: first results. *Bulletin OILB/SROP*, 23(11): 35–40.

Bhardwaj, L.N. and Gupta, M. 2005. Red stele disease of strawberry. *International Journal of Tropical Plant Diseases*, 19(1/2): 39–58.

Bhardwaj, L.N. and Sharma, S.K. 1999. *Diseases of Strawberry, Gooseberry and Raspberry. Diseases of Horticultural Crops-Fruits*. Indus Pub. Com., New Delhi, India: pp. 316–336.

Bilu, A., David, D.R., Dag, A., Shafir, S., Abu-Toamy, M. and Elad, Y. 2004. Using honeybees to deliver a biocontrol agent for the control of strawberry *Botrytis cinerea*-fruit rots. *Bulletin OILB/SROP*, 27(8): 17–21.

Blanco, C., Santos, B. De. Los., Barrau, C., Arroyo, F.T., Porras, M. and Romero, F. 2004. Relationship among concentrations of *Sphaerotheca macularis* conidia in the air, environmental conditions, and the incidence of powdery mildew in strawberry. *Plant Disease*, 88(8): 878–881.

Boff, P., Kohl, J., Jansen, M., Horsten, P.J.F.M., Plas, C.L. van-der and Gerlagh, M. 2002. Biological control of grey mold with *Ulocladium atrum* in annual strawberry crops. *Plant Disease*, 86(3): 220–224.

Borecka, H., Borecki, Z. and Millikan, D.F. 1973. Recent investigations on the control of grey mould of strawberries in Poland. *Plant Disase Report,* 57: 31–33.

Bose, S.K. 1970. Diseases of valley fruits in Kumaon (III), leaf spot disease of strawberry. *Progressive Horticulture*, 2: 33–53.

Brahm, R.U, Ueno, B and Oliveira, R. P. de 2005. Reaction of strawberry cultivars to powdery mildew under greenhouse conditions. *Revista-Brasileira-de-Fruticultura*, 27(2): 219–221.

Braun, P.G. and Hildebr, P.D. 2013. Effect of sugar alcohols, antioxidants and activators of systemically acquired resistance on severity of bacterial angular leaf spot (*Xanthomonas fragariae*) of strawberry in controlled environment conditions. *Canadian Journal of Plant Pathology*, 35(1): 20–26.

Brooks, A.N. 1931. Anthracnose of strawberry caused by *Colletotrichum fragariae*, N. Sp. *Phytopathology*, 21: 739–744.

Carisse, O. and Peyrachon, B. 1999. Influence of temperature, cultivar, and time on sporulation of *Mycosphaerella fragariae* on detached strawberry leaves. *Canadian Journal of Plant Pathology*, 21(3): 276–283.

Carisse, O., Bourgeois, G. and Duthie, J.A. 2000. Influence of temperature and leaf wetness duration on infection of strawberry leaves by *Mycosphaerella fragariae*. *Phytopathology*, 90(10): 1120–1125.

Casado Diaz, A., Encinas Villarejo, S., Santos, B. de los, Schiliro, E., Yubero Serrano, E.M., Amil-Ruiz, F., Pocovi, M.I., Pliego Alfaro, F., Dorado, G., Rey, M., Romero, F., Munoz-Blanco, J. and Caballero, J.L. 2006. Analysis of strawberry genes differentially expressed in response to *Colletotrichum* infection. *Physiologia Plantarum*, 128(4): 633–650.

Chandler, C.K. 2005. Sweet Charlie strawberry. *Journal of the American Pomological Society*, 59(2): 67.

Chandler, C.K., Albregts, E.E. and Howard, C.M. 1988. Evaluation of strawberry cultivars and advanced selections at Dover Florida, 1986–88. *Advances Strawberry Production*, 8: 19–22.

Chandler, C.K., Mertely, J.C. and Peres, N. 2006. Resistance of selected strawberry cultivars to anthracnose fruit rot and botrytis fruit rot. *Acta Horticulturae*, 708: 123–126.

Chaves, N. and Wang, A. 2004. Control of gray mold (*Botrytis cinerea*) in strawberry with *Gliocladium roseum*. *Agronomia Costarricense*, 28(2): 73–85.

Cheah, L.H. and Soterosi, J.J. 1984. Control of black fruit rot of strawberry. *Proceeding of the 37th N.Z. Weed Pest Control Conference*, 37: 160–162.

Chen Li, Tan, Gen Jia and Ding Ke Jian. 2004. Inhibiting effect of *Bacillus subtilis* on four *Botrytis cinerea*. *Journal of Fungal Research*, 2(4): 44–47.

Chen, C., Wei, Y., Fan, T., Zhang, W., Zhang, G., Liu, Q., Liu, H., Wang, J. and Zhang, L. 2014. Bio-control *Bacillus* strain TS02 for control of strawberry powdery mildew and its molecular identification. *Acta Horticulturae*, 1049: 607–611.

Chira, A., Chira, L. and Nicolae, D. 2002. Influence of the postharvest heat treatment upon the strawberry quality during storage. *Buletinul Universitatii de Stiinte Agricole si Medicina Veterinara Cluj Napoca Seria Horticultura*. 57: 131–133.

Dai, F.M., Ren, X.J and Lu, J.P 2006. First report of anthracnose fruit rot of strawberry caused by *Colletotrichum acutatum* in China. *Plant Disease*, 90(11): 1460.

Damm, U., Cannon, P.F., Woudenberg, J.H. and Crous, P.W. 2012. The *Colletotrichum acutatum* species complex. *Studies in Mycology*, 73: 37–113.

Daubeny, H.A. and Pepin, H.S. 1977. Evaluation of strawberry clones for fruit rot resistance. *Journal of the American Society for Horticultural Science*, 102: 431–435.

Daugaard, H., Sorensen, L. and Loschenkohl, B. 2003. Effect of plant spacing, nitrogen fertilisation, post-harvest defoliation and finger harrowing in the control of *Botrytis cinerea* Pers. in strawberry. *European Journal of Horticultural Science*, 68(2): 77–82.

Davalos Gonzalez, P.A., Jofre Garfias, A.E., Hernadndez Razo, A.R., Narro Sanchez, J., Castro Franco, J., Vazquez Sanchez, N. and Bujanos Muniz, R. 2006. Strawberry breeding for the Central Plateau of Mexico. *Acta Horticulturae*, 708: 547–552.

Davik, J. and Honne, B.I. 2005. Genetic variance and breeding values for resistance to a wind-borne disease [*Sphaerotheca macularis* (Wallr. ex Fr.)] in strawberry (Fragaria x ananassa Duch.) estimated by exploring mixed and spatial models and pedigree information. *TAG-Theoretical-and-Applied-Genetics*. 111(2): 256–264.

Delp, C.J. 1981. Pesticide resistance-its impact on pest and disease control. *Proceeding 1981 British Crop Protection Conference*, 3: 865–871.

Denoyes Rothan, B., Lerceteau Kohler, E., Guerin, G., Baudry, A., Molinero, V. and Navatel, J.C. 2006. Anthracnose on strawberry in France: situation and perspectives. *Acta Horticulturae*, 708: 277–280.

Domingues, R.J., Tofoli, J.G., Oliveira, S.H.F. and Garcia Junior, O. 2001. Chemical control of strawberry black flower rot (*Colletotrichum acutatum* Simmonds) under field conditions. *Arquivos do Instituto Biologico Sao Paulo*, 68(2): 37–42.

Dorozhkin, N.A. and Grisanovich, A.K. 1972. Effect of microelements on susceptibility of strawberry to white spot and grey rot. *Khimiyav Sel'Shom khozyaistve*, 10: 50–51.

Drozdowski, J. 1987. A survey of fungi associated with strawberry black root rot in commercial fields. Advance Strawberry Production, N. *America Strawberry Growers Assoc*, 6: 47–49.

Duniway, J.M. 2005. Alternatives to methyl bromide for strawberry production in California. *Acta Horticulturae*, 698: 27–32.

Eastburm, D.M. and Gubler, W.D. 1990. The detection and survival of *Colletotrichum acutatum*, a causal agent of strawberry anthracnose in soil. *Plant Disease*, 74: 161–163.

Eikemo, H. and Stensv, A. 2015. Resistance of strawberry genotypes to leather rot and crown rot caused by *Phytophthora cactorum*. *European Journal of Plant Pathology*, 143(2): 407–413.

Eikemo, H., Klemsdal, S.S., Riisberg, I., Bonants, P., Stensvand, A. and Tronsmo, A.M. 2004. Genetic variation between *Phytophthora cactorum* isolates differing in their ability to cause crown rot in strawberry. *Mycological-Research*, 108(3): 317–324.

El-Neshawy, S.M., Okasha, K.A. and Cook, L.W. 2003. Preharvest gray mold control and fungicide residues in the harvested strawberries. *Egyptian Journal of Agricultural Research*. 81(4): 1521–1534.

El-Shami, M.A., El-Desouky, S.M., Mostafa, A.M. and Rahman, T.G.A. 2004. Management of strawberry fruit rot disease by some inorganic salts. *Annals of Agricultural Science, Moshtohor*, 42(3): 1071–1078.

El-Zayat, M.M., Okashakha, K.H.A., El-Tobshy, Z.M., El-Kohli, M.M.A. and El-Neshuary, S.M. 1993. Efficiency of *Trichoderma harziamum* as a biocontrol agent against gray mould rot (*Botrytis cinerea*) of strawberry fruits. *Annals of Agricultural Science* Cairo, 38: 283–290.

Ercole, N. 1976. Verticillium wilt of strawberry, a difficult disease to control. *Informattore Agrario*, 32: 925–926.

Evenhuis, A. and Wanten, P.J. 2006. Effect of polythene tunnels and cultivars on grey mould caused by Botrytis cinerea in organically grown strawberries. *Agriculturae Conspectus Scientificus Poljoprivredna Znanstvena Smotra*, 71(4): 111–114.

Faedi, W. and Baruzzi, G. 2004. New strawberry cultivars from Italian breeding activity. *Acta Horticulturae*, 649: 81–84.

Fang, X., Phillips, D., Li, H., Sivasithamparam, K. and Barbetti, M.J. 2011. Severity of crown and root diseases of strawberry and associated fungal and oomyceties pathogens in Western Australia. *Plant Pathology*, 40: 109–119.

Fang, X., Kuo, J., You, M.P., Finnegan, P.M. and Barbetti, M.J. 2012. Comparative root colonisation of strawberry cultivars Camarosa and Festival by *Fusarium oxysporum* f. sp. *fragariae*. *Plant Soil*, 358: 75–89.

Feliziani, E., Landi, L. and Romanazzi, G. 2015. Preharvest treatments with chitosan and other alternatives to conventional fungicides to control postharvest decay of strawberry. *Carbohydrate Polymers*, 132: 111–117.

Festic, H. and Delkic, I. 1977. The reactions of some strawberry varieties to *Mycosphaerella fragariae* (Schw.) Lind at Bosanska Krajina. *Jugoslovonsko-Vocarstovo*, 10: 633–637.

Integrated Disease Management

Fouly, H.M. 2004. Effect of some micronutrients and the antioxidant salicylic acid on suppressing the infection with strawberry powdery mildew disease. *Bulletin of Faculty of Agriculture, Cairo University*, 55(3): 475–486.

Freeman, S. 2008. Management, survival strategies, and host range of *Colletotrichum acutatum* on strawberry. *HortScience*, 43: 66–68.

Freeman, S., Barbul, O., David, D.R., Nitzani, Y., Zveibil, A. and Elad, Y. 2001. *Trichoderma* spp. for biocontrol of *Colletotrichum acutatum* and *Botrytis cinerea* in strawberry. *Bulletin OILB/SROP*, 24(3): 147–150.

Freeman, S., Kolesnik, I., Barbul, O., Zveibil, A., Maymon, M., Nitzani, Y., Kirshner, B., Rav-David, D. and Elad, Y. 2002. Use of *Trichoderma* spp. for biocontrol of *Colletotrichum acutatum* (anthracnose) and *Botrytis cinerea* (grey mould) in strawberry, and study of biocontrol population survival by PCR. *Bulletin OILB/SROP*, 25(10): 167–170.

Freeman, S., Minz, D., Kolesnik, I., Barbul, O., Zveibil, A., Maymon, M., Nitzani, Y., Kirshner, B., Rav-David, D., Bilu, A., Dag, A., Shafir, S. and Elad, Y. 2004. *Trichoderma* biocontrol of *Colletotrichum acutatum* and *Botrytis cinerea* and survival in strawberry. *European Journal of Plant Pathology*, 110(4): 361–370.

Fritsch, J. 2005. Dimethyl disulfide as a new chemical potential alternative to methyl bromide in soil disinfestation in France. *Acta Horticulturae*, 698: 71–76.

Fulton, R.H. 1958. Studies on strawberry leaf spot in Michigan. *Michigan Agricultural Experiment Station Quarterly Bulletin*, 40: 581–588.

Galletta, G.J., Mass, J.L., Enns, J.M., Draper A.D., Dale, A. and Swartz, H. J. 1995a. Mohawk strawberry. *Hort Science*, 30: 631–634.

Galletta, G.J., Mass, J.L., Enns, J.M., Draper, A.D., Fiola, J.A., Scheerens, J.C., Archbold, D.D. and Ballington, J.R. 1995b. Delmarvel strawberry. *Hort Science*, 30, 1099–1103.

Garcia Gimeno, R.M., Sanz Martinez, C., Garcia Martos, J.M. and Zurera Cosano, G. 2002. Modeling *Botrytis cinerea* spores growth in carbon dioxide enriched atmospheres. *Journal of Food Science*, 67(5): 1904–1907.

Garrido, C., Carbu, M., Fernandez Acero, F.J., Vallejo, I. and Cantoral, J.M. 2009. Phylogenetic relationships and genome organisation of *Colletotrichum acutatum* causing anthracnose in strawberry. *European Journal of Plant Pathology*, 125: 397–411.

Gengotti, S., Ceredi, G. and Paoli, E-de. 2002. Anti-botrytis measures in organic and integrated strawberries. *Informatore Agrario Colletotrichum Acutatum*, 58(27): 55–58.

George, S.W. and Milholland, R.D. 1986. Growth of Phytophthora fragariae on various clarified natural media and selected antibiotics. *Plant Disease*, 70: 1100–1104.

George, S.W. and Milholland, R.D. 1987. Seasonal occurrence and recovery of *Phytophthora fragariae* in infected strawberry roots. *Hort Science*, 22: 1254–1255.

Gimenez, G. and Ballington, J.R. 2002. Inheritance of resistance to Simmonds on runners of garden strawberry and its backcrosses. *Hort Science*, 37(4): 686–690.

Gordon, T.R. 2012. A comprehensive approach to management of wilt diseases caused by *Fusarium oxysporum* and *Verticillium dahliae*. CA Strawberry Commission Annual Production Research Report, 2012–2013: 9–19.

Gordon, T.R. 2013. Management of diseases caused by Fusarium oxysporum, Verticillium dahliae and Macrophomina phaseolina. CA Strawberry Commission Annual Production Research Report, 2012–2013: 9–21.

Gourley, C.O. 1974. A comparison of benomyl, thiophanate methyl and captan for control of strawberry fruit rot. *Canadian Plant Disease Survey*, 54: 27–30.

Gu, J., Jiang Z., Tong, H., Ri Q., Wang, C.H. and Wang, W.L. 2005. Breeding report on new strawberry cultivar Hongshimei. *China Fruits*, 3: 3–5.

Guetsky, R., Shtienberg, D., Elad, Y. and Dinoor, A. 2001. Establishment, survival and activity of biocontrol agents applied as a mixture in strawberry crops. *Bulletin-OILB/SROP*, 24(3): 193–196.

Gupta, M. and Bhardwaj, L.N. 2005. Studies on *Phytophthora* spp. attacking strawberry in Himachal Pradesh. Integrated plant disease management Challenging problems in horticultural and forest pathology, Solan, India, 14–15 November 2005: 75–81.

Gupta, M., Bhardwaj, L.N. and Sharma, R.C. 2005. Biological control of red stele of strawberry with bacterial antagonists. *Acta Horticulturae*, 696: 363–366.

Harender, R. and Sharma, S.D. 2005. Integrated management of collar and root rot (*Sclerotium rolfsii*) of strawberry. *Acta Horticulturae*, 696: 375–379.

Harnendo, V. and Casada, M. 1976. The relationship between *Mycosphaerella fragariae* infection and the nutritional status of strawberry plants. *Anales de Edafologia-y-Agrobiologia*, 35: 1–2.

Harris, D.C. 1989. Control of Verticillium wilt of strawberry in Britain by chemical soil disinfestations. *Journal of Horticulture Science*, 64: 683–686.

Harrison, G.L. and Stewart, A. 1998. The effect of leaf age on susceptibility of strawberry to infection by *Mycosphaerella fragariae*. *Journal of Phytopathology*, 124(2): 112–118.

Hartung, J.S., Gouin, C.C., Lewers, K.S., Maas, J.L. and Hokanson, S. 2003. Identification of sources of resistance to bacterial angular leaf spot disease of strawberry. *Acta Horticulturae*, 626: 155–159.

Hartz, T.K., De Vay, J.E. and Elmore, C.L. 1993. Solarization is an effective disinfestation technique for strawberry production. *Hort Science*, 28: 104–106.

Hauke, K., Creemers, P., Brugmans, W. and Laer, S. van 2004. Signum, a new fungicide with interesting properties in resistance management of fungal diseases in strawberries. *Communications in Agricultural and Applied Biological Sciences*, 69(4): 743–755.

Hietaranta, T. 2006. Strawberry cultivars for the Finnish fresh market. *Journal of the American Pomological Society*, 60(1): 20–27.

Horn, N.L. 1961. Control of Botrytis rot of strawberries. *Plant Disease Report*, 45: 818–822.

Howard, C.M. 1972. A strawberry fruit rot caused by *Colletotrichum fragariae*. *Phytopathology*, 62: 600–602.

Howard, C.M. and Alberts, E.E. 1973. Cassia obtusifolia, a possible reservoir for inoculation of *Colletotrichum fragariae*. *Phytopathology*, 63: 533–534.

Hua, X.F. and Chen, X.F. 2004. Control effect of 50% kresoxim-methyl dry flow able formulation on strawberry powdery mildew. *Agricultural Science and Technology*, 4: 26–27.

Huang, Y., Zhen, W., Zhang, L., Zhang, G., Tian, L., Cheng, H., Shi, Y. and Li, S. 2005. Separation of causal organisms and biocontrol of strawberry (*Fragaria ananassa* Duch) replant disease. *Biotechnology*, 15: 74–76.

Hukkanen, A.T., Kokko, H.I., Buchala, A.J., McDougall, G.J., Stewart, D., Karenlampi, S.O. and Karjalainen, R.O. 2007. Benzothiadiazole induces the accumulation of phenolics and improves resistance to powdery mildew in strawberries. *Journal of Agricultural and Food Chemistry*, 55(5): 1862–1870.

Hutton, D. 2006. Successful management of Colletotrichum crown and stolon rot in runner production in subtropical Australia. *Acta Horticulturae*, 708: 293–298.

Igarashi, I., Monma, S., Fujino, M., Okimura, M., Okitsu, S., Takada, K. and Nii, T. 1994. Breeding of new everbearing strawberry variety Ever Berry. *Bulletin Natl. Res. Inst. Veg. Ornamental plant and Ta Series A. Vegetables and Ornamental Plants*, 9: 69–84.

Inada, M. and Yamaguchi, J.I. 2006. Occurrence of anthracnose fruit rot caused by *Colletotrichum acutatum* and *Colletotrichum gloeosporioides* on strawberry in forcing culture. *Kyushu Plant Protection Research*, 52: 11–17.

Jamieson, A.R., Nickerson, N.L., Forney, C.F., Sanford, K.A., Sanderson, K.R., Prive, J.P., Tremblay, R.J.A. and Hendrickson, P. 2004a. "Cabot" strawberry. *Hort Science*, 39(7): 1778–1780.

Jamieson, A.R., Nickerson, N.L., Sanderson, K.R., Prive, J.P., Tremblay, R.J.A. and Hendrickson, P. 2004b. Evangeline strawberry. *Hort Science*, 39(7): 1783–1784.

Jamieson, A.R., Nickerson, N.L., Sanderson, K.R., Prive, J.P., Tremblay, R.J.A. and Hendrickson, P. 2004c. "Brunswick" strawberry. *Hort Science*, 39(7): 1781–1782.

Jayamani, A. and Zeller, W. 2004. Studies on efficacy and mode of action of rhizosphere bacteria against *Phytophthora* spp. in strawberry. *Bulletin OILB/SROP*, 27(8): 261–264.

Jhooty, J.S. and Mc Kean, W.E. 1965. Studies on powdery mildew of strawberry caused by *Sphaerotheca macularis*. *Phytopathology*, 55: 281–285.

Jiang, G.H., Miao, L.M., Yang, X.F., Zhang, Y.C., Ren, H.Y. and Wang, H.R. 2014. Technology of anthracnose control on increasing strawberry plantlets. *Acta Horticulturae*, 1049: 669–671.

Joffens, W.F. 1957. Soil treatments for control of the red stele disease of strawberries. *Plant Disease Report*, 41: 415–418.

Johnson, A.W., Simpson, D.W. and Berrie, A. 2006. Hot water treatment to eliminate *Colletotrichum acutatum* from strawberry runner cuttings. *Acta Horticulturae*, 708: 255–258.

Jovaisiene, Z. 2004. *Phytophthora fragariae*, a new fungi species in Lithuania. *Botanica Lithuanica* 10(3): 233–236.

Kandybin, N.V. 2003. For activization of microbiomethod. *Zashchita i Karantin Rastenii*, 7: 13–14.

Kanto, T., Miyoshi, A., Ogawa, T., Maekawa, K. and Aino, M. 2004. Suppressive effect of potassium silicate on powdery mildew of strawberry in hydroponics. *Journal of General Plant Pathology*, 70(4): 207–211.

Kanto, T., Matsuura, K., Ogawa, T., Yamada, M., Ishiwata, M., Usami, T. and Amemiya, Y. 2014. A new UV-B lighting system controls powdery mildew of strawberry. *Acta Horticulturae*, 1049: 655–660.

Integrated Disease Management

Karabulut, O.A., Arslan, U. and Kuruoglu, G. 2004a. Control of postharvest diseases of organically grown strawberry with preharvest applications of some food additives and postharvest hot water dips. *Journal of Phytopathplogy*, 152(4): 224–228.

Karabulut, O.A., Tezcan, H., Daus, A., Cohen, L., Wiess, B. and Droby, S. 2004b. Control of preharvest and postharvest fruit rot in strawberry by *Metschnikowia fructicola*. *Biocontrol Science and Technology*, 14(5): 513–521.

Kawagishi, K., Kato, S., Ubukata, M., Abe, T., Tachikawa, S., Inagawa, Y. and Fukukawa, E. 2001. "Kentaro", a new short-day strawberry variety for cold regions. *Bulletin of Hokkaido Prefectural Agricultural Experiment Stations*, 81: 11–20.

Keyworth, W. and Benett, M. 1951. Verticillium wilt of the strawberry. *Journal of Horticulture Science*, 26: 304–316.

Khanizadeh, S., Buszard, D., Carrise, O. and Thiobdean, P.O. 1996. Joliette strawberry. *Horticulture Science*, 31: 1036–1037.

Khanizadeh, S., Cousineau, J., Deschenes, M., Levasseur, A., Carisse, O., DeEll, J., Gauthier, L. and Sullivan, J.A. 2002a. "Saint-Pierre" strawberry. *Hort Science*, 37(7): 1133–1134.

Khanizadeh, S., Cousineau, J., Deschenes, M. and Levasseur, A. 2002b. 'Roseberry' and 'Rosalyne': two new hardy, day-neutral, red-flowering strawberry cultivars. *Strawberry research to 2001 Proceedings of the 5th North American Strawberry Conference*, 42.

Khanizadeh, S., Carisse, O. and Buszard, D. 2002c. "AC Yamaska" and "AC L'Acadie" strawberries. *Strawberry Research to 2001 Proceedings of the 5th North American Strawberry Conference, 2002*, 38–40.

Khanizadeh, S., Deschenes, M., Levasseur, A., Carisse, O., Charles, M.T., Rekika, D., Tsao Rong, Yang, R., DeEll, J., Gauthier, L., Gosselin, A., Sullivan, J.A. and Davidson, C. 2005a. "Saint Laurent d'Orleans" strawberry. *Hort Science*, 40(7): 2195–2196.

Khanizadeh, S., Deschenes, M., Levasseur, A., Carisse, O., Charles, M.T., Rekika, D., Tsao-Rong Yang, R., DeEll, J., Thibeault, P., Prive, J.P., Davidson, C. and Bors, B. 2005b. "Harmonie" strawberry. *Hort Science*, 40(2): 480–481.

Khosla, K. and Kumar, J. 2005. Management of root rot and blight (Sclerotium rolfsii Sacc.) of strawberry. *Acta Horticulturae*, 696: 371–373.

Kim, M.A. and Choi, S.J. 2002. Induction of gray mold rot resistance by methyl salicylate application in strawberry fruits. *Journal of the Korean Society for Horticultural Science*, 43(1): 29–33.

Kim, J.G., Cho, C.T., Bia, T.U., Han, H.S. and Mun, B.J. 1978. Varietal resistance to strawberry to two strains of *Mycosphaerella fragariae* and chemical control of the pathogen. *Korean Journal of Plant Protection* 17: 149–154.

Kim, C.H., Seo, H.D., Cho, W.D. and Kim, S.B., 1982. Studies on varietal resistance and chemical control to the wilt of strawberry caused by *Fusarium oxysporum*. *Korean Journal of Plant Protection*, 21: 61–67.

Kim, T., Jang, W.S., Choi, J.H., Nam, M.H., Kim, W.S. and Lee, S.S. 2004. Breeding of strawberry "Maehyang" for forcing culture. Induction of gray mold rot resistance by methyl salicylate application in strawberry fruits. *Korean Journal of Horticultural Science and Technology*, 22(4): 434–437.

Koch-Dean, M., Taanami, Z. and Freeman, S. 2002. Development of strawberry cultivars resistant to anthracnose (*Colletotrichum acutatum*) in Israel. *Acta Horticulturae*, 567(1): 97–100.

Kodama, T. 1974. Characteristics of strawberry yellows disease caused by *Fusarium* and cultivar resistance. *Bulletin of Nara Agricultural Experimental Station*, 6: 68–75.

Koike, S.T. and Gordon, T.R. 2015. Management of Fusarium wilt of strawberry. *Crop Protection*, 73: 67–72.

Koike, S.T., Kirkpatrick, S.C. and Gordon, T.R. 2009. Fusarium wilt of strawberry caused by *Fusarium oxysporum* in California. *Plant Disease*, 93: 1077.

Kondakova, V. and Schuster, G. 1991. Elimination of strawberry mottle virus and strawberry crinkle virus from isolated apices of three strawberry varieties by the addition of 2,4-dioxohexahydro-1,3,5-triazine (5-azadihydrouracil) to the nutrient medium. *Journal of Phytopathology*, 132(1): 84–86.

Konno, H., Inagawa, Y., Kawagishi, K., Sawada, K., Shiozawa, K., Kato, S. and Tachikawa, S. 2001. A new short-day strawberry variety "Kita-ekubo". *Bulletin of Hokkaido Prefectural Agricultural Experiment Stations*, 81: 1–10.

Kovach, J., Petzoldt, R. and Harman, G.E. 2000. Use of honey bees and bumble bees to disseminate *Trichoderma harzianum* 1295–22 to strawberries for Botrytis control. *Biological Control*, 18: 235–242.

Laer, S van, Hauke, K., Meesters, P. and Creemers, P. 2005. Botrytis infection warnings in strawberry: reduced enhanced chemical control. *Communications in Agricultural and Applied Biological Sciences*, 70(3): 61–71.

LaMondia, J.A. and Cowles, R.S. 2005. Comparison of *Pratylenchus penetrans* infection and *Maladera castanea* feeding on strawberry root rot. *Journal of Nematology*, 37(2): 131–135.

Laugale, V. and Morocko, I. 2000. Resistance of strawberry cultivars to fungal diseases. *Bulletin OILB/SROP*, 23(11): 117–118.

Leandro, L.F.S., Gleason, M.L., Wegulo, S.N. and Nutter, F.W. Jr. 2002. *Strawberry Research to 2001 Proceedings of the 5th North American Strawberry Conference*, 2002, 74–76.

Lee, S.Y., Lee, S.B. and Kim, C.H. 2004. Biological control of powdery mildew by Q-fect WP (*Ampelomyces quisqualis* 94013) in various crops. *Bulletin OILB/SROP*, 27(8): 329–331.

Legard, D.E., Mertely, J.C., Chandler, C.K., Price, J.F. and Duval, J.D. 2002a. Management of strawberry diseases in the 21st century: chemical and cultural control of botryis fruit rot. Strawberry research to 2001. *Proceedings of the 5th North American Strawberry Conference*, 58–63.

Legard, D.E., Mertely, J.C., Xiao, C.L., Chandler, C.K., Duval, J.R. and Price, J.P. 2002b. Cultural and chemical control of Botrytis fruit rot of strawberry in annual winter production systems. *Acta Horticulturae*, 567(2): 651–654.

Legard, D.E., Mac Kenzie, S.J., Mertely, J.C., Chandler, C.K. and Peres, N.A. 2005. Development of a reduced use fungicide program for control of Botrytis fruit rot on annual winter strawberry. *Plant Disease*, 89(12): 1353–1358.

Lele, V.C. and Pathak, H.C. 1965. Leaf blight and dry stalk rot of strawberry caused by *Rhizoctonia bataticola*. *Indian Phytopathology*, 18: 117–163.

Levy, N.O., Harel, Y., Haile, Z.M., Elad, Y., Rav-David, E., Jurkevitch, E. and Katan, J. 2015. Induced resistance to foliar diseases by soil solarization and *Trichoderma harzianum*. *Plant Pathology*, 64(2): 365–374.

Lewers, K.S., Enns, J.M., Wang, S.Y., Maas, J.L., Galletta, G.J., Hokanson, S.C., Clark, J.R., Demchak, K., Funt, R.C., Garrison, S.A., Jelenkovic, G.L., Nonnecke, G.R., Probasco, P.R., Smith, B.J., Smith, B.R. and Weber, C.A. 2004. "Ovation" strawberry. *Hort Science*, 39(7): 1785–1788.

Li, Y.H., Miyawaki, Y., Matsubara, Y. and Koshikawa, K. 2006. Tolerance to anthracnose in mycorrhizal strawberry plants grown by capillary watering method. *Environment Control in Biology*, 44(4): 301–307.

Liang, L. and Lin, Z. 2014. Genetic study on resistance to powdery mildew in strawberry. *Acta Horticulturae*, 1049: 255–258.

Liang, X.M., Wang, Y.Z., Wang, P.S., Wang, J.Q. and Wu, G.B. 2004. Effect of using Luying mechanical oil emulsion for control of diseases and pests of fruit crops. *China Fruits*, 2: 7–10.

Lilja, A.T., Parikka, P.K., Paaskynkivi, E.A., Hantula, J.I., Vainio, E.J., Vartiamaki, H.A., Lemmetty, A.H. and Vestberg, M.V. 2006. *Phytophthora cactorum* and *Colletotrichum acutatum*: survival and detection. *Agriculturae Conspectus Scientificus Poljoprivredna Znanstvena Smotra*, 71(4): 121–128.

Linsley Noakes, G., Wilken, L. and Villiers, S. 2006. High density, vertical hydroponics growing system for strawberries. *Acta Horticulturae*, 708: 365–370.

Lopez-Aranda, J.M., Soria, C., Sanchez-Sevilla, J.F., Galvez, J., Medina, J.J., Arjona, A., Marsal, J.I. and Bartual, R. 2004. "Marina" strawberry. *Hort Science*, 39(7): 1776–1777.

Lopez-Aranda, J.M., Soria, C., Sanchez-Sevilla, J.F., Galvez, J., Medina, J.J., Arjona, A., Marsal, J.I. and Bartual, R. 2005. "Medina" strawberry. *Hort Science*, 40(2): 482–483.

Ludvikova, J. and Paprstein, F. 2003. Strawberry cultivar "Honeoye". *Vedecke Prace Ovocnarske*, 18: 169–171.

Maccagnani, B., Mocioni, M., Gullino, M.L. and Ladurner, E. 1999. Application of *Trichoderma harzianum* by using *Apis mellifera* as a vector for the control of grey mould of strawberry: first results. *Bulletin OILB/SROP*, 22(1): 161–164.

MacKenzie, S.J., Legard, D.E., Timmer, L.W., Chandler, C.K. and Peres, N.A. 2006. Resistance of strawberry cultivars to crown rot caused by *Colletotrichum gloeosporioides* isolates from Florida is nonspecific. *Plant Disease*, 90(8): 1091–1097.

Madden, L.V., Wilson, L.L. and Ellis, M.A. 1993. Field spread of anthracnose fruit rot of strawberry in relation to ground cover and ambient weather conditions. *Plant Disease*, 77: 861–866.

Mass, J.L. 1978. Screening for resistance to fruit rot in strawberries and red raspberries: a review. *Hort Science*, 13: 423–426.

Mass, J.L. 1984. *Compendium of Strawberry Diseases*. American Phytopathological Society, St. Paul, Minnesota: p. 60.

Matsuo, K., Suga, Y. and Jan, M. 1994. Chemical control of strawberry anthracnose caused by *Colletotrichum acutatum*. *Plant Prot.* Kyushu, 40: 17–21.

McGechan, J.K. and Fahy, P.C. 1976. Angular leaf spot of strawberry, *Xanthomonas fragariae*: first record of its occurrence in Australia and attempts to eradicate the disease. *APPS Newsletter*, 5: 57–59.

Mcintyre, J.L. and Walton, G.S. 1981. Control of strawberry anthracnose caused by *Colletotrichum acutatum*. *Proceedings of the Association for Plant Protection of Kyushu*, 40: 17–21.

Meesters, P., Meurrens, F. and Brugmans, W. 1996. Strawberry cultivation. Total soil fumigation and fumigation in strips for growing *Elsanta*. *Fruit Belge*, 64: 65–69.

Mertely, J.C., MacKenzie, S.J. and Legard, D.E. 2002. Timing of fungicide applications for *Botrytis cinerea* based on development stage of strawberry flowers and fruit. *Plant Disease*, 86(9): 1019–1024.

Meszka, B. and Bielenin, A. 2002. New fungicides for controlling grey mould in strawberry. *Zeszyty Naukowe Instytutu Sadownictwa i Kwiaciarstwa w Skierniewicach*, 10: 217–221.

Meszka, B. and Bielenin, A. 2004. Efficacy of Euparen 50 WP and Euparen M 50 WG in the control of grey mould (*Botrytis cinerea*) and reduction of the population of two-spotted spider mite (*Tetranychus urticae* Koch) on strawberries. *Bulletin OILB/SROP*, 27(4): 41–45.

Meszka, B., Abanowska, B.H. and Bielenin, A. 2000. Efficacy of Euparen 50 WP and Euparen M 50 WG in the control of grey mould (*Botrytis cinerea*) and reduction of the population of two-spotted spider mite (*Tetranychus urticae* Koch) on strawberries. *Bulletin OILB/SROP*, 23(11): 133–136.

Meszka, B., Bielenin, A. and Nowacki, J. 2004. Procymidone in control of strawberry grey mould and fungicide residues on fruits. *Folia Universitatis Agriculturae Stetinensis, Agricultura*, 96: 113–118.

Milicevic, T., Cvjetkovic, B. and Jurjevic, Z. 2002. Biology and control of the fungus *Diplocarpon earliana* (Ell. & Ev.) Wolf on strawberries. *Fragmenta Phytomedica et Herbologica*, 27(1/2): 5–13.

Milicevic, T., Ivic, D., Cvjetkovic, B. and Duralija, B. 2006. Possibilities of strawberry integrated disease management in different cultivation systems. *Agriculturae Conspectus Scientificus Poljoprivredna Znanstvena Smotra*, 71(4): 129–134.

Miller, T.C., Gubler, W.D., Laemmlen, F.F., Geng, S. and Rizzo, D.M. 2004. Potential for using *Lecanicillium lecanii* for suppression of strawberry powdery mildew. *Biocontrol Science and Technology*, 14(2): 215–220.

Mitra, S.K. 1993. Strawberries. In : S.K. Mitra, T.K. Bose and D.S. Rathore (eds.). *Temperate Fruits*. Alert and Allied Publishers, 27/3 Chakraberia Lane Calcutta, India, pp. 549–596.

Momma, N., Kobara, Y., Uematsu, S., Kita, N. and Shinmura, A. 2013. Development of biological soil disinfestations in Japan. *Applied Microbiology Biotechnology*. Online. http://dx.doi.org/10.1007/s00253-013-4826-9.

Monaco, M.E., Salazar, S.M., Aprea, A., Diaz-Ricci, D.C., Zembo, J.C. and Castagnaro, A. 2000. First report of *Colletotrichum gloeosporioides* on strawberry in Northwestern Argentina. *Plant Disease*, 84(5): 395.

Moore, J.N. 1981. "Arking", a late season, disease resistant strawberry variety. *Arkansas Farm Research*, 30: 6.

Mutisya, J.M., Sullivan, J.A., Sutton, J.C., Zheng, J. and Couling, S. 2005. Influence of leaf scorch on vegetative growth and yield of three strawberry (*Fragaria* x *ananassa* Duch.) cultivars with differing levels of resistance. *Canadian Journal of Plant Science*, 85(3): 679–686.

Nadas, A., Olmo, M. and Garcia, J.M. 2003. Growth of *Botrytis cinerea* and strawberry quality in ozone-enriched atmospheres. *Journal of Food Science*, 68(5): 1798–1802.

Nam, M.H., Jung, S.K., Ra, S.W. and Kim, H.G. 2003. Control efficacy of sodium bicarbonate alone and in mixture with polyoxyethylene sorbitanmonolaurate on powdery mildew of strawberry. *Korean Journal of Horticultural Science and Technology*, 21(2): 98–101.

Naradisorn, M., Klieber, A., Sedgley, M., Scott, E. and Able, A.J. 2006. Effect of preharvest calcium application on grey mould development and postharvest quality in strawberries. *Acta Horticulturae*, 708: 147–150.

Navarthan, S.J., Degefu, T. and Haite, M. 1991. Outbreak and new records. Ethopia: angular leaf spot of strawberry in Ethopia. *Plant Prot Bulletin*, 39: 116.

Nilsson, U., Carlson-Nilsson, U. and Svedelius, G. 2005. First report of anthracnose fruit rot caused by *Colletotrichum acutatum* on strawberry in Sweden. *Plant Disease*, 89(11): 1242.

Njoroge, S.M.C., Vallad, G.E., Park, S.Y., Kang, S., Koike, S.T., Bolda, M., Burman, P., Polonik, W. and Subbarao, K.V. 2011. Phenological and phytochemical changes correlated with differential interactions of *Verticillium dahliae* with broccoli and cauliflower. *Phytopathology*, 101: 523–534.

Nobre, S.A.M., Maffia, L.A., Mizubuti, E.S.G., Cota, L.V. and Dias, A.P.S. 2005. Selection of *Clonostachys rosea* isolates from Brazilian ecosystems effective in controlling *Botrytis cinerea*. *Biological Control*, 34(2): 132–143.

Nunes, M.C.N., Morais, A.M.M.B., Brecht, J.K., Sargent, S.A. and Bartz, J.A. 2005. Prompt cooling reduces incidence and severity of decay caused by *Botrytis cinerea* and *Rhizopus stolonifer* in strawberry. *Hort Technology*, 15(1): 153–156.

Okamoto, H., Fujii, S., Kato, K. and Yoshika, A. 1970. A new strawberry disease "Fusarium wilt". *Plant Protection*, 24: 231–235.

Okimura, M., Noguchi, Y., Mochizuki, T., Sone, K. and Kitadani, E. 2004. Breeding "Strawberry Parental Line No-2" with anthracnose resistance. *Horticultural Research Japan*, 3(3): 257–260.

Ossipov, V.V., Haukioja, E., Ossipova, S., Hanhimaki, S. and Pihlaja, K. 2001. Phenolic and phenolic-related factors as determinants of suitability of mountain birch leaves to an herbivorous insect. *Biochemical Systematics and Ecology*, 29: 223–240.

Pachauri, R.K. and Reisinger, A. (eds.). 2007. Climate change 2007. Synthesis Report. Contribution of working groups I, II and III to the Fourth Assessment Report of The Intergovernmental Panel of Climate Change. IPCC, Geneva, Switzerland, 104p.

Palmer, S., Scott, E., Stangoulis, J. and Able, A.J. 2006. The effect of foliar-applied Ca and Si on the severity of powdery mildew in two strawberry cultivars. *Acta Horticulturae*, 708: 135–139.

Pan, L., Zhang, W., Zhu, N., Mao, S. and Tu, K. 2014. Early detection and classification of pathogenic fungal disease in post-harvest strawberry fruit by electronic nose and gas chromatography–mass spectrometry. *Food Research International*, 62: 162–168.

Parikka, P. and Lemmetty, A. 2009. *Colletotrichum acutatum*: Survival in plant debris and infection of some weeds and cultivated plants. *Acta Horticulturae*, 842: 307–310.

Parikka, P. and Tuovinen, T. 2014. Plant protection challenges in strawberry production in Northern Europe. *Acta Horticulturae*, 1094: 173–179.

Particka, C.A. and Hancock, J.F. 2005. Field evaluation of strawberry genotypes for tolerance to black root rot on fumigated and non-fumigated soil. *Journal of the American Society for Horticultural Science*, 130(5): 688–693.

Parikka, P., Lemmetty, A. and Paaskynkivi, E. 2006. Survival of *Colletotrichum acutatum* in dead plant material and soil in Finland. *Acta Horticulturae*, 708: 131–134.

Paulus, A.O., Voth, V., Nelson, J. and Bowen, H. 1978. Control of Botrytis fruit rot of strawberry. *California Agriculture*, 32: 9.

Pennisi, A.M., Agosteo, G.E. and Funaro, M. 2002. Control of pre and post-harvest rot of strawberry using alternative active ingredients. Atti, Giornate fitopatologiche, Baselga di Pine, Trento, Italy, 7–11 Aprile 2002, 2: 443–448.

Peris, O.S. 1962. Studies on strawberry mildew, caused by *Sphaerotheca macularis* (Wallr. Ex Fries) Jaczcwski II: Host parasite relationships on foliage on strawberry varieties. *Annals of Applied Biology*, 50: 225–233.

Perotto, M.C., Luciani, C., Celli, M.G., Torrico, A. and Conci, V.C. 2014. First report of *Strawberry crinkle virus* in Argentina. *New Disease Reports*, 30, 5. http://dx.doi.org/10.5197/j.2044-0588.2014.030.005.

Porras, M., Barrau, C., Arroyo, F.T., Santos, B., Blanco, C. and Romero, F. 2007. Reduction of *Phytophthora cactorum* in strawberry fields by *Trichoderma* spp. and soil solarization. *Plant Disease*, 91(2): 142–146.

Porras, M., Barrau, C. and Romero, F. 2009. Influence of Trichoderma and soil solarization on strawberry yield. ISHS *Acta Horticulturae*, 842: VI International Strawberry Symposium: 75. http://www.actahort.org/book/842/842_75.htm

Porta-Puglia, A. and Mifsud, D. 2006. First record of *Colletotrichum acutatum* on strawberry in Malta. *Journal of Plant Pathology*, 88(2): 228.

Prasad, K. and Stadelbacher, G.J. 1974. Control of postharvest decay of fresh raspberries by acctaldehyde vapor. *Phytopathology*, 64: 948–951.

Prokkola, S., Kivijarvi, P. and Parikka, P. 2003. Effects of biological sprays, mulching materials, and irrigation methods on grey mould in organic strawberry production. *Acta Horticulturae*, 626: 169–175.

Ragab, M., El-Dougdoug K., Mousa S., Attia, A., Zeidan, M., Sobolev, I., Spiegel, S., Freeman, S., Tzanetakis, I.E. and Martin, R.R. 2009. Detection of strawberry viruses in Egypt. *Acta Horticulturae*, 842: 319–322.

Ramallo, C.J., Ploper, L.D., Ontivero, M., Filippone, M.P., Castagnaro, A. and Diaz-Ricci, J. 2000. First report of *Colletotrichum acutatum* on strawberry in northwestern Argentina. *Plant Disease*, 84(6): 706.

Rebollar-Alviter, A., Madden, L.V. and Ellis, M.A. 2005. Efficacy of azoxystrobin, pyraclostrobin, potassium phosphite, and mefenoxam for control of strawberry leather rot caused by *Phytophthora cactorum*. *Plant Health Progress*: 1–6. doi:10.1094/PHP-2005-0107-01-RS

Reddy, M.V.B., Belkacemi, K., Corcuff, R., Castaigne, F. and Arul, J. 2000. Effect of preharvest chitosan sprays on post-harvest infection by *Botrytis cinerea* and quality of strawberry fruit. *Postharvest Biology and Technology*, 20: 39–51.

Richardson, J., Frazier, N.W. and Sylvester, E.S. 1972. Rhabdovirus particles associated with *strawberry crinkle virus*. *Phytopathology* 62: 491–492.

Integrated Disease Management

Ritchie, P.F., Averre, C.W. and Milholland, R.D. 1993. First report of angular leaf spot, caused by *Xanthomonas fragariae*, on strawberry in North Carolina. *Plant Disease*, 77: 1263.

Rondon, S.I., Paranjpe, A.V., Cantliffe, D.J. and Price, J.F. 2004. Strawberry cultivars grown under protected structure and their susceptibility to natural infestation of the cotton aphid, *Aphis gossypii* and to powdery mildew, *Sphaerotheca maculari* f. sp. *fragarie. Acta Horticulturae*, 659(1): 357–362.

Rugienius, R., Gelvonauskiene, D., Zalunskaite, I., Pakalniskyte, J. and Stanys, V. 2006. Interspecific hybridization in genus Fragaria and evaluation of the hybrids for cold hardiness and red stele resistance using in vitro methods. *Acta Horticulturae*, 725(1): 451–456.

Saber, M.M., Sabet, K.K., Moustafa-Mahmoud, S.M. and Khafagi, I.Y.S. 2003. Evaluation of biological products, antioxidants and salts for control of strawberry fruit rots. *Egyptian Journal of Phytopathology*, 31(1/2): 31–43.

Salazar, S.M., Castagnaro, A.P., Arias, M.E., Chalfoun, N., Tonello, U. and Diaz-Ricci, J.C. 2007. Induction of a defense response in strawberry mediated by an avirulent strain of *Colletotrichum. European Journal of Plant Pathology*, 117(2): 109–122.

Sallato, B.V., Torres, R., Zoffoli, J.P. and Latorre, B.A. 2007. Effect of boscalid on postharvest decay of strawberry caused by *Botrytis cinerea* and *Rhizopus stolonifer. Spanish Journal of Agricultural Research*, 5(1): 67–78.

Schestibratov, K.A. and Dolgov, S.V. 2005. Transgenic strawberry plants expressing a thaumatin II gene demonstrate enhanced resistance to *Botrytis cinerea. Scientia Horticulturae*, 106(2): 177–189.

Schmid, A., Daniel, C. and Weibel, F.P. 2004. Effect of cultural methods on leaf spot (*Mycosphaerella fragariae* Tul.) incidence in strawberries. *Bulletin OILB/SROP*, 27(4): 55–56.

Schoen, C.D., Limpens, W., Moller, I., Groeneveld, L., Klerks M.M. and. Lindner J.L. 2001. The complete genomic sequence of *strawberry crinkle virus*, a member of the *Rhabdoviridae. Acta Horticulturae*, 656: 45–50.

Shafir, S., Dag, A., Bilu, A., Abu-Toamy, M. and Elad, Y. 2006. Honey bee dispersal of the biocontrol agent *Trichoderma harzianum* T39: effectiveness in suppressing *Botrytis cinerea* on strawberry under field conditions. *European Journal of Plant Pathology*, 116(2): 119–128.

Shalaby, O.Y.M. and Mohamed, A.A. 2005. Effect of soil mulching with different thickness of polyethylene sheets on the incidence of strawberry root rot and wilt diseases. *Annals of Agricultural Science, Moshtohor*, 43(3): 1095–1105.

Shao, X., Wang, H., Xu, F. and Cheng, S. 2013. Effects and possible mechanisms of tea tree oil vapor treatment on the main disease in postharvest strawberry fruit. *Postharvest Biology and Technology*, 77: 94–101.

Sharma, A., Bhardwaj, L.N. and Gupta M. 2005. Leather rot of strawberry and its management - a review. *Agricultural Reviews*, 26(1): 59–66.

Shennan, C., Muramoto, J., Lamers, J., Mazzola, M., Rosskopf, E., Kokalis-Burelle, N., Momma, N., Butler, D. and Kobara, Y. 2014. Anaerobic soil disinfestation for soilborne disease control in strawberry and vegetable systems: current knowledge and future directions. *Acta Horticulturae*, 1044: 165–175.

Shin, G.H., Seo, J.B., Kim, D.I., Koo, Y.S., Won, S.Y., Song, J.H. and Ko, Y.J. 2014. Environment-friendly control of major diseases at strawberry cultivation regions in Korea. *Acta Hprticulturae*, 1049: 651–654.

Shuman, J.L. and Ballington, J.R. 2003. Incidence and development of anthracnose fruit rot in seven strawberry genotypes inoculated with five concentrations of *Colletotrichum acutatum. Acta Horticulturae*, 626: 199–205.

Silva-Rosales, L., Vazquez-Sanchez, M.N., Gallegos, V., Ortiz-Castellanos, M.L. and Rivera-Bustamante, R., Davalos-González, P.A. and Jofre-Garfias, A.E. 2013. First Report of *Fragaria chiloensis* cryptic virus, *Fragariachiloensis latent virus*, Strawberry mild yellow edge virus, Strawberry necrotic shock virus, and Strawberry pallidosis associated virus in Single and Mixed Infections in Strawberry in Central Mexico. *Plant Disease*, 97(70): 1002.

Simpson, D. and Blanke, M. 1996. Calypso and Tango perpetual flowering strawberry varieties from Horticultural Research International East Malling. *Erwerbsobstbau*, 38: 55–57.

Simpson, D., Bell, J.A., Harris, D.C., Schmidt, H. and Kellaerhols, M. 1994. Breeding to fungal diseases in strawberry. *Proceeding of Eucarpia Fruit Breeding Section Meeting, Wadenswil*. Einsiedeln. Switzerland 30 August to 3 September, 1993: 63–66.

Simpson, D., Hammond, K., Lesemann, S. and Whitehouse, A. 2006. Pathogenicity of UK isolates of *Colletotrichum acutatum* and relative resistance among a range of strawberry cultivars. *Acta Horticulturae*, 708: 281–285.

Sindhan, G.S. and Roy, A.J. 1981. Screening of different varieties and selections of strawberry against leaf diseases and their control. *Indian Phytopathology*, 34: 304–306.

Singh, S.J. 1974. A ripe fruit rot of strawberry caused by *Colletotrichum fragariae*. *Indian Phytopathology*, 27: 433–434.

Singh, S.J., Sastry, K.S.M. and Sastry, K.S. 1975. Investigations on mosaic disease of cape gooseberry. *Current Science*, 44: 95–96.

Singh R., Sharma, R.R. and Goyal, R.K. 2007. Interactive effects of planting time and mulching on "Chandler" strawberry (*Fragaria* x *ananassa* Duch.). *Scientia Horticulturae*, 111(4): 344–351.

Smith, B.J. 2006. USDA-ARS strawberry anthracnose resistance breeding program. *Acta Horticulturae*, 708: 463–470.

Smith, B.J. 2013. Strawberry anthracnose: progress toward control through science. *International Journal of Fruit Science*, 13(1–2): 91–102.

Smith, B.J. and Black, L.L. 1993. In vitro activity of fungicides against *Colletotrichum fragariae*. *Acta Horticulturae*, 348: 509–512.

Sowik, I, Wawrzynczak, D and Michalczuk, L. 2004. Selection of somaclonal variants of strawberry with increased tolerance to Verticillium wilt. *Sodininkyste ir Darzininkyste*, 23(2): 164–173.

Spornberger, A., Steffek, R. and Altenburger, J. 2006. Testing of early ripening strawberry cultivars tolerant to soil-borne pathogens as alternative to "Elsanta". *Agriculturae Conspectus Scientificus Poljoprivredna Znanstvena Smotra*, 71(4): 135–139.

Stankovic, I., Ristic, D., Vucurovic, A., Milojevic, K., Nikolic, D., Krstic, B. and Bulajic, A. 2014. First report of Fusarium wilt of strawberry caused by *Fusarium oxysporum* in Serbia. *Plant Disease*, 98: 1435.

Steffek, R., Spornberger, A. and Altenburger, J. 2006. Detection of microsclerotia of *Verticillium dahliae* in soil samples and prospects to reduce the inoculum potential of the fungus in the soil. *Agriculturae Conspectus Scientificus Poljoprivredna Znanstvena Smotra*, 71(4): 145–148.

Stensvand, A., Stromeng, G.M., Langnes, R., Hjeljord, L.G. and Tronsmo, A. 2001. First report of *Colletotrichum acutatum* in strawberry in Norway. *Plant Disease*, 85(5): 558.

Stubbs, L.L. 1956. Diseases of strawberries. *Journal of Agriculture (Victoria)*, 54: 232–236.

Subbarao, K.V., Hubbard, J.C. and Koike, S.T. 1999. Evaluation of broccoli residue incorporation into field soil for Verticillium wilt control in cauliflower. *Plant Disease*, 83: 124–129.

Sundelin, T., Schiller, M., Lubeck, M., Jensen, D.F., Paaske, K. and Petersen, B.D. 2005. First report of anthracnose fruit rot caused by *Colletotrichum acutatum* on strawberry in Denmark. *Plant Disease*, 89(4): 432.

Sylla, J., Alsanius, B.W., Krüger E. and Wohanka, W. 2015. Control of *Botrytis cinerea* in strawberries by biological control agents applied as single or combined treatments. *European Journal of Plant Pathology*, 143(3): 461–471.

Terry, L.A., Joyce, D.C., Adikaram, N.K.B. and Khambay, B.P.S. 2004. Preformed antifungal compounds in strawberry fruit and flower tissues. *Postharvest Biology and Technology*, 31(2): 201–212.

Timudo Torrevilla, O.E., Everett, K.R., Waipara, N.W., Boyd-Wilson, K.S.H., Weeds, P., Langford, G.I. and Walter, M. 2005. Present status of strawberry fruit rot diseases in New Zealand. New Zealand Plant Protection, 58, *Proceedings of a Conference*, Wellington, New Zealand, 9–11 August, 2005, 74–79.

Tu, K., Chen, L. and Pan, X.J. 2006. Effects of different pre-storage heat treatments on the shelf quality and mold control of strawberry fruit. *Acta Horticulturae*, 712(2): 805–810.

Turechek, W.W., Heidenreich, C. and Pritts, M.P. 2002. First report of strawberry anthracnose (*Colletotrichum acutatum*) in strawberry fields in New York. *Plant Disease*, 86(8): 922.

Turechek, W.W., Heidenreich, M.C., Lakso, A.N. and Pritts, M.P. 2007. Estimation of the impact of leaf scorch on photosynthesis and "physiological-lesion" size in strawberry. *Canadian Journal of Plant Pathology*, 29(2): 159–165.

Turechek, W.W., Wang, S., Tiwari, G. and Peres, N.A. 2013. Investigating alternative strategies for managing bacterial angular leaf spot in strawberry nursery production. *International Journal of Fruit Science*, 13(1–2): 234–245.

Uselis, N., Valiuskaite, A. and Raudonis, L. 2006. Incidence of fungal leaf diseases and phytophagous mites in different strawberry cultivars. *Agronomy Research*, 4: 421–425.

Vaughan, E.K. 1933. Transmission of *strawberry crinkle* disease of strawberry. *Phytopathology*, 23: 738–740.

Vellicce, G.R., Diaz-Ricci, J.C., Hernandez, L. and Castagnaro, A.P. 2006. Enhanced resistance to Botrytis cinerea mediated by the transgenic expression of the chitinase gene ch5B in strawberry. *Transgenic Research*, 15(1): 57–68.

Weber, N., Veberic, R., Mikulic-Petkovsek, M., Stampar, F., Koron, D., Munda, A. and Jakopic, J. 2015. Metabolite accumulation in strawberry (Fragaria x ananassa Duch.) fruits and runners in response to *Colletotrichum nymphaeae* infection. *Physiological and Molecular Plant Pathology*, 92: 119–129.

Wilhelm, M. 1952. *Verticillium wilt and Black Root Rot of Strawberry*. California, Agriculture, 6: 8–9.

Integrated Disease Management

Wilhelm, M. 1961. Diseases of strawberry. A guide for the commercial growers. *University of California Agricultural Experiment Station Circular*, pp: 494.

Wilhelm, S. and Paulus, A.O. 1980. How soil fumigation benefits the California strawberry industry. *Plant Disease*, 64: 264–269.

Winks, B.L. and Williams, Y.N. 1965. A wilt of strawberry caused by a new form of *Fusarium oxysporum*. *Queensland Journal of Agricultural and Animal Sciences*, 22: 475–479.

Wojcik, P. and Lewandowski, M. 2003. Effect of calcium and boron sprays on yield and quality of "Elsanta" strawberry. *Journal of Plant Nutrition*, 26(3): 671–682.

Wolf, F.A. 1924. Strawberry leaf scorch. *Journal of Elisha Mitchell Scientific Society*, 39: 117–163.

Wszelaki, A.L. and Mitcham, E.J. 2003. Effect of combinations of hot water dips, biological control and controlled atmospheres for control of gray mold on harvested strawberries. *Postharvest Biology and Technology*, 27(3): 255–264.

Xu, X.M., Harris, D.C. and Berrie, A.M. 2000. Modelling infection of strawberry flowers by *Botrytis cinerea* (grey mould) using field data. The BCPC Conference: Pests and diseases, Volume 2, *Proceedings of an International Conference Held at the Brighton Hilton Metropole Hotel*, Brighton, UK, 13–16 November 2000: pp. 809–814.

Yohalem, D., Lower, K., Harvey, N. and Passey, T. 2009. *Potentilla reptans* is an alternative host for two strawberry viruses. *Journal of Phytopathology*, 157(10): 646–648.

Yokota, C. 2004. Cyflufenamid – a new fungicide against powdery mildew on various crops. *Agrochemicals Japan*, 84: 12–14.

Zhang, Z.H., Liu, Y., Gao, X.Y., Du, G.D. and Dai, H.Y. 2004. Leaf disc assay for identification of resistance of strawberry to powdery mildew and screening of fungicides. *Acta Horticulturae Sinica*, 31(4): 505–507.

Zhang, H., Zheng, X., Wang, L., Li, S. and Liu, R. 2007. Effect of yeast antagonist in combination with hot water dipson postharvest Rhizopus rot of strawberries. *Journal of Food Engineering*, 78: 281–287.

Zhang, H.Y., Wang, L., Dong, Y., Jiang, S., Cao, J. and Meng, R.J. 2007. Postharvest biological control of gray mold decay of strawberry with *Rhodotorula glutinis*. *Biological Control*, 40(2): 287–292.

Zhao, X.S., Zhen, W.C., Qi, Y.Z., Liu, X.J. and Yin, B.Z. 2009. Coordinated effects of rootautotoxic substances and *Fusarium oxysporum* Schl. f. sp. *fragariae* on the growth and replant disease of strawberry. *Frontiers of Agriculture in China*, 3: 34–39.

Zurawicz, E. and Daminikowski, J. 1995. Preliminary evaluation of productivity of several new strawberry (*Fragaria* x *ananassa* Duch.) clone and cultivars in central Poland. *Zesszyty Naukouse Instytute Sadowhictwa I Kwiaciarstwa w Skiniewicach*, 2: 5–11.

30 Integrated Insect Pest Management

Uma Shankar and D. P. Abrol

CONTENTS

30.1 Major Insect-Pests 488
    30.1.1 Sucking Insect-Pests 488
        30.1.1.1 Strawberry Aphids 488
        30.1.1.2 Strawberry Aphid, *C. fragaefolii* 488
        30.1.1.3 Melon Aphid, *A. gossypii* 488
        30.1.1.4 Green Peach Aphid, *M. persicae* 488
        30.1.1.5 Strawberry Spider Mites 490
        30.1.1.6 The Greenhouse Whitefly and *Lygus* Bug 491
        30.1.1.7 Strawberry Thrips, *Frankliniella occidentalis* (Pergande) 492
    30.1.2 Lepidopteran Insect-Pests 493
        30.1.2.1 Greasy Cut Worms 493
        30.1.2.2 Tobacco Caterpillars 493
        30.1.2.3 Strawberry Leaf Folder 494
    30.1.3 Coleopteran Insect Pests 495
        30.1.3.1 Strawberry Bud Weevil or Clipper 495
        30.1.3.2 Strawberry Crown Borer 496
        30.1.3.3 White Grubs 496
    30.1.4 Soil Arthropods 496
        30.1.4.1 Termites 496
        30.1.4.2 Snails and Slugs 497
    30.1.5 Mammalian and Frugivorous Bird Pests 498
30.2 IPM 499
30.3 Future Strategies 500
References 503

Strawberries are one of the most widely grown fruit crops in the world. They can flourish from tropical highlands to subarctic regions. Strawberry is one of the most popular soft fruits, and is cultivated in plains as well as in hills up to an elevation of 3000 m above mean sea level in humid or dry regions, widely grown in protected and open conditions in temperate and subtropical countries. Among all the different types of berries, strawberry gives the quickest return in a shortest possible period. Its heart-shaped fruit of love was mentioned by the Roman poets Virgil and Ovid in the 1st centuries BC and AD, and in England, gardeners have had been cultivating strawberries since the 16th century (Boriss et al., 2006).

Strawberry has long been associated with human society with its attractive appearance, delicacy and nutritional quality (Abrol, 1989). Over the past few years, strawberries have experienced a very high rate of growth in consumption, as a fresh fruit rich in vitamins and minerals. Further, it is regarded as one of the best natural sources of antioxidants. Besides, strawberries can be processed for making wine, jam, jelly, ice cream and soft drinks etc. However, due to the highly perishable nature of the crop, fruits need to be sold within a very short span of time to fetch premium prices. Because of its small canopy stature, fruit size, softness and dense foliage, strawberry harbours a large magnitude of insect-pests and diseases causing considerable quantitative and qualitative losses to fruits (Abrol and Kumar, 2009; Pandey et al., 2012). The entire insect-pest complexes are responsible for causing 15–40% damage in strawberry production (Sharma et al., 2008) resulting in huge losses due to a variety of diseases inflicted by diverse pathogens.

The strawberry insect-pest complex comprises of sucking pests, lepidopteran pests, soil arthropods, diseases, nematodes, birds and mammalian pests. The major insect-pests attacking strawberry are discussed in this chapter.

30.1 MAJOR INSECT-PESTS

30.1.1 SUCKING INSECT-PESTS

30.1.1.1 Strawberry Aphids

Aphids infesting strawberry fruits (Blackman and Eastop, 2000) are rarely abundant enough to cause economic damage, and their primary economic impact is as virus vectors. Aphid populations may continue to increase to damaging levels when spring temperatures are moderate with high humidity. Honeydew deposits on fruits, causing proliferation of sooty molds and the white outer covering shed by aphid nymphs sticks to the fruit. This contamination renders the affected fruits unmarketable as fresh fruit. They are capable of spreading virus diseases (especially in nursery production), and can cause significant economic losses in strawberries, if the same plantation remains in one field for several crop cycles. Although aphids are not a major problem in the field strawberries (Mossler and Nesheim, 2003), in greenhouse cultivation, the cotton aphid, *Aphis gossypii* Glover (Homoptera: Aphididae), can be a serious problem (Leclant and Deguine, 1994). This small, soft-bodied insect feeds on the underside of leaves and suck the plant sap out. The cotton aphid varies in colour and size (Watt and Hales, 1996); spring populations can be darker and may be twice the size of *yellow dwarfs* generally present in the summer (Nevo and Coll, 2001). High populations of aphids can reduce the vigour of the plant, making it susceptible to other pests. The honeydew excreted by aphids facilitates the development of black sooty mould on the substrate and causes reduction in fruit quality. Moreover, this sooty mould reduces efficiency of photosynthesis system and otherwise adversely affects the quality of the plant, causing considerable economic injury. There are three types of aphids, infesting strawberries at various crop growth stages: strawberry aphid, *Chaetosiphon fragaefolii*; melon aphid, *Aphis gossypii* and green peach aphid, *Myzus persicae*.

30.1.1.2 Strawberry Aphid, *C. fragaefolii*

It is pale green–yellowish and both adults and nymphs have transverse striations across the abdomen, and are covered with knobbed hairs that are readily seen with a hand lens. These striations and hairs are not found on any of the other aphid species infesting strawberries (Figure 30.1).

30.1.1.3 Melon Aphid, *A. gossypii*

These are small, globular with varying colour ranging from yellowish-green–greenish-black. This species is often the first to migrate into the strawberry fields each season and is the most difficult to control with insecticides.

30.1.1.4 Green Peach Aphid, *M. persicae*

The green peach aphid is green–greenish-yellow and is more streamlined than the rounded melon aphid. The peach aphid is much larger than other species and has two different forms, a pink and

FIGURE 30.1 *Chaetosiphon fragaefolii.*

FIGURE 30.2 *Myzus persicae.*

another green form. The long legs of this species give it a characteristic spiderlike appearance (Figure 30.2).

Aphids are designated as serious insect-pests in most strawberry production regions of the world due to their involvement in transmission of viruses (Maas, 1998). Selection of insecticides that minimize impact on predatory fauna in strawberry fields (Easterbrook, 1997) has been emphasized in integrated pest management (IPM) programmes for strawberry (Zalom et al., 2001). Treatments are also applied in strawberry nurseries to prevent aphid population build-up and virus spread. Judicious use of nitrogenous fertilizers may also be done to ward off the aphid populations. Cultural and biological methods of aphid management including spray of botanical insecticide soaps are acceptable in organic cultivation systems. Alternatively, row covers such as plastic tunnels or Remay-type enclosures have been observed to minimize aphid populations to below economic levels, but the costs are substantial, and the economic viability for large- or even small-scale plantings has not been very encouraging. Parasitoids and predator complexes have been observed to suppress the aphid infestation on strawberry plants. Any factor reducing parasitoids, predators or other biological control agent could result in economic damage to the crop (Kaplan and Eubanks, 2002). Natural enemies, including lady beetles, *Coccinella septempunctata* (Figure 30.3) and *Hippodamia convergens*, green lacewing, *Chrysoperla carnea* and hymenopteran wasps, *Lysiphlebuste staceipes* and *Aphidius colemani* (Howard et al., 1985; Van Driesche and Bellows, 1996; Cross et al., 2001; Kaplan and Eubanks, 2002), have been observed effective against the cotton aphid. The pink spotted lady beetle, *Coleomegilla maculate* (Coleoptera: Coccinellidae), is also known to feed on and

FIGURE 30.3 *Coccinella septempunctata.*

control aphids on greenhouse strawberry (Rondon et al., 2004a), and may be a good candidate for augmentative biological control of cotton aphid.

30.1.1.5 Strawberry Spider Mites

Spider mites are the most important pest of field- as well as greenhouse-grown strawberries worldwide (Oatman et al., 1985; Stonneveld et al., 1996; Walsh et al., 2002; Sato et al., 2004; Cloyd et al., 2006). They suck sap from the plant and cause loss of plant vigour. Among mites, the two-spotted spider mite, *Tetranychus urticae* Koch (Acari: Tetranychidae), is a major and potential pest wherever strawberries are produced (Howard et al., 1985; Price and Kring, 1991). It has a high rate of fecundity and a short developmental time at high temperatures (32°C). High populations of two-spotted spider mite can reduce foliar and floral development, thereby decreasing the quality and quantity of mature fruits (Rhodes et al., 2006). Two-spotted spider mites have been observed to be resistant to most acaricides due to their short lifecycle and high fecundity (Huffaker et al., 1969; Williams, 2000; Cross et al., 2001; Sato et al., 2004). Outbreaks of two-spotted spider mite have become more frequent over the past few decades due to increased use of pesticides in modern cultural practices (Bylemans and Meurrens, 1997). As a result, more growers are utilizing biological control as an alternative to chemical management (Huffaker et al., 1969; Escudero and Ferragut, 2005; Rhodes and Liburd, 2006; Rhodes et al., 2006). However, as of now, our common knowledge about the non-target effects of releases of biocontrol agents on beneficial arthropods in the strawberry ecosystem is very limited.

Predatory phytoseiid mites, *Phytoseiulus persimilis* have become important elements of IPM in Florida strawberry production (Decou, 1994), and these have been observed to be highly effective predators in controlling two-spotted spider mites (Zhi-Quiang and Sanderson, 1995). They have successfully been controlled in the field in some areas of Florida by the introduction of *P. persimilis* (Acari: Phitoseiidae) at 5–10% of the strawberry leaflets infestation with one or more mites (Van de Vrie and Price, 1994). Two of the most commonly used Phytoseiids are *P. persimilis* and *Neoseiulus californicus* (McMurtry and Croft, 1997; Cloyd et al., 2006) in the greenhouse. *N. californicus* (McGregor) has been observed to be more resilient than *P. persimilis*. It is a general predator, and can adapt to fluctuations in prey populations and temperature, providing consistent pest suppression (Croft et al., 1998; Greco et al., 2005; Escudero and Farragut, 2005). Croft et al. (1998), Escudero and Ferragut (2005) and Liburd et al. (2007) have demonstrated that *N. californicus* is tolerant to many insecticides and fungicides, and is able to remain viable at temperatures from as low as 10°C to as high as ≈ 32°C (Hart et al., 2002). It has been observed that *N. californicus* is able to maintain more consistent control of two-spotted spider mite populations compared with *P. persimilis* throughout the season in north Florida strawberry fields (Rhodes et al., 2006).

30.1.1.6 The Greenhouse Whitefly and *Lygus* Bug

The greenhouse whitefly, *Trialeurodes vaporariom*, and the *Lygus* bug, *Lygus hespersus*, are serious insect-pests of strawberries (Figure 30.4). Tiny whiteflies often settle on the underside of leaves and suck the sap. They can be a source for potential vectors of disease, which eventually reduces the quantity and quality of marketable strawberries. Its infestations are usually found in patches rather than over the entire field. The incidence of *Lygus* bug is recorded from flowering up to the harvest of the fruits and the continuous feeding by adults and nymphs causes "cat facing" of berries (Figures 30.5 and 30.6) and deteriorates the marketable quality (Pandey et al., 2012). Both these insect-pests are notorious for their ability to develop tolerance and/or resistance to different groups of insecticides. Neonicotinoids are a new class of insecticides with systemic properties and long residual activity against sucking insect-pests such as whiteflies (Kagabu, 1999; Yamamoto, 1999; Ware, 2000). These insecticides, such as Admire (imidacloprid), the first registered neonicotinoid followed by Platinum (thiamexthoxam) and Assail (acetamiprid), act on acetylcholine receptors in the central nervous system of insects. Thiamexthoxam and acetamiprid are considered second-generation neonicotinoids (Yamamoto et al., 1999; Palumbo et al., 2001), while the more water-soluble Venom (dinotefuran) is a third-generation neonicotinoid (Wakita et al., 2003). Since 1998, Admire, Platinum/Actara and Assail have been registered for use on several crops in the Ventura and Oxnard

FIGURE 30.4 Greenhouse whitefly.

FIGURE 30.5 *Lygus* bug.

FIGURE 30.6 Cat facing of berries.

areas of California for control of whiteflies in the greenhouse. Conventional insecticides such as chlorinated hydrocarbons, organophosphates, carbamates and pyrethroids are still being used to manage several insect pests on various crops in the Ventura and Oxnard areas. At present, the insecticides of these classes are recommended for only limited use in rotation with neonicotinoids and/or insect growth regulators (IGRs) to control whiteflies (Liu and Meister, 2001; Palumbo et al., 2001). On strawberries, foliar application of these conventional insecticides are recommended to suppress high greenhouse whitefly populations during mid- or late-season when the efficacy of Admire, applied during planting or early post-planting, diminishes.

30.1.1.7 Strawberry Thrips, *Frankliniella occidentalis* (Pergande)

Western flower-thrips (Thysanoptera: Thripidae), are most important pests of economic significance on a range of agrihorticultural crops (Kirk and Terry, 2003) causing damage directly by feeding on flowers, fruits, leaves and other plant parts (Figure 30.7), and indirectly via transmission of viruses into the host plants. It also causes russeting and withering of flowers and fruits, thereby reduced commercial value of strawberries (Nondillo et al., 2008). Strawberry thrips are thought to be pesticide-induced problems (Kirk and Terry, 2003), although pesticides are usually the main strategy used for their management (Broughton and Herron, 2009). Glasshouse and field

FIGURE 30.7 Frankliniella occidentalis.

populations of thrips resistant to several major chemical insecticide classes have been recorded in different parts of the world (Jensen, 2000). Of the insecticides used, spinosad is unique, and has been observed to be highly effective against *F. occidentalis* (Funderburk et al., 2000), and is classified as a reduced risk bio-insecticide (Sparks et al., 1995). Alternative management strategies include biological control, cultural control (e.g., removal of non-crop host plants), physical control (e.g., thrips-proof net/mesh, reflective mulches), host-plant resistance and the use of minimum-risk pesticides. These have been developed and implemented for *F. occidentalis* management in several high-value horticultural crops in glasshouses (e.g., sweet pepper, cucumber, ornamentals) and in open field conditions (e.g., sweet pepper, tomato) in Europe and the United States (Stavisky et al., 2002; Reitz et al., 2003). Among the natural enemies used for biological control of thrips, predatory mites have been observed effective in field and glasshouse conditions (van Lenteren and Woets, 1988; McMurtry and Croft, 1997).

30.1.2 Lepidopteran Insect-Pests

30.1.2.1 Greasy Cut Worms

The greasy cut worm, *Agrotis ipsilon* (Hufn.) is an important insect-pest of strawberry. The larvae of all the instars of this cutworm cause severe injury to strawberry plants (Figure 30.8). They dwell just below the ground surface and cut and eat the plants from just above or a little below the surface of the soil. These caterpillars often cut off the leaf petioles of strawberry plants or the pedicels of the young fruits, also attacking the green and full-grown fruit from below, often hollowing it out, leaving the shell of the berry (Taha, 1984; Pandey et al., 2012).

30.1.2.2 Tobacco Caterpillars

Tobacco caterpillars (Figure 30.9) are a serious problem in nurseries and greenhouses, where plants are multiplied for offseason production. Their occurrence is noticed during late summer and observed damaging the strawberry runners. Non-availability of other food sources may be the possible reason for *Spodoptera* infestation on strawberry. The adult gravid female lays eggs in batches. After hatching, from the eggs, the newly hatched larvae are responsible to completely defoliate the leaves, leaving only the midribs. *Spodoptera litura*, which completely defoliate and skeletonize the leaf area, cause reduction in the rate of growth and vigour of runners during summer months in partial–complete shade.

FIGURE 30.8 *Agrotis ipsilon* larva.

FIGURE 30.9 *Spodoptera litura* larvae.

30.1.2.3 Strawberry Leaf Folder

The strawberry leaflets are often observed folded and webbed together, wherein greenish or brownish caterpillars are found feeding inside (Figure 30.10). These leaves soon turn brown and die. In severe infestations, there is great loss of plant vigour and as a result, fruits fail to mature. The adult moths appear during February, and the females lay eggs singly on the underside of the leaves. The eggs hatch in about a week, and the larvae feed for about a month. When fully grown, they are about 1.25 cm long and vary from yellowish-green–greenish-brown with brown head. They transform to brown pupae within the folded leaves, and 10 days later, the moths emerge (Pandey et al., 2012).

Entomopathogenic nematodes in the families Steinernematidae and Heterorhabditidae are soil-inhabiting insect pathogens that possess potential as biological control agents (Kaya and Gaugler, 1993). Most biocontrol agents require days or weeks to kill their hosts, yet entomopathogenic nematodes, in association with symbiotic bacteria, kill insects in 24–48 hours. Once in the body cavity of the host insect, a symbiotic bacterium (*Xenorhabdus* for steinernematids and *Photorhabdus* for heterorhabditids) is released from the entomopathogenic nematode, multiplies rapidly and causes rapid death of the affected insect (Poinar, 1990; Adams and Nguyen, 2002).

The impact of the beneficial nematodes and two biopesticides when used alone or in combination against the greasy cut worm, *A. ipsilon* (Hufn.), in infested strawberry fields has been identified as a new approach of IPM. In laboratory conditions, larvae and pupae of *A. ipsilon* are highly susceptible to the two nematode species, *Steinernema carpocapsi* (Sc) and *Heterorhabditis bacteriophora* (Hb)

FIGURE 30.10 Strawberry leaf folder larva.

when used separately, and the percentage mortality enhances with increase in the dose of nematodes. The concentration of 100 infective juveniles (IJs) has been observed as more effective than 25 IJs for both the species of applied nematodes. The two bio pesticides, viz., spinosad and proclaim, are more effective than nematodes when used separately, spinosad being significantly more effective than proclaim. Mixing of nematodes and both spinosad or proclaim increases the efficacy and significance of mixtures both in laboratory and field conditions (Fetoh et al., 2009).

Spinosad is a fermentation product produced by one or more chemical mutants of naturally occurring actinomycetes soil bacterium *Saccharopolyspora spinose* (Boek et al., 1994). Spinosad combines the best features of synthetic chemical insecticides and naturally derived biological insect control products. It has the efficacy and broad-spectrum activity against lepidopteron pests, for example, *Ostrinianubilalis, Helicoverpa zea, Spodoptera* spp., *A. ipsilon, Trichoplusia ni, Plutella xylostella, Heliothis* spp. and *Pieris rapae*. In addition, spinosad has a very favourable safety profile towards mammals (Sparks et al., 1995) with reduced environmental risk and phytotoxicity (Harris and Maclean, 1999) and safety to beneficial insects (Schoonover and Larson, 1995; Sparks et al., 1995). Moreover, the low toxicity of spinosad towards beneficial insects allows it to be incorporated into most IPM programmes that heavily rely on predators and parasites (Bret et al., 1997).

Proclaim is a highly potent pesticide and unique foliar insecticide that controls lepidopteron pests (caterpillars and worm) in cole crops, turnip greens, leafy and fruiting vegetables. Proclaim effectively controls the larval stages of many pests at low rates and increasing the crop's value. Proclaim contains the active ingredient emamectin benzoate, a semi-synthetic second-generation avermectin pesticides. Avermectins are obtained from the naturally occurring soil bacterium, *Streptomyces avermitilis*. Proclaim can be applied by ground air or ground only, giving the growers the flexibility needed for effective IPM programmes (Jansson et al., 1997).

30.1.3 COLEOPTERAN INSECT PESTS

30.1.3.1 Strawberry Bud Weevil or Clipper

Strawberry bud weevils or clipper weevils (Figure 30.11) are a direct pest of the berries and their maximum incidence is recorded during prebloom and bloom periods. They lay eggs in the flower bud and then eat the pedicel partly off. In severe infestations, this insect may cause up to 50–60% loss of the crop. However, strawberry plants usually compensate for clipped buds and loss of yield is not often detectable. The beetle is about 3 mm long and varies from black–reddish-brown. The larva feeds almost entirely on pollen and the eggs are laid in the buds.

FIGURE 30.11 Strawberry clipper.

30.1.3.2 Strawberry Crown Borer

The grubs of this flightless weevil (Figure 30.12) tunnel downward through the crown of the plant, and by the time they are fully grown make the plants severely injured. They appear in early spring, and the females lay eggs singly in cavities eaten in the plant near the surface of the ground. There is one annual generation. Insect pathogenic nematode applications may help to reduce populations of this pest.

30.1.3.3 White Grubs

White grubs larvae cause severe injury to strawberry fields by eating the plant roots and thereby causing the plants to die all of a sudden. White grubs species include Japanese beetle, Oriental beetle, European chafer and Asiatic garden beetle, all of which also feed on foliage as adults. Japanese beetles feed on foliage during the day, whereas Asiatic garden beetles at night. There are no effective control measures to protect the roots from larvae, so the best option is to reduce white grub populations in surrounding weed hosts by installing light traps as a monitoring tool for collecting large population of adult white grubs. *Beauvaria* sp. and *Bacillus papillae* provide good control against white grub populations.

30.1.4 SOIL ARTHROPODS

Soil insects includes ants, white grubs, wireworms, root weevils, mole crickets, termites and nematodes that feed on the roots or underground part of the strawberry plants. Soil dwelling insect-pests can be controlled by application of recommended pesticides at least two weeks before transplanting of runners in the main field. Crop rotation should be followed in fields infested with nematodes, and planting materials should be raised in solarized soil beds. The best method to control nematodes in the field conditions is to follow crop rotation with nematode non-host and anti-nematode crops such as African marigold.

30.1.4.1 Termites

Termites are the greatest menace in strawberry plantations in subtropical conditions with sandy soils and a low water table. Most of the rainfed areas of Jammu, India are prone to termite infestations. Apart from this, it is a big problem when straw mulch and undecomposed farmyard manure are used for strawberry production. Its attack is especially serious (Figure 30.13) when paddy straw mulch completely covers the bed soil and developing fruits lie on top of the straw mulch (Pandey et al., 2012).

The use of black plastic mulch is also recommended to protect the crops and fruits from termite infestation. Drenching of imidacloprid @ 0.3–0.5 ml per litre of water or chlorpyriphos 10 EC @ 1 ml per litre of water is suitable to control the soil dwelling insect- pests.

FIGURE 30.12 Strawberry crown borer.

FIGURE 30.13 Termite damage in straw mulch strawberry field.

Entomopathogenic nematodes belonging to families Steinernematidae and Heterorhabditidae are soil-inhabiting insect pathogens that possess potential as biological control agents (Kaya and Gaugler, 1993) and against termites (Shankar and Kaul, 2010). It is advisable to keep the field fallow for some time and follow suitable crop rotation for few years to control nematodes in fields meant for strawberry plantation. Planting materials need to be raised in solarized soil beds. Cultivation of marigold (African marigold) is promising in suppressing the nematode population in field condition. Soil application of carbofuron 3 G @ 25 kg per ha is beneficial to suppress the nematode population in infested fields.

30.1.4.2 Snails and Slugs

The problem of snails (Figure 30.14) and slugs is sometimes faced in areas where high soil moisture conditions prevail and the weather remains humid for long periods. These are voracious leaf feeders and can completely defoliate entire plants. These pests feed on fruits near the soil and make the fruits unfit for human consumption. Slugs are soft-bodied molluscs that resemble snails but are without a shell covering. They are most active during the night as well as during cool, wet weather and feed on ripening fruit, leaving holes in the berries. Translucent silver or whitish slime trails can be observed on damaged plant parts.

Snails and slugs are controlled by the use of metaldehyde baits, which should be applied in evening hours. These poison baits are placed between rows and plants. Control of snail and slug pests during the fruit production period requires close scouting to ascertain their presence in plantations. Non-chemical pest management includes the management of mulch and water (excess of either increases slug problems), field sanitation and weed control.

FIGURE 30.14 Snails damaging leaves.

30.1.5 Mammalian and Frugivorous Bird Pests

Certain mammalian pests such as squirrels, rodents, porcupine, cattle and dogs are quite troublesome in strawberry fruit production. Besides, many species of bird's especially insectivorous birds eat agricultural insect-pests. However, some species of birds especially the red-vented bulbuls (*Pycnonotus cafer*) and yellow-vented bulbuls (e.g., *Pycnonotus goiavier*) are more problematic for strawberry crop. Bird damage may account for as much as 30% of crop loss depending upon the year and the location. Damage is most frequently caused by just a few species, including robins (e.g., *Turdus migratorius*), grackles (e.g., *Quiscalus quiscula*) and starlings (e.g., *Sturnus vulgaris*), but other songbirds may also be problematic, especially during the drought period of the year when other food resources are limited. These birds cause damage by eating the entire berry or puncturing it and/or eating a portion of the fruit (Figure 30.15), leaving it unmarketable and prone to disease infection. Birds also knock berries off bushes as they forage. Noise devices, bird-scaring devices, including reflecting ribbons, bird nets (Figure 30.16) and many other home remedies, could be tried, with fair–good success. Successful scaring tactics depend on the bird species, bird population pressure and the grower's management of the devices. Fruit growers use sonic and visual scaring devices only at critical times in the growing season, such as fruit ripening, and remove them promptly as soon as that period is over (Ames and Kuepper, 2004). Bird netting has been indicated to be the control method that worked best for the maximum number of growers. Auditory frightening devices which include bird distress calls, bottle rockets, cannons and sirens, have been mentioned as effective measures, but cannons specifically are not universally helpful. Research has shown that birds quickly habituate to these noises, so successful growers should make sure that sound should be at least 130 decibels. The sounds should cover a wide frequency range and be introduced at random

FIGURE 30.15 Fruits damaged by birds.

FIGURE 30.16 Net cover over fruits to protect from birds.

Integrated Insect Pest Management   499

intervals. The source of the sound should be moved frequently and combining visual deterrents and real danger (i.e. shooting) help reinforce the frightening aspects of the noise. According to growers, distress recordings provide the best control in the evening and early morning, which might be the most important times for bird control.

30.2   IPM

IPM is a holistic approach in which various pest management practices are implemented from plant regeneration and transplanting up to crop harvesting for keeping the pest population below economically damaging levels. Pest management has largely been accomplished through application of pesticides and that still remains a mainstay of the production process. Effective IPM depends on a combination of cultural (modifying crop production procedures to suppress problems), physical control (exclusion and hand picking), biological (use of predators, parasites and pathogens), sanitation, resistant variety and safe pesticide application methods. Biological control utilizing natural enemies of insect-pests to suppress their population is now a new area of focus to overcome pesticide triggered problems. There are many new molecules available that are substantially less toxic, more selective and more eco-friendly than the conventional pesticides, and can beneficially be exploited in pest management programmes (Pandey et al., 2012).

Reducing the risks associated with pesticides in pest management is a great challenge in insect-pest control in the 21st century. Biorational agents and approaches for pest management are a recent development for maintaining the quality and quantity of fruit production, and is the key factor for enhancing and enriching IPM strategies (Ishaaya, 2003; Horowitz and Ishaaya, 2004; Ishaaya et al., 2005) in strawberry. These approaches are based upon our enhanced awareness of existing natural resources, ecological processes, relationship between insect-pest or diseases and host as well as aetiology and physiology of the pest or disease. It tends to reduce the pesticide pressure and enhances the farmer's interest to fetch a higher price for sale of their produce. It also opens new vistas to opt for better choices as biological control and use of biorational products, which are less toxic and affect only the targeted pest. It also includes biopesticides derived from microorganisms, parasitoids, predators, botanicals and all conventional as well as non-chemical methods of pest management. It refers to the more dynamic and ecologically sound approach to IPM that considers the strawberry production unit as a part of holistic vital agro-ecosystem.

Understanding of an agro-ecosystem, determination of economic threshold level and monitoring and sampling of insect-pests are some of the prerequisites of IPM. An agro-ecosystem is a set of models having uniform pattern or cultivation of crops and exposed for the biotic and abiotic stress to act upon. Understanding the existing ecosystem and its relationship to an environment is fundamental to an IPM programme (Kaul et al., 2009). The strawberry ecosystem is quite similar to that of vegetables and many other short-duration agri-horticultural ecosystems. Although continuous flowering not only encourages but also conserves natural pest enemies in the short-duration strawberry crop, regular monitoring is essential in strawberry fields for major arthropod pests. However, it may be possible to determine pests' status and accordingly allow treatment options to be applied in the most effective way. In selecting a pesticide for control of pests in small fruits such as strawberries, there are several factors that must be considered. Degree of control desired, type of fruit harvest required for the market, type of spray used, compatibility with other pesticides and efficacy against other insect pests are some of the important factors that must be considered. An IPM programme depends on monitoring insect populations or damage in a field and using action thresholds to determine when insect control is necessary.

Monitoring of strawberry crop in field is needed given the fluctuating trends of insect-pests and allows the growers to take decisions on pest control interventions (Barnes, 1990). The monitoring procedure requires survey and surveillance of the injury caused by insect-pests. Monitoring data assists in formulating the decision for control measures whether to spray or not and thereby reduce the unnecessary pesticide applications or pressure. The area under crop should be monitored randomly regularly and at least once in a week to implement an IPM programme. A relatively small sample of plants or insects in a field can provide good estimates of the total number of insects or the amount of damage present. To be accurate, a sample must be representative of an entire field. Typically, when a field is sampled or "scouted", the scout stops at 10 locations in a field following a "V", "X" or "Z" pattern, and examines a certain number of plants or area at each location, looking for either the presence of pests or the damage they cause. Visual counts or various types of traps such as yellow or blue sticky traps could be used for detection of small and fragile insects such as aphids, thrips, whiteflies and others. Their adults are attracted to the yellow or blue colour of traps. The traps should be located at canopy level, and the number of traps will depend on the area planted and pest type. Some recommendations suggest one trap per 900 m^2 for detection and up to 10 or more per 900 m^2 for control (Rondon et al., 2004b).

30.3 FUTURE STRATEGIES

Effective control of insect pests in commercial strawberry plantations can be done through the judicious use of insecticides/pesticides combined with sanitation and sound management agronomic practices. In recent years, concerns have been raised over the effects of the overuse of chemical pesticides on the environment and human health. The continuous application of synthetic chemical insecticides has triggered several externalities including development of insect resistance, food safety issues such as food hazards, ground water contamination and ecological and biodiversity issues such as destruction of natural enemies. This disadvantage serves as a strong impetus for the development of alternative insect control measures. Biocontrol agents can serve as a potential alternative to chemicals in IPM systems. Although the adoption of biocontrol agents is strongly affected by the socioeconomic environment in which they are to be applied and by farmers' attitudes, but these sociobiological factors have, in fact, been rather poorly investigated in research and development programmes for biocontrol agents.

Moser et al. (2008) investigated about the understanding farmers' perceptions and behaviours concerning the use of biocontrol agents in strawberry production systems using IPM strategies. The survey was conducted in three regions (Italy, Israel and Germany) characterized by distinct cultural systems. The results showed that growers were more aware of the positive aspects of biocontrol agents than the negative ones, and used biocontrol agents with different levels of satisfaction. It was found that media coverage, as a source of information about biocontrol agents' positive characteristics, were the most significant factors affecting growers' confidence in biocontrol agents. They also identified the following major problem in each area: the amount of time required for monitoring in Italy, the lack of full pest control in Israel and the influence of weather on biocontrol agents' efficacy in Germany.

Among other strategies, site selection is an important activity: planting of strawberry crop close to woody orchard or forest areas or plantation crops needs to be avoided. The selection of high-quality runners free from pests and disease for transplanting is of paramount importance. It is strongly advisable to follow the clean cultivation and remove the crop residues, debris and weeds from bunds and edges as they may shelter potential insect-pests for their further multiplication. Beneficial biocontrol agents may be released to achieve the desired level of pest suppression.

It is further advised to try to follow proper cultural practices and procure the disease free planting materials/runners or recommended resistant varieties for region specific cultivation. All the weeds in and around the vicinity of strawberry fields need to be remove as they are the potential sources to harbour insect-pests. Mulching with organic materials enhances soil aggregation, water holding capacity and moderating soil temperatures, and thus reducing plant stress and also creates habitats for beneficial arthropods including predators (Figures 30.17 through 30.22) such as coccinellids, syrphid fly, green lacewings, big eyed bugs, minute pirate bugs, damsel bugs, predatory mites and various species of spiders, which are useful in suppressing the pest populations.

Integrated Insect Pest Management 501

FIGURE 30.17 Aphid parasitoids.

FIGURE 30.18 General predator spider.

FIGURE 30.19 Big eyed bugs.

Biocontrol strategies hold a great promise for organic production of strawberries for better returns and consumer acceptability. Several biocontrol agents have been described as effective against insects such as thrips (Steiner et al., 2006), aphids (Easterbrook et al., 2006), root weevils (Mahar et al., 2004), mites (Rhodes et al., 2006), nematodes and mollusc pests (Cross et al., 2001). Several biocontrol agent-based products are already commercialized and recorded throughout the

FIGURE 30.20 Green lacewing.

FIGURE 30.21 Minute pirate bug feeding on spider mites.

FIGURE 30.22 Syrphid fly laying eggs.

world. Fraulo et al. (2008) obtained the desirable effects of the predatory mite, *N. californicus* (McGregor) on arthropod community structure when released as a biological control agent for the two-spotted spider mite, *T. urticae* Koch, in north Florida strawberries.

Solomon et al. (2001) reviewed the natural enemies of the more important arthropod, nematode and mollusc pests of strawberry, and their use as biocontrol agents in northern and central Europe. Most pests of strawberry are polyphagous, and they and their natural enemies occur on other host plants (especially from Family Rosaceae) as well as on other crops. Strawberry cultivation methods, viz., protected cultivation, soil sterilization, mulching and pesticide spray schedule have profound influences on the pest and natural enemy complex, though such effects are yet to be quantified precisely. A few natural enemy groups are known to act as important natural limiting factors in

Integrated Insect Pest Management

503

pest population development in commercial strawberry plantations. Two examples are naturally occurring phytoseiid predatory mites, which regulate mite-pest populations, and predatory carabid beetles which regulate root weevil populations. Apart from the introduction of predatory phytoseiid mites to control two-spotted spider mites, biocontrol approaches are not widely used in commercial practice, because of their higher costs in comparison with conventional insecticides for commercial adoption.

In addition, research efforts need to be concentrated on those for common pests also that are currently controlled by frequent sprays of broad-spectrum insecticides, for example, aphids, blossom weevil, capsids and vine weevil. As strawberry is often grown as an annual or short-term perennial crop, exploiting natural populations of natural enemies is difficult. More effort needs to be devoted to the development of microorganism and nematode-based biocontrol agents, which can be used as biopesticides. Protected cultivation of strawberry provides more favourable conditions for exploitation of biocontrol, including the introduction of insect predators and parasites. Paranjpe and Cantliffe (2004) observed that two-spotted spider mites (*T. urticae*) and aphids (*A. gossypii*) were the two main arthropod pests during transplant production. Approximately 2000 predatory mites (*N. californicus*) were released for controlling two-spotted spider mites. About 200 ladybug larvae (*C. maculate)* and 200 nymphs of the big eyed bug (*Geocoris punctipes)* were released on strawberry plants for controlling aphids. Certain entomopathogenic nematodes can be used as biological control agents to suppress several economically important insect-pests (Shapiro-Ilan, 2004; Grewal et al., 2005).

Integrating Geographic Information System (GIS) technology into agricultural production provides predictive risk assessment on pests, pesticide use and other environmental concerns. Geographic Information System (GIS) hardware (server) and software (ArcView, ArcIMS) allows for the development of a geospatial database that shows current production areas and their interfaces, for example, determination of the number of acres that are in contact with urban development, natural areas and other agricultural production. This database can be used for real time reporting of pest pressure and weather conditions for any given geographic area, as well as any number of other "layers" as deemed appropriate, for example, soil maps, fertilizer applications, natural drainage and watercourses and irrigation maps.

REFERENCES

Abrol, D.P. 1989. Studies on ecology and behaviour of insect pollinators frequenting strawberry blossoms and their impact on yield and fruit quality. *Tropical Ecology*, 30(1): 96–100.

Abrol, D.P. and Kumar, A. 2009. Foraging activity of *Apis* species on strawberry blossoms as influenced by pesticides. *Pakistan Entomologist*, 31(1): 57–65.

Adams, B.J. and Nguyen, K.B. 2002. Taxonomy and systematics. In: R. Gaugler (ed.), *Entomopathogenic Nematology*. CABI, New York, New York: pp. 1–34.

Ames, G.K. and Kuepper, G. 2004. *Tree Fruits: Organic Production Overview*. http://attra.ncat.org/attra-pub/fruitover.html

Barnes, B.N. 1990. *FFTRI Monitoring Manual for Orchard Research Institute*. Stellenbosch, Cape Flora, South Africa: p. 21.

Blackman, R.L. and Eastop, V.F. 2000. *Aphids on the World's Crops: An Identification and Information Guide*. Wiley, Chichester, UK: p. 466.

Boek, L.D., Hang, C., Eaton, T.E., Godfrey, O.W., Michel, K.H., Nakatsukasa, W.M. and Yao, R.C. 1994. Process for producing A83543 compounds. US Patent No. 5,362,634. Assigned to Dow Elanco.

Boriss, H., Brunke, H. and Keith, M. 2006. *Commodity Profile: Strawberries*. Agricultural Issues Center, University of California. Agricultural Marketing Resource Center. http://aic.ucdavis.edu/profiles/Strawberries-2006.pdf

Bret, B.L., Larson, L.L., Schoonover, J.R., Sparks, T.C. and Thompson, G.D. 1997. Biological properties of Spinosad. *Down to Earth*, 52: 6–13.

Broughton, S. and Herron, G.A. 2009. Management of western flowerthrips, *Frankliniella occidentalis* Pergande (Thysanoptera: Thripidae) on strawberries. *General and Applied Entomology*, 38: 37–41.

504   Strawberries

Bylemans, D. and Meurrens, F. 1997. Anti-resistance strategies for the two-spotted spider mite, *Tetranychus urticae* (Acari: Tetranychidae), in strawberry culture. *Acta Horticulrurae*, 439: 869–876.

Cloyd, R.A., Galle, C.L. and Keith, S.R. 2006. Compatibility of three miticides with predatory mites *Neoseiulus californicus* Mcgregor and *Phytoseiulus persimilis* Athias Henriot (Acari: Phytoseiidae). *HortScience*, 41: 707–710.

Croft, B.A., Monetti, L.N. and Pratt, P.D. 1998. Comparative life histories and predation types: Are *Neoseiulus californicus* and *N. fallacies* (Acari: Phytoseiidae) similar type II selective predators of spider mite? *Environmental Entomology*, 27: 531–538.

Cross, J.V., Easterbrook, M.A., Crook, A.M., Crook, D., Fitzgerald, J.D., Innocenzi, P.J., Jay, C.N. and Solomon, M.G. 2001. Review: Natural enemies and biocontrol of pests of strawberry in Northern and Central Europe. *Biocontrol Science and Technology*, 11(2): 165–216.

Decou, G.C. 1994. Biological control of the two-spotted spider mite (Acarina: Tetranychidae) on commercial strawberries in Florida with *Phytoseiulus persimilis* (Acarina: Phytopseiidae). *Florida Entomologist*, 77: 33–41.

Easterbrook, M.A. 1997. A field assessment of the effects of insecticides on the beneficial fauna of strawberry. *Crop Protection*, 16: 147–152.

Easterbrook, M.A., Fitzgerald, J.D. and Solomon, M.G. 2006. Suppression of aphids on strawberry by augmentative releases of larvae of the lacewing *Chrysoperla carnea* Stephens. *Biocontrol Science and Technology*, 16: 893–900.

Escudero, L.A. and Farragut, F. 2005. Life-history of predatory mites *Neoseiulus californicus* and *Phytoseiulus persimilis* (Acari: Phytoseiidae) on four spider mites species as prey, with special reference to *Tetranychus evansi* (Acari: Tetranychidae). *Biological Control*, 32: 378–384.

Fetoh, B., El, S.A., Khaled, A.S. and El-Nagar, T.F.K. 2009. Combined effect of entomopathogenic nematodes and biopesticides to control the greasy cut worm, Agrotisipsilon (Hufn.) in the strawberry fields. *Egyptian Academic Journal of Biological Sciences*, 2(1): 227–236.

Fraulo, A.B., McSorley, R. and Liburd, O.E. 2008. Effect of the biological control agent *Neoseiulus californicus* (Acari: Phytoseiidae) on arthropod community structure in North Florida strawberry fields. *Florida Entomologist*, 91: 436–445.

Funderburk, J.E., Stavisky, J. and Olson, S. 2000. Predation of *Frankliniella occidentalis* (Thysanoptera: Thripidae) in field peppers by *Orius insidiosus* (Hemiptera: Anthocoridae). *Environmental Entomology*, 29: 376–382.

Greco, N.M., Sanchez, N.E. and Liljesthrom, G.G. 2005. *Neoseiulus californicus* (Acari:Phytoseiid) as a potential control agent of *Tetranychus urticae* (Acari: Tetranychidae): Effect of pest/predator ratio on pest abundance on strawberry. *Experimental and Applied Acarology*, 37: 57–66.

Grewal, P., Ehlers, R.U. and Shapiro-Ilan, D. (eds.). 2005. *Nematodes as Biological Control Agents*. CABI Publishing, Wallingford, UK.

Harris, B.M. and Maclean, B. 1999. Spinosad, control of lepidopterous pests in vegetable brassiscas. *Proceedings of the 52nd Plant Protection Conference, New Zealand*: 65–69.

Hart, A.J., Bale, J.S., Tullett, A.G., Worland, M.R. and Walters, K.F.A. 2002. Effects of temperature on the establishment potential for the predatory mite *Amblyseius californicus* McGregor (Acari: Phytoseiid) in the UK. *Journal of Insect Physiology*, 48: 593–599.

Horowitz, A.R. and Ishaaya, I. 2004. Biorational insecticides – mechanisms, selectivity and importance in pest management programs. In: A.R. Horowitz and I. Ishaaya (eds.). *Insect Pest Management – Field and Protected Crops*. Springer, Berlin, Germany, Heidelberg, Germany, New York, New York: pp. 1–28.

Howard, C.M., Overman, A.J., Price, J.F. and Albregts, E.E. 1985. Diseases, nematodes, mites, and insects affecting strawberries in Florida. University of Florida Institute of Food and Agricultural Sciences, Agricultural Experimental Station. *Bulletin*, 857: 52.

Huffaker, C.B., Van de Vrie, M. and McMurtry, J.A. 1969. The ecology of tetranychid mites and their natural control. *Annual Review of Entomology*, 14: 125–174.

Ishaaya, I. 2003. Introduction: Biorational insecticides – mechanism and application. *Archives of Insect Biochemistry and Physiology*, 54: 144.

Ishaaya, I., Kontsedalov, S. and Horowitz, A.R. 2005. Biorational insecticides: Mechanism and cross resistance. *Archives of Insect Biochemistry and Physiology*, 58: 192–199.

Jansson, R.K., Peterson, R.F., Mooke, P.K., Halliday, W.R. and Dybas, R.A. 1997. Development of novel soluble granule formulation of enamectin benzoate for control of lepidopterous pest. *Florida Entomologists*, 80: 425–450.

Jensen, S.E. 2000. Insecticide resistance in the western flower thrips, *Frankliniella occidentalis*. *Integrated Pest Management Reviews*, 5: 131–146.

Integrated Insect Pest Management

Kagabu, S. 1999. Studies on the synthesis and insecticidal activity of neonicotinoid compounds. *Pesticide Science*, 21: 231–239.

Kaplan, I. and Eubanks, M.D. 2002. Disruption of cotton aphid (Homoptera: Aphididae) natural enemy dynamics by red imported fire ants (Hymenoptera: Formicidae). *Environmental Entomology*, 31: 1175–1183.

Kaul, V., Shankar, U. and Khushu, M. 2009. Bio-intensive integrated pest management in fruit crop ecosystem. In: R. Peshin and A.K. Dhawan (eds.). *Integrated Pest Management: Innovation Development Process Vol-1, 2009*. Springer Science+Business Media B.V.: pp. 631–666. doi:10.1007/978-1-4020-8992-321.

Kaya, H. K. and Gaugler, R. 1993. Entomopathogenic nematodes. *Annual Review of Entomology*, 38: 181–206.

Kirk, W.D.J. and Terry, L.I. 2003. The spread of the western flower thrips *Frankliniella occidentalis* Pergande. *Agricultural and Forest Entomology*, 5: 301–310.

Leclant, F. and Deguine, J.P. 1994. Aphids. In: G.A. Mathews and J.P. Tunstall (eds.). *Insect Pests of Cotton*. CAB, Oxfordshire, UK: pp. 285–324.

Liburd, O.E., White, J.C., Rhodes, E.R. and Browdy, A.A. 2007. The residual and direct effects of reduced-risk and conventional miticides on twospotted spider mites, *Tetranychus urticae* (Acari: Tetranychidae) and predatory mites (Acari: Phytoseiidae). *Florida Entomologist*, 90: 249–257.

Liu, T.X. and Meister, C. 2001. Managing Bemisiatabaci on spring melon with insect growth regulators, entomopathogens and imidacloprid in south Texas. *Subtropical Plant Science*, 53: 44–48.

Maas, J. L. 1998. *Compendium of Strawberry Diseases*. American Phytopathological Society, St Paul, Minnesota: p. 98.

Mahar, A.N., Munir, M., Gowen, S.R., Hague, N.G.M. and Tabil, M.A. 2004. Biological control of black vine weevil *Otiorhynchus sulcatus*, Coleoptera: Curculionidae by entomopathogenic bacteria and their cell-free toxic metabolites. *Journal of Food Agriculture & Environment*, 2: 209–212.

McMurtry, J.A. and Croft, B.A. 1997. Life-styles of phytoseiid mites and their roles in biological control. *Annual Review of Entomology*, 42: 291–321.

Moser, R., Pertot, I., Elad, Y. and Raffaelli, R. 2008. Farmers' attitudes toward the use of biocontrol agents in IPM strawberry production in three countries. *Biological Control*, 47: 125–132.

Mossler, M. and Nesheim, O.N. 2003. *Florida Crop/Pest Management Profile: Strawberries*. University of Florida, IFAS Cooperative Extension Service, EDIS 129.

Nevo, E. and Coll, M. 2001. Effect of nitrogen fertilization on *Aphis gossypii* (Homoptera: Aphididae): Variation in size, color and reproduction. *Journal of Economic Entomology*, 94: 27–32.

Nondillo, A., Redaelli, L.R., Marcos, B., Pinent, S.M.J. and Gitz, R. 2008. Thermal requirements and estimate of the annual number of generations of *Frankliniella occidentalis* (pergande) (Thysanoptera: Thripidae) on strawberry crop. *Neotropical Entomology*, 37(6): 646–650.

Oatman, E.R., Badgley, M.E. and Platner, G.R. 1985. Predators of the two-spotted spider mite on strawberry. *California Agriculture*, 39(1–2): 9–12.

Palumbo, J.C., Horowitz, A.R. and Prabhaker, N. 2001. Insecticidal control and resistance management for *Bemisiatabaci*. *Crop Protection*, 20: 739–765.

Pandey, M.K., Shankar, U. and Sharma, R.M. 2012. Sustainable strawberry production in sub-tropical plains. In: D.P. Abrol and U. Shankar (eds.). *Ecologically Based Integrated Pest Management*. New India Publishing Agency, New Delhi, India, pp. 787–820.

Paranjpe, A.V. and Cantliffe, D.J. 2004. Developing a system to produce organic plug transplants for organic strawberry production. *Proceedings of the Florida State Horticultural Society*, 117: 276–282.

Poinar, Jr., G.O. 1990. Biology and taxonomy of Steinernematide and Heterorhabditidae. In: R. Gaugler and H.K. Kaya (eds.), *Entomopathogenic Nematodes in Biological Control*. CRC Press, Boca Raton, Florida: pp. 23–62.

Price, J.F. and Kring, J.B. 1991. Response of twospotted spider mite, *Tetranychus urticae* Koch, and fruit yield to new miticides and their use patterns in strawberry. *Journal of Agricultural Entomology*, 8: 83–91.

Reitz, S.R., Yearby, E.L., Funderburk, J.E., Stavisky, J., Olson, S.M. and Momol, M.T. 2003. Integrated management tactics for *Frankliniella* thrips (Thysanoptera: Thripidae) in field grown pepper. *Journal of Economic Entomology*, 96: 1201–1214.

Rhodes, E.M. and Liburd, O.E. 2006. Evaluation of predatory mites and Acramite for control of two spotted spider mites in strawberries in north central Florida. *Journal of Economic Entomology*, 99(4): 1291–298.

Rhodes, E.M., Liburd, O.E., Kelts, C., Rondon, S.I. and Francis, R.R. 2006. Comparison of single and combination treatments of *Phytoseiulus persimilis, Neoseiulus californicus*, and Acramite (bifenazate) for control of two spotted spider mites in strawberries. *Experimental and Applied Acarology*, 39: 213–225.

Rondon, S.I., Cantliffe, D.J. and Price, J.F. 2004a. The feeding behavior of the bigeyed bug, minute pirate bug, and pink spotted lady beetle relative to main strawberry pests. *Environmental Entomology*, 33(4): 1014–1019.

Rondon, S.I., Price, J.F. and Cantliffe, D.J. 2004b. An integrated pest management approach: Monitoring strawberry pests grown under protected structures. *Acta Horticulturae*, 659: 351–356.

Sato, M.E., Miyata, T., Da Silva, M., Raga, A. De Souza and Filho, M.F. 2004. Selections for fenpyroximate resistance and susceptibility, and inheritance, cross-resistance and stability of fenpyroximate resistance in *Tetranychus urticae* Koch (Acari: Tetranychidae). *Applied Entomology and Zoology*, 39(2): 293–302.

Schoonover, J.R. and Larson, L.L. 1995. Laboratory activity of spinosad on non target beneficial arthropods. *Arthropod Management Tests*, 20: 357.

Shankar, U. and Kaul, V. 2010. Entomopathogenic nematodes: A tool in IPM for vegetable crops. In: J.P. Sharma (ed.), *Quality Seed Production of Vegetable Crops*. Kalyani Publishers, Ludhiana, India: pp. 207–225.

Shapiro-Ilan, D.I. 2004. Entomopathogenic nematodes and insect management. In: J.L. Capinera (ed.), *Encyclopedia of Entomology*. Vol. 1 Kluwer Academic Publishing, Dordrecht, the Netherlands: 781–784.

Sharma, R.M., Kher, R., Dogra, J., Sood, M., Shankar, U. and Verma, V.S. 2008. Sustainable strawberry production in North Indian Plains. Abstracts published in 3rd Indian Horticulture Congress, 6–9 November 2008: 168–169.

Solomon, M.G., Jay, C.N., Innocenzi, P.J., Fitzgerald, J.D., Crook, D., Crook, A.M., Easterbrook, M.A. and Cross, J.V. 2001. Review: Natural enemies and biocontrol of pests of strawberry in Northern and Central Europe. *Biocontrol Science and Technology*, 11(2): 165–216.

Sparks, T.C., Thompson, G.D., Larson, L.L., Kirst, H.A., Jantz, O.K., Worden, T.V., Hertlien, M. B. and Busacca, J. D. 1995. Biological characteristic of the spinosyns: New naturally derived insect control agents. In: *Proceedings, Beltwide Cotton Conference*, National Cotton Council, Memphis, Tennessee: 903–907.

Stavisky, J., Funderburk, J.E., Brodbeck, B.V. and Olson, S.M. 2002. Population dynamics of *Frankliniella* spp. and tomato spotted wilt incidence as influenced by cultural management tactics in tomato. *Journal of Economic Entomology*, 95: 1216–1221.

Steiner, M.Y., Goodwin, S. and Waite, G. 2006. Getting a grip on thrips in strawberries. *Acta Horticulturae*, 708: 109–114.

Stonneveld, T., Wainwright, H. and Labuschagne, L. 1996. Development of twospotted spider mite (Acari: Tetranychidae) populations on strawberry and raspberry cultivars. *Annals of Applied Biology*, 129: 405–413.

Taha, A.M. 1984. Studies on some pests infesting strawberry plants at Qalubia governorate. M. Sc. Faculty of Agriculture, Al-Azhar University, Egypt: p. 146.

Van deVrie, M. and Price, J. 1994. Manual for biological control of two-spotted spider mite on strawberry in Florida. University of Florida, Dover Research Report, DOV:1–10.

Van Driesche, R.G.V. and Bellows, T. S. 1996. Biology and arthropod parasitoids and predators. In: *Biological Control*. Chapman and Hall, New York: 309–335. http://vvv.caes.state.ct.us/PlantPestHandbookFiles/pphS/pphstra.htm

vanLenteren, J.C. and Woets, J. 1988. Biological and integrated pest control in greenhouses. *Annual Review of Entomology*, 33: 239–269.

Wakita, T., Kinoshita, K., Yamada, E., Yasui, N., Kawahara, N., Naoi, A., Nakaya, M., Ebihara, K., Matsuno, H. and Kodaka, K. 2003. The discovery of dinotefuran: A novel neonicotinoid. *Pest Management Science*, 59(9): 1016–1022.

Walsh, D.B., Zalom, F.G., Shaw, D.V. and Larson, K.D. 2002. Yield reduction caused by twospotted spider mite feeding in an advanced-cycle strawberry breeding population. *Journal of the American Society for Horticultural Science*, 127: 230–237.

Ware, G.W. 2000. *The Pesticide Book*. 5th edn. Fresno, CA: Thompson Publishing: p. 415.

Watt, M. and Hales, D.F. 1996. Dwarf phenotype of the cotton aphid, *Aphis gossypii* Glover (Hemiptera: Aphididae). *Australian Journal Entomology*, 35: 153–159.

Williams, D. 2000. Two spotted spider mite on ornamental plants. *Agricultural Notes. State of Victoria*, Department of Primary Industries, Victoria, Australia: pp. 1–2.

Yamamoto, I. 1999. Nicotine to nicotinoids: 1962 to 1997. In: I. Yamamoto and J.E. Casida (eds.). *Nicotinoid Insecticides and the Nicotinic Acetylcholine Receptor*. Springer, Tokyo: pp. 3–27.

Zalom, F.G., Phillips, P.A., Toscano, N.C. and Udayagiri, S. 2001. *Strawberry Aphids*. University of California Department of Agriculture and Natural Resources Publication No. 3339: p. 4.

Zhi-Quiang, Z. and Sanderson, J.P. 1995. Two spotted spider mite (Acari: Tetranychidae) and *Phytoseiulus persimilis* (Acari: Phytoseiidae) on greenhouse roses: Spatial distribution and predator efficacy. *Journal of Economic Entomology*, 88: 352–357.

31 Pesticide Residues

Tarun Verma, Beena Kumari, and Kaushik Banerjee

CONTENTS

31.1 Toxicological Considerations 507
31.2 Pesticide Residues 508
    31.2.1 Maximum Residue Limit 508
31.3 Supervised Trials 510
31.4 Monitoring Trials 510
    31.4.1 Multiresidue Methods 511
    31.4.2 Results of Monitoring Studies 512
31.5 Decontamination of Pesticide Residues 513
    31.5.1 Effect of Washing 513
    31.5.2 Effect of Cooking/Boiling/Juicing 515
    31.5.3 Harvesting Precautions 515
31.6 Future Strategies 516
References 516

The term "pesticide" includes any substance or mixture of substances that is intended for preventing, destroying, attracting, repelling or controlling various pests including unwanted species of plants or animals during the production, storage, transportation, distribution and processing of foodstuffs, agricultural commodities or animal feed or that may be administered to animals for different types of ecto-parasites. On the basis of the target pests, pesticides are classified as insecticides, herbicides, fungicides, acaricides, nematicides, antibiotics etc.

According to the Food and Agricultural Organization (FAO, 1986), pesticides also include any substance or mixture of substances intended for use as a plant growth regulator, defoliant or desiccant. The term, however, excludes manures, fertilizers and antibiotics or other chemicals administered to animals to stimulate their growth or to modify their reproductive behaviour.

31.1 TOXICOLOGICAL CONSIDERATIONS

A chemical substance may evoke either or both acute and chronic toxic effects to an organism. Acute toxicity occurs shortly after exposure to a single dose of the chemical, and its symptoms are usually readily comprehensible even to nonprofessionals. The chronic effect occurs when an organism is exposed to repeated small and non-lethal doses of a potentially harmful substance. A frequently used measure of acute toxicity is the 50% lethal dose or LD_{50}. It is the amount of a poison that kills half the target organisms and is expressed in terms of milligrams poison per kilogram body weight of the experimental animal (mg kg^{-1}). LD_{50} is usually adopted as a standard for comparing the relative toxicity of various substances.

Strawberry crop is sensitive to several pests and diseases. There are many harmful insect and mite species causing injuries to strawberry roots, leaves and flowers. Red spider mite (Acarina: Tetranychidae) is often considered as a major pest (acari) of strawberry, which destroys the mesophyll cells of leaves with high population densities that can result in the plant being dwarfed or killed (Yioit and Erkylyc, 1992). Plants that somehow sustain a severe infestation of red spider mites may become severely weakened and appear stunted, dry and red. Other important pests of strawberry crop in different regions of the world include thrips, aphids, whitefly (*Trialeurodes*), *Agrotis*, *Tipula* and white grubs, cyclomein mite, strawberry mite, vivel, leaf folder, tarnish plant bug and leafhopper. Among diseases, leaf blight is very common.

Among diseases, leaf blight is very common. Application of a number of insecticides such as synthetic pyrethroids (fenvalerate, deltamethrin etc.), organophosphates (dimethoate, chlorpyrifos etc.) are advised for the management of all these pests. Similarly, for the control of diseases, carbendazim, copper oxychloride and thiram are generally used. Besides insects and diseases, strawberries are attacked by nematodes, birds and non-insect, vertebrate pests such as small animals like rabbits (Sharma and Gupta, 2004).

The introduction of pesticides in agriculture has been significantly helpful in increasing productivity of crops and has thus contributed to steadily rising food production since the Second World War (Barbash, 2006). Although the use of pesticides has led to increased yields in arable farming, due to their indiscriminate use and bioaccumulation through the food chain, they can eventually pose a risk or threat to both animal and human life. Because of their long persistence and potential threat to human health, production and use of some of the organochlorine pesticides have been prohibited in most countries. In India also, use of DDT and many other pesticides from the organochlorine group has been banned in agricultural crops since 1993. And more recently, endosulfan has also been banned for production, use and sale all over India. Organophosphorus pesticides are used as substitutes for organochlorine pesticides in many countries nowadays because they can degrade more easily in the environment (Pesticide Action Network, 2003). Indiscriminate use of pesticides can lead to insect resistance, which compels farmers to misuse/overuse the various insecticides for management of economically important, target insect-pests. Similarly, pesticides that are intended for use on one crop may be mistakenly used on others, or pesticides for public health purposes to combat insect-transmitted diseases, such as malaria or locusts, may be misused on agricultural crops (Pesticide Action Network, 1998). The uses of such substances have a great toxicological significance due to the fact that they may cause health hazards, especially to infants and children due to the immaturity of their body systems. An infant and a child's diets are usually the same as an adult, but the vulnerability to chemical compounds should be considered due to their metabolic rates, which are different, and their food and liquid intake, which is proportionally higher than an adult's. These days developed countries have started working to protect infants and children, and some maximum residues limits have been established at the limits of quantification; usually at 0.01 or 0.005 mg/kg (Gebara et al., 2011). The problem of pesticide residues in strawberry is particularly more serious because it often receives unregulated applications of pesticides and fruits are consumed in raw form.

31.2 PESTICIDE RESIDUES

Pesticide residues have been defined as any specified substance in food, agricultural commodities, animal feed, soil or water, resulting from the use of a pesticide. The term includes any derivatives of a pesticide such as conversion products, metabolites, reaction products and impurities that are of toxicological significance.

The FAO (1986) has defined them as substances that remain in or on a feed or food commodity, soil, air or water following use of a pesticide. For regulatory purposes, it includes the parent compound and any specified derivatives such as degradation and conversion products, metabolites, and impurities considered to be of toxicological significance.

31.2.1 MAXIMUM RESIDUE LIMIT

To regulate the pesticides residues at a safe level, a concept was introduced by the Joint FAO/ WHO Expert Committee on Food Additives (1955) and Codex Alimentarius Commission, which was established in 1964. The maximum residue limit of a pesticide is its maximum

Pesticide Residues

concentration in a crop or food commodity resulting from its use according to good agricultural practices. The concentration is expressed in milligram of pesticide residues per kilogram of the commodity (mg kg^{-1}).

A strawberry crop is frequently treated with various pesticides such as fungicides and insecticides to protect it from various biotic stresses causing direct damage or indirect losses to crop. And sometimes the application of pesticides may become compulsory even when the crop is close to harvest. This results in the presence of residues in the harvested fruits. The residue levels generally may remain below the maximum limits if the pesticides are used judiciously. The maximum residue limit values fixed by the Codex Alimentarius Commission for strawberry are shown in Table 31.1. For sustainable use (prevention of resistance in pests and pathogens), it is generally advised to alternate and/or combine different active ingredients of pesticide formulations. This might result in the presence of multiple residues albeit at low levels. Of the 170 strawberry samples studied in Brazil, 78.8% were observed to be frequently contaminated with pesticides viz., acephate, bifenthrin, captan, carbendazim, chlorpyrifos, chlorothalonil, lambda-cyhalothrin, diazinon, dimethoate, endosulfan, fenpropathrin, fluazinam, folpet, iprodione, methamidophos, phenthoate, procymidone and tetradifon (Gebara et al., 2011).

As strawberry is one of the most high-value crops, hence management of various pests and diseases of economic importance is indispensable. Despite the negative effects on human health, huge quantity of pesticides is used on strawberry crop (Goodwin et al., 1985; Kilmer et al., 2001; Safi et al., 2002).

TABLE 31.1

Maximum Residue Limits (MRL) of Some Important Pesticides Applied on Strawberry Crop

Pesticide	MRL (mg/kg)
Abamectin and ethoprophos	0.02
Dimethomorph	0.05
Chlorpyrifos-methyl	0.06
Cypermethrins (including alpha- and zeta-cypermethrin)	0.07
Diazinon and penconazole	0.1
Deltamethrin	0.2
Meptyldinocap[a], chlorpyrifos and glufosinate-ammonium	0.3
Fluopyram	0.4
Dinocap, imidacloprid, novaluron, acetamiprid and sulfoxaflor	0.5
Triadimenol (based on triadimenol use only) and triadimefon	0.7
Permethrin, malathion, fenarimol, myclobutanil bifenthrin, triforine, trifloxystrobin, quinoxyfen and methiocarb	1.0
Pyraclostrobin	1.5
Spirodiclofen, bifenazate, clofentezine, bromopropylate, imazalil, methoxyfenozide and cyprodinil[b]	2.0
Penthiopyrad, boscalid, cycloxydim, buprofezin, fludioxonil and pyrimethanil[b]	3.0
Chlorothalonil, folpet, tolylfluanid and dithiocarbamates	5.0
Hexythiazox	6.0
Azoxystrobin, fenbutatin oxide, iprodione, fenhexamid and dichlofluanid	10.0
Captan	15.0
Bromide Ion	30.0

Source: Codex (2013).

[a] The current dinocap Codex MRL of 0.5 mg/kg covers the use of meptyldinocap.

[b] Withdrawal recommended in 2013 by the Joint FAO/WHO Meeting on Pesticide Residue.

510   Strawberries

Therefore, considering human health, ecological parameters and social obligations, it is most pertinent to maintain food safety standards and for that to happen, the strawberry growers and other stakeholders should resort to pesticide residue analysis for facilitating production of safe fruits of strawberries fit for human consumption in the fresh state and processing into various value-added products. Pesticide residue analysis is divided into two heads, that is, supervised trials and monitoring trials. This chapter attempts to deal with the extent of contamination of strawberry fruits with pesticides residues as revealed by surveillance and compliance studies carried out globally. Finally, some kitchen processes may help in reducing the residues in strawberry to the safe limits.

31.3 SUPERVISED TRIALS

Supervised trials are those under which complete study of a pesticide, for example, its persistence, degradation and dissipation are conducted and its half-life period and safe waiting period etc. are estimated. These trials are conducted to primarily generate data on extent of pesticide residue required for registration of pesticides, setting up of maximum residue limits and finding safe waiting periods for their application. To obtain the necessary data for these purposes, the crops are grown under different agroclimatic conditions and treated with requisite pesticides under good agricultural practices. Under good agricultural practices, taking the minimum quantities necessary to achieve an adequate pest control into consideration, the pesticide treatments are in accordance with the registered or approved usage pattern. The crop commodities are then analyzed for the presence of pesticide residues.

A supervised trial conducted in China revealed that the dissipation of the pesticide cyprodinil (a pyrimidinamine fungicide) differed under different strawberry growing conditions; with the half-lives being 14.5 and 12.5 days in strawberry fruit and soil, respectively, in field conditions. In greenhouse conditions, the half-lives were 5.5 and 6.5 days, respectively indicating a much faster dissipation rate in both strawberry fruits and soil than in field conditions. The terminal pesticide residues in strawberry fruits were below the maximum residue level (5 mg kg^{-1}) as per European Union (EU) Standards at seven days after application, suggesting reasonable use of cyprodinil in different cultivating conditions. The residue data provided the reference to set the maximum residue limits of cyprodinil in strawberry in China (Liu et al., 2011).

After field treatments, the residues of the pesticides azoxystrobin (0.55 mg kg^{-1}) and pyrimethanil (2.98 mg/kg) in strawberry were below the EU maximum residue limit (2 and 5 mg/kg, respectively). The residues of another fungicide fenhexamid (2.99 mg/kg) were on an average close to the EU maximum residue limit (3 mg/kg), with some samples even exceeding the maximum residue limit. The residues of azoxystrobin and fenhexamid dissipated with a half-life of about eight days, whereas the half-life of pyrimithanil was 4.8 days (Angioni et al., 2004).

Petrosini et al. (1969) recovered 80% residue of chlorcholine chloride in strawberries by a chemical method that consisted of extraction, preliminary purification of the quaternary ammonium compounds by cation-exchange resion chromatography, further purified by adsorption with an alumina column and absorptivity measurement of a dipicrylamine-chlorcholine chloride complex at 415 nm.

31.4 MONITORING TRIALS

Pesticide residues not only threaten human health but can also cause an imbalance in ecosystem. Monitoring of commodities helps in understanding the current usage pattern and the status of pesticide residues in a commodity. On the basis of results obtained from the monitoring studies, prevention and control measures can be initiated by the Regulatory Agencies of the Government Administration before the problem becomes too serious and widespread. Monitoring of food commodities for pesticide residues also helps in our understanding if good agricultural practices have been followed "in letter and spirit" or not. Pesticide residues above

Pesticide Residues

the maximum limits are unacceptable to trade, and detection of unauthorized (without official label claim) pesticides might act as non-tariff barrier. Pesticide residues value in the marketable fruits should be within their prescribed maximum residue limits for both domestic as well as export markets.

Under monitoring studies, basket samples originating from the markets from different parts of the country as well as different countries are considered. Since several pesticides are used on strawberry crop, there are ample chances of contamination of strawberry fruits with many pesticide residues. Therefore, samples collected from the market are analyzed to check for possible contamination due to pesticide residues. Under these studies, various efficient methods have been developed for the estimation of multiresidues in strawberry fruits.

31.4.1 Multiresidue Methods

Some of the latest methods used for monitoring studies are discussed in the following.

Souza et al. (2014) demonstrated an analytical method for the determination of organophosphate pesticides (diazinon, disulfoton, parathion, chlorpyrifos, and malathion) in strawberries employing gas chromatography (GC) with flame photometric detection. Sample preparation was carried out by Quick Easy Cheap Effective Rugged and Safe (QuEChERS) method which involved extraction with acetonitrile, followed by clean up by dispersive solid-phase extraction. QuEChERS is inexpensive, fast, and easy for the separation of the analytes from the matrix. In addition, the GC method provided linear calibration curves, ranging from 0.10–1.00 µg/g, for diazinon, disulfoton, parathion, and chlorpyrifos, and 0.10–2.00 µg/g for malathion. Recovery studies yielded values ranging from 81.64 to 100%.

Fernandes et al. (2011) developed a rapid, specific, and sensitive technique based on the QuEChERS method involving clean-up using dispersive solid-phase extraction with magnesium sulphate, primary secondary amine (PSA) and C_{18} sorbents, and applied the method for the routine analysis of 14 pesticides in strawberries. The analyses were performed by three different analytical methodologies: GC with electron capture detection (ECD), mass spectrometry (MS) and tandem mass spectrometry (MS/MS). The recoveries of all the pesticides under study ranged from 46–128% (SANTE, 2015), with relative standard deviation of <15% in the concentration range of 0.005–0.25 mg/kg. The limit of detection (LOD) for all compounds met maximum residue limits applicable in Portugal for the organochlorine pesticides. A survey based study on strawberry fruits produced through organic farming and integrated pest management in Portugal during 2009–2010 revealed detection of the residues of lindane and β-endosulfan above the maximum residue limits in both organic farming and integrated pest management. Other organochlorine pesticides (aldrin, o, p'-DDT, their metabolites, and methoxychlor) were below the maximum residue limits.

Christensen et al. (2003) developed and validated a high-pressure liquid chromatography-mass spectrometry (HPLC-MS/MS) method for the analysis of three pesticides, viz., tolylfluanid, fenhexamid and pyrimethanil in strawberries. Recoveries measured at three spiking levels ranged within 85–99%. The effects of processing (ranging from rinsing to jam production) of strawberry fruits on the pesticide residue content were also investigated for all the three target fungicides applied under field conditions. Although, the fungicide kresoxim-methyl was also applied in the field, but was not detected in any of the samples investigated.

Yamazaki and Ninomiya (1998) developed a simple and rapid method for determining bitertanol residues in strawberries using liquid chromatography with fluorescence detection. Recoveries calculated at four fortified levels (0.05, 0.25, 0.5 and 1 µg/g), by the internal standard method, ranged from 92.1–99.1%, with coefficients of variation ranging from 0.3–4%. The detection limit was 0.01 µg/g. Out of 25 commercial strawberry samples analyzed for bitertanol residues, five contained them at concentrations ranging from 0.02–0.51 µg/g. Positive samples were confirmed by gas chromatography/mass spectrometry with selected ion monitoring (m/z 170 and 168).

512   Strawberries

Such monitoring studies have been reported from various parts of the world where strawberries are cultivated, traded and consumed in significant quantity but in countries such as India with comparatively fewer strawberries produced, information on results of the monitoring are lacking probably because of the fact that the produce is mostly consumed in areas nearby the sites of bulk production.

31.4.2 Results of Monitoring Studies

Out of the 400 strawberry samples analyzed (Schule and Looser, 2004) in Germany for a wide range (>250 compounds) of pesticides, 386 (97%) samples were found to contain pesticide residues. A total of 87 different pesticides were detected across the samples, with 50 samples (13%) containing residues exceeding the German or harmonized EU MRLs. Almost 90% of the samples were observed to contain multiple residues with a maximum of 14 pesticides per sample, and an average of 4.4 pesticides per sample. Due to the differences in place of origin, agroclimatic conditions, and the prevailing agricultural practices, the nature and number of pesticides detected per sample were variable across the tested lot. The researchers concluded that strawberries harvested during winter and early spring had a significantly higher amount of pesticide residues compared with the later harvested samples originated from the same country.

In UK, a total of 289 samples were tested, which included six organic specimens (Pesticide Action Network, 2006). The five most commonly reported pesticides included fenhexamid (30 samples), iprodione (25 samples), pyrimethanil (20 samples), bupirimate (13 samples) and cyprodinil (six samples). Sixty-nine per cent (198) of all the samples contained measurable concentration of residues, and this included even one organic sample. Forty-two per cent (122) of the samples had more than one pesticide residue. The maximum number of pesticides observed in one sample was six. The average number of pesticides found in non-organic samples was 1.47. Taking only those samples into consideration that do contain residues, the average number of pesticides observed was 2.12 per sample. Five samples had one residue above the legal limit and one sample had three residues above the legal limits. Residues of dicofol from UK and Spain (0.2 mg/kg and 0.5 mg/kg), kresoxim-methyl (0.09 mg/kg) from UK, penconazole (0.2 mg/kg) from Israel, carbendazim (1.5 mg/kg) from Morocco, and endosulfan (0.6 mg/kg) from Spain exceeded the maximum residue limits of 0.02, 0.05, 0.1, 0.1, and 0.1 mg/kg respectively.

Within the framework of food surveillance, Looser et al. (2006) analyzed 593 conventionally grown strawberry samples mainly originating from Germany, Spain, Italy and Morocco for pesticide residues at the Chemisches und Veterinäruntersuchungsamt, Stuttgart between 2002 and 2005, using the QuEChERS multiresidue method (Anastassiades et al., 2003) which allowed a very fast, rugged and efficient extraction of a broad spectrum (>400 compounds) of GC-and LC-amenable pesticides. The sample preparation consisted of an extraction step with acetonitrile followed by a liquid–liquid partitioning induced by the addition of salts and a clean-up employing dispersive solid-phase extraction. The QuEChERS extracts were directly subjected to determinative analysis by LC-MS (/MS), GC-MS (EI, CI, ToF), GC-ECD and GC-NPD. Pesticide residues were detected in 98% of the samples and 93% of all the strawberry samples were observed to contain multiple pesticide residues. The German or EU-harmonized maximum residue limits were exceeded by 54 strawberry samples (9%), with mepanipyrim being the most frequently violated compound. The average total concentration of pesticides per sample was 0.41 mg/kg, which was similar to the average value for fruits in general, with Italian products containing the maximum and the German ones having the minimum residue concentrations.

In Egypt, a total of 144 kg strawberry fruits and 38 soil samples, representing two types of production systems, for example, conventional and organic, were collected and screened for residues of 86 pesticide with a broad range of physico-chemical properties, viz., organochlorine pesticides (such as endosulfan, aldrin, o,p-DDD, p,p'-DDE), organophosphorus (such as pirimiphos-me and dimethoate), carbamate (such as carbaryl, propamocarb) and pyrathroids (such as deltamethrin).

Pesticide Residues

513

The fruit samples obtained from conventionally grown strawberry plots revealed the maximum amount of total pesticides (15.55 mg/kg), followed by fruit samples obtained from organically grown plots of strawberries (0.132 mg/kg). On the other hand, the total detected amount of pesticide residues in conventional soils were 6.702, 6.225 and 4.551 ppm and ranked in descending order at the soil surface, 20 cm- and 10 cm-deep soil profile, respectively. In organic soil, there were no residues in surface soil samples, but the total detected residues were 0.192 and 0.093 ppm at the 10 cm and 20 cm depths, respectively (Ghabbour et al., 2012).

Fenhexamid is a new-generation agrochemical extensively used nowadays to protect high-value crops and fruits from attack of pathogenic fungi. The development of an indirect competitive immunoassay for determination of fenhexamid residues in fruit samples was described by Esteve-Turrillas et al. (2011). After optimisation of immunoreagent concentrations, the assay, based on monoclonal antibody FHo4#27, showed high sensitivity, and selectivity, with a limit of detection for fenhexamid standards in buffer of 0.04 µg/l. The influence of several physico-chemical factors, including pH, ionic strength, solvent tolerance and matrix interferences on the analytical parameters of the standard curve was studied. For strawberry samples, a two-step extraction procedure with methanol was proposed in this study, which afforded recovery values that ranged between 103% and 116%. The limit of quantification of the developed assay was 25 µg/kg. The optimized assay was compared with a reference procedure based on liquid chromatography using incurred samples from the market, and an excellent correlation between both the methods could be established.

A method for determination of organohalogen pesticides (alpha-endosulfan, beta-endosulfan, endosulfan sulphate, lambda-cyhalothrin, procymidone and trifluralin) in strawberry by gas chromatography with electron capture detection was validated and applied in a monitoring program. The mean recoveries ranged from 74.6–115.4%, with repeatability standard deviations between 1.6% and 21% and intermediate precision between 5.9% and 21%. Detection, quantification and decision limit, and detection linearity ranged from 0.003–0.007 mg/kg, 0.005–0.013 mg/kg; 0.003–3.128 mg/kg; and 0.005–3.266 mg/kg, respectively, suggesting the method to be fit for the monitoring of organohalogen residues in strawberries (Augusto et al., 2013).

31.5 DECONTAMINATION OF PESTICIDE RESIDUES

A large proportion of total strawberry fruits produced are consumed fresh so it is desirable to make them free from pesticide residues to reduce health hazards. The processing of food commodities generally implies the transformation of the perishable fresh commodity to a value-added product that has a greater shelf life and is closer to being table ready. Fate of pesticide residues during the most common food processing techniques applied to raw agricultural commodity (RAC) in both household and industrial processing (washing, removal of the outer parts of the RAC, such as peeling, cooking and juicing), and the most important industrial processing of RAC in terms of frequent consumption of the processed food such as jam, jelly and others are assessed in the following.

31.5.1 Effect of Washing

Washing is the preliminary step in both household and commercial food preparation, and the effect of washing on the fate of the pesticide residues has been well studied and reviewed (Krol et al., 2000; Zabik et al., 2000; Kaushik et al., 2009). The most interesting conclusions of these studies are that the rinsability of a pesticide is not always correlated with its water solubility (Angioni et al., 2004; Boulaid et al., 2005; Cengiz et al., 2007) and that different pesticides may be rinsed off from the fruits by different washing procedures (Lentza-Rizos and Kokkinaki, 2002; Angioni et al., 2004; Pugliese et al., 2004). The removal of pesticides with the washing of RAC may be performed not only through the dissolution of pesticide residues in the washing water or the rinsing with chemical baths such as detergents, alkaline, acid, hypochlorite, metabisulfite salt and ozonated water (Holland et al., 1994) but also through the removal of dust or soil particles with absorbed

514   Strawberries

residues from the outer layer of the commodity (Angioni et al., 2004; Guardia Rubio et al., 2006 and 2007). Penetration is again the most dynamic process that may control the fate of a pesticide residue on fruit during washing. Furthermore, in many studies, washing was not related with the water solubility of the pesticide residue but with K_{ow} (a unitless parameter that provides a useful predictor of the other physical properties for most pesticides and other organic substances with molecular weight less than 500), reinforcing the view that partition coefficients between cuticle and washing water correlate well with the K_{ow} of pesticides (Baur et al., 1997). In consequence, the use of an appropriate detergent that has the ability to solubilise waxes may dissipate the residue present in the epicuticular wax layer of fruits (Angioni et al., 2004). Other washing agents or dipping treatments may also lead to selective removal of pesticide residues with systemic action through similar mechanisms (Cabras et al., 1998; Femenia et al., 1998). The residue still present in the fruit after washing can be ascribed to the pesticide that has penetrated into the cuticle. The possible formation of toxic byproducts during washing has also been studied by several authors (Cabrera et al., 2000; Pugliese et al., 2004; Ou-Yang et al., 2004). According to these results, typical washing with tap water is not expected to form any toxic metabolites, whereas the use of high levels of sodium hypochlorite, hydrogen peroxide and potassium permanganate in washing water as well as ozonated water could form oxons from the organophosphorus pesticides by chemical oxidation (Cabrera et al., 2000; Pugliese et al., 2004; Ou-Yang et al., 2004).

Kin and Huat (2010) used headspace solid-phase microextraction to examine the organophosphorus (diazinon, malathion, chlorpyrifos, quinalphos, profenofos) and organochlorine (chlorothalonil, α-endosulfan and β-endosulfan) pesticide residues in vegetable (cucumber) and fruit (strawberry) samples. The effects of washing by different solutions were evaluated for the reduction of organophosphorus and organochlorine pesticide residue contents. Washing treatments utilizing acetic acid, sodium carbonate, sodium chloride solutions and tap water were effective in reducing the organochlorine and organophosphorus pesticides. Acetic acid was most effective in removing residues of the investigated pesticides, with 44–70% removal, with an average of 58.4% of the residues being eliminated from the samples. This was followed by sodium carbonate with 30–50% (average: 43.1%) of residues being eliminated and sodium chloride with 23–40% (average: 30.8%) reduction of residues. Among all these methods, washing with only tap water proved least effective since it only reduced the residues by 10–20% (average: 15.5%) as a whole. The relationship between the effectiveness of washing treatments and the physico-chemical properties of the investigated pesticides were explored. Among the organophosphorus pesticides, maximum removal of residues was recorded for malathion (69.8%), followed by diazinon (69.0%), profenofos (67.8%), quinalphos (63.4%) and chlorpyrifos (61.9%). Comparatively, with similar washing treatments, in case of the organochlorine pesticides, the percentage removal of chlorothalonil (58.6%), α-endosulfan (58.3%) and β-endosulfan (57.0%) were relatively less during 30-minute washing in acetic acid (10%). This may be due to the fact that most of the organophosphorus pesticides are prone to hydrolysis and get decomposed in acidic aqueous solutions. On the other hand, the organochlorine pesticides are stable in acidic aqueous solutions. The residues of organochlorine pesticides such as chlorothalonil, α-endosulfan and β-endosulfan residues were difficult to remove due to their persistent behaviour and low water solubility. Besides, vapour pressure is also an important factor influencing the decontamination behaviour of pesticides that in turn governs the reduction of residues because many pesticides are known to distribute and dissipate in the vapour phase. A satisfactory relationship is usually observed between the vapour pressures of the pesticides and the extent of their removal through washing treatments. The pesticides with comparatively larger vapour pressure values showed a high percentage removal in response to the washing treatments.

Christensen et al. (2003) studied the effects of processing of strawberries ranging from rinsing to jam production in relation to residues of three fungicides (tolylfluanid, fenhexamid and pyrimethanil), applied under field conditions. Kresoxim-methyl was also applied in the field, but was not found in any of the samples investigated. The results from rinsing treatment showed that the residues of all the three fungicides were on an average reduced by 37% for tolylfluanid, 34% for fenhexamid, and 19% for pyrimethanil. For tolylfluanid and fenhexamid, the initial concentration significantly affected the extent of reduction or decontamination. For fenhexamid, dose could also have a minor influence on reduction. For pyrimethanil, none of the parameters had any significant influence on the extent of residue removals. Washing the fruit with tap water reduced the residues of azoxystrobin and fenhexamid but did not dislodge the pyrimethanil residues. Finally, when fruits were washed with a commercial detergent, greater amounts, for example, about 45% of azoxystrobin and pyrimethanil and 60% of fenhexamid (Angioni et al., 2004) were removed.

31.5.2 Effect of Cooking/Boiling/Juicing

Pesticide residues might be vaporized, hydrolysed and/or thermally degraded during cooking/boiling. The residue levels in juices from fruits or vegetables are generally reduced by 70–100% (Zabik et al., 2000; Rasmussen et al., 2003), and their reduction depends on the partitioning properties of the pesticide between the fruit skin, pulp and the juice (which generally contains some solids). The pulp or pomace byproducts, which often include the skin, retain a substantial proportion of lipophilic pesticide residues.

During production of strawberry jam from the fruits treated with three fungicides (tolylfluanid, fenhexamid and pyrimethanil), cooking significantly reduced tolylfluanid residues by an average of 91%. For fenhexamid (25%) and pyrimethanil (33%), a smaller reduction was recorded (Christensen et al., 2003).

Will and Kruger (1999) studied the fate of three fungicides (dichlofluanid, procymidone, and iprodione applied under field conditions) during processing of strawberry fruits to juice, wine, and jam. An untreated control was compared to the raw material treated with these fungicides according to their recommended doses and to a sample with six-fold higher application rates. The highest residue values were observed in the pomace after pressing. Residue values in readily produced juices and fruit wines were very low and did not exceed legally required maximum residue levels. Generally, processing steps such as pressing and clarification diminished fungicide residues from 50–100%. When the whole fruit was processed, as in fruit preparations or jam, the residue levels remained comparatively higher due to limited processing steps.

An experiment was conducted to evaluate the effects of some technological processes on the residual levels of imidacloprid in strawberry fruits and products. According to their half-life ($t_{1/2}$) values, strawberry fruits has been found safe for human consumption after 7.4 days of spraying. The removal ratio of imidacloprid residue ranged from 9.9 to 30.55% by washing with tap water. The average amount of imidacloprid residue in fruits, juice, and syrup under cold and hot break greatly decreased in comparison to those in unwashed strawberry fruits. Moreover, the pesticide residue decreased more in strawberry syrup under hot break than cold break. Imidacloprid residue was concentrated into jam and increased to higher levels than strawberry juice and syrup under cold and hot break (Hendawi et al., 2013).

31.5.3 Harvesting Precautions

The accumulation of pesticide residues on rubber latex gloves that are used by strawberry harvesters to protect their skin, reduces pesticide exposure and promotes food safety. In one study, gloves were observed to accumulate the residues of 16 active ingredients including azoxystrobin, bifenthrin, boscalid, captan, cyprodinil, fenhexamid, fenpropathrin, fludioxonil, hexythiazox, malathion, methomyl, naled, propiconazole, pyraclostrobin, quinoline and quinoxyfen at different times. The residue accumulation on gloves ($t_{1/2}$ 2.8 to 3.7 d) was very similar to the dissipation of dislodgeable foliar residue (DFRs) ($t_{1/2}$ 2.1 to 3 d) during the first three weeks after malathion applications. Dermal malathion dose was 0.2 mg/kg at the preharvest interval and the residues declined to trace levels during the next three months. Accumulation of malathion in gloves indicated trace surface residue availability and was used to assess the relationship between dislodgable foliar residues and potential hand exposure (Li et al., 2011).

31.6 FUTURE STRATEGIES

The following strategies are suggested to minimize the residues of pesticides in strawberries though judicious application:

- The use of relatively easily degradable pesticides should be encouraged.
- Adoption of all other alternative methods of controlling insect-pests and diseases such as natural enemies of pests (biological pest control), phyto-sanitation, and crop rotation should be encouraged.
- Use of biopesticides, which are safer to human beings and do not leave any toxic residues in crops/environment should be given due emphasis.
- Sowing of insect-pest and disease resistant varieties/hybrids should be encouraged.
- The farmers should be educated through media for the adoption of good agricultural practices.
- Regular surveillance of pesticide residues in soil, water, and food commodities as a part of national programme should be carried out in different agroclimatic regions of the country. The information thus generated will help the policy makers and researchers in regulating the use of safe pesticides. Additionally, the information will provide a correct idea of pesticidal contaminations to guide consumers for avoiding exposure to toxic residues.
- The dumping-off of industrial effluents in rivers, canals or any other water body should be prevented.
- Health costs should be taken into consideration while deciding pesticide policy of a nation.

REFERENCES

Anastassiades, M., Lehotay, S.J., Stajnbaher, D. and Schenck, F. J. 2003. Fast and easy multiresidue method employing acetonitrile extraction/partitioning and "dispersive solid-phase extraction" for the determination of pesticide residues in produce. *The Journal of AOAC International* 86(2): 412–431.

Angioni, A., Schirra, M., Garau, V.L., Melis, M., Tuberoso, C.I.G. and Cabras, P. 2004. Residues of azoxystrobin, fenhexamid and pyrimethanil in strawberry following field treatments and the effect of domestic washing. *Food Additives and Contaminants*, 21(11): 1065–1070.

Augusto, S.A., Hooper, A.E., de, S.L.A.F., de, S.S.V.C. and Gonçalves, J. R. 2013. In-house method validation and occurrence of alpha-, beta-endosulfan, endosulfan sulphate, lambda-cyhalothrin, procymidone and trifluralin residues in strawberry. *Ciência e Tecnologia de Alimentos,* 33(4): 765–775.

Barbash J.E. 2006. *The Geochemistry of Pesticides US Geological Survey.* Tacoma, Washington: pp. 2–4.

Baur P., Buchholz A. and Schonherr J. 1997. Diffusion in plant cuticles as affected by temperature and size of organic solutes: similarity and diversity among species. *Plant Cell and & Environment*, 20: 982–994.

Boulaid, M., Aguilera A., Camacho, F., Soussi, M. and Valverde, A. 2005. Effect of household processing and unit-to-unit variability of pyrifenox, pyridaben, and tralomethrin residues in tomatoes. *Journal of Agricultural and Food Chemistry*, 53(10): 4054–4058.

Cabras, P., Angioni, A., Garau, V.L., Melis, M., Pirisi, F.M., Cabitza, F. and Cubeddu, M. 1998. Pesticide residues on field-sprayed apricots and in apricot drying processes. *Journal of Agricultural and Food Chemistry*, 46: 2306–2308.

Cabrera H. A. P., Menezes H.C., Oliveira J.V. and Batista, R. F. S. 2000. Evaluation of residual levels of benomyl, methyl parathion, diuron, and vamidothion in pineapple pulp and bagasse (Smooth Cayenne). *Journal of Agricultural and Food Chemistry*, 48: 5750–5753.

Cengiz, M.F., Certel, M., Karakas, B. and Gocmen, H. 2007. Residue contents of captan and procymidone applied on tomatoes grown in greenhouses and their reduction by duration of a pre-harvest interval and post-harvest culinary applications. *Food Chemistry*, 100: 1611–1619, 0308–8146.

Christensen, H.B., Granby, K. and Rabølle, M. 2003. Processing factors and variability of pyrimethanil, fenhexamid and tolylfluanid in strawberries. *Food Additives and Contaminants*, 20(8): 728–41.

Codex. 2013. *Pesticide Residues in Food and Feed.* Codex Alimetarius International Food Standards. www.codexalimentarius.org.

Esteve-Turrillas, F.A., Abad-Fuentes, A. and Mercader, J.V. 2011. Determination of fenhexamid residues in grape must, kiwifruit, and strawberry samples by enzyme-linked immunosorbent assay. *Food Chemistry*, 124: 1727–1733.

Femenia, A., Sánchez, E.S., Simal, S. and Rossello, C. 1998. Effects of drying pretreatments on the cell wall composition of grape tissues. *Journal of Agricultural and Food Chemistry*, 46: 271–276.

Fernandes, V.C., Domingues, V.F. and Mateus N., 2011. Organochlorine pesticide residues in strawberries from Integrated Pest Management and organic farming. *Journal of Agricultural and Food Chemistry*, 59(14): 7582–7591.

Gebara, A.B., Ciscato, C.H.P., Monteiro, S.H. and Souza, G.S. 2011. Pesticide residues in some commodities: dietary risk for children. *Bulletin of Environmental Contamination and Toxicology*, 86: 506–510.

Ghabbour, S.I., Zidan, Z.H., Sobhy, H.M., Mikhail, W.Z.A. and Selim, M.T. 2012. Monitoring of pesticide residues in strawberry and soil from different farming systems in Egypt. *American-Eurasian Journal of Agriculture & Environmental Science*, 12 (2): 177–187.

Goodwin, S., Ahmad, N. and Newell, G. 1985. Dimethoate spray residues in strawberries. *Pesticide Science*, 16: 143–146.

Guardia Rubio, M., Ruiz Medina, A., Molina Diaz, A. and Canada Ayora, M.J. 2006. Influence of harvesting method and washing on the presence of pesticide residues in olives and olive oil. *Journal of Agricultural and Food Chemistry*, 54: 8538–8544.

Guardia Rubio, M., Ruiz Medina, A., Molina Diaz, A. and Canada Ayora, M.J. 2007. Multiresidue analysis of three groups of pesticides in washing waters from olive processing by solid-phase extraction-gas chromatograhy with electron capture and thermionic specific detection. *Microchemical Journal*, 85: 257–264, 0026-265X.

Hendawi, M.Y., Romeh, A.A. and Mekky, T.M. 2013. Effect of food processing on residue of imidacloprid in strawberry fruits. *Journal of Agricultural Science and Technology*, 15(5): 951–959.

Holland, P.T., Hamilton, D., Ohlin, B. and Skidmore, M.W. 1994. Effects of storage and processing on pesticide residues in plant products. *Pure and Applied Chemistry*, 66: 335–356.

Kaushik, G., Satya, S. and Naik, S. N. 2009. Food processing a tool to pesticide residue dissipation – A review. *Food Research International*, 42: 26–40.

Kilmer R., Andre A.M. and Stevens T.J. 2001. Pesticide residues and vertical integration in Florida strawberries and tomatoes. *Agribusiness*, 17: 213–226.

Kin, Ch.M. and Huat, T.G. 2010. Headspace solid-phase microextraction for the evaluation of pesticide residue contents in cucumber and strawberry after washing treatment. *Food Chemistry*, 123: 760–764.

Krol, W.J., Arsenault, T.L., Pylypiw, H.M. Jr. and Mattina, M. J. I. 2000. Reduction of pesticide residues on produce by rinsing. *Journal of Agricultural and Food Chemistry*, 48: 4666–4670.

Lentza-Rizos, C. and Kokkinaki, K. 2002. Residues of cypermethrin in field-treated grapes and raisins produced after various treatments. *Food Additives and Contaminants*, 19: 1162–1168.

Li, Y., Chen, Li, Chen, Z., Coehlo, J., Cui, Li, Liu, Yu, Lopez, T, Sankaran G. , Vega, H. Krieger, R. 2011. Glove accumulation of pesticide residues for strawberry harvester exposure assessment. *The Bulletin of Environmental Contamination and Toxicology*, 86: 615–620.

Liu, C., Wang, S., Li, Li, Ge, J., Jiang, S. and Liu, F. 2011. Dissipation and residue of cyprodinil in strawberry and soil. *The Bulletin of Environmental Contamination and Toxicology*, 86: 323–325.

Looser, N., Kostelac, D., Scherbaum, E., Anastassiades, M. and Zipper, H. 2006. Pesticide residues in strawberries sampled from the Market of the Federal State of Baden-Württemberg in the period between 2002 and 2005. *Journal für Verbraucherschutz und Lebensmittelsicherheit*, 1 (2): 135–141.

Qu-Yang, X.K., Liu, S.M. and Ying, M. 2004. Study on the mechanism of ozone reaction with parathion-methyl. *Safety and Environmental Engineering*, 11: 38–41.

Pesticide Action Network (PAN). 1998. *Pesticide Residues in Food*, Pest Management Notes. Vol. 8. UK.

Pesticide Action Network (PAN). 2003. *Organophosphate Insecticides Fact Sheet*. UK.

Pesticide Action Network (PAN). 2006. *Pesticides in Your Food*. Pesticide Action Network, UK. 1–2.

Petrosini, G., Businelli, M., Tafuri, F. and Scarponi, L. 1969. Determination of chlorcholine chloride residues in strawberries. *Analyst*, 94: 674–677.

Pugliese, P., Molto, J.C., Damiani, P., Marin, R., Cossignani, L and, Manes, J. 2004. Gas chromatographic evaluation of pesticide residue contents in nectarines after nontoxic washing treatments. *Journal of Chromatography A*. 1050: 185–191, 0021-9673.

Rasmussen, R.R., Poulsen, M.E. and Hansen, H.C.B. 2003. Distribution of multiple pesticide residues in apple segments after home processing. *Food Additives and Contaminants*. 20: 1044–1063.

Safi, J.M., Abou-Foul, N.S., El-Nahhal, Y.Z. and El-Sebae, A.H. 2002. Monitoring of pesticide residues on cucumber, tomatoes and strawberries in Gaza governorates, Palestine. *Nahrung/Food*, 46: 34–39.

SANTE. 2015. European Union Guidance document on analytical quality control and method validation procedures for pesticide residues analysis in food and feed, Document No., SANTE/11945/2015.

Schule, E. and Looser, N. 2004. Multiple pesticide residues in strawberries from the German market analysed in the period between 2000 and Spring 2004. Chemisches und Veterinäruntersuchungsamt Stuttgart, EPRW 2004.

Sharma, R. R. and Gupta, A. K. 2004. Strawberry aadhunik utpadan tekneekein. Satish Serial Publishing House, 403, Express Tower, Commercial Complex, Azadpur, Delhi–110033, India.

Souza, M. C. O., Baldim, I. M., Souza, J. C. J. da C., Boralli, V. B., Maia, P. P. and Martins, I. 2014. QUECHERS technique for the gas Chromatographic determination of Organophosphate residues in strawberries. *Analytical Letters*, 47: 1324–1333, 2014.

Will, F. and Kruger, E. 1999. Fungicide residues in strawberry processing. *Journal of Agricultural and Food Chemistry*, 47: 858–861.

Yamazaki, Y. and Ninomiya, T. 1998. Determination of bitertanol residues in strawberries by liquid chromatography with fluorescence detection and confirmation by gas chromatography/mass spectrometry. *Journal of the Association of Official Agricultural Chemists, International*, 81(6): 1252–1256.

Yioit, A. and Erkylyc, L. 1992. Studies on the chemical control of *Tetranchus cinnabarinus* Boisd. (Acarina: tetranychidae), a pest of strawberry in the east Mediterranean region of Turkey. *Crop Protection*, 11: 433–438.

Zabik, M.J., El-Hadidi, M.F.A.J., Cash, N., Zabik, M.E. and Jones, A.L. 2000. Reduction of azinphos-methyl, chlorpyrifos, esfenvalerate, and methomyl residues in processed apples. *Journal of Agricultural and Food Chemistry*, 48 (4199–4203): 0021–8561.

32 Production Economics and Marketing

Sudhakar Dwivedi and Pawan Kumar Sharma

CONTENTS

32.1 Production Economics 520
    32.1.1 Establishment Cost..520
        32.1.1.1 Operating Costs 520
        32.1.1.2 Labour and Material Costs 520
        32.1.1.3 Fixed Costs...521
    32.1.2 Picking Year Costs...521
        32.1.2.1 Labour and Material Costs 522
        32.1.2.2 Fixed Costs...523
32.2 Returns from Strawberry Production 523
    32.2.1 Cost and Return Summary 523
32.3 Economics of Annual Strawberry Production...523
32.4 Marketing..523
    32.4.1 Marketing Channels 527
        32.4.1.1 Direct Marketing Channel..527
        32.4.1.2 Retail Channel..527
        32.4.1.3 Wholesale Channel...527
    32.4.2 Price Pattern for Strawberry 527
32.5 Conclusions...529
References..529

The production and marketing of strawberries at a commercial scale form a complex business enterprise because of the large needed investment in terms of land and infrastructure with irregular and uncertain income due to the delicate nature and limited shelf life of its fruits. Its success for desired economic returns requires:

- suitable land with adequate drainage and high-tech growing environment
- access to an adequate quantity and quality of irrigation water with precise water discharging unit
- financial resources that permit a significant level of investment into infrastructure and input costs
- the ability to apply intensive management skills to the crop in a precisely timed manner
- the ability to determine market potential
- the willingness to accept risk.

It is, however, among the few fruit crops that gives quick and very high returns per unit area on capital investment, as the crop is ready for harvesting within four months of planting. Therefore, a complete assessment of the customer base and marketing opportunities needs to be completed before planting the first crop of strawberries. Both current and potential strawberry growers need production, marketing and financial information to make effective decisions about starting, expanding or leaving strawberry business. Like all business owners, their main objective should be to earn profit to make their farms financially successful. Ideally, growers should keep detailed records that they can use to estimate production, harvest and marketing costs. But in many countries, only a few growers have either the time or skill to keep detailed records. Likewise, marketing is a key factor in the success of strawberry business. Therefore, in this chapter, an attempt has been made to discuss the costs and returns of growing, harvesting and marketing of strawberries through various channels for the benefit of stakeholders.

The economics of strawberry production has been estimated from time to time by the extension wings of the Department of Agriculture or many other such establishments of many countries (Daugovish et al., 2011; Wahl et al., 2014; Wright and Ernst, 2014). Some of the researchers also analyzed the economics of strawberry processing including wine (Sharma and Joshi, 2004; Joshi et al., 2005; Sharma et al., 2009). The cost estimation of strawberry production and marketing in India is not simple because of its adaptability to subtropical and temperate regions, in plain and hilly areas, in open conditions and in high-tech greenhouses (Rowley and Drost, 2010) and with or without improved technologies. Costs and yields in different farm situations also differ due to soil type, climatic conditions and agronomic practices. Therefore, detailed planning is necessary for estimating the budget for capital expenditures, and also for the annual operating costs of strawberry production. All these factors needs to be considered if you want to develop an accurate cost of production for strawberry in Indian conditions. In this chapter, an attempt is made to fix the costs so that a generalized estimation of strawberry production costs in Indian conditions can be obtained. The capital and cash input costs associated with production of strawberry, and costs and margins associated with its marketing have been assessed separately. The same procedure can be adopted with needful modifications for working economics of production and trade of strawberry in several other regions of the world wherever cultivation of strawberry is a commercial enterprise.

32.1 PRODUCTION ECONOMICS

Strawberry is produced both as a perennial as well as an annual crop. In perennial production, the strawberry runners are maintained for three years, whereas the in case of annual production, runners are uprooted and planted freshly. The economics of perennial strawberry production is given in the following.

32.1.1 ESTABLISHMENT COST

In the establishment year, the costs of strawberry include costs incurred in various practices in the establishment of 1 ha planting of strawberry (Table 32.1).

32.1.1.1 Operating Costs
These are those costs incurred during any operation in process of production and marketing. These broadly include labour cost and material cost besides fixed costs in the estimation of cost of production and selling cost in case of the estimation of marketing cost.

32.1.1.2 Labour and Material Costs
To establish 1 ha of strawberry, 157 person-days (@ US$8 per day) are required for various activities such as furrow/bed preparation, transplanting, irrigation, straw mulching, application of farmyard manure and fertilizers and spraying of herbicides and pesticides etc.

The material costs include expenditure on all the physical items required during orchard establishment for optimum production of strawberry. Such items consist of good-quality plants (runners), organic manure, water, chemical fertilizers, pesticides and herbicides (Table 32.1).

Production Economics and Marketing

TABLE 32.1

Costs of 1 ha Strawberry Production During the Establishment Year

S. No.	Item of Cost	Units/ha	Price (US$ Per Unit)	Amount (US$)
A.	**Operating costs**			
1.	**Labour costs**			
a.	Furrow/bed preparation	60	8	480.00
b.	Application of farmyard manure and fertilizers	15	8	120.00
c.	Transplanting	55	8	440.00
d.	Mulching	10	8	80.00
e.	Spraying of herbicides and pesticides	5	8	40.00
f.	Irrigation	12	8	96.00
2.	**Material costs**			
a.	Plants	60,000	0.15	9,000.00
b.	**Manures and fertilizers**			
	Farmyard manure	50 tonnes	15	750.00
	Nitrogen	60 kg	12	720.00
	Phosphorus	80 kg	32	2560.00
	Potassium	20 kg	4	80.00
	Zinc	10 kg	43	430.00
	Sulphur	4 kg	13	52.00
c.	**Herbicides**			
	Broad leaf	–	–	22.00
	Wild oats	–	–	8.00
d.	Insecticides	–	–	15.00
e.	Fungicides	–	–	10.00
f.	Mulching with polythene	–	–	215.00
g.	Miscellaneous	–	500	50.00
3.	Repairs and maintenance	–	200	500.00
4.	Interest on operating costs	14%	–	2,193.52
Total operating costs				17,861.52
B.	**Fixed costs**			
1.	Installation of microirrigation unit	1	2,500.00	2,500.00
3.	Interest on fixed cost	12%	–	300.00
Total Fixed Costs				2,800.00
C.	**Total cost per hectare**	–	–	20,661.52

32.1.1.3 Fixed Costs

These costs do not vary with output, and will always be there even if there is absence of any farm activity. These are the investments made in farm durables having long gestation periods. Farm machinery, equipment, irrigation structures and buildings are examples of fixed costs. In the estimation of establishment cost, the full cost of the establishment of microirrigation infrastructure is considered, as it is the major investment incurred in the strawberry plantation.

32.1.2 PICKING YEAR COSTS

The various costs incurred on 1 ha plantings in the picking years of strawberry is calculated for three consecutive years, as shown in Table 32.2. Strawberry is an intensively managed crop that

TABLE 32.2

Picking Year Costs and Returns of Strawberry Production (in US$ Per Hectare)

S. No.	Item of Cost	Picking Year Costs			Total
		First Year	Second Year	Third Year	
A.	**Operating costs**				
1.	**Labour costs**				
a.	Mulching	27.27	35.45	40.91	103.64
b.	Field operations	54.55	49.09	49.09	152.73
c.	Application of manures and fertilizers	19.09	19.09	19.09	57.27
d.	Spraying of herbicides and pesticides	19.09	16.36	19.09	54.55
e.	Irrigation	32.73	32.73	32.73	98.18
2.	**Material costs**				
a.	Fertilizers				
	Nitrogen	23.64	20.00	16.36	60.00
	Phosphorus	67.27	63.64	58.18	189.09
	Potassium	2.91	2.91	2.91	8.73
	Zinc	36.36	32.73	30.91	100.00
	Sulphur	8.18	8.18	8.18	24.55
b.	Herbicides				
	Broad leaf	21.82	21.82	21.82	65.45
	Wild oats	10.55	10.55	10.55	31.64
c.	Insecticides	21.82	18.18	23.64	63.64
d.	Fungicides	9.45	7.27	10.55	27.27
e.	Mulch (polythene)	218.18	218.18	218.18	654.55
f.	Miscellaneous	9.09	10.91	12.73	32.73
3.	Repairs and maintenance	3.64	5.45	7.27	16.36
4.	Interest on operating costs @ 14%	61.62	59.80	61.15	182.56
	Total operating costs	647.26	632.34	643.34	1922.94
B.	**Fixed costs**				
1.	Land revenue	0.64	0.64	0.64	1.91
2.	Depreciation				
a.	Building	36.36	36.36	36.36	109.09
b.	Other machinery and equipment	29.09	29.09	29.09	87.27
3.	Estimated rental value	118.18	120.00	121.82	360.00
4.	Interest on fixed cost @ 12%	22.11	22.33	22.55	66.98
	Total fixed costs	206.38	208.42	210.45	625.25
C.	Total cost per hectare	853.64	840.76	853.79	2548.19

requires timely intercultural operations. Yield and quality of fruits obtained depend, to a great extent, upon timely management practices of the previous year, as well as care and maintenance before picking.

32.1.2.1 Labour and Material Costs

In the first picking year of strawberry, US$152.73 is required for hiring labourers for carrying out different operations in 1 ha of land. Labour is required for doing various activities such as mulching, field operations, application of manures and fertilizers, spraying of herbicides and pesticides, and irrigation. For the second and third year, more labour is required to remove mulch applied in the previous year and for fresh mulching. On the other hand, the labour requirement on various field operations and irrigation decline in subsequent years.

Production Economics and Marketing

The material cost includes expenditure on fertilizers (120 kg nitrogen, 180 kg phosphorus and 20 kg potash), cost of herbicides and pesticides and miscellaneous costs incurred during picking years of strawberry.

32.1.2.2 Fixed Costs

This includes land revenue, estimated rental value, investment on installation of micro-irrigation unit and depreciation on building and machinery/equipment. Estimated rental value of a farmer's own land was estimated to be US$118.18, US$120 and US$121.82 per hectare for the first, second and third year, respectively. Depreciations of equipment were included, which were used at different intervals at various stages, other than the microirrigation unit. The depreciation of the microirrigation unit was excluded because its full value (US$2500) has been charged during the assessment of establishment year costs. The depreciation was calculated using the following formula:

$$\text{Depreciation} = \frac{\text{Original value} - \text{Salvage value}}{\text{Useful life}(3 \text{ years})}$$

The combined depreciation of building and machinery/equipment was estimated to be US$196.36 per hectare. The interest on fixed costs was charged at the rate of 12%.

32.2 RETURNS FROM STRAWBERRY PRODUCTION

The strawberry is a short-duration crop that, if planted in September–October, starts giving yield in February–March. It continues to give yield up to the third year and thereafter it needs to be replanted. The average yield per year of strawberry was worked out be 67 quintals per hectare. The yield decreases from 70 quintals per hectare in the first year to 66 quintals in the second year to 64 quintals per hectare in the third year. The average price of strawberry was US$200 per quintal. Thus gross return could be worked out to US$15,000, US$14,200 and US$13,800 per hectare for the first, second and third year, respectively, including the returns from sale of runners.

32.2.1 COST AND RETURN SUMMARY

The cost and return summary of strawberry is shown in Table 32.3. The cost of establishing a 1 ha strawberry plantation was estimated to be US$18,452.85, whereas the total cost for three years including picking cost worked out to be US$7313.37. The total net return for three years from a 1 ha strawberry plantation was US$13,033.98, which means an annual net return of US$4344.66. Table 32.4 indicates the estimated monthly cash cost incurred on a 1 ha strawberry plantation with an annual production of 67 quintals per hectare. This would help the farmers in assessing their financial requirements for every month and desired credit need in case of insufficient income.

32.3 ECONOMICS OF ANNUAL STRAWBERRY PRODUCTION

The economics of annual strawberry production has been presented in Table 32.5. The total operating cost for 1 ha strawberry production was estimated to be US$6932.78, whereas the fixed cost was US$212.05. The gross and net incomes obtained from one hectare annual strawberry production, including fruits and runners, were US$15,000 and US$7855.17, respectively with C:B ratio of 1:1.1.

32.4 MARKETING

Fruits and vegetables are usually more difficult to market than other agriculture produce such as cereals and pulses. There are ready markets available daily or weekly for grain and livestock produce in almost all parts of the country, but there are few similar markets for perishable horticultural

TABLE 32.3
Cost Components and Returns of Three Years Strawberry Plantation (Amount in US$ Per Hectare)

Item of Cost	Establishment Costs	Picking Costs	Annual Costs
A. Operating costs			
Labour	1256.00	466.37	574.12
Planting material	9000.00	–	3000.00
Manure and fertilizers	4592.00	382.37	1658.12
Herbicides	30.00	97.09	127.09
Insecticides	15.00	63.64	26.21
Fungicides	10.00	27.27	12.42
Straw	215.00	654.55	289.85
Repair and maintenance	200.00	900.00	366.00
Miscellaneous	50.00	32.73	27.58
Interest on operating cost @14%	2151.52	367.36	851.39
Total operating cost	17519.52	2991.38	6932.78
B. Fixed costs			
Investment	2500.00	–	833.33
Land revenue	–	1.91	0.64
Depreciation			
– Building	–	109.09	36.36
– Machinery and equipment	–	87.27	29.09
Estimated rental value		360	120.00
Interest on fixed cost @ 12%	100	22.33	40.77
Total fixed cost	933.33	208.42	380.58
Total cost of production	18452.85	3199.80	7313.37

Returns from strawberry production	First Year	Second Year	Third Year
Production (quintal per hectare)	70	66	64
Sale of fruits	14000.00	13200.00	12800.00
(@ US$200 per quintal)			
Sale of runners	1000.00	1000.00	1000.00
(20,000 @ US$0.05 per plant)			
Total income per hectare	15000.00	14200.00	13800.00
Total return/hectare for three years		42000.00	
Net return/hectare for three years		13033.98	
Annual total return/hectare		9655.34	
Annual net return per hectare		4344.66	

commodities. Most of the commodities are produced in abundance, and long-established market channels may be discouraging for small-scale or new producers. Practically, a producer may need several years to establish a marketing network. Efficient marketing of agricultural commodities in general and fruit crops in particular plays an important role in safeguarding interest of producers as well as consumers. An efficient marketing system is one that optimizes the net returns from a given transaction and helps in expanding the market in respect of grading, packing, transportation etc. As all these functions are important determinants of the price a product fetches in market, intensive care has to be ensured at every step. Any carelessness at any stage in the marketing channel will result in poor prices and would miss the target of higher net returns. Marketing of strawberry

Production Economics and Marketing

TABLE 32.4

Monthly Operating Cost of Strawberry Production in US$ Per Hectare (Year: 2016/17)

Yield: 67 quintals/hectare

Expenditure	Jan	Feb	Mar	Apr	May	Jun	Jul	Aug	Sep	Oct	Nov	Dec	Total
Labour costs													
Field operations	11.00	41.00	12.00	15.00	14.00	51.00	18.00	8.00	9.00	56.00	8.00	9.00	252.00
Mulching	0.00	0.00	0.00	36.00	0.00	0.00	0.00	0.00	0.00	0.00	0.00	24.00	60.00
Application of manures and fertilizers	0.00	10.00	0.00	0.00	0.00	0.00	0.00	0.00	0.00	76.00	0.00	36.00	122.00
Spraying of herbicides and pesticides	0.00	15.00	0.00	0.00	0.00	0.00	0.00	0.00	0.00	46.00	0.00	22.00	83.00
Irrigation	22.00	14.00	0.00	0.00	0.00	13.00	8.00	0.00	0.00	0.00	0.00	0.00	57.00
Total labour costs	33.00	80.00	12.00	51.00	14.00	64.00	26.00	8.00	9.00	178.00	8.00	91.00	574.00
Material costs													
Manures and fertilizers	0.00	182.00	0.00	0.00	0.00	0.00	0.00	0.00	0.00	1610.00	0.00	710.00	2502.00
Pesticides and fungicides	0.00	72.00	0.00	0.00	0.00	0.00	0.00	0.00	0.00	174.00	0.00	94.00	340.00
Mulching	0.00	0.00	0.00	360.00	0.00	0.00	0.00	0.00	0.00	0.00	0.00	156.00	516.00
Total material costs	0.00	254.00	0.00	360.00	0.00	0.00	0.00	0.00	0.00	1784.00	0.00	960.00	3358.00
Total operating cost	33.00	334.00	12.00	411.00	14.00	64.00	26.00	8.00	9.00	1962.00	8.00	1051.00	3932.00
Land revenue	0.05	0.05	0.05	0.05	0.05	0.05	0.05	0.05	0.05	0.05	0.05	0.05	0.64
Office expenses including rent	10.00	10.00	10.00	10.00	10.00	10.00	10.00	10.00	10.00	10.00	10.00	10.00	120.00
Investment repairs	5.45	5.45	5.45	5.45	5.45	5.45	5.45	5.45	5.45	5.45	5.45	5.45	65.45
Cash overhead cost	15.50	15.50	15.50	15.50	15.50	15.50	15.50	15.50	15.50	15.50	15.50	15.50	186.09
Total cash costs/hectare	48.50	349.50	27.50	426.50	29.50	79.50	41.50	23.50	24.50	1977.50	23.50	1066.50	4118.09
Total cash costs/quintal	0.69	4.99	0.39	6.09	0.42	1.14	0.59	0.34	0.35	28.25	0.34	15.24	58.83

TABLE 32.5

Costs and Returns from Annual Strawberry Production (in US$ Per Hectare)

Particulars	Amount
Costs	
Operating cost	
Labour	574.12
Planting material	3000
Manure and fertilizers	1658.12
Herbicides	127.09
Insecticides	26.21
Fungicides	12.42
Straw	289.85
Repair and maintenance	366
Miscellaneous	27.58
Interest on operating cost @14%	851.39
Total operating costs	6932.78
Fixed cost	
Land revenue	0.64
Depreciation on building	36.36
Depreciation on machinery and equipment	29.09
Estimated rental value	120
Interest on fixed cost @ 12%	25.96
Total fixed costs	212.05
Total costs	7144.83
Returns	
Production (quintal per hectare)	70.00
Sale of fruits	14000.00
(@ US$200 per quintal)	
Sale of runners	1000.00
(20,000 @ US$0.05 per plant)	
Total returns	15000.00
Net returns	7855.17
Output–input ratio	2.10
C:B ratio	1.10

encompasses all the business activities involved in the flow of fruits and fruit products from producers to consumers. Since strawberry is a highly perishable fruit, it needs a better and assured marketing infrastructure. The fruit loses quality from immediately after harvesting. Immediate marketing to ensure freshness is a most desirable feature of locally grown produce. Produce not sold immediately needs to be stored properly to maintain its appearance, flavour and quality. Harvesting at the right time, cooling to minimize field heat and scientific storage help to retain fruit quality during transportation and marketing.

The scope of the strawberry business shows more promise for big and small producers than for midsized producers. The large-scale producer can afford the big equipment needed for production and marketing of strawberry. Small-scale producers can use smaller equipment, often hand operated, and family labour to substitute for the other equipment. A midsized producer is less efficient and often can't economically justify the purchase of needed equipment or substitute workers for equipment.

32.4.1 Marketing Channels

The marketing system for strawberry is somewhat more complex than for other fruits. In India, there are very few marketing channels available for strawberry. However, in general, there are three different marketing channels utilized by strawberry growers in India (Figure 32.1). The degree of importance of each has been difficult to determine since individual producers utilize a mixture of channels. The main characteristics of these channels are described in the following.

32.4.1.1 Direct Marketing Channel

Marketing directly to consumers includes farm gate selling in which the producers sell strawberries outside their farm. It is one of the best channels of selling for small growers as consumers get fresh produce and farmer does not require any large investment in transport, packing and marketing.

32.4.1.2 Retail Channel

This is a form of indirect marketing in which farmers deal with intermediaries such as retail shops and supermarkets. The majority of produce is sold through this channel utilizing packaging in plastic containers.

32.4.1.3 Wholesale Channel

This channel is used by the growers who have a large market surplus. The wholesalers sell the produce to retailers, through which it reaches to the ultimate consumer. Two types of strawberry packaging (individual punnets and punnet-filled cartons) are available in Indian markets.

32.4.2 Price Pattern for Strawberry

The price pattern in the marketing of strawberry in the three following channels is shown in Table 32.6.

1. Producer – Consumer
2. Producer – Retailer/Hoteliers/Super markets – Retailer – Consumer
3. Producer – Wholesaler – Retailer – Consumer

In direct marketing, the strawberry grower got a net price of US$372 per quintal. Their selling price was US$464 per quintal after adding marketing costs, which included picking, packing and loading/unloading charges. US$464 was also the consumer's purchase price in marketing channel 1.

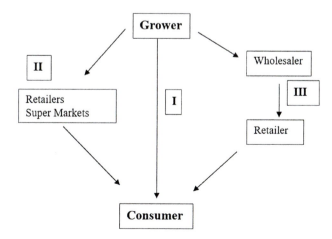

FIGURE 32.1 Marketing channels of strawberry.

528 Strawberries

TABLE 32.6
Price Spread Analysis of Strawberry in Different Marketing Channels (US$ Per Quintal)

Particulars	Direct Marketing 1		Retailer/Super Markets 2		Wholesalers 3	
	US$	%	US$	%	US$	%
			Grower's			
1. Net price	372.00	80.17	302.00	64.67	285.00	59.50
2. Marketing costs						
– Picking cost	16.00	3.45	25.00	5.35	18.00	3.76
– Packing cost	34.00	7.33	42.00	8.99	0.00	0.00
– Loading/Unloading	25.00	5.39	26.00	5.57	0.00	0.00
– Transportation cost	17.00	3.66	42.00	8.99	0.00	0.00
3. Selling price	464.00	100.00	437.00	93.58	303.00	63.26
			Wholesaler's			
1. Purchase price	–	–	–	–	303.00	63.26
2. Marketing cost						
– Unloading	–	–	–	–	56.00	11.69
– Spoilage and wastage	–	–	–	–	48.00	10.02
3. Margin	–	–	–	–	42.00	8.77
4. Selling price	–	–	–	–	449.00	93.74
			Retailer's			
1. Purchase price	–	–	437.00	93.58	449.00	93.74
2. Marketing cost						
– Transportation cost	–	–	4.00	0.86	4.00	0.84
– Shop charges	–	–	6.00	1.28	6.00	1.25
– Spoilage and wastage	–	–	8.00	1.71	8.00	1.67
– Miscellaneous	–	–	2.00	0.43	2.00	0.42
3. Margin	–	–	10.00	2.14	10.00	2.09
Consumer's purchase price	464.00	100.00	467.00	100.00	479.00	100.00

Source: Personal Communications.

In marketing channel 2, the grower gets a net price of US$302 per quintal. The retailer in this case spent US$155 as various marketing expenses and earn a margin of US$10 per quintal. In marketing channel 3, the grower's selling price was US$303 per quintal, which included their net received price of US$285 and marketing expenses of US$18. The wholesaler spent US$122 and earns US$42 per quintal. The consumer's purchase price in this case was US$479, which also includes retailer's expenses and margin of US$20 and US$10 per quintal, respectively.

The marketing efficiency given by Acharya and Aggarwal (2001) has been calculated in Table 32.7, which shows that channel 1 (4.04) was the most efficient marketing channel for strawberry growers, followed by channel 2 (1.83) and channel 3 (1.47). In direct marketing of strawberries, growers have to pay out 20% of the consumers' price for marketing functions, which includes picking, packing and loading. There was no marketing margin because of the direct sale to the consumer. This practice is quite prevalent in tourist places such as Jammu and Kashmir, where the farmers sell their produce to the tourists outside their farms on the roads. The consumers also preferred to buy strawberry directly from the growers during their tour because of its freshness.

The quantity sold, however, through channel 1 and channel 2 is always very small due to coverage of a small area. This also reflects that the growers could not afford the marketing through channel 3, either due to low surplus or lack of resources. Growers with large surplus have to market their produce through channel 3. This also requires good infrastructure besides an airconditioned vehicle for safe transportation of strawberry to the market.

TABLE 32.7
Marketing Efficiency of Different Channels

Particulars	1		2		3	
	US$	%	US$	%	US$	%
Net price received by farmer	372.00	80.17	302.00	64.67	285.00	59.50
Marketing margin	0	0	10	2.14	52.00	10.86
Marketing cost	92.00	19.83	155	33.19	142.00	29.65
Marketing efficiency[a]	4.04		1.83		1.47	

[a]
$$MME = \frac{NFP}{TMC + MM}$$
where MME = measurement of marketing efficiency

NFP = net price received by the farmers
TMC = total marketing cost
MM = total net margin of intermediaries.

32.5 CONCLUSIONS

Small-scale strawberry growers have immense opportunities to sell their produce at on-farm markets, organized farm markets, locally owned supermarkets and locally owned fruit markets. However, selling the produce to any market, and especially to local supermarkets or fruit and vegetable markets, needs good communication between grower and buyer. A producer must know about what, when and how much the buyer can use, while the buyer needs to understand what is available and when, as he has to keep the shelves stocked. Retailers generally double the price paid to account for shrinkage and spoilage.

Farmers can obtain a premium price when using plastic containers for packaging. There are financial implications but strawberry needs some marketing expenses for effective marketing. Besides, own counters of strawberry growers associations along national highways are also one of the potential marketing channels if supported by the government agencies.

Note: The facts and figures given in this chapter are only indicative, accordingly readers/stakeholders wisdom is equally important in dealing with socioeconomic issues such as the production and trade of strawberries in India or elsewhere in the world.

REFERENCES

Acharya, S.S. and Agarwal, N.L. 2001. *Agricultural Marketing in India.* Oxford and IBH Publishing Co., New Delhi, India.

Daugovish, O., Klonsky, K.M. and De Moura, R.L. 2011. *Sample Costs to Produce Strawberries in South Coast Region– Ventura County Oxnard Plain.* University of California Cooperative Extension, ST-SC-11-2.

Joshi, V.K., Sharma, S. and Bhushan, S. 2005. Effect of method of preparation and cultivar on the quality of strawberry wine. *Act Aliment*, 34(4): 339–355.

Rowley, D.B. and Drost, D. 2010. High tunnel strawberry production. *Horticulture.* https://extension.usu.edu/files/publications/publication/Horticulture_HighTunnels_2010-01pr.pdf; accessed online on 10-09-2016.

Sharma, S. and Joshi, V.K. 2004. Flavour profiling of Strawberry wine by quantitative descriptive analysis technique. *Journal of Food and Science Technology*, 41(1): 22–26.

Sharma, S., Joshi, V.K. and Bhushan, S. 2009. An overview on Strawberry [Fragaria × ananssa (Weston) Duchesne ex Rozier] wine production technology, composition, maturation and quality evaluation. *Natural Product Radiance*, 8(4): 356–365.

Wahl, C., Liegel, L. and Seavert, C.F. 2014. Strawberry economics: comparing the costs and returns of establishing and producing fresh and processed market June bearing strawberries in a perennial matted row system to day-neutrals in a perennial hill, plasticulture system, in the Willamette valley. Oregon State University Extension Service, AEB 0052.

Wright, S. and Ernst, M. 2014. *Plasticulture Strawberries.* University of Kentucky, Cooperative Extension Service. https://www.uky.edu/Ag/CCD/introsheets/plasticulturestrawberry.pdf. Accessed online on 10 September 2016.

33 Challenges, Potential and Future Strategies

K. Kumar

CONTENTS

33.1 Challenges.. 531
33.2 Potential 532
33.3 Future Strategies 535
References... 536

Cultivated strawberries, are "man-made" and one of the most recently domesticated fruit crops in the world. They have witnessed unprecedented popularity, leading to a manifold increase in production and usage all over the world. The United States and China collectively contribute approximately two-thirds of the world's total production. Other producers that contribute significantly to world strawberry production are Turkey, Spain, Egypt and Mexico. After bananas, apples, grapes and melons, strawberries rank high in world production. Its attractive fruits with a distinct flavour, rich in antioxidants and minerals, have made it a preferred fruit for human consumption from health benefit points of view. Its use both as fresh and processed and value-added products have firmly placed the strawberry sector as a global industry worth many billions of dollars. It is a source of employment to millions of households across the globe.

Cultivated strawberries are among the few fruit crops that have undergone extensive genetic improvement through controlled breeding efforts. In addition, cultivation and management practices including post-harvest handling, processing and trade have witnessed a great deal of transformation and the industry has emerged as one of the most sophisticated and modern enterprises. No fruit crop has similarly witnessed the evolution of thousands of artificially bred cultivars like cultivated strawberries except perhaps grapes and peaches to some extent.

33.1 CHALLENGES

Strawberry production systems depend heavily on environmental factors, which significantly influence various plant growth, fruit yield and quality attributes. The ideal growing climate for strawberries is one where they are exposed to warm, sunny days and cool nights. The strawberry sector is expanding worldwide. Consequently, strawberry production both for fresh and processed products is increasing at a global level. The strawberry sector gains further impetus from the consumer shift towards greater choices for healthier food, and growth in emerging markets far away from the areas of production. But adverse weather conditions, mainly excessive soil moisture and low temperatures during night hours, badly affect production, market prices, trade and consumption of strawberries.

Despite the fact that a lot has been achieved in terms of high remunerations through modern-day strawberry production systems largely driven by advanced technical knowhow, threats and challenges still remain to sustainable strawberry production at global level. Among these factors, some of the most important are declining availability of land suitable for successful strawberry cultivation, the scarcity and high cost of good-quality planting materials, non-uniformity of bare-rooted transplants often infected with pests and diseases, high soil pH, low soil fertility, ever-decreasing availability of water for irrigation, dependence on ethyl bromide for soil fumigation, frost injury due to unpredictable winters, capturing the early-season market, seasonality barriers restricting availability of varieties ideal for extended availability of marketable fruits, severity of pests and diseases necessitating excessive use of pesticide and consequently leading to food and environmental safety issues due to pesticide residues, lack of knowledge of appropriate planting and crop management practices when compared with technological advances taking place in this sector, low awareness of the huge potential of the crop and available market opportunities, etc. The goal of combining technological advancements with high input use efficiency to harness the optimum potential is likely to remain a formidable for scientists and other stakeholders associated with the strawberry sector.

33.2 POTENTIAL

Categorically, short-day and day-neutral strawberry cultivars are being grown all over the world, but different cultivars are adapted to different agroclimatic conditions. Largely, those varieties that have a dormant phase are well adapted to cooler climates and account for major areas in the world with significant commercial strawberry production. Short-day strawberries form flower buds in the field during late summer and early autumn and the minimum number of short days necessary to induce flower bud initiation varies from as low as one week to as high as more than three weeks. However, experimental findings from more recent research studies indicate that temperature also plays a more vital role than had been understood earlier (Hollender et al., 2011) for adaptation of strawberries to particular agroclimatic conditions.

Apart from the range of strawberry cultivars under cultivation, diverse cultivation technologies are being followed in different countries and in varied agroclimatic zones throughout the world. The strawberry production areas and their respective yields vary considerably, largely due to production methods. In protected cropping under plastic or in greenhouse conditions, fruit yields are as high as 60–70 tonnes/ha, while in an open field cultivation system, fruit yield can be about 20 tonnes/ha. The maximum fruit yield of 56 tonnes/ha has been recorded in the United States where a large proportion of the crop is intensively produced under plastic tunnels.

Strawberries are either grown in open conditions or indoors in polytunnels/greenhouses. There are two ways of open field cultivation of strawberries, the hill system and the matted row system (Martin and Tepe, 2014). In hill systems, a plastic mulch is often used on raised beds (15–25 cm high) with drip irrigation. In such a system, day-neutral strawberry varieties are usually grown, and the first-season flowers are removed to enhance plant vigour through promotion of vegetative shoot and root growth.

Regular irrigation is necessary and if water available is insufficient, it is detrimental to the survival of plants (Carroll et al., 2015). The plants are allowed to grow for one or two seasons depending upon the situation, and mother plants are replaced by new ones. In a matted row system, strawberries are grown on flat beds with assured drip or sprinkler/fogger irrigation. Depending upon plant health and yield potential, short-day or June-bearing strawberries are routinely grown for three–five seasons in matted row systems. As with day-neutrals, the first flowers are removed, although the practice continues for a full season with short-day strawberries. In either case, maintaining adequate soil moisture during the period of flowering is of utmost importance for achieving optimum fruit quality and yield. Stolons are usually allowed to grow to a certain extent before their shifting towards the middle of the row. At the end of the season, growers replant in the rows by removing the initially planted mother plants to promote new daughter plants from the stolon, or they are used for large-scale multiplication by nurseries.

Strawberry growers in many parts of the world use plastic tunnels or greenhouses for indoor cultivation of strawberries. Low tunnels are built over strawberry rows to extend the fruiting season, besides providing protection from the environmental exigencies and also minimizing the severity of various pests and diseases. Recent studies also indicate a high potential for low tunnels to extend the fruiting season of day-neutral strawberries into late autumn or early winter (Petran et al., 2015).

High tunnels are also used widely for strawberry production, especially in Europe (Hancock and Simpson, 1995). High tunnels are costlier and much more difficult to construct than low tunnels, but under the suitable climatic conditions, they can be used for round-the-year production with appropriate technological interventions. Like low tunnels, temperature inside the high tunnels also rises, which helps in protecting the strawberries from vagaries of extreme weather conditions prevailing outside (Rowley 2010) in cool climatic regions. The polytunnels, in fact, provide almost similar conditions to those experienced in the greenhouse, but at a relatively low cost, so they are helpful in minimizing the difficulties of constructing large infrastructure.

Still, some growers around the world do use greenhouses for strawberry production. In Japan, growers plant strawberries on a coir/peat media mixture on elevated cultivation benches inside a greenhouse. It is a kind of automatic turn-key system wherein irrigation, nutrition, humidity and temperature control are fully computerized. Planting material is either procured from outdoor nurseries or runners are produced inside by placing the newly emerging stolons in small pots placed just below the elevated benches. This simple intervention saves the growers a good amount of money and also the risks of using infested material are avoided. The Japanese have been growing strawberries in protected conditions since 1910, by simply putting plastic covering over strawberries planted in concrete blocks placed on a hillside, although this could be considered primitive given the high-tech structures prevalent today (Oda and Kawata, 1993).

Among the recent advancements in indoor strawberry cultivation, the more modern greenhouse production system allows for round-the-year production, and the example of Minnesota model has immense potential to realize this (Rubinstein, 2015). In Europe, Dutch strawberry growers lead in greenhouse production. Dutch and Belgian growers use cultivars with dormancies that can be overcome using interruption lighting during night hours, thus avoiding the need to chill their plants and facilitating round-the-year production (Lieten, 2013). Use of growing media in protected environments is catching up in view of restrictions imposed on soil fumigants such as methyl bromide, chloropicrin etc. Soilless media such as coco-peat, perlite, rockwool and several others are being promoted but demand sound quality, yield and economic analysis. Efforts are on to standardise these alternative mixtures in open field environments too to avoid soil fumigants in strawberry cultivation (Wang et al., 2016). Coconut coir (100%) has shown initial promise in this regard, leading to comparable yields. European growers also tend to use a coir/peat mixture and stagger their plantings to get round-the-year production.

Harvesting of strawberries is done when berries are up to three-quarters coloured to store well or when fully ripe if they are to be sold sooner. Berries intended for value addition should be picked usually later. After harvest, the fruit are precooled to have a post-harvest life of five–nine days depending on how much ripe they were at the time of harvest and how they are stored in terms of temperature and other conditions.

Propagation via seed is quite rare, as the varying ploidy levels of different species can complicate sexual reproduction; asexual propagation is cheaper and straight forward. Japanese researchers are attempting sexual propagation as a possible method of eliminating viruses or other pathogens (Mauro et al., 2007). Standardization of the sexual method and its conversion into a grower-friendly approach may be a boon for subtropical strawberry growers, where annual procurement of planting materials at a higher rate is the major limitation in expanding this crop despite high production per unit area. Nowadays, strawberry propagation using plugs has become more popular (Hokanson et al., 2004). Plugs are grown using the runner tips or stolonate crowns from mother plants, grown either in the field or in a greenhouse (Rowley et al., 2010). The runners are harvested and immediately rooted in cell trays or used in tissue culture before transplanting into cell trays. With proper management, plugs become ready for transplanting within one month. Plugs can be mechanically planted more easily and require less water for root establishment, and do not suffer any root damage on being taken out from cells, unlike bare-rooted plants.

Modern strawberry breeding programmes are aimed to evolve genotypically superior varieties with resistance/tolerance to economically important pests and diseases, to excess or deficit soil moisture stress and to excessively hot or cold temperatures, better flowering habit and fruit quality and yield and amenability to mechanical harvesting practices.

Some of the outstanding achievements during the recent past are the development of everbearing and day-neutral varieties, those with exceptionally good fruit quality, longer shelf life etc. The recent development of white-skinned strawberries in Japan has opened a new vista for growers subject to consumer acceptance of a change from red-fruited varieties. Strawberry, being a rich source of phytonutrients, warrants development of genetically improved genotypes fortified with health and nutritional benefits (Mezzetti et al., 2016). Although significant research on developing improved varieties has been carried out and notable achievements made, modern biotechnological tools such as gene transfer and gene editing could beneficially be used in supplementing well-planned conventional breeding programmes for developing trait-specific and unique varieties.

Of late, organic farming of strawberries is being seen as an alternative production system due to environmental concerns and considerations on health related benefits. There is an absolute and immediate need to put in place vigorous cultivars with high degree of tolerance to economically important pests and diseases without compromising on typical fruit characteristics and yield of strawberries under the organic production system. The strawberry breeding programmes in many leading strawberry-producing countries such as Germany are expected to yield new cultivars amenable to organic production with low input costs besides being tolerant to root diseases. So far, as a result of sustained breeding efforts, three clones, viz., P-7189, P-8043 and P-8071, have been observed to be suitable for organic production and also exhibited tolerance to *Verticillium* wilt (Weissinger et al., 2014). Success in pollination has been recorded, and the proportion of fully pollinated flowers have been reported to be higher in organic farming due to pollinator abundance than in conventional farms and this difference becomes more evident a couple of years after conversion to organic farming. It has been suggested that conversion to organic farming may rapidly increase success in pollination, and hence benefit the ecosystem services through improved crop pollination with respect to both quantity and quality of fruits produced (Andersson et al., 2012).

Fruit yield in strawberry is very sensitive to weather conditions as environmental temperature and rainfall determine the intensity of all physiological processes. Climate change is posing increased risks for strawberry production. Extreme weather, such as unusually high and low temperatures, could cause significant losses in fruit yield. Strawberry fruit development is characterized by an uneven distribution of strawberry yield over the production season, which, in conjunction with the highly perishable fruit, often results in large fluctuations in price of fruits and result in volatile marketing.

Indoor production of strawberry is at risk, particularly due to requiring a lot of water. Hydroponic or aquaponic systems offer a possible solution but will need sustained efforts to develop technologies applicable to strawberries. Researchers from the University of Arizona and the University of Arkansas initiated working on fertigation techniques for hydroponic strawberries, based on Japanese practices (Kubota and Kroggel, 2013). Other indoor growing methods have used drip irrigation systems but often require multiple waterings and fertilizer applications per day (Lantz et al., 2010). Hydroponics could provide a way to significantly reduce water requirements in the climate change scenario that sees continually receding precipitation and water table globally. While hydroponic strawberry production occurs in several places around the world, aquaponic strawberry production is quite rare, and resources for growers are limited. Roosta and Afsharipoor (2012) advocated for the use of fish tank waste instead of fertigation. Both aquaponic or hydroponic systems make possible growing of day-neutral strawberry cultivars, and accordingly greenhouse conditions can be improvised. Especially, the cost of intensive lighting can be reduced as day-neutral strawberries have proven to be insensitive to photoperiod (Stewart and Folta, 2010). However, requisite temperature regime would be needed for proper growth and development in greenhouses. While a greenhouse is unlikely to reach 30°C during winter, ensuring greenhouse temperatures lower than 30°C during summer could be difficult. Therefore, evolving a cultivar having better tolerance to high day temperatures would be a welcome step for growers to overcome the problem of loss of fruit yield due to hot summers. Such a venture would pave the way for growing strawberries yielding fruits over an extended period during spring and autumn. Adjusting the minimum and maximum temperature conditions under which these new cultivars can produce flowers and fruits will help make round-the-year strawberry production a viable enterprise. As the demand for local food increases, round-the-year strawberry production is a niche area of research and development. Already, research is in progress in this direction.

While promoting indoor production, it is not necessary to alter the physical appearance and flavour of strawberries. However, a new strawberry cv. Pineberry has recently shown some popularity in Europe. Pineberry has the white colour of *F. chiloensis*, the wild Chilean strawberry, and is slightly smaller with a different flavour profile and sweetness (Osbourne, 2014). While interest in the Pineberry has so far been limited in the United States, it could be a cultivar to explore to increase interest in the strawberry industry. While developing a new, ideal, indoor strawberry cultivar, researchers should consider the Pineberry as well. Finally, the cultivar best suited for indoor production should have standard fruit size with relatively less variability. While most cultivars have a range of fruit size and shape, developing a standard would make indoor cultivation with mechanized intercultural operations possible and make the system economically more feasible. Several studies have already explored the possibility of using robots for fruit harvesting in raised-trough and indoor conditions primarily using the size and colour of fruits to determine stage of ripeness (Qingchun et al., 2012). In the UK, it has been reported that Green has developed a strawberry harvester that can perform picking of fruits faster than humans (King, 2017). Harper Adams University is considering setting up a spin-off company to commercialize this technology.

33.3 FUTURE STRATEGIES

Nevertheless, beyond the foregoing discussion, the future strategic research priorities in strawberry should concentrate on developing technical knowhow for efficient resource use, low-cost open/indoor cultivation, round-the-year production technologies, amenability to mechanization, suitability for processing, disease resistant and consumer-friendly cultivars with sustainable production for enhanced returns.

In areas where open cultivation is the common practice, there is the need to collect as much practical and scientific information as possible, which could be handy in preparing the first draft of the step-by-step transition from field to greenhouse cultivation of strawberries (Paparozzi, 2013). Further research is required to develop cultivars having characteristic good fruit size and reduced runner production (Yao et al., 2015) behaviour, for example, the cultivar "Florida Fortuna", with its ability to produce very early in the season, besides maintaining constant more fruit size and quality throughout the season despite a heavy crop load, is of great value for future breeding programmes (Tsormpatsidis et al., 2017). Biological control of economically important pests and protected cultivation can enable the growers to market the berries with branding as pesticide free.

It is seriously felt that no single measure will suffice to meet the challenges ahead. Rather, a multifaceted approach will be required, one that integrates advances in disease resistance through breeding with closer attention to the factors that influence the survival, activity and spread of pathogen populations in soil, and those that address the issues relating to labour and other input costs. Higher levels of genetic resistance to diseases and pests in strawberry cultivars can be achieved through breeding on one hand, and making available low-cost structures for protected cultivation coupled with judicious planning and market intelligence may facilitate addressing many issues concerning strawberry sector.

Of late, rooftop gardening and vertical gardening has become much-sought-after subjects both from environmental and economic points of view, more especially in and around large cities. Strawberry is a crop which can suffice both beauty and utility aspects of formal gardening on roof top or vertical gardening in residential or office premises. Breeding of varieties, cultivation techniques including customized fertilizer formulations and engineering interventions for providing

536 Strawberries

structures for these cultivation practices and many other issues related with rooftop or vertical gardening could not only be helpful in enhancing strawberry production but a source of employment generation in and around large cities. Research on devising technologies for utilization of city/municipal waste in strawberry cultivation could be an eyeopener for many other sectors.

Carbon footprints are growing concerns in all the energy intensive and excess water requiring production systems. These influence trade and policy matters. Our greater understanding of the technologies and practices of strawberry production management systems with minimum carbon footprints could serve as a guiding principles for many other sectors.

REFERENCES

Andersson, G.K.S., Rundlof, M. and Smith, H.G. 2012. Organic farming improves pollination success in strawberries. *PLoS ONE*, 7: e31599. doi:10.1371/ journal.pone.0031599

Carroll, J., Pritts, M. and Heidenreich, C. 2015. *Organic and IPM Guide for Strawberries*. Cornell University Cooperative Extension Publication, New York State Integrated Pest Magement Program. Ithaca, New York.

Hancock, J.F. and Simpson, D. 1995. Methods of extending the strawberry season in Europe. *HortTechnology*, 5: 286–290.

Hokanson, S.C., Takeda, F., Enns J. and Black, B.L. 2004. Influence of plant storage duration on strawberry runner tip viability and field performance. *HortScience*, 39: 1596–1600.

Hollender, C., Geretz, A., Slovin, J.P. and Liu, Z. 2011. Flower and early fruit development in a diploid strawberry, *Fragaria vesca. Planta*, 235: 1123–1139.

King, A. 2017. The future of agriculture. *Nature*, 544: 21–23.

Kubota, C. and Kroggel, M. 2013. Hydroponic strawberry information website. University of Arizona Controlled Environment Agriculture Center Publication. Website. http://cals.arizona.edu/strawberry/Hydroponic_Strawberry_Information_Website/Welcome.html

Lantz, W., Swartz, H., Demchack, K. and Frick, S. 2010. *Season-Long Strawberry Production with Ever Bearers for Northeastern Producers*. University of Maryland Publication, Lake Park, MD. EB 401.

Lieten, P. 2013. Advances in strawberry substrate culture during the last twenty years in the Netherlands and Belgium. *International Journal of Fruit Science*, 13: 84–90.

Martin, E. and Tepe, E. 2014. *Cold Climate Strawberry Farming. Department of Horticultural Science e-book*, University of Minnesota, St. Paul, Minnesota.

Mauro, T., Ito, Y., Ishikawa, M., Kamisoyama, D., Ohmameuda, M. and Ogura, Y. 2007. About the practical use of the seed propagation type strawberry (*F. x ananassa* Duch.) cultivation system (using June-bearing cultivar). *HortResearch*, 61: 73–78.

Mezzetti, B., Balducci, F., Capocasc, F., Zhong, C.F., Cappelletti, R., Di Vittori, L., Mazzoni, L., Giampieri, F. and Battino, M. 2016. Breeding strategy for higher phytochemicals content and claim It: Is it possible? *International Journal of Fruit Science*, 16 (S1): 194–206.

Oda, Y. and Kawata, K. 1993. Cement block wall strawberry cultivation of strawberry in Japan. *Acta Horticulturae*, 348: 219–226.

Osbourne, L. 2014. Rescued from extinction: The bizarre "pineberry" that's soon to grace our supermarket shelves. *Daily Mail*, online edition. www.dailymail.co.uk

Paparozzi, E. T. 2013. The Challenges of growing strawberries in the greenhouse. *HortTecnology*, 23: 800–802.

Petran, A., et al. 2015. *Low Tunnel Strawberry Blog*. University of Minnesota Commercial Fruit Publication. http://fruit.cfans.umn.edu/category/ strawberries/low-tunnel-strawberry

Qingchun, F., Xiu, W., Wengang, Z., Quan, Q. and Kai, J. 2012. A new strawberry harvesting robot for elevated-trough culture. *International Journal of Agricultural and Biological Engineering*, 5: 1–8.

Roosta, H.R. and Afsharipoor, S. 2012. Effects of different cultivation media on vegetative growth, ecophysiological traits and nutrient concentration in strawberry under hydroponic and aquaponic cultivation systems. *Advances in Environmental Biology*, 6: 543–555.

Rowley, D. 2010. Season extension of strawberry and raspberry production using high tunnels. Master's Thesis, Utah State University. http://digitalcommons. usu.edu/etd/716/.

Rowley, D., Black, B. and Drost, D. 2010. *Strawberry Plug Plant Production. Utah State University Cooperative Extension*, Horticulture Publication, Logan, UT.

Rubinstein, J. 2015. *Fragaria x ananassa: Past, Present and Future Production of the Modern Strawberry.* Retrieved from the University of Minnesota Digital Conservancy, http://hdl.handle.net/11299/175838.

Challenges, Potential and Future Strategies

Stewart, P. and Folta, K. 2010. A review of photoperiodic flowering research in strawberry (*Fragaria* spp.). *Critical Reviews in Plant Sciences*, 29(1): 1–13.

Tsormpatsidis, E., Vysini, E. and Papanikolopoulos, T. 2017. Challenges and opportunities for early strawberry production in Greece: incorporating new cultivars in the production system. *Acta Horticulturae*, 1156: 915–920.

Wang, D., Gabriel, M.Z., Legard, D. and Sjulin, T. 2016. Characteristics of growing mixes and application for open-field production of strawberry. *Scientia Horticulturae*, 198: 294–303.

Weissinger, H., Flachowsky, H. and Spornberger, A. 2014. New strawberry genotypes tested for organic production on a *Verticillium*-infested site. *Horticultural Science* (Prague), 41: 167–174.

Yao S., Guldan, S., Flynn, R. and Ochoa, C. 2015. Challenges of strawberry production in high-pH soil at high elevation in the Southwestern United States. *HortScience*, 50: 254–258.

Index

A

Abscisic acid, 342, 343, 349
Acadie (Glooscap × Guardian), 85
Adenosine triphosphate (ATP), 214
Adina (88-042-35 × Parker), 85
Adria (Gramda × Miss), 85
Advanced berry ripening, 368
Aerated flow technique, 379
Aeration, 378, 379
Aeroponics system, 379
Agrobacterium-mediated gene transfer mechanism, 131–135
Agro-ecosystem, 499
Agrotis ipsilon, 493, 494
Aguedilla (Camarosa × 67.35), 86
Albinism, 446–447
Alessandra (Director P. Wallbaum × Pocahontas), 86
Alice, 86
Alinta (88-011-30 × Chandler), 86
Amelia, 86
AMF, *see* Arbuscular mycorrhizal fungi
2,4-D amine (Formula 40, Amine 4), 284
Amino acids, 340
Anaerobic soil disinfestation (ASD), 176
Anagallis arvensis, 271
Anaheim (Irvine × Cal 85.92-602), 86
Androgenesis, 125
Annapolis [K74-5(Micmac × Raritan) × Earligow], 86–87
Anther and pollen culture, 125–126
Anthracnose, 456–458
Anti-transpirant, 256–257
Aphis gossypii, 488
Apis cerana, 325, 327
Apis mellifera, 325, 327, 328
Aquaculture, 374
Arbuscular mycorrhizal fungi (AMF), 220
Ariel, 87
ASD, *see* Anaerobic soil disinfestation
ATP, *see* Adenosine triphosphate
Auxins, 303, 342, 349, 352
Available water, 231

B

Baotong (Benihomalei × Camarosa), 87
Barak (Tamir × Aro 730), 87
Barrel and pyramid system, 199
Bed renovation, 276–277
Bell raugex (77-36 × Morioka 19) × (Morioka 19), 87
Belrubi (Pocahontas × Redcoat), 87, 88
Benicia (Palomar × Cal 0.18-601), 87, 88
Big eyed bugs, 501, 503
Biochemical markers (isozymes), 142–143
Biofumigation, 176
Biological control, weed management, 280

Biotic and abiotic stresses, 253–255
    drought, 253–254
    flooding, 254
    pathogens, 255
    salinity, 254
    weeds, 254
Black root rot, 461–462
Blomidon [K72-4 × {Micmac × (Guardsman × Tioga)}], 87–88
Bob (1501 × 1242), 88
Bohemia strawberry *(Fragaria major sterilisseubohemica),* 2, 37
Bolero (Redgauntlet, Gorella, Cordinal and Selva), 88
Boron, 215
Botrytis cinerea, 326, 413, 416, 419, 424, 464, 466
Bountiful (Linn × Totem), 88
Breeding and improvement, 49–72
    cytogenetics, 58–61
    floral biology, 62
    genetics, 52–54
    history and evolution, 50–51
    methods, 62–66
    objectives, 66–72
    quantitative genetics, 55–58
Breeding methods, 62–66
    apomixis, 66
    hybridization, 63–66
        intergeneric, 65–66
        interspecific, 64–65
    introductions and selection, 63
    mutagenesis, 66
Breeding objectives, 66–72
    adaptability to soil factors, 66–67
    environmental factors, 67
    plant characters, 67–72
        disease resistance, 70–72
        fruiting habit, 68
        fruit size, 68–69
        pest resistance, 69–70
        quality, 68–69, 70
        ripening time and fruit development, 68
        stolon production, 67–68
        variegation, 68
        yield, 68–69
Bronzing, 450
Brunswick (Cavendish × Honeoye), 88–89
Bur clover *(Medicago polymorpha),* 11, 269

C

Cabot (K87-5 × K86-19), 89
Calcium, 214
Calcium chloride (CaCl$_2$), 413, 414, 449
Calypso (Rapella × Selva), 89
Camarosa (Douglas × Cal 85.218-605), 89
Cambridge Favourite [(Etter seedling × Avant tout) × Blackmore], 90

539

540 Index

Cambridge prizewinner (Early Cambridge × Howard 17), 89–90
Cambridge Rival (Dorsette × Early Cambridge), 90
Cannabis sativa, 271
Capillary water, 231
Capitola (Cal 75.121-101 × Rarker), 90
Capri (Civri-30 × R6R1-26), 90–91
Carbon footprints, 536
Carisma (Oso Grande × Vilanova), 91
Carmine (Rose Linda × FL 93-53), 91
Carthamus oxyacantha, 270
Cat facing, berries, 491, 492
Catskill (Marshall × Premier), 91
Cavalier (Valentine × Sparkle), 91–92
Cellular organelles, 338
Cell wall constituents, 337–338
Chaetosiphon fragaefolii, 488, 489
Chambly (Sparkle × Honeoye), 92
Chandler (Douglas × Cal 72.361-105), 92
Chemical control, weed management, 281
Chilean strawberry *(F. chiloensis),* 4–5, 38
Chilling treatment, 310
Chlorine dioxide (ClO_2), 418–419
Chunxing (Chunxiang × Haiguan Zaohony), 92
Chunxu (Chunxiang and a Polish cultivar), 93
Chutney
    ingredients, 436
    preservation tips, 436
Citric acid, 339, 433
CKs, *see* Cytokinins
Clethodim (Arrow®, Prism®, Select®), 283
Climatic requirements, 161–165
    light, 164
    ozone exposure, 164–165
    photoperiod, 163, 164
    temperature, 161–163, 164
Coccinella septempunctata, 489, 490
Cocklebur *(Xanthium strumarium),* 272
Coconut coir, 533
Coir, media, 173
Coleopteran insect pests
    strawberry bud weevil, 495
    strawberry crown borer, 496
    white grubs, 496
Colletotrichum acutatum, 456, 457, 458
Colletotrichum fragariae, 456, 457
Colletotrichum gloeosporioides, 456, 458, 459
Colletotrichum nymphaeae, 457, 458
Coloured mulch, 294
Colour effects, 294
Combridge vigour, 90
Contact herbicide, 281
Container and hydroponic planting, 200
Copper, 216
Cristina (Selection AN01.211.51), 92
Critical moisture point, *see* Wilting coefficient
Crop improvement and varietal wealth, 5–6
Crown gall *(A. tumefaciens),* 131, 133
Crown rot *(Phytophthora cactorum),* 175
Crusader [Auchincruive SDLG. 50 × 5 {SDLG. 6M17 (Temple selfed × SDLG.(Aberdeen open)} × Cambridge Vigour], 93
Cuesta [Seascape × Cal 83.25-2 (Fern × Parker)], 93
Cultivars

country-specific suitability, 404
    high quality and traits, country-specific, 406
    high-yielding, country-specific, 405
Cultivated strawberries, 531
Cultural controls, weed management, 278–279
    crop competition, 279
    crop rotation, 278–279
    mulches, 279
    sanitation and prevention, 278
    site selection, 278
Cupcake (102213 × 1581), 93
Cycocel, 351
Cyperus spp, 272
Cyprodinil, 510
Cytogenetics, 58–61
    chromosome doubling, 61
    chromosome number, 61
Cytokinins (CKs), 304, 342, 349, 351, 388

D

Dange (Venta × Redgauntlet), 93
Day-neutral plants, 308
Day-neutral strawberries, 532
DCPA (Dacthal®), 283
Deblossoming, 361
Decision support systems, 252–253
Deep flow technique, 378–379
Defoliation, 360
Delia (Honeoye × ISF 80-52-1), 93
Dely (T2-6 × A20–17), 93
Dichasial cyme/dichasium, 34
Dichasium, 301
Donna (Fern seeding × Douglas), 94
Douglas (Tufts × 64-57-108 (Tioga × Sequoia)), 94
Dried calyx, 450
Drip irrigation technique, 380–381
Drip/trickle irrigation, 245–252
    precision irrigation system design, 248–252
    variation factors, water uniformity, 247–248
Dris Straw Eighteen (91J302 × 26H165), 94
Dris Straw Nineteen (Driscoll Atlantis × 43j313), 95
Dris Straw Twenty (2K297 × Driscoll Ojai), 95
Dris Straw Twenty Eight (95L299 × 251M27), 96
Dris Straw Twenty Five (18L33 × 192M122), 96
Dris Straw Twenty Four (3M44 × 50L174), 95
Dris Straw Twenty One (13H377 × 587L48), 95
Dris Straw Twenty Seven (Dris Straw Eight × 10L297), 96
Dris Straw Twenty Six (18L33 × 193M68), 96
Dris Straw Twenty Three (1M16 × 87K286), 95
Dris Straw Twenty Two (Dris Straw Three × 50L206), 95

E

Earlibrite (Rosa Linda × FL 90-38), 96
Earlidawn (Midland × Tennessee Shipper), 96
Early Bommel (New Dutch Seedling No. 69-16), 97
Electrofusion, 123–124
Elegance (EM834 × EM1033), 97
Elista [(Jucunda × Self) × US-3763], 97
Elvira (Vola × Gorella), 97
Enzed Levin (Orion) (84-1-38 × 84-8-443), 97
Erie (Sparkle × Premier), 97
Escapes, weeds, 274

Index

EST, *see* Expressed sequence tags
Ethylene, 343, 349, 350
Ethylene evolution rate, 341
Etna (Marlate × Belrubi), 98
Evangeline (K 88-4 × NYUS 119), 98
Everbearers/long-day plants, 307
Everest (Irvine × Evita), 98
Evita [Chandler × (Gorella × Brighton)], 98
Expressed sequence tags (EST), 147

F

FAO, *see* Food and Agricultural Organization
Farmyard manure (FYM), 213, 216, 220
Fengguan 1 (Toyonoka × Tochitome), 98
Fenhexamid, 510, 512, 513, 515
Fern (Tufts × Cal 69.63-103), 98, 99
Ferrara (MD. US4083 × Holiday), 98
Fertigation, 218–219, 250
Fertilizer use efficiency (FUE), 218
Festival (Rosa Linda × Oso Grande), 98–99
Field propagation, runners, 180–182
    cold-stored plants (frigo plants), 181–182
    fresh bare-rooted plants, 180–181
    waiting bed plants, 181, 182
Filaree *(Erodium cicutarium)*, 11, 269, 277
Fineleaf fumitory/ Indian fumitory *(Fumaria parviflora)*, 272
FL 05-107 OR Winterstar™ (Florida Radiance × Earlibrite), 99
Flair (Flevo 00-24-7 × Flevo 00-08-4), 99
Flame defoliation, 360
Flavorfest (B759 × B786), 99–100
Floral biology, 62
Florence [Tioga × (Redgauhtlet × (Weltguard × Gorella)] × (Providence × Self), 100
Florida Radiance (Winter Dawn × FL 99-35), 100
Flower biology, 321
Flowering, strawberry species, 12
    genetic constitution varieties
        day-neutrals, 308
        everbearers/long-day plants, 307
        flowering habits, 309
        junebearer/short-day plants, 306
        remontant strawberries, 308
    light influences, 309–310
    light × temperature conditions, 310–312
    molecular insights, 313–315
    photoperiod, 310
    physiology
        auxins, 303
        cytokinines, 304
        gibberellins, 303
        nutrition, 305–306
        primary metabolites, 304–305
    plant age, 313
    season and environmental factors, 312
    temperature conditions, 310
Flumloxazin (Chateau®), 284
F. nilgerrensis, 52, 53, 68
Fog feed technique, 381
Foliar nutrition, 217–218
Food and Agricultural Organization (FAO), 507, 508
Foreign strawberry *(Fragaria major pannonica)*, 2, 37

Frankliniella occidentalis, 492–493
Free water, 231
Frost and disease control, 367–368
Frozen strawberries
    flowsheet, 438
    ingredients, 438
    preservation tips, 438
Fruit development
    achenes, 341–342
    cellular changes
        cellular organelles, 338
        cell wall constituents, 337–338
    climate changes, 341
    compositional changes
        amino acids, 340
        enzymes, 338
        organic acids, 339
        phenolic substances, 339–340
        pigments, 338
        sugars, 339
        volatile constituents, 340
    fruit position, inflorescence, 342
    growth, 336
    irrigation and nutrition, 341
    physiological changes
        ethylene evolution rate, 341
        respiration rate, 340–341
    plant bioregulators, 342–343
    ripening, 336–337
Fruiting, strawberry species, 12–13
Fruit quality, 23–24
    genetic diversity, 25–26
Fruit yields, 532, 534
FUE, *see* Fertilizer use efficiency
Functional genomics, 151–153
Furrow irrigation, 244
Fusarium wilt, 468–469
F. viridis, 4, 39, 52, 53, 59
FYM, *see* Farmyard manure

G

GA₃, *see* Gibberellic acid
Galletta (NCH 87-22 × Earliglow), 100
Gardena (Addie × Pajaro), 100
GAs, *see* Gibberellins
Gas chromatography (GC), 511
GC, *see* Gas chromatography
GCA, *see* General combining ability
GCV, *see* Genotypic coefficient of variation
Gea [R66/60{(Senga Sengana × Valentine) × Surprise des Halls} × MD US38], 100
Gem (Brilliant and Superfection), 100
General combining ability (GCA), 56, 57, 58, 71
Genetic mapping, 147–151
    genome mapping, 151
    linkage mapping, 147–149
    QTL mapping, 149–150
Genetic transformation, 131–135
Genome mapping, 72, 151
Genotypic coefficient of variation (GCV), 56
Geographic Information System (GIS), 503
Gibberellic acid (GA₃), 350, 352
Gibberellins (GAs), 303, 342, 349

542
Index

Ginza (1929 × 1902), 101
GIS, *see* Geographic Information System
Glory (Nelson. PS-1269 × PS-2286), 101
Gorella (Juspa × US-3763), 101
Grading, strawberries, 413
Gravitational water, 231
Greasy cut worm, 493
Greenhouses, 533
Greenhouse whitefly, 491–492
Grenadier (Valentine × Fairfax), 101
Grey mould (*Botrytis cinerea*), 220, 296
Grey mould rot, 424, 464–468
Growing medium, 173
Gurdsman (Claribel × Sparkle), 101
Gynogenesis, 125

H

Hairy root (*A. rhizogenes*), 131
Haruyoi (Harunoka × Hokowase), 102
Harvesting
  manual harvesting, 399–400
  mechanical harvesting
    night-time picking assistance, 400
    picking machine, 400–401
    picking robot, 401
Hedge row system, 199
Herbicidal injury, 282–283
  precautions, 283
Herriot (NYUS299 × Winona), 102
High-pressure liquid chromatography-mass spectrometry
    (HPLC-MS/MS), 511
Hill row system, 197
Honeoye (Vibrant × Holiday), 102
Honey bees; *see also Apis cerana; Apis mellifera*
  floral activity, 327–328
  as pollinators, 325, 326, 327
Hongjia, 102
Hongshimei (Zhangji × Ducra), 102
Hongxiutianxiang (SEL. 06-37-2) (Camarosa ×
    Benihoppe), 102
Horsetail (*Equisetum arvense*), 285–286
Horseweed (*Conyza canadensis*), 272
Howard 17, 50
HPLC-MS/MS, *see* High-pressure liquid chromatography-
    mass spectrometry
Huxley (Ettersburg 80), 102
Hydro-cooling method, 412
Hydroponic strawberries, 534
Hydroponic system, 374, 378, 379
Hygroscopic water, 231

I

Independence [Orus 850-48(Linn × Orus 3727) ×
    Orus 750-1 (Totem × Orus 3746)], 103
Indian goosegrass (*Eleusine indica*), 273
Indoor strawberry cultivation, 533
Induction, ovary culture, 125
Inflorescence, 33
Inorganic mulches, 291–293
  black polyethylene mulch, 291–292
  infrared-transmitting mulches, 293
  red-coloured polyethylene mulch, 293

white, coextruded white-on-black/silver reflecting
    mulch, 293
white polyethylene mulch, 292–293
Integrated disease management
  anthracnose, 456–458
  black root rot, 461–462
  Fusarium wilt, 468–469
  grey mould rot, 464–468
  leaf blight, 472
  leaf scorch, 472–473
  leaf spots
    angular leaf spot, 471–472
    mycosphaerella leaf spot, 469–471
    Pestaliopsis leaf spot, 471
    yellow leaf spot, 471
  leather rot, 463–464
  powdery mildew, 454–456
  red stele, 462–463
  rhizopus rot, 464
  Verticillium wilt, 459–461
  viruses, 473–474
Integrated insect pest management
  coleopteran insect pests
    strawberry bud weevil, 495
    strawberry crown borer, 496
    white grubs, 496
  future strategies, 500–503
  integrated pest management (IPM), 499–500
  lepidopteran insect-pests
    greasy cut worm, 493
    strawberry leaf folder, 494–495
    tobacco caterpillars, 493
  mammalian and frugivorous bird pests, 498–499
  soil arthropods
    snails and dugs, 497
    termites, 496–497
  sucking insect-pests
    greenhouse whitefly, 491–492
    green peach aphid (*Myzus persicae*), 488–490
    Lygus bug, 491–492
    melon aphid (*Aphis gossypii*), 488
    spider mites, 490
    strawberry aphids (*Chaetosiphon fragaefolii*),
        488, 489
    strawberry thrips (*Frankliniella occidentalis*),
        492–493
Integrated pest management (IPM), 499–500
Intercropping, 362
In vitro culture, 121
IPM, *see* Integrated pest management
Iprodione, 467, 512, 515
Iron, 215–216
Irrigation, 171
Irrigation scheduling factors, 232–244
  amount of, 235–236
  bed capacity, 240
  climatic parameters, 241
  criteria for, 233
  critical time of, 234–235
  depth of, 234
  frequency of, 234
  growing conditions effect, 237–238
  plant parameters, 240–241, 242
  soil parameters, 238–240

Index **543**

stages of, 235
time of, 236
varietal response to, 236–237
Irrigation systems, 244–252
drip irrigation, 245–252
furrow irrigation, 244
pressurized/micro-irrigation systems, 244–245
sprinkler irrigation, 245

J

Jams
ingredients, 434
preservation tips, 434
Jewel (NY1221 × Holiday), 103
Jingchunxiang {(01-12-15′ (advanced selection) ×
Kinuama)} SYN. SEL. 06-54-2, 103
Jingchunxiang {(01-12-15 (advanced selection) ×
Benihoppe)} SYN. SEL. 06-56-6, 103
Jingyuxiang (Camarosa × Benihoppe) SYN. SEL.
06-37-8, 103
Jingyuxiang (Camarosa × Benihoppe) SYN. SEL.
06-37-10, 103
Joliette (Jewel × SJ 85189), 104
Joly (T2-6 × A20-17), 103
Juice
ingredients, 433
preservation tips, 433
Jukhyang (Redpearl × Maehyang), 104
Junebearer/short-day plants, 306
June-bearing strawberries, *see* Short-day strawberries
June yellows, 448

K

Kaorino (Line 0028401 × Line 0023001), 104
Kentaro (Kita-Ekubo × Toyonoka), 104
Kimberley (Gorella × Chandler), 104
Kita-Ekubo [59 KOU-13-37(Aito × Morioko 19)
× Reiko], 105
Kyowa technique, *see* Aerated flow technique

L

Laguna (Irvine × Cal 85.92-602), 105
Lambada [IVT76013 (Silvetta × Holiday) × IVT74112
(Karina × Primella)], 105
Laurel (Allstar × Cavendish), 105
L'authentique Orleans (AC-L'acadie × Joliette), 105
LDPE, *see* Low-density polyethylene
Leaf spots
angular leaf spot, 471–472
mycosphaerella leaf spot, 469–471
Pestaliopsis leaf spot, 471
yellow leaf spot, 471
Leather rot, 463–464
Lepidopteran insect-pests
greasy cut worm, 493
strawberry leaf folder, 494–495
tobacco caterpillars, 493
Limit of detection (LOD), 511
Lincoln (Donna × 85-24-1), 105
Linear low density polyethylene (LLDPE), 291
Linkage mapping, 147–149

Little mallow *(Malva parviflora)*, 11, 269, 277
LLDPE, *see* Linear low density polyethylene
LOD, *see* Limit of detection
Louise (Ettersburg 80 Selfed), 105
Low-density polyethylene (LDPE), 421
Low-temperature injury, 448
Lygus bug, 491–492

M

Maehyang (Trochinomine × Akihime), 106
Magnesium, 214
Mahabaleshwar Strawberry, 4
Malformed berries, 445–446
Malwina (Unnamed Seedling × Sophie), 106
Mammalian and frugivorous bird pests, 498–499
Manganese, 216
MAP, *see* Modified atmosphere packaging
Mari (Pocahontas × Lihama), 106
Marina (Cartuno × Camarosa), 106
Marker assisted selection (MAS), 150
Marketing, 523–529
direct marketing channel, 527
efficiency, different channels, 529
price pattern, 527–528
retail channel, 527
wholesale channel, 527
Marsh cupweed *(Gnaphalium uliginosum)*, 270
Marvel (PS-1269 × PS-2280), 106
MAS, *see* Marker assisted selection
Matted row system, 198–199
Maturity and harvesting, 411–412
Maximum residue limits (MRL), 508–510
MCP, *see* 1-Methylcyclopropene
Mechanical harvesting
night-time picking assistance, 400
picking machine, 400–401
picking robot, 401
Mechanical suppression, weed control, 279–280
hand weeding, 280
mowing, 280
tillage, 279–280
MeJA, *see* Methyl jasmonate
Melilotus indica, 270
Meristem/tissue culture stages, 126–127
culture initiation, 126
explants, identification and preparation, 126
multiplication, 126
plantlets acclimatization, 127
rooting of shoots, 126–127
Merit (PS-2880 × PS-4630), 106
Merton Princess, 106
1-Methylcyclopropene (MCP), 416
Methyl jasmonate (MeJA), 417
Micro-cuttings, 127
Micropropagation, 7, 122–123, 184–187
acclimatization, 187
bioreactors for, 187
Microsatellite markers (SSRs), 146–147
Milsei (Parker × Chandler), 106–107
Mira (Scoot × Honeoye), 107
Missionary, 107
MNUS 210 [Earligrow × MNUS 52] SYN. Winona TM, 107
Modified atmosphere packaging (MAP), 420

544

Mojave (Palomar × Cal 1.57-601), 107
Molecular markers, 72, 128, 141, 143–147
Morphological markers, 142
    AFLP, 145
    ISSR, 145–146
    microsatellite markers (SSRs), 146–147
    RAPD, 143–145
    RFLP, 143
MRL, *see* Maximum residue limits
Mulching process, 10, 255–256, 289–297, 294–297
    colour effects, 294
    fruit yield and quality, 295–296
    growth and earliness, 294–295
    ideal mulch materials, 256
    inorganic mulches, 291–293
    mulching effects, 294–297
    mulching period, 297
    mulch removal, 297
    organic mulches, 255, 290–291
    plant protection, 296
    plastic mulches, 255–256
    soil moisture conservation, 295
    weed management, 297
Muriate of potash (KCl) fertilizer, 10
Musky strawberry *(Fragaria moschata)*, 2, 37
Myzus persicae, 488–490

N

NAA, *see* Naphthalene acetic acid
Nabila (Ventanax Q6Q8-26), 107–108
Naphthalene acetic acid (NAA), 350, 352, 353
Napropamide (Devrinol®), 283
Nectar production, 323–324
Neoseiulus californicus, 490, 502
Niigata S3 (KEI 812 (Seed Parent) × Asuka-Ruby), 108
Ningyu (Sachinoka × Akihime), 108
Nitrogen fertilizer, 210, 212–213, 305
Non-selective herbicide, 281
Nubbins/button berries, 445–446
Nutlets (achenes), 2
Nutrient film technique, 380
Nutrient solution
    management
        electrical conductivity (EC) management, 391–392
        oxygenation, 392–393
        pH regulation, 391
        temperature control, 392
    preparation
        chemical compounds/fertilizers, 383
        composition, 382, 383–385
        water quality, 382–383
Nutrition, 209–222
    fertigation, 218–219
    nutrients, 210–216
        boron, 215
        calcium, 214
        copper, 216
        iron, 215–216
        magnesium, 214
        manganese, 216
        nitrogen, 210, 212–213
        phosphorus, 213
        potassium, 213

        recommendations, 216–217
        zinc, 214–215
    organic production, 219–222

O

Octoploid species, flower types, 2
Oka (K 75-13 × Kentville), 108
Ontaria, 109
Open field cultivation, 532
Opines, 133
Organic mulches, 290–291
Organochlorine pesticides, 508, 512, 514
Organophosphorus pesticides, 508, 512, 514
Oso Grande [Parker (U.S. Plant PAT. NO. 5263) × Cal 77.3-603], 109
Oxalis corniculata, 271
Oxalis *syn.* wood-sorrel *(Oxalis stricta)*, 285
Ozone exposure, 164–165

P

Packaging, strawberries, 420–421
Pajaro (Cal 63.7-101 × Sequoia), 109
PAL, *see* Phenylalanine ammonialyase
Palomar (Camino Real × Ventana), 109
Paraquat (Gramoxone Inteon®), 284
Parthenium hysterophorus, 271
Parthenocarpy, 352
Pavana, 109
PCV, *see* Phenotypic coefficient of variation
Pectin methylesterase, 338
Pedicel, 301, 302, 432
Peduncle, 301, 302
Pegasus (Redgauntlet × Gorella) SYN. ES608, 110
Pelargonic acid (Scythe®), 284
Pendimethalin (Prowl H20®), 284
Pesticide residues
    decontamination
        cooking/boiling/juicing effect, 515
        harvesting precautions, 515
        washing effect, 513–514
    future strategies, 516
    maximum residue limit, 508–510
    monitoring trials, 510–511
        monitoring studies result, 512–513
        multiresidue methods, 511–512
    supervised trials, 510
    toxicological considerations, 507–508
Phenotypic coefficient of variation (PCV), 56
Phenylalanine ammonialyase (PAL), 338
Phosphorus, 213
Photoperiod, 163
Phyllody, 447
Physiological disorders
    albinism, 446–447
    bronzing, 450
    dried calyx, 450
    june yellows, 448
    low-temperature injury, 448
    malformed berries, 445–446
    nubbins/button berries, 445–446
    phyllody, 447
    salts injury, 449–450

Index

545

sunscald, 448
   tip burn, 448–449
   uneven ripening, 449
Phytophthora cactorum, 377, 463, 464
Phytoseiulus persimilis, 490
Picking assistant, 400; *see also* Mechanical harvesting,
      picking machine
Picking robots
   components, 401
   limitation, 401
Pick your own (PYO) harvesting, 399
Pinching, 359–360
Pineberry, 535
Pink Beauty and Pretty Beauty (Kinuama ×
      Pink Panda), 110
Pink Panda (*F. Comarum* Hybrid), 110
Pinnacle (Laguna × Orus 1267-250), 110
Planasa 02-32 (94-020 × 9719), 110
Plant bioregulators, 342–343, 350, 352
Plant bio-regulators uses, 349–353
   earliness, 352
   flowering, 351–352
   fruit quality, 353
   parthenocarpy, 352
   plant growth, 350–351
   runner production, 351
   yield, 352–353
Plant characters, 67–72
   disease resistance, 70–72
   fruiting habit, 68
   fruit size, 68–69
   pest resistance, 69–70
   quality, 68–69, 70
   ripening time and fruit development, 68
   stolon production, 67–68
   variegation, 68
   yield, 68–69
Planting, strawberry, 193–205
   considerations, 193–194
   density, 203–205
   plan, 193, 196–200
   system, 197–200
   time, 201–203
Planting system, 197–200
   barrel and pyramid system, 199
   container and hydroponic planting, 200
   hedge row system, 199
   hill row, 197
   matted row, 198–199
   ribbon row, 199
   spaced row, 199
   strawberry plasticulture planting, 198
   vertical system, 199–200
Plant thinning, 361
Plant tissue testing, 210, 211, 212
Plant type, 194–196
Plasmodesmata, 123
Pollen production, 323–324
Pollination
   honey bees, floral activity, 327–328
   management practices, 328
   modes of, 324
   requirements, 321–328
      nectar production, 323–324

pollen production, 323–324
   pollinators, 324–327
   stigma receptivity, 323–324
Pollinators, 324–327
Polyethylene, 369, 391
Polyvinyl chloride (PVC), 245
Post-emergence herbicide, 281
Post-harvest defoliation, 360
Post-harvest treatments
   biocontrol, 419
   calcium role, 413–414
   chlorine dioxide (ClO_2), 418–419
   decay control, biocontrol agents, 420
   diseases, 423–424
   heat treatments, 416
   1-methylcyclopropene, 416
   oxygen, 416–417
   ozone, 417
   plant bioregulators, 417
   quality and shelf life improvement, 415
   titanium dioxide (TiO_2), 419
   ultrasound technology, 417–418
   ultraviolet irradiation, 418
   waxing/edible coating, 414
Potassium, 213
Powdery mildew, 454–456
Preconditioning, 360
Precooling, 412–413
Predatory mites, 502, 503
Pre-emergence herbicide, 281
Premier (Crescent × Howard No 1), 111
Preplant herbicide, 281
Pressurized/micro-irrigation systems, 244–245
Production economics
   annual strawberry production, 523, 526
   cost and return summary, 523, 524, 525
   establishment cost, 520–521
      fixed cost, 521
      labour and material cost, 520
      operating cost, 520
   picking year costs, 521–523
      fixed costs, 523
      labour and material costs, 522–523
Prompt cooling, 423
Propagation methods, 179–187
   sexual propagation, 179
   vegetative propagation, 179–187
      factors affecting runner production, 183–184
      field propagation, 180–182
      micropropagation, 184–187
      substrate culture, 182–183
Protected cultivation, 365–370
   advantages
      advanced berry ripening, 368
      frost and disease control, 367–368
      plant growth, 368
      runner production, 369, 370
      yield and quality improvement, 368–369
   cultivars for, 367
Protoplast culture, 123–125
Protoplast fusion techniques
   electrofusion, 123–124
   polyethylene glycol (PEG)-mediated fusion, 123
Puree

546   Index

flowsheet, 438
ingredients, 438
preservation tips, 438
PVC, *see* Polyvinyl chloride
PYO, *see* Pick your own harvesting
Pyrimethanil, 510, 515

Q

QTL, *see* Quantitative Trait Loci
QTL mapping, 149–150
Quality planting material types, 7
Quantitative genetics, 55–58
coefficient of variation (CV), 55–56
combining ability, 56–58
correlations, 58, 59
genetic advance (GA), 55–56
heritability (h^2), 55–56
Quantitative Trait Loci (QTL), 6
Quick Easy Cheap Effective Rugged and Safe, 511, 512

R

Raised bed solarization, 464
Rania (Ventanax Q6Q8-26), 111
Ratooning, 362
Rebecka [Fern × *F. Vesca*) × *F. Ananassa* F 861502], 111
Receptacle, 34
Red Coat (Sparkle × Valentine), 111
Redgauntlet (New Jersey 1051 × Auch. Climax), 111
Red Rich (Red-Glo and Hagerstrom's Everbearing), 111
Red spider mite, 507
Red stele, 462–463
Regeneration, ovary culture, 125
Relative irrigation supply (RIS), 248, 249
Remontant strawberries, 308
Renovation (renewing the planting), 361–362
Residual herbicide, 281
Respiration rate, 340–341, 423
Rhizopus rot, 424, 464
Rhizopus stolonifer, 419, 424, 464
Ribbon row system, 199
RIS, *see* Relative irrigation supply
Romina (Selection AN99, 78, 51), 111–112
Root-inducing plasmids *(pRi)*, 133
Root mist technique, 381
Rosalyne [Fern × (SJ9661-1 × Pink Panda)], 112
Roseberry [(Raritan × K74-12) × (SJ9616-1 × Pink Panda)], 112
Rosettes, 275
Rosie (Honeoye × Forli), 112
Roxana [Surprise Des Halles (Market Surprise) × Senga Sengana, 112

S

Sable (Veestar × Cavendish), 112
Sabrina (9719 × 94-020), 112
Saint-laurent D'orleans [L Acadie × (SJ 8916 × Pink Panda)], 113
Saint Pierre (Chandler × Jewel), 112–113
Salicylic acid, 417
Salla (Addie × Pajaro), 113
Santaclara (Carisma × Plasirfre), 113

Sauce
flowsheet, 437
ingredients, 437
preservation tips, 437
Saulene (Shuksan × Senga Sengana), 113
SCA, *see* Specific combining ability
Scouting, weed, 273–274
SCV, *see* Strawberry crinkle virus
SDCD, *see* Strawberry dried calyx disorder
Seasonal flowering locus (SFL), 313, 314
Selective herbicide, 281
Selva (Cal 70.3-117 × Cal 71.98-605), 113
Senga Sengana (Markee × Sieger), 113
Sethoxydim (Poast®), 283
Sexual propagation, 179
SFL, *see* Seasonal flowering locus
Shimei 3 (183-2 × All Star), 113
Shimei 4 (Hokowase × Shimei 1), 113
Shimei-7 (Tochiotome × Allstar), 114
Short-day strawberries, 532
Single sequence repeat (SSR) markers, 129
Site selection, 169–170
Soil arthropods
snails and dugs, 497
termites, 496–497
Soil biosolarization, 7
Soil fertility, 172–173
Soil fumigants, 533
Soil fumigation, 277
Soilless culture
advantages
crop per year increment, 376
labour requirement reduction, 375
nutrient solution, precise use, 375
productivity enhancement, 375
root environment control, 375
soil problem suitability, 376
substrates sterilization practices, 375
water economy, 375
factors influencing
anion competition, 390
electrical conductivity (EC), nutrient solution, 386
nitrate update and concentration, 389
nutrient medium composition, 383–385
pH, nutrient solution, 385
reaction (pH), nutrient solution, 389
root aeration, 390
salinity, 390
substrate, 390
temperature, nutrient solution, 387–388
limitations
disease infections, high risk, 376
diverse technical skill requirements, 376
high initial investment, 376
nutrient solution
management, 390–393
preparation, 382–383
substrates, 381–382
suitable cultivars, 381
system, 376–378
techniques
aerated flow technique, 379
aeroponics system, 379
deep flow technique, 378–379

Index 547

drip irrigation technique, 380–381
ebb and flow technique/flood and drain technique, 378
fog feed technique, 381
nutrient film technique, 380
root mist technique, 381
Soilless media, 533
Soil moisture constants, water management, 231–232
available soil moisture, 232
evapotranspiration, 232
field capacity, 231
maximum water holding capacity, 231
transpiration, water loss, 232
water requirement, 232
wilting point, 231–232
Soil preparation, 173–176
Soil solarization, 173, 278
Soil type, 170–171
Somaclonal variation, 127–129
Somatic embryogenesis, 127
Sophie (NY 1261 × Holiday), 114
Spaced row system, 199
Sparkle (Fairfax × Aberdeen), 114
Special cultural practices
deblossoming, 361
defoliation, 360
intercropping, 362
pinching, 359–360
plant thinning, 361
preconditioning, 360
ratooning, 362
renovation (renewing the planting), 361–362
Specific combining ability (SCA), 57
Sphaerotheca macularis, 454
Spider mites, 490, 502, 503
Spodoptera litura, 493, 494
Sprinkler irrigation, 245
Squash and syrup
flowsheet, 434
ingredients, 433
preservation tips, 433
Stigma receptivity, 323–324
Storage conditions, 421–424
CO_2 injury, 423
modified and controlled atmosphere, 422–423
Strawberries *(Fragaria vesca); see also* individual entries
botany and taxonomic status, 2–3
composition, 23, 24
crop improvement and varietal wealth, 5–6
diseases, 15
economics, 14
epilogue, 15–16
evolution and distribution, 3–5
flower, 321, 322
growing environments, 6–7
handling, 13
harvesting, 13
historical background, 1–2
medicinal use, 26–28
origin, taxonomy and distribution, 37–46
plant protection, 14–15
processing, 13
production technology, 7–11
protected cultivation, 11
yield, 14

Strawberry breeding programmes, 534
Strawberry clipper, 495
Strawberry crinkle virus (SCV), 473
Strawberry dried calyx disorder (SDCD), 450
Strawberry Festival (Rosa Linda × Oso Grande), 114
Strawberry leaf folder, 494–495
Strawberry mild yellow edge (SMYE) virus, 130
Strawberry plasticulture planting, 198
Strawberry production systems
challenges, 531–532
Strawberry propagation, 533
Structures and functions, 31–35
crown, 32–33
flower, 33–34
fruit, 34–35
leaves, 33
roots, 32
runners, 32, 33
Substrate culture, 182–183
plug plants, 182–183
tray plants, 183
Sucking insect-pests
greenhouse whitefly, 491–492
green peach aphid *(Myzus persicae),* 488–490
Lygus bug, 491–492
melon aphid *(Aphis gossypii),* 488
spider mites, 490
strawberry aphids *(Chaetosiphon fragaefolii),* 488, 489
strawberry thrips *(Frankliniella occidentalis),* 492–493
Sucrose, 304, 339
Summer Ruby (57K104 × 572I16), 114
Sunscald, 448
Surecrop (Fairland × Maryland-U.S.1972), 114
Suspension culture, 125, 187
Sveva (EM483 × 87.734.3), 114
Sweet Ann (4A28 × 10B131), 115
Sweet Charlie (FL 80-456 × Pajaro), 115
Sweet clover *(Melilotus officinalis),* 11, 269, 277
Sweet 'n' sour pickle
flowsheet, 437
ingredients, 436–437
preservation tips, 436
Systemic herbicide, 281

T

Talisman (New Jersey × Auch. Climax), 115
Tango (Rapella × Selva), 115
Templer (Seedling Auch. 11 × Cambridge Vigour), 115
Temporary immersion bioreactors (TIB), 187
Terbacil (Sinbar®), 283
Terunoka (Hokowase × Donner), 115
Tetranychus urticae, 490
Thermal controls, 278
TIB, *see* Temporary immersion bioreactors
Tillamook (Cuesta × Puget Reliance), 115–116
Tioga [Fresno × Torrey (lassen × Cal 42.8-16)], 116
Tip burn, 448–449
Tissue culture, 121–135
applications, 127–135
biotic/abiotic stress tolerance, *in vitro* selection, 130–131

548 Index

genetic transformation, 131–135
somaclonal variation, 127–129
virus elimination, 129–130
techniques, 122–127
anther and pollen culture, 125–126
meristem culture, 126–127
micro-cuttings, 127
micropropagation, 122–123
protoplast culture, 123–125
somatic embryogenesis, 127
Titanium dioxide (TiO_2), 419
Toadflax *(Linaria maroccana),* 285
Tobacco caterpillars, 493
Tochiotome (Kurume 49 × Tochinomine), 116
Tolylfluanid, 514, 515
Torrey (Lassen × Cal 42.8.16), 116
Total soluble solids (TSS), 339, 353
Totipotency, 122
Trichoderma harzianum, 455, 466, 471
Tricky weeds, 286
TSS, *see* Total soluble solids
Tumour-inducing plasmids *(pTi),* 131, 133

U

Unavailable water, 231
Uneven ripening, 449
United States Department of Agriculture (USDA), 413
USDA, *see* United States Department of Agriculture

V

Value addition
chutney, 436
frozen strawberries, 438
jam, 434
preprocessing considerations, 432
preserve, 434–435
puree, 438
quality processing, 440–441
sauce, 437
squash/syrup, 433–434
strawberry juice, 433
strawberry wine, 439
sweet 'n' sour pickle, 436–437
Vegetative propagation, 179–187
Vertical system, 199–200
Verticillium wilt, 57, 70, 71, 175, 459–461
Vibrant [(SDBL101 × EM881) × (Rosie × Onebor)], 116
Vinegar and essential oil herbicides, 280
Virus elimination, 129–130
Volatile constituents, 340

W

Water economy, 375
Water management, 230–261
biotic and abiotic stresses, 253–255
drought, 253–254
flooding, 254
pathogens, 255
salinity, 254
weeds, 254
classification, soil water/moisture, 231

decision support systems, 252–253
efficient irrigation system, indicators, 257
high crop water use efficiency, methods, 257
irrigation scheduling, 232–244
amount of, 235–236
bed capacity, 240
climatic parameters, 241
criteria for, 233
critical time of, 234–235
depth of, 234
frequency of, 234
growing conditions effect, 237–238
plant parameters, 240–241, 242
soil parameters, 238–240
stages of, 235
time of, 236
varietal response to, 236–237
irrigation systems, 244–252
role in plants, 230–231
soil moisture constants, 231–232
available soil moisture, 232
evapotranspiration, 232
field capacity, 231
maximum water holding capacity, 231
transpiration, water loss, 232
water requirement, 232
wilting point, 231–232
water quality, 257–259
water saving practices, 255–257
anti-transpirant, 256–257
mulching, 255–256
Water quality, 257–259
Water requirement, 241, 243–244
Weed biology, 275–276
Weed identification, 273–274
Weed management, 269–286
chemical weed control options, 283–285
components, 273–281
herbicidal injury, 282–283
herbicide formulation, 281
Weed threshold and action levels, 276
Wild strawberry, 31, 39, 66, 185, 468
Wilting coefficient, 232
Wine
flowsheet, 439
ingredients, 439
preservation tips, 439
Winterstar™ (Florida Radiance × Earlibrite) SYN. FL 05-107, 116–117
Wood strawberry *(F. vesca),* 2, 37

Y

Yamaska (Pandora × Bogota), 117
Yellow nut sedge *(Cyperus esculentus),* 285
Yield and varietal performance, 403–406
Yueli (Benihoppe × Sachinoka), 117
Yuezhu (Akihime × Toyonoka), 117

Z

Zarina (1007 × 880), 117
Zinc, 214–215
Zuoheqingxiang, 117